FLORA OF WEST TROPICAL AFRICA

FLORA

OF

WEST TROPICAL AFRICA

ALL TERRITORIES IN WEST AFRICA SOUTH
OF LATITUDE 18°N. AND TO THE WEST OF
LAKE CHAD, AND FERNANDO PO

FIRST EDITION BY

J. HUTCHINSON, LL.D., F.R.S., V.M.H., F.L.S.

FORMERLY KEEPER OF THE MUSEUMS OF BOTANY, ROYAL BOTANIC GARDENS, KEW

AND

J. M. DALZIEL, M.D., B.Sc., F.L.S.

FORMERLY OF THE WEST AFRICAN MEDICAL SERVICE, AND
ASSISTANT FOR WEST AFRICA, ROYAL BOTANIC GARDENS, KEW

SECOND EDITION

REVISION EDITED BY

F. N. HEPPER, B.Sc., F.L.S.

PRINCIPAL SCIENTIFIC OFFICER, ROYAL BOTANIC GARDENS, KEW

PREPARED AND REVISED AT THE HERBARIUM,
ROYAL BOTANIC GARDENS, KEW, UNDER THE
SUPERVISION OF THE DIRECTOR

VOL. III PART 2

2ND MAY, 1972

ISBN 0 85592 018 1

PUBLISHED ON BEHALF OF THE GOVERNMENTS
OF NIGERIA, GHANA, SIERRA LEONE AND
THE GAMBIA

BY THE

CROWN AGENTS FOR OVERSEA GOVERNMENTS
AND ADMINISTRATIONS

MILLBANK, LONDON. S.W.1

Price £1.70

Printed in Great Britain by
The Whitefriars Press Ltd., London and Tonbridge.

200. JUNCACEAE

By F. N. Hepper

Perennial or annual herbs. Leaves mostly in a basal tuft, grass-like, linear or filiform, sheathing at the base or reduced to a sheath, sheaths open or closed. Flowers hermaphrodite or unisexual, usually very small. Perianth-segments 6, in 2 whorls, or rarely only 3, usually glumaceous. Stamens 6 or 3, free; anthers 2-celled, basifixed, opening lengthwise; pollen in tetrads. Ovary superior, 1-celled or 3-celled; styles and stigmas 1 or 3. Ovules ascending or parietal, 3 or more. Fruit a dry capsule. Seeds sometimes tailed, with a small straight embryo in the middle of endosperm.

World-wide distribution, mostly in temperate and cold or montane regions, often in damp places.

Leaves flat, hairy on the margin with a closed basal sheath; ovary 1-celled, with 3 sub-basal ovules.. .. .. 1. **Luzula**
Leaves subterete, glabrous, with an open basal sheath; ovary subcompletely 3-celled, with several ovules in each cell .. .. 2. **Juncus**

1. LUZULA DC., Fl. Fr., ed. 3, 3: 158 (1805); F.T.A. 8: 95 (1901). *Nom. cons.*

Roots fibrous; leaves in a basal rosette, linear-lanceolate to elongate-linear, acute, up to 20 cm. long and 5 mm. broad, about 7-nerved, ciliate when young, at length glabrous; inflorescence capitate or clusters of flowers pendunculate; leafy bracts subtending the inflorescence densely villous inside towards the base; floral bracts

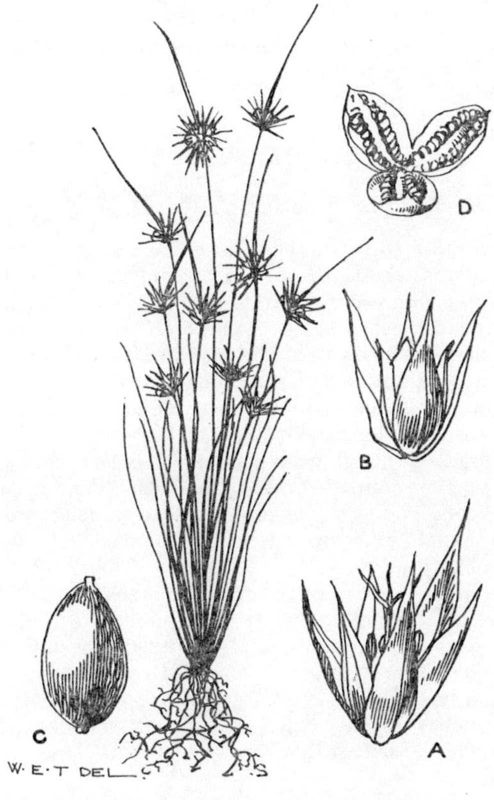

Fig. 406—JUNCUS CAPITATUS *Weig.* (JUNCACEAE).
A, flower. B, fruit. C, seed. D, open capsule.

acutely acuminate, long-ciliate, as long as the perianth-segments, the latter very dark brown, acuminate; capsule ovoid-trigonous, mucronate, 2 mm. long *campestris* var. *mannii*

L. campestris (*Linn.*) *DC.* Fl. Fr., ed. 3, 3: 161 (1805). Widely distributed, principally in the N. Temperate Zone, with one variety in our area.

L. campestris var. **mannii** *Buchenau* in Engl., Bot. Jahrb. 12: 159 (1890), and in Engl., Pflanzenr. 4, 36: 87 (1906); F.T.A. 8: 96; Hedberg in Symb. Bot. Ups. 25, 1: 63 (1957); Carter in Kew Bull. 19: 515 (1965). Small grass-like perennial; in montane grassland.
W. Cam.: Cam. Mt., 10,000–13,300 ft. (Dec., Jan.) *Mann* 2108! *Johnston* 48! *Maitland* 1248! *Hinds* C17! *Morton* GC 6904! **F. Po:** Clarence Peak, 7,000 ft. to summit (Apr.) *Mann* 658! 1462! *Guinea* 2868! Also in Congo (Kivu) and Uganda (Mt. Elgon).

2. JUNCUS Linn., Sp. Pl. 325 (1753); F.T.A. 8: 92 (1901).

Leaves about 3 cm. long in a basal rosette, linear-filiform, acute; flowers 4–8 in a pedunculate cluster; subtending bract subulate-linear, about 2 cm. long; floral bracts ovate, subulate-acuminate; perianth-segments acuminate, membranous, with recurved tips; stamens 3; capsule ovoid-trigonous, brown, much shorter than the perianth; seeds ellipsoid, not tailed .. .. 1. *capitatus*

Leaves about 30–40 cm. long, few, basal, terete, acute; stems terete, wiry, terminated by a cyme of small flowers subtended by a subulate terete leaf about 10 cm. long which expands into a membranous wing at the base about 3 cm. long; bracts lanceolate, chaffy, about 5 mm. long; perianth-segments lanceolate, acute, pale green; stamens 6; capsule as long as or longer than the perianth; seeds shortly tailed at one end 2. *maritimus*

1. **J. capitatus** *Weig.* Obs. 28 (1772); F.T.A. 8: 95; Buchenau in Engl., Pflanzenr. 4, 36: 256 (1906). A small annual rush about 9 cm. high; at high altitudes.
W. Cam.: Cam. Mt., 7,000 ft. (fr. Dec.) *Mann* 2094! Also in Ethiopia, N. Africa, Atlantic islands and S. Europe.

2. **J. rigidus** *Desf.* Fl. Atlant. 1: 312 (1798); Snogerup in Fl. Iranica 75: 4 (1971). *J. arabicus* (Asch. & Buch.) Adamson (1935). *J. maritimus* of F.T.A. 8: 93; F.W.T.A., ed. 1, 2: 464; Berhaut, Fl. Sén. ed. 2, 372. A tufted perennial herb with inflorescences about 60 cm. high and very acute leaf apices; in salt marshes and salt water pools by oases.
Sen.: Cayor *Leprieur*! Tibeghun *Chipp* 94! Gandhiolle *Heudelot* 511! St. Louis *Berhaut* 1351! Baïbani *Trochain* 2352! Salty places in the Sahara, tropical and southern Africa and the Middle East.

201. CYPERACEAE

By Miss S. S. Hooper and Miss D. M. Napper

Annuals or caespitose (tufted), rhizomatous, occasionally tuberiferous perennials (*Microdracoides* is pseudo-arborescent); underground stems bearing scales which grade into the culm leaves; culms generally leafy only at, or towards the base and generally unbranched below the inflorescence, solid, triquetrous, round, flattened or 5-angled; leaves with a distinct cylindrical sheath, closed (except in *Coleochloa*), generally without a ligule, prolonged at the apex on the side opposite the blade (antiligule) in species of *Scleria* and *Afrotrilepis*, and a generally linear blade (reduced to an apiculus in *Eleocharis* and elsewhere), not articulated with the sheath as in grasses, though deciduous in *Coleochloa*. Flowers small, simple, often consisting only of stamens or pistil or both within a subtending glume, sometimes with accompanying hypogynous bristles, hairs or squamellae, arranged in bisexual or unisexual (*Microdracoides, Carex*, some *Sclerieae*) spikelets. Spikelets sometimes solitary (always so in *Eleocharis*), more often aggregated into capitula, spikes or glomerules which in turn may be solitary or variously arranged, often in 1–3 times compound, unequally rayed umbels, subtended by 1–several, more or less leaf-like, bracts. Stamens 1–numerous, often 3, anthers basifixed with two pollen sacs often with sterile tips and crest, opening lengthwise by a slit, often protandrous. Ovary superior with one erect anatropous ovule, and generally 2–3 branched style. Fruit indehiscent, nut-like (achene) generally lenticular or trigonous. Seed erect with a small embryo and abundant mealy or fleshy endosperm.

A large family distributed throughout the world, mostly dominant in damp or marshy places, especially in temperate or cold regions.

Key to the Genera

By Miss S. S. Hooper

- Plants either male or female, with a trunk-like base and short lanceolate leaves set in a close spiral .. .. 23. **Microdracoides**
- Plants bisexual:
  - Flowers bisexual, enclosed by glumes, and arranged in two rows on the rhachilla of the more or less distinctly flattened spikelet; usually all glumes above the prophyll fertile:
    - Spikelets falling entire from the rhachis at maturity, leaving the subtending bract and prophyll surrounding the rhachilla scar, often very numerous, sessile and crowded on short rhachides, generally composed of a small number of glumes:
      - Stigmas 3; achene more or less distinctly trigonous:
        - Spikelets numerous, clustered in a lobed *Kyllinga*-like head, each with a single achene partly sheathed in a thickened section of the rhachilla; plant with long creeping rhizomes .. .. 3. **Remirea**
        - Spikelets generally with 2 to several achenes, if one then this not sheathed in the thickened rhachilla, (sometimes glumes numerous and then distinguishable from *Cyperus* only at maturity) .. .. 2. **Mariscus**
      - Stigmas 2; achene laterally (that is, in the plane of the rhachilla) flattened; spikelets often composed of 2–3 glumes folded within one another and maturing only 1–2 achenes .. .. 6. **Kyllinga**
    - Spikelets at maturity breaking into several pieces or remaining attached to the rhachis whilst the lower mature glumes and achenes are shed, exposing the more or less deeply pitted rhachillas, generally composed of a large number of glumes:
      - Rhachilla of spikelet breaking into segments on maturity, these composed of rhachilla, glume and achene .. .. 4. **Torulinium**
      - Rhachilla of spikelet remaining entire at maturity, often bare below due to loss of glumes and achenes:
        - Stigmas 3, occasionally 2 or 1, achene usually trigonous .. 1. **Cyperus**
        - Stigmas 2, achene laterally flattened .. .. 5. **Pycreus**
  - Flowers unisexual or bisexual in a rounded spikelet (sometimes flattened at the base but then with the lowest two glumes empty):
    - Flowers (at least some) bisexual with stamens and ovary within the same subtending glume:
      - Spikelets of numerous flowers spirally arranged at least at the top of the spikelet, generally producing many achenes, only the lowermost 0–2 sterile:
        - Ovary and stamens enclosed in modified scale(s) (squamellae) within the glume:
          - Ovary and stamens enclosed or enwrapped by a squamella more conspicuous than the subtending glume .. .. 13. **Ascolepis**
          - Ovary and stamens enclosed in two membranous squamellae within the more conspicuous glume .. .. 14. **Lipocarpha***
        - Ovary and stamens not enclosed in modified scales:
          - Achene with a terminal appendage (beak) formed by the thickened and persistent style base:
            - Plants without well developed leaf-blades, sheaths ending in a small point; flowering stems generally numerous and ending in a solitary spikelet, sometimes sterile; bristles often present round the ovary:
              - Achene more or less truncate bearing a distinct style-base 8. **Eleocharis**
              - Style-base very small; submerged aquatic with numerous sterile stems in successive whorls .. .. 9. **Websteria**
            - Plants with at least some leaves present; sheaths often with long hairs at the mouth and frequently long hairs also present at the base of the bracts 10. **Bulbostylis**
          - Achene without a distinct appendage at maturity:
            - Base of the style not enlarged; bristles, at the base of achene, present or absent:
              - Glumes hairy; laminate bristles often present .. .. 12. **Fuirena**
              - Glumes generally glabrous; bristles absent or simple .. 7. **Scirpus**
            - Base of the style enlarged but generally falling with the rest of the style; bristles absent .. .. 11. **Fimbristylis**
      - Spikelets of few fertile flowers accompanied by 1—several empty glumes:
        - Plants large with a branched inflorescence of many partial umbels and saw-like edges of the leaves; styles 3; achene narrower above but not with a distinct beak; rare .. .. 16. **Cladium**
        - Plants variable but without saw-like edges to the leaves and with a distinct beak on the achene; styles 2 .. .. 15. **Rhynchospora**

* *Scirpus isolepis*, a rare slender tufted annual with 1–2 pseudolateral dark spikelets also sometimes has 1–2 delicate membranous scales investing the achene.

Flowers unisexual, without ovary and stamens enclosed within the same glume:
Ovary completely enclosed in a flask-shaped utricle .. .. 24. **Carex**
Ovary not completely enclosed in a utricle:
Ovary with a boat-shaped glume, generally ciliate on the keel, on either side, containing one stamen each; achene ovate or obovate, inflated or wrinkled in the upper part:
Inflorescence umbellate .. .. 18. **Hypolytrum**
Inflorescence capitate .. .. 17. **Mapania**
Ovary not accompanied by two boat-shaped laterally placed glumes:
Achene accompanied by numerous long bristles:
Leaf-sheaths open on one side (as in grasses) with a ring of ligular hairs; blade deciduous .. .. 21. **Coleochloa**
Leaf-sheaths closed and prolonged at the mouth on the side opposite the blade 22. **Afrotrilepis**
Achene not accompanied by numerous long bristles:
Achene with two 3-lobed glumes close beneath it .. .. 20. **Diplacrum**
Achene with a simple glume, sometimes with a 3-lobed disk or ciliatecu pule at its base .. .. 19. **Scleria**

1. CYPERUS Linn., Sp. Pl. 44 (1753); F.T.A. 8: 310–377 (1901); Kükenthal in Engl., Pflanzenr., Cyperaceae—Scirpoideae—Cypereae 41–315 (1935–36). *Juncellus* (Griseb.) C.B.Cl. in Hook. f., Fl. Brit. Ind. 6: 594 (1893); F.T.A. 8: 306 (1901); F.W.T.A., ed. 1, 489 (1936); Kükenthal in Engler l.c. 315–326.

By Miss S. S. Hooper

Stems up to 2 m. high, leafless, pithy, rounded with transverse septa showing in the dry state as rings; bracts broadly lanceolate up to 1 cm. long; spikelets 10–25 × 1–1·25 mm., pale brownish, in short rounded spikes .. .. .. 7. *articulatus*
Stems without transverse septa, even when dry:
Rays of the inflorescence very numerous (50 or more), slender, equal, or almost so, often sterile; spikelets pale to golden brown in short cylindrical spikes 1. *papyrus*
Rays of the inflorescence not very numerous and subequal:
*Spikelets arranged along the rhachis in more or less elongated spikes (to p. 282):
Spikes cylindrical or oblong at least twice as long as wide; spikelets in flower not exceeding 1·5 cm. long:
Spikelets very numerous and crowded together so as to obscure the rhachis:
Spikelets and bracts erect or suberect at maturity:
Stem-bases swollen into a woody pseudobulb; glumes up to 1·5 mm. long 26a. *tonkinensis* subsp. *baikiei*
Stem bases hardly thickened; sheaths yellow; glume 2 mm. long; rare 25. *eleusinoides*
Spikelets spreading:
Glumes about 1 mm. long with a distinct green keel and recurved mucro; rays sometimes very short or absent; bracts 5–7 mm. wide .. 5. *imbricatus*
Glumes with a short straight mucro; rays well-developed:
Glumes with a broad rounded keel, margins inrolled at maturity to reveal the dorso-ventrally flattened, pale achene; stigmas and stamens generally 2 6. *alopecuroides*
Glumes with a narrow keel; spikelets golden, 3–5 mm. long, 10–14 flowered; stigmas 3; achene trigonous .. .. 4. *dives*
Spikelets not so numerous and crowded as to obscure the rhachis, often erecto-patent to erect:
Glumes black with a green keel .. .. 24. *atroviridis*
Glumes straw-coloured, silvery, yellow or golden-brown:
Annual; glumes broadly elliptical, obtuse with a membranous margin, golden-sided, nearly equalled by the elliptical flattened achene .. .. 28. *iria*
Perennial; glumes longer than the achene:
Spikelets narrow, less than 1 mm. wide, erecto-patent:
Spikelets golden or brownish with a pale central stripe, almost terete; glumes acute and minutely apiculate .. .. 2. *digitatus* subsp. *auricomus*
Plants up to 1 m. high with leaves and bracts to 1·5 cm. wide; inflorescence umbellate; rays smooth 2a. *digitatus* subsp. *auricomus* var. *auricomus*
Plants up to 3 m. high with leaves and bracts to 3 cm. wide; inflorescence bisumbellate; rays scabrid 2b. *digitatus* subsp. *auricomus* var. *bruntii*
Spikelets uniformly golden-brown, flattened with a zigzag rhachilla; plant large, with inflorescence rays to 35 cm. long .. .. 23. *koyaliensis*
Spikelets more than 1 mm. wide; glumes mucronate:
Glumes numerous, less than 2 mm. long with a clear green keel; rhachilla straight .. .. 3. *exaltatus*
Glumes few, more than 2 mm. long, golden; rhachilla zigzag 19. *permacer*

Spikes obovoid, hemispherical or fan-shaped:
Base of the stem swollen and enclosed with the lower leaf-bases in hard, brown scales, often with bulbils in the leaf-axils surrounded by fibrous-edged scales; leaves numerous, clustered, nearly equalling stems; spikelets brown 15. *bulbosus*
Base of the stem not enclosed in brown scales, swollen or not:
Rhachis of the spike hairy and more or less flexuose; glumes with a conspicuous white membranous margin:
Spikelets 3–5 mm. wide; rhachis of spike almost straight, often short; glumes about 3·5 mm. long .. .. 30. *incompressus*
Spikelets up to 3 mm. wide; plant with slender rhizomes, sometimes bearing tubers; stems sharply angled; rhachis of spike elongate, flexuose; bracts long acuminate:
Glumes straight at maturity; plant with slender rhizomes bearing tubers; leaves and bracts up to 8 mm. wide, gradually narrowed into a slender tip 9. *procerus*
Glumes arching away from the rhachilla at maturity; stems and sheaths tinged purple; leaves and bracts up to 18 mm. wide, suddenly narrowed into a slender tip .. .. 8. *latifolius*
Rhachis of the spike not hairy:
Plant bearing slender rhizomes ending in tubers:
Rhizomes rather soft, with pale scales; tubers globular to ovoid, smooth or sometimes hairy when old; stems thickened in the lower part by numerous leaf-sheaths but not swollen; spikelets in short spikes, generally golden or yellowish, 1–2 cm. long, with an irregular outline at maturity; glumes obtuse, sides with raised nerves, only the margins nerveless and membranous 14. *esculentus*
Rhizomes somewhat woody with dark scales, soon becoming fibrous; tubers covered with dark fibres and persisting as hard swollen bases to the stems; spikelets often in fascicles; glumes with a wide nerveless side:
Spikelets red-brown to dark red-purple, glumes 2·5–4 mm. long:
Glumes elliptical, about 3 mm. long, keel somewhat incurved towards the subacute tip; spikelets about 2 mm. wide, acute .. 12. *rotundus*
Glumes lanceolate, about 4 mm. long, gradually narrowed to an acute, membranous, slightly recurved tip; spikelets 1–1·5 mm. wide with long, slender, sometimes flexuose tips .. .. .. 13. *tuberosus*
Spikelets pale brown or silvery; glumes 1·5–2 mm. long, numerous, closely and regularly imbricate:
Glumes 2 mm. long, obtuse, not spreading; spikes lax to dense spike-letted .. .. 11. *maculatus*
Glumes 1–1·5 mm. long, slightly spreading at the tip, obtuse and apiculate; spikes dense spikeletted .. .. 26a. *tonkinensis* subsp. *baikiei*
Plant rhizomatous or not but not bearing scaly tubers:
Glumes, at least some, with a dark red-brown spot on the side near the base; plant caespitose; stem often yellow-green, at least when dry; rhachilla narrow, black, zig-zag.. .. .. 16. *sphacelatus*
Glumes without red-brown spots; stems not yellow-green below:
Spikelets narrow, less than 1·5 mm. wide:
Glumes less than 2 mm. long with obtuse tips; rhachilla internodes over 1 mm. long .. .. 22. *distans*
Glumes 2 mm. long or more; rhachilla internodes less than half the length of the glume:
Tips of the glumes mucronulate, sides with clear raised red nerves and broad pale membranous margins .. .. 27. *congensis*
Tips of the glumes obtuse to acute, nerves slender, obscure, margin narrowly membranous .. .. 10. *fenzelianus*
Spikelets 1·5 mm. wide or more:
Spikes with distant spikelets, the lower spreading or reflexed at maturity; plant with slender, about 2 mm. wide, curved rhizome; glumes pale with a pink or reddish tinge, the apex emarginate and shortly mucronate 17. *dilatatus*
Spikes with distant or clustered spikelets; erect or spreading at maturity; glumes yellow, golden or brown with green mid-rib:
Glumes very large (5 × 4 mm.), deltoid with spreading tips; plants up to 150 cm. high; stems smooth, sharply triangular .. 21. *baoulensis*
Glumes smaller (3–4 mm. long), ovate-lanceolate .. 18. *tenuiculmis*
Glumes clearly two coloured, streaked with dark brown on a golden back-ground .. .. 18c. *tenuiculmis* var. *guineensis*
Glumes brown or golden, concolorous:
Glumes shortly mucronate .. 18a. *tenuiculmis* var. *tenuiculmis*

Glumes with a distinct mucro; stem below the inflorescence often scabrid 18b. *tenuiculmis* var. *schweinfurthianus*

*Spikelets digitately or very shortly spicately arranged in a central sessile head, with or without additional unbranched, rayed spikes, or spikelets few, paired or solitary:

Plant with elliptical, dark green, white-or purple-veined leaves and bracts; rays stem-like, triangular or flattened, flexuose, longer than the stems, sometimes recurved and proliferating, bearing 1–4 oblong, flattened spikelets about 10 × 5 mm. .. .. 38. *fertilis*

Plant not having elliptical leaves and bracts and long stem-like rays:

†Rayed spikes present (to p. 283):

Spikelets large, at least 10 × 2·5 mm.:

Achene flattened, lenticular or with one long and two short sides; spikelets broadly elliptical in section; glumes rounded on back with a broad green keel, falling to reveal a deep, narrow rhachilla:

Spikelets linear, obtuse; glumes pale, often spotted with dark colour and crossed by two raised lines on the sides; stigmas generally 2 31. *pustulatus*

Spikelets lanceolate, acute; glumes golden or brown; stigmas 3:

Sides of glume almost nerveless, apex acute.. .. .. 32. *podocarpus*

Sides of glume finely nerved except for a membranous margin:

Glumes 4–5 mm. long, subobtuse and mucronulate .. 54. *aucheri*

Glumes 2·5–3·5 mm. long:

Spikelets strongly compressed; glumes purple-brown, acuminate 53. *jeminicus*

Spikelets subterete; glumes pale .. .. .. 52. *conglomeratus*

Achene trigonous, with equal sides:

Spikelet subquadrangular in section, greenish; bracts numerous, leafy, spreading; glumes apiculate .. .. 20. *zollingeri*

Spikelet flattened, narrowly elliptical to elliptical in section:

Glumes aristate, with a three-nerved keel and pale or chestnut-brown sides 49. *cuspidatus*

Glumes obtuse or apiculate:

Glumes with a sharp green keel prolonged into an apiculus:

Glumes lanceolate, not overlapping, persistent and spreading at achene fall, green or golden; achene narrowly elliptical, little shorter than the body of the glume .. .. 42. *reduncus*

Glumes ovate, overlapping, deciduous, pale or chestnut-brown; achene obovate or broadly elliptical, about half as long as the glume 29. *compressus*

Glumes with a rounded keel and retuse membranous apex 30. *incompressus*

Spikelets small, less than 1 cm. long and 2·5 mm. wide, and numerous:

Large cultivated or escaped, leafless plants with a whorl of numerous subequal bracts, sometimes sterile or with spherical clusters of golden-brown spikelets 33. *alternifolius*

Plants not leafless and with many subequal bracts:

Glumes black or dark purple on the sides, elliptical with a short green recurved apiculus; inflorescence diffuse with lax secondary rays; achene slightly shorter than the glume .. .. 40. *dichroöstachyus*

Glumes green or pale to dark brown:

Achene equalling the glume in length and exposed by the spreading of the glume at maturity:

Inflorescence pseudolateral with an erect bract continuing the line of the stem; rays short, up to 5 cm. long; spikes spherical; glumes rounded on the back 43. *submicrolepis*

Inflorescence with several spreading or reflexed leafy bracts:

Glumes 0·5 mm. long, obovate, obtuse with a winged keel; spikelets ovate, flattened .. .. 41. *difformis*

Glumes 1·5 mm. long, broadly ovate, acute and shortly apiculate, keel not winged; spikelets linear, few-flowered, sometimes proliferating 37a. *diffusus* subsp. *buchholzii*

Achene shorter than the glumes and not exposed at maturity:

Plant annual:

Glumes retuse with a long recurved point from the three-nerved keel and pale or chestnut-brown sides .. .. 49. *cuspidatus*

Glumes apiculate, shortly pointed:

Spikelets narrowly linear, uniformly pale to deep golden-brown; glumes 1·0–1·25 mm. long .. .. 50. *amabilis*

Spikelets lanceolate or linear-elliptic, greenish white sometimes tinged with gold; glumes 1·5 mm. or more long:

Spikelets generally exceeding 2 mm. wide; glume with a sharp erect tip and nerved sides; achene broadly elliptical, sharply trigonous, almost smooth, more than half the length of the glume .. 29. *compressus*

Spikelets not exceeding 2 mm. wide; glumes with nerveless sides; achene globose, stipitate, puncticulate, up to half as long as the glume:

Spikelets generally 2 or 3 together; glumes at maturity spreading, with a recurved mucro .. .. 45. *tenuispica*

Spikelets mostly solitary; glumes at maturity erect and deciduous 44. *remotispicatus*

Plant perennial:

Bracts 6–10, up to 2·5 cm. wide; inflorescence diffuse with secondary and tertiary rays spreading or reflexed in fruit; spikes brown, hemispherical, 3·5–6 mm. across; glumes recurved, apiculate with clearly nerved sides, the lower ones spinulose-ciliate on the keel .. .. 36. *renschii*

Bracts 1–5:

Plant with an elongated rhizome, 3-winged stems and bladeless or shortly bladed leaves; bracts 1–2, short, trigonous, the lower erect or nearly so; spikelets greenish, golden or pale brown; anthers 0·65–0·9 mm. long 47. *denudatus*

Plant caespitose or with a short rhizome, sometimes with well-developed leaves:

Side of glume with several nerves; plants 90–150 cm. high with leafy stems; spikelets red-brown; glumes lanceolate, tapering to a recurved apiculus .. .. 35. *mannii*

Side of glume with a single nerve; generally smaller plants:

Leaves all basal, spreading, with hard red-brown sheaths and inrolled margins; inflorescence compact with rays up to only a few cm. long, bearing crowded spikelets; glumes shining, red-brown to nearly black .. .. 48. *tenax*

Leaves lax or wanting; rays well developed, bearing few-spikeletted spikes:

Glumes 2·5–3·0 mm. long, lanceolate and apiculate 34. *pseudoleptocladus*

Glumes 1·5–2·0 mm. long, acute and cuspidate .. 46. *haspan*

†Rayed spikes normally absent, though 1 or 2 may occasionally be developed; spikelets normally sessile in a solitary head:

Style simple, without stigmatic branches; slender plants, stems generally not exceeding 25 cm. with a tiny, pseudo-bulbous base; achene turbinate:

Spikelets broadly elliptical 4–6 mm. long in a compact head, brownish at maturity; achene about 0·5 mm. long .. .. 64. *meeboldii*

Spikelets narrowly elliptical or lanceolate 8–18 mm. long in a terminal or pseudo-lateral cluster, pale even at maturity; achene about 1 mm. long:

Spikelets narrowly elliptical, radiating; sides of the glumes without nerves; achene with rounded angles .. .. 63. *clavinux*

Spikelets lanceolate, erect or curved upwards; glumes nerved on the sides; achene with acute angles .. .. 65. *lateriticus*

Style with two or three branches:

Leaves reduced to bladeless sheaths; spikelets few, large, about 15 × 5 mm., ovate with numerous overlapping, acuminate glumes .. 62. *nudicaulis*

Leaves, at least some, with blades:

Inflorescence of few, white, turgid, subacute spikelets in a pseudolateral cluster, subtended by an erect, stem-like bract .. .. .. 67. *laevigatus*

Inflorescence not pseudolateral:

Plant perennial with knotted, fibre-covered rhizome, purple sheaths and a head of large, whitish spikelets .. .. 39. *mapanioides*

Plant annual or perennial without knotted, fibre-covered rhizomes:

Spikelets small, up to 5 × 1·5 mm., congested in a *Kyllinga*-like head; plant slender:

Inflorescence surrounded by conspicuous, leafy bracts with widened whitish bases; glumes subdistichously arranged, overlapping, with a green apiculus; achene flattened .. 66a. *michelianus* subsp. *pygmaeus*

Inflorescence bracts scarcely widened at the base; glumes distichous, not overlapping, obtuse, pale; achene trigonous .. .. 55. *pulchellus*

Spikelets 2 mm. or more wide, few and digitate, or if numerous and congested then plants robust:

Stems about 1 mm. wide, not or scarcely thickened at the base; plant annual; glumes green or golden with a green, spreading acumen 29. *compressus*

Stems either more than 1 mm. wide or markedly thickened and pseudo-bulbous at the base, often surrounded by the remains of old leaf-sheaths;

Stems robust, leafy but not pseudobulbous at the base; leaves flat, coriaceous:
Inflorescence hemispherical, up to 2 cm. across; lowermost leaves represented by acute scales, gradually lengthening into blades up the stem; bracts short, spreading, rigid .. .. 56. *angolensis*
Inflorescence globose or subglobose, 3–5 cm. across; lower leaf-blades not gradually reduced; bracts long, reflexed .. .. 51. *maritimus*
Stems pseudobulbous at the base, thickened by the remains of old leaf-sheaths:
Glumes lanceolate, 6–8 mm. long, white; robust creeping plants up to 90 cm. high; leaves about 4 mm. wide; bracts spreading; spikelets much flattened, somewhat confluent .. .. .. 61. *ledermannii*
Glumes ovate or broadly ovate with a boat-shaped tip, white, yellow or pink; spikelets little flattened, discrete:
Robust plants with few, large, yellowish spikelets up to 4·5 × 1 cm.; leaves up to 6 mm. wide .. .. .. 57. *karlschumannii*
Plants of moderate size with several whitish or pinkish spikelets up to 30 × 8 mm.:
Rhizome composed of a series of contiguous, short, swollen stem-bases clad in membranous sheaths; leaves flat, up to 4 mm. wide; spikelets at maturity oblong, 15–30 × 6–8 mm., flattened; glumes boat-shaped, spreading, shining white with green keel 58. *margaritaceus*
Stem-bases elongated, flask-shaped, clad in hard remains of leaf-sheaths; leaves capillary, up to 1 mm. wide; spikelets at maturity broadly ovate, 6–12 cm. × 5–7 mm.:
Bracts 3–5, several times longer than the spikelets 60. *tisserantii*
Bracts 0–3, shorter to slightly longer than the spikelets 59. *nduru*

1. **C. papyrus** *Linn.* Sp. Pl. 47 (1753); F.T.A. 8: 374; Kük. in Engl., Pflanzenr. Cyper. 45 (1935); Berhaut, Fl. Sén. ed. 2, 379 (1967). *C. papyrus* Linn. subsp. *antiquorum* (Willd.) Chiov. in Mém. R. Instit. Bot. Modena 1: 73, t.2 (1931); Kük. l.c. 47. Stems spongy, up to 5 m. high with characteristic mop-like inflorescences of very numerous primary, often sterile rays; forming rafts in open water and fringing lakes.
Sen.: Dakar *Berhaut* 1399! **Dah.:** Porto-Novo (Jan.) *Chev.* 22714! **N. Nig.:** Maiduguri (Mar.) *Tuley* 1116! Baga Seyoram, L. Chad (Oct.) *Golding* 23! **S. Nig.:** Lagos *Barter* 2153! Ejirin (Apr.) *Ejiofor & Horne* FHI 29400! Epe (Feb.) *Onochie* FHI 20690! Also reported by Kükenthal from Guin. and Iv. C. and from Lib. by Dinklage; throughout tropical and S. Africa and Madagascar and cultivated elsewhere.

2. **C. digitatus** *Roxb.* Hort. Beng. 81 (1814), and Fl. Ind. 1: 209 (1820); F.T.A. 8: 372; Kük. l.c. 55. The typical subspecies occurs in Eastern Asia, and central and S. America.
C. digitatus subsp. **auricomus** (Spreng.) Kük. in Bot. Notis. 1934: 65 (1934).

2a. **C. digitatus** subsp. **auricomus** (Spreng.) Kük. var. **auricomus.** *C. auricomus* Sieb. ex Spreng., Syst. 1: 230 (1824); F.T.A. 8: 373; F.W.T.A., ed. 1, 2: 483; Cherm. in Arch. Bot. Caen 7, Mém. 4: 16 (1936); Berhaut, Fl. Sén. ed. 2, 382 (1967). Characterized by the lax-spikeletted, cylindrical, digitately-arranged spikes of spreading spikelets with a central golden stripe; shallow water and swamps.
Maur.: N'Diago (Oct.) *Adam* 19214! **Sen.:** Walo country (Dec.) *Roger*! Bignona (Sept.) *Adam* 18109! Gouloumbou (July) *Adam* 14871! Tambacounda (May) *Trochain* 3553! Galam *Heudelot* 322! **Gam.:** *Saunders* 14! (July) *Ruxton* 70! Wallikunda (Aug., Sept.) *Macluskie* 3! 24! **Mali:** Bamako (May) *Duong*! San (June) *Chev.* 1057! Macina (July) *Roberty* 2560! Diafarabé to Roundé (May) *Lean* 27! Gao (Jan.) *de Wailly* 4953! **Guin.:** Kapatchez, Katako R. (Nov.) *Jac.-Fél.* 7290! **U. Volta:** Banzon *Duverger* in *Hb. Raynal* 6183! **Ghana:** Nangodi to Bawku (May) *Morton* GC 7483! Port Tamale (Mar.) *Morton* 8759! Yeji (Apr.) *Hall* VBS 1227! **Niger:** Kouroumé, Yatakala to Tillabery (July) *Rogeon* 502! Tillabery to Niamey (July) *Rogeon* 521! **N. Nig.:** Nupe *Barter* 1565! Anara F.R., Zaria (May) *Keay* FHI 19197! Giwa Fadama, Zaria (May) *Kershaw* 900310! Sokoto Prov. *Dalz.* 548! Maska, Katsina Prov. (June) *Keay* FHI 25902! Sherifuri, Bauchi Dist. (May) *Thornewill* 164! Throughout tropical Africa and in Egypt and S. Africa. The typical subspecies occurs in eastern Asia, and central and S. America.

2b. **C. digitatus** subsp. **auricomus** var. **bruntii** *Hooper* in Kew Bull. 26: 577 (1972). Similar to the preceding variety in spikelet structure but larger in all its parts with stems up to 3 m. high, leaves and bracts up to 3 cm. wide and a bisumbellate inflorescence.
N. Nig.: Ibaji Ojoko F.R., Kabba (May) *Latilo* FHI 47653! **W. Cam.:** Babungo, Ndop Plain, 3,800 ft. (June) *Brunt* 1146!

3. **C. exaltatus** *Retz.* Obs. 5: 11 (1789); F.T.A. 8: 370; Kük. l.c. 64. *C. dives* of F.W.T.A., ed. 1: 483. Robust plant up to 1·5 m. high, with obtuse, cylindrical, lax-spikeletted, golden, rayed spikes; near open water.
Sen.: Ourossogui to Matam (Dec.) *Naegelé* 13183! **Iv.C.:** Téhini (Aug.) *de Wilde* 781! Ferkessédougou (May) *Aké Assi* 4798! Marabadiassa to Gottoro (July) *Chev.* 22026! **U. Volta:** Koupéla to Ouagadougou (July–Aug.) *Chev.* 24582! Namenda, White Volta R. (June) *Scholz* 311! **Ghana:** Nwereme (May) *Irvine* 2497! Tumu (May) *Morton* GC 7505! Zuarungu (Dec.) *Williams* 859! Kakumdu (Sept.) *Hall* 2625! Accra to Ada (Jan.) *Morton* GC 8248! **N. Nig.:** Abinsi (June) *Dalz.* 819! Maska, Katsina Prov. (June) *Keay* FHI 25901! Bauchi to Dindima (Dec.) *Haines* 85! Katagum *Dalz.* 239! Gurum, Vogel Peak area (Nov.). *Hepper* 1258! **S. Nig.:** Lagos *W. MacGregor* 119! Ogun Crossing, Lagos (Apr.) *Haines* 52! Ilesha (Apr.) *Haines* 54! Pantropical.

4. **C. dives** *Del.* Descr. Egypte Hist. Nat. 2: 5, t.4 fig. 3 (1813); Kük. l.c. 68; Cherm. in Arch. Bot. Caen 7, Mém. 4: 17 (1936); Berhaut, Fl. Sén. ed. 2, 382 (1967). *C. exaltatus* Retz. var. *dives* (Del.) C.B. Cl. in J. Linn. Soc. 21: 187 (1884); F.T.A. 8: 370. Distinguished from *C. exaltatus* Retz. by the shorter, congested spikelets and from *C. alopecuroides* by the keeled glumes and three-sided achene; marshes.
Sen.: Mboro (Apr.) *Leprieur*! Sangalkam (July) *Pitot*! Kayar (Dec.) *Pitot*! Tivaouane (Apr.) *Trochain* 3110! Sangako, nr. Toubacouta (Oct.) *Monod* 8566! Throughout tropical Africa, in Madagascar and in Egypt and the near East.

5. **C. imbricatus** *Retz.* Observ. 5: 12 (1789); Kük. l.c. 69; Nelmes & Baldwin in Amer. J. Bot. 39: 379 (1952); Berhaut, Fl. Sén. ed. 2, 383 (1967). *C. radiatus* Vahl, Enum. Pl. 2: 369 (1806); F.T.A. 8: 369; Cherm. in Arch. Bot. Caen 7, Mém. 4: 16 (1936). *C. imbricatus* Retz. var. *capitatus* (Boeck.) Kük. l.c. 70; Berhaut l.c. 383. Plant to 120 cm. high and generally less robust than *C. exaltatus* Retz. with dense-spikeletted, cylindrical spikes in digitate clusters; edge of open water.
Sen.: Senegal R. *Gay*! Sangalkam (July) *Berhaut* 1123! Coular *Trochain* 3404! Galam *Heudelot* 211! Niokolo-Koba *Adam* 14297! **Mali:** Koulikoro (June) *Roberty* 3694! Niger R., 10 miles E. of Macina (May)

Lean 19! Mopti (June) *Lean* 35! Timbuktu (Sept.) *Chudeau*! Gao to Bossobougou (July) *de Wailly* 5092! **Guin.:** Dabola (Sept.) *Pitot*! Kouroussa (July) *Pobéguin* 415! Macenta (Apr.) *Pitot*! **S.L.:** Mano (May) *Deighton* 2898! Gbap (Apr.) *Deighton* 4298! Gbundapi (Mar.) *A. S. Macdonald* 2! Daru (Apr.) *Deighton* 3164! Pandobu (Apr.) *Deighton* 3926! **Lib.:** Webo, Cavally R. (Apr.) *Dinklage* 2649! **Iv. C.:** Soubré (Apr.) *Leeuwenberg* 3222! **U. Volta:** Red Volta R., nr. Po (June) *Scholz* 309! **Ghana:** Bjury, Black Volta R. (July) *Chipp* 503! Yeji (Apr.) *Hall* VBS 1253! Kete Krachi to Oti R. (May) *Morton* GC 7295! Onyinan Lake, nr. Winneba (June) *Hall* 3069! Kukrukui lagoon, Kete Krachi (Apr.) *Hall* VBS 1300! **Niger:** Tillabery to Niamey (July) *Rogeon* 527! **N. Nig.:** Shagunu (July) *Cook* 307! Birnin Gwari (Apr.) *Meikle* 1378! Nupe *Barter* 1127! Lokoja (Apr.) *Haines* 55! Abinsi (July) *Dalz.* 816! **S. Nig.:** Ogun R. Crossing, Lagos (May) *Haines* 46! Ikoyi Plains (Apr.) *Dalz.* 1296! Ibadan (Apr., May) *Meikle* 1417! *Guile* 133! Shasha F.R. (Apr.) *Gledhill* 1003! **W. Cam.:** Ikom to Mamfe (Apr.) *Morton* GC 7139! Pantropical. The rays of the inflorescence may be very short or absent, producing the var. *capitatus*.

6. **C. alopecuroides** *Rottb.* Descr. Pl. Rar. Programm. 20 (1772), and Descr. et Icon. 38, t.8 fig. 2 (1773); Kük. l.c. 71; Cherm. in Arch. Bot. Caen 7, Mém. 4: 17 (1936). *Juncellus alopecuroides* (Rottb.) C.B. Cl. in Hook. f., Fl. Brit. Ind. 6: 595 (1893); F.T.A. 8: 307; F.W.T.A., ed. 1, 2: 489; Berhaut, Fl. Sén. ed. 2, 382 (1967). Somewhat less robust than *C. dives* Del. and with the edges of the flattened achenes exposed in the ripe spikelet; in or near open water.
Maur.: Nioudjéria, Tagant (Apr., Nov.) *Monod* 109! *Chudeau*! **Sen.:** M'Boro, St. Yago Is. *Brunner* 64! M'Bane, L. Guiers (Oct.) *Adam* 19239! Tanma L. (Jan.) *Pitot*! Dakar (Jan.–Apr.) *Chev.* 15792! Sangalkam (June) *Adam* 1532! Diourbel (Mar.) *Chev.* 44114! **Mali:** Laougna (Jan.) *Davey* 164! Karseni (Apr.) *Davey* 265! Faguibine L. (July) *Leclerq*! Diré (Mar.) *Chev.* 43888 *bis*! Gao (Mar.) *de Wailly* 5344! **U. Volta:** Ouagadougou (July) *Scholz* 318! **Ghana:** Winneba (June) *Hall* 3066! Weija (July) *Hall* 3231! Accra (July) *Irvine* 2787! Dawa (Dec.) *Adams* 3616! Accra to Ada (June) *Morton* GC 9267! **Niger:** Niamey to Kolo (Jan.) *Chev.* 43160! **N. Nig.:** Gotomo, Birnin Kebbi to Argungu, Sokoto Prov. (July) *Latilo* FHI 62633! Sokoto (July) *Dalz.* 466! Kalkala (May) *Golding* 81! SW. of L. Chad (Dec.) *Gaillard*! Also reported from Dah. by Chermezon; throughout tropical and S. Africa, Madagascar, India and Malaysia.

7. **C. articulatus** *Linn.* Sp. Pl. 44 (1753); F.T.A. 8: 356; Kük. l.c. 77; Cherm. in Arch. Bot. Caen 7, Mém. 4: 16 (1936); Nelmes & Baldwin in Amer. J. Bot. 39: 379 (1952); Berhaut, Fl. Sén. ed. 2, 379 (1967). The only species of *Cyperus* to have a round leafless pithy stem which on drying appears articulated; rays slender bearing clusters of cigar-shaped spikelets; in or near open water.
Maur.: Mondjema (Nov.) *Chudeau*! Moudjéria (Apr.) *Monod* 107! **Sen.:** Dakar (Mar.) *Adam* 17678! St. Louis (July) *Adam* 1565! M'bidjem (Oct.) *Trochain* 622! Richard Tol (May) *Döllinger* 58! Bignona, Casamance (Sept.) *Adam* 18106! **Gam.:** (Aug.) *Ruxton* 90! Kuntama (July) *Fox* 156! **Mali:** Forécariah (Nov.) *Schnell* 2077! Goundam (Nov.) *Roberty* 3203! *Chev.* 2284 *bis*! **Guin.:** Kouyeya, Coyah (June) *Chillou* 1536! Timbo (July) *Pobéguin* 739! Kindia *Jac.-Fél.* 136! Semou *Jac.-Fél.* 7405! **S. L.:** Kitchom (Jan.) *Deighton* 929! Makump (Dec.) *Glanville* 405! Bagroo R. *Mann* 894! Kania, Njala (Oct.) *Deighton* 1417! Mano Salija (Nov.) *Deighton* 325! **Lib.:** Fisherman's L., Grand Cape Mount Dist. (Jan., Dec.) *Harley* 1367! *Baldwin* 10884! Bodje, Montserrado Dist. (Dec.) *Adam* 16300! Nyaake, Webo Dist. (June) *Baldwin* 6144! **Iv. C.:** Cavally (Aug.) *Chev.* 19848! Baoulé (June) *Chev.* 22187! Flambo Lagoon (Feb.) *Aké Assi* 7327! **U. Volta:** Foulasso *Duverger* in *Hb. Raynal* 6161! **Ghana:** Cape Coast (June) *Hall* 3128! Koforidua (Nov.) *Morton* 6067! Sakumo Lagoon, Tema (Sept.) *Ankrah* GC 20552! Weija, Accra (Oct.) *Irvine* 4804! Keta (June) *Thorold* C.B. 98! **Togo Rep.:** Lomé *Warnecke* 114! *Mahoux* 2042! **N. Nig.:** Acharane, Anka Dist. (June) *Daramola* FHI 38033! Lokoja *Barter*! Abinsi *Dalz.* 825! Kadaura, Kano Dist. (Sept.) *Lely* 597! L. Chad (Dec.) *Elliott* 157! **S. Nig.:** Ogun River F.R., Lagos (Mar.) *Hepper* 2260! Ibadan to Lagos (Mar.) *Killick* 214! Ogun Crossing, Lagos (May) *Haines* 38! Badagry (Aug.) *Onochie* FHI 33463! Ikorodu, Lagos Lagoon (Dec.) *Onochie* FHI 32024! **W. Cam.:** *Diestel* 44! **F. Po:** (June) *Barter*! Tropical and subtropical Africa and America, extending to India.

8. **C. latifolius** *Poir.* in Lam., Encycl. 7: 268 (1806); F.T.A. 8: 351; Kük. l.c. 87, t.11. Robust plants to about 120 cm. high with broad leaves and bracts, characterized by branched spikes of numerous spikelets with glumes arcuate at maturity; swamps.
W. Cam.: Ndop Plain (Apr.–June) *Brunt* 312! 1133! 1144! 1157! Mbengwi to Mankon (Apr.) *Brunt* 1112! Throughout tropical and S. Africa and in Madagascar.

9. **C. procerus** *Rottb.* Descr. et Icon. 29 (1773). *C. pilosus* of F.T.A. 8: 351. *C. procerus* Rottb. var. *stenanthus* Kük. in Engl. Pflanzenr. Cyper. 92 (1935). Stoloniferous sedge to about 120 cm. high with elongated rhizomes, winged triangular stems, long acuminate bracts, and a small umbel of flattened spikelets with membranous-edged, brown glumes; marshy grassland, rice farms and near open water.
Mali: Niger 'delta' *Duong* 1944! Diafarabé (Aug.) *Lean* 98! **U. Volta:** Banfora (Sept.) *Adam* 15379! (July) *Aké Assi* 10730! Mossi (July) *Chev.* 24585! Ougarou to Nassougou (June) *Scholz* 312! **Ghana:** Bawku (Aug.) *Hall* 2066! Bawku to Nakpanduri (Aug.) *Hall* CC 539! Tamale to Bolga (Aug.) *Hall* CC 424! Winneba plains (July) *Hall* 3206! Ada to Accra (June) *Morton* GC 9273! **Togo Rep.:** Coblékové (May) *Mahoux* 535! 2031! **N. Nig.:** Shagunu, Ilorin Prov. (July) *Cook* 443! Giwa Fadama, Zaria Dist. (Aug.) *Kershaw* 900413! Katsina to Jibya (Aug.) *J. Hall* 436! Kaura Namoda (July) *Latilo* FHI 62595! Also in E. Cameroun, north-east Africa, India, Malesia and Australia.

10. **C. fenzelianus** *Steud.* Syn. Pl. Glum. 2: 33 (1855); F.T.A. 8: 368; J. & A. Raynal in Adansonia sér. 2, 7: 317 (1967). *C. longus* Linn. var. *pallidus* Boeck. in Linnaea 36: 280 (1870); Kük. l.c. 100. Perennial to 1 m. with slender rhizomes, leafy stems, long, gradually acuminate, leafy bracts and slender, rather few-spikeletted rays; marshes and edge of open water.
Sen.: *Heudelot* 523! Richard Toll (June) *Adam* 1419! M'bidgem (Nov.) *J. & A. Raynal* 6557! L. Guiers (June) *Roger*! Thillé to Boubakar (Dec.) *Naegelé* 13130! **Mali:** Bagoundié, nr. Gao (Mar.) *de Wailly* 5346! 5347! **N. Nig.:** Ngala *Gwynn* 116! Kalkala, nr. L. Chad (May) *Golding* 79! Northern parts of tropical Africa extending to India.
[The typical European form of *C. longus* Linn. seems not to extend to W. Africa, though the boundary between it and *C. fenzelianus* Steud. requires further study.]

11. **C. maculatus** *Boeck.* in Peters, Reise Mossamb. Bot. 2: 539 (1864); F.T.A. 8: 363, partly; Kük. l.c. 103; Berhaut, Fl. Sén. ed. 2, 385 (1967). *C. heudelotii* C.B. Cl. in F.T.A. 8: 364 (1901). *C. maculatus* Boeck. var. *contractus* Cherm. in Bull. Soc. Bot. Fr. 85: 368 (1938). Rhizomatous perennial with slender, leafy stems from hard, round tubers; leaves glaucous, slender; spikelets neat, pale, sometimes dark-spotted; sand banks and sandy ground near rivers.
Maur.: L. Rkiz (Oct., Nov.) *Adam* 21825! *Sadio* 300! **Sen.:** M'boro, Cayor (May) *Leprieur*! Taouee, L. Guiers (June) *Adam* 1403! Podor (June) *Adam* 1424! Tambacounda (Apr., June) *Berhaut* 1506! *Adam* 14299! **Mali:** Sotuba (Feb.) *J. & A. Raynal* 5462! Macina (May) *Lean* 20! Diafarabé (May) *Lean* 29! Timbuktu (Aug.) *Lean* 70! Gao to Bossobougou (July) *de Wailly* 5094! **Guin.:** Dabola (Apr.) *Pitot*! Dabola to Faranah (Apr.) *Adam* 4533! Faranah (Apr.) *Pitot*! **S. L.:** Makump (May) *Deighton* 1710! 1719! 1720! Njala (Feb.) *Deighton* 1573! **Iv. C.:** Abiati, lower Comoé R. (Mar.) *Chev.* 17533! **U. Volta:** Namenda, Tenkodogo (June) *Scholz* 316! **Ghana:** Saboba, Dagomba Dist. (Mar.) *Hepper & Morton* A 3125! Damongo to Tamale (Mar.) *Morton* GC 8766! Yeji (Apr.) *Hall* VBS 1238! Chindiri, Daka R. (Apr.) *Hall* VBS 1273! Kete Krachi (May) *Morton* GC 7181! **Dah.:** Savé to Agouagon (May) *Chev.* 23592! **Niger:** Tillabery *Rogeon* 515! **N. Nig.:** Sokoto (July) *Dalz.* 474! Shagunu (July) *Cook* 308! Nupe *Barter* 1571! Lokoja (Apr.) *Haines* 49! Abinsi (Mar.) *Dalz.* 811! Bama, Dikwa Div. (Dec.) *McClintock* 98! **S. Nig.:** Etemi, Omo R. (Mar.) *Jones & Onochie* FHI 16691! Olokemeji (Mar.) *Meikle & Keay* 1243! In E. Cameroun, throughout tropical Africa and in Madagascar.

12. **C. rotundus** *Linn.* Sp. Pl. 45 (1753); F.T.A. 8: 364; Kük. l.c. 107, t.13 (including subsp. *retzii* (Nees) Kük.); Cherm. in Arch. Bot. Caen 7, Mém. 4: 15 (1936); Berhaut, Fl. Sén. ed. 2, 387 (1967). Common

weed with slender, elongated rhizomes, rounded, fibre-covered tubers at the stem bases and large spikelets, characteristically dark red-brown, in short spikes; cultivated and other damp places.
Maur.: *Naegelé*! Akjoujt (Oct.) *Adam* 19465! Tamchakett (Aug.) *Monod* 2020! Oued Initi, W. of Oualata (Sept.) *Rossetti* 61/240*B*! **Sen.:** *Heudelot* 545! Lagbar, nr. Linguére (Sept.) *Adam* 12373! Dakar (June) *Adam* 1374! Darou, L. Tanma (June) *J. & A. Raynal* 5992! Oussouyé (Oct.) *Adam* 18314! **Mali:** Kayes (Feb.) *Duong*! Bamako (June) *Waterlot* 1192! Souinkoura (May) *Roberty* 2131! Dioura (Jan.) *Davey* 249! Gao to Berra (Aug.) *de Wailly* 5106! **Guin.:** Conakry (May) *Pitot*! **S. L.:** *T. Vogel* 76! Freetown (Oct.) *Deighton* 1773! Mt. Aureol, Freetown (June) *Gledhill* 603! Njala (June) *Deighton* 3012! Bonthe (Nov.) *Deighton* 2260! **U. Volta:** Farakoba, Bobo-Dioulasso Dist. (June) *Scholz* 322*a*! Ouagadougou to Laï (Aug.) *Chev.* 24691! Ouahigouya (July) *Scholz* 319! **Ghana:** Tamale to Damongo (Mar.) *Morton* GC 8765! Yapei (Mar.) *Adams* 3927! Nsawam to Achimota (Nov.) *Boughey*! Achimota (Nov.) *Irvine* 1521! Ada (June) *Morton* GC 9242! **Niger:** Niamey (July) *Rogeon* 529! Toukounous (Sept.) *Koechlin* 6455! *Bartha* 29! Dijdija (Sept.) *Koechlin* 6535! Dan Issa to Maradi (Aug.) *Hall* 435! **N. Nig.:** Edozhigi, nr. Bida (June) *Vaillant* 2797! Zaria (June) *Kershaw* 900350! Sokoto *Dalz.* 467! Katagum *Dalz.* 244! Maiduguri (June) *Clayton* 1084! **S. Nig.:** Lagos (July) *Haines* 58! Ibadan (Mar.) *Lowe* 2183! Pantropical and subtropical.

13. **C. tuberosus** *Rottb.* Descr. Pl. Nov. Programm. 18 (1772), and Descr. et Icon. 28, t.7 fig. 1 (1773); F.T.A. 8: 368. *C. rotundus* Linn. subsp. *tuberosus* (Rottb.) Kük. in Engl., Pflanzenr. Cyper. 113 (1935). Similar to *C. rotundus* Linn. but with somewhat laxer spikes of more numerous acute spikelets of longer, lanceolate, acute glumes; damp grassy and waste places.
Mali: Benenigeni, San (June) *Lean* 46! Djenné to Soufara (July) *Chev.* 1148! **Guin.:** Youkounkoun *Adam* 14830! **Ghana:** Tumu-Lawra boundary (May) *Morton* 7592! Akuse (July) *Hall* 3170! Nungua (Apr.) *Rose Innes* GC 30107! Oda to Tefli Ferry (Jan.) *Morton* GC 8262! **N. Nig.:** Nupe *Barter* 853! Dumbi, Kaduna Rd., Zaria Prov. (Apr.) *Kershaw* 900292! Kuffena, Zaria Prov. (June) *Kershaw* 900363! Lemme (May) *Lely* 149! **S. Nig.:** Lagos (July) *Kershaw* 900649! Ibadan (Apr.) *Lowe* 959! Oyo to Iseyin (Apr.) *Keay* FHI 37843! Gbongan, nr. Ife (Apr.) *Haines* 208! Port Harcourt (Sept.) *Stubbings* 42! Throughout Old World tropics.

14. **C. esculentus** *Linn.* Sp. Pl. 45 (1753); F.T.A. 8: 355; Kük. l.c. 116; Cherm. in Arch. Bot. Caen 7, Mém. 4: 13 (1936); Berhaut, Fl. Sén. ed. 2, 385 (1967). Generally distinguished from the other slenderly rhizomatous, tuber-bearing species by its golden, rather blunt spikelets, 1–2 cm. long and nerved sides to the glumes; weed in cultivated and waste places.
Maur.: Akjoujt (Oct.) *Adams* 19554! **Sen.:** Dagana *Heudelot* 518! Almadies (Aug.) *Broadbent* 23! Hann (Aug.) *Berhaut* 903! Mt. Roland, nr. Thiès (Aug.) *Broadbent* 52! Diaroumé, Casamance (Sept.) *Adam* 18176! **Gam.:** (July) *Ruxton* 14! Yundum Agric. Sta., nr. Bathurst (Aug.) *Ashrif* 38! Genieri (July) *Fox* 123! Wallikunda (Aug.) *Macluskie* 2! **Mali:** Kita Mts. (July) *Jaeger* 38! Bamako (June) *Waterlot* 1203! Sansanding (June) *Chev.* 950! Macina (July) *Roberty* 2542! Gao to Goura (Aug.) *de Wailly* 5122! **Guin.:** Conakry (May) *Paroisse* 223! Kindia *Jac.-Fél* 105! Kouroussa (July) *Pobéguin* 404! Mamou (Aug.) *Adam* 7944! **S. L.:** *T. Vogel* 31! Musaia (July) *Miszewski* 41! Konta to Kukuna (June) *Gledhill* 608! Torma (Mar.) *Bakshi* 78! Freetown (June) *Gledhill* 602! **Iv. C.:** Bouaké (July) *Chev.* 22133! **U. Volta:** Banfora (July) *Maire*! **Ghana:** Takoradi (May) *Morton* 2106! Cape Coast (May) *Hall* 1774! Ada (June) *Morton* GC 9241! Kwahu Tafo to Mankrong (Apr.) *Morton* A670! Yendi dam (Apr.) *Morton* 9073! **Togo Rep.:** Woumea (June) *Mahoux* 2090! Blitta (June) *Mahoux* 2094! **Dah.:** Djougou (June) *Chev.* 23939! **Niger:** 20 mls. from Eliki (Aug.) *Barry* 59! **N. Nig.:** Jebba (Apr.–May) *Kamphorst* 190! Anara F.R., Zaria (May) *Keay* FHI 22876! Bindawa, Katsina Prov. (Aug.) *Grove* 17! Bununun Kasa, Bauchi Prov. (June) *G. V. Summerhayes* 2! Maiduguri, Bornu (June) *Clayton* 1095! **S. Nig.:** Oke Iho to Iseyin, Oyo Prov. (Apr.) *Keay & Polunin* FHI 37840! Ibadan to Abeokuta (Mar.) *Schlechter* 13017! **W. Cam.:** Bambui, Bamenda (May) *Pedder* 18! Ndop to Bamenda (Dec.) *Boughey* GC 11383! Pantropical and in Madagascar and the Mediterranean region.

15. **C. bulbosus** *Vahl* Enum. Pl. 2: 342 (1806); F.T.A. 8: 352; Kük. l.c. 125; Cherm. l.c. 12; Berhaut, Fl. Sén. ed. 2, 337 (1967). *C. polyphylla* Vahl, Enum. Pl. 2: 317 (1806). Characterized by the scales surrounding the swollen base of the stem, numerous, slender, basal leaves nearly equalling the stems, and the dark red-brown spikelets generally in a single spike; dune slacks and damp sandy places.
Maur.: Tiguent to Nouakchott (Oct.) *Adam* 19396! **Sen.:** Makana, St. Louis (Sept.) *Fotius* 165! Almadies, nr. Dakar (Aug.) *Broadbent* 18! Mbao, Cape Verde (Aug.) *Pitot*! Hann (Aug.) *Berhaut* 2682! Malika (Aug.) *Broadbent* 44! Kaolak *Berhaut* 1028! **Guin.:** *teste* Cherm. **Ghana:** Korle Lagoon, Accra (July) *Hall* 3185! Labadi (Mar.) *Morton* GC 6657! New Ningo (May) *Morton* A 1680! Tema (June) *Hall* 2596! Keta (Aug.) *Akpabla*! **Niger:** Agadez *Chopard & Villiers*! Irabellaben, Baguezane Mts. *Chopard & Villiers*! **N. Nig.:** Kauwa, Bornu Dist. (Oct.) *Golding* 14*b*! North tropical Africa, through Arabia to N. Australia.

16. **C. sphacelatus** *Rottb.* Descr. Pl. Nov. Programm. 21 (1772), and Descr. et Icon. 26 (1773); F.T.A. 8: 346; Kük. l.c. 129; Cherm. l.c. 14; Berhaut, Fl. Sén. 2, 387 (1967). *C. locuples* C.B. Cl. in F.T.A. 8: 362 (1901). Generally characterized by the dark red-purple spot at the lower edge of the glume and the golden-green colour of the lower sheaths but both may be absent; weed of disturbed ground and in damp grassy places.
Maur.: L. Rkiz (Sept.) *Sadio* 302! **Sen.:** *Naegelé*! Kolda, Casamance *Raynal* 7805! **Mali:** Sindou (May) *Chev.* 873! Macina (Nov., Dec.) *Duong* 1310! **Guin.:** Faranah (Apr.) *Pitot*! Kissidougou to Gueckédou (Apr.) *Pitot*! Macenta (Apr.) *Pitot*! Sérédou to Nzérékoré (Apr.) *Pitot*! **S. L.:** Rokupr (Mar.) *Jordan* 16! Bonthe (Nov.) *Deighton* 2261! Njala (July) *Pyne* 174! Gboyama (Apr.) *Deighton* 1600! Tombo (Jan.) *Deighton* 997! **Lib.:** Bushrod I. (Apr.) *Barker* 1303! Monrovia (May) *Baldwin* 5805! Grand Bassa *Ansell*! Ganta, Sanokwele Dist. (May) *Harley* 947! Nimba Mts. (Mar.) *Adam* 21076! **Iv. C.:** Issia, Daloa to Abidjan (Dec.) *Boughey* 13592! Sassandra R., 14 km. from Sassandra (Apr.) *Leeuwenberg* 4008! Dabou (Dec.) *Giovannetti* 450! Adiopodoumé (Nov.) *Leeuwenberg* 2099! Abidjan (Dec.) *Pitot*! **U. Volta:** Banfora *Adam* 15389! **Ghana:** Burufo Spring, Lawra (Apr.) *Adams* 4047! Damongo (Mar.) *Morton* GC 8711! Kumasi (Nov.) *Bannerman-Bruce* C. D. A 4422! Atwabo *Fishlock* 14 of 1931! Kpeshi lagoon, Labadi (May) *Adams* 2731! **Togo Rep.:** Sokodé *Mahoux* 2064! Lomé *Warnecke* 129! **Dah.:** Grand Popo (May) *Roberty* 1570! Sémé (Sept.) *Adjanohoun* 479! **N. Nig.:** Sokoto *Dalz.* 463! Lokoja *Richardson* 11 (Apr.) *Haines* 47! Wushishi (June) *Clayton* 1104! **S. Nig.:** Lagos (Jan.–Mar.) *Haines* 92! *Hepper* 2264! Ibadan (Oct.) *Hambler* 73! Okomu F. R., Benin (Feb.) *Brenan* 9156! Ajagbododu (July) *Wright* 48! **W. Cam.:** Mamfe (Dec.) *Migeod* 273! Mankon (Apr.) *Brunt* 1120! Bota, nr. Victoria (Aug.) *Chuml* 124! **F. Po:** *T. Vogel* 13! (June) *Barter*! Throughout tropical Africa, and in India, Malesia and tropical America.

17. **C. dilatatus** *Schum. & Thonn.* Beskr. Guin. Pl. 38 (1827); F.T.A. 8: 375. *C. gracilinux* C.B. Cl. in J. Linn. Soc. 21: 162 (1884); F.T.A. 8: 362; F.W.T.A., ed. 1, 2: 485; Kük. l.c. 131; Cherm. in Arch. Bot. Caen 7, Mém. 4: 14 (1936); Berhaut, Fl. Sén. ed. 2, 385 (1967). *C. gracilinux* var. *platyphylla* C.B. Cl. in Mém. Soc. Bot. Fr. 2, 8: 27 (1907). Plant with many curved slender rhizomes, and generally neat, acute, flat spikelets with red-tinged lanceolate glumes; damp grassland and as a weed in cultivation.
Gam.: Yundum *Frith* 130! **Sen.:** Sébikotane (Sept.) *Pitot*! Sédhiou *Chev.* 2448. Kaolack to Guinguinéo (Aug.) *Trochain* 144! Richard-Toll *Berhaut* 1576. **Mali:** Baguineda (Aug.) *Roberty* 2665! Bamako *Adam* 14978! 14999! 15000! 15001! **Guin.:** Youkounkoun (Oct.) *Pitot*! Labé (June) *Adam* 14491! Kouroussa (July) *Pobéguin* 423! Kindia *Jac.-Fél.* 106! N'zérékoré to Beyla (May) *Pitot*! **S. L.:** Hastings (Aug.) *Melville & Hooker* 196! Konta (Aug.) *Deighton* 1242! Pendembu (July) *Thomas* 811! Koinadugu (Aug.) *Haswell* 237! Sefadu (June) *Fisher* 20! **Lib.:** Duport (Nov.) *Linder* 1447! Nimba Mts. (May) *Leeuwenberg & Voorhoeve* 4606*b*! **Iv. C.:** Téhini (Aug.) *de Wilde* 777! Mankono (Oct.) *Aké Assi* 7098! Adiopodoumé (July) *Boughey* 14498! **U. Volta:** Ouagadougou *Adam* 15458! (Apr.) *Nongonierma* 55!

Ghana: Yendi (Dec.) *Adams & Akpabla* 4092! Tamale (Mar.) *Morton* GC 8811! Ejura (Dec.) *Morton* GC 9728! Cape Coast (Apr.) *Hall* 1335! Accra (July) *Dalz.* 173! **Togo Rep.:** Bafilo (May) *Roberty* 1455! **N. Nig.:** Shagunu (July) *Cook* 310! Jebba to Mokwa (Aug.) *J. Hall* 296! Kaduna to Keffi (June) *Kershaw* 900625! Bauchi (Aug.) *Lely* P458! Yola (June) *Dalz.* 260! **S. Nig.:** Lagos (Oct.) *Haines* 44! Ifaki, Ondo Prov. (June) *Clayton* 1118! Mamu River F.R., Onitsha Prov. (May) *Lowe* 1008! Sapoba (Nov.) *Meikle* 586! Obudu Plateau *Tuley* 971! Throughout tropical Africa.

18. **C. tenuiculmis** *Boeck.* in Linnaea 36: 286 (1870); Kern in Reinwardtia 3: 28–31 (1954). *C. zollingeri* of F.T.A. 8: 360, partly; F.W.T.A., ed. 1, 2: 483; Kük. l.c. 133; Cherm. l.c. 14; not of Steud.

18a. **C. tenuiculmis** *Boeck.* var. **tenuiculmis.** Slender perennial with a short swollen stem-base and a large inflorescence with slender erect bracts and rays bearing a sparse spike of angular, yellowish spikelets with a zigzag rhachilla; grassland, swamps and sandy waste places.
Sen.: Sangalkam (June) *Berhaut* 1574! Djembering (Sept.) *Broadbent* 133! Bignona (Sept.) *Adam* 18065! Ziguinchor (Sept.) *Broadbent* 80! Kabrousse (Oct.) *Adam* 18372! **Mali:** Kita Mts. (Aug.) *Jaeger* 27! **Guin.:** Conakry (Oct.) *Adam* 12619! Friguiagbé (May, July) *Boismaré* 269! *Chillou* 2675! Hollandé Tossékré (Oct.) *Adam* 12724! Macenta (Apr.) *Pitot*! **S. L.:** Lumley to York (Aug.) *Harvey* 77! Bumban (Aug.) *Deighton* 1291! Mabum (Aug.) *Thomas* 1542! Njala (July) *Deighton* 745! Gbap (June) *Adames* 55! **Lib.:** Monrovia (May, Sept.) *Baldwin* 5820! *Wrigley & Melville* 710! Sinkor (June) *Barker* 1372! Nimba Mts. (July) *Leeuwenberg & Voorhoeve* 4606a! Sangwin (Mar.) *Baldwin* 11330! **Ghana:** Atwabo *Fishlock* 16 of 1931! Ayikuma to Doyum (May) *Hall & Bowling* GC 38513! Eikwe (Aug.) *Hall* 3374! **W. Cam.:** Bum, Bamenda, 5,000 ft. (May) *Maitland* 13a! Bali (Apr.) *Brunt* 1103! Throughout the Old World tropics.

18b. **C. tenuiculmis** var. **schweinfurthianus** (*Boeck.*) *Hooper* in Kew Bull. 26: 578 (1972). *C. zollingeri* Steud. var. *schweinfurthianus* (Boeck.) Kük. in Engl., Pflanzenr. Cyper. 134 (1935). *C. schweinfurthianus* Boeck. in Flora 62: 553 (1879); F.T.A. 8: 361; F.W.T.A., ed. 1, 2: 483; Cherm. in Arch. Bot. Caen 7, Mém. 4: 14 (1936). Differing from the typical *var.* in having the stem scabrid beneath the inflorescence and dense spikes of golden, angular spikelets; grassland, savanna and waste places.
Sen.: Kolda, Casamance (Oct.) *Adam* 18548! **Mali:** Kita *Duong* 407! (July) *Jaeger* 16! 23! Bamako to Ouagadougou *Adam* 15087! Ségou (Aug.) *Roberty* 3753! **Guin.:** Sambaïlo (Sept., Oct.) *Pitot*! Friguiagbé (Sept.) *Chillou* 630! Kissidougou to Guékédou (Apr.) *Pitot*! Nzérékoré (Apr.) *Pitot*! **S. L.:** Musaia (July) *Morton & Jarr* SL 2118a! Koinadugu (Aug.) *Haswell* 239! **Iv. C.:** Foro-foro Forest, 25 km. N. of Bouaké (Sept.) *Oldeman* 405! Bouaké (May) *de Wilde* 35! Cosrou (May) *Leeuwenberg* 4249! Dabou (Nov.) *Leeuwenberg* 1952! Adiopodoumé (Dec.) *Koechlin* 2328! **U. Volta:** 18 km. N. of Banfora (June) *Leeuwenberg* 4353! **Ghana:** Ejura (Aug.) *Andoh* 5039! Essiama (Oct.) *Hall* 2664! Abura, Cape Coast (Apr.) *Hall* 2291! Krobo Mt., Accra Plains (May) *Hall* 2317! Amedzofe (May) *Morton* GC 7220! **Togo Rep.:** *Kling* 71. Sokodé *Schilling* 15c. *Kersting* 104. **Dah.:** Savalu *Chev.* 23682. **N. Nig.:** Ahmadu Bello University, Zaria (Aug.) *Kershaw* 900412! Panshanu, Bauchi Prov. (Aug.) *Lawlor & Hall* 367! Minna (Jan.) *Keay* FHI 37322! Badeggi, Niger Prov. (June) *Onochie* FHI 18682! Ibaji Ojoko N.A.F.R., Kabba Prov. (June) *Latilo* FHI 47752! **S. Nig.:** Shaki (May) *Lagos Govt.* 61! Ilora turn, 2 miles S. of Oyo (July) *Keay & Stanfield* FHI 46318! Lagos *Millen* 107! Ifaki, Ondo Prov. (June) *Clayton* 1130! 10 miles E. Benin Rd., Owo Prov. (Apr.) *Haines* 56! **W. Cam.:** Ndop (May) *Brunt* 492! Throughout tropical Africa.

18c. **C. tenuiculmis** var. **guineensis** (*Nelmes*) *Hooper* in Kew Bull. 26: 583 (1972). *C. guineensis* Nelmes in Kew Bull. 6: 165 (1951); Nelmes & Baldwin in Amer. J. Bot. 39: 381, f. 64–69 (1952). Similar to var. *schweinfurthianus* but characterized by the spreading inflorescence and dark brown coloration of the sides of the glumes; montane grassland.
Guin.: Chutes de la Sala (Oct.) *Pitot*! Dalaba *Monod*! **S. L.:** Kamabai (Aug.) *Deighton* 1232! Perankonko, Loma Mts. (Sept.) *Jaeger* 1766! **Lib.:** Vonjama (Oct.) *Baldwin* 9901! Nimba Mts., 4,000–4,500 ft. (Sept.–Jan.) *Adam* 20076! 20539! *Adames* 507! 716! **Iv. C.:** Petit Tonkoui (Sept.) *Schnell*! **Ghana:** Tonogo (Nov.) *Morton* A 3476! Amedzofe (Nov.) *Morton* 9393! **S. Nig.:** Orosun Peak, Idanre Hills (Oct.) *Keay* FHI 22677! Obudu Plateau, 5,000 ft. (Aug.) *Kershaw* 900680! **W. Cam.:** Bamenda *Dept. of Agric.* 6! Also in E. Cameroun.

19. **C. permacer** *C.B. Cl.* in Mém. Bull. Soc. Bot. Fr. 2, 8: 27 (1907). *C. zollingeri* Steud. var. *permacer* (C.B. Cl.) Kük. in Engl., Pflanzenr. Cyper. 134 (1935). Slenderly rhizomatous perennial similar in habit to *C. tenuiculmis* but with cylindrical or oblong spikes of short spikelets with distinctly apiculate glumes; grassland, river banks and other damp places.
Guin.: Kandiara R. (Oct.) *Pitot*! La Sala (Oct.) *Pitot*! Timbi to Pita (Oct.) *Adam* 12554! Télimélé to Pita (Oct.) *Pitot*! Nimba Mts. (Sept.) *Schnell* 3592! **S. L.:** Port Loko (Oct.) *Jordan* 808! Batkanu (Sept.) *Glanville* 352! Koinadugu Dist. (Aug.) *Haswell* 233! Loma Mts. (Nov.) *Morton* SL 2694! Tingi Mts. (Dec.) *Morton & Gledhill* SL 3043! **Ghana:** Banda (Aug.) *Hall* 2034! **S. Nig.:** Olokemeji F.R. to Eruwa Rd. turning (Aug.) *Keay* FHI 37171!

20. **C. zollingeri** *Steud.* Synops. Cyper. 17 (1855), not of C.B. Cl.; Kern in Reinwardtia 3: 28 (1954). *C. rubroviridis* Cherm. in Bull. Soc. Bot. Fr. 66: 350 (1920); Kük. l.c. 135. *C. sphacelatus* Rottb. var. *tenuior* C.B. Cl. in F.T.A. 8: 347 (1901); F.W.T.A., ed. 1, 2: 485. *C. zollingeri* var. *parva* C.B. Cl. in F.T.A. 8: 361 (1901). Rare, caespitose annual about 20 cm. high with flaccid leaves and bracts and few, subquadrangular greenish spikelets; grassland.
Sen.: *Huard*! Ngalene *Heudelot* 485! **Ghana:** Kwahu Tafo (June) *Hossain & Enti* GC 35498! **Togo Rep.:** *Mahoux* 2064. Throughout tropical Africa, Madagascar, Malaysia and N. Australia.

21. **C. baoulensis** *Kük.* in Engl., Pflanzenr. Cyper. 467 (1936). *Pycreus baoulensis* A. Chev., Bot. 1: 695 (1920), name only. *Mariscus baoulensis* Hutch., F.W.T.A., ed. 1, 2: 486, English descr. only. Plant up to 1·2 m. tall with few, long rays, characterized by the very large, triangular, light brown glumes, spreading at maturity; marshy grassland.
Iv. C.: Tiégouakro to Kodiokoffi *Chev.* 22336! **Ghana:** Pong Tamale (Aug., Sept.) *Hall* CC 414! *Yamoah* SLUS 336! Tamale (Oct.) *Baldwin* 13560! 10 mls. N. of Ejura (Aug.) *Hall* CC 306! Oyibo (July) *Hall & Jenik* 3689! Ayikuma to Doyum (May) *Hall & Bowling* GC 38512!

22. **C. distans** *Linn. f.* Suppl. 103 (1781); F.T.A. 8: 349; Kük. l.c. 137; Berhaut, Fl. Sén. ed. 2, 383 (1967). Characterized by the slender, brown spikelets of small, distant glumes in a spreading compound umbel; in cultivated and waste ground and, generally damp, grassland.
Sen.: Dakar (June) *Adam* 1500! L. Tanma (June) *Adam* 17705! Sangalkam (July) *Pitot*! Thiès (Nov.) *Berhaut* 3989! Richard Toll (Oct.) *Berhaut* 2646! **Mali:** Kitaba, Kita Mts. (Oct.) *Jaeger* 76! **Guin.:** Conakry (Oct.) *Maclaud* 40! Kindia *Jac.-Fél.* 155! Fouta Djalon *Jac.-Fél.* 502! Faranah (Apr.) *Pitot*! Nzérékoré (May) *Pitot*! **S. L.:** Newton (June) *Deighton* 3008! Mayombo, nr. Roruks (July) *Deighton* 4677! Makene (Aug.) *Deighton* 1265! Mabonto (Aug.) *Deighton* 1395! Bumbuna (Oct.) *Thomas* 3339! **Lib.:** Zokatown, Tappita Dist. (Nov.) *Adam* 16139! Ganta, Sanokwele Dist. (Sept.) *Harley* 1011! Nimba Mts. (July) *Leeuwenberg & Voorhoeve* 4606! **Iv. C.:** Bouaké (May) *de Wilde* 65! Banco forest (Nov.) *Bégué* 88! Abidjan (Dec.) *Pitot*! Adiopodoumé (July, Dec.) *Boughey* 14502! *Leeuwenberg* 2126! **Ghana:** Medea to Kotoku (Apr.) *Morton* GC 7151! Hohoe (Apr.) *Morton* GC 9186! Nungua (May) *Rose Innes* GC 30115! Accra (July) *Irvine* 705! Prampram (Nov.) *Morton* GC 6153! **Dah.:** Cabolé (June) *Annet* 75! **N. Nig.:** Ilesha (Apr.) *Haines* 50! Abinsi (Nov.) *Dalz.* 820! **S. Nig.:** Umudike (July) *Ariwaodo* ARS 1278! Ibadan (Apr.) *Meikle* 1473! Shasha F.R. (Apr.) *Gledhill* 1001! Ikotekpene (June) *Guile*! Calabar *Maitland* 11! **W. Cam.:** Bota (Aug.) *Chuml* 126! Batibo (Apr.) *Brunt* 1058! Buea (Nov.) *Migeod* 59! Ndop, Bamenda (Dec.) *Boughey* GC 11381! Bamenda, 5,000 ft. (Jan.) *Migeod* 308! **F. Po:** Moka, 4,000 ft. (Dec.) *Boughey* 108! Pantropical.

23. **C. koyaliensis** *Cherm.* Arch. Bot. Caen 7, Mém. 3: 7 (1936); J. Raynal in Adansonia sér. 2, 6: 390 (1966); *C. immanis* Nelmes in Kew Bull. 6: 164 (1951); Nelmes & Baldwin in Amer. J. Bot. 39: 379, t.58–63 (1952). Tall "papyrus"-like sedge with slender drooping rays, those of the second or third order bearing clusters of slender spikelets; marshes.

S. L.: Masadou to Loma Mts. (Aug.) *Jaeger* 897! **Guin.:** Sambadougou to Boria (Jan.) *Chev.* 20553! Béro massif (June) *Adams* 5422! **Lib.:** Vonjama (Oct.) *Baldwin* 12044! **Iv. C.:** Disandougou to Nianouépleu (May) *Chev.* 21520! **N. Nig.:** Jos to Vom (Dec.) *Harris* 96! Jos *Batten-Poole* 390! (Aug.) *J. Hall* 604! Gurum, nr. Vogel Peak (Nov.) *Hepper* 1259! Also in E. Cameroun and C. African Republic.

[A specimen (*de Wit* 971) from the Ivory Coast is close to the S. American *C. prolixus* Kunth having spikelets 2·5 mm. broad and mucronate glumes 3 mm. long. It may be introduced.]

24. **C. atroviridis** *C.B. Cl.* in F.T.A. 8: 359 (1901); Kük. l.c. 141. *C. aterrimus* Steud. var. *atroviridis* (C.B. Cl.) Kük. l.c. 142. *C. aterrimus* of F.T.A. 8: 358; Kük. l.c. 141; not of Steud. Tufted sedge to **1**·3 m. high characterized by the secondary rays of the inflorescence (when present) spreading or deflexed and the elongated, subcylindric spikes of black and green slender spikelets; damp places.

W. Cam.: Mann's Spring, Cam. Mt. (Dec.) *Boughey* GC 10634! Bamenda, 8,000 ft. (Feb.) *Migeod* 492! Oku, 6,000 ft. (June) *Brunt* 634! Bafut-Ngemba F.R., 6,200 ft. (July) *Brunt* 835! **F. Po:** 8,500 ft. (Apr.) *Mann* 1466! Throughout tropical Africa.

25. **C. eleusinoides** *Kunth* Enum. Pl. 2: 39 (1837); Kük. l.c. 144; F.T.A. 8: 350. Stout, medium-sized plant with sulphur-yellow sheaths and congested spikes with mucronate, recurved glumes; damp places.

Niger: Myriah (Aug.) *P. de Fabrègues* 173! Also in north-east and east Africa, Madagascar, southern Asia and Australia.

26. **C. tonkinensis** *C.B. Cl.* in Kew Bull. add. ser. 8: 9 (1908). The typical variety comes from S.E. Asia.

26a. **C. tonkinensis** var. **baikiei** (*C.B. Cl.*) *Hooper* in Kew Bull. 26: 577 (1972). *C. baikiei* C.B. Cl. ex Kük. in Fedde, Rep. 21: 325 (1925); F.W.T.A., ed. 1, 2: 483; Kük. in Engl., Pflanzenr. Cyper. 148, t.17 (1935); Nelmes & Baldwin in Amer. J. Bot. 39: 381 (1952). *C. maculatus* of F.T.A. 8: 363, partly. *C. kottensis* Cherm. in Arch. Bot. Caen 4, Mém. 7: 23 (1931). Caespitose plant with a knotted mass of pseudobulbous, swollen stem-bases and a narrow inflorescence with clusters of erect, silvery, linear spikelets of numerous tiny glumes; sand-banks in rivers.

Sen.: Niokolo-Koba, Tambacounda (June, Dec.) *J. & A. Raynal* 6935! *Adam* 14300! Geneto (May) *Naegelé*! Vorouli, Gambia R. (Apr.) *Nongonierma*! **Mali:** San, Bani R. (Feb.) *Roberty* 3373! Ségou (Feb.) *Roberty* 3385! 3646! Mgomi, Mopti (June) *Lean* 34! **Guin.:** Dabola (Apr.) *Pitot*! Bissikrima (June) *Pobéguin* 1060! **S. L.:** Baiima (May) *Deighton* 5771! Bumban to Port Lokko (Apr.) *Sc. Elliot* 5676! Samaia (May) *Thomas* 214! Makump (May) *Deighton* 1711! **Lib.:** Baila, Sanokwele Dist. (May) *Harley* 1416! **Ghana:** Yeji, R. Volta (June) *Deighton* 713! Zabzugu to Yendi (Apr.) *Morton* GC 9070! **Dah.:** Savé to Agouagon (May) *Chev.* 23595! **N. Nig.:** Gurara Falls, Niger R. (Dec.) *Haines* 81! **S. Nig.:** Niger R. bank *Baikie*! Also in Chad, C. African Rep. and Congo.

27. **C. congensis** *C.B. Cl.* in Dur. & Schinz, Études Fl. Congo 1: 285 (1896); Kük. l.c. 115; J. & A. Raynal in Adansonia sér. 2, 7: 316 (1967). *C. eleusinoides* Kunth var. *dinklageanus* Kük. l.c. 145. *C. distans* Linn. f. var. *mucronatus* Berhaut, Fl. Sén. ed. 1, 231 (1954), and ed. 2, 383 (1967), French descr. only. Resembling *C. distans* L.f. but with longer, more crowded and spreading glumes in a generally more congested inflorescence, variable; river banks and similar damp places.

Sen.: Kayar *J. & A. Raynal* 6043! 6060! 6451! Kanouméri, Kédougou (July) *Fotius* K250! **S. L.:** Magbema, nr. Kambia (July) *Jordan* 463! Makump (May) *Deighton* 1716! Sumbuya to Serabu (Apr.) *Deighton* 1672! Kambia (June) *Gledhill* 618! **Lib.:** (May) *Harley* 1808! Cavally R., Webo Dist. (Apr.) *Dinklage* 2652! Also in Gabon and the Congo.

28. **C. iria** *Linn.* Sp. Pl. 45 (1753); F.T.A. 8: 346; Kük. l.c.: 150; Cherm. in Arch. Bot. Caen 7, Mém. 4: 12 (1936); Berhaut, Fl. Sén. ed. 2: 376, 385 (1967). Annual up to 75 cm. high, with obtuse, golden, very compressed spikelets in oblong, erect spikes; edge of open water.

Sen.: Djoudj, Senegal R. delta (Oct.) *Adam* 21743! Richard Toll (Sept., Oct.) *Leprieur* 1825! *Berhaut* 1416! Djolof (Aug.) *Naegelé* 11748! Bakel *Heudelot* 312! 329! **Gam.:** (July) *Ruxton* 23! **Mali:** Niger "delta" *Duong*! Dioura (Sept.) *Davey* 102! Faguibine to Bouna *Chev.* 2430! Gao to Goura (Sept.) *de Wailly* 4818! **U. Volta:** Ouagadougou to Lai (Aug.) *Chev.* 24693! 24703! **Ghana:** Natinga, Nakpanduri to Bawku (Aug.) *Hall* CC 534! **N. Nig.:** Damaturu, W. Bornu (Aug.) *De Leeuw* 1069! Old world tropics and subtropics extending into U.S.A.

29. **C. compressus** *Linn.* Sp. Pl. 46 (1753), emend. Dandy in Exell, Cat. S. Tomé 357 (1944); F.T.A. 8: 347; Kük. l.c. 156; Berhaut, Fl. Sén. ed. 2, 376, 380 (1967). Annual from 4–60 cm. high, with 1–6 digitate clusters of obtuse, compressed spikelets up to 3 cm. long with apiculate, green-keeled glumes; weed of waste places, often on sandy soil.

Sen.: Labgal (Oct.) *Naegelé* 12875! Sebikotane, Cap Verde (Oct.) *Pitot*! Oussouye to Ziguinchor (Sept.) *Broadbent* 114! Velingara (Oct.) *Adam* 18574! Tambacounda (Nov.) *Berhaut* 2681! **Guin.:** Conakry (Aug.) *Martine* 37! 48! 63! *Chillou* 3356! **S. L.:** Newton (Nov.) *Deighton* 1438! Mambolo (Sept.) *Jordan* 86! Mahera (Nov.) *Deighton* 5676! Njala (Sept.) *Deighton* 2082! Bonthe (Nov.) *Deighton* 2262! **Lib.:** Buchanan, Grand Bassa Dist. (Nov.) *Adam* 16034/1! Sinoe (Nov.) *Dinklage* 2338! Gbau, Sanokwele Dist. (Sept.) *Baldwin* 9451! Zwedru, Tchien Dist. (Aug.) *Baldwin* 7044! **Iv. C.:** Bingerville (Dec.) *Chev.* 16087! **Ghana:** Krisan, Axim (Sept.) *Cudjoe*! Abura (Mar.) *Hall* 2230! Cape Coast (May–June) *Hall* 1474! *Morton* A3668! Keta (Sept.) *Hall* 3417! **Togo Rep.:** Lomé (May) *Mahoux* 2022! Agblowé (Sept.) *Mahoux* 125! **N. Nig.:** Bacita (Nov.) *Hepper* 3909! **S. Nig.:** Jedda, Lagos (Dec.) *Haines* 83! Ijora, Lagos (July) *Haines* 39! Ajagbodudu, Benin (July) *Wright* 46! Warri (July) *Kershaw* 900678! Pantropical.

30. **C. incompressus** *C.B. Cl.* in F.T.A. 8: 348 (1901); Kük. l.c. 160. *C. trachycladus* Cherm. var. *multiflorus* of Berhaut, Fl. Sén. ed. 2, 384 (1967). *C. procerus* Rottb. var. *vanderystii* Kük. l.c. 92. Annual or perennial sedge 40–100 cm. high with wide, soft, triangular stems, long rays and spicately arranged large, obtuse, golden or pink tinged spikelets with obtuse membranous-edged glumes; rice fields, river-banks and near water.

Sen.: Karang (Nov.) *Berhaut* 1791! Sédhiou, Casamance (Oct.) *Adam* 18437! **Guin.:** Koba (Nov.) *Jac.-Fél.* 7216! **S. L.:** Kambia (Dec.) *Sc. Elliot* 4373! Gbap, Nongoba (Oct.) *Jordan* 619! Also in the western Congo and northern Madagascar.

31. **C. pustulatus** *Vahl* Enum. Pl. 2: 341 (1806); Kük. l.c. 161 (incl. var. *djalonis* (Chev.) Kük.). *Juncellus pustulatus* (Vahl) C.B. Cl. in Dur. & Schinz, Consp. Fl. Afr. 5: 546 (1895); F.T.A. 8: 307; F.W.T.A., ed. 1, 2: 489; Berhaut, Fl. Sén. ed. 2, 377, 381 (1967). Tufted annual 30–100 cm. high with few, rather unequal rays bearing clusters of oblong, often dark-spotted spikelets of broadly ovate, obtuse glumes; marshes and near open water.

Sen.: Fayil *Berhaut* 3932! Kolda (Oct.) *Adam* 18547! Ouassadou (Sept.) *Berhaut* 2677! Gouloumbo to Tambacounda (Oct.) *Pitot*! Niokolo-Koba (Oct.) *Adam* 15644! **Gam.:** (July) *Ruxton* 21a! Wallikunda (Oct.) *Macluskie* 32! **Mali:** *Scäetta* 3329! Koulikoro (Apr.) *Chev.* 2474! Macina (Nov.) *Duong*! **Guin.:** Sambailo (Oct.) *Pitot*! Seriba *Adam* 14803! Fouta Djalon *Chev.* 18613! Kouroussa (Aug.) *Pobéguin* 399! Lola to Nzo *Chev.* 20978! **S. L.:** Kambia (Aug.) *Harvey* 24b! Sendugu (June) *N. W. Thomas* 604! Njala (Aug.) *Deighton* 1836! Tisana (Nov.) *Deighton* 2321! Lunsar (Aug.) *Adames* 123! **Iv. C.:** Ferkessédougou (Nov.) *Leeuwenberg* 2038! Mankono *Aké Assi* 4808! Toumodi (Oct.) *Pobéguin* 81! Singrobo (Dec.) *Adjanohoun & Aké Assi* 4766! Baoulé Kodiokoffi *Chev.* 22333! **U. Volta:** Bedia to Barbotrom *Hb. ALF*! Banfora (Sept.) *Jaeger* 5245! Ouagadougou (Oct.) *Nongonierma* 201! **Ghana:** Burufo Spring, nr. Lawra (Apr.) *Adams* 4041! Gambaga (May) *Morton* GC 7464! Banda (Aug.) *Hall* 1961! Iture, nr. Elmina (Aug.) *Hall* 2528! Kpeshi lagoon (May) *Adams* 4288! **Togo Rep.:** Koumea (June) *Mahoux* 2089! Bassari to Sokodé (Aug.) *Mahoux*! **N. Nig.:** Gawu Hills, Gwari (Aug.) *Onochie* FHI

35932! Shagunu (July) *Cook* 325! Kufena Hill, Zaria (Sept.) *Clayton* 1322! Jos (July) *Lawlor & Hall* 16! Nupe *Barter* 1563! Benue R., Yola (Sept.) *Kennedy* FHI 7287! Damaturu, W. Bornu (Aug.) *De Leeuw* 1070! **S. Nig.:** (Jan.) *Dodd* 390! Shaki, nr. Lagos *Macgregor* 164! Ado Rock, Ado-Awaiye, Ibadan Dist. (May) *Onochie* FHI 31215! Throughout tropical Africa.

[Very slender, perhaps depauperate specimens corresponding to the var. *debilis* Kük. occur on the margins of lateritic hard pans in Sierra Leone (*Jordan* 884 etc.) and have 1–3 spikelets only.]

32. **C. podocarpus** *Boeck.* in Flora 62: 551 (1879); F.T.A. 8: 333; C.B. Cl., Illustr. Cyper. t.10 (1909); Kük. l.c. 162; Berhaut, Fl. Sén. ed. 2, 377, 381 (1967). Slender annual, variable in size, with few, narrow leaves, an erect bract and 1–3 digitate clusters of acute, terete spikelets with broad-keeled glumes on a wide rhachilla, curved at maturity; marshes and beside pools on rock outcrops.

Sen.: Sangalkam *Berhaut* 2669! Ouassadou *Berhaut* 1566! Kanéméré (Sept.) *Fotius* K492! Tambacounda *Berhaut* 3183! Tambacounda to Gouloumbo (Oct.) *Pitot*! **Mali:** Kita (July) *Jaeger* 14! Bamako *Waterlot* 1291! Bougouni *Scaëtta* in *Hb. Roberty* 3130! Ségou (Sept.) *Roberty* 2707! Boré (July) *Demange* 31/1957! **Guin.:** Seriba *Adam* 14802! Télimelé (Oct.) *Pitot*! Medina to Nyvelande *Adam* 14624! Kouroussa (Aug.) *Pobéguin* 400! Kankan to Bougouni *Scaëtta* 3165! **S. L.:** Kambia (Aug.) *Harvey* 24c! **Iv. C.:** Ouango–Fitini (July) *Aké Assi* 4828! Bouna Reserve, Kakpin (June) *Aké Assi* 10301! **U. Volta:** Sikasso to Bobo-Dioulasso *Adam* 15104! Ouagadougou to Ouahigouya (Aug.) *Chev.* 24720! **Ghana:** Burufo (Sept.) *Hall* CC 727! Lawra (Sept.) *Hall* CC 703! **N. Nig.:** (Aug.) *Lely* P464! 80 mls. N. of Gusau, Sokoto Prov. (July) *De Leeuw* 912! Bindawa, Katsina Prov. (Aug.) *Grove* 42! Kan Gimi, Zaria Prov. (May) *Keay* FHI 22897! Anara F.R., Zaria Prov. (July) *Clayton* 1227! 1258! Also in central and north-east tropical Africa.

[Very variable in size, having a depauperate form with one to a few spikelets similar to the var. *debilis* of the previous species from which it can be distinguished by the much more acute, often curved, spikelets.]

33. **C. alternifolius** *Linn.* Mant. 2: 28 (1771). *C. flabelliformis* Rottb., Descr. et Icon. 42, t.12 f.2 (1773); F.T.A. 8: 336. Robust, leafless, triangular stems to 2 m. high with numerous, leafy bracts and subequal rays bearing clusters of elliptical, acute, shining brown spikelets; cultivated.

Sen.: Dakar (Apr.) *Adam* 1113! 1114! **S. L.:** Njala (Nov.) *Deighton* 3785! **Ghana:** Achimota (Mar.) *Akpabla* 2007! Aburi (Apr.) *Irvine* 2041! **W. Cam.:** Mouth of Limba R. (Mar.) *Brunt* 1041! Introduced; native in Madagascar, east tropical and S. Africa and introduced in other parts of the tropics.

34. **C. pseudoleptocladus** *Kük.* in Fedde, Rep. 29: 196 (1931). *C. deckenii* of F.T.A. 8: 342 partly, not of Boeck. Similar to *C. haspan L.* but with somewhat longer, darker spikelets with longer apiculate glumes; near water.

W. Cam.: Musake, Buea, 6,000 ft. (Nov.) *Migeod* 127! L. Bambuluwe to Bamuko (July) *Daramola* FHI 41582! Pinyin, 5,500 ft. (Apr.) *Brunt* 1097! Also in east tropical Africa.

35. **C. mannii** *C.B. Cl.* in F.T.A. 8: 341 (1901). *C. baronii* C.B. Cl. var. *mannii* (C.B. Cl.) Kük. in Engl., Pflanzenr. Cyper. 202 (1936). Leafy perennial to 150 cm. high with an irregular, thrice-branched umbel bearing small clusters of brown spikelets, with paler, less clearly apiculate glumes and fewer bracts than the preceding species; montane forest.

S. L.: Bintumane Peak, Loma Mts. (Oct.–Jan.) *T. S. Jones* 97! 131! *Jaeger* 306! 495! Da-Oulen Peak, Loma Mts. (Aug.) *Jaeger* 1346! **Lib.:** Nimba Mts. (Dec.) *Adam* 20077! **N. Nig.:** Bellel, Mambila Plateau (Jan.) *Hepper* 1693! **W. Cam.:** Cam. Mt., 7,000 ft. (Jan.) *Mann* 1358! Mann's Spring, 6,650 ft. (Feb., Mar.) *Adams* GC 11740! *E. W. Jones* 9524! Buea, 5,000 ft. (Jan.) *Maitland* 347! Bambui (Dec.) *Boughey* GC 10778! 4,000 to 7,000 ft. *Mann* 320! 1479! **F. Po:** Biao, 6,400 ft. (Sept.) *Wrigley & Melville* 474! El Pico, 6,000 ft. (Dec.) *Boughey* 125! Moka (Dec.) *Boughey* 64!

[Perhaps to be equated with one of the east African species of this section of the genus.]

36. **C. renschii** *Boeck.* in Flora 65: 11 (1882); F.T.A. 8: 345; Kük. l.c. 206. Robust perennial with a much-branched, lax inflorescence with broad bracts and very numerous small, round clusters of few-flowered dark spikelets with recurved tips to the glumes; secondary or gallery forest, by streams.

Guin.: Sébori, Fouta Djalon (Oct.) *Adam* 12674! **S. L.:** Yetaya (Sept.) *Thomas* 2444! **Lib.:** Kitoma, Sanokwele Dist. (Mar.) *Adam* 16810! **Iv. C.:** Sassandra to Beyo (Jan.) *Leeuwenberg* 2551! Issia, Daloa to Abidjan (Dec.) *Boughey* GC 13612*b*! **Ghana:** *Burton*! Wenchi to Techiman, Ashanti (Mar.) *Morton* GC 8556! Kwahu (Apr.) *Johnson* 689! **S. Nig.:** Obudu Plateau, 5,000 ft. (Apr.) *Haines* 201! Sonkwala, Obudu Div. (Dec.) *Savory & Keay* FHI 25185! **W. Cam.:** Cam. Mt., 2–3,000 ft. (Dec.) *Mann* 2103! Wum (Jan.) *Brunt* 925! Kumbo, 5,500 ft. (Feb.) *Hepper* 1977! Ashong, Batibo, 5,100 ft. (Apr.) *Brunt* 1090! Ndop (Dec.) *Adams* GC 11058! Throughout tropical Africa.

37. **C. diffusus** *Vahl* Enum. Pl. 2: 321 (1806). The typical subspecies occurs in India and south-east Asia.

37a. **C. diffusus** subsp. **buchholzii** (*Boeck.*) *Kük.* in Engl., Pflanzenr. Cyper. 210 (1936). *C. buchholzii* Boeck., Beitr. Kenntn. Cyper. 1: 3 (1883). *C. diffusus* of F.T.A. 8: 343; F.W.T.A., ed. 1, 2: 483. Leafy, caespitose perennial 25–50 cm. high with purple leaf-sheaths, 4–8 leafy bracts and a 2–3 times umbellate inflorescence with small clusters of few-flowered spikelets with spreading glumes; path sides and other openings in forest.

Guin.: Kindia *Jac.-Fél.* 144! Benty (June) *Chillou* 2652! Macenta (Apr.) *Pitot*! Ziama (Apr.) *Adam* 4790! Nimba (Dec.) *Adam* 7374! **S. L.:** Koinadugu Dist. (Aug.) *Haswell* 203! Rokupr (Sept.) *Jordan* 313! Makump (July) *Thomas* 994! Njala (July) *Deighton* 748! Male R. Ferry, Segbwema to Kenema (July) *Deighton* 3751! **Lib.:** *Harley* 1407! Tappita (Aug.) *Baldwin* 9081! Baila, Sanokwele Dist. (May) *Harley* 1407! Webo on Cavally R. (Apr.) *Dinklage* 2650! Nimba, 1,100 m. (May) *Adam* 25680! **Iv. C.:** Mankono (July) *Chev.* 22003! Sinfra (Dec.) *Aké Assi* 7215! Issia (Aug.) *Boughey* 14679! Sassandra (Apr.) *Leeuwenberg* 4011! Lamé (Nov.) *Leeuwenberg* 1903! Adiopodoumé (Mar.) *Bernardi* 8569! **Ghana:** Ejura Scarp (Mar.) *Adams* 2444! Cape Coast (June) *Hall* 1476! Mpraeso Scarp, Kwahu (Dec.) *Adams* 5176! Aburi (June, July) *Johnson* 195! *Morton*! Volta River F.R. (Nov.) *Morton* 6065! **Togo Rep.:** Misahöhe *Baumann* 222. **S. Nig.:** Ijaiye F.R., Oyo Dist. (May) *Onochie* FHI 19170! Lagos *Millen* 182! Omo F.R., Ijebu Dist. (Apr.) *Onochie* FHI 15537! Okelifi, Ondo Dist. (Nov.) *Onochie* FHI 34227! Owo (Apr.) *Haines* 41! Calabar to Atimbo (Feb.) *Onyeachusim & Latilo* FHI 48173! **W. Cam.:** *Kingsley*! Mamfe to Victoria (Apr.) *Morton* GC 7122! Bimbia Rubber Plantation, Victoria (Aug.) *Chuml* 159! Mbalange F.R., Kumba (Jan.) *Binuyo & Daramola* FHI 35472! **F. Po:** *T. Vogel*! Also in E. Cameroun, central and east tropical Africa.

[Not uncommonly proliferous, the achenes being replaced by green shoots giving the inflorescence a witches'-broom appearance.]

38. **C. fertilis** *Boeck.* in Engl., Bot. Jahrb. 5: 90 (1884); F.T.A. 8: 341; C.B. Cl., Illustr. Cyper. t.12 (1909); Kük. l.c. 227. Distinctive small sedge with short stems, generally less than 15 cm., usually much exceeded by the flaccid rays subtended by broad, leaf-like bracts; river-banks, track-sides and other damp places in forest.

Lib.: Truo, Sinoe Dist. (Mar.) *Baldwin* 11439! Mt. Tokadeh Peak, Nimba Mts. (June) *Adam* 21439! Yratoke, Webo Dist. (July) *Baldwin* 6237! **Iv. C.:** Oubi to Kéeta, Cavally R. (July) *Chev.* 19315! Davo to Soubré, Sassandra R. (May) *Chev.* 17986! N'zida (Dec.) *Mangenot, Miège & Aké Assi* 1741! Davo R., E. of Beyo (Jan.–Feb.) *Leeuwenberg* 2607! 2847! Sékré, 15 km. E. of Béréby (Nov.) *Oldeman* 540! **Ghana:** Tanoso (Feb.) *Bowling* GC 36645! **S. Nig.:** Epe (Aug.) *J.B. Gillett* 15263! Ajebandele to Akilla, Ijebu Ode Dist. (Apr.) *Tamajong* FHI 16838! Orisunbare, Omo F.R. (Apr.) *Jones & Onochie* FHI 17245! Sapoba F.R. (Oct.) *Ujor* FHI 23924! Umondealichi F.R., Itu Dist. (May) *Onochie* FHI 33229! **W. Cam.:** Likomba (Dec.) *Mildbraed* 10772! Banga, S. Bakundu F.R. (Mar.) *Brenan* 9285! Maikoko to Mbalange, S. Bakundu (Jan.) *Binuyo & Daramola* FHI 35063! S. Bakundu F.R. (Apr.) *Daramola* FHI 29825! Also in E. Cameroun, central and western Africa to Angola.

39. **C. mapanioides** *C.B. Cl.* in F.T.A. 8: 340 (1901); Kük. in Engl., Pflanzenr. Cyper. 230, incl. var. *major* (Boeck.) Kük. (1936). *C. dichromenaeformis* Kunth var. *major* Boeck.—F.W.T.A., ed. 1, 2 : 482. Leafy plant up to 70 cm., having broad bracts surrounding a head of sessile, pale spikelets with ovate, acute, ciliate glumes; forest and cultivated land.
Guin.: Gangan (May) *Roberty* 17788! Kindia *Jac.-Fél.* 143! Tonta *Desmeules & Jac.-Fél.* 41! Dalaba (Oct.) *Chev.* 18826! Labé (June) *Adam* 14598! **S. L.:** Sinkunia (July) *Morton & Jarr* SL 2141! Gberia Fotombu (Oct.) *Small* 413! **Iv. C.:** Gro, Bouaké Dist. (Aug.) *Aké Assi* 6832! Flansobly Forest (July) *Aké Assi* 9065! **Togo Rep.:** (July) *Mahoux* 2063! **N. Nig.:** Acharane, Kabba Prov. (June) *Daramola* FHI 38035! Dogon Kurmi F.R., Plateau Prov. (July) *Okafor & Binuyo* FHI 47520! **S. Nig.:** Abakaliki (June) *Nwauzo in Hb. Tuley* 793! Umudike (July) *Nwauzo in Hb. Tuley* 789! Ewohimi to Idumuje, Benin Prov. (Aug.) *Onochie* FHI 33271! **W. Cam.:** Bamenda, 5,000 ft. (May) *Maitland* 12a! Throughout tropical Africa.

40. **C. dichroöstachyus** *Hochst. ex A. Rich.* Tent. Fl. Abyss. 2: 481 (1851); F.T.A. 8: 331; Kük. l.c. 233. Leafy plant with a thin rhizome and very small blackish and green spikelets; wet ground.
W. Cam.: Madele, Wum Dist. (June) *Daramola* FHI 41098! Cheddar Gorge, Bamenda (Mar.) *Morton* K188! Bamenda, 5,000 ft. (May) *Brunt* 1138! Bambuluwe, nr. Santa (July) *Bumpus* July 46/Bam/28! Throughout tropical Africa, Madagascar and Asia.

41. **C. difformis** *Linn.* Cent. Pl. 2: 6 (1756), and in Amoen. Acad. 4: 302 (1760); F.T.A. 8: 330; Kük. l.c. 237, t.27, incl. var. *subdecompositus* Kük. (1936); Cherm. in Arch. Bot. Caen 7, Mém. 4: 8 (1936); Berhaut, Fl. Sén. ed. 2, 380 (1967). Annual up to 90 cm. high with soft, triangular stems, leafy bracts, and spherical clusters of very small, neat, pale spikelets; rice fields, roadside ditches and other damp places.
Sen.: Séléty, Casamance (Sept.) *Adam* 18215! Messira (Nov.) *Berhaut* 2674! Messira to Kaolack (Aug.) *Trochain* 4007! Dakar (May) *Baldwin* 5715! Richard-Toll (Oct.) *Berhaut* 2673! **Gam.:** (Aug.) *Ruxton* 26! 125! Wallikunda *Macluskie* 32a! **Mali:** Kati (Jan.) *Chev.* 181! Bamako (Mar.) *Waterlot* 1099! Koutiala (Feb.) *Roberty* 3411! Sikasso (Feb.) *Roberty* 3448! Gao (Sept.) *de Wailly* 4809! **Guin.:** Friguiagbé (June) *Chillou* 532! Mamou (May) *Pitot*! Faranah (Apr.) *Pitot*! Kouroussa (Jan.) *Pobéguin* 625! Koba *Jac.-Fél.* 7242! **S. L.:** Mahera (Oct.) *Glanville* 417! Rokupr (Sept.) *Jordan* 90! Falaba (Mar.) *Sc. Elliot* 5158! Pujehun to Gardohun (Apr.) *Deighton* 1642! Gola (Feb.) *Bakshi* 32! **Lib.:** Monrovia (July) *Kunkel* 65! Salala (Aug.) *Voorhoeve* 398! Gbanga (Sept.) *Linder* 707! Tappita (Aug.) *Baldwin* 9096! Ganta (May) *Harley* 951! **Iv. C.:** Bouaké (May) *de Wilde* 91! Baoulé (Sept.) *Pobéguin* 102! Sinfra (Dec.) *Aké Assi* 7217! Songon Agban to Dabou (Feb.) *Leeuwenberg* 2641! **U. Volta:** Ouagadougou (Oct.) *Nongonierma* 210! **Ghana:** Burufo Spring, Lawra (Apr.) *Adams* GC 4051! Damongo (Dec.) *Morton* GC 25050! Ejura (Dec.) *Morton* GC 9577! Kumasi (Mar.) *Hall* 2506! Kpong (Mar.) *Asiedu-Gyadu* GC 20553! **Togo Rep.:** (May–June) *Mahoux* 2019! 2086! **Dah.:** Savalou (June) *Annet* 10! Massé to Kétou (Feb.) *Chev.* 23011! **Niger:** Zinder (Feb.) *Chev.* 43640! Toukounous (Sept.) *Koechlin* 6515! **N. Nig.:** Sokoto (Aug.) *Dalz.* 460! Argungu, Sokoto Prov. (July) *Latilo* FHI 62755! Kuffena, Zaria Prov. (Jan.) *Kershaw* 90020! Shendam, Plateau Prov. (Nov.) *Clayton* 1462! Dikwa, Bornu Prov. (Oct.) *Johnston* N53! **S. Nig.:** Ilesha (Apr.) *Haines* 42! Olokemeji (Nov.) *Onochie* FHI 8121! Oyo to Iseyin (Oct.) *Onochie* FHI 34921! Ibadan (Feb.) *Brenan* 8951! Lagos *Dawodu* 37! Pantropical and subtropical.

42. **C. reduncus** *Hochst. ex Boeck.* in Linnaea 35: 580 (1868); F.T.A. 8: 329; C.B. Cl., Ill. Cyper. t.9 (1909); Kük. l.c.: 240; Meneses in Garcia de Orta 4: 560 (1956). Weak annual up to 30 cm. high with soft, flat leaves and bracts and broad yellow-green spikelets with conspicuously spreading glumes ending in recurved green tips; marshy savanna, rice fields and other wet places.
Sen.: *Heudelot* 172! Kolda (Oct.) *Adam* 18545! Ouassadou (Nov.) *Berhaut* 1517! Niokolo-Koba (Nov.) *Adam* 16992! Moulessi, Tambacounda to Kédougou (Dec.) *J. & A. Raynal* 6744! **Mali:** Nafadié (Dec.) *Chev.* 2427! Goundaka (Dec.) *Monod*! Fafa, Gao (Aug.) *de Wailly* 5370! **Port. G.:** Bafata (Dec.) *Esp. Santo* 2854. **Guin.:** Koulikoro (Oct.) *Chev.* 2456! Kouroussa *Pobéguin* 620 (partly)! Beyla to Sinko *Adam* 7234! **Iv. C.:** Maramana, Boundiali (Nov.) *Aké Assi* 8273! **Ghana:** Kamba, Lawra (Nov.) *Harris*! Tumu (Dec.) *Adams & Akpabla* 4369! Pong-Tamale (Nov.) *Akpabla* 383! Ejura (Dec.) *Morton* GC 9575! Kumasi (Oct.) *Baldwin* 13461! **Togo Rep.:** Sokodé *Kersting* 259. **N. Nig.:** Jebba (Dec.) *Haines* 65! Kabba (Oct.) *Parsons* 31! Gwari Hill, Minna (Dec.) *Meikle* 707! Samaru (Sept.) *Kershaw* 900426! Bauchi Plateau (Oct.) *Lely* P787! Gurum, Vogel Peak (Nov.) *Hepper* 1243! **S. Nig.:** Abakaliki (Nov.) *Tuley* 4! Ogurude, Cross R. (Jan.) *Holland* 257! Also in central, north-east and east Africa.

43. **C. submicrolepis** *Kük.* in Engl., Pflanzenr. Cyper. 241 (1936). *C. microlepis* Boeck. in Flora 62: 551 (1879); F.T.A. 8: 330; F.W.T.A., ed. 1, 2: 483; Berhaut, Fl. Sén. ed. 2, 380 (1967); not of Baker. Tufted annual with purple roots close to *C. difformis* Linn. but with narrower glumes and more flattened achene; edge of pools chiefly on laterite.
Sen.: *Naegelé!* Tambacounda (Sept.) *Berhaut* 3286! Tambacounda to Gouloumbo (Oct.) *Pitot*! Ouassadou *Berhaut* 1567! 3036! **Mali:** Kita (Sept.) *Jaeger* 52! Bamako *Adam* 14951! Koulikoro *Chev.* 2469! Bamako to Ségou (Sept.) *Jaeger* 5111! **Guin.:** Thianguel-Bori (Oct.) *Pitot*! Ditinn (Oct.) *Pitot*! **Iv. C.:** Bouna Reserve, Kakpin (June) *Aké Assi* 10229! **U. Volta:** Sikasso to Ouagadougou *Adam* 15442! **Ghana:** Burufo (Sept.) *Hall* CC 744! Lawra (Sept.) *Hall* CC 717! Kampaha (June) *Adams* 710! Nakpanduri to Bawku (Aug.) *Hall* CC 536! Damongo scarp (Sept.) *Hall* CC 846! **N. Nig.:** Jebba *Barter*! Shagunu (July) *C. D. K. Cook* 322! 323! Idon, Kaduna to Keffi (Aug.) *Kershaw* 900710! Anara F.R., Zaria (July) *Keay* FHI 25965! Throughout tropical Africa.

44. **C. remotispicatus** *Hooper* in Kew Bull. 26: 577 (1972). Resembling *C. tenuispica* but distinguished by the solitary greenish spikelets which remain narrow at maturity; damp places, generally on disturbed ground.
Sen.: Velingara (Oct.) *Adam* 18582! **Iv. C.:** Aboakouamékro (Nov.) *Aké Assi* 10351! **Ghana:** nr. Zolo-Kpuita (Nov.) *Morton* GC 9392! Achimota (July, Aug.) *Adams* 4890! 4903! Damongo (Dec.) *Morton* GC 25059! **N. Nig.:** Kainji (Nov.) *Hepper* 3900! **S. Nig.:** Ezillo (Oct.) *Ariwaodo* in *Hb. Tuley* 958!

45. **C. tenuispica** *Steud.* Syn. Pl. Glum. 2: 11 (1855); Kük. l.c. 245; Cherm. in Arch. Bot. Caen 7, Mém. 4: 8 (1936); Berhaut, Fl. Sén. ed. 2, 381 (1967). *C. flavidus* of F.T.A. 8: 333, not of Retz. (1789). Weak annual 10–50 cm. high with flaccid leaves, a diffuse inflorescence with rays often as long as the weak angular stems, and neat elliptical spikelets often 2 or 3 together; rice fields, marshy savanna and other damp places.
Maur.: L. Rkiz (Sept.) *Sadio* 298! **Sen.:** Fatick *Maymard*! Richard-Toll (Feb.) *Döllinger* 14! Mbidjem (Nov.) *J. & A. Raynal* 6563! Niokolo-Koba *Adam* 15726! Kolda, Casamance (Oct.) *Adam* 18549! Bakel *Heudelot* 325! **Mali:** Kita Mts. (Oct.) *Jaeger* 89! Kara (Dec.) *Davey* 187! Fafa, Gao (Mar.) *de Wailly* 5372! **Port. G.:** Buruntuma (Dec.) *Pereira* 216! **Guin.:** Sambaïlo (Oct.) *Pitot*! Kiffaya to Seriba (Oct.) *Pitot*! Koba (Nov.) *Jac.-Fel.* 7241! Thiangeul bori (Oct.) *Pitot*! Mamou to Senegal frontier *Adam* 7937! **S. L.:** Lumley Beach, Freetown (Sept.) *Harvey* 136! Rokupr (Sept.) *Jordan* 316! Ribi R. (Feb.) *Morton & Gledhill* SL 1719! Njala (Nov.) *Morton & Jarr* SL 3923! Sandra Tenraron, nr. Mateboi (Nov.) *Jordan* 969! **Lib.:** nr. Monrovia (Aug.) *Naumann* 25. **Iv. C.:** N'Douci to Toumodi (Aug.) *de Wit* 7345! Aboakouamékro (Nov.) *Aké Assi* 10350! Adiopodoumé (July) *Boughey* GC 14507! **Ghana:** Tumu (Dec.) *Adams & Akpabla* 4371! Sakogu, Gambaga Scarp (Dec.) *Adams* 4245! Tamale (Aug.) *Irvine* 4583! Tafo (July) *Irvine* 4973! Accra to Aburi (Oct.) *Johnson* 1003! **Togo Rep.:** Lilikopé *Mahoux* 110! Atakpamé to Lomé *Mahoux* 2005! **N. Nig.:** Rofia, nr. Yelwa *C. D. K. Cook*! Badeggi, nr. Bida (Jan.) *Vaillant* 2778! Kaduna to Keffi, Zaria Prov. (Aug.) *Kershaw* 900703! Panshanu, Bauchi Prov. (Aug.) *Lawlor & Hall* 619! Damaturu, W. Bornu (Sept.) *De Leeuw* 1212! Old world tropics.

[The typical form of this species from India, Australia and elsewhere in the Far East has spikelets less than 1 mm. wide. Many of the West African collections have larger spikelets (1·5 mm. wide) and may represent a distinct taxon but the position requires further investigation.]

46. **C. haspan** *Linn.* Sp. Pl. 45 (1753); C.B. Cl. in J. Linn. Soc. 21: 119 (1884); F.T.A. 8: 332; Kük. in Engl., Pflanzenr. Cyper. 247, incl. var. *flaccidissimus* Kük. (1936); Cherm. in Arch. Bot. Caen 7, Mém. 4: 8 (1936); Berhaut, Fl. Sén. ed. 2, 379, 381 (1967). Tufted perennial with weak angular stems, purple sheaths, two unequal bracts, and a spreading inflorescence with near linear spikelets in groups of 1–3; open, moist places.
Sen.: Dakar (Jan.) *Adam* 409! M'bidjem to Cayar (Apr.) *Trochain* 3174! Mont Roland, Thiès (Mar.) *de Wailly* 4505! Niokolo-Koba *Adam* 15914! Diouloulou, Casamance (Sept.) *Adam* 18206! **Gam.:** (Aug.) *Ruxton* 99! 121! **Mali:** Koulouba *Duong*! Bamako (July) *Waterlot* 1213! Sotuba (Nov.) *J. Raynal* 5165! Bamako to Ségou (Apr.) *Rogeon* 35! Macina (July) *Pitot*! **Guin.:** Kindia *Jac.-Fél.* 103! Labé (Apr.) *Pitot*! Mali (Dec.) *Schnell* 2357! Macenta (Apr.) *Pitot*! Nzérékoré (Apr.) *Pitot*! **S. L.:** Lunsar (Aug.) *Adames* 126! Rokupr (Mar.) *Jordan* 12! Bintumane Peak, 5,400 ft. (Jan.) *T. S. Jones* 75! Njala (Sept.) *Deighton* 2098! Mano Salija (Nov.) *Deighton* 324! **Lib.:** Marshall, Montserrado Dist. (Feb.) *Baldwin* 11050! Tawata, Boporo Dist. (Nov.) *Baldwin* 10314! Gbanga *Linder* 698! Vonjama (Oct.) *Baldwin* 9931! Yéképa, Nimba Mts. (Jan.) *Adam* 20694! **Iv. C.:** Bouaflé to Sinfra (Dec.) *Aké Assi* 7227! Tiégouakro to Kodiokoffi (Aug.) *Chev.* 22335! Baoulé (Sept.) *Pobéguin* 85! Songon Agban to Dabou (Feb.) *Leeuwenberg* 2643! Anguédédou Forest, nr. Abidjan (Dec.) *Leeuwenberg* 2289! **U. Volta:** Toussiana waterfall, Bobo-Dioulasso (June) *Scholz* 308! **Ghana:** Burufo (Dec.) *Adams & Akpabla* 4392! Damongo (Mar.) *Morton* GC 8701! Bimbila, Dagomba Dist. (Mar.) *Hepper & Morton* A3098! Ejura (Dec.) *Morton* GC 9717! Nungua (Nov.) *Adams* 5380! **Togo Rep.:** *Baumann* 191! Lama Kara *Mahoux* 2072! **Dah.:** Porto Novo (Sept.) *Adjanohoun* 270! **N. Nig.:** Sokoto (Aug.) *Dalz.* 459! Katsina to Daura (Apr.) *Meikle* 1358! Minna (Dec.) *Meikle* 718! Jos Plateau *Batten-Poole* 399! Mayo Selbe, Gashaka Dist. (Jan.) *Hepper* 1640! **S. Nig.:** Lagos *W. MacGregor* 26! Ifaki, Ondo Prov. (June) *Clayton* 1123! Ajagbodudu, Benin Dist. (Aug.) *Wright* 54! Nun R. (Aug.) *T. Vogel* 13! Obudu, 5,000 ft. (Apr.) *Haines* 100! **W. Cam.:** Madele, Wum Dist. (June) *Daramola* FHI 41099! Jakiri, Bamenda, 6,500 ft. (Feb.) *Hepper* 2717! Bamenda, 5,000 ft. *Maitland*! Mankon to Mbengwi (Apr.) *Brunt* 1104! Pantropical and subtropical.

47. **C. denudatus** *Linn. f.* Suppl. 102 (1781); F.T.A. 8: 338; Kük. l.c. 254. Plant with purple, creeping rhizome, reduced leaves, and bract shorter than the slender inflorescence rays; marshy grassland.
Sen.: L. Tanma (June) *Adam* 17691! Sangalkam (June) *Berhaut* 2658! **Mali:** Bamako (Sept., Dec.) *Adam* 11287! 15337! San (June) *Chev.* 1050! *Lean* 44! Dogo (May) *Davey* 110! **S. L.:** Luti (Aug.) *Jordan* 86! L. Yomboro, nr. Konta (Apr.) *Hepper* 2662! Kambia (June) *Gledhill* 616! **Ghana:** Velere R., Wa to Lawra (Sept.) *Hall* CC 620! E. of Wa (Apr.) *Morton* A3319! Sing to Boli (June) *Adams* GC 740! Nasia, Tamale to Bolgatanga (Apr.) *Adams* 4150! Doyum, Accra Plains (July) *Hall* 3722! **Niger:** Maouri, S. Dallol (Jan.–Feb.) *Virgo* 73! **N. Nig.:** Nupe *Barter*! Giwa Fadama, Funtua, Zaria Prov. (Aug.) *Kershaw* 900407! Badeggi, nr. Bida (June) *Onochie* FHI 18688! Gwari, Gawu to Abuja (Aug.) *Onochie* FHI 35711! Shendam (Nov.) *Clayton* 1459! **S. Nig.:** Abbeokuta *Irving*! Central and east tropical Africa, Madagascar and S. Africa.

48. **C. tenax** *Boeck.* in Linnaea 35: 504 (1868); F.T.A. 8: 334; Kük. l.c. 259, incl. var. *monroviensis* (Boeck.) Kük. (1936); Nelmes & Baldwin in Amer. J. Bot. 39: 383 (1952). Densely tufted plant with hard, persistent leaf-sheaths and a compact inflorescence of short rays bearing spherical spikes of dark, obtuse spikelets; sandy places.
S. L.: Lumley (May) *Deighton* 2681! Tisana, Bonthe I. (Nov.) *Deighton* 2296! Gbap (June, Oct.) *Adames* 56! *Jordan* 587! Messima (May) *Jordan* 2110! **Lib.:** Monrovia (Nov.) *Linder* 1433! Robertsfield, Grand Bassa Dist. (Mar.) *Harley* 2118! Sasstown, Sinoe Dist. (Mar.) *Baldwin* 11611! Harper, Maryland Dist. (June) *Baldwin* 5978! **Dah.:** Cotonou (Mar.) *Morton* GC 6670! Sémé, E. of Cotonou (Feb.) *Raynal* 13510! **S. Nig.:** Lagos I. *Dalz.* 1114! Throughout tropical Africa, excluding the north-east, and extending into S. Africa.
[Sierra Leone and Liberian gatherings are characterized by rather short glumes and spikelets and correspond to the var. *monroviensis*.]

49. **C. cuspidatus** *Kunth* in Humb., Bonpl. & Kunth, Nov. Gen. et Spec. 1: 204 (1816); Kük. l.c. 261, incl. f. *angustifolius* (Nees) Kük. (1936); Berhaut, Fl. Sén. ed. 2, 380 (1967). *C. uncinatus* C.B. Cl. (1895)–F.T.A. 8: 328; F.W.T.A., ed. 1, 2: 482; not of Poir (1806). ? *C. recurvus* Vahl, Enum. Pl. 2: 310 (1806). Slender annual 4–15 cm. high with large spikelets in rounded clusters, having glumes red-brown at maturity and with a distinct recurved tip; in sandy soil, on rocks, in turf, generally damp places.
Sen.: Mbao (Nov.) *Pitot*! Oussouye (Sept.) *Broadbent* 110! 111*a*! Elinkine (Oct.) *Adam* 18260! Velingara (Oct.) *Adam* 18562/2! **Mali:** Kita Massif (July) *Jaeger* 22! **Port. G.:** Bissau, Antula (Oct.) *Esp. Santo* 2558. **Guin.:** Youkounkoun (Oct.) *Pitot*! Boké (Nov.) *Jac.-Fél.* 7303! Conakry (Oct.) *Maclaud* 20! Kindia *Jac.-Fél.* 134! Friguiagbé (Aug.) *Chillou* 686! **S. L.:** Newton (Nov.) *Deighton* 1441! Rokupr (July) *Jordan* 56! Makump (Aug.) *Deighton* 1388! Mayosso (Aug.) *Thomas* 1426! Koinadugu (Aug.) *Haswell* 258! **Lib.:** Robertsport, Grand Cape Mt. Dist. (Dec.) *Baldwin* 10917! Gbau, Sanokwele Dist. (Sept.) *Baldwin* 9450! Zwedru, Tchien Dist. (Aug.) *Baldwin* 7089! **Iv. C.:** Danané to Kouanhoulé (Apr.) *Chev.* 21184! Issia rock (May, Dec.) *Leeuwenberg* 4133! *Boughey* GC 13591! **U. Volta:** Ouagadougou (Oct.) *Nongonierma* 203! **Ghana:** Tumu (May) *Morton* GC 7533! Kampala, nr. Wa (June) *Adams* 867! Damongo Scarp (Sept.) *Hall* CC 868! Awiebo (Aug.) *Hall* 3393! Kpeshi Lagoon, Labadi (May) *Adams* 2730! **Togo Rep.:** Sokodé (July) *Mahoux* 2053! **N. Nig.:** (Aug.) *Lely* P459! Gawu hills, Gwari (Aug.) *Onochie* FHI 35918! Anara FR., Zaria (May) *Keay* FHI 22881! Izom to Abuja (Aug.) *Onochie* FHI 35738! Nupe *Barter* 1569! **S. Nig.:** Idanre to Aweba, Akure Dist. (Jan.) *Brenan & Keay* 8687! Ado Rock, Ibadan (May) *Onochie* FHI 31209! Lagos *W. MacGregor* 19! Obiaruku (July) *Kershaw* 900655! Calabar (Sept.) *Keay* FHI 22579! Pantropical.

50. **C. amabilis** *Vahl* Enum. 2: 318 (1806); F.T.A. 8: 327; Kük. l.c. 265; Cherm. in Arch. Bot. Caen 7, Mém. 4: 11 (1936); Berhaut, Fl. Sén. ed. 2, 381 (1967). Annual generally 15–20 cm. high with purple-streaked sheaths and semicircular clusters of golden brown spikelets about 1 cm. long; open savanna, cultivated or waste ground, often on sandy soil.
Sen.: Mbambey *Chev.* 33813! Camberéné (Sept.) *Berhaut* 2668! Gorom (Sept.) *Berhaut* 3520! Sangalkam (Sept.) *Trochain* 435! Ziguinchor (Sept.) *Broadbent* 69! **Gam.:** (Aug.) *Ruxton* 154! Yundum Agric. Stn. *Ashrif* 37! **Mali:** Kita (Oct.) *Jaeger* 66! Koulikoro *Vuillet* 208! Ségou (Sept.) *Chev.* 2481! Bamako (Sept.) *Adam* 14346! Boré (Aug.) *Demange* 27/1957! **Guin.:** Friguiagbé (Aug.) *Chillou* 2763! Kouroussa (July) *Pobéguin* 409! Sambailo (Oct.) *Pitot*! Youkounkoun (Oct.) *Pitot*! **S. L.:** Musaia (Sept.) *Small* 237! **Iv. C.:** Dihabo to Bouaké (July) *Chev.* 22065! Baoulé (Oct.) *Pobéguin*! Tehini, Bouna Reserve (Aug.) *de Wilde* 771! Mankono (June) *Chev.* 21949! **U. Volta:** Banfora to Ferkessédougou (June) *Leeuwenberg* 4426! Ouagadougou (Oct.) *Nongonierma* 200! **Ghana:** Damongo (July) *Andoh* FH 5192! Ejura (Dec.) *Darko* 755! Tamale (Oct.) *Baldwin* 13544! Ada to Accra (May) *Adams* 2732! Achimota (July) *Adams* 4893! **Togo Rep.:** Lama Wara Nturi (June) *Mahoux* 2070! Lomé (July) *Mahoux* 128. 2037! **Dah.:** Savalou (June) *Hb. E. Annet* 109! Bassa-Zoumé (May) *Hb. Annet* 88! **Niger:** Dosso (Oct.) *Monod* 7156! Niamey *Terraz* (ex *Hb. Cherm.*) 1930! (Aug.) *Vaillant* 35! Filingué, Toukounous *Bartha* 88! Wadi Dallol *L. P. White* 29! **N. Nig.:** Nupe *Baikie*! Patti Lokoja (Sept.) *T. Vogel* 193! Sokoto (Sept., Nov.) *Moiser* 134! *Dalz.* 462! Katsina (Oct.) *Clayton* 1334! Allaguerno, Bornu Prov. (Sept.) *De Leeuw* 1220*a*! Gurum, nr. Vogel Peak (Nov.) *Hepper* 1233! **S. Nig.:** Lagos *W. MacGregor* 31! Eruwa to Lanlate, Ibadan (Oct.) *Savory & Keay* FHI 25292! Olokemeji (Nov.) *Guile*! Pantropical and subtropical.

51. **C. maritimus** *Poir.* in Lam., Encycl. 7: 240 (1806); F.T.A. 8: 326; Kük. l.c. 269, incl. vars. *crassipes* (Vahl) C.B. Cl., *gracilescens* Kük. and *subtilis* Kük. (1936); Berhaut, Fl. Sén. ed. 2, 375 (1967). Strand

plant with a thick, oblique rhizome, glaucous leathery leaves, and a generally large capitate inflorescence of large, acute, only slightly compressed, pale brownish spikelets; sandy places.

Sen.: Djembering (Sept.) *Broadbent* 176! Dakar *Monaud*! M'bidjem (Oct.) *Thierry* 86! Malika (Aug.) *Broadbent* 28! Wallo country (Jan.) *Roger* 69! **Port. G.:** S. Domingos (Dec.) *Esp. Santo* 785. **Gam.:** Bathurst (Nov.) *J. C. Gardiner* Gambia 1! **S. L.:** Lumley (Aug.) *Melville & Hooker* 246! *Jordan* 777! Toke (Jan.) *Morton* SL 618! Yele, Turtle Is. (Nov.) *Deighton* 2308! Yungeru (Jan.) *A. W. Thomas* 7297! **Lib.:** Robertsport (Dec.) *Baldwin* 10912! Cape Mount (Nov.) *Dinklage* 2246! Monrovia (June–Dec.) *Baldwin* 5940! 10503! *Kunkel* 76! **Iv. C.:** Grand Bassam to Azuretti (Oct.) *de Wilde* 1010! Grand Bassam *Giovannetti* 985! (Mar., Dec.) *Leeuwenberg* 3133! *de Wilde* 974! SE. frontier beach (July) *Chipp* 281! **Ghana:** Cape Coast (July) *Hall* 3203! Winneba (June) *Boughey* GC 1311! Teshie (Nov.) *Morton* GC 6251! Mokwe Lagoon, Tema (June) *Rose Innes* GC 30357! Ada (June) *Thorold* CB 96! **Togo Rep.:** Lomé *Warnecke* 138! (May, Aug.) *Mahoux* s.n.! 7! **Dah.:** Cotonou (Apr., Nov.) *Chev.* 2366! 23473! *Poisson* 126! **S. Nig.:** Badagry (Aug.) *Onochie* FHI 33490! Lagos I. *Dalz.* 1108! Lagos (June–Dec.) *Onochie* FHI 23307! *Stubbings* 13! *Dalz.* 1295! Tropical Africa, Madagascar and S. Africa.

[Most West African specimens correspond to the var. *crassipes* (Vahl) C.B. Cl., having a large capitate inflorescence.]

52. **C. conglomeratus** *Rottb.* Descr. Pl. Nov. Programm. 16 (1772), and Descr. et Icon. 21, t.15 f.7 (1773); F.T.A. 8: 324, partly; F.W.T.A., ed. 1, 2: 482, partly; Kük. l.c. 272, t.30. *C. adansonii* C.B. Cl. (1895)–F.T.A. 8: 335. Tufted plant with woolly roots, pale, papery leaf sheaths, stiff leaves and bracts and spherical clusters of generally pale spikelets; sandy places.

Maur.: Rosso (Oct.) *Adam* 19375! Oujat, E. of Aratane (Aug.) *Rossetti* 61/229! **Sen.:** *Adanson*! L. Tanma (Jan.) *Pitot*! Dakar *Adam* 14889! **Mali:** Timbuktu (July) *Chev.* 1235! Kabarah to Day (Aug.) *Chev.* 1346! Hombori (Feb.) *Aké Assi* 11476! Extending through north-east Africa, and the Near East to India.

53. **C. jeminicus** *Rottb.* Descr. Pl. Nov. Programm. 24 (1772). *C. conglomeratus* Rottb. var. *multiculmis* (Boeck.) Kük. in Engl., Pflanzenr. Cyper. 274, incl. f. *excisus* (Boeck.) Kük. (1936). *C. cruentus* Rottb. var. *excisus* C.B. Cl. in F.T.A. 8: 325 (1901); Cherm. in Arch. Bot. Caen 7, Mém. 4: 9 (1936). *C. effusus* of F.W.T.A., ed. 1, 2; 482; Berhaut, Fl. Sén. ed. 2, 376 (1967). Tufted plant with narrow wiry roots, glaucous leaves, varying in width, and digitate clusters of spikelets about 3 cm. across; glumes lanceolate, acute with a recurved mucro becoming purple-brown; sandy places.

Maur.: Mederdra to Tiguent (Oct.) *Adam* 19350! Rkiz (Oct.) *Adam* 19323! **Sen.:** *Heudelot* 517! St. Louis to Thiés (Aug.) *J. & A. Raynal* 6271! Leybar, St. Louis (Oct.) *Adam* 19209! Malika (Aug.) *Broadbent* 33! **Niger:** Toukounous, Filingué *Bartha* 59! Also in north-east Africa and Arabia.

54. **C. aucheri** *Jaub. & Spach* Ill. Pl. Orient 2: 1, t.101 (1844). *C. conglomeratus* Rottb. var. *aucheri* (Jaub. & Spach) C.B. Cl. in Dur. et Schinz, Consp. Fl. Afr. 5: 554 (1895); F.W.T.A., ed. 1, 2: 482. Similar to *C. conglomeratus* Rottb. in habit but with a broader spikelet and winged achene; sandy places.

Niger: Agadem *E. Vogel!*

[The relationship between this specimen and typical examples of the species from the Persian Gulf requires further study.]

55. **C. pulchellus** *R.Br.* Prodr. Fl. Nov. Holland 213 (1810); Kern in Reinwardtia 3: 39–46, fig. 5 (1954). *C. leucocephalus* of F.T.A. 8: 323; F.W.T.A., ed. 1, 2: 481; Cherm. in Arch. Bot. Caen 7, Mém. 4: 9 (1936); Azancot de Meneses in Garcia de Orta 4: 559 (1956); not of Retz. (1789). Slender herb up to 40 cm. high from a fibrously thickened base, with a white, rounded head of numerous, small spikelets; damp places.

Sen.: Somone (Aug.) *Leprieur*! Kaolack (Sept.) *Berhaut* 2684! Tambacounda (Sept.) *Berhaut* 3270! Guermalale (Sept.) *Trochain* 4494! Ouassadou (Sept.) *Berhaut* 1814! **Gam.:** (Aug.) *Ruxton* 147! Beguirmi *Vaillant* 298! **Mali:** Klela (May) *Demange* 2250! **Port. G.:** Bafata (Sept.) *Esp. Santo* 2749! **Guin.:** Kiffaya to Seriba (Oct.) *Pitot*! Thiangeul Bori (Oct.) *Pitot*! **U. Volta:** Mossi, nr. Koupéla (July) *Chev.* 24587! **N. Nig.:** Damaturu, Bornu Prov. (Aug.) *De Leeuw* 1073! Katagum Dist. *Dalz.* 240! Yola (Aug.) *Dalz.* 267! Throughout the Old World tropics.

[Very close, except in the distichous arrangement of the glumes, to *Scirpus microcephalus* (Steud.) Dandy.]

56. **C. angolensis** *Boeck.* in Flora 63: 435 (1880); F.T.A. 8: 321; Kük. l.c. 281. Perennial from a stout creeping rhizome with firm, rough, grey-green leaves up to 1 cm. wide, erect stems to 1 m. high and a hemispherical white head up to 2 cm. across of coarse *Mariscus*-like spikelets subtended by spreading bracts; anthers large, yellow; montane savanna, conspicuous after burning.

S. L.: Loma Mts. (Feb.–Aug.) *Jaeger* 6818! 9548! *Gledhill* 357! *Morton* SL 3641! **Ghana:** Banda, Ashanti (Mar.) *Morton* GC 8634! **N. Nig.:** Vom, Jos Plateau *Dent Young* 259! Bida, Bornu Dist. (June) *Vaillant* 876! Kakara, Mambila Plateau (Jan.) *Hepper* 1777! Mai Idoanu to Ngoroje, Mambila Plateau (Feb.) *Latilo & Daramola* FHI 34466! Vogel Peak, 5,500 ft. (Mar.) *J. Hall* 1084! Also in E. Cameroun and throughout tropical Africa.

57. **C. karlschumannii** *C.B. Cl.* in Kew. Bull., Add. Ser. 8: 5 (1908). *C. margaritaceus* Vahl var. *karlschumannii* (C.B. Cl.) Kük. in Engl., Pflanzenr. Cyper. 285 (1936). Stiffly erect perennial to 120 cm. high with purple sheaths, glaucous leaves and 1–5 broad, yellow spikelets with hard, boat-shaped spreading glumes; savanna.

Mali: Bamako to Ouagadougou *Adam* 15234! 15237! **Iv. C.:** Nansian to Kong (Nov.) *Aké Assi* 9732! Bouna (Sept.) *Aké Assi* 6513! **U. Volta:** Banfora to Bobo-Dioulasso (June) *Leeuwenberg* 4414! **Ghana:** Damongo Scarp F.R. (Nov.) *Rose Innes* GC 30840! Banda (Apr., Aug., Nov.) *Morton* A3280! *Hall* 1975! *Harris*! Jaketi to Anyaboni (Oct.) *Morton*! **Togo Rep.:** Sokodé-Basari *Kersting* 659! **N. Nig.:** Idon, Kaduna to Keffi (July) *Kershaw* 900626! Jos Plateau *Batten-Poole* 395! Naraguta F.R. (July, Aug.) *Keay* FHI 20174! 4,000 ft. *Lawlor & Hall* FHI 46589!

[The material cited from N. Nig. may represent a distinct variety differing most clearly in the leaves being flat and up to 6 mm. wide.]

58. **C. margaritaceus** *Vahl* Enum. Pl. 2: 307 (1806); F.T.A. 8: 321; Kük. l.c. 284, incl. var. *pseudoniveus* (Boeck.) C.B.Cl.; Berhaut, Fl. Sén. ed. 2, 375 (1967). Tufted plant with contiguous stem bases thickened by numerous leaf-sheaths, and with a capitate inflorescence of large spikelets, whitish with a green edge formed by the flattened glume tips; sandy ground.

Sen.: Cambéréne (Sept.) *Pitot*! Dakar (Sept.) *Adam* 1887! Djembering (Sept.) *Broadbent* 134! Abémé, Casamance (Sept.) *Adam* 18185! Elinkine, Casamance (Oct.) *Adam* 18271! **Gam.:** *Skues*! **Port. G.:** Varela, S. Domingos (Aug.) *Esp. Santo* 3090! **Guin.:** Sambaïlo (Oct.) *Pitot*! **U. Volta:** Takalédougou (July) *Aké Assi* 4822! **Ghana:** Gambaga (Apr.) *Akpabla*! Ejura (Apr.) *Akpabla* 36! Elmina (Mar.) *Hall* 1311! Accra to Ada (May) *Adams* 357! Nungua, nr. Accra (May) *Boughey* GC 10416! **Togo Rep.:** Lomé *Warnecke* 127! **Dah.:** *Poisson*! **N. Nig.:** Sokoto (June) *Dalz.* 556! Zungeru (Sept.) *Dalz.* 275! Shagunu (Aug.) *Cook* 499! Nupe *Barter* 620! Quorra = Niger R. (Sept.) *T. Vogel* 19! **S. Nig.:** Lagos I. *Dalz.* 1117! Orle F.R., Kukuruku (Aug.) *Onochie* FHI 33294! Milliken Hill, Enugu (Sept.) *Tuley* 882! Iva Valley F.R., Enugu (July) *Daramola* FHI 55163! Enugu to Agbani (May) *Onochie & Awa* (May) FHI 35821! Throughout tropical Africa and extending into S. Africa.

59. **C. nduru** *Cherm.* in Arch. Bot. Caen 4, Mém. 7: 18 (1931). *C. margaritaceus* Vahl var. *nduru* (Cherm.) Kük. in Bot. Notis. 1934: 67 (1934). Tufted perennial with clustered stems 30–70 cm. high, thickened below by bladeless, purple, hard sheaths producing a bulb-like appearance, few very short leaves and bracts and a white head of 1–6 broad spikelets 7–10 mm. long; montane grassland subject to burning.

Guin.: Nionssomoridou to Beyla (Mar.) *Chev.* 20862! **S. L.:** Loma Mts. (Dec.–Feb.) *Jaeger* 4181! *Morton* SL 436! *Gledhill* 353! **Ghana:** Shiare (Dec.) *Jenik & Hall* CC 1142! Maliato, Togo Plateau (Dec.) *Morton* A3836! **N. Nig.:** Kakara, Mambila Plateau (Jan.) *Hepper* 1784! **W. Cam.:** Ndop (Mar.) *Brunt* 29! Ndop to Kumbo (Dec.) *Boughey* GC 11158*b*! Also in central and east tropical Africa.

60. **C. tisserantii** *Cherm.* in Arch. Bot. Caen 4, Mém. 7: 18 (1931). *C. margaritaceus* Vahl var. *tisserantii* (Cherm.) Kük. in Engl., Pflanzenr. Cyper. 285 (1936). *C. compactus* Lam. var. *tenerior* C.B. Cl. in F.T.A. 8: 321, partly. Similar to *C. margaritaceus* Vahl but with a longer, pseudobulbous base, narrower leaves and bracts and glumes without a distinct flattened tip; savanna.
Sen.: Tambacounda *Adam* 14338! Kanoumeri, Kédougou (June) *Fotius* K137! **Iv. C.:** Kongasso to Séguéla (Mar.) *Aké Assi* 9903! Tiégouakro to Kodiokoffi (Aug.) *Chev.* 22343! Assakra (Dec.) *Aké Assi* 4767! Lamto (Dec.) *Aké Assi* 9805! Dabou (Nov.) *Leeuwenberg* 1958! **Ghana:** Trumiso, Boku to Wenchi (Mar.) *Morton* GC 8588! Damongo to Sawla (Apr.) *Adams* 3980! **Togo Rep.:** (June) *Kling* 55! **N. Nig.:** Niger R. *Baikie*! Aguji, nr. Ilorin *Thornton*! Kachia to Kaduna (Feb.) *Meikle* 1201! Zaria (May) *Kershaw* 900366! Vogel Peak, 3,000 ft. (Mar.) *J. Hall* 1080! Also in E. Cameroun.

61. **C. ledermannii** (*Kük.*) *Hooper* in Kew Bull. 26: 578 (1972). *C. obtusiflorus* Vahl var. *ledermannii* Kük. in Engl., Pflanzenr. Cyper. 287 (1936). *C. compactus* of F.T.A. 8: 319, partly. *C. obtusiflorus* of F.W.T.A., ed. 1, 2: 481. Perennial 30–90 cm. high with clustered stem-bases thickened by leaf-sheaths and a head of numerous large, flattened, pinkish-white, congested spikelets of numerous long, narrow, folded glumes; savanna or open woodland.
Dah.: Atacora Mt. *Chev.* 24140! **N. Nig.:** Kaduna to Jos (May) *Clayton* 1034! Anara F.R., Zaria Prov. (May) *Keay* FHI 22958! Samaru, Zaria Prov. (July) *C. B. Taylor* 18! Jos Plateau (May) *Lely* 291! Panshanu, Bauchi Prov. (Aug.) *Lawlor & Hall* 383! Also in E. Cameroun and ? east tropical Africa.

62. **C. nudicaulis** *Poir.* in Lam., Encycl. 7: 240 (1806); F.T.A. 8: 316; C.B. Cl., Ill. Cyper. t.7 (1909); Kük. l.c. 291; Cherm. in Arch. Bot. Caen 7, Mém. 4: 7 (1936). Tufted perennial with slender, leafless stems and a capitate inflorescence of few, broad, lanceolate spikelets of numerous, closely imbricated, yellow-green to brown glumes, sometimes proliferating; floating in mats, on mud or in swamps.
Sen.: Ndenout (Mar.) *Dollinger* 41! Mbao to Dakar *Duong*! Sangalkam *Berhaut* 1620! Mt. Rolland, Thiès *de Wailly* 4553! M'boro (Apr.) *Leprieur* 29! **Mali:** Dendela (Mar.) *Chev.* 628! Sabalibougou (Mar.) *J. Raynal* 5613! Niger delta *Duong*! Saroféré (July) *Lean* 58! **Guin.:** Kapatchez *Jac.-Fél.* 7292! **S. L.:** Rhombe (Mar., July) *Adames* 147! *Jackson* in *Hb. Jordan* 479! Yomboro, Konta (Apr.) *Hepper* 2657! **Lib.:** Cape Palmas (Apr.) *Dinklage* 2679! **Iv. C.:** Bouna Wild-life Reserve, Ouangofétini (Feb.) *de Wilde* 3462! Agnéby (Mar.) *Aké Assi* 4784! Dabou *Giovannetti* 438! **Ghana:** Akotokyir, Cape Coast (Feb.) *Hall* 2187! Amisano (Dec.) *Hall* 1244! Yeji (Apr.) *Hall* VBS 1294! **N. Nig.:** Nupe *Barter* 1568! Matyoro (Oct.) *Thornewill* 136! Kuffena, Zaria Dist. (Jan.) *Kershaw* 900019! Yankara Game Reserve, Bauchi Dist. (Dec.) *Haines* 79! **S. Nig.:** Ajagbododu, Benin (Aug.) *Wright* 52! 52*a*! Sapoba (Feb.) *Onochie* FHI 31945! Jamieson R. (Jan., Sept.) *Brenan* 8941! *Onochie* FHI 34273! Agege, Owo R. (Nov.) *Haines* 229! **W. Cam.:** Baba, Ndop Plain (Mar.) *Brunt* 125! Also in central and east tropical Africa and in Madagascar.

63. **C. clavinux** *C.B. Cl.* in F.T.A. 8: 319 (1901); Kük. l.c. 304; J. Raynal in Adansonia, sér. 2, 6: 306 (1966). *C. monostigma* C.B. Cl. in Mém. Soc. Bot. Fr. 2, 8: 26 (1907); Kük. l.c. 281. *C. meeboldii* var. *gigas* Berhaut in Bull. Soc. Bot. Fr. 100: 174 (1953). Similar to *C. meeboldii* Kük. but with fewer, larger, pale spikelets in the head; seasonally damp sandy ground.
N. Nig.: Bornu *T. Vogel* 64! 65! Also in Chad, and south east Africa.

64. **C. meeboldii** *Kük.* in Fedde, Rep. 18: 345 (1922); Cherm. in Arch. Bot. Caen 7, Mém. 4: 11 (1936); Berhaut, Fl. Sén. ed. 2, 375 (1967); J. Raynal in Adansonia, sér. 2, 6: 304 (1966). Slender leafy plants with slightly swollen stem-bases and heads about 1 cm. across, of broadly elliptical spikelets, brown when ripe; achene turbinate; clayey soils at edge of temporary water.
Sen.: St. Louis (Sept.) *Trochain* 4712*b*! Ndiakher, nr. St. Louis (Aug.) *J. & A. Raynal* 6296! Ouassadou (Nov.) *Berhaut* 1584! Rhaddar, nr. Linguère (Nov.) *Adam* 12353! **Niger:** Toukounous *Bartha* 89*a*! Also in Chad, Kenya, Tanzania and peninsular India.

65. **C. lateriticus** *J. Raynal* in Adansonia, sér. 2, 6: 308 (1966). Plant of similar size and habit to *C. meeboldii* but with the inflorescence a pseudolateral cluster of pale lanceolate spikes, similar to those of *Juncellus laevigatus*; edge of temporary water.
Sen.: Maël, nr. Tambacounda (Sept.) *Berhaut* 1722! Perhaps overlooked elsewhere.

66. **C. michelianus** (*Linn.*) *Link* Hort. Bot. Berol. 1: 303 (1827).

66a. **C. michelianus** subsp. **pygmaeus** (*Rottb.*) *Aschers & Graebn.* Synops. Mitteleur. Fl. 2, 2: 273 (1904); Kük. l.c. 312. *Cyperus pygmaeus* Rottb., Descr. et Icon. 20, t.14, fig. 5 (1773); Cherm. in Arch. Bot. Caen 7, Mém. 4: 11 (1936). *Juncellus pygmaeus* (Rottb.) C.B. Cl. in Hook., Fl. Brit. Ind. 6: 596 (1893); F.T.A. 8: 308; F.W.T.A., ed. 1, 2: 489. *Scirpus occultus* C.B. Cl. in Mém. Soc. Bot. Fr. 2, 8: 28 (1907). Tufted annual 2–25 cm. high with conspicuous leafy bracts and a dense *Kyllinga*-like head of small whitish-green spikelets; muddy margins of watercourses.
Sen.: R. Senegal at Boki Diavé *Trochain* 2880! Tamboukané (Dec.) *Chev.* 2443! Ourosogui to Matam (Dec., Feb.) *Naegelé* 12127! Matam (Nov.) *Berhaut* 1414! Galam *Heudelot* 387! **Mali:** Gao (June) *de Wailly* 5071! Langana (Jan.) *Chev.* 265! Néguébabougou, nr. Bamako (Jan.) *J. & A. Raynal* 5445*c*! Sotuba (Jan.) *J. & A. Raynal* 5460! San (Feb.) *Roberty* 3380! **Ghana:** Nungua (Apr.) *Hall* GC 37174! Avilor lagoon, Battor (Nov.) *Hall* 2735! Old World tropics and subtropics to Mediterranean region and temperate Asia.

67. **C. laevigatus** *Linn.* Mant. Pl. 2: 179 (1771); Kük. in Engl., Pflanzenr. Cyper. 321 (1936). *Juncellus laevigatus* (Linn.) C.B. Cl. in Hook. f., Fl. Br. Ind. 6: 596 (1893); F.T.A. 8: 308; F.W.T.A., ed. 1, 2: 489; Berhaut, Fl. Sén. ed. 1, 228 and ed. 2, 379. A creeping, rush-like sedge, varying much in size but constantly with an erect bract and a capitate, pseudo-lateral inflorescence of sessile, white spikelets; damp sandy places, particularly on the margins of brackish pools.
Maur.: Taoujafête, Tagant (Aug.) *Monod* 1658! **Sen.:** L. Rhetba (Nov.) *Pitot*! L. Youi (Nov.) *Pitot*! Malika (Aug.) *Broadbent* 48! Malika to Yombel (Nov.) *Pitot*! St. Louis *Brunner* 18! **Mali:** Timbuktu (Sept.) *Chev.* 1236! **Niger:** Zinder (Dec.) *Vaillant* 2595! Niamey (Oct.) *Hagerup* 554! Bilma *Le Coeur* 1! 71! **N. Nig.:** L. Chad, Bornu (Dec.) *Elliott* 158! Throughout tropical and subtropical regions.

Insufficiently known, doubtful and excluded species

C. bulamensis *Steud.* Syn. Pl. Glum. 2: 19 (1855); F.T.A. 8: 375; Kük. l.c. 136. Portuguese Guinea, Bulama Islands, collector?
C. callistus *Ridl.* reported from Dahomey (*Newton* 10) by Kükenthal is perhaps *C. esculentus* Linn.
C. cancellatus *Ridl.* is an Angolan species of the *C. haspan* Linn. alliance not recorded from West Africa, other than by a Maitland specimen reputedly from W. Cameroon, with few obtuse oblong-elliptic spikelets 4–7 × 1·5–2·0 mm., borne 1–4 together on a slender plant.
C. corymbosus *Rottb.* reported from Togo (*Kersting* 440) by Kükenthal, is perhaps *C. fenzelianus* Steud.
C. foliaceus *C.B. Cl.*—*Warnecke* 388 is given by Clarke as from Togo, but as from Kenya by Kükenthal.
C. glaber *Linn.* Kükenthal cites a specimen of *Afzelius* from Sierra Leone, but this southern European and western Asiatic species does not seem to have been collected subsequently.
C. hamulosus *M. Bieb.*—see *Mariscus hamulosus*
C. sp. Sierra Leone, Sefadu (Dec.) *Deighton* 3574! Belongs to the *C. tenuiculmis* group of species with a narrow sinuous rhachilla and subquadrangular spikelets up to 4 cm. long.

2. MARISCUS Vahl, Enum. Pl. 2: 372 (1806), *nom. cons.* *Mariscus* Gaertn. (1788)—F.T.A. 8: 377 (1901); F.W.T.A., ed. 1, 2: 485. *Cyperus* Linn. subgen. *Mariscus* (Gaertn.) C.B.Cl. (1884)—Kükenthal in Engl., Pflanzenr. Cyperaceae—Scirpoideae—Cypereae 402 (1936).

By Miss S. S. Hooper

Annual or slender caespitose perennial; glumes aristate or clearly apiculate:
- Annuals with a spicy smell; glumes aristate:
  - Spikelets flattened with many glumes, aristas long, recurved .. 1. *squarrosus*
  - Spikelets subspiral with few glumes .. .. 2. *hamulosus*
- Slender perennial; spikelets terete or angular with few, obtuse, clearly apiculate glumes .. .. 3. *soyauxii*

Perennial; glumes acute or acuminate:
- Robust caespitose plant; stem bases thickened with numerous leaf-sheaths; leaves glaucous, stiff, septate-nodulose (when dry) with denticulate cutting edges; primary rays thick, bearing ovoid or shortly cylindrical spikes in a pyramidal cluster; spikelets terete, reddish, about 5 × 2 mm. .. .. 4. *ligularis*
- Plant slender or robust; leaves smooth to slightly scabrid on the margins:
  - Plant rhizomatous; rhizome horizontal, about 8 mm. across, covered with purple to black scales; stems leafless, winged, up to 120 cm. high; inflorescence bisumbellate, rather strict; glumes spreading at maturity .. .. 5. *socialis*
  - Plant caespitose or shortly rhizomatous with stems leafy below:
    - Stems clavately thickened with papery white, later brown, fibrous sheaths; inflorescence capitate; head oblong or hemispherical; spikelets about 5 mm. long 6. *dubius*
    - Stem-bases not clavately thickened; sheaths herbaceous or coriaceous, green or purple:
      - Leaf-bearing portion of the stem elongated, woody, sometimes branched; leaves firm, septate-nodulose beneath (dry); inflorescence rays bearing a single cylindrical greenish-white spike, up to 6 × 2 cm. .. .. 7. *tomaiophyllus*
      - Leaves basal; stem above ground not woody and branched; inflorescence not simply umbellate with very large cylindrical spikes:
        - Inflorescence up to 35 cm. across, either forming a compound umbel with secondary rays, or primary rays bearing clustered or distinctly branched spikes:
          - Inflorescence spreading; spikelets 2–3 cm. long when mature, remote and patent; glumes rounded on the back with a broad keel; rhachilla becoming thickened and breaking into segments with one achene (see *Torulinium odoratum* p. 297)
          - Inflorescence and spikelets smaller; glumes more or less distinctly keeled; rhachilla not breaking into pieces:
            - Spikelets pale to red-brown .. .. 8. *longibracteatus*
            - Spikelets dark red-brown to black .. .. 9. *keniensis*
        - Inflorescence capitate or simply umbellate with rays bearing simple or occasionally shortly branched spikes; spikelets not exceeding 1·5 cm. long:
          - Spikelets 10–15 mm. long, linear with slender acute tips which become curved or sinuous as the spikelet matures:
            - Glumes 2·4–2·8 mm. long; achene 1·4–1·85 mm. long:
              - Spikes cylindrical or oblong, spikelets remote 10a. *flabelliformis* var. *flabelliformis*
              - Spikes ovoid, about 1·5 cm. across 10b. *flabelliformis* var. *aximensis*
            - Glumes 3·4–4·5 mm. long; achene 1·9–2·6 mm. long:
              - Leaves and bracts up to 4·5–5·5 mm. broad, leaf-sheaths papery; rays up to 11 cm. long; spikes cylindrical; spikelets pale with green keel; glume tips spreading in fruit (W. Cam.) .. .. 11. *foliosus*
              - Leaves and bracts 2·5–3·5 mm. broad; rays up to 2·5 cm. long; spikes hemispherical; spikelets crowded, reddish or golden with green keel; glume tips appressed even in fruit .. .. 12. *luridus*
          - Spikelets not exceeding 5 mm. long, more or less terete, one or sometimes 2–3-flowered .. .. 13. *alternifolius*

1. M. squarrosus (*Linn.*) *C.B. Cl.* in Hook. f., Fl. Brit. Ind. 6: 623 (1893), as to name only. *Cyperus squarrosus* Linn., Cent. Pl. 2, 6 (1756); Kern in Blumea 10: 642 (1960). *Mariscus aristatus* (Rottb.) Cherm. in Bull. Soc. Bot. Fr. 85: 366 (1938). *Cyperus aristatus* Rottb., Descr. et Icon. 23, t. 6 fig. 1 (1773); F.T.A. 8: 348; F.W.T.A., ed. 1, 2: 483; Kük. in Engl., Pflanzenr. Cyper. 502 (1936). *C. aristatus* var. *semiglobosus* Kük. l.c. 504; Cherm. in Arch. Bot. Caen 7, Mém. 4: 12 (1936). Annual weed with golden or reddish hemispherical or ovoid spikes and soft leaves; widespread in open ground.
Maur.: Traya (Jan.) *Naegelé*! **Sen.:** *Leprieur*! *Heudelot* 328! Hann (Aug.) *Berhaut* 2655! Diouloulou (Sept.) *Adam* 18219! Wallo (Dec., Jan.) *Roger* 68! **Gam.:** Kuntaur (Aug.) *Ruxton* 163! Kaiaaf, Genieri (July) *Fox* 169! Yundum *Ashrif* 36! **Mali:** Fodébougou, Kita (Oct.) *Jaeger* 59! 65! Sotuba (Oct.) *Raynal* 5033! Bore (Aug.) *Demange* 29/1957! **Guin.:** Conakry (Aug.) *Martine* 51! Tamba to Sambaïlo (Sept.) *Pitot*! **Iv. C.:** Garango *Prost!* Téhini, Bouna Reserve (Aug.) *de Wilde* 770! **U. Volta:** Koupéla to Ouagadougou (July, Aug.) *Chev.* 24583! Ouagadougou (Oct.) *Nongonierma* 205! **Ghana:** Tumu (May) *Morton* GC 7546! Yendi (Dec.) *Adams* 4091! Nsawkaw (Aug.) *Hall* 1955! Labadi (Mar.) *Morton* GC 6658! Achimota (May) *Morton* A879! **Niger:** Filingué, Toukounous *Bartha* 89*b*! **N. Nig.:** Sokoto *Dalz.* 452! Samaru, Zaria (Aug.) *Clayton* 1326! Nupe *Barter* 1570! Yola (Sept.) *Dalz.* 268! Gurum, Vogel Peak area (Nov.) *Hepper* 1232! **S. Nig.:** Ado Rock (Oct.) *Savory & Keay* FHI 25441! Lagos *W. MacGregor*! Igboora (Oct.) *Haines* 216! Widespread in tropical regions, extending in the New World to Canada and Chile.

2. M. hamulosus (*M. Bieb.*) *Hooper* in Kew Bull. 26: 578 (1972). *Cyperus hamulosus* M. Bieb., Fl. Taur.-cauc. 1: 35 (1808); J. Raynal in Adansonia sér. 2, 6: 581–588 (1967). *Scirpus hamulosus* (M. Bieb.) Stev. in

Mém. Soc. Nat. Moscou 5: 356 (1817). *S. lugardii* C.B. Cl. in F.T.A. 8: 458 (1902). Slender annual 3–15 cm. high with few, purple sheathed leaves and up to 6 sessile or stalked, ovate, starry spikes; distinguished from the previous species by the shorter-pointed, sub-spiral glumes and disk at the base of the achene; sandy soil subject to occasional flooding.

Sen.: Kayar, Cape Verde (Oct.) *J. & A. Raynal* 6430! 6935! **Mali:** Gao to Berra (Mar.) *de Wailly* 5012! Dogo (Apr.) *Demange* 2071! **N. Nig.:** Maiduguri, Bornu (Dec.) *McClintock* 117! In subdesert zones in Africa, Europe and western Asia.

3. **M. soyauxii** (*Boeck.*) *C.B. Cl.* in Dur. et Schinz, Consp. Fl. Afr. 5: 593 (1895); F.T.A. 8: 393. *Cyperus soyauxii* Boeck. in Engl., Bot. Jahrb. 5: 501 (1884); Kük. l.c. 496 (1936). Small leafy weed often with a single echinate head of turgid spikelets; disturbed ground.

Sen.: Sangalkam (June, Oct., Nov.) *Berhaut* 1622! 2620! 5453! **Iv. C.:** Dabou *Giovannetti* 456! 469! Adiopodoumé (June) *Aké Assi* 6065! **Ghana:** Kwahu (Apr.) *Johnson* 663! Mankrong (Apr.) *Morton* A604! Aburi (Nov.) *Johnson* 1038! Achimota (Oct.) *Morton* 6011! Adaiso (Nov.) *Morton* 6146! **Togo Rep.:** Aledjo, Atacora Mts. *Villiers*! **S. Nig.:** Ikoyi Plains (June) *Dalz.* 1431! Yaba (May) *Haines* 32! Otta (May) *Haines* 210! Ibadan *West*! Ijaiye F.R. Oyo (May) *Daggash* FHI 21997! From Senegal to Gabon.

[A form of this species appearing very different from the typical has been found in Ghana at the foot of the Atewa Hills (*Hall* 3461). The inflorescence is anthelate with the rays about 1 cm. long and the spikelets numerous and distinct.]

4. **M. ligularis** (*Linn.*) *Urb.* Symb. Antill. 2: 165 (1900); Nelmes & Baldwin in Amer. Journ. Bot. 39: 385, t. 70–79 (1952). *Cyperus ligularis* Linn., Pl. Jamaic. Pugill. 5 (1759), and Syst. Nat. ed. 10, 2: 867 (1759); Kük. l.c. 474. *Mariscus rufus* Kunth, Nov. Gen. et Spec. 1: 216, t. 67 (1816); F.T.A. 8: 396. A coarse leafy plant with congested reddish heads of small terete spikelets; spikelets often hypertrophy after egg deposition by a gall-fly, attaining 10–12 mm. in length; swamps and marshes, generally near sea.

Sen.: *Farmar* 57! *Heudelot* 442! Dakar (May) *Baldwin* 5722! Mbao (Aug.) *Broadbent* 11! Toubacouta (Oct.) *Monod* 8480! **Gam.:** Kuntaur (July) *Ruxton* 2! **Guin.:** Golea, Conakry (Aug.–Sept.) *de Wit* 51! *Martine* 113! Koba (Nov.) *Jac.-Fél.* 7219! Manéa, Falikouré (June) *Chillou* 2609! 2625! **S. L.:** Mambolo (Jan.) *Deighton* 1006! Lumley (June) *Tindall* 5! Aberdeen (Aug.) *Melville & Hooker* 126! Christineville (Aug.) *Deighton* 2045! Mahera, Kafu Bulom (Oct.) *Glanville* 416! **Lib.:** Bushrod I. (Feb.) *Baldwin* 11075! *Dinklage* 2877! Robertsport (Dec.) *Baldwin* 10890! Monrovia (May–Sept.) *Barker* 1305! *Baldwin* 9211! Cess R. (Mar.) *Baldwin* 11259! **Iv. C.:** Sassandra (June) *de Wilde* 187! Dabou *Giovannetti* 44! Abouabou Forest, Abidjan to Grand Bassam (Jan.) *Leeuwenberg* 2380! Banco (Dec.) *Bégué* 152! Grand Bassam *Giovannetti* 306! **Ghana:** Atwabo *Fishlock* 28 of 1931! Sakumo Lagoon, Tema (Sept.) *Rose Innes* GC 31274! Mile 13, Ada road *Adams* 3320! Keta (Mar., June) *Thorold* 99! *Morton* GC 6532! **Togo Rep.:** Lomé *Warnecke* 136! *Duffour* 6478! **Dah.:** *Newton*! **S. Nig.:** Lagos *Millen* 219! Ikoyi Plains (Mar.) *Dalz.* 1305! Yaba (Apr.) *Haines* 31! Ogun River F.R. (Mar.) *Hepper* 2254! Brass (Dec.) *Daramola* FHI 46933! **W. Cam.:** Victoria (July) *Maitland* 31! **F. Po:** (June) *Barter*! Senegal to the Congo, Mascarenes, tropical America and Andes.

5. **M. socialis** (*C.B. Cl.*) *Hooper* in Kew Bull. 26: 578 (1972). *Cyperus socialis* C.B. Cl. in F.T.A. 8: 351 (1901); Kük. l.c. 87 & 633. *Mariscus trinervis* C.B. Cl. in F.T.A. 8: 399 (1902). *M. pseudo-pilosus* C.B. Cl. in F.T.A. 8: 402 (1902); Nelmes & Baldwin in Amer. J. Bot. 39: 385 (1952). Stems up to 2 m. high, bracts conspicuous, inflorescence rays only about 10 cm. long; swamps.

Guin.: N'Zérékoré, Nimba (Sept.) *Adam* 6313! **Lib.:** Jabroke, Webo Dist. (July) *Baldwin* 6686! Monroviatown, Tchien Dist. (Aug.) *Baldwin* 8016! Gbanga (Sept.) *Linder* 484! Sanokwele Dist. (Dec.) *Adam* 16515! Ganta (Mar.) *Harley* 994! Nimba Mts. (Sept.) *Adames* 532! **Iv. C.:** Sampleu to Ganhoué (Apr.) *Chev.* 21150! Danané to Nzo (Aug.) *Mangenot & Aké Assi* 4833! Extends to Angola and Uganda.

6. **M. dubius** (*Rottb.*) *C. E. C. Fischer* in Gamble, Fl. Madras 1644 (1931). *Cyperus dubius* Rottb., Descr. et Icon. 20, t. 4, fig. 5 (1773); Kük. l.c. 563. *C. kyllingioides* Vahl and *C. coloratus* Vahl, Enum. Pl. 2: 312 (1806). *Mariscus coloratus* (Vahl) Nees in Linnaea 9: 286 (1835); F.T.A. 8: 381; F.W.T.A., ed. 1, 2: 485. Herb to 30 cm. high with bulb-like stem-bases and oblong or hemispherical golden or greenish heads; widespread as a weed, and in natural habitats on rocks.

Guin.: Nzérékoré (Oct.) *Baldwin* 9693! **Lib.:** Soplima, Vonjama Dist. (Nov.) *Baldwin* 10033! **Iv. C.:** Séguéla (Aug.) *Boughey* GC 18451! Bouaké (May) *de Wilde* 130! Bécédi, 45 km. NNE of Dabou (Aug.) *de Wilde* 684! **Ghana:** Atwabo *Fishlock* 27 of 1931! Achimota (Oct.) *Morton* 6009! *Irvine* 5014! Oblogo (May) *Adams* 273! Ajena (May) *Morton*! Keta (Mar.) *Morton* GC 6521! **Dah.:** towards 10°N *de Gironcourt* 199! **N. Nig.:** Jebba *Barter*! Gwari, Sokoto Prov. (Aug.) *Onochie* FHI 35933! Jos Plateau (July) *Lawlor & Hall* FHI 46502! Panshanu Pass (Aug.) *Lawlor & Hall* 268! 558! **S. Nig.:** Lagos (Jan., July) *Haines* 40! *Dalz.* 1429! Mt. Orosun, Idanre (Jan.) *Brenan* 8650! Owo F.R. (July) *Onochie* FHI 33249! Akure (June) *Swarbrick* 2697! Tropical Africa and Asia.

[Specimens from Mann's Spring on the Cameroon Mountain (*Brenan* 9380!, *Morton* GC 7067!) with the leaf-sheaths forming an unusually large bulbous base about 3 cm. across and having very small spikelets may represent a form resulting from burning.]

7. **M. tomaiophyllus** (*K. Schum.*) *C.B. Cl.* in F.T.A. 8: 392 (1902). *Cyperus tomaiophyllus* K. Schum. in Engl., Pfl. Ost-Afr. C: 122 (1895), incl. var. *crassistolon* Kük. l.c. 429, and var. *laxiflorus* Kük. l.c. 430 (1936). Robust leafy sedge with decumbent woody branches; montane grassland and forest.

S. Nig.: Obudu Plateau, 5,000 ft. (Apr.) *Haines* 203! *Tuley* 968! **W. Cam.:** Cam. Mt., 5,000 ft., *Maitland*! Bafut-Ngemba F.R. (Feb.) *Hepper* 2721! Bamenda Nkwe to Santa (May) *Daramola* FHI 41193! Ashong, nr. Batibo, 5,100 ft. (Apr.) *Brunt* 1089! Oku, 6,000 ft. (June) *Brunt* 611! **F. Po:** Moka to Biao, 5,000 ft. (Sept.) *Melville* 472! Throughout tropical Africa.

[The West African material has a distinctive appearance due to the possession of a long, thick, often branched leafy aerial portion of the stem generally absent in East African material; further study is needed to prove whether it forms a distinct taxon.]

8. **M. longibracteatus** *Cherm.* in Bull. Mus. Hist. Nat. Paris 25: 407 (1919). *Cyperus longibracteatus* (Cherm.) Kük. in Fedde, Rep. 26: 250 (1929); Kük. in Engl., Pflanzenr. Cyper. 413 (1936). *Cyperus distans* of F.T.A. 8: 349 partly, not of Linn. f. Tufted plant up to 120 cm. high with a spreading inflorescence similar to that of *Cyperus distans* Linn. f. which is laxer with flexuose spikelets shedding their glumes; and to *Torulinium odoratum* (Linn.) Hooper which is a more robust plant with mucronate tipped glumes and the rhachilla breaking into segments when ripe; swampy ground.

Sen.: Sédhiou, Casamance (May) *Nongonierma* 525! **Mali:** Kléla, plain of Lotio (June) *Demange* 2875! **Guin.:** *Heudelot* 788! Koba (Nov.) *Jac.-Fél.* 7368! Dabola (Apr.) *Pitot*! Farana to Banian (Apr.) *Pitot*! Kissidougou to Guekédou (Apr.) *Pitot*! **S. L.:** *Thomas* 9937! Mahera (Oct.) *Glanville* 414! Kambia (Dec.) *Deighton* 819! Musaia (Sept.) *Small* 244! Yetaya (Sept.) *Thomas* 2440! Gboyama to Bendu *Deighton* 1603! **Lib.:** Firestone Plantations, Dukwai R. (Oct.–Nov.) *Cooper* 27! Kakatown *Whyte*! Sanokwele (Jan., Sept.) *Harley* 1747! *Baldwin* 9547! Gretown, Tchien Dist. (July) *Baldwin* 6784! **Iv. C.:** Man (Aug.) *Boughey* GC 18410! Guitry (Dec.) *Boughey* GC 13557! Sinfra (Dec.) *Aké Assi* 7231! **Ghana:** *Burton*! Simpa, nr. Tarkwa (Feb.) *Kinloch* 3243! Kumasi (Nov.) *Bannerman-Bruce* in Hb. *Adams* 4437! Biakpa to Vane (Nov.) *Morton* A 1660! Tono, Navrongo (Aug.) *Irvine* 5178! **S. Nig.:** Ibadan (Sept.) *Okafor* FHI 57170! Obiakuru, Warri (July) *Kershaw* 900659! Umuahia (Feb.) *Vaillant* 2758! Adani (Nov.–Mar.) *Tuley* 16! 70! Obudu, 5,000 ft. (Aug.) *Kershaw* 900685! **W. Cam.:** Victoria (Jan.) *Maitland* 320! Mamfe to Ikom (Apr.) *Morton* GC 7134! **F. Po:** *Mann* 121! *T. Vogel* 218! Throughout tropical Africa.

9. **M. keniensis** (Kük.) Hooper in Kew Bull. 26: 579 (1972). *Cyperus keniensis* Kük. in Notizbl. Bot. Gart. Mus. Berlin 9: 306 (1925), as *keniaeensis*. Similar to *M. longibracteatus* but with more compact spikes of very dark spikelets; damp grassland or marshes.

N. Nig.: Jos Plateau (Aug.) *Lely* P492! Upper tributary of Delimi River, nr. Naraguta (July) *Lawlor & Hall* 26! FHI 46645! Naraguta (June) *Lely* 329! Maisamari (Aug.) *De Leeuw* 1736! **S. Nig.:** Obudu Plateau, 5,500 ft. *Tuley* 969! **W. Cam.:** Cheddar Gorge, Bamenda Dist. (Mar.) *Morton* K189! Bambui Farm, Bamenda (May) *Pedder* 22! Santa to L. Bambulue, Bamenda (Mar.) *Morton* K30! Ashong, 5,600 ft. (Apr.) *Brunt* 1088! Bamessi, Ndop Plain (Apr.) *Brunt* 311! Throughout tropical Africa.

10. **M. flabelliformis** *Kunth* in H.B.K., Nov. Gen. et Sp. Pl. 1: 215 (1816); F.T.A. 8: 397. *Cyperus flabelliformis* Spreng., Linn. Syst. Veg. 1: 228 (1825), not of Rottb. (1773). *C. tenuis* Swartz, Prodr. Veg. Ind. Occ. 20 (1788); Kük. in Engl., Pflanzenr. Cyper. 416 (1936), incl. var. *grandiceps* Kük., not *Mariscus tenuis* C.B. Cl. (1900).

10a. **M. flabelliformis** *Kunth* var. **flabelliformis.** A densely caespitose plant with purple sheaths and spikes of distant, spreading, very acute and flexuose tipped spikelets; open ground, sometimes as a weed of cultivation.
Sen.: Ouassadou (Sept.) *Berhaut* 1102! **Mali:** Doniena (Oct.) *Demange* 3124! **S. L.:** Mano Bonjema (May) *Jordan* 2116! Pujehun (Apr.) *Deighton* 1653! Kenema (Apr.) *Jordan* 2029! Daru (Apr.) *Deighton* 3163! **Lib.:** Paynesville (Sept.) *van Dillewijn* 95! Ganta (May, Nov.) *Harley* 946! 1722! Gretown (July) *Baldwin* 6772! **Iv. C.:** Guitry (Dec.) *Boughey* GC 13523! Adiopodoumé (Nov.) *Leeuwenberg* 2098! Abidjan (Dec.) *Pitot*! **Ghana:** Assuantsi (July) *Irvine* 5115! Takoradi to Achimota (Jan.) *Morton* GC 6368! Anyinam (June) *Deighton* 3402! Kpong (June) *Irvine* 4925! Amedzofe (June) *Boughey* GC 13044! **Togo Rep.:** Palimé (June) *Mahoux* 2096! **S. Nig.:** Lagos *MacGregor* 97! Olokemeji (Oct.) *Hepper & Keay* 936! Ibadan (Apr.) *Keay* FHI 37657! Sapoba (Nov.) *Meikle* 597! Cross R. *Johnston*! **W. Cam.:** Mamfe (Dec.) *Migeod* 273! Mamfe to Ikom (Apr.) *Morton* GC 7134! **F. Po:** Nioko (Dec.) *Monod* 10335! Santa Isabel *Swarbrick* 2917! Also in Chad and tropical America.

10b. **M. flabelliformis** var. **aximensis** (*C.B. Cl.*) *Hooper* in Kew Bull. 26: 579 (1972). *M. aximensis* C.B. Cl. in F.T.A. 8: 398 (1902). *Cyperus tenuis* Swartz var. *aximensis* (C.B. Cl.) Kük. in Engl., Pflanzenr. Cyper. 418 (1936). Differs from the typical variety only in details of the spikelet; damp places.
Ghana: Axim (May) *Gürich* 37!

11. **M. foliosus** *C.B. Cl.* in F.T.A. 8: 399 (1902). *Cyperus chermezonianus* Robyns & Tournay, Fl. Sperm. Parc Nat. Albert 3: 246 (1955). Tufted sedge characterized by long glumes and broad leaves; clearings in upland forest and disturbed ground.
W. Cam.: Buea, 3,000 ft. (Nov.) *Migeod* 103! Musake Camp, Cam. Mt., 5,000 ft. (Jan.) *Maitland* 1288! Cam. Mt., 5,900 ft. (Dec.) *Adams* GC 11774! Bafut-Ngemba F.R., 7,000 ft. (Feb.) *Hepper* 2196! Throughout tropical Africa.
[Perhaps to be equated, as by Kükenthal, with *C. luteus* Boeck.; also similar to the Brazilian *M. palustris* Schrad.]

12. **M. luridus** *C.B. Cl.* in F.T.A. 8: 399 (1902). *Cyperus tenuis* Swartz var. *luridus* (C.B. Cl.) Kük. l.c. 418 (1936). Similar in habit to *M. flabelliformis* but spikes hemispherical to spherical, golden to reddish; river banks.
Sen.: Ouassadou (Nov.) *Berhaut* 2656! **Guin.:** Kouroussa (July) *Pobéguin* 412! **S. L.:** Njala (July) *Deighton* 747! **Lib.:** Webbo to Cavally (Apr.) *Dinklage* 2651! **Iv. C.:** Sassandra (Apr.) *Leeuwenberg* 4009! Gansé (June) *Aké Assi* 10223! **Ghana:** Otisu ferry, Oti R. (Apr.) *Morton* GC 9158! Mankrong (Apr.) *Morton* A573! Kpong to Akuse (May) *Morton* A3679! Dawhenya, nr. Prampram (Apr.) *Hall* 2256! Mankessim (May) *Hall* 2970! **Dah.:** Savé to Agouagon (May) *Chev.* 23601! Also in E. Cameroun, Chad, C. African Republic and the Congo.
[Differs in only minor characters from *M. palustris* Schrad. and *M. lancastriensis* Porter from Brazil.]

13. **M. alternifolius** *Vahl* Enum. Pl. 2: 376 (1806), not *Cyperus alternifolius* Linn. *Kyllinga umbellata* Rottb., Descr. et Ic. Pl. Nov. 15 t. 4, f. 2 (1773), nom. illeg. *Mariscus umbellatus* Vahl l.c.: 376 (1806), nom. illeg.; F.T.A. 8: 390; F.W.T.A., ed. 1, 2: 486. *M. nossibeensis* Steud., Syn. Pl. Glum. 2: 63 (1855); F.T.A. 8: 391. *Cyperus subumbellatus* Kük. in Engl., Pflanzenr. Cyper. 523 (1936), nom illeg., incl. var. *subglobosus* Kük. and var. *thomensis* (C.B. Cl.) Kük. Tufted plants with a short woody rhizome more or less composed of swollen stem bases; leaf sheaths purple; inflorescence variable, rays more or less well developed and bearing spikes of more or less crowded small greenish or reddish one or two-flowered spikelets; common in damp grassy places.
Sen.: Mbao, Cape Verde (Aug.) *Pitot*! Sangalkam (June) *J. & A. Raynal* 5949! Kanoumeri, Kédougou (July) *Fotius* K153! M'bidjem (July) *Thierry* 12! Ziguinchor (Sept.) *Broadbent* 81! **Gam.:** *Ruxton* 47! Genieri (July) *Fox* 139! **Mali:** Kita Massif (July–Oct.) *Jaeger* 5! 25! 29! 37! 86! 121! 5759! 5761! Sikasso (July) *Duong*! Soufarasso (May) *Chev.* 811! Bamako (July–Aug.) *Waterlot* 1540! *Roberty* 2668! **Guin.:** Conakry (July) *de Wit* 7324! Kouroussa (July) *Pobéguin* 417! Friguiagbé (Aug.) *Chillou* 2702! 2705! Farana to Kissidougou (Apr.) *Pitot*! Farana (Mar.) *Sc. Elliot* 5361*a*! *b*! Nzérékoré to Beyla (May) *Pitot*! **S. L.:** Freetown (Aug.) *T. Vogel*! *Melville & Hooker* 102! Mateboi (July) *Small* 143! Sendugu (June) *Thomas* 598! Musaia (Apr.) *Deighton* 5461! Loma Mts., foothills—5,600 ft. (July) *Jaeger* 6783! *Bakshi* 243! **Lib.:** Monrovia (May–July) *Dinklage* 2182! *Baldwin* 5806! 5807! *Kunkel* 93! Gbanga (Jan.) *Daniel* 93! Buchanan (Nov.) *Adam* 16050! 16051! Nimba (June, July) *Adam* 21581! *Leeuwenberg & Voorhoeve* 4605! Sinoe Basin *Whyte*! **Iv. C.:** Adiopodoumé (Oct.) *Leeuwenberg* 1713! Mbaso (July) *Oldeman* 211! Téhini, 40 km. E. of Ouangofétini (Aug.) *de Wilde* 831! R. Nzi, N'Douci Reg. (Aug.) *De Wit* 7332! Divo, SE. of Gagnoa (Aug.) *Boughey* 14654! **Ghana:** Achimota (June) *Irvine* 5012! Akroful *Cummins* 29! Damongo (July) *Andoh* FH 5189! Folifoli, Afram Plains (Aug.) *Hall* 194! Wenchi to Techiman (Mar.) *Morton* GC 8542! **Togo Rep.:** Misahöhe (Mar.) *Baumann* 134! Atacora Mts. (June) *Villiers*! Bafilo Mt. (May) *Roberty* 1441! Palimé (Mar.) *Stage* 7! **Dah.:** Porto Novo (June) *Roberty* 1722! Grand Popo (May) *Roberty* 1553! **N. Nig.:** Nupe *Barter* 1583! Zungeru (Aug.) *Dalz.* 256! Sanga River F.R., Jemaa Div. (May) *E. W. Jones* 56! Jos Plateau (Sept.) *Lawlor & Hall* 662! Abinsi (May) *Dalz.* 812! **S. Nig.:** Lagos *Dalz.* 1115! Ogun River F.R. (Mar.) *Hepper* 2265! Ibadan (Apr.) *Meikle* 1475! Benin (Jan.) *Brenan* 8749! Obudu Plateau, 5,400 ft. (Apr.) *Haines* 99! Ikom (Jan.) *Rosevear* 2/31! **W. Cam.:** Likomba (Dec.) *Mildbr.* 10787! Buea *Lehmbach* 150! *Preuss* 753! Bamenda (Jan.) *Migeod* 368! Bamessi, Ndop Plain (Mar.) *Brunt* 257! Oku, 6,350 ft. (June) *Brunt* 565! **F. Po:** *Barter* 1584! Balachá (Jan.) *Guinea* 1529! S. Isabel *Swarbrick* 2902! Pantropical.
[Very variable in the size of the spikes, length of rays and crowding of the spikelets; a common form with subglobose or shortly obovoid spikes was described as var. *subglobosus* Kük. but is not clearly separable from specimens with laxer more cylindrical spikes (var. *thomensis* (C.B. Cl.) Kük.) which in turn run into the large form with more or less branched spikes described as *M. nossibeensis* Steud.]

Imperfectly known species

M. sp. A. A robust plant 120 cm. high with flattened golden-brown spikelets, 10–15 × 1·25 mm., in cylindrical spikes on rays up to 12 cm. long; resembling *Mariscus congestus* Vahl but with a more expanded inflorescence and also to be compared, when better known, with *Mariscus strigosus* (Linn.) C.B. Cl. from N., central, and east S. America.
S. L.: Koinadugu, Northern Prov. (Aug.) *Haswell* 245!

M. sp. B. Plant resembling *Mariscus alternifolius* Vahl but differing in the slender rhizomes covered with dark, eventually fibrous, scales, and the broadly elliptical, equal-sided achene.
Gam.: ?Kuntaur (July) *Ruxton* 45!

M. sp. C. A slender species about 25 cm. high, with a short, swollen rhizome and distinctive whitish, *Kyllinga*-like head of 1-flowered spikelets; in upland grassland.
N. Nig.: nr. radio masts, Shere Mts., 5,000—5,500 ft. (July) *J. Hall* 1940!

3. REMIREA Aubl., Hist. Pl. Guian. Fr. 1: 44 (1775); F.T.A. 8: 485 (1902).

By Miss S. S. Hooper

Rhizome slender, branched, long-creeping; shoots short, with a cluster of lanceolate, stiff, spreading, harsh leaves, and a terminal, nearly sessile head of one to several sessile spikes surrounded by leaf-like bracts; spikelets with 3 or 4 sterile and one fertile glume; achene when ripe surrounded by a corky sheath formed from the apex of the rhachilla .. .. *maritima*

R. maritima *Aubl.* Hist. Pl. Guian. Fr. 1: 45, t. 16 (1775); F.T.A. 8: 486. *Cyperus pedunculatus* (R. Br.) Kern in Acta Bot. 7: 798 (1958). Strand plant with shoots arising singly or in small clusters from a long rhizome; sandy places.
S. L.: Mahera (Oct.) *Jordan* 371! Lungi shore (Apr.) *Hepper* 2671! Aberdeen to Lumley (Aug.) *Melville & Hooker* 153! No. 2 River Beach, Colony (July) *Small* 158! Yele, Turtle Is. (Nov.) *Deighton* 2309! **Lib.:** Robertsport (Dec.) *Baldwin* 10913! Monrovia (May, Nov.) *Baldwin* 5803! *Linder* 1436! Cestos River (May) *Dinklage* 1950! Greenville (Apr.) *de Wilde* 3796! **Iv. C.:** Abouabou Forest, Abidjan to Grand Bassam (Jan.) *Leeuwenberg* 2377! Grand Bassam to Azuretti (Oct.) *de Wilde* 1017! Grand Bassam (Sept., Oct.) *de Wilde* 545! *Giovannetti* 283! *Oldeman* 423! **Ghana:** Axim (Mar.) *Morton* A416! Cape Coast (July) *Hall* 3204! Teshi (Nov.) *Irvine* 801! Ada (June) *Thorold* CB 107! Old Ningo, E. Lagoon (Jan.) *Morton* 6352! **Dah.:** *Le Testu* 231! Cotonou (Mar., Apr.) *Keay* FHI 37577! *Chev.* 23474! Grand Popo (May) *Roberty* 1561! **S. Nig.:** Badagri (Apr.) *Haines* 25! Tarqua Is. (May) *Ward* 5! Victoria Beach, Lagos (Feb.) *Richards* 5083! Lighthouse Beach, Lagos (Sept.) *Stubbings* 8! Nun River (Sept.) *Mann* 534! Coastal zones throughout the tropics.

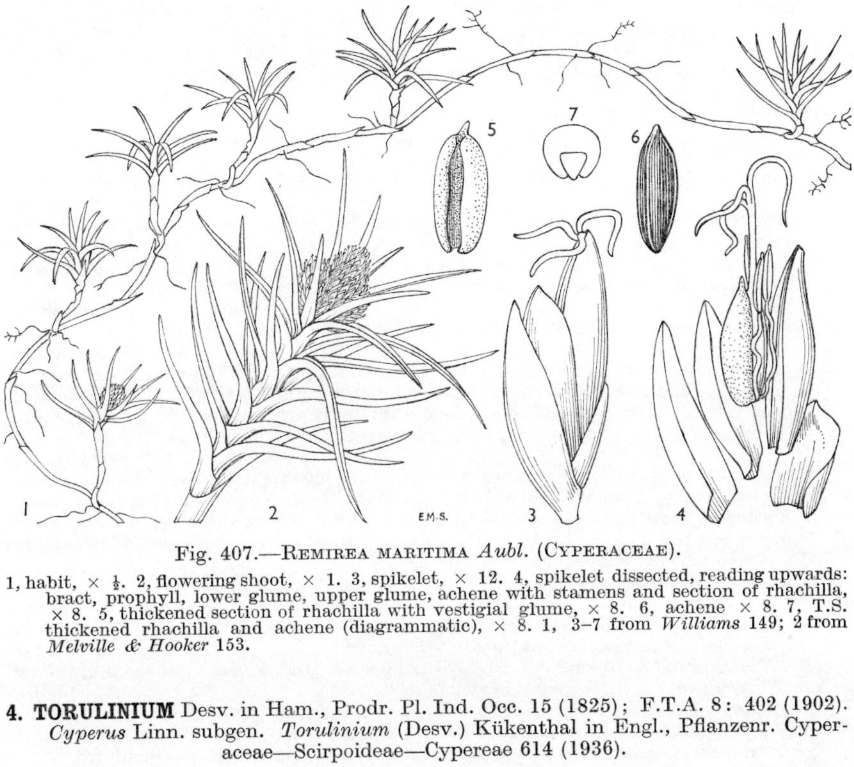

Fig. 407.—Remirea maritima *Aubl.* (Cyperaceae).

1, habit, × ½. 2, flowering shoot, × 1. 3, spikelet, × 12. 4, spikelet dissected, reading upwards: bract, prophyll, lower glume, upper glume, achene with stamens and section of rhachilla, × 8. 5, thickened section of rhachilla with vestigial glume, × 8. 6, achene × 8. 7, T.S. thickened rhachilla and achene (diagrammatic), × 8. 1, 3–7 from *Williams* 149; 2 from *Melville & Hooker* 153.

4. TORULINIUM Desv. in Ham., Prodr. Pl. Ind. Occ. 15 (1825); F.T.A. 8: 402 (1902). *Cyperus* Linn. subgen. *Torulinium* (Desv.) Kükenthal in Engl., Pflanzenr. Cyperaceae—Scirpoideae—Cypereae 614 (1936).

By Miss S. S. Hooper

Robust perennial to 2·5 m. tall, inflorescence a compound umbel up to 30 cm. across; bracts numerous about 2 cm. broad, scabrid beneath; spikelets terete up to 2·5 cm. long, patent; rhachilla disarticulating above each glume at maturity; stigmas 3 *odoratum*

T. odoratum (*Linn.*) *Hooper* in Kew Bull. 26: 579 (1972). *Cyperus odoratus* Linn., Sp. Pl. 46 (1753). *Torulinium confertum* Ham., Prodr. Pl. Ind. Occ. 15 (1825); F.T.A. 8: 403; F.W.T.A., ed. 1, 2: 486. *Cyperus ferax* L. C. Rich. in Act. Soc. Hist. Nat. Paris 1: 106 (1792); Kük. l.c. 615. Like the larger species of *Mariscus* in habit but distinguished when immature by the broad rounded green keel to the glumes, and when mature by the thickened rhachilla breaking into segments; near surface water.
Sen.: Tivaouane (Nov.) *Pitot*! Thiès (May) *Pitot*! Kayar (Mar.) *Pitot*! Sangalkam (June) *J. & A. Raynal* 5961! Hann (Sept.) *Pitot*! **Iv. C.:** Agnéby (Feb.) *Aké Assi* 6963! **Ghana:** Tano River, 75 km. W. of Kumasi (Dec.) *Oldeman* 824! Bibiani (Dec.) *Adams* 2032! Brimsu, Cape Coast (Mar., Sept.) *Hall* 328! 2618! Akotokyir, Cape Coast (June) *Hall* 3115! Adaiso to Kade (Jan.) *Morton* 8346! Nsawam to Suhum

(Nov.) *Morton* GC 7996! **Dah.:** Athiémé (Feb.) *J. & A. Raynal* 13522! **S. Nig.:** Ibadan (Oct.) *Onochie* FHI 34335! Lagos *Dawodu* 213! (June, Nov.) *Haines* 35! *Millen* 11! Pantropical; particularly New World.

5. PYCREUS P. Beauv., Fl. Oware 2: 48, (1816); F.T.A. 8: 288 (1901). *Cyperus* Linn. subgen. *Pycreus* (P. Beauv.) C.B.Cl.—Kükenthal in Engl., Pflanzenr. Cyperaceae —Scirpoideae—Cypereae 326 (1936).

By Miss S. S. Hooper

Aquatic or semi-aquatic plants growing in, or at the edge of, open water:
Weak creeping or floating plants with slender stems and leaves and few spikelets:
Inflorescence with rays:
Spikelets borne singly on the rays of the inflorescence, pale .. 1. *waillyi*
Spikelets 2–5 on each ray, brown or tinged brown .. .. 2. *demangei*
Inflorescence of 1–2 sessile spikelets .. .. 3. *felicis*
Stout semi-aquatic plants more or less decumbent at the base but with robust stems up to 4 mm. across, thickened by yellowish leaf sheaths below; inflorescence an umbel of spreading subequal rays bearing short spikes of red-brown spikelets; glumes obtuse and boat-shaped at the apex:
Leaves lanceolate, borne at intervals up the stem, stiff, spreading, about 10 cm. long; sheaths with a membranous, convex, often wrinkled and red-dotted area at the mouth of the sheath opposite the blade insertion; bracts $\frac{3}{4}$–$1\frac{1}{2}$ times as long as the rays; spikelets lanceolate 6–10 × 2–2·5 mm. .. .. 23. *mundtii*
Leaves linear, flat, long acuminate, up to 60 cm. long; sheath concave at the mouth, splitting early; longest bract 2–3 times as long as the rays; spikelets elliptical to oblong, about 15 × 4 mm. .. .. 4. *nitidus*
Land plants, not normally submerged or floating:
Spikelets white or greenish-white even at maturity, borne in a capitate cluster:
Leaves capillary, not more than 1 mm. broad; culms thickened below by brown fibrillose remains of leaf-sheaths; spikelets few, 3–4 mm. broad, subacute 7. *scaëttae*
Leaves flat, 1–5 mm. broad; spikelets numerous (more than 15), about 2·5 mm. broad, acute:
Robust plant, stem about 2·5 mm. broad; spikelets up to 30 mm. long, in a short, branched, bracteate spike; glumes lanceolate; achene broadly elliptical 9. *cataractarum*
Stem 1·5–2 mm. broad; spikelets up to 15 mm. long, sessile or nearly so; glumes ovate; achene obovate .. .. 8. *smithianus*
Spikelets yellow, brown, red or black, at least at maturity; inflorescence various:
Glumes with a distinct white membranous margin and broad rounded green keel; plant robust, with a spreading anthelate inflorescence; spikes compound; spikelets distant; glumes brown, spreading to reveal the black achene 17. *macrostachyos*
Glumes with no white membranous margin or if a narrow one then slender annuals:
Spikelets ovate, 2–4 mm. long, 8–12 flowered, in small compound spikes; stems clustered, shortly woody at the base; leaves flat; glumes black with conspicuous green keels .. .. 6. *elegantulus*
Spikelets elliptical, lanceolate or linear, more than 4 mm. long:
Glumes with an acute green keel excurrent in a mucro:
Spikelets up to 5 per stem, solitary in the axils of bracts arranged in a simple spike, broadly elliptical, about 10 × 4 mm. .. 26. *divulsus* subsp. *africanus*
Spikelets numerous in an anthelate inflorescence:
Weak annual 5–40 cm. high; spikelets linear 4–16 × 1 mm.; glumes lax, spreading to reveal the wide, pale rhachilla .. .. .. 22. *pumilus*
Perennial, culms arising from a short broad rhizome covered with black fibres; spikelets lanceolate, about 10–15 × 3 mm., compressed, brown, channelled; achene obovate or suborbicular, white muriculate, obtuse or apiculate 21. *acuticarinatus*
Glumes acute, obtuse or obtuse and apiculate:
Inflorescence a digitate or very shortly spicate cluster of spikelets, sometimes either reduced to 1 or 2 spikelets or enlarged by the addition of a lateral rayed spike:
Caespitose perennial; culms arising from a slender yellow rhizome; spikelets linear-lanceolate, 10–16 × 2·5–3 mm., yellow; glumes obtuse with a narrow golden sinuate margin; achene chestnut, puncticulate, apiculate 10. *lanceolatus*
Annual with fibrous roots:
Spikelets linear up to 1·5 mm. broad; leaves capillary:
Glumes black with a narrow white membranous margin; spikelets less than 1 mm. broad; achene muriculate .. .. 15. *melas*

Glumes golden to dark brown; spikelets 1–1·5 mm. broad; achene puncticulate .. .. 16. *capillifolius*
Spikelets 1·5 mm. broad or more:
Spikelets dark reddish black with a narrow membranous margin, lanceolate, about 10 × 2·5 mm.; glumes with a narrow insertion exposing the wide straight rhachilla; leaves narrow, inrolled; longest bract erect; achene obovate or suborbicular, black, puncticulate .. .. 25. *atrorubidus*
Spikelets straw-coloured to brown:
Spikelets usually up to 5 per stem, broadly elliptical to oblong-lanceolate, 3·5–5 mm. broad:
Spikelets 1–2 per stem, strongly compressed; glumes golden, tinged and tipped dark red (rare) .. .. 24. *pauper*
Spikelets generally more than 2 per stem, somewhat compressed, obtuse; glumes pale, becoming golden at maturity
12b. *pseudodiaphanus* var. *occidentalis*
Spikelets usually more than 5, narrowly elliptical to linear, 1·5–2·5 mm. broad:
Spikelets numerous, linear, about 1·5 mm. broad .. 27. *fallaciosus*
Spikelets narrowly elliptical, 2–2·5 mm. broad .. .. 28. *flavescens*
Inflorescence with several unequally rayed spikes; rays sometimes very short:
Spikelets very dark red-brown to black; perennials; stems 40–100 cm. high; glumes lanceolate, imbricate:
Leaves up to 4·5 mm. broad, flat; spikelets 12–20 mm. long with about 30 glumes; achene black, narrowly elliptical; stamens 2, anthers about 0·5 mm. long .. .. 13. *nuerensis*
Leaves about 2 mm. broad, inrolled; spikelets 10–14 mm. long with about 20 glumes; achene brown; stamens 3, anthers about 1 mm. long 14. *testui*
Spikelets pale to dark brown:
Rhizomatous perennial; culms erect, 50–100 cm., sharply triangular; leaf-sheaths tinged red-purple; bracts sharply keeled, spinulose-serrulate on the margin; rays stiff, up to 10 cm.; spikelets golden, elliptical, 10–15 × 4 mm., channelled .. .. 5. *unioloides*
Annuals, or perennials without the above combination of characters:
Inflorescence lax, spreading, forming $\frac{1}{3}$–$\frac{1}{2}$ height of the plant, with capillary rays up to 12 cm. long; spikelets lanceolate, golden to brown, compressed; glumes obovate, obtuse and minutely apiculate, spreading to reveal the discoid black achene; rhachilla markedly zig-zag .. .. 18. *pelophilus*
Inflorescence not forming more than $\frac{1}{4}$ height of plant; rhachilla not markedly zig-zag:
Achene narrowly elliptical to oblanceolate, distinctly papillate with papillae in longitudinal rows; spikelets acute, pale becoming more or less tinged reddish-brown; glumes elliptical, regularly imbricate giving the spikelet a neat " plaited " appearance:
Spikelets more or less erect in dense fascicles; rays generally up to 3 cm. long, sometimes almost wanting; stamens generally 2, anthers about 0·4 mm. long .. .. .. 19a. *polystachyos* var. *polystachyos*
Spikelets spreading in lax branched spikes, rays often exceeding 3 cm.; glumes often tinged chestnut-brown; stamens generally 3:
Spikelets about 1·0 cm. long by 1·5 mm. broad; glumes 1·5 mm. long; anthers about 0·8 mm. long .. 19b. *polystachyos* var. *laxiflorus*
Spikelets up to 2·5 cm. long by 2 mm. broad; glumes 2 mm. long; anthers 1·0–1·2 mm. long.. .. .. 20. *ferrugineus*
Achene broadly elliptical, obovoid or spherical:
Achene obovoid, slightly compressed, beakless, puncticulate; robust erect plant to 70 cm. high; spikelets golden, tinged brown, 10–18 × 3–4 mm., channelled; stamens 3 .. .. 11. *mortonii*
Achene not puncticulate:
Achene compressed with a surface pattern of vertically much elongated cells in irregular transverse bands separated by raised ridges:
Inflorescence with well developed rays and lax-spikeletted spikes; spikelets acute; glumes broadly elliptical, generally tinged red-brown, very obtuse, 2·5–3 mm. long .. .. 29. *intermedius*
Inflorescence with generally short rays, often reduced to a single spike; spikes more or less dense-spikeletted; glumes broadly ovate, generally golden, 2–2·5 mm. long .. .. 28. *flavescens*
Achene subspherical with a surface pattern of hexagonal cells, not much elongated vertically with a greyish white horizontal muriculation
12. *pseudodiaphanus*
Spikelets 2–3 mm. broad; glumes 1·6–2·1 mm. long; achene 0·7–0·95 × 0·6–0·9 mm. .. .. 12a. *pseudodiaphanus* var. *pseudodiaphanus*

Spikelets 3·5–5·0 mm. broad; glumes 2·4–2·7 mm. long; achene 0·9–1·25 × 0·9–1·05 mm. .. 12b. *pseudodiaphanus* var. *occidentalis*

1. **P. waillyi** *Cherm.* in Bull. Soc. Bot. Fr. 85: 366 (1938). Submerged or floating with an inflorescence of spikelets borne singly; in shallow water.
Mali: Dogo (Mar.) *Davey* 6! Gao to Bena (Feb.) *de Wailly* 5340! Also in Zambia.
2. **P. demangei** *J. Raynal* in Kew Bull. 23: 314 (1969). Similar to *P. waillyi* Cherm. in its floating habit but having very slender (filiform) stems and leaves and clustered spikelets; flood plains.
Mali: Dogo (Feb.) *Demange* 1589! 3114! (Jan., Mar.) *Davey* 168! 575! Sarédina, Debo (Sept.) *Demange* in *J. & A. Raynal* 11297! Also in Chad.
3. **P. felicis** *J. Raynal* in Kew Bull. 26: 568 (1972). Slender herb with straggling branched stems, setaceous leaves and 1–2 sessile pseudolateral spikelets, 5–12 × 3 mm.; near surface water.
Guin.: Fouta Djalon *Jac.-Fél.* 7413 *b*!
4. **P. nitidus** (*Lam.*) *J. Raynal* in Kew Bull. 23: 314 (1969). *Cyperus nitens* Lam., Ill. Gen. 1: 145 (1791). *Pycreus lanceus* (Thunb.) Turrill in Kew Bull. 1925: 67 (1925). *Cyperus lanceus* Thunb., Prodr. Pl. Cap. 18 (1794); Kük. l.c. 333. *Pycreus umbrosus* Nees in Linnaea 10: 130 (1836); F.T.A. 8: 303. Robust plant, with culms up to about 60 cm. and a yellow to dark red-brown inflorescence; in or near permanent water.
Sen.: Hann (Apr.) *Chev.* 25711! Niakoulrab (May) *Berhaut* 661! 5181! Malika to Yombeul (Nov.) *Pitot*! Berr Tialane (May) *J. & A. Raynal* 5892! Lompoul (Mar.) *J. & A. Raynal* 13653! **Dah.:** Cadjehoun, Cotonou (Feb.) *J. & A. Raynal* 13528! **Niger:** Guidimouni (July) *P. de Fabrègues* 524! 1892! **N. Nig.:** Matyoro (Feb) *Thornewill* 158! Baga Seyoram, L. Chad (Mar.) *Gwynn* 112! **W. Cam.:** Babeni, Ndop Plain (Apr.) *Brunt* 372! Africa south of the Sahara and Madagascar; also in Palestine.
5. **P. unioloides** (*R. Br.*) *Urb.* Symb. Antill. 2: 164 (1900). *Cyperus unioloides* R. Br., Prodr. Fl. Nov. Holl.: 216 (1810); Kük. l.c. 338. *Pycreus angulatus* (Nees) Nees in Linnaea 9: 283 (1835); C.B. Cl., Illustr. Cyper. t. 4 (1909); F.T.A. 8: 305; F.W.T.A., ed. 1, 2: 490. Erect plant, woody at the base with an inflorescence of large golden spikelets; swamps or moist places in savanna.
Gam.: (Aug.) *Ruxton* 82! **Mali:** Klela, Sikasso (Aug.) *Demange* 2906! **Iv. C.:** Kodiokoffi, Baoulé *Pobéguin* 109! *Chev.* 22323! **U. Volta:** Banfora (June) *Leeuwenberg* 4358! **Ghana:** Sampa, N.W. Ashanti (Apr.) *Morton* A3268! **S. Nig.:** Shaki Dist. (Aug.) *Latilo & Taylor* FHI 43398! Pantropical.
6. **P. elegantulus** (*Steud.*) *C.B. Cl.* in Dur. & Schinz, Consp. Fl. Afr. 5: 538 (1895). *Cyperus elegantulus* Steud. in Flora 25: 583 (1842); Kük. l.c. 342. Plant with a compact anthelate inflorescence of small black strongly compressed spikelets with green keels, subtended by long bracts; in swamps.
N. Nig.: Mambila Plateau (Mar.) *J. Hall* 1642! **W. Cam.:** Bamenda, 5,000 ft. (May) *Brunt* 1137! Bafut-Ngemba F.R., Bamenda (May) *Daramola* FHI 41183! **F. Po:** 8,500 ft. (Apr.) *Mann* 1470! Throughout tropical and S. Africa.
[Closely approaching, especially in the Fernando Po specimen, the central and S. American species *P. niger* (Ruiz & Pavon) Cufodontis.]
7. **P. scaëttae** *Cherm.* in Bull. Jard. Bot. Brux. 13: 278 (1935). *Cyperus fibrillosus* Kük. var. *scaëttae* (Cherm.) Kük. l.c. 348 (1936). Small tufted perennial with a white head of spikelets; montane grassland liable to burning.
W. Cam.: Bafut-Ngemba F.R., Bamenda, 5,500–7,000 ft. (Mar.) *Hepper* 2241! *Coombe* 213! Lake Bambulue, Bamenda (Apr.) *Morton* K. 268! Also in E. Cameroun and Congo.
8. **P. smithianus** (*Ridley*) *C.B. Cl.* in Dur. et Schinz, Consp. Fl. Afr. 5: 542 (1895); F.T.A. 8: 301. *Cyperus smithianus* Ridley in J. Bot. 22: 15 (1884); Kük. l.c. 349. Similar to *P. scaëttae* but larger, with grass-like leaves and a white head of numerous spikelets; river banks and other damp open habitats.
Guin.: Friguiagbé (June) *Chillou* 531! Macenta (Apr.) *Pitot*! Dabola (Apr.) *Pitot*! Madina Tossekré (Oct.) *Adam* 12514! **S. L.:** Falaba (Apr.) *Deighton* 1645! **U. Volta:** Banfora (July) *Aké Assi* 10737! **N. Nig.:** Ropp, Plateau Prov. (Apr.) *J. Hall* 1114! Mai Samare, Mambila Plateau (Mar.) *J. Hall* 1629! **S. Nig.:** Obudu Plateau, 5,000–5,500 ft. *Tuley* 972! (Apr., Aug.) *Haines* 232! *Kershaw* 900684! **F. Po:** Moka (Sept., Dec.) *Melville* 680! *Boughey* 111! Throughout tropical Africa.
9. **P. cataractarum** *C.B. Cl.* in Engl., Bot. Jahrb. 38: 132 (1906). More robust than the preceding species with a less contracted inflorescence of larger spikelets; river banks.
S. Nig.: R. Nwop, Boje enclave (May) *Jones & Onochie* FHI 18609! **W. Cam.:** Mamfe to Ikom (Apr.) *Morton* GC 7140! Falls of R. Lobo, Kribi (July) *J. Hall* 1866! Also from E. Cameroun to Congo.
10. **P. lanceolatus** (*Poir.*) *C.B.Cl.* in Dur. & Schinz, Consp. Fl. Afr. 5: 538 (1895), partly. *Cyperus lanceolatus* Poir. in Lam., Encycl. 7: 245 (1806); Kük. l.c. 349. *Pycreus propinquus* Nees in Mart., Fl. Bras. 2, 1: 7 (1842); F.T.A. 8: 300, partly; Chev. Bot. 696. Loosely tufted perennial, stems generally with a single cluster of golden spikelets; riverbanks, marshes and other damp habitats.
Sen.: (Nov.) *Fotius* K.713! Niokolo-Koba (June) *Adam* 14256! Badi (Sept.) *Berhaut* 3054! **Gam.:** Albreda (June) *Leprieur*! **Mali:** Balasoko, Bamako to Koulouba (Dec.) *Monod*! Koulouba *Duong* 1503! Sébéla (Jan.) *Roberty* 3305! Ségou (May) *Roberty* 3652! **Guin.:** Friguiagbé (May) *Schnell* 5626 *bis*! Hollandé Tossekré (Oct.) *Adam* 12725! Daralabe, Fouta Djalon (Jan.) *Langdale-Brown* 2573! Kouroussa (Jan.) *Pobéguin* 626! Koba (Dec.) *Jac.-Fél.* 7367! **S.L.:** Waterloo (Aug.) *Melville & Hooker* 298! Mahera (Nov.) *Deighton* 5666! Robump, nr. Rokupr (Mar.) *Jordan* 14! Kombile (Oct.) *Small* 452! Bwedu (Apr.) *Deighton* 3156! **Lib.:** Kailahun, Kolahun Dist. (Nov.) *Baldwin* 10128! Vonjama (Oct.) *Baldwin* 9930! Gbanga (Sept.) *Linder* 494! Ganta, Sanokwele Dist. (May) *Harley* 1703! Nimba Mts. (Dec.) *Adam* 20204! **Iv. C.:** Sanlo to Kakpin (Nov.) *Aké Assi* 9276! Bouaké (May) *de Wilde* 90! Brobo (Sept.) *de Wilde* 939! **U. Volta:** Darsalami, nr. Robo (Dec.) *Bille* 3351! Bekui (Dec.) *Bille* 3381! **Ghana:** Burufo, nr. Lawra (Sept.) *Hall* CC 730! Banda (Aug.) *Hall* 1982! Assin Bereku *West-Skinn* 68! Tiasi, Afram Plains (Aug.) *Hall* CC 166! Kukrukui Lagoon, Kete Krachi (Apr.) *Hall* VBS 005! **Togo Rep.:** Dzolokpuita (Nov.) *Morton* GC 9355! Sansanné-Mango (Oct.) *Bille* TG 362! **Dah.:** Agouagon (Apr.) *Chev.* 23495! **N. Nig.:** Minna (Dec.) *Meikle* 717! Kufena, Zaria Prov. (Jan.) *Kershaw* 900015! Wushishi, Niger Prov. (June) *Clayton* 1105! Richa, Jos Plateau (Sept.) *Lawlor & Hall* 663! **S. Nig.:** Ilesha (Apr.) *Haines* 36! Oyo to Iseyin (Oct.) *Onochie* FHI 34920! Ibadan *Hambler* 116! Iva stream, Udi Dist. (Sept.) *Onochie* FHI 34087! Obudu (Aug.) *Kershaw* 900683! **W. Cam.:** Madele, Wum Dist. (June) Daramola FHI 41100! Jakiri, Bamenda, 6,500 ft. (Feb.) *Hepper* 2068! Throughout tropical Africa and Madagascar and in central and S. America.
11. **P. mortonii** *Hooper* in Kew Bull. 23: 313 (1969). Similar to *P. unioloides* (R. Br.) Urb. but annual, altogether smaller, and without spinulose-scabrid leaf and bract-tips; wet savanna.
Sen.: Niokolo-Koba (Nov.) *Adam* 17143! Siminti (Jan.) *Adam* s.n.! Kanèméré (Nov.) *Fotius* K.689! **Ghana:** Pong Tamale (Dec.) *Morton* GC 9832! Damongo (Dec.) *Morton* GC 25060! **N. Nig.:** Gampu (Dec.) *Peter & Tuley* 28! Igbete-Kishi (Mar.) *Gledhill* 731! Also in E. Cameroun, Chad and Uganda.
12. **P. pseudodiaphanus** *Hooper* in Kew Bull. 23: 313 (1969). Similar in habit to *P. mortonii* but generally smaller and with a less compressed, greyish-white often iridescent achene; damp places.
12a. **P. pseudodiaphanus** *Hooper* var. **pseudodiaphanus**
Sen.: Niokolo-Koba (Nov.) *Adam* 17079/15! Velingara (Nov.) *Boudet* 4607! 4616! 4621! **Mali:** Sotuba (Sept.) *Adam* 15329! 15330! 15345! (Oct.) *Raynal* 5014! Sikasso (Oct.) *Demange* 2713! Kouoro *Demange* 3528! **U. Volta:** Comoé R. (Sept.) *Adam* 15262! 15269! Banfora *Jaeger* 6556! 6557! **Ghana:** Ejura scarp (Dec.) *Morton* GC 9621! Sawla *Hall* CC 806! Damongo to Yapei (Sept.) *Hall* CC 881! Tamale (Oct.) *Baldwin* 13564! Nakpanduri to Bawku (Aug.) *Hall* CC 537, partly! **N. Nig.:** Idon, Kaduna to Keffi, Zaria Prov. (Aug.) *Kershaw* 900702! 900707! E. of Bida (Dec.) *Haines* 95! Shagunu, Ilorin (Oct.) *J. Hall* 1451!

12b. **P. pseudodiaphanus** var. **occidentalis** Hooper l.c. 313 (1969). Spikelets wider and achene larger than in the typical variety.
Sen.: Oussouye (Oct.) *Adam* 18311! 18337! Djembering (Oct.) *Adam* 18343! Abémé (Sept.) *Adam* 18183! Kabrousse (Oct.) *Adam* 18304! Niokolo-Koba *Adam* 17079/15. **Guin.:** Kandiara Plateau (Oct.) *Pitot*! Kaiserville to Youkounkoun (Oct.) *Pitot*! **Ghana:** Kpandai (July) *Hall* GC 39508!

13. **P. nuerensis** (*Boeck.*) *Hooper* in Kew Bull. 26: 579 (1972). *Cyperus nuerensis* Boeck. in Flora 62: 555 (1879). *Cyperus globosus* All. var. *nuerensis* (Boeck.) Kük. l.c. 356, partly (1936). *Pycreus polystachyos* (Rottb.) P. Beauv. var. *sanguineus* Kük. l.c. 371 (1936). *P. globosus* (All.) Rchb. var. *nilagiricus* of F.T.A. 8: 299, partly. Distinguished from the other dark brown or black spikeletted species by the broader flat leaves and long linear spikelets; swampy ground up to 5,800 ft.
Guin.: Milo (Apr.) *Adam* 7959! Loffa (Feb.) *Adam* 3760! **S.L.:** Koinadugu Dist. (Aug.) *Haswell* 248*a*! 252! Loma Mts. *Jaeger* 2001! 7027! **Iv. C.:** Mankono (June) *Chev.* 21928! **N. Nig.:** Jos Plateau (Aug.) *Lawlor & Hall* FHI 46642! Banyo R. (Apr.) *J. Hall* 1772! Also in Central Africa and the Sudan, probably also extending through east Africa to Rhodesia.

14. **P. testui** *Cherm.* in Arch. Bot. Caen 4, Mém. 7: 13 (1931). *Cyperus laxespicatus* Kük. var. *testui* (Cherm.) Kük. l.c. 332 (1936). *Pycreus globosus* (All.) Rchb. var. *nilagiricus* of F.W.T.A., ed. 1, 2: 490. Erect sedge with inrolled leaves and narrowly elliptical blackish spikelets; wet savanna.
Sen.: Badi to Fourou (Apr.) *Nongonierma* 441! **Mali:** Sikasso (May, July) *Demange* 2249! 2888! **Guin.:** Dabola to Farana (Apr.) *Pitot*! Farana *Sc. Elliot* 5334! (Apr.) *Pitot*! Timbi (June) *Adam* 14642! Madina Tossékré (Mar.) *Adam* 11612! **Iv. C.:** Tehini, Bouna Dist. (May) *Aké Assi* 4802! Nassian (June) *Aké Assi* 10218! **Ghana:** Banda, Wenchi Dist. (Dec.) *Morton* GC 24204! Nwereme, N.W. Ashanti (May) *Irvine* 2474! Sampa, N.W. Ashanti (Apr.) *Morton* A3266! Ejura (Aug.) *Hall* CC 308! **S. Nig.:** Old Oyo F.R., Igboho (Feb.) *Keay* FHI 23446! **W. Cam.:** Baba, Ndop Plain (Mar.) *Brunt* 123! 124! Throughout tropical Africa.

15. **P. melas** (*Ridl.*) *C.B. Cl.* in Dur. & Schinz, Consp. Fl. Afr. 5: 538 (1895). *Cyperus melas* Ridl. in Trans. Linn. Soc. ser. 2, Bot. 2: 127 (1884); F.T.A. 8: 302; Kük. l.c. 357. Elegant, slender, tufted annual, similar to *P. capillifolius* but the spikelets darker and narrower; wet flushes on rocky hillsides.
Ghana: Kintampo (Sept.) *Hall* CC 924! **N. Nig.:** Kainji (Nov.) *Hepper* 3878! Kaduna to Keffi, Zaria Prov. (Aug.) *Kershaw* 900701! Panshanu, Bauchi Prov. (Aug.) *Lawlor & Hall* 620! Widespread in tropical Africa.

16. **P. capillifolius** (*A. Rich.*) *C.B. Cl.* in Dur. & Schinz, Consp. Fl. Afr. 5: 535 (1895); F.T.A. 8: 300. *Cyperus capillifolius* A. Rich., Tent. Fl. Abyss. 2: 475 (1851); Kük. l.c. 357, t. 42. *C. afzelii* Boeck. in Linnaea 35: 475 (1868). Small tufted plants with pseudolateral sessile spikes of radiating spikelets; rice nurseries, marshy grassland, stream edges and other wet habitats.
Sen.: *Afzelius* 45! Sédhiou, Casamance (Oct.) *Adam* 18448! Niokolo-Koba (Dec.) *J. & A. Raynal* 6800! Fafakonou to Medina Yoro Foula (Sept.) *Boudet* 3851! Ouassadou (Sept.) *Berhaut* 1435! **Mali:** Kita Mts. (Sept.) *Jaeger* 62! 64! 73! Bamako (Nov.) *Duong*! Koulikoro (Oct.) *Chev.* 2478! **Guin.:** Youkounkoun (Oct.) *Pitot*! Sébori (Oct.) *Adam* 12677! Seriba (Oct.) *Adam* 12691! Kinkou to Pita (Oct.) *Pitot*! Friguiagbé *Chillou* 598! **S. L.:** Wellington (Sept.) *Harvey* 116! Kambia (July) *Jordan* 467! Bumban (Sept.) *Thomas* 2016! Koinadugu Dist. (Aug.) *Haswell* 257! Bintimani Mt., 5,600 ft. (July) *Bakshi* 294! **Lib.:** Kailahun (Nov.) *Baldwin* 10143! 12048! Nimba Mts. (Dec.) *Adam* 16473! **U. Volta:** Banfora (Sept.) *Jaeger* 6558! *Adam* 15395! **Ghana:** Burufo, nr. Lawra (Sept.) *Hall* CC 736! Kwahu Tafo (Aug., Dec.) *Harris*! *Hall* CC 83! Berekum to Sampa (Dec.) *Adams* 5279! **N. Nig.:** (Aug.) *Lely* P465! P470! Kabba (Oct.) *Parsons* 33! Kufena, Zaria (Aug.) *Kershaw* 900691! Kaduna to Keffi, Zaria (Aug.) *Kershaw* 900709! Panshanu, Bauchi Dist. (Aug.) *Lawlor & Hall* 560! Throughout tropical Africa and Madagascar; also in Brazil.

17. **P. macrostachyos** (*Lam.*) *J. Raynal* in Kew Bull. 23: 314 (1969). *Cyperus macrostachyos* Lam., Ill. Gen. 1: 147 (1791). *Pycreus tremulus* (Poir.) C.B. Cl. in Dur & Schinz, Consp. Fl. Afr. 5: 542 (1895); F.T.A. 8: 306; Chev. Bot. 969. *Cyperus tremulus* Poir. in Lam., Encycl. 7: 264 (1806); Kük. l.c. 361. *Pycreus albomarginatus* Nees in Mart., Fl. Bras. 2, 1: 9 (1842). Rather robust annual with few, flaccid leaves but a conspicuous spreading inflorescence; in or near open water and other damp places.
Sen.: *Roger* 30! 39! *Heudelot* 537! Dagana (Aug.) *Naegelé* 11750! Oussouye to Ziguinchor (Sept.) *Broadbent* 113! Velingara (Oct.) *Adam* 18577! **Gam.:** (Aug., Sept.) *Macluskie* 15! *Ruxton* 87! 118! Bansang *Duke* 10! **Mali:** Kita Mts. (Aug.) *Jaeger* 18! Soninkoro, Ségou (Sept.) *Roberty* N2686! Dioura (Sept.) *Davey* 70! San (June, Sept.) *Chev.* 1090! 2484! Keribane (Oct.) *Davey* 85 = C.I.P.A.S. 486! **Guin.:** Friguiagbé (Sept.) *Chillou* 2121! Ditinn (Oct.) *Pitot*! Kandiara Plateau (Oct.) *Pitot*! Thianguel -Bori (Oct.) *Pitot*! Sambailo (Oct.) *Pitot*! **S. L.:** Kambia (July) *Jordan* 464! **U. Volta:** Djibasso (Nov.) *Bille* 3290! Ouagadougou (Aug.) *Nongonierma* 80! **Ghana:** Burufu, Lawra (May) *Morton* 7668! Amanma, Cape Coast (June) *Hall* 3143! Yeji (Apr., Aug.) *Hall* VBS 1269! CC 388! Pwalugu (Sept.) *Hall* CC 582! Bawku (Aug.) *Hall* 2070! **Niger:** Toukounous (Sept.) *Koechlin* 6618! nr. Niamey (Sept.) *Cremers* 964! Niamey to Filingué (Sept.) *Cremers* 949! Say *Koechlin* 6393! **N. Nig.:** Shagunu, N. Ilorin Prov. (July–Aug.) *C. D. K. Cook* 426! Kufena, Zaria Prov. (Aug.) *Kershaw* 900690! Naraguta F.R., Jos Plateau (July) *Lawlor & Hall* FHI 46538! Panshanu Pass, Bauchi Dist. (Aug.) *Lawlor & Hall* 218! Katagum *Dalz.* 241! Gajibo, Bornu Prov. (Nov.) *H. B. Johnston* N55! Pantropical.

18. **P. pelophilus** (*Ridl.*) *C.B. Cl.* in Dur. & Schinz, Consp. Fl. Afr. 5: 540 (1895); F.T.A. 8: 298. *Cyperus pelophilus* Ridl. in Trans. Linn. Soc. ser. 2, Bot. 2: 129 (1884). Annual with flaccid stems, few leaves and a large spreading inflorescence of rather few golden, lanceolate spikelets; moist sandy places near water.
Guin.: *Brun* 718! Friguiagbé (June, July) *Chillou* 2681! 3048! Scattered throughout tropical Africa and Madagascar.

19. **P. polystachyos** (*Rottb.*) *P. Beauv.* Fl. Oware 2: 48, t. 86 (1816); F.T.A. 8: 296. *Cyperus polystachyos* Rottb., Descr. Pl. Rar. Programm. 21 (1772); Kük. l.c. 367. *Pycreus odoratus* Urb., Symb. Antill. 2: 164 (1900); F.W.T.A., ed. 1, 2: 490, not *Cyperus odoratus* Linn. (1753).

19a. **P. polystachyos** (*Rottb.*) *P. Beauv.* var. **Polystachyos.** Tufted, with a rosette of basal leaves shorter than the culms and with a capitate or radiate inflorescence of numerous clustered, erecto-patent spikelets having a characteristically neat "plaited" appearance; dunes, rice-swamps and other damp places in the coastal region.
Sen.: Dakar (Aug., Sept.) *Adam* 1795! 1920! Hann (Sept.) *Trochain* 467! M'boro (Aug.) *Trochain* 4241! Cambérène (Oct.) *Berhaut* 1050! **Lib.:** Robertsport, Grand Cape Mount Dist. (Dec.) *Baldwin* 10901! Monrovia (June) *Baldwin* 5860! Sasstown, Sinoe Dist. (Mar.) *Baldwin* 11607! Cape Palmas *Ansell*! Harper, Maryland Dist. (June) *Baldwin* 5977! **Iv. C.:** Grand Bassam to Azuretti (Aug.) *des Abbayes* 383! Canal d'Assini (July) *Oldeman* 155! Cavally to Bliéron (Aug.) *Chev.* 19913! Baoulé (Sept.) *Pobéguin*! **Ghana:** Atwabo *Fishlock* 15 of 1931! Weija (Oct.) *Irvine* 4801! Keta (June) *Thorold* CB 100! Nungua (Apr.) *Rose Innes* GC 30108! Prampram (Nov.) *Morton* 5077! **Togo Rep.:** Lomé *Warnecke* 123! (Sept.) *Mahoux* 130! **Dah.:** Ouidah (Aug.) *Adjanohoun* 236! **Niger:** Watameze (Nov.) *P. de Fabrègues* 1304! Dallol Boboye, Niamey (Sept.) *Cremers* 965! **N. Nig.:** Sokoto (July) *Dalz.* 471! Marachi (Aug.) *J. Hall* 398! **S. Nig.:** Badagry (Aug.) *Onochie* FHI 34256! Ogun River F.R., Lagos (Mar.) *Hepper* 2247! Ibadan (Apr.) *Meikle* 1415! Ajagbodudu, Benin (Aug.) *Wright* 55! Nun R. (Aug.) *T. Vogel* 11! Tropical and subtropical zones of both hemispheres.

19b. **P. polystachyos** var. **laxiflorus** (*Benth.*) *C.B. Cl.* in Hook. f., Fl. Brit. Ind. 6: 592 (1893); Cherm. in Arch. Bot. Caen 7, Mém. 4: 6 (1936). *Cyperus polystachyos* Rottb. var. *laxiflorus* Benth., Fl. Austral. 7: 261 (1878). Spikelets remote and spreading, generally red-brown; rice-swamps on edge of tidal area and marshy places.
Sen.: *Roger* 39! St. Louis *Heudelot* 498! Sangalkam (Sept.) *Berhaut* 1573! Fatick (Sept.) *Adam* 18026!

Oussouye (Sept.) *Broadbent* 119! **Gam.:** (July, Aug.) *Ruxton* 13! 123! **Guin.:** Koba (Nov.) *Jac.-Fél.* 7282! Ile de Kabak (Dec.) *Jac.-Fél.* 7384! Samou (Dec.) *Jac.-Fél.* 7409! **S. L.:** Lunsar (Aug.) *Adames* 124! Kichom (July) *Deighton* 3759! Rokupr (Aug.) *Deighton* 3023! Kalangba (Apr.) *Glanville* 216! Njama (June) *Deighton* 5958! Old World tropics.

20. **P. ferrugineus** (*Poir.*) *C.B. Cl.* in Hook. f., Fl. Brit. Ind. 6: 593 (1893). *Cyperus ferrugineus* Poir. in Lam., Encycl. 7: 261 (1806). Similar to 19b but more robust; in similar habitats but less widespread.
Sen.: Baïla Bignona (Sept.) *Adam* 18115! Kabrousse (Oct.) *Adam* 18295, partly!
[This Madagascan and S. African species appears to have been introduced into West Africa. The relationship of West African plants to *Pycreus polystachyos* requires investigation.]

21. **P. acuticarinatus** (*Kük.*) *Cherm.* in Bull. Soc. Bot. Fr. 81: 262 (1934). *Cyperus acuticarinatus* Kük. in Fedde, Rep. 12: 93 (1913). *C. macranthus* Boeck. var. *mucronatus* (Kunth) Kük. f. *acuticarinatus* Kük. in Engl., Pflanzenr. Cyper. 389 (1936). *C. crustaceus* Raymond in Garcia de Orta 11: 374 (1963). *Pycreus angulatus* of F.W.T.A., ed. 1, 2: 490. Caespitose from a woody base, spikelets golden-brown and characterized by the acute and mucronate glumes; wet grassland in savanna region.
Sen.: Selikekemé (Sept., Oct.) *Boudet* 3670! 3970! Diulayel (Sept.) *Boudet* 3670! **Mali:** Sotuba (Dec.) *Duong*! *Adam* 11414! 15332! **Port. G.:** Boruntuma (Dec.) *Pereira* 2285! **Guin.:** Youkounkoun (Oct.) *Pitot*! **Iv. C.:** Tiegouakro to Kodiokoffi, Baoulé (Aug.) *Chev.* 22333 *bis*! Pacoubo (Aug.) *Scaëtta* 3037! Foro-foro Forest, 25 km. N. of Bouaké (Sept.) *Oldeman* 404! Toumodi (Oct.) *Pobéguin* 72! **Ghana:** Wa to Lawra (Sept.) *Hall* CC 610! Banda (Aug.) *Hall* 1984! Ejura (Aug., Dec.) *Andoh* 5057! Ayafie to Atafie (Dec.) *Adams* 4678! Tiasi, Afram Plains (Aug.) *Hall* CC 165! **Togo:** *Mahoux* 2060! **S. Nig.:** Ezillo, Ogoja Dist. (Oct.) *Ariwaodo* in Hb. *Tuley* 925! Also in E. Cameroun.

22. **P. pumilus** (*Linn.*) *Nees* in Linnaea 9: 283 (1834); Berhaut, Fl. Sén. ed. 2, 384 (1967). *Cyperus pumilus* Linn., Cent. Pl. 2: 6 (1756). *C. pumilus* Linn. var. *patens* (Vahl) Kük. l.c. 378. Weak annual 5–40 cm. high with linear yellow-brown spikelets; wet places, often on cultivated or otherwise disturbed sandy soil.
Sen.: *Heudelot* 331! Velingara (Oct.) *Adam* 18590! Kabrousse (Oct.) *Adam* 18303! Elinkine (Oct.) *Adam* 18270! Kolda (Oct.) *Adam* 18542! **Mali:** Kita (Oct.) *Jaeger* 81! Bamako (Sept.) *Adam* 14960! Gorinnta, Dogo (Apr.) *Davey* 556! Gao (Mar., Apr.) *de Wailly* 4687! 5373! **Guin.:** Sambaïlo (Oct.) *Pitot*! Dielila (Oct.) *Pobéguin* 1880! **S. L.:** Kutr (Jan.) *Jordan* 4! Rokupr (Aug.) *Jordan* 491! Tauyia (Sept.) *Jordan* 89! Bunbuna (Aug.) *Deighton* 1402! Yakala (Sept.) *Thomas* 2376! **Iv. C.:** Bouaké (May) *de Wilde* 73! Mankoro (Sept.) *Nozeran*! Aboakouamékro (Nov.) *Aké Assi* 10340! **U. Volta:** Banfora (Sept.) *Adam* 15406! **Ghana:** Tumu (May) *Morton* GC 7536! Navrongo (Oct.) *Rose Innes* GC 31090! Tamale (Oct.) *Baldwin* 13577! Mankrong (Aug.) *Hall* CC 218! Labadi, Accra (Aug.) *Hall* 3281! **Togo Rep.:** Palimé (Sept.) *Mahoux* (111) 2006! Dapongo (Oct., Nov.) *Bille* TG 8! **Dah.:** Ouécé (Nov.) *Annet* 52! **Niger:** Niamey (Sept.) *Cremers* 955! **N. Nig.:** Kufena hill, Zaria (Sept., Oct.) *Clayton* 1325! *Kershaw* 900416! 900427! Panshanu, Bauchi Prov. (Aug.) *Lawlor & Hall* 389! Gurum, nr. Vogel Peak (Nov.) *Hepper* 1283! **S. Nig.:** Olokemeji F.R. to Eruwa (Aug.) *Keay* FHI 37170! Lagos *W. MacGregor* 9! (Sept.) *Stubbings* 19! Badagry (Aug.) *Onochie* FHI 33494! Igboora (Oct.) *Haines* 218! Old World Tropics, extending to the Himalayas and central America.
[The African and Madagascan taxon has been distinguished as *Cyperus patens* Vahl. Kükenthal regards it as a variety, distinguished by having muticous or shortly mucronate glumes, spreading at maturity; but plants with these characteristics occur throughout the range of the species. African specimens have generally longer and broader glumes than Indian but the ranges overlap.]

23. **P. mundtii** *Nees* in Linnaea 10: 131 (1836); F.T.A. 8: 294. *Cyperus mundtii* (Nees) Kunth, Enum. Pl. 2: 17 (1837); Kük. in Engl., Pflanzenr. Cyper. 380 (1936). A semi-aquatic plant with a generally submerged creeping rhizome, stiff leaves and pale to dark brown clusters of spikelets; stream and lake margins in or near open water.
Maur.: Moudjeria (Nov.) *Chudeau*! **Sen.:** M'bidjem (Nov.) *Pitot*! Kayar (Mar.) *Pitot*! Sangalkam (May) *J. & A. Raynal* 5860! Dagoudane to Pikine (Apr.) *J. & A. Raynal* 5715 *ter*! Berr Tialane (May) *J. & A. Raynal* 5893! **Guin.:** Koba *Jac.-Fél.* 7206! **Ghana:** Samreboi, W. Region (Feb.) *Hall* 2485! Accra to Winneba (Apr.) *Hall* 1874! Amisano, nr. Elmina (Mar.) *Hall* 1309! Kumasi (Mar.) *Adam* 19068! **Dah.:** Sémé Forest (Mar.) *Aké Assi* 11153! **N. Nig.:** Sokoto (July) *Dalz.* 465! Vom, Plateau Prov. (Apr., Dec.) *Morton* K326! *Haines* 71! Mallamfatori, Chad Dist. *Jackson* FHI 59168! Gwoza, Dikwa Dist. (Dec.) *McClintock* 69! Vogel Peak, Sardauna Prov. (Dec.) *Hepper* 1528! Mediterranean region, and throughout Africa, and in Madagascar.

24. **P. pauper** (*A. Rich.*) *C.B. Cl.* in Dur. & Schinz, Consp. Fl. Afr. 5: 540 (1895). *Cyperus pauper* Hochst. ex A. Rich., Tent. Fl. Abyss. 2: 478 (1851); Kük. l.c. 394. Slender tufted annual with few large elliptical spikelets; boggy ground.
N. Nig.: Kufena, Zaria Prov. (Aug.) *Kershaw* 900698! Also in E. Cameroun, Central African Republic, Ethiopia and SE tropical Africa.

25. **P. atrorubidus** *Nelmes* in Kew Bull. 6: 320 (1952). *Cyperus atrorubidus* (Nelmes) Raymond in Nat. Canad. 91: 127 (1964). More or less caespitose plants with dark red-black spikelets in a sessile pseudo-lateral cluster; damp places on mountains to 5,000 ft.
Guin.: Tossékré, Labé (Oct.) *Adam* 12732! Nzérékoré (Oct.) *Jac.-Fél.* 1923! **S. L.:** Bintumane Peak (Oct.-Nov.) *Jaeger* 400! 7886! **Lib.:** Nimba Mts. (Aug.) *Schnell* 6241! **Iv. C.:** Nimba Mt., 4,500 ft. (Aug.) *Boughey* GC 18164! Also in SE Africa.

26. **P. divulsus** Ridl. subsp. **africanus** *Hooper* in Kew Bull. 26: 579 (1972). Unique in its few large spikelets each sessile in the axil of a bract and arranged in a spike; open grassland.
S. L.: Freetown (Nov.) *Deighton* 1868! **N. Nig.:** Farin Rua, Plateau Prov. (Aug.) *J. Hall* 652! Also in E. Cameroun, Ethiopia and Zambia; subsp. *divulsus* comes from Madagascar.

27. **P. fallaciosus** *Cherm.* in Arch. Bot. Caen 7, Mém. 4: 7 (1936). *Cyperus fallaciosus* (Cherm.) Raymond in Nat. Canad. 91: 129 (1964). Differing from *P. flavescens* in the narrower spikelet and narrower brown achene; damp places.
Sen.: Massadala *Trochain* 3538! 3545! **Mali:** Djenné (July) *Chev.* 1141! **Guin.:** Dabola (Apr.) *Unknown Collector*! **Ghana:** Oti R., Zabzugu (Apr.) *Morton* GC 9057! Yeji (Apr.) *Hall* VBS 1279! Chindiri (Apr.) *Hall* VBS 1292! Kete Krachi (Apr., July) *Hall* VBS 1284! *Ansah-Emmim* VBS 46! **Niger:** Mirria (July) *P. de Fabrègues* 2706! Guidiguir *P. de Fabrègues* 1731! **N. Nig.:** Nupe *Barter* 1567! Also in Central African Republic.
[Very close to *P. flavescens* (from which it may not be specifically distinct) but reminiscent of *P. polystachyos* in the size and crowding of the spikelets.]

28. **P. flavescens** (*Linn.*) *Reichenb.* Fl. Germ. Excurs. 72 (1830); F.T.A. 8: 290. *Cyperus flavescens* Linn., Sp. Pl. 46 (1753); Kük. l.c. 298. Similar to *P. lanceolatus* but generally not exceeding 30 cm. high, without a rhizome and often with shorter spikelets; wet places.
Sen.: Kolda (Feb.) *Adam & Kerharo* 19754! **Mali:** Niafounké (July) *Demange* 2201! Kolongotomo (Apr.) *Demange* 2233! Bamako (Mar.) *Demange* 2228! **Guin.:** Sériba (Oct.) *Adam* 12695! Madina Tossékré (Oct.) *Adam* 12513! 12523! Kouroussa (July) *Pobéguin* 420! **S. L.:** York (Sept.) *Melville & Hooker* 407! Rokupr (Sept.) *Jordan* 118! **Ghana:** Tumu dam (Mar.) *Morton* GC 8855! Natinga, Nakpanduri to Bawku (Aug.) *Hall* CC 537! **Togo Rep.:** Lama-Kara *Mahoux* 2058! **Niger:** Guidiguir (June) *P. de Fabrègues* 1731! Zinder (Feb.) *Chev.* 43642! **N. Nig.:** Bida (Dec.) *Haines* 95! Kufena, Zaria Prov. (Sept.) *Kershaw* 900417! Jos (Dec.) *Haines* 80! Bauchi to Dindima (Dec.) *Haines* 78! Katsina to Daura (Feb.) *Meikle* 1217! **W. Cam.:** Bamessi, Ndop Plain (Apr.) *Brunt* 310! Pantropical and in temperate regions extending to central Europe, Himalayas, Canada and southern Brazil.

29. **P. intermedius** *C.B. Cl.* in F.T.A. 8: 290 (1901); J. & A. Raynal in Adansonia sér. 2, 7: 320 (1967). *Cyperus intermedius* Steud. in Flora 25: 581 (1842), and Syn. Pl. Glum. 2: 5 (1855); not of Guss. (1832). *C. subintermedius* Kük. l.c. 390 (1936). Similar to *P. flavescens*, but a generally taller plant with a more expanded inflorescence of acute spikelets; damp places.
Sen.: Badène, Casamance (Jan.) *J. & A. Raynal* 7958! Mare Diamowel, Niokolo-Koba (Dec.) *J. & A. Raynal* 6889! Scattered throughout tropical Africa and in Madagascar.

6. KYLLINGA Rottb., Descr. Ic. Nov. Pl. 12 (1773); F.T.A. 8: 268; *nom. cons. Cyperus* L. subgen. *Kyllinga* (Rottb.) Valck.-Sur.—Kükenthal in Engl., Pflanzenr. Cyperaceae—Scirpoideae—Cypereae 566 (1936).

By Miss S. S. Hooper

Inflorescence of several subequal rays; spikelets few, remote; plant annual 1. *debilis*
Inflorescence capitate, with one or more short spikes bearing crowded spikelets:
Annual with wing of glume having lobes about 1 mm. long; achene suborbicular 2. *squamulata*
Perennial, or annual not having a lobed wing to the glume:
Plant with tuberous stem bases and thread-like slender rhizomes; spikelet with at least 4 opaque, white flowering glumes .. .. 3. *bulbosa*
Plant not having many slender rhizomes; spikelet with 2 or 3 flowering glumes:
Central (or only) spike of inflorescence cylindrical, white, greenish-white or pink:
Plant with a slender stem, slightly swollen at the base; centre spike about 4 × 3 mm., pinkish with 1–2 small lateral spikes; glumes scarcely exceeding achene 18. *tisserantii*
Plant with more or less well developed rhizome; glumes exceeding achene:
Plant with a thick, often vertical stem covered with fibrous remains of leaf-bases; keel of glume generally smooth, thickened .. .. .. 17. *appendiculata*
Plant with a thin, horizontal rhizome; keel of glume often ciliate, green 4. *odorata*
Central (or only) spike of inflorescence spherical, ovoid or ellipsoidal, rarely cylindrical and then golden:
Plant caespitose without a conspicuous rhizome; flowering stems crowded:
All heads of a single spherical spike:
Spike 10–12 mm. wide, reddish when dry; keel of glume with a wide, finely ciliate wing; leaves 2–4 mm. broad .. .. 5. *nigritana*
Spike 8–10 mm. wide, white even when dry; keel with an inconspicuous wing; leaves often decaying into black fibres .. .. .. 6. *echinata*
At least some heads of more than one spike:
Spikes more or less confluent into a reddish hemispherical head, 7–10 mm. across, partially enveloped by the broad bases of the bracts; leaves glaucous, lower leaf-sheaths and roots purple; achene obovate; in saline swamps 7. *robusta*
Spikes discrete, not reddish:
Leaf-sheaths usually entire, pale crimson, not fibrous; glumes equal, with white, weakly nerved sides, a conspicuous green ciliate keel and recurved apiculus .. .. 8. *pumila*
Leaf-sheaths usually becoming fibrous; stem bases often thickened; upper of two flowering glumes smaller than lower, distinctly stipitate, tips of glumes not recurved:
Keel of glume conspicuously winged and ciliate; achene broadly elliptical, little exceeded by the glumes .. .. 9. *welwitschii*
Keel of glume smooth or sparingly ciliate; achene elliptical or ovate generally much exceeded by the glumes .. .. 10. *tenuifolia*
Plant rhizomatous:
Inflorescence of a single spherical spike, 8–12 mm. across, subtended by 3 short, often very short, bracts; spikelets elliptical, somewhat blunt; stems spongy 2–3 mm. wide, tinged purple below; generally lower, and often all, leaves bladeless; rhizome aromatic, scales cinnamon-coloured .. 11. *peruviana*
Inflorescence of one or more cylindrical, hemispherical or ovoid spikes; spikelets acute, often with spreading tips to the glumes; stems leafy:
Spike solitary, golden with green keels; spikelets large (about 4–5 mm.) with acuminate, spreading tips:
Rhizome elongate, generally flexuose and slender when dry, loosely invested with elliptical, apiculate, scarious, pink-tinged scales; leaves and bracts about 5 mm. broad, flaccid, acuminate, bracts erecto-patent or patent; (F. Po, W. Cam.) .. .. 12. *elatior*
Rhizome thick with ovate, acute scales; bracts deflexed at maturity; spike large (about 10–12 × 6–8 mm.); (N. Nig.) .. .. .. 13. *melanosperma*
Spikes one or more, white, greenish-white or yellow-green:
Rhizome thin, wiry, generally elongated between the flowering stems, closely invested with dark scales:

Head small, 4–5 mm. across, subtended by three stiff bracts; glumes white with green keels .. .. 14. *brevifolia*
Head 5–10 mm. across, subtended by 4 long, flaccid bracts; glumes membranous below, opaque and spongy above, red dotted .. .. 15. *nemoralis*
Rhizome thicker, generally bearing a row of contiguous flowering stems:
Rhizome horizontal covered with more or less persistent, ovate, acute, dark margined scales .. .. 16. *erecta*
Bracts generally 3, short to fairly long; spike rarely more than one, ovoid, greenish-white or yellow; glumes smooth or very sparsely ciliate on the keel .. .. 16a. *erecta* var. *erecta*
Bracts 4–7:
Bracts 4–5; spike solitary, spherical, white; glumes smooth or somewhat ciliate .. .. 16b. *erecta* var. *africana*
Bracts 4–7, long; spikes several, confluent, greenish white or yellow-green; glumes ciliate on the keel .. .. 16c. *erecta* var. *polyphylla*
Rhizome short, often vertical:
Rhizome covered with reddish scales or their fibrous remains; keel of glume generally smooth, thickened, opaque .. .. .. 17. *appendiculata*
Rhizome not covered with scales; keel of glume smooth or ciliate, not thickened .. .. 4. *odorata*

1. **K. debilis** *C.B. Cl.* in Mém. Soc. Bot. Fr. 2, 8: 26 (1907); Berhaut, Fl. Sén. ed. 2, 376. *Cyperus leptorhachis* Mattf. & Kük. in Engl., Pflanzenr. Cyper. 595, fig. 63 A–D (1936). A weak caespitose annual 5 to 30 cm. tall, with a *Mariscus*-like inflorescence but broadly obovate one-fruited spikelets; in dry, rocky places. **Sen.:** Camberené (Nov.) *Pitot*! Tambacounda *Berhaut* 620! 3181. Tambacounda to Sambaïlo (Sept.) *Pitot*! **Mali:** Gulo Kourou (Aug.) *Jaeger* 8! Kita (Sept.) *Jaeger* 93! Bamako (Sept.) *Adam* 15034! Koulikoro (Oct.) *Chev.* 2455! **Guin.:** *Scaëtta* 3362! **Iv. C.:** Téhini, Bouna Reserve (Aug.) *de Wilde* 689! **U. Volta:** Sindou (Sept.) *Adam* 15199! Gaoua (Oct.) *Rose Innes* GC 31661*a*! **Togo Rep.:** Mango *Mahoux* 2131! **N. Nig.:** Anara F.R., Zaria (July) *Clayton* 1257! Dumbi, Zaria (Sept.) *Kershaw* 900424! Also in Central African Republic.

2. **K. squamulata** *Thonn. ex Vahl* Enum. Pl. 2: 381 (1806); F.T.A. 8: 270 (excl. syn. *K. dipsacoides* Schumach.); Nelmes & Baldwin, Amer. J. Bot. 39: 389 (1952); Berhaut, Fl. Sén. ed. 2, 365. *Cyperus metzii* (Steud.) Mattf. & Kük. l.c. 612, fig. 64 J–K (1936). A weak, leafy annual bearing pyramidal, spinulose heads of few, large spikelets above the broad bract bases; a widespread weed of sandy soil. **Maur.:** Traya (Jan.) *Naegelé*! **Sen.:** Sebikotane (Oct.) *Pitot*! M'bao (Oct.) *Giovanetti*! Oussouye to Ziguinchor (Sept.) *Broadbent* 112! Velingara (Oct.) *Adam* 18585! Djembering (Sept.) *Broadbent* 135! **Gam.:** *Saunders* 125! *Hayes* 568! Kuntaur (Aug.) *Ruxton* 155! Yundum Agric. Stn. *Ashrif* 39! **Mali:** Kita (Sept.) *Jaeger* 88! Fodébougou (Oct.) *Jaeger* 67! Baroneli (Aug.) *Roberty* 2649! San (Sept.) *Chev.* 2485! Boré (Aug.) *Demange* 30/1957! **Port. G.:** Bissao (Aug.) *Dinklage* 3093! **Guin.:** Inguiagbé (Aug.) *Chillou* 2703! Télimélé (Oct.) *Pitot*! Youkounkoun (Oct.) *Pitot*! Kouroussa (Aug.) *Pobéguin* 397! **S. L.:** Rokupr (July) *Jordan* 480! Mambolo (Sept.) *Jordan* 85! Rogbasa, Bikolo–Makene (Aug.) *Deighton* 1270! Njala (July) *Deighton* 749! **Lib.:** *Dinklage* 2743. **Iv. C.:** Mankono (Oct.) *Aké Assi* 7097! **Ghana:** Cape Coast (June) *Hall* 1499! 3116! Nsawkaw, Wenchi (Aug.) *Hall* 1954! Abetifi to Pepease (Apr.) *Morton* A758! Gambaga (Apr.) *Akpabla* 76! Yendi (Dec.) *Adams* 4090! Achimota *Akpabla* 2003! **Togo Rep.:** Sokode (June) *Mahoux* 2089! *Kersting* 232. Mangu *Mellin* 103. Kumea *Mahoux* 2082! **N. Nig.:** Sokoto (Aug.) *Dalz.* 451! Nupe *Baikie*! Damaturu, Bornu Prov. (Aug.) *De Leeuw* 1083! **S. Nig.:** Lagos *W. MacGregor* 12! Ibadan (Aug.) *Keay* FHI 37143! Auchi (Apr.) *Haines* 204! Throughout tropical Africa, also in India and America.

3. **K. bulbosa** *P. Beauv.* Fl. Oware 1: 11, t. 8 fig. 1 (1805). *K. macrocephala* A. Rich., Tent. Fl. Abyss. 2: 491 (1851); F.T.A. 8: 286; C.B. Cl., Ill. Cyper. t. 2, fig. 7–9 (1909). *K. albiceps* (Ridley) C.B. Cl. ex Rendle, Cat. Welw. 2, 1: 106 (1899); F.T.A. 8: 286; F.W.T.A., ed. 1, 2: 487. *Cyperus richardii* Steud., Syn. Pl. Glum. 2: 8 (1855) and var. *angustior* Kük. in Engl., Pflanzenr. Cyper. 570 (1936). A generally small, slenderly rhizomatous and tubiferous perennial with white heads appearing yellow during flowering because of the numerous anthers; in grass on tracks and other disturbed habitats.
Mali: San (June) *Chev.* 1035! **Guin.:** Madina Tossékré (Oct.) *Adam* 12525! **S. L.:** Mapotolon (Sept.) *Jordan* 918! Freetown (Oct.) *Deighton* 2148! Lumley (Aug.) *Melville & Hooker* 258! Mano (June) *Deighton* 6075! Yakala (Sept.) *Thomas* 2382! **U. Volta:** Banfora *Duong* 1317! **Ghana:** Tamale (Mar.) *Morton* GC 8817! **N. Nig.:** Sokoto *Dalz.* 453! Acharane, Ankpa Dist. (June) *Daramola* FHI 38036! **S. Nig.:** Lagos Is. (July) *Haines* 8! Sapoba *Kennedy* 2642! Foriku, Ondo Dist. (July) *Onochie* FHI 33421! Obiaruku, Warri (July) *Kershaw* 900658! Calabar (May) *Holland* 50! Obokum, nr. Ikom *Swarbrick* 2998! **W. Cam.:** Bambui, 5,000 ft. (July) *Bumpus* Bam/22! Ndop Plain, 3,800 ft. (May) *Brunt* 434! Mamfe (Apr.) *Morton* K 322! Throughout tropical Africa.

4. **K. odorata** *Vahl* Enum. Pl. 2: 382 (1806); C.B. Cl., Ill. Cyper. t. 2, fig. 3–5 (1909); Nelmes & Baldwin in Am. J. Bot. 39: 387 (1952); Berhaut, Fl. Sén. ed. 2, 366. *K. cylindrica* Nees in Wight, Contrib. Bot. India 91 (1834); F.T.A. 8: 282; Berhaut, Fl. Sén. ed. 2, 366. *Cyperus sesquiflorus* (Torr.) Mattf. & Kük. l.c. 591 (1936), incl. var. *cylindricus* (Nees) Kük. and f. *spinulosus* Kük. Tufted or with a short, slender rhizome, leafy; spikes white, generally several with the central spike cylindrical and large; spikelets ovate, glumes clearly nerved; general in damp habitats.
Guin.: Friguiagbé (July) *Chillou* 590! Kinkou to Pita (Oct.) *Pitot*! Guekédou (Apr.) *Pitot*! Sérédou to N'zérékoré (Apr.) *Pitot*! Macenta (Apr.) *Pitot*! **S. L.:** Lumley (Aug.) *Melville & Hooker* 259! Rokupr, Magbema (Nov.) *Jordan* 700! Musaia (Sept.) *Small* 273! Mayombo, nr. Roruks (July) *Deighton* 4678! Tombo (Jan.) *Deighton* 994! Madina (Aug.) *Jordan* 908! **Lib.:** Monrovia (June–Sept.) *Dinklage* 2228! 2859! Buchanan (Dec.) *Adam* 16522! Bassa Cove *Ansell*! **Iv. C.:** *Boughey* GC 14595! Man (Aug.) *Boughey* X 418! Issia (Dec.) *Boughey* GC 13595! Téhini, Bouna Reserve (Aug.) *de Wilde* 826! **Ghana:** Wenchi Ferry (May) *Morton* GC 7234! Ejura (Aug.) *Hall* CC 353! Kwahu Tafo to Mankrong (Apr.) *Morton* A669! Kete Krachi (Apr.) *Morton* GC 9135! Amedzofe Hill (May) *Morton* GC 7222! **N. Nig.:** Panshanu Pass, Bauchi Prov. (Aug.) *Lawlor & Hall* 292! **S. Nig.:** Igboora, Oyo Prov. (Oct.) *Haines* 224! Ajasse, SE. Ilorin *Clayton* 1110! Oka, Owo Dist. (Apr.) *Haines* 4*a*! Obudu Plateau, Ogoja Prov., 5,500 ft. (Aug.) *Stone* 42! **W. Cam.:** Cam. Mt., 8,000–10,000 ft. *H. H. Johnston*! Bambui Farm, Bamenda (May) *Pedder* 20! **F. Po:** Santa Isabel *Swarbrick* 2901! Throughout tropical Africa and America, also in Madagascar.

5. **K. nigritana** *C.B. Cl.* in F.T.A. 8: 272 (1902). *K. ascolepidioides* Cherm. in Arch. Bot. Caen 4, Mém. 7: 4 (1931). ? *K. alata* Nees in Linnaea 10: 139 (1835), as *allata*; F.W.T.A., ed. 1, 2: 487; Aké Assi, Contrib. 2: 71. *Cyperus cristatus* (Kunth) Mattf. & Kük. var. *nigritanus* (C.B. Cl.) Kük. l.c. 610 (1936) (as to W. African specimens only). A tufted perennial, with aromatic roots, leaves up to 25 cm. long and a spherical head of large, winged spikelets; in savanna.

Iv. C.: Foro-foro Forest, 25 km. N. of Bouaké (Sept.) *Oldeman* 416! **Ghana:** Bole (May) *Morton* GC 7693! Banda (Apr., Aug.) *Morton* A3292! *Hall* 1993! Kintampo (Aug.) *Hall* 2056! **N. Nig.:** Damaturu, Bornu Prov. (Aug.) *De Leeuw* 1072! Nupe *Barter* 1588! Also in the Central African Republic; and this or closely related species of the *K. alba* Nees complex throughout tropical Africa.

6. **K. echinata** *Hooper* in Kew Bull. 26: 580 (1972). Tufted plant with capillary stems to 40 cm. high, leaf-sheaths readily splitting into black fibres, capillary leaves and bracts and rounded heads of long-tipped white spikelets; damp sandy ground.
Ghana: Adamso, NW. Ashanti (Nov.) *Harris*! Kaleo Hill, nr. Wa (Apr.) *Adams* 4006! Damongo (Mar.–July) *Morton* GC 8664! *Adams* 953! Damongo to L. Kpirri (Mar.) *Adams* 3961!

7. **K. robusta** *Boeck.* in Linnaea 35: 409 (1863). *K. caespitosa* Nees var. *major* Nees in Mart., Fl. Brasil. 2, 1: 12 (1842). *Cyperus densicaespitosus* Mattf. & Kük. var. *major* (Nees) Kük. l.c. 599 (1936). A strongly caespitose perennial with a reddish, irregularly hemispherical head partially enveloped by the broad bract-bases; confined to brackish swamps.
Sen.: Kayar (Mar.) *Pitot*! **Gam.:** Pakali Nding *Raynal* 7978! **Guin.:** Ouassou (Nov.) *Jac.-Fél.* 7252! **S. L.:** Rokupr (Mar.) *Jordan* 414! Tumbu, Little Scarcies R. (Apr.) *Glanville* 207! Kitchom (Jan.) *Deighton* 948! Talil salt area (Jan.) *Deighton* 3549! **Lib.:** Stockton creek, Monrovia (May) *Dinklage* 2725! Buchanan (Dec.) *Adam* 16034! **Iv. C.:** Sassandra (July) *Aké Assi* 10690! **Ghana:** Ankobra River, Axim (Mar.) *Morton* GC 8472! **S. Nig.:** Ikorodu, Lagos Lagoon (Dec.) *Onochie* FHI 32022! Ikoyi, Lagos (Jan., July) *Jones* FHI 19426! *Dalz.* 1430! Abonnema (Nov.) *Anderson* 1414! **F. Po:** (Oct.) *T. Vogel* 11! Also in E. Cameroun, the Congo and in Brazil.
[The type (*Sellow* Brazil) has leaves longer than the flowering stems and coriaceous and thus differs slightly from the W. African material here cited in which the herbaceous leaves are shorter than or equal to the flowering stems.]

8. **K. pumila** *Michx.* Fl. Bor.-Amer. 1: 28 (1803); F.T.A. 8: 281; Nelmes & Baldwin in Amer. J. Bot. 39: 387 (1952); Berhaut, Fl. Sén. ed. 2, 368. *Cyperus densicaespitosus* Mattf. & Kük. l.c. 597, incl. var. *rigidulus* (Steud.) Kük. (1936); Raymond in Garcia de Orta 11: 374 (1963). Caespitose with numerous, small, green and white, 3–4 spiked, pyramidal heads; on margins of streams, rivers and swamps and similar moist places.
Sen.: L. Tanma (Jan.) *Pitot*! Sangalkam (Jan.) *Pitot*! **Port G.:** Bafata (Jan.) *Raimundo & Guerra* 896. **Mali:** Fodebougou (July) *Jaeger* 28! Kita (Oct.) *Jaeger* 78! 90! 5756! Sotuba (Nov.–Dec.) *J. & A. Raynal* 5124! 5242! **Guin.:** Friguiagbé (Apr.) *Chillou* 1248! Mamou (May) *Pitot*! Koba (Nov.) *Jac.-Fél.* 7243! 7261! Faranah (Apr.) *Pitot*! Kissidougou to Guékédou (Apr.) *Pitot*! **S. L.:** Rokupr (Oct.) *Jordan* 579! Makump (May) *Deighton* 1712! Newton (Nov.) *Deighton* 1465! Njala (Sept.) *Deighton* 2097! Bumban (Aug.) *Deighton* 1293! **Lib.:** (Feb.) *Harley* 1802! Monrovia (July, Aug.) *Naumann* 16! *Kunkel* 89! Nimba (Dec.) *Adam* 20074! Gretown, Tchien Dist. (July) *Baldwin* 6773! **Iv. C.:** Bouaké (May) *de Wilde* 71! Adiopodoumé (July) *Boughey* GC 14513! Yapo Forest, Abidjan (July) *Boughey* GC 14591! Anguedédou Forest, Abidjan (Nov.) *Leeuwenberg* 1879! **Ghana:** Assin-Prasu *West-Skinn* 67! Kumasi (Nov.) *Adams* 4427! Burufo, Lawra (May) *Morton* 7643! Gambaga (May) *Morton* 7443! Kade (June) *Irvine* 4853! Aframso, nr. Mampong (Apr.) *Hall* 3030! **N. Nig.:** Lemme (May) *Lely* 155! Lokoja *Richardson*! Jemaa (Feb.) *McClintock* 192! Mayo Selbe, Gashaka Dist. (Jan.) *Hepper* 1638! **S. Nig.:** Ibadan (Jan., Feb.) *Meikle* 996! Sapoba *Kennedy* 2712! *Brenan* 8977! Ogbomosho (Dec.) *Haines* 64! Ifaki (June) *Clayton* 1126! Owo to Benin (Apr.) *Haines* 2! Nsukka (Jan.) *Okigbo* 145! **W. Cam.:** Ndop, 3,500 ft. (Nov., Dec.) *S. Gillett* 11! *Boughey* GC 11384! GC 11389! Mamfe (Apr.) *Morton* GC 7136! **F. Po:** *T. Vogel* 219! Moka, 4,000 ft. (Dec.) *Boughey* 36! Throughout tropical Africa and temperate and tropical America.

9. **K. welwitschii** *Ridley* in Trans. Linn. Soc. Bot. ser. 2, 2: 147 (1884). *K. triceps* Rottb. var. *ciliata* Boeck. in Peters, Reise Mossamb. Bot. 535 (1864); Kük. l.c. 579 (1936); F.T.A. 8: 281. *K. blepharinota* Hochst. ex Boeck., *nom. illeg.*; Berhaut, Fl. Sén. ed. 2, 365. *K. controversa* Steud. var. *subexalata* C.B. Cl. in F.T.A. 8: 271 (1902). *K. controversa* of F.W.T.A., ed. 1, 2: 487; Aké Assi, Contrib. 2: 71. *Cyperus controversus* (Steud.) Mattf. & Kük. var. *subexalatus* (C.B. Cl.) Kük. l.c. 612 (1936). Similar to *K. tenuifolia* in habit but distinguished by the clearly ciliate keel to the glume and the broader achene; sandy ground in savanna, etc.
Maur.: Rosso (Oct.) *Adam* 19299! Tanfott, Tagant (Aug.) *Monod* 1916! **Sen.:** *Heudelot* 399! *Naegelé*! Dahra-Djolof (Sept.) *Fotius* 166! *Davey* 011! Bouta (Feb.) *Davey* 36! **Iv. C.:** Séguela, fide *Aké Assi* l.c. **Niger:** Filingué, Toukounous *Bartha* 91! Nord Gouré (Aug.) *P. de Fabrègues* 683! **N. Nig.:** Sokoto (Aug.) *Dalz.* 455! Bindawa F.R., Katsina (July) *Grove* 3! Damaturu, Bornu Prov. (Aug.) *De Leeuw* 1101! Also in E. Cameroun, Chad and throughout tropical Africa.

10. **K. tenuifolia** *Steud.* Syn. Pl. Glum. 2: 69 (1855). *K. triceps* Rottb., Descr. et Ic. 14, t. 4, fig. 6 (1773), *nom. illeg.*; F.T.A. 8: 280; F.W.T.A., ed. 1, 2: 487; Berhaut, Fl. Sén. ed. 2, 368. *Cyperus triceps* (Rottb.) Endl., Cat. Hort. Vindob. 1: 94 (1842); Kük. l.c. 578 (1936). Caespitose; stems thickened at base with old leaf-sheaths; leaves gradually acuminate; inflorescence rounded-pyramidal, bracts spreading; damp places in savanna, etc.
Sen.: Sebikotane (Sept.) *Pitot*! Djembering (Sept.) *Broadbent* 136! Niokolo-Koba (July) *Adam* 14356! **Mali:** Kita (July–Oct.) *Duong* 431! *Jaeger* 2950! Bamako (Sept.) *Waterlot* 1393! *Jaeger* 5082! Segou (Aug.) *Roberty*! **Guin.:** Friguiagbé (Aug.) *Chillou*! Mamou (May) *Pitot*! Sérédou (Apr.) *Pitot*! **Iv. C.:** Issia Rock (May–Aug.) *Leeuwenberg* 4136! *Boughey* 14702! **Ghana:** Tono, nr. Navrongo (Aug.) *Irvine* 4661! Salaga (Apr.) *Morton* A3908! Ejura, Ashanti (Dec.) *Morton* GC 9709! Tiasi, Afram Plains (Aug.) *Hall* CC 164! Agomeda (July) *Adams* 4872! **Togo Rep.:** Agudewe, Sokode (Aug.) *Schilling* 27a! **Dah.:** Agouagon (May) *Chev.* 23529! **N. Nig.:** Nupe *Barter* 1588! Samaru, Zaria Prov. (July) *C. B. Taylor* 21! Naraguta F.R., Jos Plateau (July) *Lawlor & Hall* FHI 46590! Damaturu, Bornu Prov. (Aug.) *De Leeuw* 1115! Yola (July) *Dalz.* 265! Abinsi (May) *Dalz.* 824! **S. Nig.:** Olokemeji (Oct.) *Hepper & Keay* 935! Throughout the Old Word tropics.
[The following specimens, distinguished by the small inflorescence and smooth keel to the glume, may represent a distinct taxon:— **Mali:** Kita Mt. (July) *Jaeger* 21! 44! **Ghana:** Wenchi to Techiman (Mar.) *Morton* GC 8551! Gambaga *Hall* 781!]

11. **K. peruviana** *Lam.* Encycl. 3: 366 (1789); F.T.A. 8: 278; Nelmes & Baldwin in Amer. J. Bot. 39: 389, t. 80–87 (1952); Berhaut, Fl. Sén. ed. 2, 364. *Cyperus peruvianus* (Lam.) F. N. Williams in Bull. Herb. Boiss. 2, sér. 7: 90 (1907); Kük. l.c. 586. Rhizomatous with a distinctive spherical, white inflorescence, very short bracts and often leafless stems; dunes, tidal swamps, foreshores.
Sen.: *Heudelot* 150! *Sieber* 5! *Roger*! Kayar (Mar.) *Pitot*! Diouloulou (Sept.) *Adam* 18177! Djembering (Sept.) *Broadbent* 170! **Gam.:** *Skues*! **Guin.:** Iles Tristao *Jac.-Fél.* 7316! **S. L.:** Aberdeen (Aug.) *Melville & Hooker* 136! Lumley (May) *Marmo* 206! Samarank, Rokel R. (Mar.) *Glanville* 209! Yele, Turtle Is. (Nov.) *Deighton* 2377! **Lib.:** Robertsport (Dec.) *Baldwin* 10904! Bushrod Is. (Feb.) *Baldwin* 11073! Grand Bassa *Ansell*! Cape Palmas (Dec.) *Dinklage* 2360! Filoke, Webo Dist. (July) *Baldwin* 6667! **Ghana:** Half Assinie (July) *Chipp* 268! Atwabo *Fishlock* 22 (of 1931)! Axim (Dec.) *Johnson* 1002! Shama (May) *Hall* 2977! Keta (Mar.) *Morton* GC 6522! **S. Nig.:** Badagry (Aug.) *Onochie* FHI 33486! Onikan, Lagos (June) *Onochie* FHI 23309! Lagos Is. *Dalz.* 1113! Central Prov. *Rosevear* M.21! Nun R. (Aug.) *Mann* 455! Brass *Barter* 63! **W. Cam.:** Bibundi R. (May) *Domke* 711! Throughout coastal regions of W. Africa to the Congo and in central and tropical S. America.

12. **K. elatior** *Kunth* Enum. Pl. 2: 135 (1837); F.T.A. 8: 275. *Cyperus aromaticus* (Ridl.) Mattf. & Kük. var. *elatior* (Kunth) Kük. l.c. 582 (1936). Rhizome creeping; head cylindrical, bracts very long; in grassland and on forest paths.
W. Cam.: Cam. Mt. (Jan.) *Dunlap* 60! Buea, 3,000–4,000 ft. (Nov.–Mar.) *Migeod* 67! *Maitland* 329!

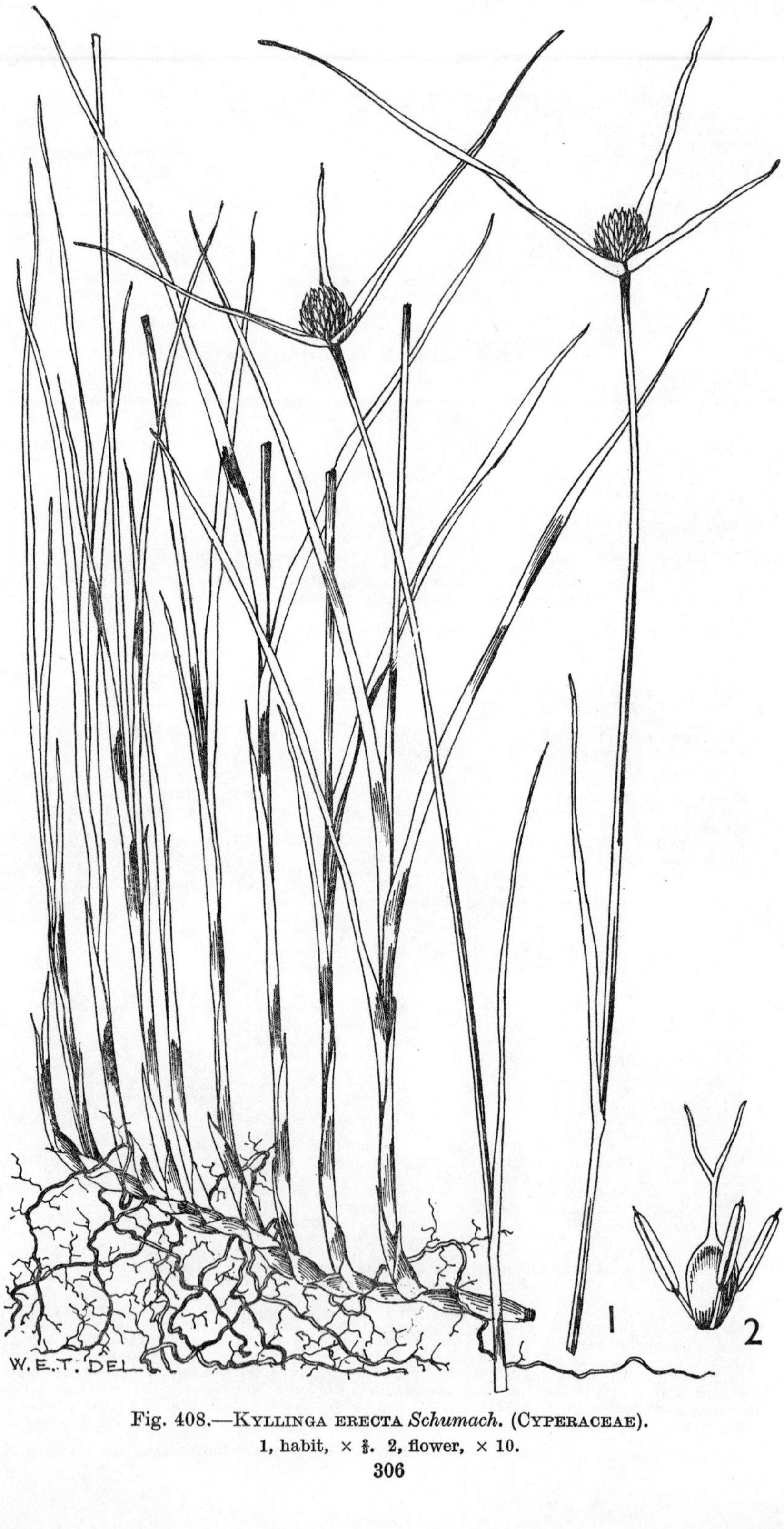

Fig. 408.—Kyllinga erecta *Schumach.* (Cyperaceae).
1, habit, × $\frac{2}{3}$. 2, flower, × 10.

Morton GC 6745! Bamessi, Ndop Plain (Apr.) *Brunt* 320! **F. Po:** Moka, 4,000 ft. (Dec.) *Boughey* 114! Also in E. Cameroun and throughout tropical and S. Africa.

13. **K. melanosperma** *Nees* in Wight, Contrib. Bot. Ind. 91 (1834); F.T.A. 8: 277. *Cyperus melanospermus* (Nees) Valck.-Sur., Het Gesl. Cyp. in Mal. Archip. 50 (1898). Rhizome thick, creeping; spike large, cylindrical, golden; damp places.
N. Nig.: Kuna, Pankshin Dist. (June) *Okafor* FHI 59252! Jos Plateau (Apr.) *Lely* P227! From Nigeria throughout tropical and S. Africa, also in Madagascar and the Old World tropics.
[This species is recorded from the flora area by Kükenthal and the above cited specimens agree with Indian specimens in inflorescence characters; however, both are incomplete and might represent large specimens of the closely allied *K. elatior*.]

14. **K. brevifolia** *Rottb.* Descr. et Icon. 13, t. 4, fig. 3 (1773); F.T.A. 8: 273; C.B. Cl., Ill. Cyper. t. 1, fig. 1–4 (1909). *K. colorata* of F.W.T.A., ed. 1, 2: 487. *Cyperus brevifolius* (Rottb.) Hassk., Catal. Bogor. 24 (1844); Kük. l.c. 600 (1936). Similar to *K. erecta* in the inflorescence but having an elongated slender rhizome; in damp places.
Lib.: Bushrod I., Monserrado Dist. (Nov.) *Winne* 557! Monrovia (June) *Dinklage* 2233! White Plains (May) *Dinklage* 2705! **S. Nig.:** *T. Vogel* 37! **W. Cam.:** Ekona (Aug.) *Chuml* CDG 374! Tropical and subtropical regions of the Old and New Worlds.

15. **K. nemoralis** (*Forst.*) *Dandy ex Hutch.* in F.W.T.A., ed. 1, 2: 487 (1936). *Thyrocephalon nemorale* J.R. & G. Forst., Char. Gen. Pl. 130 (1776). *K. monocephala* Rottb., Descr. et Icon. 13, t. 4, fig. 4 (1773), partly, excl. syn., nom. illeg.; F.T.A. 8: 272; C.B. Cl., Illustr. Cyper. t. 2, fig. 1–2 (1909). *Cyperus kyllingia* Endl., Cat. Hort. Acad. Vindob. 1: 94 (1842); Kük. l.c. 606, tab. 64 C–D. Rhizome slender, creeping, bearing stems at intervals, bracts and leaves long, flaccid, nut black; a weed of secondary forest and disturbed habitats.
Guin.: Macenta to Kérouané (Nov.) *Adam* 6970! **Lib.:** *Linder*! Monrovia (May–June) *Dinklage* 2181! 2713! Cape Palmas *Schönlein*! **Iv. C.:** Niégré Forest (Sept.) *Nozeran*! Sassandra (May) *Chev.* 17919! (Mar.) *Bernardi* 8542! **Ghana:** Deposo, Takoradi (June) *Hall* 1497! Kumasi (Nov.) *Adams* 4426! Mampong (Aug.) *Irvine* 5157! Suhum, Okroase to Amanase (June) *Adams* 387! Afram Mankrong (Apr.) *Morton* A772! Aburi (Oct.) *Morton* GC 7871*b*! **Togo Rep.:** Sokode *Schilling* 14*a*! **S. Nig.:** Ajagbodudu, Benin Div. (Aug.) *Wright* 51! Sapele, Sapoba Div. *Kennedy* 1422! **W. Cam.:** Victoria (Dec.) *Maitland* 712! Likomba Plantation (Dec.) *Mildbr.* 10790! Buea (Jan.–Mar.) *Maitland* 330! *Morton* GC 7052! Pantropical.

16. **K. erecta** *Schumach.* Beskr. Guin. Pl. 42 (1828); F.T.A. 8: 274, partly; Berhaut, Fl. Sén. ed. 2, 366.

16a. **K. erecta** *Schumach.* var. **erecta**—*Cyperus erectus* (Schumach.) Mattf. & Kük. l.c. 588 (1936), incl. var. *pleiocarpus* Kük. *Kyllinga pungens* of F.W.T.A., ed. 1, 2: 487, not of Link. Rhizome producing a row of contiguous stems; head ovoid greenish-yellow, bracts variable in length, generally three; in damp grassy places.
Sen.: Noflaye (July) *Adam* 12251! Sangalkam *Berhaut* (Aug.) 365! (May) 5147! (Nov.) 5664! L. Tanma *Adam* 17710! **Mali:** Bamako (Aug.) *Waterlot* 1274! **Iv. C.:** Baoulé, Toumodi *Pobéguin* 77! Pacoubou Tiassalé *Scaëtta* 3047! Abidjan (Oct.) *de Wilde* 1105! **U. Volta:** Fö to Koudougou (June) *Chev.* 954! **Ghana:** Bole (Aug.) *Irvine* 5183! Winneba (June) *Hall* 3134! Accra to Mankessum (Aug.) *Hall* 2527! Achimota (July) *Morton* A1014! Kete Krachi (Dec.) *Morton* GC 6297! **Togo Rep.:** Paratau, Sokodé (May) *Büttner* 626! *Mahoux* 2143! Lomé *Warnecke* 139! **Dah.:** Agouagon (May) *Chev.* 23497! Dassa Zoumé (May) *Annet* 86! Ahozon (Feb.) *J. & A. Raynal* 13524! Cotonou (June) *Poisson* B2–12! (Apr.) *Chev.* 4457! (July) 4473! **N. Nig.:** Samaru, Zaria (July) *C.B. Taylor* 15! Obiaruku (July) *Kershaw* 900660! Lokoja *Richardson* 6! **S. Nig.:** Badagry (Aug.) *Onochie* FHI 33459! Lagos *Dalz.* 1111! Ilora (July) *Gillett* 15187! Aba (Mar.) *Coombe* 162! Throughout tropical and S. Africa and in the Mascarene Islands.

16b. **K. erecta** *Schumach.* var. **africana** (*Kük.*) *Hooper* in Kew Bull. 26: 580 (1972). *Cyperus obtusatus* (Presl) Mattf. & Kük. var. *africanus* Kük. l.c. 586 (1936). *Kyllinga pungens* of F.T.A. 8: 277, not of Link. Robust rhizomatous perennial, differing from the typical var. in having a white hemispherical head and more than three bracts; damp grassland among thickets.
Mali: Kleba *Demange* 2135! Kimparana *Demange* 2202! **Iv. C.:** Bouroukourou (Dec.) *Chev.* 16576! **Ghana:** Nkuntrodu, Cape Coast (June) *Hall* 3058! Accra (Apr.) *Ménager*! Accra to Mankessim (May–Aug.) *Hall* 2524! 3065! **Togo Rep.:** Lomé (Sept.) *Mahoux* 106! **Dah.:** Athiéme *J. & A. Raynal* 13520! Cotonou (Mar.) *Morton* GC 6677! **N. Nig.:** Nupe *Barter* 1586 partly! **S. Nig.:** Lagos (Jan.–Oct.) *Haines* 5! 7! 84! Also in the Congo, E. Cameroun, the Central African Republic and probably elsewhere in tropical Africa.

16c. **K. erecta** *Schumach.* var. **polyphylla** (*Kunth*) *Hooper* l.c. 580 (1972). *K. polyphylla* Willd. ex Kunth, Enum. Pl. 2: 134 (1837); F.T.A. 8: 276. *K. aromatica* (Ridley) Mattf. & Kük. l.c. 581 (1936). ? *K. senegalensis* C.B. Cl. in F.T.A. 8: 276 (1902); F.W.T.A., ed. 1, 2: 487. *Cyperus aromaticus* (Ridl.) Mattf. & Kük. l.c. (1936). Similar to var. *erecta* but the head composed of several, more or less confluent, spikes with more than three bracts and distinctly ciliate keels to the glumes; damp grassland.
Ghana: Kumasi (Aug.) *Hall* 2521! Skumo Lagoon, Accra (Nov.) *Morton* GC 6134! Keta *Morton* 6137! **N. Nig.:** Badeggi (Jan.) *Vaillant* 2776! **S. Nig.:** Olokomeji F.R. (Aug.) *Keay* FHI 37163! Ijora, Lagos (July) *Haines* 3! Okumu F.R., Benin (Dec.) *Brenan* 8624! Obiaruku, Warri (July) *Kershaw* 900654! Burutu (Sept.) *Parsons* 3! Throughout tropical and S. Africa, also in Madagascar and the Mascarenes.

17. **K. appendiculata** *K. Schum.* in Engl., Bot. Jahrb. 24: 338 t. 4, fig. A–G (1898), fig. poor. *K. cylindrica* Nees var. *appendiculata* (K. Schum.) C.B. Cl. and var. *major* C.B. Cl. in F.T.A. 8: 283 (1902). *Cyperus sesquiflorus* (Torr.) Mattf. & Kük. var. *major* (C.B. Cl.) Kük. l.c. 594 (1936). Rhizome short, scaly; leaves long, often equalling stems; head pyramidal with an elongated central spike; alpine grassland.
W. Cam.: Cam. Mt., 7,000 ft. (Nov.) *Mann* 2104! Mann's Springs, 7,500 ft. (Apr.) *Morton* GC 7092! Buea *Preuss* 923! Mba Kokeka, Bamenda, 7,000 ft. (Mar.) *Richards* 5325! Bamenda to Santa (July) *Daramola* FHI 41564! **F. Po:** 7,500 ft. (Apr.) *Mann* 1474! Clarence Peak, 8,500 ft. (Dec.) *Mann* 659! Also in E. Cameroun and Ethiopia.

18. **K. tisserantii** *Cherm.* in Arch. Bot. Caen 4, Mém. 7: 6 (1931). Similar to small specimens of *K. odorata* but with reddish spikelets less than 2 mm. long, and glumes scarcely exceeding the spherical nut; damp places.
S. L.: Koinadugu Dist. (Aug.) *Haswell* 231! **Iv. C.:** Mont Doura, Koualé (May) *Chev.* 21728! **N. Nig.:** Naraguta (June) *Lely* 324! **W. Cam.:** *fide* Kük. l.c. Also in the Central African Republic.

7. SCIRPUS Linn., Sp. Pl. 1: 47 (1753); F.T.A. 8: 446 (1902).

By Miss S. S. Hooper

Perennial with an elongated rhizome and tuberous stem-bases; stems stoutly triangular, bearing several spirally arranged, broadly laminate leaves; bracts two or more, leafy, all spreading or one erect; inflorescence of one to several large, golden or rufous spikelets; glumes scarious, bifid at the membranous apex with a recurved arista; achene large with retrorsely scabrid bristles:

Inflorescence of 1–6 broadly ovate, golden spikelets, 15–20 × 10 mm. style bifid; achene lenticular .. .. 1. *grandispicus*

Inflorescence of numerous, elliptic to cylindric rufous spikelets 15–30 × 5 mm.; style trifid; achene trigonous .. .. 2. *maritimus*
Annual or perennial without the above combination of characters:
Flowering stems bearing one or two spikelets; if two then spikelets not exceeding 4 mm. long:
Spikelet solitary, 5–8 × 4 mm., obtuse, rufous or brownish black, terete; glumes scarious, persistent; stems and leaves filiform .. 3a. *angolensis* var. *brizaeformis*
Spikelets not exceeding 4 × 2 mm.:
Spikelets always solitary, terminal, acute, without a conspicuous bract; water plant with elongated, flaccid, leafy stems .. .. 5. *fluitans*
Spikelets 1 or 2, apparently lateral with an erect and stem-like bract; glumes dark brown with a green or pale tip:
Style trifid; achene longitudinally ridged .. .. .. 6. *setaceus*
Style bifid; achene papillate:
Transparent scale(s) present within the broadly obovate, obtuse and mucronulate glume; sterile tip to anther connective large.. .. .. 7. *isolepis*
Transparent scale absent; glume elliptic or oblanceolate, acute and apiculate; sterile tip to connective small .. .. 8. *micranthus*
Flowering stems either with more than 2 spikelets or with 2 spikelets more than 4 mm. long:
Spikelets sessile in a single apparently lateral (pseudolateral) cluster subtended by a stem-like, generally erect bract:
Flowers with simple retrorsely-scabrid bristles; rhizomatous perennial; bract sharply trigonous, sometimes deflexed at fruiting .. .. 10. *mucronatus*
Flowers without bristles; bracts not sharply trigonous:
Stems and bracts septate with transverse septa and more or less pronounced constrictions visible externally:
Stems about 5 mm. broad, spongy above, narrower below, with well-marked constrictions at the septa; spikelets 8–10 mm. long; glumes acute, about 5 mm. long, rusty brown with a broad green keel; achene smooth, black 11. *articulatus*
Stems not exceeding 3 mm. broad; glumes less than 5 mm. long, five-ranked; achene undulately ridged, black or brown:
Stems about 2·5 mm. wide; glumes about as long as wide, very concave and obtuse, producing a terete, turgid spikelet; achene 1·0–1·7 mm. long 12. *jacobii*
Stems not exceeding 1·5 mm. wide; glumes longer than wide, chaffy, clearly five-ranked, persistent and spreading after maturity producing a cylindrical spikelet; achene 0·85–1·15 mm. long .. .. .. 13. *roylei*
Stems and bracts not septate; stigmas 2:
Small annual with filiform stems and leaves; spikelets few, dark red-brown with paler tips, 1·5–3 × 1 mm., obtuse .. .. 8. *micranthus*
Stems, and leaves when present, not filiform; spikelets greenish, yellowish or straw-coloured with dark tips, more than 3 × 1 mm., acute or acuminate; basal flowers usually present:
Bracts very long, exceeding stems; sheaths bladeless; spikelets greenish or yellowish, generally 2, often curved, 10–15 × 2–3 mm. .. 14. *oxyjulos*
Bracts not exceeding stems; sheaths with a short or long blade; spikelets greenish with dark tips, 5–10 × 2 mm., acute .. .. 15. *uninodis*
Spikelets not in a single, pseudolateral cluster:
Inflorescence terminal; bracts several, spreading or reflexed:
Annual; spikelets several, crowded in a single greenish head; glumes with a long recurved arista .. .. 9. *kernii*
Perennial; glumes without a long recurved arista:
Leaves distichous; shoots flattened; sheaths tough, often red; spikelets distinct, generally not exceeding 10 in number, very dark brown or black, in a contracted panicle or capitate cluster .. .. 4. *bulbostylidoides*
Leaves not distichous; spikelets crowded, numerous, white or green and brown:
Inflorescence a single white head up to 8 mm. across; plant slender; bracts spreading or reflexed .. .. 22. *microcephalus*
Inflorescence a lobed green or brown head often with additional shortly stalked heads forming a flattish corymb; bracts several, spreading; glumes white ciliate on the margin and with a strong green apiculus .. 23. *cubensis*
Inflorescence pseudolateral, with an erect bract continuing the line of the stem; spikelets in one sessile and one to many stalked clusters or spikes, or sometimes solitary:
Flowers with laminate, pectinate bristles surrounding the achene:
Inflorescence lax; spikelets on long pedicels; glumes with a broad pale membranous only slightly ciliate margin, stems up to 10 mm. broad, obtusely trigonous above, often tinged yellow below; achene about 2 × 1·5 mm. 18. *litoralis*

Inflorescence rather dense; spikelets on short curved pedicels; glumes with a very narrow membranous, distinctly ciliate margin; stems up to 15 mm. broad, terete except immediately beneath the inflorescence, spongy; achene about 2·5 × 1·5 mm. .. .. 19. *pterolepis*

Flowers with simple bristles or bristles absent:

Rush-like plants with a creeping rhizome and spongy, terete stems up to 2 m. high, bearing bladeless sheaths and a corymbose inflorescence:

Bract with a sheathing base and blunt tip, shorter than the, generally several, rays; glumes generally pale below with dark reddish tips; anther 1·5 mm. long; achene surface smooth .. .. .. 20. *brachyceras*

Bract stem-like, generally exceeding the contracted inflorescence; anther 0·5 mm. long; achene surface transversely ridged .. .. 21. sp. *A*

Tufted plants less than 1 m. high with or without bladed sheaths; spikelets either solitary or in one sessile and one to few stalked clusters or spikes:

Spikelets solitary, one sessile and 3–4 stalked 3a. *angolensis* var. *brizaeformis*

Spikelets clustered:

Achene with three equal sides and angles; leaves without blades; margin of glume smooth; anther 0·5 mm. long with a small, smooth connective-tip 16. *aureiglumis*

Achene with one or both sides flattened and only two acute angles; uppermost sheath sometimes producing a long blade; margin of glume minutely ciliate; anther connective-tip bristly:

Stigmas 2; anther 0·8–1·5 mm. long; achene lenticular, 1·25–1·5 × 1–1·2 mm., transversely undulate but not ridged .. .. 15. *uninodis*

Stigmas 3; anther 0·25–0·6 mm. long; achene with one flat and one convex side, 0·95–1·2 × 0·7–1·1 mm., rugulose and ridged at least in the upper half .. .. 17. *lateriflorus*

1. **S. grandispicus** (*Steud.*) *Berhaut* in Bull. Soc. Bot. Fr. 100: 176 (1953), and Fl. Sén. ed. 2, 361, 369 (1967). *Isolepis grandispica* Steud., Syn. Pl. Cyp. 318 (1855). Resembling *S. maritimus* in its leafy stems and membranous, bifid, aristate glumes but with few broadly ovate spikelets, and lenticular achenes; depressions behind dunes near coast.
Sen.: Hann (June) *Adam* 12262! 17664! *J. & A. Raynal* 5965! Lac Retba (Apr.) *Adam* 17627! *Naegelé* 8107! Mboro (May) *Adam* 17660!

2. **S. maritimus** *Linn.* Sp. Pl. 1: 51 (1753); F.T.A. 8: 455; Berhaut, Fl. Sén. ed. 2, 371 (1967). Creeping perennial with leafy stems up to 1 m. high, bearing a bracteate corymb of large brown spikelets; saline and semi-saline marshes.
Maur.: *Naegelé*! *Monod* 7068! **Sen.:** Yankoba (June) *J. & A. Raynal* 5925! Richard-Toll *Couey* 6513! Walo (May) *Roger*! Lac Tanma (July) *Berhaut* 1101! M'bidjem *Pitot*! **Mali:** Mare de Buonankodjie (Feb.) *Davey* 56/CIPAS 444! Toubel (May) *Davey* 71! Tiouki to Selingourou (Feb.) *Davey* 22! Diré (Aug.) *Lean* 62! Gao (Apr.) *de Wailly* 5382! Throughout tropical and temperate regions.

3. **S. angolensis** *C.B. Cl.* in F.T.A. 8: 448 (1902). Tropical Africa.

3a. **S. angolensis** var. **brizaeformis** (*Hutch.*) *Hooper* in Kew Bull. 26: 580 (1972). *S. briziformis* Hutch. in F.W.T.A., ed. 1, 2: 466 (1936): Berhaut, Fl. Sén. ed. 2, 364 (1967). *Eriocaulon spadiceum* Lam., Ill. 1: 214 (1792). *Scirpus spadiceus* (Lam.) Boeck. in Linnaea 36: 493 (1870); F.T.A. 8: 448; not of Linn. *S. ustulatus* Podl. in Mitt. Bot. 4: 117 (1961). Slender tufted annual with stems 5–35 cm. high and reddish-black, papery-glumed spikelets; on damp rocks, in lateritic pans and in damp, grassy places.
Sen.: Niokolo-Koba (Dec.) *J. & A. Raynal* 6796! Tambacounda to Gouloumbo (Oct.) *Pitot*! Oussouye, Casamance (Nov.) *Berhaut* 6496! Kédougou (Nov.) *Adam* 20015! Kanéméré (Sept.) *Fotius* K 493! **Mali:** Kita Mts., 600 m. (Sept.) *Jaeger* 92! 5766! Niani d'Koulou Mts., Kita country (Oct.) *Jaeger* 2976! Bamako to Koulouba (Dec.) *Monod*! Sikasso to Bobo-Dioulasso (Sept.) *Jaeger* 5159! **Port. G.:** Madina de Boé, Gabú (Dec.) *Esp. Santo* 2849! **Guin.:** Friguiagbé (Aug.) *Chillou* 613! Conakry (Oct.) *Adam* 12622! Thianguel-Bori (Oct.) *Pitot*! Mali (Dec.) *Schnell* 2405! Macenta (Oct.) *Baldwin* 9788! **S. L.:** Fourah Bay College, Freetown (Oct.) *Morton* SL8! Roboli, nr. Rokupr (Nov.) *Jordan* 168! Falaba (Sept.) *Small* 308! Tombo (Jan.) *Deighton* 999! Sankan Biriwa, 6,080 ft. (Jan.) *Cole* 140! **Lib.:** Genne-Loffa, Kolahun Dist. (Nov.) *Baldwin* 10086! Kailahun, Kolahun Dist. (Nov.) *Baldwin* 10149! Nimba Mts., 1,800 ft. (Oct., Dec.) *Adam* 702! 20145! **Iv. C.:** Man *Porterés*! Séguélo, Odienné Region (Oct.) *Aké Assi* 6610! 9694! Yérébodi to Sanlo (Nov.) *Aké Assi* 9272! **U. Volta:** Banfora (Sept.) *Jaeger* 6584! Sikasso to Bobo-Dioulasso (Sept.) *Adam* 15103! **Ghana:** Kwahu Tafo (Aug.) *Hall* CC 130! Sampa (Dec.) *Morton* A1065! **N. Nig.:** Kogigiri, Jos Plateau, 4,100 ft. (Oct.) *Hepper* 1067! Panshanu, Bauchi Prov. (Aug.) *Lawlor & Hall* 618! Agbaja, Kabba (Nov.) *J. Hall* 808! **S. Nig.:** Mt. Orosun, Idanre, Akure Dist. (Oct., Jan.) *Keay* FHI 22598! *Brenan et al.* 8642! Iseyin rocks (Oct.) *Jackson & Etukudo* 1251069! **W. Cam.:** Mamfe (Mar.) *Guile* 122! Also in E. Cameroun, Central African Rep., Uganda, Congo and Zambia.

4. **S. bulbostylidoides** *Hooper* in Kew Bull. 26: 581 (1972). A tufted or shortly rhizomatous perennial with flattened orange-red shoots, sharp needle-like leaves bearing long hairs at the top of the sheath and a small number of large spikelets similar to those of *S. angolensis* var. *brizaeformis* with black or red-black papery glumes; marshy montane grassland.
Guin.: Nzo, Nimba Mts. (July) *Schnell* 1494! Nzérékoré (Oct.) *Jac.-Fél.* 1922! Nimba Mts. (Feb.) *Schnell* 482! 933! Bereguizi, Macenta (Aug.) *Adam* 6126! **S. L.:** Loma Mt. *Jaeger* 395! 1043! *T. S. Jones* 106! Kponkponto to Waia, 2,000 ft. (July) *Marmo* 208! Tingi Mts., 4,000 ft. (Dec.) *Morton & Gledhill* SL 2953! **Iv. C.:** Nimba Mts., 4,500 ft. (Aug.) *Boughey* GC 18165! Mt. Tonkoui (Oct.) *Aké Assi* 5760! (Aug.) *Boughey* X279!

5. **S. fluitans** *Linn.* Sp. Pl. 1: 48 (1753); F.T.A. 8: 449. Aquatic or semi-aquatic with weak floating or trailing many-noded stems; in or near water.
Mali: Gao to Berra (Mar.) *de Wailly* 5012*b*! **N. Nig.:** Nguroje, Mambila Plateau (Mar.) *J. Hall* 1630! **W. Cam.:** Bamenda, c. 5,000 ft. (Mar.) *Coombe* 228! Bali, 4,100 ft. (Apr.) *Brunt* 1056! Also in the Congo and sporadically in east and S. Africa.

[Until the West African plant has been collected in fruit it is impossible to say the extent to which it differs from European and Asian *S. fluitans*, but it certainly does so in some respects, as the glumes and anthers appear to be shorter.]

6. **S. setaceus** *Linn.* Mant. 2: 321 (1771); F.T.A. 8: 450. *Isolepis setaceus* (Linn.) R. Br., Prod. 222 (1810). Annual with filiform leaves and stems up to 10 cm. and a single dark, ovate, acute spikelet; damp grassland.

W. Cam.: Cam. Mt., Mann's Spring, c. 9,500 ft. (Dec.) *Morton* K865! c. 8,000 ft. (Apr.) *Morton* GC 7065! Above Buea, 9,100 ft. (Apr.) *Morton* GC 6866! Mountains of east Africa to S. Africa; also Europe and Australia.

7. **S. isolepis** (*Nees*) *Boeck.* in Linnaea 36: 498 (1870); F.T.A. 8: 459; Berhaut, Fl. Sén. ed. 2, 360 (1967). *Hemicarpha isolepis* Nees in Edinb. New Phil. Journ. 17: 263 (1834). Annual with filiform leaves and stems to 15 cm., with one or two dark, elongate-ovate, obtuse, *Lipocarpha*-like spikelets; damp grassland.
Sen.: *Leprieur* 9. *Perrottet*. **Ghana:** Yapei Bridge, Tamale (Nov.) *Hall & Enti* GC 35730! **S. Nig.:** Agbaja (Feb.) *J. Hall* 829! In India and SE. Asia and scattered throughout tropical and S. Africa.

8. **S. micranthus** *Vahl* Enum. 2: 254 (1806); F.T.A. 8: 459; Berhaut, Fl. Sén. ed. 2, 361 (1967). Resembling *S. isolepis* but having a small cluster of pseudolateral spikelets and oblanceolate, not broadly obovate, glumes.
Sen.: Mbao *Berhaut* 1186. Kayar (Oct.) *J. & A. Raynal* 6432! Also in Angola and throughout America.

9. **S. kernii** *Raymond* in Nat. Canad. 86: 230 (1959); J. Raynal in Adansonia sér. 2, 8: 95 (1968). *S. squarrosus* of F.T.A. 8: 458, partly; F.W.T.A., ed. 1, 2: 466; Berhaut, Fl. Sén. ed. 2, 368 (1967). Stems 2–25 cm. long; somewhat stouter than *Ascolepis dipsacoides* which it resembles in the head of spikelets with squarrose glumes; damp ground or marshes.
Sen.: (Oct.) *Fotius* K569! Fatick (Dec.) *J. & A. Raynal* 6708 *bis*! Kaolak (Nov.) *Berhaut* 2622! Niokolo-Koba (Oct.) *Adam* 15561! 17018! *Berhaut* 4692! Vélingara (Oct.) *Adam* 18563! **Mali:** Kita Mts. (Nov.) *Jaeger* 3514! Sotuba (Oct.) *Raynal* 5028! **U. Volta:** Ouagadougou (Oct.) *Nongonierma* 202! **Ghana:** Tumu, Northern Prov. (Mar., May) *Morton* GC 7547! GC 8859! Tamale to Lumbaga (Dec.) *Morton* GC 6265! Yapei Bridge (Nov.) *Hall & Enti* GC 35841! 5 miles E. of Yendi (Sept.) *Goodall* GC 15846! **Togo Rep.:** Dapango (Oct.–Nov.) *Bille* TG 10! **Niger:** Niamey (Sept.) *Cremers* 952! **N. Nig.:** Shagunu, Ilorin (Oct.) *J. Hall* 1462! **S. Nig.:** Igboora, Oyo Dist. (Oct.) *Haines* 219! Agbaja (Nov.) *J. Hall* 832! Also in E. Cameroun and eastwards to Ethiopia, about L. Tanganyika and in India (Mysore).

10. **S. mucronatus** *Linn.* Sp. Pl. 1: 50 (1753); F.T.A. 8: 454. Leafless rhizomatous perennial to 1·5 m. characterized by the sharply triangular stems and short stiff triangular bracts; in swamps or open water.
Guin.: Daralabé, Fouta Djalon (Jan.) *Langdale-Brown* 2575! (Apr.) *Chev.* 12341! Labé (Apr.) *Pitot*! Pita (Dec.) *Pobéguin* 2191! Mamou (May) *Pitot*! **N. Nig.:** Mandaga, Mambila Plateau, 6,000 ft. (Jan.) *Latilo & Daramola* FHI 29000! *J. Hall* 1692! **W. Cam.:** Bamenda (Mar.) *Morton* K159! Also in E. Cameroun, central, east and south Africa and widespread in the temperate and tropical regions of the Old World.

11. **S. articulatus** *Linn.* Sp. Pl. 1: 47 (1753); F.T.A. 8: 453. A rush-like plant with spongy rounded septate stems and a pseudolateral cluster of large acute spikelets; in or near standing water.
Ghana: Accra (Apr.) *Hall* 3458! Accra to Ada (Dec.) *Morton* 6324! Dawa, nr. Ada (Dec.) *Adams* 3618! **Niger:** Zinder (Oct.) *Hagerup* 588! **N. Nig.:** Katagum *Dalz.* 242! Fadama, Dikwa Dist. (Nov.) *Davey* FHI 27105! Bama, Dikwa Dist. (Dec.) *McClintock* 115! Also in E. Cameroun, Chad, from the Sudan to the Cape, in Madagascar and intermittently throughout the Old World tropics. There is a *Leprieur* specimen marked as from the Senegal River, but the species has not been recently recorded from Senegal.
[The typical Indian plant has an obtuse and minutely apiculate glume whilst the glume is acute and apiculate or acuminate in specimens from Africa.]

12. **S. jacobii** *C. E. C. Fischer* in Kew Bull. 1931: 103. *S. praelongatus* of authors, not of Poir. (1804); F.W.T.A., ed. 1, 2: 466; Berhaut, Fl. Sén. ed. 2, 361 (1967). *Isolepis senegalensis* Hochst. ex Steud., Syn. Pl. Cyp. 96 (1855), not *Scirpus senegalensis* Lam. (1791). Similar to the preceding species in the septate stems and pseudolateral inflorescence but much more slender with smaller, catkin-like, often very obtuse spikelets; edge of standing water, marshes, temporary pools on laterite.
Maur.: Aleg (Sept.) *Baron* 57 Aleg! **Sen.:** *Heudelot* 319! L. Tanma (Nov.) *Pitot*! Richard Tol (Feb.) *Dollinger* 13! Ourosogui to Matam (Dec.) *Naegelé* 13162! Tambacounda *Adam* 14325 *bis*! **Mali:** Kita (Aug.) *Jaeger* 45! Bamako to Ségou (Sept.) *Jaeger* 5119! Macina (Nov., Dec.) *Duong*! Gao (Sept.) *de Wailly* 4817! South Koubita (Feb.) *Davey* 61 (*C.I.P.A.S.* 449)! **Guin.:** Sambailo to Youkounkoun (Oct.) *Pitot*! Kandiara plateau (Oct.) *Pitot*! **Iv. C.:** Ouango-fitini (July) *Aké Assi* 4827! Bassawa (Nov.) *Aké Assi* 9734! **U. Volta:** Ouagadougou (Oct.) *Adam* 12744! **Ghana:** Nasia, Bolgatanga to Tamale (Apr.) *Adams* 4151! Pong Tamale (Aug.) *Hall* CC 406! Bole (Sept.) *Irvine* 4753! Folifoli, Afram Plains (Aug.) *Hall* CC 173! Achimota, nr. Accra (Nov.) *Irvine* 1523! **N. Nig.:** Kufena, Zaria Prov. (Sept.) *Kershaw* 900421! Anara F.R., Igabi Dist. (July) *Keay* FHI 25964! Panshanu, Bauchi Prov. (Aug.) *Lawlor & Hall* 622! 15 miles S. of Damaturu, Bornu Prov. (Aug.) *De Leeuw* 1137*a*! Logomani, N. Bornu (Nov.) *Johnston* N88! Throughout tropical and S. Africa and in India.

13. **S. roylei** (*Nees*) *Parker* in Duthie, Fl. Upper Gangetic Plain 3: 361 (1929). *Isolepis roylei* Nees in Wight, Contrib. Bot. Ind. 107 (1834). *Scirpus quinquefarius* Buch.-Ham. ex Boeck. in Linnaea 36: 701 (1870); F.T.A. 8: 454; Berhaut, Fl. Sén. ed. 2, 1361 (1967). Similar to the preceding species but with narrower stems generally longer than the bracts and brown papery spreading glumes; in shallow water or swampy grassland.
Sen.: *Naegelé*! San, M'boro (Jan.) *Leprieur*! Oussouye, Casamance (Oct.) *Adam* 18286! Kaolak (Nov.) *Berhaut* 1578! Ross (Feb.) *Berhaut* 2630! **Mali:** S. Koubita (Feb.) *C.I.P.A.S.* 450/062! **Ghana:** Busunu, Damongo to Yapei (Sept.) *Hall* CC 898! **N. Nig.:** 8 miles N. of Damaturu, Bornu Prov. (Nov.) *De Leeuw* 1337! Also in Mauritania, Chad and Congo, intermittently through east Africa and Angola to S.W. Africa and in India.
[The African collections are characterized by smaller glumes and spikelets than the Indian.]

14. **S. oxyjulos** *Hooper* in Kew Bull. 26: 581 (1972). Tufted rush-like plant 15–35 cm. high with spikelets tapering to an acute point and long prominent erect bracts; rock pools or wet flushes on inselbergs.
Guin.: Thianguel-Bori (Oct.) *Pitot*! Telimélé (Oct.) *Pitot*! **S. L.:** Kambia to Kukuna (Dec.) *Morton & Gledhill* SL 27! **N. Nig.:** Anara F.R. (Oct.) *Keay* FHI 20127! Kufena Hill, Zaria, 2,300 ft. (Sept.) *Clayton* 1323! *Kershaw* 420! Dumbi inselberg, Zaria (Sept.) *Kershaw* 422! Also in E. Cameroun, C. African Rep. and Sudan.

15. **S. uninodis** (*Del.*) *Boiss.* Fl. Or. 5: 380 (1884). *Isolepis uninodis* Del., Fl. Égypte 8, t. 6 fig. 1 (1813). *Scirpus supinus* Linn. var. *uninodis* C.B. Cl. in Hook. f., Fl. Brit. Ind. 6: 656 (1894); F.T.A. 8: 453. Tufted annual, generally less than 30 cm. high with a generally sessile cluster of few pseudolateral spikelets with dark red-tipped glumes; edge of lakes and swamps.
Sen.: *Naegelé* s.n.! 4994! Dakar *Adam* 17655! **Mali:** Gao to Berra (Mar.) *de Wailly* 5006! **N. Nig.:** L. Chad, Bornu Prov. (Mar.) *Jackson* 2574! Also in Egypt and Sudan.

16. **S. aureiglumis** *Hooper* in Kew Bull. 26: 581 (1972). *Schoenus junceus* Willd., Phyt. 1: 2, t. 1, 4 (1794), not *Scirpus junceus* Forst. f. (1786). *Scirpus supinus* of F.T.A. 8: 452, partly. Tufted annual up to 60 cm. with soft rounded stems and bracts, leaves reduced to sheaths and spikelets often golden at maturity; margin of ponds and swamps.
Sen.: Walo *Leprieur*! Kolda to Salikenié *Trochain* 1330! **Guin.:** Koba *Jac.-Fél.* 7248! **Iv. C.:** Néro-Mer, nr. Béréby (Nov.) *Oldeman* 520! Adiopodoumé (July) *Boughey* 14505! **Ghana:** Cape Coast (Oct.) *Hall* 2533! Accra (Mar.) *Irvine* 2803! Accra to Ada, Mile 30 (May) *Boughey* 14224! Prampram (Nov.) *Morton* GC 6154! Asubsuare, nr. Akuse (Aug.) *Hall* 3306! **Togo Rep.:** Lomé (June, Sept.) *Mahoux* 544! 2020! 2032! **N. Nig.:** Bama, Dikwa Div. (Dec.) *McClintock* 87*a*! **S. Nig.:** Galarige (Oct.) *Golding* 30! Also in tropical east Africa.

17. **S. lateriflorus** *Gmel.* Syst. Veg. 1: 127 (1791); Kern in Blumea, suppl. 4: 164 (1958). *S. erectus* of Berhaut, Fl. Sén. ed. 2, 361 (1967). Similar to the preceding species, but generally somewhat larger and distinguished by the stigma number and achene shape; in similar habitats.
Sen.: Mbao (Jan.) *Berhaut* 1327! **Mali:** Sotuba (Feb.) *J. & A. Raynal* 5489! Dogo (Mar., May) *C.I.P.A.S.*

573! *Davey* 111! **Ghana:** Busunu, Damongo to Yapei (Sept.) *Hall* CC 880! Yapei Bridge (Nov.) *Hall & Enti* GC 35910! **Dah.:** Cotonou (Mar.) *Morton* GC 6676! **S. Nig.:** Ikoyi Plains, Lagos (Nov.) *Dalz.* 1307! Also in tropical east Africa, Madagascar and India.

18. **S. litoralis** *Schrad.* Fl. Germ. 1: 142, t. 5 f. 7 (1806); F.T.A. 8: 456; Berhaut, Fl. Sén. ed. 2, 369 (1967). Perennial with rush-like stems to 1·5 m. high and a loose pseudolateral inflorescence of reddish brown spikelets about 1 cm. long; in or near open saline water.
Sen.: *Heudelot* 533! Sor, nr. St. Louis (Aug.) *J. & A. Raynal* 6284! Fatick *Maymard*! Diouloulou, Casamance (Sept.) *Adam* 18217! Bignona, Casamance (Sept.) *Adam* 18118! **Guin.:** Koba to Rokiti *Jac.-Fél.* 7230! Samou *Jac.-Fél.* 7396! Monchon *Jac.-Fél.* 7359! **S. L.:** Mapotolon (Sept.) *Jordan* 923! **Ghana:** Weija (June) *Hall* 3102! Keta (Sept.) *Hall* 3414! Labadi-Teshi lagoon (Apr.) *Irvine* 2762! Sakumu Lagoon (July) *Irvine* 2782! Temperate and tropical regions of the Old World.

19. **S. pterolepis** (*Nees*) *Kunth* Enum. 2: 166 (1837); Berhaut, Fl. Sén. ed. 2, 369 (1967). *Malacochaete pterolepis* Nees in Linnaea 9: 292 (1835). *Pterolepis scirpoides* Schrad. in Goett., Gel. Anz. St. 208: 2071 (1821), and Anal. Fl. Cap. 1 Cyper. 50 (1832), not *Scirpus scirpoides* (Michx.) Koyama (1958). *S. lacustris* of F.W.T.A., ed. 1, 467, and Adam, Not. Syst. 16: 317 (1960). Stems tufted from a woody rhizome, up to 120 cm. high, terete, spongy; spikelets brown, elliptical, paired or solitary; near open water.
Sen.: Mbao (July) *J. & A. Raynal* 6067! Malika (Jan.) *Naegelé* 31! Malika to Yombeul (Nov.) *Pitot*! L. Youi (Nov.) *Pitot*! Sebikotane (July) *Pitot*! Also in Chad, Gabon, Congo and east and S. Africa.

20. **S. brachyceras** *Hochst. ex A. Rich.* Tent. Fl. Abyss. 2: 496 (1851); Berhaut, Fl. Sén. ed. 2, 369 (1967). *S. corymbosus* Heyne ex Roth (1821); F.T.A. 8: 455; not of Linn. (1760), nor of Forsk. (1775). Tufted, rush-like plant, to 2 m. high with a spreading pseudolateral inflorescence of rounded clusters of straw-coloured or red-brown, acute spikelets; in or near standing water.
Mali: Macina (Nov.) *Monod*! Tenenkan (Apr.) *Roberty* 3792! Tiouki to Selingourou (Feb,) *Davey* 24! Dogo (May) *Davey* 125! Gao to Bagoundié (Feb.) *de Wailly* 4965! **N. Nig.:** Vom, Plateau Prov. (Dec., Feb.) *Haines* 73! *McClintock* 223! Jos, Plateau Prov. (Nov.) *Keay* FHI 21441! Kurra Gashish, Pankshin Dist. (June) *Okafor* FHI 59254! Ngel Nyaki, Mambila Dist, 5,400 ft. (Jan.) *Hepper* 1717! **W. Cam.:** Lakom, Bamenda, 6,000 ft. (June) *Maitland* 1546! Jakiri, Bamenda, 5,800 ft. (Feb.) *Hepper* 1960! **F. Po:** L. Biao (Sept.) *Wrigley & Melville* 483! Tropical and S. Africa and Madagascar.

21. **S. sp. A.** Similar to *S. brachyceras* but with a contracted inflorescence of dark spikelets exceeded by the erect stem-like bract and a transversely ridged achene; in or near water.
N. Nig.: Jos (Aug.) *Lely* P485!
[Probably belonging to a species of the *S. confusus* N.E. Br., *S. rogersii* C.B. Cl. alliance from East Africa, but too incomplete a specimen to allow certain identification.]

22. **S. microcephalus** (*Steud.*) *Dandy* in F. W. Andrews, Fl. Sudan 3: 366 (1956). *Kyllinga microcephala* Steud. in Flora 25: 597 (1842). *Scirpus kyllingioides* (A. Rich.) Boeck. in Linnaea 36: 733 (1870); F.T.A. 8: 457; Berhaut, Fl. Sén. ed. 2, 365 (1967). A slender, tufted or shortly rhizomatous plant with swollen stem-bases covered by brownish fibres; the spherical white head is very similar to that of *Cyperus pulchellus* R. Br.; open woodland or savanna.
Sen.: Dakar (Aug.) *Adam* 1656! Ndjarno, Dahra to Linguéré (Aug.) *J. & A. Raynal* 7221! Koungheul (Aug.) *Trochain* 179! Niokolo-Koba (July) *Adam* 14118! Kanouméri, Kedougou (July) *Fotius* K216! K218! **Mali:** Klela (May) *Demange* 3361! Koura (July) *Roberty* 2405! nr. Koupéla, Mossi (July) *Chev.* 24587! **Guin.:** Iles Tristao, *Paroisse* 24! Youkounkoun (Oct.) *Pitot*! Friguiagbé *Chillou* 381! Ténémadou, Macenta *Adam* 4223! Tougué (June) *Adam* 14745! **Ghana:** Saboba, nr. Yendi (June) *Harris*! **Niger:** (Sept.) *P. de Fabrègues* 2128! **N. Nig.:** Gusau, Sokoto Dist. (Sept.) *De Leeuw* 1574! Bindawa Village F.R., Katsina Dist. (July) *Grove* 3a! Sange River F.R., Zaria Prov. (May) *E. W. Jones* 105! Anara F.R., Zaria Prov. (May) *Keay* FHI 25774! Jos Plateau (Aug.) *Lely* P462! Throughout tropical Africa and in India.

23. **S. cubensis** *Poeppig & Kunth* in Kunth, Enum. 2: 172 (1837); F.T.A. 8: 451; Berhaut, Fl. Sén. ed. 2, 366, 371 (1967). Leafy plants up to 50 cm. high with slender rhizomes, trigonous stems and very long spreading bracts; in permanent marsh or in or near permanent water, sometimes floating.
Sen.: Lac de Guiers (Oct.) *Adam* 19246! Ile du Diable, Casamance (Oct.) *Adam* 18438! Kolda, Casamance (Oct.) *Adam* 18455! Sédhiou, Casamance (July) *Berhaut* 6129! Diaroumé, Casamance (Oct.) *Adam* 1839! **Mali:** Niger delta *Duong*! L. Korienza (Apr.) *Duong* 2132! Macina (Mar.) *Roberty* 3617! Saraféré (July) *Lean* 56! Gao (Apr.) *de Wailly* 5014! 5022! **U. Volta:** L. Tengréla, Banfora (July) *Aké Assi* 4821! **Ghana:** Akotokyir, nr. Cape Coast (May) *Hall* 2294! Mankessim (Feb.) *Hall* 2202! Densu R., Accra to Winneba (Apr.) *Morton* A4157! Weija, nr. Accra (Apr.) *Hall* 1883! Nungua, nr. Accra (Apr.) *Morton* A4172! **Dah.:** L. Aziri to Zagnanado (Feb.) *Chev.* 23055! **Niger:** Niamey (May) *P. de Fabrègues* 1561! Diffa (May) *P. de Fabrègues* 2494! **N. Nig.:** Nupe *Barter* 1068! Yankari Game Reserve, Bauchi Province (Dec.) *Haines* 72! 20 miles S. of Maiduguri (Mar.) *Tuley* 1113! **S. Nig.:** Sangwabe, Ilorin Province (Mar.) *J. Hall* 1025! Throughout tropical Africa and America.

Doubtful species.

S. verruculosus (*Nees*) *Steud.* in Flora 12: 145 (1829); F.W.T.A., ed. 1, 2: 466; Berhaut, Fl. Sén. ed. 1, 215, 217 (1954), and ed. 2: 361 (1967). The two records for the occurrence of this S. African species in the Flora area are doubtful; Leprieur's specimen in the Paris herbarium does not, as normally, bear an original label and Berhaut's specimen is not present in Paris.

8. ELEOCHARIS R. Br., Prod. Fl. Nov. Holl. 224 (1810); F.T.A. 8: 404 (1902). *Heleocharis* of many authors.

By Miss S. S. Hooper

Plants aquatic or semiaquatic with weak, often floating capillary stems, often many sterile; spikelets small, more or less flattened with few glumes:
Plants aquatic, submerged but for the flowering stems; sterile stems borne in clusters; spikelets acuminate, composed of up to 8 lanceolate glumes; achene with an acuminate style-base as tall as or taller than broad:
Spikelets composed of 2–3 glumes producing 1 achene .. .. 1. *naumanniana*
Spikelets 4–5 mm. long; floating stems clearly septate
1a. *naumanniana* var. *naumanniana*
Spikelets 1·5–3 mm. long; floating stems not, or only obscurely, septate
1b. *naumanniana* var. *caillei*
Spikelets composed of 6–8 glumes producing 2–4 achenes; stems not septate
2. *subtilissima*
Plants semi-aquatic; spikelets 1·5–2 × 0·75–1 mm., elliptical, acute; glumes obtuse; achene with a clear quadrate surface pattern and depressed style-base broader than **tall** .. .. 3. *brainii*

Plants normally terrestrial with most stems erect and fertile; spikelets terete, not acuminate, producing many achenes:
- Spikelets ovate, oblong or spherical, up to $1\frac{1}{2}$ times as long as broad; plants small to medium-sized:
  - Achene trigonous; stigmas 3:
    - Stems capillary; spikelets small, few-flowered:
      - Achene about 0·75 × 0·5 mm. with a clear surface pattern and depressed style-base; stems often submerged .. .. 3. *brainii*
      - Achene with a faint or no surface patterning:
        - Achene 1 mm. long; glumes reddish .. .. 4. *trilophus*
        - Achene 0·65–0·7 mm. long; glumes pale .. .. 5. *setifolia*
    - Stems flattened, ridged, 1–1·5 mm. thick, 10–25 cm. high, arising in tufts from a more or less distinct rhizome; spikelets ovate, about 5 mm. long; glumes obtuse with a wide membranous edge and distinct keel; achene with three clear angular ridges .. .. 6. *complanata*
  - Achene lenticular; stigmas 2:
    - Achene dark red or black:
      - Stems about 1 mm. thick; basal sheaths conspicuous, purple below, white above; spikelets broadly ovoid, about 5 mm. long, subglobular when young; achene about 1 mm. long, bristles pink, longer than the achene or absent 7. *geniculata*
      - Stems less than 0·5 mm. thick; spikelets ovoid or elliptic about 2 mm. long; achene about 0·5 mm. long; bristles white, shorter than the achene or absent 8. *atropurpurea*
    - Achene greenish yellow; spikelets narrowly ovoid or elliptical, greenish; achene about 1 mm. long with white bristles .. .. 9. *deightonii*
- Spikelets cylindrical, twice or more as long as broad; generally at least 2 cm. long; plants large:
  - Stems terete, spongy, septate, up to 1 cm. thick; sheaths greyish or purple, shining, papery; spikelet not exceeding stem in width; glumes with a single prominent nerve; surface of the achene almost smooth .. .. 10. *dulcis*
  - Stems angular, or if terete not exceeding 5 mm. thick, not spongy and septate:
    - Margin of the glume with a distinct dark band; glume 2 × 1·25 mm.; achene turbinate with quadrate surface pitting .. .. .. 11. *decoriglumis*
    - Margin of the glume membranous, sometimes dotted or streaked with red; achene longitudinally striate, the surface pitting horizontally elongated and arranged in vertical rows:
      - Achene markedly constricted above with a neck narrower than the style-base; style-base acute, taller than broad; glumes generally oblong and clearly nerved; stems sharply trigonous with concave faces .. .. .. 12. *acutangula*
      - Achene slightly constricted above with a wide neck surrounding the style-base:
        - Glumes close-packed, orbicular to obovate, obscurely nerved; striations on the achene fine; generally in brackish coastal swamps .. .. 13. *mutata*
        - Glumes oblong, clearly several-nerved; achene clearly striate; stems slender, generally terete or obscurely angled:
          - Achene 2·3–2·8 mm. long (including style-base) with bristles as long as itself; outer layer of surface cells transparent; style-base as tall or taller than broad .. .. 14. *variegata*
          - Achene 1·7–2·1 mm. long with bristles much shorter; style-base broader than tall .. .. 15. *nupeënsis*

1. **E. naumanniana** *Boeck.* in Engl., Bot. Jahrb. 5: 92 (1884); F.T.A. 8: 411; Svenson in Rhodora 34: 253, t. 461, fig. 6 (1932).

1a. **E. naumanniana** *Boeck.* var. **naumanniana**—Nelmes and Baldwin in Amer. J. Bot. 39: 373 (1952). *E. testui* of Berhaut, Fl. Sén. ed. 2, 11, 360, 363. Generally aquatic, forming mats of numerous slender clustered stems, sometimes terrestrial and sterile; in or near open water.
Sen.: Diantème (Oct.) *Kerharo & Adam* s.n. (H.F.M.P. (Dakar) 1561)! Nioro du Rip (Nov.) *Berhaut* 1802! Oussouye, Casamance (Mar.) *Berhaut* 7229! Tambana, Casamance (Feb.) *Chev.* 2448! **Guin.:** Friguiagbé (Aug.) *Chillou* 1649! Mankouta (Nov.) *Jac.-Fél.* 7364! Koba *Jac.-Fél.* 7371! **S. L.:** Peninsula (Sept.) *Melville & Hooker* 471! Newton (Nov.) *Deighton* 5627! Kasanko, Maforki (Oct.) *Jordan* 814! Rokupr (Mar., Apr.) *Jordan* 21! 205! **Lib.:** Bushrod I., Montserrado Dist. (Aug.) *Baldwin* 13021! Monrovia *Naumann* 20! (Jan.) *Dinklage* 2403! 20 miles East of Monrovia (Aug.) *Baldwin* 13049*a*! **Iv. C.:** Abouabou, Grand Bassam (Mar.) *J. & A. Raynal* 13649! Kodioboué, nr. Bianoua (Apr.) *Guillaumet* 1886! **Ghana:** Beyin, Axim Dist. (July) *Hall* GC 35536! **S. Nig.:** Sapoba (Sept.) *Onochie* FHI 34304! Nekede, nr. Owerri (Feb.) *Killick* 309! Also in C. African Republic, Gabon and Congo.

1b. **E. naumanniana** var. **caillei** (*Hutch. ex Nelmes*) *Hooper* in Kew Bull. 26: 582 (1972). *E. caillei* Hutch. ex Nelmes in Kew Bull. 6: 165 (1951); F.W.T.A., ed. 1, 2: 468; Nelmes and Baldwin in Amer. J. Bot. 39: 375, tt. 14–25 (1952). Differing from the typical variety in the more slender stems and smaller spikelets.
Mali: Macina (Nov.) *Monod*! **Guin.:** Kouria (Nov.) *Caille* in *Hb. Chev.* 14957! Friguiagbé (July, Nov.) *Chillou* 593! 894! Manéa (Dec.) *Chillou* 2322! **S. L.:** Kambia (Nov.) *Morton & Gledhill* SL 92! **Lib.:** Monrovia (Nov.) *Dinklage* 2748! 10 miles to 20 miles East of Monrovia (Aug.) *Baldwin* 13009! 13049*b*!

2. **E. subtilissima** *Nelmes* in Kew Bull. 6: 422 (1952); Nelmes and Baldwin in Amer. J. Bot. 39: 375, tt. 26–39 (1952). Differing from the foregoing aquatic species in producing more than one achene per spikelet and having a three-lobed style-base; in or at the margin of open water.
Mali: Macina (Nov.) *Duong*! **Guin.:** Siguiri (Nov.) *Jac.-Fél.* 1355! Friguiagbé (Nov.) *Chillou* 835! 893!

S. L.: Mafufui, nr. Mateboi (Nov.) *Jordan* 838! 839! **Lib.:** Monrovia (Nov.) *Dinklage* 3272! Also in the Congo.

[Very close to the Indian *E. retroflexa* (Poir.) Urb. (*E. chaetaria* Roem. & Schult.) from which it differs only in the slender habit, frequent presence of sterile stems, the narrower more acute glumes and somewhat smaller achene and style-base.]

3. **E. brainii** *Svenson* in Rhodora 39: 251, t. 461, fig. 4 (1937); J. & A. Raynal in Adansonia sér. 2, 7: 318 (1967). Semi-aquatic annual with numerous slender stems and very small elliptical spikelets; flooded savanna.
Sen.: Badi to Siminti (Dec.) *J. & A. Raynal* 6997*b*! **Guin.:** Kiffaya to Sériba (Oct.) *Pitot*! **Ghana:** Busunu, Damongo to Yapei (Sept.) *Hall* CC 891! Yendi to Zalazugu (Dec.) *Morton* A1463! **N. Nig.:** Basita, nr. Jebba, Ilorin Prov. (Feb.) *Clayton* 1508 (FHI 42003)! Also in E. Cameroun, C. African Rep. and east and south-east Africa.

4. **E. trilophus** *C.B. Cl.* in Fl. Trop. Afr. 8: 409 (1902); Svenson in Rhodora 34: 253, t. 462, fig. 10 (1932). Small tufted annual up to 10 cm. high with red-purple sheaths and elliptical red-purple glumed spikelets about 2·5 × 1 mm.; ground subject to flooding.
Sen.: *Roger* 113*a*! Podor to Guédé (Dec.) *Martine* 35! Thillé to Boubacar (Dec.) *Naegelé* 13124! Ndioum (Dec.) *Naegelé* 13105! Also in E. Cameroun and Congo.

[Perhaps conspecific with the American *E. minima* Kunth from which it differs only in the somewhat larger glumes and achene.]

5. **E. setifolia** (*A. Rich.*) *J. Raynal* in Adansonia sér. 2, 7: 318 (1967). *Isolepis setifolia* A. Rich., Tent. Fl. Abyss. 2: 498 (1851). *Eleocharis schweinfurthiana* Boeck. in Flora 62: 562 (1879); F.W.T.A., ed. 1, 2: 468. *E. microcarpa* of C.B. Cl. in F.T.A. 8: 410. Slender, tufted plant with neat, elliptical spikelets 2–3 mm. long of pale to golden membranous glumes; damp places on rocks and in grassland.
Sen.: (Sept.) *Fotius* 399! Tambacounda (Sept.) *Berhaut* 1538! Tambacounda to Sambailo (Sept.) *Pitot*! Sambailo (Oct.) *Pitot*! **Mali:** Ouolo Kourou Mountain (Aug.) *Jaeger* 7! Bamako to Ségou (Sept.) *Jaeger* 5110! Bamako (Sept.) *Adam* 14961! **Guin.:** Youkounkoun (Oct.) *Pitot*! Kiffaya to Seriba (Oct.) *Pitot*! Touqué (June) *Adam* 14745! foot of Mt. Nimba (Aug.) *Schnell* 3426*b*! Nimba (Sept.) *Adam* 6266! **S. L.:** Kambia (Sept.) *Jordan* 311! **Iv. C.:** Séguéla (Oct.) *Aké Assi* 6670! **U. Volta:** Ouahigouya to Koro (Aug.) *Chev.* 24806! **Ghana:** Burufu Plateau, nr. Lawra (July) *Adams* 921! Nasia, Bolgatanga to Tamale (Apr.) *Adams* 4156! Nakpanduri (Aug.) *Hall* CC 502! Busunu, Damongo to Yapei (Sept.) *Hall* CC 887! Ejura (Aug.) *Hall* CC 331! **N. Nig.:** Panshanu Pass, Bauchi Prov. (Aug.) *Lawlor & Hall* 442! Anara F.R., Zaria Prov. (July) *Clayton* 1228! Also in Chad, Central African Republic and throughout tropical Africa.

[Differs little from the American *E. microcarpa* Torrey.]

6. **E. complanata** *Boeck.* in Flora 62: 562 (1879); F.T.A. 8: 409. *E. anceps* Ridley in Trans. Linn. Soc. Lond., Ser. 2, Bot. 2: 148 (1884); F.T.A. 8: 410; Svenson in Rhodora 34: 253, t. 462, fig. 11 (1932). Medium sized, densely tufted annual with ovoid, pale or purplish spikelets about 5 mm. long; rice fields, pools and other marshy places.
Sen.: Badi (Sept.) *Berhaut* 1687! Niokolo-Koba (Nov.) *Adam* 16971! Dialacoto (Dec.) *J. & A. Raynal* 6773!Banharé (Dec.) *J. & A. Raynal* 6947! Velingara (Oct.) *Adam* 18588! **Mali:** Kita (Sept.) *Jaeger* 2659! Balassoko, Bamako (Nov., Dec.) *Duong*! *Monod*! Koulikoro (Oct.) *Chev.* 2454! **S. L.:** Kamalu (Jan.) *Deighton* 4578! **Iv. C.:** Toumodi (Aug.) *Chev.* 22376! Singrobo (Dec.) *Aké Assi* 7245! Tiégouakro to Kodiokoffi (Aug.) *Chev.* 22351! Sanlo to Kakpin (Nov.) *Aké Assi* 9293! **Ghana:** Busunu, Damongo to Yapei (Sept.) *Hall*) CC 886! Banda, Wenchi Dist. (Dec.) *Morton* GC 25099! Yeji (Apr.) *Hall* VBS 275! Kete Krachi (Dec.)*Adams* 4574! Anum to Asikuma (Sept.) *Rose Innes* GC 30389! **Togo Rep.:** Sansanné, Mango region (Oct, Nov.) *Bille* TG 263! **Dah.:** Cotonou (Mar.) *Morton* GC 6675! Ouèçé (Nov.) *Annet* 43! Agouagon (May) *Chev.* 23496! **Niger:** Bandé *P. de Fabrègues* 2253! Magaria *P. de Fabrègues* 2662! **N. Nig.:** Maska Dam, Funtua (Jan.) *Kershaw* 900028! Idon, Kaduna to Keffi (Aug.) *Kershaw* 900699! Bauchi (Dec.) *Haines* 59! Jebba (Feb.) *Meikle* 1188! Nupe *Barter* 1574! **S. Nig.:** Ifaki, Ondo Prov. (June) *Clayton* 1125! Olokemeji to Eruwa, Egba Dist. (Nov.) *Daramola* FHI 43847! Lagos (Oct.) *Dalz.* 1297! Also in E. Cameroun and throughout tropical Africa.

7. **E. geniculata** (*Linn.*) *Roem. & Schult.* Syst. Veg. 2: 150 (1817), emend.; Svenson in Rhodora 34: 259, t. 463, fig. 1 (1932) and 41: 50 (1939); Berhaut, Fl. Sén. ed. 2, 364 (1967). *Scirpus geniculatus* Linn., Sp. Pl. 1: 48 (1753), partly. *Eleocharis capitata* R. Br., Prodr. 225 (1810); F.T.A. 8: 407. *E. caribaea* (Rottb.) S. F. Blake in Rhodora 20: 24 (1918); F.W.T.A. ed 1, 2: 468; Berhaut, Fl. Sén. ed. 1, 217 (1954). Medium sized, closely tufted plant up to 40 cm. high with brownish, broadly ovate spikelets; brackish marshes, edges of mangrove swamps.
Sen.: St. Louis (Oct.) *Adam* 19204! L. Youi (Nov.) *Pitot*! Tivaouane (Nov.) *Pitot*! Malika (Jan.) *Naegelé* 29! Casamance (Sept.) *Adam* 18122! **Gam.:** St. Mary's I. (Dec.) *Heudelot* 549! **Guin.:** Is. Tristao (Nov.) *Jac.-Fél.* 7311! Coyah to Conakry (June) *Chillou* 1547! Kabak (Dec.) *Jac.-Fél.* 7331! 7376! Manéa (May) *Chillou* 2086! **S. L.:** Royof, Samu Chiefdom *Jordan* 387! Kayinti, Loko Masama Chiefdom (Mar.) *Glanville* 214! Mabange Railway Bridge, Ribi R. (Feb.) *Morton & Gledhill* SL 1724! Bagroo R. (Apr.) *Mann* 89! Mano Salija (Dec.) *Deighton* 281! **Lib.:** Cape Mount (Mar.) *Dinklage* 2489! Monrovia (May, Dec.) *Baldwin* 5832! *Kunkel* 235! Fishtown, Grand Bassa (Aug.) *Dinklage* 1705! Cape Palmas (July) *T. Vogel* 7! **Ghana:** Sakumo Lagoon, Weija (Jan.) *Morton* GC 8287! Keta (Sept.) *Hall* 3418! **S. Nig.:** Lagos *W. MacGregor* 337! (Jan., July) *Baldwin* 14014! *Haines* 22! Nembe, Niger Delta (Mar.) *Anderson* 1359! Abonnema (Nov.) *Anderson* 1415! **F. Po:** Moka *Guinea* 2426! Also in N. Mauritania (Adrar); pantropical, extending to S. Africa and southern S. America.

8. **E. atropurpurea** (*Retz.*) *Presl* Rel. Haenk. 1: 196 (1828); F.T.A. 8: 407; Svenson in Rhodora 31: 227, t. 49 (1929); Berhaut, Fl. Sén. ed. 2, 364 (1967). *Scirpus atropurpureus* Retz., Obs. 5: 14 (1789). *Eleocharis atropurpureus* var. *minor* Kunth, Enum. Pl. 2: 151 (1837). *Isolepis dichroa* Steud., Syn. Pl. Glum. 2: 91 (1855). Slender tufted annual with clusters of capillary stems up to 15 cm. high; spikelets ovate, obtuse, with green-keeled glumes and black or red achenes; damp grassy places.
Maur.: Letfotar (Oct.) *Chudeau*! **Sen.:** (Jan.) *Roger* 113! Podor to Guedé (Dec.) *Martine* 35! Bargny (Oct.) *Berhaut* 5525 *bis*! Richard-Toll (Feb.) *Berhaut* 1185! Casamance (Oct.) *Adam* 18349! **Mali:** Dogo (May) *Davey* 114! Buonankodjie, nr. Koubita (Nov.) *Davey* 549! Bourem to Bamba (Mar.) *Chev.* 43796! Bamba to Timbuctoo (May) *Chev.*! Gao to Goura (Apr.) *de Wailly* 4684! **Guin.:** Kiffaya to Sériba (Oct.) *Pitot*! **Ghana:** Tamale to Bolgatanga (Aug.) *Hall* CC 436! **Niger:** (Oct.) *P. de Fabrègues* 2568! **N. Nig.:** Yola, Adamawa Prov. (Nov.) *Hepper* 1204! Also in E. Cameroun; pantropical.

9. **E. deightonii** *Hooper* in Kew Bull. 26: 582 (1972). Tufted annual 6–25 cm. high with flattened stems and pale ovoid spikelets 3–4 mm. long; rice swamps.
Guin.: Kabak (Dec.) *Jac.-Fél.* 7379! Koba (Nov.) *Jac.-Fél.* 7250! **S. L.:** Rokupr (Aug., Sept.) *Deighton* 4361! *Jordan* 117!

10. **E. dulcis** (*Burm. f.*) *Henschel* Vita Rumph. 186 (1833); Merrill, Interpret. Herb. Amb. 104 (1917); Svenson in Rhodora 31: 158 t. 16 (1929). *Andropogon dulce* Burm. f., Fl. Ind. 219 (1768). *Eleocharis plantaginea* of F.T.A. 8: 405; F.W.T.A., ed. 1, 2: 467. The Chinese Water Chestnut, a large plant with soft, septate stems; in or near surface water.
Sen.: Casamance (Oct.) *Adam* 18417! Niokolo-Koba (Feb., Mar.) *Adam* 15938! 17443! 17964! Niombato (Nov.) *Berhaut* 1804! **Gam.:** (Aug.) *Ruxton* 76! 86! **Mali:** Macina (Nov., Dec.) *Duong* B18! 1231! Tenenkou (Sept.) *Lean* 104! Bourem I. (July, Aug.) *Chev.* 1234! Gao (Jan.) *Chev.* 43089! Ouatagouna, Ansongo (Sept.) *Hagerup* 432! **Guin.:** Youkounkoun (Oct.) *Pitot*! Koba (Nov.) *Jac.-Fél.* 7235! **S. L.:** Rosino, Great Scarcies (Mar., July) *Jordan* 40! *Deighton* 4120! Freetown to Magburaka (Dec.) *Morton & Gledhill* SL 275! Njala (July) *Deighton* 1165! Erimakuna, Falaba (Mar.) *Sc. Elliot* 4453*a*! **Lib.:** Bushrod I.

(Apr.) *Barker* 1302! **U. Volta:** Banfora L. (July) *Duong*! *Le Gall*! (Sept.) *Adam* 15386! Tengréla L., Banfora (July) *Aké Assi* 4761! 4861! **Ghana:** Akotokyir, Cape Coast (Feb.) *Hall* 2188! Cape Coast (June) *Hall* 3129! Atiavi to Keta (Feb.) *Hall* CC 1220! **S. Nig.:** Sapoba, Benin Dist. (Feb.) *Onochie* FHI 31940! Old World tropics, China and Japan.

11. **E. decoriglumis** *Berhaut* in Bull. Soc. Bot. France 100: 174 (1953), and Fl. Sén. ed. 2: 363 (1967). *E. fistulosa* Link var. *micrantha* Cherm. in Arch. Bot. Caen 7, Mém. 4: 25 (1936), partly. Similar in size and habit to *E. acutangula* but with a regular dark margin to the obtuse glume; marshes.
Sen.: *Perrottet* 839! Niakhéné (Nov.) *Trochain* 948! Linguéré *Roberty* 10068! Oualo dist. (Mar.) *Leprieur*! **Mali:** Fori (Oct.) *Wintrebert* in *Demange* 487! **Niger:** Kissambana to Hamdara (Oct.) *P. de Fabrègues* 2231! Also in E. Cameroun and Chad.

12. **E. acutangula** (*Roxb.*) *Schult.* in Roem. et Schult., Mant. 2: 91 (1824). *Scirpus acutangulus* Roxb., Fl. Ind. (ed. Wall.) 1: 216 (1820). *Eleocharis fistulosa* of F.T.A. 8: 406 partly, and F.W.T.A., ed. 1, 2: 468. *Scirpus fistulosus* Poir. (1804), not of Forsk. (1786). Tufted stoloniferous perennial with greenish cylindrical spikes; beside ponds and streams and in swamps.
Maur.: L. Rkiz (Sept.) *Sadio* 292! Néma (Oct.) *Davey* 488! **Sen.:** *Heudelot* 320! *Roberty* 10149! Bignona (Feb.) *Chev.* 2432! Kaolak (Nov.) *Berhaut* 1824! Diohine (Nov.) *Berhaut* 3979! **Gam.:** (Aug.) *Ruxton* 89! **Mali:** *Duong*! Kita (Oct.) *Jaeger* 55! 5752! Gorinnta, nr. Dogo (Apr.) *Davey* 555! **Guin.:** La Kolenté (Nov.) *Chillou* 950! Manéa (Dec.) *Chillou* 2320! Koya (Mar.) *Chillou* 2548! Koba (Nov.) *Jac.-Fél.* 7239! Faranah (Apr.) *Pitot*! **S. L.:** Erimakuna (Mar.) *Sc. Elliot* 4453b! Rokupr (Mar.) *Jordan* 20! Pendembu (July) *N. W. Thomas* 839! Ronietta (Nov.) *N. W. Thomas* 5605! Njala (July) *Deighton* 5944! **Lib.:** Ganta, Sanokwelle Dist. (Dec.) *Harley* 781! Nimba (Mar., Oct.) *Adam* 21082! *Adames* 704! **Iv. C.:** Mossou (Aug.) *Schnell* 6549! Séguéla to Mankono (Nov.) *J. J. F. E. de Wilde* 945! Grand Bassam (Sept.) *de Wit* 7398/1218! Abidjan (Oct.) *W. J. J. O. de Wilde* 1109! Béréby (Nov.) *Oldeman* 537! **Ghana:** Gambaga to Walewale (Dec.) *Adams* 4201! Nasia, Bolgatanga to Tamale (Apr.) *Adams* 4149! Yeji (Aug.) *Hall* CC 370! Dawa, Accra Plains (July) *Hall* 3246! Amisano marsh, Cape Coast (Sept.) *Hall* 2380! **Niger:** Doutchi to Dosso (Aug.) *P. de Fabrègues* 2483! Magaria (July) *P. de Fabrègues* 2489! Dallol Boboye, Niamey (Sept.) *Cremers* 970! **N. Nig.:** Bida to Abuja (Dec.) *Haines* 60! Badeggi, nr. Bida (Jan.) *Vaillant* 2775! Vom, Plateau Prov. (Feb.) *McClintock* 227! Giwa Fadama, Zaria Prov. (Jan.) *Kershaw* 900027! **S. Nig.:** Uke to Nobi, Onitsha Dist. (May) *Onochie* FHI 35794! Pantropical.

13. **E. mutata** (*Linn.*) *Roem. & Schult.* Syst. 2: 155 (1817); Svenson in Rhodora 31: 133, t. 188, fig. 8 (1929); F.W.T.A., ed. 1, 2: 467. *Scirpus mutatus* Linn., Pugill. Jamaic. Pl., Amoen. Acad. 5: 391 (1759), and Sp. Pl. ed. 2, 1: 71 (1762); *Eleocharis fistulosa* of F.T.A. 8: 406, partly. Similar to *E. acutangula* but the glumes broader and more rounded; generally in brackish coastal swamps.
Sen.: St. Louis (Oct.) *Adam* 19205! L. Youi, Cape Vert (Apr.) *Pitot*! Fatick *Maymard*! Kaolak (Oct.) *Berhaut* 3794! Oussouye (Sept.) *Broadbent* 116! **Guin.:** Coyah (Nov.) *Chillou* 2246! Koba (Nov.) *Jac.-Fél.* 7285! **S. L.:** Rokupr (July) *Jordan* 42! Benkia, Port Loko Creek (Mar.) *Glanville* 211! Aberdeen (July) *Small* 121! Maswari (Nov.) *Deighton* 5636! Bagroo R. (Apr.) *Mann* 893! **Lib.:** Monrovia (Sept., Oct.) *Wrigley & Melville* 713! *Adam* 16532! Sinkor, Monrovia (Nov.) *Kunkel* 550! Duport (Nov.) *Linder* 1444! Buchanan (Nov.) *Adam* 16040! **Ghana:** Elmina (May) *Morton* A895! Duakor, nr. Cape Coast (Oct.) *Hall* 2384! Weija (June) *Hall* 3099! Ada (Dec.) *Harris*! **S. Nig.:** Lagos I. *Barter* 2234! Lagos *W. MacGregor* 327! Abonnema, Rivers Prov. (Nov.) *Anderson* 1411! Also in West Indies, Central and northern South America.

14. **E. variegata** (*Poir.*) *Presl* in Oken, Isis 21: 269 (1828); Kunth, Enum. Pl. 2: 153 (1837); Svenson in Rhodora 31: 156 (1929), and 41: 8, t. 537 fig. 4 (1939); J. & A. Raynal in Adansonia Sér. 2, 7: 319 (1967). *Scirpus variegatus* Poir., Encycl. 6: 6: 749 (1804). Generally more slender than the preceding two species having glumes like *E. acutangula* and a wide mouthed achene like *E. mutata*; in similar habitats.
Sen.: Toubakouta (Jan.) *J. Raynal* 7928. **Guin.:** Friguiagbé (Sept.) *Chillou* 697! Télimélé (Feb.) *Roberty* 10769! **S. L.:** Kuputu (Sept.) *Jordan* 342! Rhombe swamp (Mar.) *Adames* 149! Serabu to Mowoto (Apr.) *Deighton* 1687! nr. Buedu, Kissi Tungu (Sept.) *Jordan* 532! **Lib.:** Devilbush, nr. Paynesville, Montserrado Dist. (Aug.) *Leeuwenberg & Voorhoeve* 4901! Ganta, Sanokwele Dist. (Sept.) *Baldwin* 12042! **Iv. C.:** *fide* Raynal *l.c.* **Ghana:** Kukrukui Creek, N. of Kete Krachi (July) *Adom-Boako & Ansah-Emmim* VBS 109! **Dah.:** *fide* Raynal *l.c.* **S. Nig.:** Nigerian Inst. for Oil Palm Research, 17 m. N.W. of Benin (May) *Killick* 230! *J. Hall* 925! Intermittently throughout the tropics.

15. **E. nupeënsis** *Hutch.* in F.W.T.A., ed. 1, 2: 457 (1936); Svenson in Rhodora 41: 6 (1939) partly; Berhaut, Fl. Sén. ed. 2: 363 (1967); J. & A. Raynal in Adansonia sér. 2, 7: 318 (1967). *E. mitrata* (Griseb.) C.B. Cl. var. *africana* C.B. Cl. in F.T.A. 8: 406 (1902). Similar to slender specimens of *E. acutangula* from which it may be distinguished on the terete stem and on achene characters; similar habitats.
Sen.: Niokolo-Koba (Dec.) *J. & A. Raynal* 6883; 6998. **S. L.:** Kambia (Sept.) *Jordan* 306! Mange, Bure (Nov.) *Jordan* 819! Kasanko, Maforki (Oct.) *Jordan* 818! **Iv. C.:** Singrobo (Dec.) *Aké Assi* 4739! **Ghana:** Nyankpala, nr. Tamale (Dec.) *Adams* 4192! Atebubu (Nov). *Hall & Duodu* GC 42127! **N. Nig.:** Nupe *Barter* 1040! Also in Chad.

[A taxon of doubtful status whose relationship to the West Indian *E. mitrata* (Griseb.) C.B. Cl. and to *E. variegata* and *E. acutangula* needs further study.]

9. WEBSTERIA S. H. Wright in Bull. Torr. Bot. Club 14: 135 (1887).

By Miss S. S. Hooper

Submerged aquatic with long slender weak stems producing tufts of increasingly short and slender, finally capillary, radiating stems of the first second and sometimes third order, surrounded at the base by membranous purple sheaths; spikelets emergent, linear, acute, about 1 cm. long, of two large purple and green glumes surrounding a single achene; anthers 3, large, about 4 mm. long; achene obovate with a distinct beak bearing the slightly enlarged style-base .. .. .. *confervoides*

W. confervoides (*Poir.*) *Hooper* in Kew Bull. 26: 582 (1972). *Scirpus confervoides* Poir., Encyc. Méth. 755 (1804). *Rhynchospora ruppioides* Benth. in Hook. f., Icon. Pl. 14, t. 1344 (1881). A green, floating, alga-like plant, most closely resembling *Eleocharis naumanniana* in the single-flowered, soft, flattened spikelet; in running water.

Iv. C.: Moossou (Dec.) *Hallé & Guillaumet*! Abouabou (Mar.) *J. & A. Raynal* 13650! **S. Nig.:** Sapoba (Sept.) *Onochie* in FHI 34310! The genus, which is probably monotypic, also occurs in the Congo, Tanzania, Zambia, Madagascar, Ceylon and intermittently in the New World tropics.

10. BULBOSTYLIS Kunth, *nom. cons**; C.B. Cl. in Hook.f., Fl. Brit. Ind. 6: 651 (1893); F.T.A. 8: 427 (1902).

By Miss S. S. Hooper

Spikelets sessile, inflorescence capitate:
Spikelets large, 10–15 mm. long, up to 10 in the head; plants with a clearly woody base:
Spikelets flattened, the glumes distichous or nearly so; leaves reduced to bladeless or shortly bladed, pilose, green sheaths bearded at the mouth; stems sharply angular or ridged; glumes brown to dark red, long acuminate, apiculate; achene obovate 1. *pilosa*
Spikelets terete; sheaths glabrous or puberulent, bright red; glumes black or blackish, obtuse or apiculate; achene elliptical .. .. .. 2. *erratica*
Spikelets small, less than 10 mm. and generally less than 5 mm. long, generally numerous in a spherical or hemispherical head; plants tufted, rarely somewhat woody:
Inflorescence greenish- or golden-brown; leaves and bracts capillary; leaf-sheaths pale, membranous; glumes hairy, boat-shaped with a clear green keel prolonged into a recurved apiculus .. .. 3. *barbata*
Inflorescences black or red-brown; basal sheaths reddish:
Inflorescence spherical, reddish brown when mature; long crimped brown or white hairs at mouth of sheath and on bracts; spikelets distinctly flattened, ovate, few-flowered; glumes hairy, more or less distichous:
Slender annual up to 20 cm. high; leaves capillary; inflorescence about 5 mm. across, occasionally up to 8 mm.; glumes lanceolate, acuminate, ciliate on the margin; achene about 0·4 × 0·3 mm. .. .. .. 4. *fimbristyloides*
Perennial up to 50 cm. high; inflorescence about 10 mm. across; glumes ovate, acute, slightly ciliate on the margin; achene 0·85–1·0 × 0·6–0·7 mm. 5. *laniceps*
Inflorescence hemispherical, black or dark brown; densely tufted plants with filiform leaves; spikelets terete, angular or slightly flattened, more or less cylindrical:
Stems beneath inflorescence hispid or tomentose with robust upwardly directed hairs; spikelets angular, distinct; glumes broadly ovate, boat-shaped with a short recurved apiculus; achene broadly elliptic, yellow to brown with a pale broad, generally obtuse, style-base 0·2–0·3 mm. across and slightly rugulose sides .. .. 6. *scabricaulis*
Stems beneath inflorescence smooth or somewhat scabrid; spikelets crowded; achene obovoid, trigonous with a small conical style-base and papillate sides:
Stems robust, about 1·5 mm. thick; sheaths brown, conspicuously hairy at mouth; glumes lanceolate, at least the lower with a straight or nearly straight keel .. .. 7. *metralis*
Stems usually not exceeding 1·0 mm. thick; sheaths not conspicuous; glumes ovate with a distinct curved apiculate keel .. .. .. 8. *filamentosa*
Spikelets pedicellate, solitary or few to numerous in a spreading inflorescence:
Spikelets large, about 8 × 3 mm., solitary, brown; glumes obtuse, hairy, ciliate on the margin .. .. 9. *bodardii*
Spikelets smaller than above, more than one (rarely reduced to one):
Plant with a rhizome composed of swollen stem-bases and covered with brown acute scales; sheaths red; leaf-blades reduced; spikelets 1–5, black 10. *oritrephes*
Plant tufted, without a rhizome:
Spikelets less than ten-flowered, in a simple umbel; bracts leafy, one at least up to twice as long as the peduncles; stems and numerous leaves generally hairy; sheaths red, often disintegrating into a mass of red-brown fibres; achene 1·25–1·5 mm. long; style-base small, easily detached .. .. .. 11. *lanifera*
Spikelets numerous in a spreading panicle, or if few, then without the above combination of characters:
Perennial, plant tufted; bracts with conspicuous dark bases and short green tips; spikelets 5–6 × 2 mm., reddish-black; glumes acute with a single slender nerve and pale margins; achene surface transversely ridged .. 12. *guineensis*
Annual or apparently so; bracts without conspicuous dark bases:
Inflorescence contracted with most of the spikelets on short pedicels; glumes with a well-marked keel excurrent in a sharp acumen 16b. *densa* var. *cameroonensis*
Inflorescence diffuse with most of the spikelets on spreading pedicels longer than themselves; glumes obtuse or mucronate, with the keel not or hardly excurrent:
Upper spikelets broadly ovate with several lanceolate, aristate, white-fringed

* Although a conserved name in Nomina Conservanda of the Code of Botanical Nomenclature (1961) with the authority Kunth, Enum. Pl. Cyp. 205 (1837) it was actually described by C. B. Clarke—see S. S. Hooper in Taxon 17: 446 (1968).

persistent bracteoles at the base; glume often with a dark spot towards the tip .. .. 13. *coleotricha*

Upper spikelets elliptical to ovate with a single, usually inconspicuous empty bracteole at the base:

Stem tomentose; sheaths pinkish; bracts red-brown, shortly hairy; achene obovate to obcordate, transversely undulate, sharply trigonous 14. *congolensis*

Stems glabrous or sparsely and shortly hairy:

Bracts fringed with long white hairs (sometimes only sparsely so in small specimens); glumes glabrous, 3-nerved with a membranous margin; achene with quadrate surface cells raised in the centre to give a papillate appearance, the papillae being arranged more or less in vertical rows 15. *abortiva*

Bracts not fringed; generally small plants less than 20 cm. high; glumes hairy; achene with vertically elongated surface cells, papillate in the centre and raised to produce a transversely undulate appearance of the achene surface:

Spikelets ovate to broadly elliptic; glumes broadly ovate, clearly keeled 16a. *densa* var. *densa*

Spikelets narrowly elliptic to cylindric; glumes lanceolate, appressed, weakly keeled .. .. 17. *pusilla*

1. **B. pilosa** (*Willd.*) *Cherm.* in Bull. Soc. Bot. Fr. 81: 266 (1934), and 82: 341 (1935); Bodard in Ann. Fac. Sci. Dakar 9: 61 (1963). *Schoenus pilosus* Willd., Phyt. 1: 3 (1794). *Fimbristylis africana* C.B. Cl. in Dur. & Schinz, Consp. Fl. Afr. 5: 601 (1895). *Bulbostylis aphyllanthoides* (Ridley) C.B. Cl. in Dur. & Schinz, Consp. Fl. Afr. 5: 611 (1895); F.T.A. 8: 436; F.W.T.A., ed. 1, 2: 477. *Fimbristylis aphyllanthoides* Ridley in Trans. Linn. Soc., ser. 2, 2: 151 (1883). Tufted perennial with a woody rhizome and a head of up to 10 large flattened spikelets; savanna woodland or grassland.
Sen.: Kanouméri, nr. Kédougou (July) *Fotius* K233! **Mali:** Sikasso, bridge on the Farako (May) *Demange* 2239! **Iv. C.:** N'Douci, N'zi R. (Aug.) *de Wit* 7338/982! Dabou (Nov.) *Roberty* 12505! *Giovannetti* 441! Kodiokoffi to Kondroko, N. Baoulé (Aug.) *Chev.* 22361! Aquien *Scaetta* 3012! **Ghana:** Beyin, nr. Axim (July) *Hall* GC35538! Achimota (Feb.) *Morton* GC 25375! Legon (Oct.) *Adams* 3408! Tema (June) *Boughey* 13009! Aburi, Accra plains (Aug.) *Deighton* 3410! **Togo Rep.:** Bafilo (May) *Roberty* 1457! Agomé *Schlechter* 12967! Klouto (June) *Villiers*! 35 km. S. of Palimé (Mar.) *Stage* 8! Lomé *Warnecke* 124! 125! *Stage* 15! **Dah.:** Ouidah (Sept.) *Adjanohoun* 329a! Ahozon, 11 km. E. of Ouidah (Feb.) *Raynal* 13525! Porto Novo *Adjanohoun* 348! Abomey (Feb.) *Chev.* 23205b! **N. Nig.:** Jebba (Apr.-May) *Kamphorst* 185! 200! Nupe *Barter* 491! Ibaji Ojoko N.A. F.R., Igala Dist. (June) *Latilo* FHI 47670! **S. Nig.:** Oyo to Iseyin (July) *J. B. Gillett* 15202! Lagos (July) *Haines* 19! Ubiaja Clan F.R., Ishan Dist. (Apr.) *Charter* FHI 39423! Onitsha (Mar.) *Killick* 83! Enugu (Mar.) *Morton* GC 6681! Jesse, Warri Prov. (Sept.) *Butler-Cole* 14! Tropical Africa.

2. **B. erratica** (*Hook. f.*) *C.B. Cl.* in F.T.A. 8: 434 (1902). *Schoenus erraticus* Hook. f. in J. Linn. Soc. 6: 22 (1861). Tufted plant with lurid red colouring at the base and large black spikelets in a single head; montane grassland.
S. L.: Plateau S. of Camp 2, Loma Mts. (June) *Morton* SL 3563! **N. Nig.:** Gangirwal, Mambila Plateau, 7,500–8,000 ft. (Nov.) *Tuley* 1999! **W. Cam.:** Bafut-Ngemba F.R., Bamenda Dist. (Feb., May) *Hepper* 2127! *Daramola* FHI 41176! Cam. Mt., 8,000 to 11,000 ft. (Jan., Apr.) *Mann* 1344! *Maitland*1268 ! *Morton* GC 6874! **F. Po:** Clarence Peak, 8,500–9,000 ft. (Apr., Dec.) *Mann* 635! 1472!

3. **B. barbata** (*Rottb.*) *C.B. Cl.* in Hook. f., Fl. Brit. Ind. 6: 651 (1893); F.W.T.A., ed. 1, 2: 477, excluding syn. *Scirpus barbatus* Rottb., Ic. et Descr. 52 t. 17 f. 4 (1773). Tufted annual 5–20 cm. high with acute radiating spikelets in a hemispherical head; weed, particularly on sandy soil.
Sen.: *Sieber*! *Heudelot* 441! Dakar (Apr.) *Adam* 1097! Hann (Sept.) *Pitot*! Mbao (Oct.) *Pitot*! Nord Saveigne (Sept.) *Audru* 2641! **Gam.:** *Saunders* 38! Genieri (July) *Fox* 124! **Mali:** Ségou (Aug.) *Roberty* 3776! Macina (Nov., Dec.) *Duong* 1311! Bore (Aug.) *Demange* 26/1957! 28/1957! Gao (Aug.) *de Wailly* 5161! **Lib.:** Harper, Maryland Dist. (June) *Baldwin* 6001! **Iv. C.:** Bouaké (June) *Leeuwenberg* 4276! Bouaké to Diabo (July) *Chev.* 20065 *bis*! Pakoubo (Aug.) *Roberty* 3068(52)! Adiopodoumé (Dec.) *Leewenberg* 24! 2128! Grand Bassam (Dec.) *Giovannetti* 290! **U. Volta:** Banfora to Ferkessédougou (June) *Leewenberg* 4430! **Ghana:** Dedora Tankara F.R., Navrongo Dist. (June) *Andoh* FH 5916! Bamboi Ferry (May) *Morton* GC 7705! Ayikuma to Shai Hills (Dec.) *Morton* GC 8071! Cape Coast (May) *Hall* 2592! Accra *T. Vogel*! **Togo Rep.:** *Baumann* 538! Lomé *Warnecke* 122! **Dah.:** Agouagon (Apr., May) *Chev.* 23492! Cotonou (Mar.) *Morton* GC 6669! Savalou (July) *Annet* 109b! **Niger:** Yatacala to Kokoro (July) *Rogeon* 510! Toukounous *Bartha* 98! 99! Niamey (Oct.) *Hagerup* 498! Dosso (Dec.) *Monod* 7159! **N. Nig.:** Sokoto Prov. *Dalz.* 458! Shagunu (July) *Cook* 384! Jebba (Apr.) *Meikle* 1389! Lokoja (Apr.) *Haines* 17! Dawo, Vogel Peak (Nov.) *Hepper* 1407! **S. Nig.:** Oke Iho to Iseyin (Apr.) *Keay & Polunin* FHI 37841! Arakanga F.R., Abeokuta (May) *Keay & Onochie* FHI 37009! Lagos *Dalz.* 1109! Eruwa to Lanlate, Ibadan Dist. (Oct.) *Keay & Savory* FHI 25290! Nsugbe, nr. Onitsha (Aug.) *Maggs* 138! Widespread in the Old World tropics.

4. **B. fimbristyloides** *C.B. Cl.* in Mém. Soc. Bot. Fr. 2, 8: 28 (1907); Bodard l.c. 61. *B. cyrtathera* Cherm. in Arch. Bot. Caen 4, Mém. 7: 35 (1931). Similar to the following species but smaller in all its parts; damp places.
Mali: Klela (Oct.) *Demange* 3128! 3129! Kita (Sept.) *Jaeger* 2776! Koulikoro (Oct.) *Chev.* 2457! 2462! **Guin.:** Madina Kossekré (Oct.) *Adam* 12550! **U. Volta:** Sindoukoroni (Sept.) *Adam* 15188/3! Also in Chad, C. African Republic and Zambia; probably undercollected.

5. **B. laniceps** *C.B. Cl. ex Dur. & Schinz* Mém. Acad. Roy. Belg. 53: 306 (1896); Bodard l.c. 78. Densely tufted plant with fine leaves and spherical inflorescences which are white-woolly when young; grassland.
Guin.: Friguiagbé (Apr., May) *Chillou* 211! 2586! Mamou (May) *Pitot*! Labé *Adam* 14552! Beyla (Mar.) *Chev.* 20878! **S. L.:** Gbap (June) *Adames* 54! Malal (Apr.) *Bakshi* 144! Kambia (Apr.) *Hepper* 2636! Mapema (Feb.) *Glanville* 159! Loma Mts. (Aug.) *Jaeger* 6981! **Lib.:** Sinkor (June) *Barker* 1368! Painesville (Feb.) *Harley* 1859! Grand Bassa *T. Vogel* 92a! Sangwin (Mar.) *Baldwin* 11323! Pandamai (Mar.) *Bequaert* 117! **Iv. C.:** Nimba (Aug.) *Boughey* GC 18144! **Ghana:** Divinim, Berekum to Sampa, Ashanti (Dec.) *Adams* 5292! Kintampo to Techiman (Apr.) *Hossain & Enti* GC 36787! Sege to Batter, Accra Plains (Apr.) *Morton* A3206! Wuropong, Togo Plateau (May) *Morton* A3933! Amedzofe, 2,600 ft. (May) *Morton* GC 7208! **Togo Rep.:** Sowodi (Aug.) *Mahoux* 2139! **N. Nig.:** Vogel Peak, Sardauna Prov. *J. Hall* 1079! **W. Cam.:** Bamenda *Richards*! Also from E. Cameroun and Gabon to the Congo, Angola and Zambia.

6. **B. scabricaulis** *Cherm.* in Bull. Soc. Bot. Fr. 68: 419 (1922). *B. cardiocarpa* (Ridl.) C.B. Cl. var. *holubii* C.B. Cl. in F.T.A. 8: 434 (1902). *B. filamentosa* of C.B. Cl. in F.T.A. 8: 433, and of F.W.T.A., ed. 1, 2: 477, partly. *Fimbristylis collina* Ridl. in Trans. Linn. Soc. ser. 2, 2: 154 (1884), not *B. collina* (Kunth)

C.B. Cl. *B. filamentosa* (Vahl) Kunth var. *scabricaulis* (Cherm.) Bodard in Ann. Fac. Sci. Dakar 9: 68 (1963). Close to *B. filamentosa* and *B. metralis* but distinguished by the larger more distinct spikelets and stem below the inflorescence scabrid hairy; moist places.
Sen.: Dakar *Adam* 16938! Tambacounda (Sept.) *Berhaut* 3133! **Mali:** Kita (July) *Jaeger* 16! 24! 33! Fodébougou to Kroukoto, Kita Dist. (July) *Jaeger* 26! Bamako *Adam* 15142! (Aug.) *Roberty* 2605! **Guin.:** Kouroussa (July) *Pobéguin* 424! Faranah to Banian (Apr.) *Pitot*! Kissidougou to Kowdou (June) *Martine* 328! Kissidougou to Guéckédou (Apr.) *Pitot*! **S. L.:** Mando (Sept.) *Jordan* 97! Koinadugu (Aug.) *Haswell* 205! Falaba (July) *Morton & Jarr* SL2167! Kenema (Apr.) *Deighton* 3134! Loma Mts. (Sept.) *Jaeger* 1919! **Lib.:** Yékèpa Grassfield, Nimba Mts. (June) *Adam* 21559! **Iv. C.:** Tafiré (June) *Leeuwenberg* 4439! Téhini (Aug.) *Oldeman* 319! Toumodi (Aug.) *Boughey* GC 18549! **U. Volta:** Barmoussa (June) *Scholz* 315! **Ghana:** Tumu (May) *Morton* GC 7527! Wa (June) *Adams* 812! Banda (Apr.) *Morton* A3284! Ejura Scarp (Mar.) *Adams* 2452! **Togo Rep.:** Sokodé Basari *Baumann* 3834! **N. Nig.:** Kan Gimi, Anara F.R. (May) *Keay* FHI 22872! Shagunu, Ilorin Prov. (July) *Cook* 301! Panshanu, Bauchi Prov. (Aug.) *Lawlor & Hall* 84! Kaduna to Keffi (June) *Kershaw* 900613! Kudaru (May) *Clayton* 1032! Samaru, Zaria Prov. (July) *Clayton* 1244! **S. Nig.:** Shaki, Oyo Prov. (May) *Wanstall* A9! Owo (Apr.) *Haines* 29! Also in central, east and south-east Africa and Madagascar.

7. **B. metralis** *Cherm.* in Arch. Bot. Caen 4, Mém. 7: 34 (1931). *B. cardiocarpa* and *B. filamentosa* of F.W.T.A., ed. 1, 2: 477, partly. Distinguished from *B. filamentosa* by the dark, clearly white-fringed leaf-sheaths and the lanceolate glumes; montane grassland or savanna.
Mali: Fodébougou to Kroukoto (Aug.) *Jaeger* 5751! **Guin.:** Manea (June) *Chillou* 2617! Friguiagbé (July) *Chillou* 1636! Sala Falls (Oct.) *Pitot*! Nzérékoré to Beyla (May) *Pitot*! Nimba Mt. (Apr.) *Pitot*! **S. L.:** Mabonto to Bumbuna (Aug.) *Deighton* 1400! Binkolo, nr. Makeni (Sept.) *Harvey* 121! Kundukonko, Loma Mts. (Aug.) *Jaeger* 1089! Rowala (July) *Thomas* 1112! Wonkifu (July) *Jordan* 45! **Iv. C.:** Nimba Mts., 2,500–5,000 ft. (Aug.) *Boughey* GC 18083! Bandama Forest, Cosrou (May) *Leeuwenberg* 4250*b*! Kodiokoffi to Kondrokro (Aug.) *Chev.* 22364! Moossou (May) *Aké Assi* 4300! **Ghana:** Gemi Hill, Amedzofe (June) *Morton* A3427! (May) *Morton* GC 7212! Wuropong (May) *Morton* A3935! Shiare to Chilinga, Kete Krachi (May) *Morton* A3983! Beyin (July) *Hall* GC 35534! **N. Nig.:** Ilorin to Awtun (June) *Clayton* 1157! **S. Nig.:** Iva Valley F.R., Enugu Dist. (July) *Daramola* FHI 55156! Emene, Enugu (July) *Tuley* 803! Enugu (Mar.) *Morton* GC 6682! Also in central and east tropical Africa.

8. **B. filamentosa** (*Vahl*) *C.B. Cl.* in Dur. & Schinz, Consp. Fl. Afr. 5: 613 (1895). *Scirpus filamentosus* Vahl, Enum. 2: 262 (1806). Similar to *B. scabricaulis* but distinguished by the smaller achene and smooth or slightly scabrid stems; moist places.
Iv. C.: Grand Bassam *T. Vogel* 92*b*! Bouna, Ouangofétini *de Wilde* 786! **Ghana:** Kete Krachi (Apr.) *Morton* GC 9151! Kwahu Tafo to Mankrong (Apr.) *Morton* A691! Krobo Mt. (Oct.) *Morton* GC 7744! Accra (July) *Dalz.* 174! Kpong (Apr.) *Gyadu* GC 20560! **N. Nig.:** Samaru (Aug.) *Keay* FHI 28007! Naraguta F.R., Jos (July) *Lawlor & Hall* 146! Biu, Bornu Prov. (Sept.) *Kennedy* FHI 8002! Panshanu, Bauchi Prov. (Aug.) *Lawlor & Hall* 565! 387!

9. **B. bodardii** *Hooper* in Kew Bull. 26: 582 (1972). *B. zambesica* C.B. Cl. var. *occidentalis* Bodard in Bull. Soc. Bot. Fr. 108: 309 (1961); J. & A. Raynal in Adansonia sér. 2, 7: 316 (1967). Distinguished by the single large red-brown spikelet with obtuse white-fringed glumes; wet places, sometimes on laterite.
Sen.: Niokolo-Koba *Adam* 14196! 17894! 18956! *Bodard* 3148! *J. & A. Raynal* 6918. **Guin.:** Dalaba–Diaguissa Plateau (Feb.) *des Abbayes* 825! Mamou (May) *Pitot*!

10. **B. oritrephes** (*Ridley*) *C.B. Cl.* in Trans. Linn. Soc. ser. 2, 4: 54 (1894). *Fimbristylis oritrephes* Ridley in Trans. Linn. Soc., ser. 2, 2: 155 (1884). *Bulbostylis trichobasis* of F.W.T.A., ed. 1, 2: 477. *B. clarkeana* Hutch. ex Bodard in Bull. Soc. Bot. Fr. 108: 308 (1961); F.W.T.A., ed. 1, 2: 477. Characterised by the seriate rhizome composed of swollen stem bases covered by reddish scales and the few dark spikelets with, when flowering, very conspicuous anthers; upland grassland, often on bare rock.
Guin.: Sita (Apr.) *Pitot*! Labé *Adam* 14643! Thianguel-Bori (Apr.) *Pitot* 25! 41! Erimakuna (Mar.) *Sc. Elliot* 5244! Boula (Apr.) *Adam* 12062, partly! **S. L.:** Loma Mts. (Feb., Mar.) *Morton & Gledhill* SL 1097! *Jaeger* 4188! 4288! Bintumane Peak, Loma Mts. (June, Aug.) *Jaeger* 986! *Morton* SL 3578! **N. Nig.:** Kakara, Adamawa Prov. (Jan.) *Hepper* 1783! nr. Njawai (Nov.) *Tuley* 1899! Chappal Waddi, Gotel Mts. (Apr.) *J. Hall* 1824! **S. Nig.:** Obudu (Apr.) *Haines* 230! **W. Cam.:** Bambalang, Ndop Plain (Mar.) *Brunt* 1033! Bamenda (Jan.) *Daramola* FHI 40601! *Migeod* 318! Bafut-Ngemba F.R., Bamenda (Mar.) *Coombe* 216! *Brunt* 1035! Mayo Binka, N.E. Nkambe (Feb.) *Hepper* 1885! Throughout tropical and South Africa but distribution uncertain because of confusion with the allied *B. trichobasis* (Baker) C.B. Cl.

[A gathering from Bamenda (*Richards* 5310) seems to suggest the possibility of hybridisation between this species and *B. erratica*.]

11. **B. lanifera** (*Boeck.*) *Kük.* in Peter in Fedde, Rep. 40, 1: 418 (1938). *Scirpus laniferus* Boeck. in Linnaea 36: 768 (1870). *Bulbostylis coleotricha* (A. Rich.) C.B. Cl. var. *lanifera* (Boeck.) C.B. Cl. in Dur. & Schinz, Consp. Fl. Afr. 5: 613 (1895), partly; F.T.A. 8: 442. *B. togoensis* Cherm. in Bull. Soc. Bot. Fr. 81: 267 (1934). Tufted plant with narrow leaves often nearly as long as the stems and a small inflorescence of a few dark spikelets; savanna woodland.
Sen.: Ouassadou (Dec.) *Berhaut* 1876! Sangalkam (Dec.) *Berhaut* 2617! **Iv. C.:** Baoulé, Toumodi (Oct.) *Pobéguin* 78! Toumodi (Nov.) *Boudet* 1419! Séguéla (Apr.) *Leeuwenberg* 3270! **U. Volta:** Barmoussa (June) *Scholz* 303*a*! **Ghana:** Zian to Wahabu, Wa Dist. (July) *Adams* 875! Banda (Aug.) *Hall* 1989! Gambaga (May) *Morton* GC 7370! 7387! Accra Plains (Aug.) *Gyadu* SLUS 139! **Togo Rep.:** *Baumann*! Pagouda (July) *Mahoux* 2116! Kandé *Mahoux* 2128! **N. Nig.:** Shagunu, Ilorin Prov. (July) *Cook* 303! Nupe *Barter* 1030! Jebba (Apr., May) *Kamphorst* 138! 139! Lokoja (Apr.) *Haines* 18! **S. Nig.:** Lagos to Abeokuta (Sept.) *Killick* 252! Also in the Congo and Chad.

12. **B. guineensis** *Cherm. ex Bodard* in Bull. Soc. Bot. Fr. 108: 308 (1961), and in Ann. Fac. Sci. Dakar 9: 62 (1963). More robust than the other dark-spikeletted species with a somewhat woody base and neat elliptical spikelets.
Guin.: Mt. Nimba *Schnell* 3686*a*! (Sept.) *Scaëtta* 3258*b*!

13. **B. coleotricha** (*A. Rich.*) *C.B. Cl.* in Dur. & Schinz, Consp. Fl. Afr. 5: 613 (1895); Bodard l.c. 74. *Fimbristylis coleotricha* Hochst. ex A. Rich., Tent. Fl. Abyss. 2: 506 (1851). *Bulbostylis miegei* Bodard in Bull. Soc. Bot. Fr. 108: 307 (1961). Annual up to 40 cm. high, often less, with filiform leaves and few, relatively large, rounded, dark-tipped spikelets; damp flushes on granite or ironstone rocks, sandy open grassland and as roadside weed.
Sen.: Casamance (Oct.) *Adam* 18607! Niokolo-Koba (Dec.) *J. & A. Raynal* 6942! Boulou Kounda (Dec.) *J. & A. Raynal* 6757! Kanouméri (Aug.) *Fotius* K221! Sebikotane (Sept.) *Pitot*! **Mali:** Bafoulabé (Sept.) *Duong*! Kita (July, Oct.) *Jaeger* 15! 79! Bamako (Sept.) *Jaeger* 5067! Lafaya, Koulikoro (May–June) *Duong* 12! Ségou (Aug.) *Roberty* 2642! **Guin.:** La Kolenté (Aug.) *Chillou* 1671! Sambailo (Oct.) *Pitot*! Kouroussa (July) *Pobéguin* 418! 408! Sita, Sériba (Oct.) *Adam* 12690! **S. L.:** Brookfields, Freetown (Aug.) *Jordan* 71! Port Loko (May) *Jordan* 872! Kambia (Aug.) *Jordan* 298! Batkanu (July) *Small* 148! Kamabai nr. Makeni (Sept.) *Harvey* 91! **Iv. C.:** Nimba (Aug.) *Boughey* GC 18143! Mankono (July) *Aké Assi* 4755! **Ghana:** Bole (Apr.) *Morton* A3293! Damongo F.R. (Nov.) *Rose Innes* GC 30839! Burufu Plateau, nr. Lawra (July) *Adams* 925! Gambaga (Aug.) *Hall* 2071! Kete Krachi (May) *Morton* GC 7173! **Togo Rep.:** Sokodé *Schilling* 43! *Duffour* 7023! **Niger:** Kirtachi Seybou (Nov.) *Hepper* 3847! **N. Nig.:** Nupe *Barter* 531! Sokoto (Aug.) *Dalz.* 457! Birnin Gwari Dist., Zaria Prov. (Aug.) *Keay* FHI 25996! Panshanu Pass, Bauchi Prov. (Aug.) *Lawlor & Hall* 552*a*! Yola (July) *Dalz.* 259! **S. Nig.:** Iseyin (Aug.) *J. B. Gillett* 15407! Igboora (June) *Haines* 234! Igbete to Kishi (Mar.) *Gledhill* 724! Also in E. Cameroun, central and east tropical Africa.

14. **B. congolensis** *De Wild.* Pl. Bequaert. 4: 194 (1927); Bodard l.c. 65. *B. holotricha* Peter in Fedde, Rep. Beih. 40, 1 Anhang 127 (1936); Nelmes & Baldwin in Amer. J. Bot. 39: 368 (1952). *B. polytricha* Cherm. in Arch. Bot. Caen 4, Mém. 7: 39 (1931). Tufted annual with papery leaf-sheaths and distinctly pedicellate spikelets; granite out-crops, lateritic flats and wet flushes.
Mali: Kita Mts. (Oct.) *Jaeger* 57! Bamako (Sept.) *Waterlot* 1349! Koulikoro (Oct.) *Chev.* 2476! **Guin.:** Télémelé to Pita (Oct.) *Pitot*! Hollandé-Tossekré (Oct.) *Adam* 12727! Macenta (Oct.) *Baldwin* 9786! Nzérékoré (Apr., Oct.) *Pitot*! *Baldwin* 9694! **S. L.:** Waterloo (Aug.) *Melville & Hooker* 333! Binkolo (Aug.) *Thomas* 1771! Sefadu (Sept.) *Adames* 269! Bendugu to Kulafaga (Sept.) *Deighton* 5153! Loma Mts. (Aug.) *Jaeger* 7223! **Lib.:** Palilah, Gbanga Dist. (Aug.) *Baldwin* 9154! 12040! Nimba Mts. (June, Aug.) *Adam* 21582! *Adames* 460! 20 miles E. of Zwedru, Tchien Dist. (Aug.) *Baldwin* 7057! **Iv. C.:** *Scaëtta* 3258! 3297! Man (June) *Portères* 21 *bis*! Orodougou, Sifié to Séguéla (June) *Chev.* 21829! **Ghana:** Berekum to Sampa (Dec.) *Adams* 5295! **N. Nig.:** Kuffena, Zaria Dist. (Aug.) *Kershaw* 900694! Jos (Aug.) *Keay* FHI 20200! Naraguta F.R., Jos Plateau (July) *Lawlor & Hall* 45560! Panshanu, Bauchi Prov. (Aug.) *Lawlor & Hall* 334! 397! **S. Nig.:** Ado Rock, Oyo Dist. (Oct.) *Savory & Keay* FHI 25298! Iseyin Rock (Aug.) *Onyeachusim* FHI 59186! Fashola Rock, Oyo Dist. (Nov.) *Gledhill* 680! Oyo to Iseyin (July) *Gillett* 15199! Igboora (Oct.) *Haines* 221! Throughout tropical Africa.

15. **B. abortiva** (*Steud.*) *C.B. Cl.* in Dur. & Schinz, Consp. Fl. Afr. 5: 610 (1895); Bodard l.c. 74. *Fimbristylis abortiva* Steud., Syn. Pl. Glum. 2: 111 (1855). *Scirpus schweinfurthianus* Boeck. in Linnaea 36: 758 (1870). Readily distinguished from *B. congolensis* and large forms of *B. densa* which it resembles in size, by the ciliate-edged bracts and glabrous glumes; grassland, and cultivated and waste places.
Sen.: Souriel (Sept.) *Diallo* 247! Fajakouon to Medina Yoro Foula *Boudet* 3828! Dulayel (Sept.) *Boudet* 3540! Ouassadou (Nov.) *Berhaut* 899! Tambacounda (Sept.) *Berhaut* 3184! **Mali:** Kita (Sept., Oct.) *Jaeger* 3! or 52! 5747! Bamako (Sept.) *Jaeger* 5058! *Adam* 14972! 15032! **S. L.:** Lengekoro (July) *Glanville* 219! Koinadugu (Aug.) *Haswell* 229! Loma Mts. (Sept.) *Jaeger* 1911! Kondembaia (Aug.) *Jordan* 492! **U. Volta:** Pama to Diabiga (Aug.) *Scholz* 149! Banfora (Sept.) *Adam* 15159! Sindoukoroni (Sept.) *Adam* 15188! **Ghana:** Burufo, S. of Lawra (Sept.) *Irvine* 4722! Tongo Hill, nr. Bolgatanga (Aug.) *Hall* CC 570! Tamale (Aug.) *Irvine* 4535! Banda (Aug.) *Hall* 1990! **Togo Rep.:** *Büttner* 123! Sokodé (July) *Duffour* 7022! **N. Nig.:** Badeggi, nr. Bida (Sept.) *Vaillant* 2739! Kaduna to Keffi (June) *Kershaw* 900611! Anara F.R., Zaria Prov. (July) *Keay* FHI 25950! Jos (Aug.) *Keay* FHI 20186! Panshanu Pass (Aug.) *Lawlor & Hall* 203! **S. Nig.:** Lagos to Abeokuta (Sept.) *Killick* 251! Umudike (Aug.) *Nwanzo* in *Hb. Tuley* 773! Mamu River F.R., Onitsha Prov. (Sept.) *Onochie* FHI 34078! Igboora (June) *Haines* 233! Throughout tropical Africa and Madagascar.

16. **B. densa** (*Wall.*) *Hand.-Mazz.* in Karsten & Schenk, Vegetationsb. 20, 7: 16 (1930). *Scirpus densus* Wall. in Roxb., Fl. Ind. ed. 1, 1: 231 (1820). *Bulbostylis capillaris* (Linn.) C.B. Cl. var. *trifida* (Nees) C.B. Cl. in Hook. f., Fl. Brit. Ind. 6: 652 (1893). *B. trifida* (Nees) Nelmes in Kew Bull. 5: 209 (1950). *Isolepis trifida* Nees in Wight, Contrib. Bot. Ind. 108 (1834).

16a. **B. densa** (*Wall.*) *Hand.-Mazz.* var. **densa.** Very variable in size and inflorescence but always with clearly pedicellate spikelets.
Sen.: Tambacounda (Sept.) *Berhaut* 3285! **Mali:** Kita Mts., summit (Sept.) *Jaeger* 46! 61! **Guin.:** Friguiagbé (Apr., May, Aug.) *Chillou* 215! 517! 610 *bis*! Labé *Adam* 14562! Labé to Gaoual *Adam* 14811! **S. L.:** Hastings (Aug.) *Melville & Hooker* 194! Rokupr (July) *Jordan* 458! Falaba (July) *Morton & Jarr* SL 2094! Bintumane Peak, Loma Mts. (Oct.) *Jaeger* 195! Mokunde, nr. Njala (June) *Pyne* 173! **S. Nig.:** Obudu Plateau Rd., 4,500 ft. (Nov.) *Tuley* 1020! North of Owo F.R., Ondo Prov. (July) *Onochie* FHI 33248! **W. Cam.:** Cam. Mt., 6,000–10,000 ft. (Dec.) *Mann* 2093! Mann's Spring, Cam. Mt. (Mar.) *Richards* 9399*a*! Onyanga, Cam. Mt., 8,100 ft. (Jan.) *Steele* 76! Bagango to Bafut-Ngemba, Bamenda Dist. (July) *Daramola* FHI 43206! Bum, Bamenda Dist. (June) *Maitland* 1759! **F. Po:** *Milne*! 8,500 ft. (Jan.) *Mann* 1471! Clarence Peak (Dec.) *Mann* 660! nr. Biao (Sept.) *Melville* 466! Throughout the tropical region of the Old World.

16b. **B. densa** var. **cameroonensis** (*C.B. Cl.*) *Hooper* in Kew Bull. 26: 583 (1972). *B. puberula* (Poir.) C.B. Cl. var. *cameroonensis* C.B. Cl. in Dur. et Schinz, Consp. Fl. Afr. 5: 615 (1895), name only. Differing from the rest of the species in Africa in having spikelets on short pedicels producing a compact inflorescence and glumes with a distinctly excurrent nerve; grassland.
Cam.: Cam. Mt., 6,000–1,000 ft. (Dec.) *Mann* 1360*b*! 2093*b*!
[A variable species, the lowland form of which has a number of small spikelets with glumes about 1·5 mm. long, whilst specimens from the Nimba Mts. and Cameroun Mt. have a few larger spikelets with dark glumes 2 mm. long.]

17. **B. pusilla** (*A. Rich.*) *C.B. Cl.* in Dur. et Schinz, Consp. Fl. Afr. 5: 615 (1895); F.T.A. 8: 440; Bodard l.c. 72. *Fimbristylis pusilla* Hochst. ex A. Rich., Tent. Fl. Abyss. 2: 506 (1851). *Scirpus hochstetteri* Boeck. in Linnaea 36: 739 (1870). *Bulbostylis capillaris* of F.W.T.A., ed. 1, 2: 477, partly. Slender annual with capillary leaves and few, narrow, brown spikelets; moist grassy places often on rock outcrops.
Mali: Kita (July–Sept.) *Duong* 430! *Jaeger* 4! 17! **Guin.:** Friguiagbé (July) *Chillou* 1635! Labé *Adam* 14640! **S. L.:** Pendembu (July) *Thomas* 775*b*! **U. Volta:** Sindoukorony *Adam* 15191! Banfora to Toussiana (Sept.) *Jaeger* 6561! Ouagadougou to Ouahigouya (Aug.) *Chev.* 24722 *ter*! **Ghana:** Nakpanduri (Aug.) *Hall* CC 519! **N. Nig.:** Nupe *Barter* 1579! Kufena, Zaria Prov. (Sept.) *Clayton* 1324! Naraguta F.R., Jos (July) *Lawlor & Hall* 143! Panshanu, Bauchi Prov. (Aug.) *Lawlor & Hall* 241! 551! Also in Chad, C. African Rep. and Ethiopia.

11. FIMBRISTYLIS Vahl, Enum. 2: 285 (1806); F.T.A. 8: 411 (1902). *Abildgaardia* Vahl, Enum. 2: 296 (1806).

By Miss D. M. Napper

Achene oblong-cylindric, smooth; small slender annual up to 10 cm. high; inflorescence of 1–3 ovoid spikelets on long peduncles and a central sessile one; spikelets 2–4 mm. long, 2 mm. wide; glumes with long awns, crowded .. 1. *dipsacea*
Achene orbicular to obpyriform, trigonous or biconvex, the surface variously ornamented or smooth:
Stigmas 2; style flat, usually ciliate; achene biconvex; glumes always spirally arranged:
Style-base with a fringe of long hairs pendant over the upper part of the achene; slender annual up to 15 cm. high with numerous green to buff narrowly oblong spikelets up to 6 mm. long, 1·5–2 mm. wide; glumes scarious with a conspicuous recurved awn .. .. 2. *squarrosa*
Style-base not fringed:
Achene striate with 5–9 ribs, obpyriform; style compressed, bearded above:
Spikelets linear-oblong 1·2–1·6 mm. wide; small annual with generous inflorescence

of brown spikelets; glumes boat-shaped and sharply keeled with 1 vein in the keel .. .. 3. *bisumbellata*

Spikelets ovoid to lanceolate, 1·5–4·5 mm. wide; a very variable perennial or annual with large spreading inflorescence; glumes convex, rounded on the back with about 3 veins visible in the green keel area:

Spikelets 3–6 mm. long and not above 3 mm. wide; annual or perennial with culms 2·5 mm. or more wide at the base and soft leaves; glumes 2–2·4 mm. long .. .. 4a. *dichotoma* var. *dichotoma*

Spikelets very large and somewhat inflated, 6–10 mm. long and 3–4·5 mm. wide; stout rhizomatous perennial with fibrous base and few to many spikelets; glumes greyish to light brown, up to 3·5 mm. long 4b. *dichotoma* var. *laxa*

Achene tuberculate, smooth or faintly striolate with numerous faint lines (about 15) usually almost circular in outline, rarely obovate; style compressed and usually bearded or terete and glabrous:

Glumes glabrous or with a sparse scattering of short hairs:

Spikelets 1·5–2·8 mm. wide, never hairy; rhachilla very narrow, 0·5–0·6 mm. wide; achenes never granular:

Spikelets oblong-lanceolate, 1·8–3 mm. wide, 7–12 or more per inflorescence; achene faintly striolate to almost smooth or with a few tubercules, usually dark when ripe .. .. 4c. *dichotoma* var. *pluristriata*

Spikelets lanceolate 1·5–2·5 mm. wide, 1–5 or, rarely, up to 10 per inflorescence; slender annuals with rigidly erect solitary or tufted stems and narrow leaves not over 1·5 mm. wide:

Achenes not striate, smooth with scattered tubercules especially on the thickened margins and near the apex, surface cells thin-walled, elongate 5. *alboviridis*

Achenes faintly pluristriate with up to 15 ribs, rarely with tubercules, surface cells thick-walled, square or almost so .. .. .. 6. *striolata*

Spikelets 2·8–4 mm. wide, glabrous or sparingly hairy; rhachilla 0·8–1 mm. wide; achenes smooth, striolate or granular:

Annual or perennial up to 75 cm. high; glumes broadly ovate, obtuse, mucronate:

Glumes with conspicuous dark chestnut brown or black lateral patches, otherwise pale green, veins crowded into the keel; spikelets usually 3–9 per inflorescence .. .. 7. *pilosa*

Glumes pale greenish cream throughout, veins fairly evenly spaced across the back; spikelets solitary, rarely 2–3 per inflorescence .. 8. *schoenoides*

Stout perennials 25–60 cm. high, or more; rhizomes thick with bulbous stem-bases coated with fibrous sheath remains; glumes broadly lanceolate, acute or obtuse, mucronate:

Glumes acute, dark chestnut with a dark grey-green excurrent keel:

Pedicels straight or almost so, erect; achenes finely punctate, black at maturity .. .. 10. *nigritana*

Pedicels long, flexuous, mostly curving outwards and downwards; achenes finely granular, dark brown at maturity .. .. .. 11. *barteri*

Glumes obtuse, light yellowish brown or ferruginous, with an obscure keel scarcely excurrent .. .. 13. *chevalieri*

Glumes with a very dense and conspicuous grey pubescent patch in the upper third; stout perennial .. .. 9. *ferruginea*

Stigmas 3; style trigonous, bearded or glabrous; achene trigonous, glumes spirally arranged or distichous:

Glumes spirally arranged:

Inflorescence of 2–many pedicellate spikelets:

Cauline leaves with well developed leaf-blades; stems more or less terete or strongly compressed and 2-angled:

Achene with transverse wavy ridges or undulations; leaves filiform or very narrow; stem ribbed at the top; plant usually with scattered fine long white hairs:

Spikelets 3–5 mm. long, numerous; glumes ferruginous with a long mucro, pubescent .. .. 24. *cioniana*

Spikelets 5–8 mm. long, rarely more than 12 to an inflorescence and usually many fewer; glumes acute or shortly mucronate, pubescent or glabrescent:

Achenes obpyriform, yellowish, undulations not very pronounced or the achene almost smooth:

Glumes 2–3·5 mm. long, mucronate or acute, dark chestnut with a green keel or rarely ferruginous:

Achene with 10–12 transverse undulations; slender annual; glumes dark, acute, 2–2·5 mm. long .. 25c. *hispidula* subsp. *senegalensis*

Achene with 6–8 transverse undulations; annual or perennial; glumes dark or ferruginous shortly mucronate, 2·5–3·5 mm. long 25a. *hispidula* subsp. *hispidula*

Glumes 3·8–4·5 mm. long, acute, ferruginous, rarely dark chestnut with a green keel, pubescent; rigid perennial 25b. *hispidula* subsp. *brachyphylla*

Achenes elliptic-ovoid, white, with narrow-crested transverse ridges; annual or perennial; glumes 3·8–4·5 mm. long, acute, ferruginous with dark brown close to the green keel, glabrous .. .. 26. *tisserantii*

Achene tuberculate or vertically ribbed; leaves filiform or flat; plant never with scattered fine long white hairs:

Stem ribbed or terete at the top; spikelets small, more or less oblong to narrowly lanceolate:

Perennial with swollen creeping rhizomatous bases with a thick coating of old leaf-sheath bases:

Plants stout; leaves 1–4 mm. wide; inflorescence with numerous spikelets (12 or more); glumes dark, acutely keeled, acute .. 12. *scabrida*

Plants slender; leaves not over 0·5 mm. wide; inflorescence with 2–6 brown spikelets; glumes brown, acutely keeled, obtuse 14. *schweinfurthiana*

Annual, or perennial with a slender rhizome; leaf-blades filiform or up to 1 mm. wide; spikelets ovoid, greenish white or tinged with light chestnut, few; glumes sub-acute with or without a short mucro .. 15. *debilis*

Stem compressed, flattened and acutely 2-keeled at the top:

Slender densely leafy annual with small, linear-lanceolate spikelets up to 1·2 mm. wide; rhachilla narrow, less than 0·6 mm. wide; glumes ferruginous, 1–1·5 mm. long .. .. 16. *thonningiana*

Densely leafy annual with spikelets 1·5–2·5 mm. wide; rhachilla 0·7–1 mm. wide; glumes 2–2·8 mm. long .. .. 17. *complanata*

Cauline leaves reduced to the sheath only or with a very short blade; basal leaves well developed or absent; stems strongly 4–5 angled at the top:

Spikelets small and very numerous, globose; slender or robust leafy annual 18. *littoralis*

Spikelets ovoid to lanceolate; tufted perennials:

Basal leaves few to many, always with a well-developed blade 19. *quinquangularis*

Basal leaves absent .. .. 20. *aphylla*

Inflorescence of clusters of numerous subsessile spikelets.. .. 21. *obtusifolia*

Glumes distichous:

Inflorescence a solitary pale yellowish green spikelet; achene trigonous, granular-tuberculate .. .. 22. *ovata*

Inflorescence of 2–5 spikelets; achene smooth or almost so with an exceptionally long stipitate base .. .. 23. *triflora*

1. **F. dipsacea** (*Rottb.*) *C.B. Cl.* in Hook. f., Fl. Brit. Ind. 6: 635 (1893); F.T.A. 8: 413. *Scirpus dipsaceus* Rottb., Desc. & Ic. 56, t. 12, f. 1 (1773). Minute ephemeral herb with small green spikelets not over 4 mm. long; glumes scarious with the green midvein excurrent in a long awn; sandy places and swamps, also on laterite outcrops.
Mali: L. Korienza (Apr.) *Duong* 2148! 2152! Gao to Bagoundié (June) *de Wailly* 5061! **S. L.:** Kambia (May) *Jordan* 866! Rokupr (Apr.) *Adames* 170a! Widespread in tropical regions.

2. **F. squarrosa** *Vahl* Enum. 2: 289 (1805); F.T.A. 8: 413; Berhaut, Fl. Sén. ed. 2, 370. Small annual herb with green to buff-coloured oblong spikelets, glumes scarious with a conspicuous green midvein excurrent in a long recurved awn; in sandy river-beds and swamps.
Sen.: Sangalkam (June) *Berhaut* 5266! Kayar (Apr.) *Naegelé* 8086! Ourosogui to Matam (Feb.) *Naegelé* 12117! Dougar (Jan.) *Berhaut* 5652! Niokolo-Koba (Feb.) *Adam* 17539! **Mali:** Néguébabougou, nr. Bamako (Jan.) *J. & A. Raynal* 5420! Douna (Feb.) *Roberty* 3366! Djenné to Sofara (July) *Chev.* 1149! **S. L.:** Rokupr (Apr.) *Adames* 170! **N. Nig.:** Nupe *Barter* 1580! Gurara Falls, Niger Prov. (Mar.) *Stanfield*! Abinsi (May) *Dalz.* 822! **W. Cam.:** Bambalang, Ndop Plain (Mar.) *Brunt* 1030! Widespread in tropical regions.

3. **F. bisumbellata** (*Forsk.*) *Bubani* Dodecanthea 30 (1850); Aké Assi, Contrib. 2: 262 (1964); Berhaut, Fl. Sén. ed. 2, 371. *Scirpus bisumbellatus* Forsk., Fl. Aeg.-Arab. 15 (1775). *Fimbristylis dichotoma* Vahl, Enum 2: 287 (1805), as to description and citations only; F.T.A. 8: 414. Annual herb to 30 cm. high with numerous short leaves; inflorescence of many linear-lanceolate to cylindric light brown spikelets; damp grassy places.
Sen.: Niokolo-Koba *Berhaut* 1607. **Mali:** Ségou (Apr., May) *Roberty* 2056! 2065! Djenné (June) *Chev.* 1119! Goundam (Nov.) *Roberty* 3194! L. Faguibine (Oct.) *Monod* 4058! **Iv. C.:** Ferkessédougou *Aké Assi* IA 5623. **Ghana:** Tamale to Damongo (Mar.) *Morton* GC 8771! Zuarungu (Dec.) *Adams & Akpabla* 4302! Bawku to Bolgatanga (Apr.) *Morton* 8959! **N. Nig.:** *Moiser*! Nupe *Barter* 1209! Sokoto (July) *Dalz.* 461! Malamfaturi, N.E. Bornu (Oct.) *De Leeuw* 2068! Widespread in tropical regions.

4. **F. dichotoma** (*Linn.*) *Vahl* Enum. 2: 287 (1805), as to name only; Berhaut, Fl. Sén. ed. 2, 371, 373. *Scirpus dichotomus* Linn., Sp. Pl. ed. 2, 74 (1762). *S. diphyllus* Retz., Obs. 5: 15 (1788). *Fimbristylis diphylla* (Retz.) Vahl, Enum. 2: 289 (1805); F.T.A. 8: 415.

4a. **F. dichotoma** (*Linn.*) *Vahl* var. **dichotoma.** Tufted perennial or annual up to 1 m. high but very variable in stature with wide scarious-margined basal leaf-sheaths and numerous glabrous or sparingly hairy leaves. Inflorescence of numerous light brown to chestnut narrowly ovate or lanceolate spikelets 3·5–6 mm. long, rarely longer, with shallowly convex green-keeled glumes; damp grassland and savanna, stream banks and swamps.
Sen.: Sangalkam (May) *Berhaut* 7583! Kolda (Oct.) *Adam* 18472! Geneto, Gambia R., (May) *Naegelé* 11756! Badi (Sept.) *Berhaut* 3056! Kanouméri, Kédougou to Bandiafassi (July) *Fotius* K273! **Gam.:** Yundum (Dec.) *Austin* 19! **Mali:** (May) *Chev.* 784! Bamako (May) *Duong* 2287! Sotuba (Oct.) *Raynal* 5011! Djenné (July) *Chev.* 1140! Gao to Bossobougou (July) *de Wailly* 5090! **Guin.:** Friguiagbé (Apr., May) *Chillou* 345! *Boismaré* 246! Dalaba *Monod*! Kouroussa (July) *Pobéguin* 414! Farana (Apr.) *Pitot*! **S. L.:** Rokupr (Feb.) *Deighton* 5362! Njala (Oct.) *Deighton* 1353! Koinadugu (Aug.) *Haswell* 240! Perankonko, Loma Mts. (Sept.) *Jaeger* 1768! Kenema (Nov.) *Deighton* 5877! **Lib.:** Kailahun, Kolahun Dist. (Nov.) *Baldwin* 10129! Monrovia (May) *Baldwin* 5835! Gbanga (Sept.) *Linder* 512! Ganta,

Sanokwele Dist. (Nov.) *Harley* 1717! Nyaake (Webo), Webo Dist. (June) *Baldwin* 6154! **Iv. C.:** Tai to Tabou (Aug.) *Boughey* GC 14951! Divo Forest, Gagnoa (Aug.) *Boughey* GC 14659! Mankono (Aug.) *Boughey* GC 18509! Brafouédi (Apr.) *Leeuwenberg* 3338! Grand Bassam (Sept.) *de Wit* 7310! **U. Volta:** Sikasso to Bobo-Dioulasso (Sept.) *Jaeger* 5173! Banfora (June) *Leeuwenberg* 4362! Farakoba (June) *Scholz* 306! 307! Ouagadougou (Aug.) *Nongonierma* 93! **Ghana:** Axim (Mar.) *Morton* GC 6591! Burufo, nr. Lawra (Apr.) *Adams* 4058! Bawku (Aug.) *Hall* 2067! Tamale (Dec.) *Adams & Akpabla* 4178! Accra *Morton* 6112! **N. Nig.:** Sokoto (July) *Dalz.* 473! Katsina to Daura (Apr.) *Meikle* 1356! Minna, Niger Prov. (Jan.) *Keay* FHI 37323! Hoss, Jos Plateau (Aug.) *Keay* FHI 12695! Oturkpo, Benue Prov. (Oct.) *Kennedy* 8021! nr. Vogel Peak, Sardauna Prov. (Nov.) *Hepper* 1385! **S. Nig.:** Ibadan (Feb.) *Brenan* 8979! Owo F.R., Ondo Prov. (July) *Onochie* FHI 33246! Ajagbodudu (Aug.) *Wright* 56! Aboh *Barter* 131! Abakaliki (Nov.) *Tuley* 3! **W. Cam.:** Malende to Muyuka (Mar.) *Brenan* 9327! Mencham R. (Feb.) *Brunt* 953! Mbaw Plain (May) *Brunt* 490! Mamfe (Mar., Dec.) *Migeod* 261! *Onochie et al.* FHI 30890! **F. Po:** (Oct.) *T. Vogel* 12! S. Isabel *Swarbrick* 2918! Moka (Sept.) *Melville* 610! In all tropical areas of the world.

4b. **F. dichotoma** var. **laxa** (*Vahl*) *Napper* in Kew Bull. 25: 436 (1971). *F. laxa* Vahl, Enum. 2: 292 (1805). *F. diphylla* (Retz.) Vahl var. *laxa* (Vahl) E. G. Camus in Lecomte, Fl. Gén. Indochine 7: 104 (1912). *F. dichotoma* of F.T.A. 8: 414, in small part. A much stouter perennial than the type variety, with the sheath-bases becoming hard and the stem-bases frequently surrounded by fibres; inflorescence with few to many light brown to greyish spikelets 6–9 mm. long, 3–4·5 mm. wide and usually plumper than the species; swampy, damp grasslands and plantations.
Sen.: Kolda (Oct.) *Adam* 18472! **Guin.:** Friguiagbé (Apr., May) *Chillou* 136! 447! Farana (Mar.) *Sc. Elliot* 5355! **S. L.:** Sulima (May) *Deighton* 5781! Yonibana (Jan.) *Glanville* 133! Sendugu (June) *Thomas* 601! Samaia (May) *Thomas* 235! **S. Nig.:** Igbetti to Ogbomosho (Aug.) *Stanfield*! Almost as widely spread as the type variety but much less frequent in occurrence in Africa, though in parts of the east it may be the more abundant.

4c. **F. dichotoma** var. **pluristriata** (*C.B. Cl.*) *Napper* in Kew Bull. 25: 437 (1971). *F. diphylla* (Retz.) Vahl var. *pluristriata* C.B. Cl. in Hook. f., Fl. Brit. Ind. 6: 637 (1893). *F. podocarpa* Nees in Wight, Contrib. 98 (1834); Berhaut, Fl. Sén. ed. 2, 371. *F. pluristriata* (C.B. Cl.) Berhaut in Bull. Soc. Bot. Fr. 101: 376 (1955), illegitimate name. An annual differing from similar states of the species by the almost orbicular achene with numerous faint striae, otherwise identical; damp grassy places, stream-banks, etc.
Gam.: (Aug.) *Ruxton* 80! **Sen.:** M'Bao (Nov.) *Adam* 2311! Oussouyé (Oct.) *Kerharo & Adam* 1564! Casamance (Oct.) *Adam* 18587! **Mali:** Kita (Sept.) *Jaeger* 71! 74! Samanko, nr. Bamako (Jan.) *J. & A. Raynal* 5418! **Guin.:** Koba (Nov.) *Jac.-Fél.* 7251! **S. L.:** Waterloo (Aug.) *Melville & Hooker* 297! Rokupr (July) *Jordan* 51! Konta (Sept.) *Jordan* 106! Loma Mts. (Aug.) *Jaeger* 7224! Madina (Aug.) *Jordan* 911! **Ghana:** Oti R. nr. Kpandai (Aug.) *Hall* GC 38730! Yendi (Dec.) *Adams & Akpabla* 4094! Lumbaga to Tamale (Dec.) *Morton* GC 6266! Dedoro to Tankara (Dec.) *Adams & Akpabla* 4330! **N. Nig.:** Birnin Gwari (Apr.) *Meikle* 1383! Naraguta F.R., Jos (July) *Lawlor & Hall* 107! Damaturu, Bornu Prov. (Sept.) *De Leeuw* 1211*b*! **S. Nig.:** Okelifi, Ondo Dist. (Nov.) *Onochie* FHI 34225! Widespread with the type variety and frequently not separated from it.
[The taxonomy of this widespread and very variable species poses many problems. The varieties defined here seem a natural division of the West African material though a few anomalous intermediates have been observed. Much work remains to be done on a wider basis especially in field observation and experimentation before a proper understanding of the variability can be obtained.]

5. **F. alboviridis** *C.B. Cl.* in Fl. Brit Ind. 6: 638 (1893); Napper in Kew Bull. 25: 437 (1971). *F. diphylla* (Retz.) Vahl var. *tuberculata* Cherm. in Bot. Arch. Caen 4, Mém. 7: 31 (1931). *F. pluristriata* (C.B. Cl.) Berhaut var. *tuberculata* (Cherm.) Berhaut in Bull. Soc. Bot. Fr. 101: 376 (1955), invalidly published. *F. podocarpa* Nees var. *tuberculata* (Cherm.) Berhaut, Fl. Sén. ed 2, 371, name only. Slender narrow-leaved annual differing from *F. dichotoma* in the more rigid habit, the fewer spikelets, the smaller glumes marked with dark chestnut brown and the very characteristic suborbicular achene having few to several vertical rows of cells lacking the raised ribs, but with a thick tuberculate rim and scattered tubercules on the sides; swampy places and river banks, also damp grassland and rocky places.
Sen.: *Berhaut* 630; 2606. **Guin.:** Friguiagbé (Aug.) *Chillou* 610! Manéa (June) *Chillou* 2631! foot of Nimba Mts. *Schnell* 3428! Nzérékoré *Adam* 6265! **S. L.:** Freetown (Sept.) *Hepper* 923! Martin (Jan.) *Deighton* 3545! Konta (Sept.) *Jordan* 108! Port Loko to Mange (May) *Jordan* 871! Kambia (July) *Jordan* 469! **Ghana:** Burufo, nr. Lawra (July) *Adams* 928! *Hall* CC 761! Pepease to Kwahu (May) *Hall* 3660! Kwahu Tafo (Apr.) *Hall* 3017! 3018! **N. Nig.:** Kan Gimi, Anara F.R. (May) *Keay* FHI 22947! **S. Nig.:** Idanre, Ondo Prov. (Aug.) *Jones* FHI 20723! Mt. Orosun, Idanre Hills (Oct.) *Keay* FHI 22594! Occurs widely to India and Java though rarely recorded in central and north-eastern Africa.

6. **F. striolata** *Napper* in Kew Bull. 25: 438 (1971). Scarcely differing from *F. alboviridis* except by the achene, though the habit is usually rather stouter than commonly recorded for the other; an extreme, very slender form with solitary spikelets is common in some areas; seasonally swampy places, rice fields and grassland.
Sen.: Fatick (Dec.) *J. & A. Raynal* 6713! Diohine (Oct.) *Berhaut* 3941! Kaolack (Oct.) *Berhaut* 3802! Tambacounda (Sept.) *Berhaut* 3308! Casamance (Sept.) *Adam* 18121! **Guin.:** La Kolenté, Botocoli Rd. (Nov.) *Chillou* 1017! **S. L.:** Mateboi (Nov.) *Jordan* 830! Kabala (Dec.) *Deighton* 4409!

7. **F. pilosa** *Vahl* Enum. 2: 290 (1805); F.T.A. 8: 416. Perennial, rarely annual, with narrow leaves 0·5–1·7 mm. wide; inflorescence with few, 3–9 spikelets, glumes with dark chestnut to black patches and green keel; spikelets larger than *F. dichotoma* and the rhachilla broader; damp or swampy places in grassland and savanna.
Sen.: (Sept.) *Fotius* K394! Ziguinchor (Sept., Oct.) *Adam* 18225! *Broadbent* 82! **Guin.:** Bandakure, S.W. of Pita (Mar.) *Adames* 287! Kouroussa (July) *Pobéguin* 419! **S. L.:** Mayumbo (Aug.) *Jordan* 916! Tabe (Sept.) *Deighton* 3041! Koinadugu (Aug.) *Haswell* 226! Musaia (Sept.) *Small* 238! Sintunia (July) *Morton & Jarr* SL 2088! **Iv. C.:** N. of N'Douci, N'Zi R. (Aug.) *de Wit* 7384! N'Douci to Toumodi (Aug.) *de Wit* 7336! 7337! Gonokrom to Techikrom (Dec.) *Adams* 2973! **U. Volta:** Ouagadougou (Aug.) *Nongonierma* 82! **Ghana:** Wa (Sept.) *Hall* CC 599! Tamale (Nov.) *Williams* 441! Ejura (Aug.) *Andoh* 5055! Kpong (May) *Gyadu* GC 20561! Accra (July) *Irvine* 718! **N. Nig.:** Kurni, Kaduna to Keffi (June) *Kershaw* 900614! Plateau Prov. (Aug.) *Lely* P451! Vodni, Pankshin Div. *Saunders* 17! Panshanu, Bauchi Prov. (Aug.) *Lawlor & Hall* 554! Gawu to Abuja, Niger Prov. (Aug.) *Onochie* FHI 35448! **S. Nig.:** Igbete to Kishi, Old Oyo Game Reserve (Nov.) *Gledhill* 750! Apete, Ibadan (May) *Guile*! Igboora (Oct.) *Haines* 220! Owo (Apr.) *Haines* 202! Ikare, nr. Owo (Apr.) *Haines* 15! Also in the Congo Republic and Uganda.

8. **F. schoenoides** (*Retz.*) *Vahl* Enum. 2: 286 (1805); Berhaut, Fl. Sén. ed 2, 373. *Scirpus schoenoides* Retz., Obs. 5: 14 (1788). Narrow-leaved annual or perennial with solitary, or rarely paired, spikelets; swampy places in grassland.
Sen.: Diohine (Oct.) *Berhaut* 1711! Oussouyé (Oct.) *Adam* 18297! *Kerharo & Adam* 1563! Casamance (Oct.) *Adam* 18151! 18373! Niassia (Nov.) *Berhaut* 6655! **S. L.:** Martin (Jan.) *Deighton* 3544! Bakepet, Mambolo (Dec.) *Jordan* 735! **Ghana:** Burufo, nr. Lawra (Sept.) *Hall* CC 746! Widespread in India.

9. **F. ferruginea** (*Linn.*) *Vahl* Enum. 2: 291 (1805); F.T.A. 8: 417; Berhaut, Fl. Sén. ed. 2, 361, 370, 371, 373. *Scirpus ferrugineus* Linn., Sp. Pl. ed. 2, 74 (1762). Stout tufted erect perennial 45–75 cm. high with narrow leaves and small dense panicles of few to many spikelets and bracts of very variable length; spikelets lanceolate, grey and rust coloured; brackish swamps and damp grassland.
Sen.: Hann (Mar.) *Berhaut* 7022! Thiès (June) *Berhaut* 2612! Bamboulan, nr. Toubacouta (Oct.) *Monod*

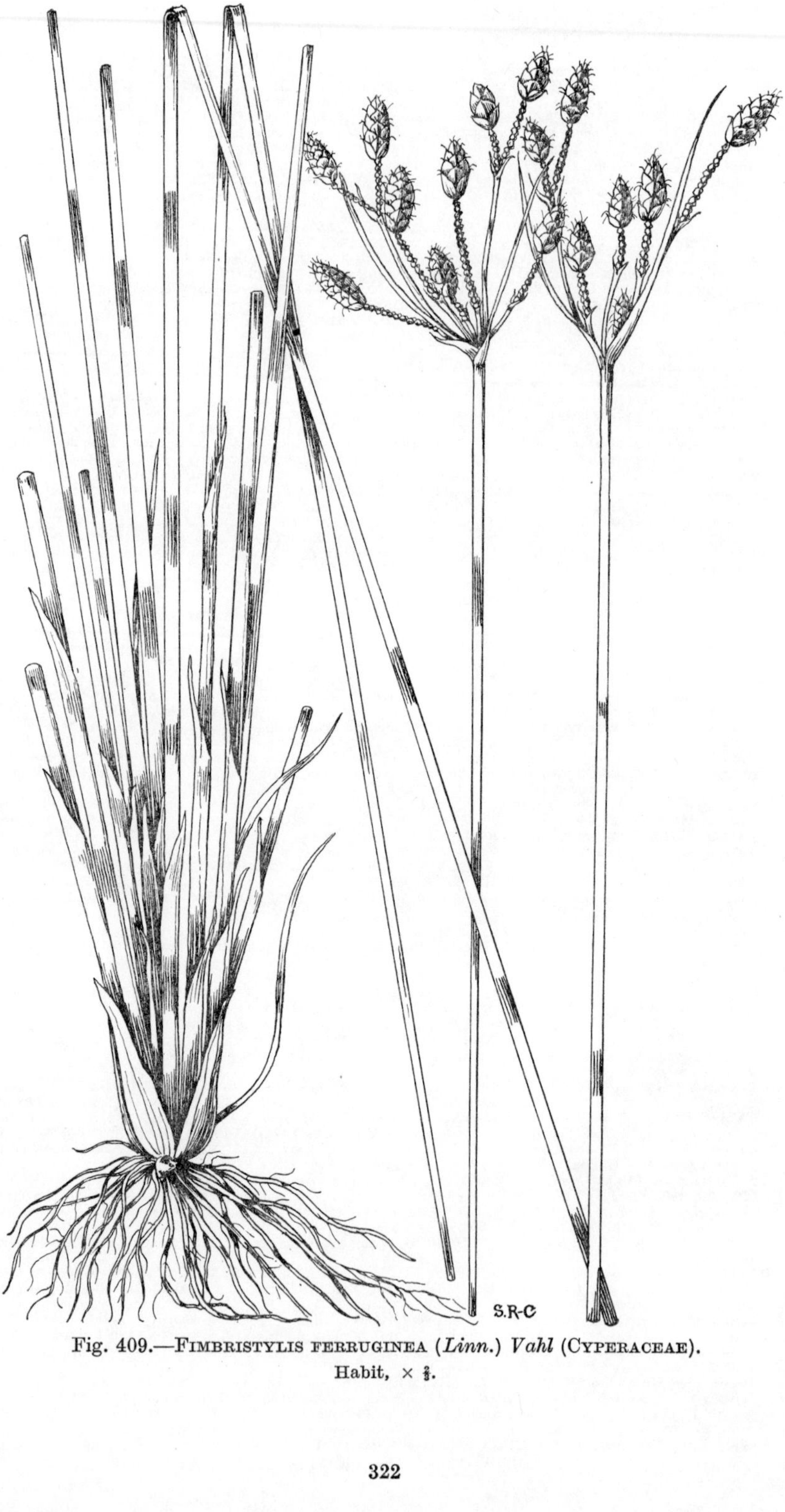

Fig. 409.—FIMBRISTYLIS FERRUGINEA (*Linn.*) *Vahl* (CYPERACEAE).
Habit, × ⅔.

8589! M'boro to Fatick (July) *Trochain* 247! Oussouyé (Sept.) *Broadbent* 115! **Gam.:** (July) *Ruxton* 5! Genieri to Massembe (July) *Fox* 121! **Guin.:** Conakry (Sept.) *Martine* 114! Golea, nr. Conakry (July) *de Wit* 7295! Koba *Jac.-Fél.* 7218! **S. L.:** Aberdeen (July) *Small* 114! Maswari, nr. Newton (Oct.) *Pyne* 129! Talil (Jan.) *Deighton* 3548! Rokai (Mar.) *Jordan* 386! Rokel (Jan.) *Jordan* 6! **Lib.:** Grand Bassa (July, Aug.) *T. Vogel* 55! *Dinklage* 1706! Cape Palmas (Dec.) *Dinklage* 2362! **Ghana:** Keta (Mar.–June) *Thorold* 102! 103! *Morton* A2188! GC 6529! Ada (June) *Morton* GC 9253! **Niger:** Niamey (Oct.) *Hagerup* 555! Tamgak Wadi, Aïr Mts. (Mar.) *Bradley* 122! **S. Nig.:** Ijora, Lagos (June) *Haines* 14! Ikoyi, Lagos (July) *Jones* FHI 19424! Lagos *Dawodu* 343! Nun R. (Sept.) *Mann* 531! *T. Vogel* 36! Widespread in all parts of the tropics.

10. **F. nigritana** *C.B.Cl.* in F.T.A. 8: 418 (1902). Stout perennial with leaves 3–4 mm. wide and dark spikelets up to 12 mm. long with finely punctate black achenes and acute glumes similar to those of *F. scabrida*.
N. Nig.: Nupe *Barter* 623!

11. **F. barteri** *Boeck.* in Linnaea 37: 33 (1871). Differs from *F. scabrida* in the much coarser habit, the widely spreading inflorescence with ovate-elliptic spikelets on long lax pedicels and the biconvex, smoother larger achene usually with a 2-fid style, though the lowest fertile florets of the spikelet may have 3-fid styles despite a biconvex achene; grassland and close to dams.
Ghana: Tumu (May) *Morton* GC 7541! **N. Nig.:** Nupe *Barter* 1578! Maska, Katsina Dist. (June) *Keay* FHI 25899!

12. **F. scabrida** *Schumach.* Beskr. Guin. Pl. 32 (1827); F.T.A. 8: 422; Aké Assi, Contrib. 263 (1964). *F. muriculata* Benth. in Fl. Nigrit. 554 (1849) (*F. muricata* Walp., Ann. 3: 688 (1854), and *F. muricatula* Steud., Syn. Pl. Glum. 2: 113 (1855)—both orthographic errors for *muriculata*.) *F. mucronata* Boeck. in Linnaea 37: 41 (1871). *F. pachystachys* Cherm. in Bull. Soc. Bot. Fr. 68: 418 (1921). *F. tenera* of C.B. Cl. in F.T.A. 8: 420, partly, and of Hutch. in F.W.T.A., ed. 1, 2: 476; not of Schult. Plant very variable in size with stem-bases forming coarse matted fibrous tufts and panicles of lanceolate-cylindric spikelets with acute glumes; damp grassland, savanna and rocky outcrops.
S. L.: Musaia (July) *Deighton* 4824! Koinadugu (Aug.) *Haswell* 208! **Iv. C.:** Séguéla (Apr.) *Leeuwenberg* 3253! Tafiré (June) *Leeuwenberg* 4440! **Ghana:** Sakogu to Nakpanduri (Aug.) *Hall* CC 462! Kete Krachi (May) *Morton* GC 7175! Ejura (Apr.) *Williams* 234! Accra *Don*! **Togo Rep.:** Bafilo (May) *Roberty* 1426! Lomé *Stage* 16! **N. Nig.:** Bonu Kurmi, Niger Prov. (June) *E. W. Jones* 170! Mando F.R., Birnin Gwari Dist. (June) *Keay* FHI 25853! Kan Gimi, Anara F.R. (May) *Keay* FHI 22862! **S. Nig.:** Wasimi Rd (May) "*Lagos Govt.*" 50! Iseyin (May) *Guile* A8! Oyo (Apr.) *Guile*! Ishan Dist. (Apr.) *Daramola* FHI 31261! Owo to Ikare (Apr.) *Haines* 9! Also eastwards to the Congo.

13. **F. chevalieri** *Kük.* in Bull. Mus. Hist. Nat. Paris sér. 2, 3: 547 (1931). Tufted perennial similar in habit to *F. scabrida* but with much paler ovate-elliptic larger spikelets; in savanna.
Sen.: Geneto to Kourientiun (May) *Naegelé*! Kanouméri, Kédougou (June) *Fotius* K123! **Guin.:** Bioumahame to Siguin (May) *Chev.* 25897! Mamou (May) *Pitot*!

14. **F. schweinfurthiana** *Boeck.* in Flora 37: 565 (1879); F.T.A. 8: 420. Small slender-stemmed plant with narrow leaves and thick perennial tufts of old culm- and leaf-bases, spikelets only 2–5, brown; rocky outcrops, shallow soils on seasonally swampy ground, seepage areas.
Mali: Kita (July) *Jaeger* 34! 39! **Guin.:** Kindia (May) *Jac.-Fél.* 1635! **S. L.:** Waterloo (May) *Deighton* 2649! Kukuna (Apr.) *Hepper* 2645! Loma Mts. (June) *Morton* SL 3565! Also across to the Sudan.

15 **F. debilis** *Steud.* Syn. Pl. Glum. 2: 109 (1855); Napper in Kew Bull. 25: 438 (1971). *F. tenera* of F.T.A. 8: 420, partly; Berhaut, Fl. Sén. ed. 2, 372; not of Schult. Slender annual up to 30 cm. high or perennial with a creeping rhizome, rarely the culm-bases surrounded with fibrous sheath-remnants; savanna and scrub but especially in and near pools on laterite outcrops.
Sen.: *Heudelot* 324! Ross-Bétio (Aug.) *Audru* 2159! *Berhaut* 372! Rufisque *Leprieur*. Kaolak to Fatick *Trochain* 4052. **Gam.:** (Aug.) *Ruxton* 148! **Mali:** Diondiou (June) *Chev.* 1010. Katibougou (July) *Kaden* 67/506. **S. L.:** Port Loko (Oct.) *Jordan* 810! **Ghana:** Kwahu Tafo (Dec.) *Harris*! Berekum to Sampa (Dec.) *Adams* 5277! Nsemre F.R., nr. Borku (Dec.) *Adams* 5225! Tumu (May) *Morton* 7414! Burufo, nr. Lawra (Sept.) *Hall* CC 762! **Niger:** Zinder (July) *P. de Fabrègues* 2705. **N. Nig.:** Bindawa Village F.R., Katsina Prov. (Aug.) *Grove* 2!

16. **F. thonningiana** *Boeck.* in Linnaea 38: 395 (1874); F.T.A. 8: 426. Leafy annual differing from *F. complanata* in the more slender annual habit, the lower growth and the much smaller spikelets; in damp grassland and wet exposed places.
S. L.: Rokupr (Apr.) *Jordan* 203! Mano Bonjema (Apr.) *Deighton* 3706! **Lib.:** Cape Mount (Mar.) *Dinklage* 2487! Monrovia (Feb.) *Dinklage* 2757! Mnanulu, Webo Dist. (June) *Baldwin* 6051! **Iv. C.:** Banco Forest, nr. Abidjan (Dec., Feb.) *Aké Assi* 4744! *Boughey* GC 13501! Yapo Forest, nr. Abidjan (July) *Boughey* GC 14590! **Ghana:** Essiama (May) *Morton* A2123! Axim (Mar.) *Morton* GC 8476! Nsuaem, Agona to Tarkwa (Aug.) *Hall* 3325! Ankasa F.R. (Apr.) *Hall & Enti* GC 38329! Kpandai (Aug.) *Hall* GC 38740! **S. Nig.:** Ibadan (Apr., Oct.) *Guile* A3! A17!

17. **F. complanata** (*Retz.*) *Link* Hort. Berol. 1: 292 (1833); F.T.A. 8: 422. *Scirpus complanatus* Retz., Obs. 5: 14 (1789). Tufted leafy perennial with broadly linear leaf-blades rounded at the tip and compressed linear stems bearing contracted panicles with elongated golden brown spikelets; in rocky places, rare.
Lib.: Monrovia (Nov.) *Barker* 1466! Gbau, Sanokwele Dist. (Sept.) *Baldwin* 9437! **N. Nig.:** Vom (Dec.) *Haines* 82! Occurring in most tropical regions.

18. **F. littoralis** *Gaud.* in Freyc. Voy. Bot.: 413 (1826); Napper in Kew Bull. 25: 439 (1971). *F. miliacea* Vahl, Enum. 2: 287 (1805); F.T.A. 8: 421; Aké Assi, Contrib. 262 (1964); Berhaut, Fl. Sén. ed. 2, 373 (1967); not *Scirpus miliaceus* Linn. Tufted leafy annual with numerous panicles of very small globose spikelets; in marshy grassland, rice-fields, edge of dams, etc.
Sen.: *Heudelot* 212! Casamance (Oct.) *Adam* 18432! **Iv. C.:** Ayamé (May) *Aké Assi* 4747! **Ghana:** Takoradi (Mar.) *Hall* 2571! Cape Coast (Mar.) *Hall* 1060! Kade (June) *Irvine* 4845! Kusu lagoon, nr. Akuse (Feb.) *Hall* CC 1213! Kete Krachi (Dec.) *Adams* 4561! **N. Nig.:** Jebba (Dec.) *Hagerup* 741! Edozhigi, nr. Bida (June) *Vaillant* 2659! Bida to Badeggi (Aug.) *Onochie* FHI 35420! **S. Nig.:** Adani (Nov.) *Tuley* 17! Widespread in eastern tropical regions, rare in Africa and possibly introduced with cultivated rice.

[Blake and Kern in their studies of *F. miliacea* (Linn.) Vahl have shown that this name belongs to a taxon other than the one usually recognised as such. *F. littoralis* is the correct name for the plant previously known as *F. miliacea* (Linn.) Vahl.]

19. **F. quinquangularis** (*Vahl*) *Kunth* Enum. 2: 229 (1837); Aké Assi, Contrib. 262 (1964), partly; Berhaut, Fl. Sén. ed. 2, 373. *Scirpus quinquangularis* Vahl, Enum. 2: 279 (1805). Tufted annual similar to *F. littoralis* but with lanceolate elliptic small spikelets; in rice-fields, marshy grassland, etc.
Sen.: Niokolo-Koba (Nov.) *Adam* 19926! Moulessi, Tambacounda to Kedougou (Dec.) *J. & A. Raynal* 6745! **N. Nig.:** Edozhigi, Bida (Apr.–Oct.) *Vaillant* 2740! 2741! 2792! Biliri, S. of Gombe (Nov.) *J. Lowe* 1516! **S. Nig.:** Adani (Nov.) *Tuley* 19! Widespread in all tropical regions.

20. **F. aphylla** *Steud.* Syn. Pl. Glum. 2: 114 (1855); Napper in Kew Bull. 25: 440 (1971). *F. quinquangularis* of F.T.A. 8: 421; F.W.T.A., ed. 1, 2: 476; Aké Assi, Contrib. 262 (1964), partly. *F. testui* Cherm. in Arch. Bot. Caen 4, Mém. 7: 33 (1931). Stout tufted perennial with the leaf-blades much reduced, scarcely reaching 1 cm. long, and acutely quadrangular almost winged stems up to 75 cm. high, widely spreading panicles often broader than long with light chestnut elliptic-lanceolate spikelets and short bracts; swamps and very damp ground in savanna and near rivers, etc.
S. L.: Waterloo (Aug.) *Melville & Hooker* 274! Mahera (Oct.) *Glanville* 415! Rokon (July) *Jordan* 483! Boajibu (Mar.) *Deighton* 4091! Koinadugu (Aug.) *Haswell* 260! Sahn (Apr.) *Deighton* 1686! **Ghana:** Sampa, NW. Ashanti (Apr.) *Morton* A3264! **S. Nig.:** Ikorodu (Mar.) *Gledhill* 629! Eastwards to the Congo.

21. **F. obtusifolia** *Kunth* Enum. 2: 240 (1837); F.T.A. 8: 423; Berhaut, Fl. Sén. ed. 2, 373. Tufted rhizomatous perennial with rather rigid leaves and with crowded clusters of dark beige spikelets with very small glumes; sea-shore, on sand, mud or among rocks.
Sen.: *Roger*! Mboro (Apr.) *Pitot*! Niaga, L. Retba (July) *J. & A. Raynal* 6079! Hann (July) *Naegelé*! Diouloulou, Casamance (Sept.) *Adam* 18149! **S. L.:** Toke, Peninsular (Jan.) *Morton* SL 610! Freetown (Sept.) *Milne-Redhead*! *Harvey* 144! Yele, Turtle Is. (Nov.) *Deighton* 2307! King Tom (Oct.) *Deighton* 2157! **Lib.:** Monrovia (June, July) *Adam* 16571! *Dinklage* 2824! Grand Bassa (Apr.) *T. Vogel* 70! Cape Palmas (June, Nov.) *Baldwin* 5968! *Dinklage* 2348! **Ghana:** Axim (Mar.) *Morton* GC 8481! Elmina (Nov.) *Boughey* GC 10315! Sakumo lagoon, nr. Tema (May) *Adams* 258! Prampram (Nov.) *Boughey* GC14307! Keta (Mar.) *Morton* GC 6531! **Togo Rep.:** Lomé *Warnecke* 130! **S. Nig.:** Lagos (Jan., Oct.) *Haines* 13! *Stubbings* 80! *Killick* 85! Nun R. *T. Vogel* 34! (Sept.) *Mann* 532!
[This species is part of the world-wide *F. cymosa* R. Br. complex of strand and atoll plants whose inter-relationships are not entirely clear. This has made it preferable in the present work to retain the name currently accepted in West Africa, pending a world-wide study of the taxonomy and nomenclature of the group.]

22. **F. ovata** (*Burm. f.*) *Kern* in Blumea 15: 126 (1967). *Carex ovata* Burm. f., Fl. Ind. 194 (1768). *Cyperus monostachyos* Linn., Mant. 180 (1771). *Fimbristylis monostachya* (Linn.) Hassk., Pl. Jav. Rar. 61 (1848); F.T.A. 8: 424; F.W.T.A., ed. 1, 2: 476. Slender tufted rhizomatous perennial with crowded culms swollen at the base, filiform or narrowly linear leaves and solitary pale green or yellowish-buff spikelets 1 cm. long; in grassland and savanna.
Iv. C.: 40 km. NE. of N'Douci (Aug.) *de Wit* 7312! **Ghana:** Bole (Aug.) *Irvine* 5182! Kete Krachi (May) *Morton* GC 7177! Achimota F.R. (Apr.) *Morton* GC 9164! Nungua (Apr.) *Rose Innes* GC 30109! Kpong (May) *Rose Innes* GC 31231! **N. Nig.:** Rinjim Mukr (May) *Lely* 221! Samaru, Zaria Prov. (June, July) *Keay* FHI 25907! *C. B. Taylor* 23! Zaria (June) *Kershaw* 900371! Shika, Zaria Prov. (July) *Magaji* 15! Widespread in the tropics.

23. **F. triflora** (*Linn.*) *K. Schum.* in Pflanzenw. Ost-Afrika, C: 124 (1895). *Cyperus triflorus* Linn., Mant. 180 (1771). *Abildgaardia tristachya* Vahl, Enum. 2: 297 (1805). *A. lanceolata* Schumach., Beskr. Guin. Pl. 33 (1827). *Fimbristylis tristachya* (Vahl) Thwaites, Enum. Pl. Zeyl. 434 (1864); F.T.A. 8: 424; not of R. Br. Stout rhizomatous perennial up to 60 cm. high with crowded culms and inflorescences of 2–3 large yellowish buff spikelets up to 17 mm.long and 3–4·5 mm. wide; grasslands, especially near the coast.
Ghana: Essiama (May) *Morton* A2130! Nungua (May) *Rose Innes* GC 31249! Kpong (Mar.) *Johnson* 1043! Ada (Jan.) *de Wit & Morton* A2973! Keta (June) *Thorold* C.B. 106! **Togo Rep.:** Lomé *Warnecke* 140! Widespread in the tropics.

24. **F. cioniana** *Savi* in Mém. Valdarn. 3: 98 (1842); F.T.A. 8: 420; Berhaut, Fl. Sén. ed. 2, 370. *F. hispidula* (Vahl) Kunth var. *cioniana* (Savi) Boeck. in Linnaea 37: 28 (1871). Small tufted annual usually about 15 cm. high, rarely more; sandy grassland and river banks.
Sen.: *Naegelé*! Vorouli, Gambia R. (Apr.) *Nongonierma* 470! Niokolo-Koba to Kédougou (Mar.) *Adam* 18834! **Mali:** Kayes *Duong*! Ségou (May) *Roberty* 2067! San (June) *Chev.* 1087! L. Debo (Nov.) *Duong* 1355! Gao to Kokoromme (Apr.) *de Wailly* 5388! **Guin.:** Niger R. (May) *Chev.* 25873! **S. L.:** Magbema, nr. Kambia (Mar.) *Jordan* 1058! Laminaiya (Apr.) *Thomas* 141! Makump (May) *Deighton* 1703! Njala (Apr.) *Deighton* 4769! Gbundapi (Mar.) *MacDonald* 1! **Ghana:** Senchi, Volta R. (Apr.) *Irvine* 2868! Otisu, Oti R. (May) *Morton* 7299! Chindiri, Daka R. (Apr.) *Hall* VBS 1298! Tamale Reservoir (Mar.) *Morton* GC 8805! **N. Nig.:** Nupe *Barter* 127! 1211! Jebba (Dec.) *Meikle* 854! **S. Nig.:** Owena R., nr. Benin (Apr.) *Haines* 205! Aboh *Barter* 132! Widespread in most tropical and warm temperate regions.

25. **F. hispidula** (*Vahl*) *Kunth* Enum. 2: 227 (1837); Boeck. in Linnaea 37: 27 (1871). *Scirpus hispidulus* Vahl, Enum. 2: 276 (1805). *Isolepis exilis* Kunth in H.B.K., Nov. Gen. 1: 224 (1815). *Fimbristylis exilis* (Kunth) Roem. & Schult., Syst. Veg. 2: 98 (1817); F.T.A. 8: 418; F.W.T.A., ed. 1, 2: 476; Berhaut, Fl. Sén. ed. 2, 371, 372.

25a. **F. hispidula** subsp. **hispidula.** Densely tufted annual, sometimes persisting, with few spikelets; glumes mucronate, glabrous or rarely puberulous to shortly pubescent, dark chestnut on the sides and with a green keel; achenes yellowish white; grassland and savanna on sandy soils.
Sen.: Kasango (Aug.) *Trochain* 198! Velor (Nov.) *Berhaut* 4026! Oussouye (Sept.) *Broadbent* 111! Ziguinchor (Sept.) *Broadbent* 70! **Gam.:** *Saunders* 36! (July, Aug.) *Ruxton* 48! 136! Yundum *Ashrif* 40! Genieri to Kaiaaf (July) *Fox* 161! **Mali:** Kita Mts. (July) *Jaeger* 31! 5753! 144 km. W. of Ségou (Aug.) *Roberty* 3748! Boré (Aug.) *Demange* 25/1957! Sotuba (Oct.) *J. Raynal* 5040! Bamako (Aug.) *Roberty* 3353! **Guin.:** Friguiagbé (May) *Chillou* 468! 516! Mamou (May) *Pitot*! Kouroussa (July) *Pobéguin* 406! 411! Youkounkoun (Oct.) *Pitot*! **Lib.:** Porobush, Duport (Oct.) *van Harten* 175! **Ghana:** Berekum to Sampa (Apr.) *Morton* A3233! Atonso (Oct.) *Baldwin* 13488! Tamale (Oct.) *Baldwin* 13576! Burufo, nr. Lawra (Nov.) *Harris*! Gambaga (May) *Morton* GC 7456! **Togo Rep.:** Lomé *Warnecke* 121! **N. Nig.:** Nupe *Barter* 1576! Lokoja (Sept.) *Parsons* 34! Kuffena, Zaria (June) *Kershaw* 900358! Naraguta F.R., Jos Plateau (July) *Lawlor & Hall* FHI 46533! Yola (Sept.) *Kennedy* 7292*a*! Damaturu, Bornu Prov. (Sept.) *De Leeuw* 1211*a*! **S. Nig.:** Oke Iho to Iseyin (Apr.) *Keay & Polunin* FHI 37839! Oyo (May) *Wanstall* A6! Ibadan (May) *Guile* 118! Owo F.R. (July) *Onochie* FHI 33247! Aboh *Barter* 356! **W. Cam.:** Cameroon R. (Feb.) *Mann* 778!

25b. **F. hispidula** subsp. **brachyphylla** (*Cherm.*) *Napper* in Kew Bull. 25: 440 (1971). *F. exilis* (Kunth) Roem. & Schult. var. *brachyphylla* Cherm. in Arch. Bot. Caen 4, Mém. 7: 32 (1931). *F. hensii* C.B. Cl. in F.T.A. 8: 419. Tufted perennial, usually much stouter than subsp. *hispidula* with scattered long hairs in the inflorescence, conspicuously paler spikelets having glumes without a mucro and similar achenes though sometimes the surface cells are scarcely raised; seasonally damp places in savanna, also in grassland, usually in sandy soils, coconut plantations and farmland.
Sen.: *Naegelé*! L. Rhetba (Nov.) *Pitot*! Dagoudane (Oct.) *J. & A. Raynal* 6495! Diouloulou, Casamance (Sept.) *Adam* 18184! **Guin.:** Friguiagbé (May, July) *Chillou* 493! 1637! 1638! Boké (Nov.) *Jac.-Fél.* 7305! Youkounkoun (Oct.) *Pitot*! **S. L.:** Tisana, Bonthe I., (Nov.) *Deighton* 2302! Aberdeen (Aug.) *Melville & Hooker* 139! Roboli, nr. Rokupr (Nov.) *Jordan* 166! Koinadugu (Aug.) *Haswell* 224! Mano Salija (Dec.) *Deighton* 285! **Lib.:** Bushrod I., Monrovia (Feb.) *Baldwin* 11068! Monrovia (May) *Baldwin* 5813! Sinkor (June) *Barker* 1373! Paynesville (Sept.) *van Dillewijn* 97! Tappita (Aug.) *Baldwin* 9065! Grand Bassa *Ansell*! **Iv. C.:** Azaguié *Adam* 6547! Aguien *Scaëtta* 3014! Dabou (Nov.) *Leeuwenberg* 1950! *Giovannetti* 422! Bouaké (May) *de Wilde* 57! **Ghana:** Axim (Mar.) *Morton* GC 6585! Legon Hill (Oct.) *Adams* 3366! Keta (Mar.) *Morton* GC 6525! Kpong (June) *Irvine* 4926! Amedzofe (Oct.) *Morton* 6037! **Togo Rep.:** Lomé *Warnecke* 126! **Dah.:** Cotonou *Debeaux* 331! (Nov.) *Poisson*! Porto Novo (Sept.) *Adjanohoun* 261! **N. Nig.:** Ilorin to Jebba (Apr.) *Wanstall* A5! Lokoja *Richardson* 5! Badeggi to Lapai (June) *E. W. Jones* 195! Lapai to Paiko (Aug.) *Onochie* FHI 35134! Plateau Prov. (Aug.) *Lely* P453! **S. Nig.:** Lagos (Oct.) *Stubbings* 79! Abeokuta to Ibadan (Mar.) *Schlechter* 12347! Oke Iho to Iseyin (Apr.) *Keay & Polunin* FHI 37834! Udi (Dec.) *Onyeagocha* FHI 7771! Calabar (Mar.) *Brenan* 9208! **W. Cam.:** Bali (Apr.) *Brunt* 1053! Basenako, Bamenda (May) *Maitland* 14*a*! Widespread in tropical and south-west Africa.

25c. **F. hispidula** subsp. **senegalensis** (*Cherm.*) *Napper* in Kew Bull. 25: 440 (1971). *Isolepis perrotetii* Steud., Syn. Pl. Glum. 2: 318 (1855). *Fimbristylis exilis* (Kunth) Roem. & Schult. var. *senegalensis* Cherm. in Arch. Bot. Caen 7, Mém. 4: 23 (1936); Berhaut, Fl. Sén. ed. 2, 371. *F. exilis* of F.T.A. 8: 418, partly; F.W.T.A., ed. 1, 2: 476, partly. Annual up to 22 cm. high, rarely more, differing from slender forms of subsp. *hispidula* in the more widely spreading inflorescence, the more pubescent glumes without a mucro and the more numerous undulations on the achene; sandy soils and as a weed.
Maur.: Mederdra to Tiguent (Oct.) *Adam* 19354! Dahr de Néma (Dec.) *Rossetti* 61/181! **Sen.:** *Sieber*!

Heudelot 330! (May) *Roger* 114! Hann (Sept.) *J. & A. Raynal* 6314! St. Louis (Oct.) *Berhaut* 2662! Dakar (Sept.) *Adam* 1856! Djembering (Sept.) *Broadbent* 153! Diourbel (Aug.) *Trochain* 5! **Mali:** Kita Mts. (Oct.) *Jaeger* 57! Dioura (July) *Davey* 10! Timbuktu (Aug.) *Lean* 77! *Hagerup* 258! Gao to Korogoussou *de Wailly* 4734! **Guin.:** Soumbalako to Boulivel (Sept.) *Chev.* 18667! Macenta *Adam* 6118! **Ghana:** Yeji (Aug.) *Hall* CC 381! Kete Krachi (May) *Morton* GC 7174! Keta (July) *Hall* 3261! Ada (June) *Morton* GC 9245! **Niger:** Toukounous *Bartha* 56! 97! **N. Nig.:** Sokoto (Sept., Oct.) *Moiser* 132! *Dalz.* 456! Katagum *Dalz.* 243! Kauwa (Oct.) *Golding* 17!

26. **F. tisserantii** *Cherm.* in Arch. Bot. Caen 4, Mém. 7: 32 (1931); Napper in Kew Bull. 25: 440 (1971). *F. exilis* (Kunth) Roem. & Schult. var. *rufescens* Cherm. in Bull. Soc. Bot. Fr. 81: 266 (1934). Tufted glabrous annual with large spikelets similar to those of *F. hispidula* subsp. *brachyphylla* having glabrous or puberulous glumes without a mucro; an ellipsoid not oblanceolate achene white before maturity and sharply ridged; swampy grassland and savanna.
Guin.: Friguiagbé (Aug.) *Chillou* 2112! **Iv. C.:** Foro-foro Forest, 25 km. N. of Bouaké (Sept.) *Oldeman* 407! **Ghana:** Ejura (Aug.) *Hall* CC 327! Yeji (Aug.) *Hall* CC 383! Oti River, nr. Kpandai (Aug.) *Hall* GC 38768! Tamale to Bolgatanga (Aug.) *Hall* CC 435! **Togo Rep.:** Sokodé (Aug.) *Mahoux* 2169!
[West African plants of this species have rather smaller spikelets than those from the Central African and Congo Republics, but it is impossible to decide until further material becomes available and the range of variation can be more accurately assessed whether one or two subspecies are involved.]

12. FUIRENA Rottb., Descr. et Ic. Pl. 70 (1773); F.T.A. 8: 461 (1902).

By Miss S. S. Hooper

Rhizomatous perennial; stems and sheaths conspicuously 5-angled; leaves with 5 main nerves, upper exceeding the internodes in length; three of the hypogynous bristles with an obovate, apiculate lamina as long as the achene .. .. 1. *umbellata*
Annual or caespitose perennial; stems and sheaths triangular, terete or inconspicuously 5-angled; leaves with one or several main nerves:
Stems and sheaths triangular; leaves stiff, keeled, one-nerved; tips of the glumes erect:
Spikelets ovoid, 4–5 mm. wide .. .. 5. sp. aff. *pubescens*
Spikelets narrowly ellipsoid, 2–3 mm. wide .. .. .. 2. *stricta*
Spikelets angular, glumes more or less obviously 5 ranked 2a. *stricta* var. *stricta*
Spikelets terete, generally greenish .. .. 2b. *stricta* var. *chlorocarpa*
Stems and sheaths terete or obscurely 5-angled; leaves soft, many-nerved, hairy; tips of the glumes conspicuous, spreading; hypogynous bristles laminate or absent:
Ligule brown, smooth or shortly hairy; spikelets about 3 mm. wide; tips of the glumes about 1·5 mm. long; three hypogynous bristles with a quadrate 3-nerved lamina; achene 0·8–1·25 mm. long .. .. 3. *ciliaris*
Ligule transparent, ciliate; spikelets about 2 mm. wide; cilia of the glume about 1·0 mm. long; 0–3 hypogynous bristles with a small rectangular lamina less than half as long as broad; achene 0·4–0·65 mm. long .. .. 4. *leptostachya*

1. **F. umbellata** *Rottb.* Descr. et Ic. Pl. 70, t. 19, f. 3 (1773); F.T.A. 8: 466; Berhaut, Fl. Sén. ed. 2, 359 (1967). *F. glomerata* Lam., Ill. 1: 150, t. 39 (1791). *F. seriata* C.B. Cl. in Mém. Soc. Bot. Fr. 2, 8: 28 (1907). *F. mahouxii* Cherm. in Bull. Soc. Bot. Fr. 81: 264 (1934). Generally robust, often with the lowermost internode of the flowering stems becoming swollen; common on river banks and other moist places.
Sen.: L. Youi (Nov.) *Pitot*! Sangalkam (Jan.) *Berhaut* 2634! Niokolo-Koba *Adam* 15955! Santamba, nr. Toubacouta (Oct.) *Monod* 8528! Kolda, Casamance (May) *Rivière*! **Gam.:** (Aug.) *Ruxton* 85! **Mali:** Sekoro (Jan.) *Chev.* 251! Koulouba, nr. Bamako (Nov.) *Jaeger* 77! Sotuba (Nov.) *J. & A. Raynal* 5163! Kabala (July) *Roberty* 2439! Sikasso (Feb.) *Roberty* 3439! **Guin.:** Sambaïlo (Oct.) *Pitot*! Friguiagbé (May) *Boismaré* 240! Dalaba (Nov.) *des Abbayes* 837! Kouroussa (Aug.) *Pobéguin* 403! Nzérékoré to Beyla (May) *Pitot*! **S. L.:** Mano Salija (Dec.) *Deighton* 461! Tisana, Bonthe I. (Nov.) *Deighton* 2320! Njala (July) *Deighton* 1164! Kumrabai (Dec.) *Thomas* 6918! Koinadugu Dist. (Aug.) *Haswell* 253! Mt. Loma (Sept.) *Jaeger* 2027! **Lib.:** Robertsport, Grand Cape Mount Dist. (Dec.) *Baldwin* 10905! Zigida, Vonjama Dist. (Oct.) *Baldwin* 9996! Palilah, Gbanga Dist. (Aug.) *Baldwin* 9164! Ganta, Sanokwele Dist. (Oct.) *Harley* 731! Harper, Maryland Dist. (June) *Baldwin* 5997! **Iv. C.:** Banco forest *Bégué* 130! Grand Bassam (Sept.) *de Wit* 7395! Songon Agban to Dabou (Feb.) *Leeuwenberg* 2640! Orumbo-Boka Mt., Toumodi (Dec.) *Boughey* A13! Ferkessédougou (Nov.) *Leeuwenberg* 2019! **U. Volta:** Comoé falls (Sept.) *Adam* 15247! Bouassi to Kouéré (July) *Scholz* 314! **Ghana:** Atwabo *Fishlock* 33 of 1931! Ejura (Aug.) *Hall* CC 303! Kete Krachi (Apr.) *Morton* GC 9110! Yendi to Tamale (Dec.) *Adams & Akpabla* 4125! Sawla to Wa (Oct.) *Rose Innes* GC 32396! Gambaga *Morton*! **Togo Rep.:** Palime *Stage* 83! Lomé (June) *Mahoux* 2020! Atakpamé (Aug.) *Mahoux* 2162! Sokodé *Mahoux* 2164! Sansanné-Mango (Oct., Nov.) *Bille* TG 363! **Dah.:** Cabolé (June) *Annet* 64! Porto Novo (Jan., Sept.) *Chev.* 22711! *Adjanohoun* 355! **N. Nig.:** Abinsi (June) *Dalz.* 817! Lokoja (Apr.) *Haines* 24! Bida to Badeggi (Aug.) *Onochie* FHI 35422! Nupe *Barter* 418! Naraguta, Jos Plateau (Aug.) *Lawlor & Hall* FHI 46643! Sokoto *Moiser*! Gurum, Vogel Peak, Sardauna Prov. (Nov.) *Hepper* 1263! **S. Nig.:** Lagos (Jan.) *Haines* 91! Oyo to Ilora (July) *Keay & Stanfield* FHI 46313! Ajagbododu, Benin (Aug.) *Wright* 53! Sapoba (July) *Kershaw* 900404! Abonnema, Niger Delta (Nov.) *Anderson* 1412! Calabar (June) *Tuley* 69! **W. Cam.:** Cameroon R. (Feb.) *Mann* 777! Bamenda (Feb.) *Daramola* FHI 40490! Nkom to Wum, Bamenda (July) *Ujor* FHI 29299! Pantropical.
[The type of *Fuirena seriata* C.B. Cl. has the hypogynous scales pale and apparently one-nerved whilst those of *F. umbellata* have 3 nerves and are generally brown.]

2. **F. stricta** *Steud.* Syn. Pl. Glum. 2: 128 (1855); F.T.A. 8: 465; Berhaut, Fl. Sén. ed. 2, 359 (1967).

2a. **F. stricta** *Steud.* var. **stricta.** Tufted, slender plants with a paniculate inflorescence of rather few, five-angled spikelets in small clusters; swamps, ricefields and wet savanna.
Sen.: Bignona (Feb.) *Trochain* 1470! Niokolo-Koba (Nov., Dec.) *Adam* 17090! *J. & A. Raynal* 6886! Badi (Jan.) *Berhaut* 4388! Dienoudiella (May) *Trochain* 3540! **Mali:** Madina (Dec.) *J. & A. Raynal* 5340! Bamako (Mar., Nov.) *Chev.* 44059! *Duong*! Bamako to Koulouba (Dec.) *Monod*! Koulouba *Duong*! **Guin.:** Friguiagbé (Feb.) *Chillou* 2454! Dabola to Faranah (Apr.) *Pitot*! Kouroussa (Jan.) *Pobéguin* 622! Sambaïlo (Oct.) *Pitot*! Koba (Nov.) *Jac.-Fél.* 7365! **S. L.:** Kambia (Jan.) *Deighton* 926! Kouta (Aug.) *Deighton* 1240! Koinadugu (Aug.) *Haswell* 247! Bintumane Peak, 5,400 ft. (Jan.) *T. S. Jones* 77! Kenema (Sept.) *Jordan* 516! **Lib.:** Siatown (Dec.) *Adam* 16335a! Ganta (Nov.) *Harley* 1707! Siaple, Sanokwele Dist. (Sept.) *Baldwin* 9465! Mt. Nimba (Jan.) *Adam* 20525! **Iv. C.:** Bouaké (Sept.) *de Wilde* 966! Lomo (Sept.) *Aké Assi* 5737! Nassian to Ferkessédougou (July) *Aké Assi* 4820! **U. Volta:** Dar Salami (Dec.) *Bille* 3351b! **Ghana:** Banda, Wenchi Dist. (Dec.) *Morton* GC 25208! Kintampo (Sept.)

Hall CC 916! Damongo Scarp (Dec.) *Morton* GC 9982! Nalerigu, Gambaga (June) *Harris*! Kukrukui lagoon, nr. Kete Krachi (Apr.) *Hall* VBS 1214! **Togo Rep.:** Sokodé (July) *Mahoux* 2099! **Niger:** Magaria (Dec.) *Dumont* in *Hb. P. de Fabrègues* 2661! **N. Nig.:** Jebba (Dec.) *Meikle* 849! Nupe *Barter* 1562! Badeggi, nr. Bida (Dec.) *Haines* 68! Bida (Mar.) *Meikle* 1321! Kuffena, Zaria Prov. (Jan.) *Kershaw* 900017! Sara Hills, Jos Plateau (Oct.) *Hepper* 1121! **S. Nig.:** Iva stream, Udi Dist. (Sept.) *Onochie* FHI 34088! Uke to Nobi, Onitsha Dist. (May) *Onochie* FHI 33799! Throughout tropical Africa and in Madagascar.

2b. **F. stricta** var. **chlorocarpa** (*Ridley*) *Kük.* in Fries & Fries in Notizbl. Bot. Gart. Berl. 9: 310 (1925). *F. chlorocarpa* Ridley in Trans. Linn. Soc. ser. 2, 2: 159 (1884). Differs from the typical variety only in the somewhat shorter, terete spikelets; damp ground near water.
N. Nig.: Vom, Plateau Prov. (Dec.) *Haines* 70! Kakara, Mambila Dist. (Jan.) *Hepper* 1772! **W. Cam.:** Cheddar Gorge, Bamenda (Mar.) *Morton* K187! Bafut-Ngemba F.R., Bamenda (June) *Daramola* FHI 41557! Jakiri, 5,000–6,500 ft. (Feb., June) *Hepper* 2062! *Brunt* 534! Pinyin, 5,500 ft. (Apr.) *Brunt* 1096! Tropical and S. Africa.

3. **F. ciliaris** (*Linn.*) *Roxb.* Fl. Ind. 1: 184 (1820); Berhaut, Fl. Sén. ed. 2, 360 (1967). *Scirpus ciliaris* Linn., Mant 1: 182 (1767); Rottb., Descr. et Ic. Pl. 55 t. 17, f. 1 (1773). *Fuirena glomerata* of F.T.A. 8: 465; F.W.T.A. ed. 1, 2: 470; Berhaut, Fl. Sén. ed. 1, 215 (1954); not of Lam. Lax, hairy, caespitose plant; tips of the glumes conspicuous, spreading; rice fields and other moist depressions.
Sen.: Sangalkam (Sept.) *Berhaut* 2632! Dakar (May) *Baldwin* 5741! Mbidjem (Nov.) *J. & A. Raynal* 6555! Fatick (Dec.) *Maymard*! Oussouye (Oct.) *Kerharo & Adam* 1568! Carabane (Jan.) *Chev.* 2433! **Guin.:** Mankoutan *Jac.-Fél.* 7352! Kabuk (Dec.) *Jac.-Fél.* 7380! Conakry *Maclaud*! **S. L.:** Bakepet, Mambolo (Dec.) *Jordan* 734! Mahela, Samu Dist. (Dec.) *Sc. Elliot* 3982! Tisana, Bonthe I. (Nov.) *Deighton* 2298! Malpaki, Massaboi (Aug.) *Jordan* 502! Tombo (Jan.) *Deighton* 993! **Iv. C.:** Néro-Mer, nr. Béreby (Nov.) *Oldeman* 497! Grand Bassam (Feb.) *Scaëtta*! **Ghana:** Accra *Brown* 365! Nungua (July) *Ankrah & Rose Innes* GC 30024! Achimota to Nsawam (Nov.) *Boughey*! Mile 35 on Ada road (May) *Adams* 2742! Adidome to Ho (May) *Morton* A2050! **Togo Rep.:** Lomé *Warnecke* 228! Tsecké (Sept.) *Mahoux* 542! **Niger:** Matameye (Nov.) *P. de Fabrègues* 1302! **N. Nig.:** *Ward* 88! Kano (Sept.) *Hagerup* 668! Katsina (Nov.) *Latilo* FHI 43769! **S. Nig.:** Badagry (Aug.) *Onochie* FHI 33475! Lagos *Barter* 2224! (Jan.) *Haines* 90! Ibadan (July) *Keay* FHI 37769*a*! Old World tropics.

4. **F. leptostachya** *Oliv.* in Trans. Linn. Soc. 29: 168, t. 108B (1875); F.T.A. 8: 466. *F. moiseri* Turrill in Kew Bull. 1925: 71 (1925). *F. pygmaea* Ridley in Trans. Linn. Soc. ser. 2, 2: 160 (1884). Hairy annual similar to the preceding species, but with smaller spikelets and a very small lamina to the inner hypogynous bristles which may be partially or entirely wanting; rice fields and other damp places, uncommon.
Mali: Madina (Dec.) *J. & A. Raynal* 5310 *bis*! 5344! Bamako (Nov.) *Waterlot* 1446! Dinderesso forest (Nov.) *Neu* 123! **U. Volta:** Ouagadougou (Nov.) *Aubréville* 2630! **N. Nig.:** Fodama, Sokoto Dist. (Dec.) *Moiser* 157! Shendam, Plateau Prov. (Nov.) *Clayton* 1466! Bauchi to Dindima (Dec.) *Haines* 88! Isanlu Mopo, Kabba Prov. (Nov.) *Latilo* FHI 61327! Scattered throughout tropical Africa.
[Differs from the Indian *F. trilobites* C.B. Cl. only in the smaller achene.]

5. **F.** sp. aff. **pubescens** (*Poir.*) *Kunth* Enum. Pl. 2: 182 (1837); F.T.A. 8: 463. *Carex pubescens* Poir., Voy. Barb. 2: 254 (1789). Plants grey-green with grey hairy glumes in few ovoid spikelets; moist ground.
N. Nig.: Kano, Korel (Apr.) *Onwudinjoh* FHI 22362!
[Very similar in habit and spikelet to *F. pubescens* which occurs in the Mediterranean region, throughout Africa and in India, but distinguished by the presence of hypogynous bristles, which are sometimes narrowly laminate.]

13. ASCOLEPIS Nees ex Steud., Syn. Pl. Cyp. 105 (1855), *nom. cons.*; F.T.A. 8: 473 (1902).

By Miss S. S. Hooper

Annual with an inflorescence of several spikelets forming a lobed *Kyllinga*-like head; stems slender, up to 25 cm. high:
- Hypogynous scale enclosing the achene, with its upper margin inflated, white or red, and with a short postical tip .. .. 1. *pusilla*
- Hypogynous scale flat, not enclosing the achene, with an acumen as long as the rest of the scale; an antical seta is persistent on the rhachilla after glume fall 2. *dipsacoides*

Annual with a globular, hemispherical or discoid inflorescence of a single spikelet, or perennial:
- Inflorescence of about 4 spikelets in a pyramidal head; plant shortly rhizomatous; sheaths reddish, not fibrous .. .. 5. *brasiliensis*
- Inflorescence of a single spikelet:
  - Inflorescence 25–40 mm. diam.; hypogynous scale about 15 mm. long 6. *elata*
  - Inflorescence not exceeding 12 mm. diam.:
    - Hypogynous scale with a subcylindric tip and narrow base; stems slender, about 0·5 mm. diam., often thickened at the base with pale fibrous sheaths 3. *protea*
    - Hypogynous scale with a flattened tip and winged base; stems about 1 mm. diam., clustered; sheaths becoming black and fibrous .. .. 4. *capensis*

1. **A. pusilla** *Ridley* in Trans. Linn. Soc., ser. 2, 2: 164 (1884); F.T.A. 8: 476; Berhaut, Fl. Sén. ed. 2, 369 (1967); J. Raynal in Adansonia sér. 2, 8: 98 (1968). Slender plant with a greenish inflorescence and smooth outline to the spikelets; rice fields and other marshy ground.
Sen.: Kanémééré (Nov.) *Fotius* K688! Niokolo-Koba (Nov.) *Adam* 16970! 17079–20! **Mali:** Sotuba (Dec.) *J. & A. Raynal* 5236. Farako (Nov.) *Demange* 3276! **Guin.:** Kandiara R. (Oct.) *Pitot*! Télimélé (Oct.) *Pitot*! Thianguel-Bori (Oct.) *Pitot*! Ditinn (Oct.) *Pitot*! **Ghana:** Yapei Bridge (Nov.) *Hall & Enti* GC 35737! Ejura (Dec.) *Morton* GC 9700! Kwahu Tafo (Dec.) *T. M. Harris*! **N. Nig.:** Kabba (Oct.) *Parsons* 32! Nupe *Barter* 761*a*! Gurum, Sardauna Prov. (Nov.) *Hepper* 1240! **S. Nig.:** Igbete to Kishi, Oyo Prov. *Gledhill* 734*c*, partly! Also in E. Cameroun, Chad, C. African Rep., and from Tanzania to Rhodesia and ? introduced in Madagascar (*teste* Raynal).

2. **A. dipsacoides** (*Schum.*) *J. Raynal* in Adansonia sér. 2, 8: 99 (1968). *Kyllinga dipsacoides* Schumach., Beskr. Guin. Pl. 61 (1827). *Ascolepis setigera* Hutch., F.W.T.A., ed. 1, 2: 474 (1936), name only. Resembling *A. pusilla* but spikelets with spreading tips to the scales; also resembling *Scirpus kernii*; damp hollows, often in cultivated land.
Sen.: Niokolo-Koba (Oct.) *Adam* 15944*a*. **Iv. C.:** Bouaké to Bamoro (Dec.) *Pitot*. Aboakouamékro (Nov.) *Aké Assi* 10353! **Ghana:** Damongo Scarp (Sept.) *Hall* CC 867! Lumbaga to Tamale (Dec.) *Morton* GC 6265! Tamale (Dec.) *Morton* GC 6219! Kintampo (Sept.) *Hall* CC 925! Nungua, Accra (July) *Hall* 3249! **Dah.:** Ouécé (Nov.) *Annet* 53. **N. Nig.:** Jebba (Dec.) *Hagerup* 710! Nupe *Barter* 761*b*! Kaduna to Keffi,

Zaria Prov. (Aug.) *Kershaw* 900700! Yola (Nov.) *Hepper* 1212! **S. Nig.:** Igbete to Kishi, Oyo Prov. *Gledhill* 734c, partly! Also from E. Cameroun; our plant is subsp. *dipsacoides*, the other subspecies occurs in Thailand and Vietnam.

3. **A. protea** *Welw.* in Trans. Linn. Soc. 27: 75 (1869); F.T.A. 8: 474. *A. eriocauloides* of Berhaut, Fl. Sén. ed. 2, 365 (1967). Plant with white, often curved, scales in a radiate head about 6–8 mm. across; on granite outcrops, in short grass.
Sen.: Tambacounda to Sambaïlo (Sept.) *Pitot*! Badi (Sept.) *Berhaut* 3070! Kanéméré (July) *Fotius* K217! Damaradji (Sept.) *Audru* 1245! **Mali:** Kita Mts. (Oct.) *Jaeger* 2882! Bamako *Adam* 15157! 15252! *Garnier*! Sikasso (Oct.) *Demange* 2721! **Guin.:** Sambaïlo (Sept.) *Pitot*! Youkounkoun (Oct.) *Pitot*! Friguiagbé (July) *Chillou* 585! Macenta to Kankan *Roberty* 3124 (204)! Nzérékoré, Nimba Mt. *Adam* 7891! **S. L.:** Warantamba, Gberia Fotombu (Oct.) *Small* 367! Koinadugu Dist. *Haswell* 272a! Lengekoro (July) *Glanville* 220! Summit of Bintumane Peak (June, Aug.) *Morton & Gledhill* SL 3575! *Jaeger* 1047! **Iv. C.:** Séguéla (Aug.) *Boughey* GC 18485b! Mankono (June) *Chev.* 21884! Mt. Bourou, nr. Koualé (May) *Fleury* in *Hb. Chev.* 21729! **U. Volta:** Banfora (Sept.) *Jaeger* 5194! Barmoussa (Sept.) *Scholz* 323! **Ghana:** Burufo, nr. Lawra (Sept.) *Hall* CC 745! Yagha, Lawra to Wa (Sept.) *Hall* CC 630! Tono, nr. Navrongo (Aug.) *Irvine* 4660! **Togo Rep.:** Atakpamé (Aug.) *Mahoux* 2172! **N. Nig.:** Shagunu, nr. Bussa (July) *C. D. K. Cook* 409! Sokoto Prov. *Dalz.* 573! Anara F.R., Zaria Prov. (July) *Keay* FHI 25970! Panshanu, Bauchi Prov. (Aug.) *Lawlor & Hall* 557! Abuja to Izom, Niger Prov. (Aug.) *Onochie* FHI 35739! **W. Cam.:** Bamenda, 6,000 ft. (June) *Maitland* 1726! Occurring, with its varieties, throughout tropical Africa.
[A gathering from Bamenda (*Ujor* FHI 30226), a stouter plant with heads up to 15 mm. diam., may represent a distinct variety of this variable species.]

4. **A. capensis** (*Kunth*) *Ridley* in Trans. Linn. Soc. ser. 2, 2: 164 (1884); F.T.A. 8: 477. *Platylepis capensis* Kunth, Enum. 2: 269 (1837). Inflorescence similar in size to *A. protea*, but with a more solid appearance due to the broader scales; wet ground, generally amongst grasses.
Mali: *Duong* 1732! Sabalibougou, Mt. Irémakdio (Mar.) *J. & A. Raynal* 5618! Sotuba (Dec.) *Adam* 11440! **Guin.:** Timbi (June) *Adam* 14644! **Iv. C.:** *Bégué* 129! Sourougban (Apr.) *Aké Assi* 8741! confluence of R. Sassandra and R. Bafing, nr. Daton (May) *Chev.* 21760b! Haut Sassandra and Haut Cavally *Portères* 707! **U. Volta:** 18 km. N. of Banfora (May) *Leeuwenberg* 4352! **Ghana:** Sampa (Dec.) *Morton* A2626! Banda, Wenchi Dist. (Dec.) *Morton* 25203! Ejura (Aug.) *Hall* CC 319! Kintampo (Dec.) *Morton* A1201! Bimbila (Mar.) *Hepper & Morton* A3094! **N. Nig.:** Kabba (Oct.) *Parsons* 24! Nupe *Barter*! Abinsi (Sept.) *Dalz.* 814! Vom, Jos Plateau (Feb.) *McClintock* 209! Kakara, Mambila Plateau (Jan.) *Hepper* 1773! **S. Nig.:** Old Oyo F.R., 8 miles E. of Igboho, Oyo Dist. (Feb.) *Keay* FHI 23445! Obudu (Apr.) *Haines* 98! Throughout tropical and S. Africa.

5. **A. brasiliensis** (*Kunth*) *Benth. ex C.B. Cl.* in Dur. & Schinz, Consp. Fl. Afr. 5: 651 (1895); F.T.A. 8: 478; Berhaut, Fl. Sén. ed. 2, 368 (1967). *Platylepis brasiliensis* Kunth, Enum. Pl. 2: 269 (1837). Similar in habit to, but somewhat stouter than, *A. capensis* and in West Africa distinguished by a lobed inflorescence; in marshland.
Sen.: Niokolo-Koba (Feb.) *Adam* 17567! Badi (Jan., Dec.) *Berhaut* 1856! 4735! **Guin.:** Konkouré (Mar.) *Pitot* 291! Mamou (May) *Pitot*! Dabola to Faranah (Apr.) *Pitot*! Hérémakono (Mar.) *Sc. Elliot* 5243! **Iv.C.:** Baoulé (Sept.) *Pobéguin* 110! **N. Nig.:** Mayo Ndaga (Apr.) *J. Hall* 1761! Kontagora to Auna (Jan.) *Meikle* 1041! Nupe *Barter* 1587! **W. Cam.:** Bamenda to Ndop, 4,500 ft. (Dec.) *Adams* GC 11342! In Madagascar and S. America.

6. **A. elata** *Welw.* in Trans. Linn. Soc. 27: 79 (1869); F.T.A. 8: 476. Plant with white radiate heads about 35 mm. across; in swampy grassland.
N. Nig.: Abinsi *Dalz.* 813! Naraguta F.R., Jos Plateau (Aug.) *Keay* FHI 20064! Panyam, 4,500 ft. (July) *Lely* 411! Ibi, Muri Prov. (July) *Hepburn* 30! Kilba country, Yola Prov. (Aug.) *Dalz.* 238! Extending over much of tropical Africa from Sudan to Rhodesia.

14. LIPOCARPHA R.Br. in Tuckey, Narr. Exp. Congo 459 (1819), *nom. cons.*; F.T.A. 8: 468 (1902).

By Miss S. S. Hooper

Spikelets pale even when mature; glumes whitish, sometimes dark towards the base:
  Spikelets confluent, 3–6 in the echinate head; glume gradually narrowed into a long blunt-tipped, spreading arista; plant with creeping rhizome .. 1. *albiceps*
  Spikelets discrete, 5–16 in the head; glume with a short blunt, triangular tip; plant densely caespitose .. .. 2. *chinensis*
Spikelets dark or becoming so at maturity; glumes red, brown or black, sometimes with pale tips:
  Perennial, shortly rhizomatous:
    Spikelets 2·5–3·0 mm. broad, black; glume with a short straight arista 3a. *atra* var. *atra*
    Spikelets 3·5–5 mm. broad, brown; glume with a pale spreading or recurved arista 3b. *atra* var. *barteri*
  Annual, sometimes robust; spikelets less than 4 mm. broad:
    Glumes clearly aristate, with a pale arista ½–1 times as long as the uniformly dark body of the glume; rhachilla of the spikelet 0·5–0·75 mm. broad; achene elliptical somewhat compressed .. .. 5. *nana*
    Glumes with a pale apiculus or short acumen not more than ⅓ as long as the generally bicolored glume; spikelets terete:
      Achene broadly obovate, trigonous; spikelets obtuse with a tendency to become cylindrical; glumes very concave; rhachilla 0·75–1 mm. broad with the persistent glume bases surrounding the scars conspicuous .. .. 4. *prieuriana*
        Glume truncate, with a small triangular mucro 4a. *prieuriana* var. *prieuriana*
        Glume with a broad pale arista .. .. 4b. *prieuriana* var. *crassicuspis*
      Achene elliptical, somewhat flattened; spikelets ovoid; rhachilla 0·5–0·75 mm. wide with the persistent glume bases small so that the scars form a regular spiral pattern .. .. 6. *sphacelata*

1. **L. albiceps** *Ridl.* in Trans. Linn. Soc. ser. 2, 2: 163 (1884); F.T.A. 8: 471. Sometimes, at least, having a distinct storage rhizome; damp places in grassland or open woodland.
Sen.: Niokolo-Koba (July) *Adam* 14206! 14260! 14266! **Mali:** Bamako (July) *Waterlot* 1188! Sotuba (Sept.) *Adam* 15346! Sikasso (July) *Demange* 2860! Katibougou (July) *Kaden* 67/252! **Iv. C.:** Bouna Reserve, Téhini (Aug.) *de Wilde* 707! Boka to Lenguera (June) *Boudet* 2825! **U. Volta:** Ouagadougou *Lefebvre* 18! **Ghana:** Wa to Lawra (Sept.) *Hall* CC 612! Tumu (May) *Morton* GC 7528! Lumbaga (Dec.) *Morton* GC 6246! Pong Tamale (Aug.) *Hall* CC 412! Kintampo (Aug.) *Hall* 2055*b*! **Togo Rep.:** Bafilo (May) *Roberty* 1429! Atakpame (Aug.) *Mahoux* 2165! Sokodé (May) *Mahoux* 2015! **N. Nig.:** Jebba (Apr.–May) *Kamphorst* 232! Shagunu, N. Ilorin Prov. (July) *Cook* 416! Gwari, Niger Prov. (Aug.) *Onochie* FHI 35444! Anara F.R., Zaria Prov. (May) *Keay* FHI 22861! Samaru, Zaria Prov. (July) *Taylor* 14! Also in the Congo and from Uganda to Rhodesia.

2. **L. chinensis** (*Osb.*) *Kern* in Blumea, Suppl. 4: 167 (1958). *Scirpus chinensis* Osb., Dagb. Ostind. Resa 220 (1757). *Lipocarpha argentea* (Vahl) R. Br. in Tuckey, Narr. Exp. Congo 459 (1819); F.T.A. 8: 469. *L. senegalensis* (Lam.) Th. & Hél. Dur., Syllog. Fl. Congol. 619 (1909); F.W.T.A., ed. 1, 2: 471; Berhaut, Fl. Sén. ed. 2, 369 (1967). Tufted, with stout flowering stems; the commonest species; beside streams and in other damp places.
Sen.: *Naegelé*! Niokolo-Koba (Nov., Dec.) *Adam* 17093! *J. & A. Raynal* 6887! Casamance (Aug.) *Berhaut* 6334! Badi (Jan.) *Berhaut* 4404! **Mali:** Boulouli *Duong* 569! Tabacco (Jan.) *Chev.* 141! Bamako (Mar.) *Chev.* 44089! Sotuba (Dec.) *J. & A. Raynal* 5250 *ter*! Kobala (July) *Roberty* 2452! **Guin.:** Youkounkoun (Oct.) *Pitot*! Friguiagbé to Bingaya (May) *Boismaré* 244! Fouta Djalon (Jan.) *Langdale-Brown* 2572! Dabola to Faranah (Apr.) *Pitot*! Macenta to Kankan *Roberty* 3125/206! **S. L.:** Newton (Nov.) *Deighton* 1473! Koinadugu Dist. (Aug.) *Haswell* 246! Kuntu, nr. Rokupr (Jan.) *Deighton* 909! Bintumane Peak (Aug.) *Jaeger* 1034! Kenema (Nov.) *Deighton* 420! **Lib.:** Kailahun, Kolahun Dist. (Nov.) *Baldwin* 10127! Tawata, Boporo Dist. (Nov.) *Baldwin* 10315! Grand Bassa (Nov.) *Dinklage* 2297! Blazié (Nov.) *Adam* 16110! Harper, Maryland Dist. (June) *Baldwin* 5989! **Iv. C.:** Mankono (June) *Fleury* in *Hb. Chev.* 21892! Baoulé (Sept.) *Pobéguin* 87! Yapo (Oct.) *Roberty* 15355! Téké Forest, 25 km. N. of Abidjan (Oct.) *Leeuwenberg* 1857! **U. Volta:** Bekoui (Dec.) *Bille* 3378! Bobo-Dioulasso (Nov.) *Bille* 3326! **Ghana:** Burufo, nr. Lawra (Apr., Sept.) *Adams* 4039! *Hall* CC 758! Sakogu, nr. Gambaga (Aug.) *Hall* CC 443! Gambaga (May) *Morton* GC 7442! Kukrukui Lagoon, nr. Kete Krachi (Apr.) *Hall* VBS 1216! **Togo Rep.:** Lama-Kara (June) *Mahoux* 2051! **Niger:** Sassoumbouroum *P. de Fabrègues*! Matameye (Nov.) *P. de Fabrègues* 1305! **N. Nig.:** Sokoto (July) *Dalz.* 454! Minna (Dec.) *Meikle* 719! Sara Hills, Jos Plateau (Oct.) *Hepper* 1120! Kurra Gorge, Pankshin Dist. (June) *Okafor* FHI 50227! Gurum, Vogel Peak, Sardauna Prov. (Nov.) *Hepper* 1254! **S. Nig.:** Shaki (May) *Lagos Govt.* 2! Ilesha (Apr.) *Haines* 20! Uke to Nobi, Onitsha Dist. (May) *Onochie* FHI 35803! **W. Cam.:** Madele, Wum Dist. (June) *Daramola* FHI 41095! Jakiri, 5,000–6,500 ft., Bamenda (Feb., June) *Hepper* 2064! *Brunt* 535! *Daramola* FHI 40492! Throughout tropical and S. Africa and from India to China and Australia.

3. **L. atra** *Ridl.* in Trans. Linn. Soc. ser. 2, 2: 162 (1884).

3a. **L. atra** *Ridl.* var. **atra.** Characterized by the rhizome, the yellowish stems much exceeding the leaves, and the dark, echinate spikelets which are smaller than those in the variety following.
Sen.: Badi (Sept.) *Berhaut* 3066! **U. Volta:** Lac de Tengréla (July) *Aké Assi* 10751! Also in E. Cameroun, Central African Rep., the Congo and SE. tropical Africa.

3b. **L. atra** var. **barteri** (*C.B. Cl.*) *J. Raynal* in Adansonia sér. 2, 7: 85 (1967). *L. barteri* C.B. Cl. in F.T.A. 8: 472 (1902). *Kyllinga baoulensis* A. Chev., Bot. 698 (1920), name only. Spikelets larger than those of other W. African species with conspicuous pale glume tips; wet patches in savanna.
Iv. C.: Sémien to Kanébly (June) *Aké Assi* 9077! Bouaké (Jan.) *de Wit* 7372/549! Dimbokro *Scaëtta* 3078 (89)! **Ghana:** Burufo, nr. Lawra (May) *Morton* GC 7641! Sampa, N.W. Ashanti (Apr.) *Morton* A3267! Banda (Aug.) *Hall* 1995! Ejura (Dec.) *Morton* GC 9715! Tiasi, Afram Plains (Aug.) *Hall* CC 163! **Togo Rep.:** Lama-Kara (June) *Mahoux* 2052! **N. Nig.:** Nupe *Barter* 1585! Lokoja to Okom (Apr.) *Haines* 21! **S. Nig.:** Igbetta, (Nov.) *Stanfield*! **W. Cam.:** Bafia (Apr.) *Eldin* 68! Also in E. Cameroun.

4. **L. prieuriana** *Steud.* Syn. Pl. Glum. 2: 130 (1855); F.T.A. 8: 471; Berhaut, Fl. Sén. ed. 2, 369 (1967).

4a. **L. prieuriana** *Steud.* var. **prieuriana.** Characterized by the terete, ovate, obtuse spikelets; damp places in savanna.
Sen.: St. Louis (Oct.) *Berhaut* 813! Fatick (Oct.) *J. & A. Raynal* 7741! Niokolo-Koba (Nov.) *Adam* 17079. 21! Gambia R., E. Senegal (Sept.) *Fotius* K436! Velingara, Casamance (Oct.) *Adam* 18597! **Mali:** Kita (Sept.) *Jaeger* 2533! Balasoko, Bamako to Koulouba (Dec.) *Monod*! Koulikoro (Nov.) *Chev?.* 2459! **Guin.:** Youkounkoun (Oct.) *Pitot*! Gaoual to Kiffaya (Oct.) *Pitot*! Kiffaya to Seriba (Oct.) *Adam* 12705! **U. Volta:** Banfora (Nov.) *Jaeger* 6543! 6565! **Ghana:** Burufo (Dec.) *Adams & Akpabla* 4388! Lumbaga, nr. Tamale (Dec.) *Morton* GC 6243! Tamale (Oct., Dec.) *Baldwin* 13563! *Adams & Akpabla* 4156! Yapei Bridge, Tamale Port (Nov.) *Hall & Enti* GC 35738! **Togo Rep.:** Dapongo (Oct., Nov.) *Bille* TG 75! **N. Nig.:** Kano (Nov.) *Hagerup* 667! Agbaja (Nov.) *J. Hall* 838! Also in Ethiopia and SE. tropical Africa.

4b. **L. prieuriana** var. **crassicuspis** *J. Raynal* in Adansonia sér. 2, 7: 86, fig. 2 (1967). Differing from the typical and commoner variety only in the development of the glume tip into a broad green arista; brackish areas.
Sen.: Fatick (Dec.) *J. & A. Raynal* 6711! Kaolack *Adam* 8241!

5. **L. nana** (*A. Rich.*) *Cherm.* in Bull. Soc. Bot. Fr. 71: 142 (1924). *Fuirena nana* A. Rich., Tent. Fl. Abyss. 2: 497 (1851). *Lipocarpha pulcherrima* Ridl. in Trans. Linn. Soc. ser. 2, 2: 162 (1884); F.T.A. 8: 473. Spikelets dark with green aristate tips; rare or perhaps overlooked amongst other ephemeral species in damp situations.
Guin.: Sébori, nr. Kanéa (Oct.) *Adam* 12676! Ditinn (Oct.) *Pitot*! **N. Nig.:** Jos Plateau (Aug.) *Lely* P471! **W. Cam.:** Nkambe, Bamenda (Sept.) *Ujor* FHI 30229! Also in the Congo and from Ethiopia to the Transvaal.

6. **L. sphacelata** (*Vahl*) *Kunth* Enum. Pl. 2: 267 (1837). *Hypaelyptum sphacelatum* Vahl, Enum. 2: 283 (1806) *Lipocarpha triceps* (Lam.) Nees (1834)—F.T.A. 8: 470; F.W.T.A., ed. 1, 2: 471; Berhaut, Fl. Sén. ed. 2' 369 (1967). *Kyllinga triceps* Lam. (1791), not of Rottb. (1773). Distinguished from *L. prieuriana* by the smaller, less obtuse spikelets and conspicuous, if short, glume tips; roadsides and other sandy brackish places.
Sen.: Niokolo-Koba (Oct.) *Adam* 15840! Oussouye (Nov.) *Berhaut* 6516! Kabrousse (Oct.) *Adam* 18308! Kolda (Oct.) *Adam* 18551! Velingara (Oct.) *Adam* 18567! **Mali:** Kita (Aug.) *Jaeger* 16! Bamako *Adam* 15041! Balossoko, Bamako to Koulouba (Nov., Dec.) *Duong*! *Monod*! Koulouba (Apr.) *Rogeon* 3! **Guin.:** Youkounkoun (Oct.) *Pitot*! Bombi Bouron (Oct.) *Pitot*! Boké (Nov.) *Jac.-Fél.* 7308! Conakry (Sept.) *Martine* 112 in *Hb. Chillou* 2814! Friguiagbé *Chillou* 598*b*! **S. L.:** Bonthe (Nov.) *Deighton* 2267! Mattru to Gbangbama (Nov.) *Deighton* 2338! Newton (Nov.) *Deighton* 1472! Roboli, nr. Rokupr (Nov.) *Jordan* 167! Koinadugu Dist. (Aug.) *Haswell* 262! Kimadugu, Loma Mts. (Aug.) *Jaeger* 7264! **Lib.:** Cape Palmas *Ansell*! **Iv. C.:** Mankono (Oct.) *Aké Assi* 6645! Bouaké (Jan.) *de Wit* 7272/549! Bouaké to Damoro (Dec.) *Pitot*! Assakra (Aug.) *de Wit* 7388/1082! Adiopodoumé (July) *Boughey* GC 14503! **U. Volta:** Banfora (Sept.) *Adam* 15395*b*! Ouagadougou (Oct.) *Nongonierma* 209! **Ghana:** Tumu Dam (Mar.) *Morton* GC 8856! Banda (Aug.) *Hall* 1981! Ejura Scarp (Dec.) *Morton* GC 9574! Kwahu Tafo (Dec.) *Harris*! Achimota (July) *Adams* 4894! **Togo Rep.:** Agbelonde (Apr., Sept.) *Mahoux* 119! 2024! Lamakara (June) *Mahoux* 2051*b*! **Dah.:** Ouèré (Nov.) *Annet* 49! **N. Nig.:** Nupe *Barter* 786! Jebba (Dec.) *Haines* 67! Minna (Dec.) *Meikle* 724! Panshanu, Bauchi Prov. (Aug.) *Lawlor & Hall* 563! Jimeta, Adamawa Prov. (Nov.) *Hepper* 1213! **S. Nig.:** Badagry (Aug.) *Onochie* FHI 33495! Lagos *W. Macgregor* 24! Olokemeji F.R. to Eruwa (Aug.) *Keay* FHI 37165! Okelifi, Ondo Dist. (Nov.) *Onochie* FHI 34221! Igboora, Ondo Dist. (Oct.) *Haines* 215! Also in the Congo, Madagascar, India and Thailand.

15. RHYNCHOSPORA Vahl, Enum. 2: 229 (1806) *nom. et orth. cons.*; F.T.A. 8: 478 (1902); Kükenthal in Engl., Bot. Jahrb. 74: 375 (1949), 75: 90 (1950), 75: 273 (1951).

By Miss S. S. Hooper

Spikelets sessile and crowded together into one or more hemispherical or spherical heads:
Inflorescence a single, terminal, hemispherical head, surrounded by stiff, erect or spreading bracts; caespitose annual or perennial .. .. 1. *rubra*
Bristles shorter than, or equal to, the achene; beak short 1a. *rubra* subsp. *africana*
Bristles longer than the very hispidulous achene; beak long 1b. *rubra* subsp. *senegalensis*
Inflorescence of one to a dozen (generally 3–4) spherical or subspherical greenish-brown heads; rhizomatous perennial .. .. .. 2. *holoschoenoides*
Spikelets distinct and more or less clearly pedicellate:
Inflorescence a single, simple, terminal corymb of 3–10 large white or whitish spikelets; perennial with a narrow, creeping, scaly rhizome; stigmas 2 .. 5. *candida*
Inflorescence not as above:
Robust, leafy perennial with a branched inflorescence and leafy bracts; spikelets 8–10 mm. long; style undivided:
Spikelets in large clusters on stout, spreading pedicels; sheaths tough, leaves distinctly keeled, often more than 1 cm. wide .. .. 3. *corymbosa*
Spikelets 1 to a few together in small clusters on long, slender, erect pedicels; sheaths soft, rounded on the back; leaves less than 1 cm. wide .. 4. *triflora*
Annual or slender perennial; spikelets up to 8 mm. long:
Spikelets solitary, pedicellate, 5–8 mm. long; inflorescence confined to upper part of stem; achene about as long as broad, rugulose, style-base tissue extending down the sides of the achene:
Pedicels stout, erect or spreading; spikelets ovate, acuminate; rhachilla straight; leaves up to 4 mm. wide, flat .. .. 6. *eximia*
Pedicels slender, capillary; spikelets elliptical; rhachilla strongly flexuose; leaves up to 1·5 mm. wide, canaliculate 7a. *gracillima* subsp. *subquadrata*
Spikelets 2–5 mm. long, in small clusters:
Spikelets dark brown; partial inflorescence rather narrow; style-base triangular, not decurrent on the achene .. .. 8. *rugosa*
Spikelets pale straw-coloured or yellow-brown; inflorescence generally occupying more than half of stem:
Spikelets straw-coloured, very small, 2–2·5 mm. long with only one achene; style-base lunate, upper edge of achene with two distinct humps 11a. *tenerrima* subsp. *microcarpa*
Spikelets 2·5–5 mm. long; style-base not lunate:
Style-base small, inconspicuous; achene surface not rugulose, generally with two lateral basal brown patches and a central dark stripe; spikelets 2·5–3·5 mm. long.. .. .. 9. *brevirostris*
Style-base apiculate, decurrent on the rugulose achene, without basal patches; spikelets 4–5 mm. long .. .. 10. *perrieri*

1. **R. rubra** (*Lour.*) *Makino* in Bot. Mag. Tokyo 17: 180 (1903). *R. rubra* subsp. *rubra* does not occur in West Africa.

1a. **R. rubra** subsp. **africana** *J. Raynal* in J. & A. Raynal in Adansonia, sér. 2, 7: 323 (1967). *R. wallichiana* of F.T.A. 8: 478, and of F.W.T.A., ed. 1, 2: 470. *R. minor* of Nelmes in Kew Bull. 11: 533 (1957); Berhaut, Fl. Sén. ed. 2, 368 (1967); not of (Nees) Schnee (1944). *R. parva* of Aké Assi, Contrib. 258 (1963). Tufted plant about 15–50 cm. high, with linear grassy leaves and a small yellowish inflorescence; moist savanna or grassland, often on sand.
Sen.: Lyndiane (Sept.) *Adam* 12487. Toubakouta, Casamance (Jan.) *J. Raynal* 7933. **S. L.:** Waterloo (Aug.) *Melville & Hooker* 296! Manoh (Apr.) *Dinklage* 2516! Mano Salija (Nov.) *Deighton* 341! **Lib.:** Robertsport, Grand Cape Mount Dist. (Dec.) *Baldwin* 10919! Monrovia (Aug.) *Baldwin* 9175! Paynesville (Sept.) *Leeuwenberg & Voorhoeve* 4904! Buchanan (Dec.) *Adam* 16426! Grand Bassa (July) *T. Vogel* 105! **Iv. C.:** Moossou (Mar.) *J. & A. Raynal* 13579! *Aké Assi* 5622! Grand Bassam (Sept.) *de Wit* 7377! 7387! **Ghana:** Eikwe (Mar.) *Rose Innes* GC 30883! Essiama (May, Oct.) *Morton* A2155! 8534! *Hall* 2665! **Dah.:** Sémé, nr. Cotonou (Feb.) *J. & A. Raynal* 13545! Also occurring in Rio Muni and Gabon and in coastal regions of SE. Africa.

1b. **R. rubra** subsp. **senegalensis** *J. Raynal* in J. & A. Raynal in Adansonia sér. 2, 7: 323, pl. 4 (1967). Resembling subsp. *africana* but perhaps shorter; brackish dune slacks.
Sen.: Fatick (Dec.) *J. & A. Raynal* 6712! (Oct.) *J. & A. Raynal* 7742. Lyndiane (Oct.) *Adam* 11020. Also in Madagascar, widely in tropical Asia, Polynesia and northern Australia.

2. **R. holoschoenoides** (*L. C. Rich.*) *Herter* in Revis. Sudamer. Bot. 9: 157 (1953); Berhaut, Fl. Sén. ed. 2, 365 (1967). *Schoenus holoschoenoides* L. C. Rich. in Act. Soc. Hist. Nat. Paris 1: 106 (1792). *Rhynchospora cyperoides* (Swartz) Mart. in Denkschr. Acad. Wiss. Muench. 6: 149 (1816–17); F.T.A. 8: 479; F.W.T.A., ed. 1, 2: 470; Berhaut, Fl. Sén. ed. 1, 218 (1954). *R. mauritii* Steud., Syn. Pl. Glum. 2: 149 (1855); Robinson in Kirkia 1: 34 (1961); Aké Assi, Contrib. 257 (1964). A generally stout, erect plant often about 1 m. high, but sometimes much dwarfed, with an inflorescence somewhat resembling *Scirpus holoschoenus* Linn.; in marshes, swamps and damp grassland, generally near surface water which may be brackish.
Sen.: Malika to Yombeul (Nov.) *Pitot*! Déni Biram Ndao, L. Retba (July) *J. & A. Raynal* 6086! 6670. Niakoulourab (Dec.) *Chev.* 2451! Niassia, Casamance (Nov.) *Berhaut* 6653! **Guin.:** fide *Aké Assi* l.c. **S. L.:** York (Nov.) *Morton & Gledhill* SL 1453*a*! Samarank, Rokel R. (Mar.) *Glanville* 210! Rhombe,

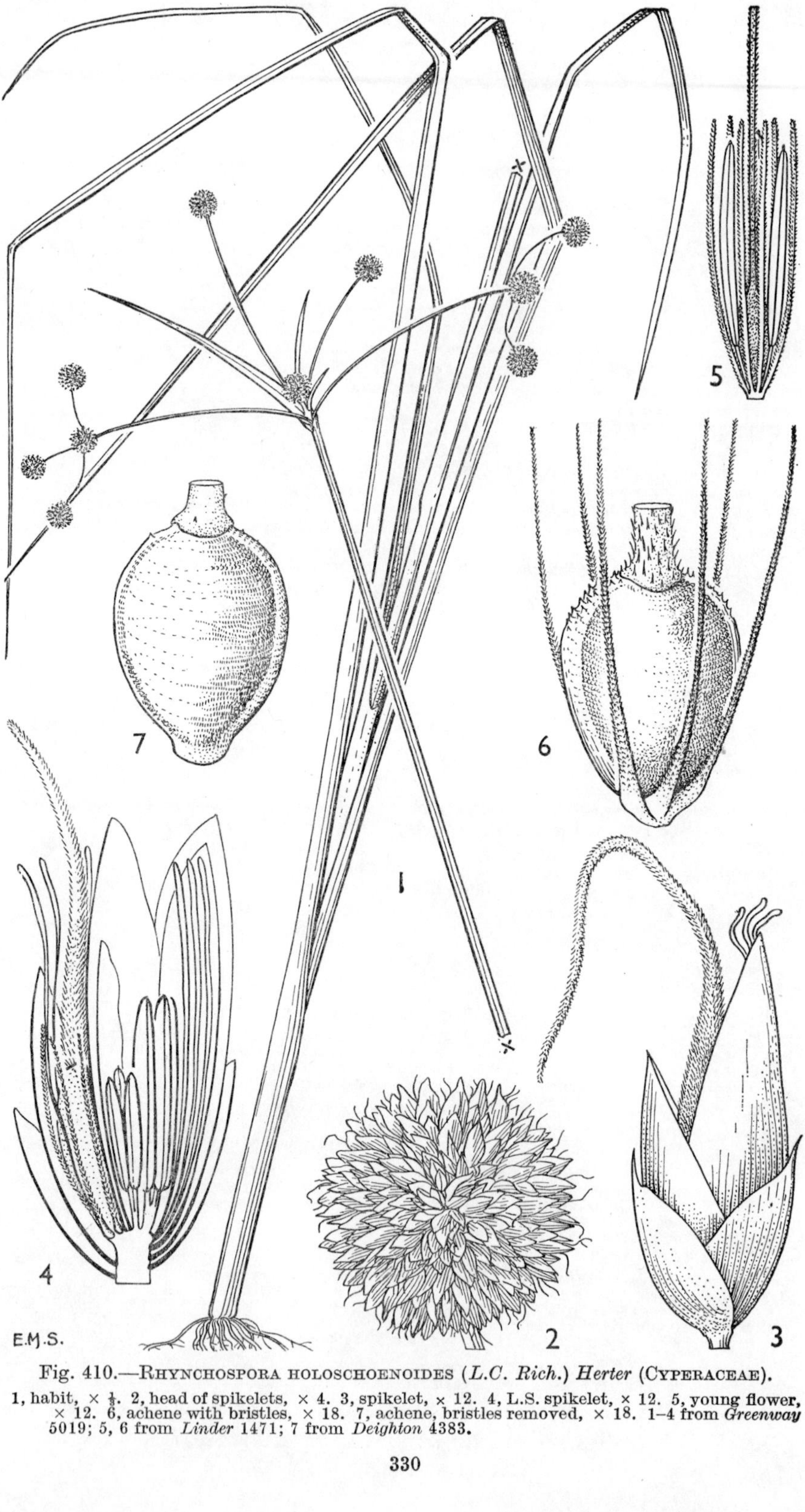

Fig. 410.—RHYNCHOSPORA HOLOSCHOENOIDES (*L.C. Rich.*) *Herter* (CYPERACEAE).

1, habit, × ⅓. 2, head of spikelets, × 4. 3, spikelet, × 12. 4, L.S. spikelet, × 12. 5, young flower, × 12. 6, achene with bristles, × 18. 7, achene, bristles removed, × 18. 1–4 from *Greenway* 5019; 5, 6 from *Linder* 1471; 7 from *Deighton* 4383.

Loko, Massama (July) *D. R. E. Jackson* in *Jordan* 478! Rosino, Samu (Oct.) *Jordan* 795! Mano Salija (Dec.) *Deighton* 283! **Lib.:** Duport, Montserrado Dist. (Nov.) *Linder* 1471! Monrovia (Dec.) *Baldwin* 10499! Johnsonville, Montserrado Dist. (Feb.) *Dinklage* 2966! Upper Mesurado Swamp, Monrovia (Sept.) *Künkel* 57! Sasstown, Sinoe Dist. (Mar.) *Baldwin* 11608! **Iv. C.:** Audouin, nr. Abidjan (Mar.) *J. & A. Raynal* 13635! Grand Bassam (Sept.) *de Wit* 7401! Grand Bassam to Abidjan (July) *de Wilde* 619! Assinie (Aug.) *Nozeran*! **Ghana:** Beyin (Feb.) *Hall* 2919! Atwabo *Fishlock* 35 of 1931! Essiama (Feb.–Oct.) *Morton* A1627! A2149! *Hall* 2663! **Niger:** Bangaza, nr. Dungass (Dec.) *Dumont* in *P. de Fabrègues* 2660! Throughout tropical and S. Africa, in Madagascar and tropical America.

3. **R. corymbosa** (*Linn.*) *Britt.* in Trans. New York Acad. Sci. 11: 84 (1892); F.T.A. 8: 480; F.W.T.A., ed. 1, 2: 468, partly; Kük. in Engl., Bot. Jahrb. 74: 410 (1949); Berhaut, Fl. Sén. ed. 2, 359; Robinson in Kirkia 1: 35 (1961). *Scirpus corymbosus* Linn., Amoen. Acad. 4: 303 (1760). *Rhynchospora aurea* Vahl, Enum. 2: 229 (1806). *R. corymbosa* var. *asperula* (Nees) Kük. l.c. 416 (1949) and var. *grandispiculosa* Kük., l.c. 417 (1949). A robust erect sedge with rasp-edged leaves and a large inflorescence of brown spikelets in flattish heads; swamps, sometimes in or near water in high forest.
Sen.: Kayor *Leprieur*! Linguière *Roberty*! Koulountou (Apr.) *Adam* 17981! Casamance (Apr.) *Fidèle*! Badi to Fourou (Apr.) *Nongonierma* 443! **Mali:** Kabala, E. Bamako (July) *Roberty* 2447! Farakoba Valley, Bamako (Jan.) *Duong* 2081! Dendela (Mar.) *Chev.* 627! Koutiala (Feb.) *Roberty* 3403! **Gam.:** Albreda (June) *Heudelot* 90! Cahout *Heudelot* 343! **Guin.:** Conakry *Maclaud*! Mamou (Mar.) *Dalz.* 8431! Kouroussa (Aug.) *Pobéguin* 402! Macenta (Jan.) *Schnell* 131! Nzérékoré (Oct.) *Baldwin* 9710! **S. L.:** Regent (Dec.) *Sc. Elliot* 3861! Ronietta (Nov.) *Thomas* 5600! Rokupr (Sept.) *Jordan* 114! Koinadugu (Aug.) *Haswell* 244! N. Gola Forest (Feb.) *Bakshi* 43! **Lib.:** Kailahun, Kolahun Dist. (Nov.) *Baldwin* 10123! Zigida, Vonjama Dist. (Oct.) *Baldwin* 10001! Timbo, Grand Bassa Dist. (Mar.) *Baldwin* 11218! Yah R., Sanokwele Dist. (Sept.) *Adames* 547! Mnanulu, Webo Dist. (June) *Baldwin* 6032! **Iv. C.:** Dinderesso (Nov.) *Néa* 138! Man (Aug.) *Boughey* GC 18411! Bliéron, Cavally R. mouth (Aug.) *Chev.* 19913! Alépé (Feb.–Mar.) *Chev.* 17483! Port Bouet (Oct.) *Nongonierma*! **U. Volta:** Bekoni (Dec.) *Bille* 3359! **Ghana:** Pamu-Berekum F.R. (Sept.) *Vigne* 2488! Bimbila (Mar.) *Hepper & Morton* A3105! Koforidua (Nov.) *Morton* 6088! Kumasi (Feb.) *Darko* 515! Nungwa (May) *Irvine* 4841! **Togo Rep.:** Agoué *Ménager*! Sokodé (Aug.) *Mahoux* 2137! **Dah.:** Porto Novo (Jan.) *Chev.* 22715! **N. Nig.:** *Thornewill*! Jos Plateau (Dec.) *Monod* 9701! Vom (Feb.) *McClintock* 225! Kurra, Pankshin Dist. (June) *Okafor* FHI 59250! Nguroje, Mambila Dist. (Jan.) *Hepper* 1741! **S. Nig.:** Lagos (June) *Dalz.* 1306! Ibadan (Nov.) *Newberry & Etim* 174! Ondo (Feb.) *Jones & Onochie* FHI 16993! Iva Valley F.R., Enugu (July) *Daramola* FHI 55152! Calabar (June) *Tuley* 68! **W. Cam.:** Cameroon R. (Jan.) *Mann* 776! Mbiami, 6,000 ft. (June) *Brunt* 753! Nsop, Mbaw Plain (May) *Brunt* 488! Bamenda (Feb.) *Daramola* FHI 40494! Jakiri, Bamenda, 6,500 ft. (Feb.) *Hepper* 2075! Pantropical.

4. **R. triflora** *Vahl* Enum. Pl. 2: 232 (1806); Kük. in Engl., Bot. Jahrb. 74: 426 (1949); Robinson in Kirkia 1: 35 (1961); Aké Assi, Contrib. 258, t. 27 (1964); Berhaut, Fl. Sén. ed. 2, 359 (1967). *R. corymbosa* of F.W.T.A., ed. 1: 468, partly. More slender than the previous species, with the spikelets fewer and very acute; marshy grassland.
Sen.: Niokolo-Koba (Dec.) *J. & A. Raynal* 6885! Casamance (Aug.–Nov.) *Adam* 18507! *Berhaut* 6331! 6457! **Mali:** Bamako (Nov.) *Duong* 2116! Sotuba *Adam* 15333! (Dec.) *Duong*! Kléla (Oct.) *Demange* 2722! **Guin.:** Sambaïlo (Oct.) *Pitot*! Fouta Djalon *Desmeules* 12! **S. L.:** Maforki, nr. Mankara (Oct.) *Jordan* 813! Mange, Bure Makonte (Oct., Dec.) *Jordan* 586! 732! Materboi, Karene Dist. (July) *Glanville* 217! Medina (July) *Deighton* 4339! **Iv. C.:** Sifié (Oct.) *Aké Assi* 6057! Grand Lahou (Jan.) *Aké Assi* 4775! Tiégouakro to Kodio Koffi (Aug.) *Chev.* 22339! Baoulé Dist. *Pobéguin* 108! **Ghana:** Wa to Lawra (Sept.) *Hall* CC 611! Bungari, Banda Dist. (Dec.) *Morton* 25077! Ejura (Aug.) *Hall* CC 302! Ayafie to Atafie (Dec.) *Adams* 4674! Bimbila (Dec.) *Morton* GC 6274! **N. Nig.:** Nupe *Barter* 1564! **S. Nig.:** Uromi to Auchi, Benin (July) *Lowe* 1758! Also in the C. African Rep., E. Cameroun, Chad and intermittently throughout the tropics.

5. **R. candida** (*Nees*) *Boeck.* in Linnaea 37: 605 (1873); F.T.A. 8: 481; Kük. in Engl., Bot. Jahrb. 75: 175 (1950); Berhaut, Fl. Sén. ed. 2, 356 (1967); Robinson in Kirkia 1: 34 (1961). *Psilocarya candida* Nees in Mart., Fl. Bras. 2, 1: 117 (1842). Slender plant, about 60 cm. high, with distinctive large white spikelets; in grassland liable to flooding.
Sen.: Badi, Gambia R. Basin (Jan.) *Berhaut* 1479! **Mali:** *Duong*! Klela (May) *Demange* 2234! **Port. G.:** Contubo el Canhamine (Sept.) *Esp. Santo* 2758! Ingore to Barro (Aug.) *Esp. Santo* 3085! **Guin.:** Friguiagbé (Dec.) *Chillou* 1096! Kindia (Oct.) *Pobéguin* 1366! Labé to Guessèwolé *Adam* 14645! Kouroussa (Apr.) *Pobéguin* 698! Kissidougou (July) *Martine* 3312! **S. L.:** Bakelimre (June) *Adames* 232! Kichom (July) *Adames* 213! Mateboi (June) *Adames* 267! Dimbaia (Sept.) *Adames* 249! Kenema (June) *Jordan* 2125! **Lib.:** Robertsport, Grand Cape Mount Dist. (Dec.) *Baldwin* 10962! **Iv. C.:** Sémien to Kanébly (July) *Aké Assi* 9072! **U. Volta:** Bouassi to Kouéré, nr. Banfora (July) *Scholz* 313! **Dah.:** Porto Novo (Sept.) *Adjanohoun* 351! **N. Nig.:** Maska Dam, Zaria Prov. (May) *Kershaw* 900319! Samaru Dam (July) *Oche* MRK 21! **S. Nig.:** Lagos *Barter* 20178! Scattered throughout tropical Africa, the Mascarene Islands and tropical S. America.

6. **R. eximia** (*Nees*) *Boeck.* in Linnaea 37: 601 (1873); Kük. in Engl., Bot. Jahrb. 75: 181 (1950); Berhaut, Fl. Sén. ed. 2, 359 (1967). *Spermodon eximius* Nees in Seem., Voy. Herald Bot. 222 (1857). *Rhynchospora schroederi* K. Schum. ex C.B. Cl. in Engl., Bot. Jahrb. 38: 135 (1906); F.W.T.A., ed. 1, 2: 470. *R. testui* Cherm. var. *pleiantha* Cherm. in Bull. Soc. Bot. Fr. 81: 267 (1934); Berhaut, Fl. Sén. ed. 1, 215 (1954). Leafy annual with brown spikelets on long but rather robust pedicels; marshes and wet flushes.
Sen.: Badi (Dec.) *Berhaut* 1810! Niokolo-Koba (Oct., Nov.) *Adam* 15716! 15838! 17094! Kanémére (Sept.) *Fotius* K443! **Mali:** (Apr.) *Duong*! Kita Mts. (Oct.) *Jaeger* 54! 80! Sotuba (Oct.) *J. Raynal* 5008 *ter*! 5029! **Guin.:** Télimélé to Pita (Oct.) *Pitot*! Ditinn (Oct.) *Pitot*! Macenta to Kérouané *Adam* 7035! Koundara (Dec.) *Berhaut* 4180! Kouroussa (Dec.) *Pobéguin* 541*b*! **S. L.:** Kambia (Dec.) *Deighton* 818! Rokupr (Nov.) *Jordan* 175! Mabonto to Bunbuna (Aug.) *Deighton* 1401! Jokibu (Nov.) *Deighton* 3777! Kenema (Nov.) *Deighton* 5875! **Iv. C.:** *Bégué* 91! Séguélo (Oct.) *Aké Assi* 6612! Dabakala to Toupé (Nov.) *Aké Assi* 9245! **Ghana:** Burufo (Dec.) *Adams & Akpabla* 4392! Tamale (Dec.) *Adams & Akpabla* 4167! Yapei Bridge (Nov.) *Hall & Enti* GC 35735! **Togo Rep.:** Sokodé to Bassari *Schroeder* 147! Bassari (Aug.) *Mahoux* 2179! **N. Nig.:** Kabba (Oct.) *Parsons* 27! Kufena Hill, Zaria Dist. (Sept.) *Clayton* 1304! Kaduna to Keffi, Zaria Prov. (Aug.) *Kershaw* 900708! Panshanu Pass, Bauchi Prov. (Aug.) *J. Hall* 505! Also in E. Cameroun, C. African Rep., Malawi and tropical America.

7. **R. gracillima** *Thw.* Enum. Pl. Zeyl. 435 (1864). *R. gracillima* subsp. *gracillima* occurs in Asia.

7a. **R. gracillima** *Thw.* subsp. **subquadrata** (*Cherm.*) *J. Raynal* in Adansonia sér. 2, 7: 321 (1967); Berhaut, Fl. Sén. ed. 2, 360 (1967). *R. subquadrata* Cherm. in Bull. Soc. Bot. Fr. 69: 720 (1922); Kük. in Engl., Bot. Jahrb. 75: 274 (1950); Robinson in Kirkia 1: 40 (1961). *R. testui* Cherm. in Arch. Bot. Caen 4, Mém. 7: 42 (1931). Slender erect plants to 80 cm. high, characterized by the slender spreading branches of the inflorescence; edges of streams, lakes and other marshy areas.
Sen.: Niokolo-Koba (Dec.) *J. & A. Raynal* 6916! (Nov.) *Adam* 17091! Badi (Dec.–Jan.) *Berhaut* 1171! 4740! **Mali:** Sotuba (Dec.) *Adam* 1438! **Guin.:** Madina Tossekré (Oct.) *Adam* 12541! **Lib.:** (Sept.) *Harley*! Upper Mesurado Swamp, nr. Monrovia (Sept.) *Künkel* 58! Buchanan, Grand Bassa Dist. (Nov.) *Adam* 16016! **Iv. C.:** Moossou (Nov.) *Bodard*! (Jan.) *Aké Assi* 6888! Grand Bassam (Sept.) *de Wit* 7334! Baoulé Nord, Manikro to Tiégouakro (Aug.) *Chev.* 22316! **U. Volta:** Ouagadougou (Oct.) *Charreau* in *Hb. Adam* 12742! **Ghana:** Burufo, nr. Lawra (Sept.) *Hall* CC 757! Sakogu (Aug.) *Hall* CC 457! Kintampo (Sept.) *Hall* CC 913! Ejura (Dec.) *Morton* GC 9487! Essiama (May) *Morton* 2151! **Togo Rep.:** Lama-Kara to Sokodé (July) *Mahoux* 2114! Bafilo Sokodé (Aug.) *Mahoux* 2155! **N. Nig.:** Jos (Aug., Dec.) *Haines* 97! *J. Hall* 602! *Lely* P481! Also in E. Cameroun, the Congo, Uganda, Tanzania, Zambia, Rhodesia and Madagascar.

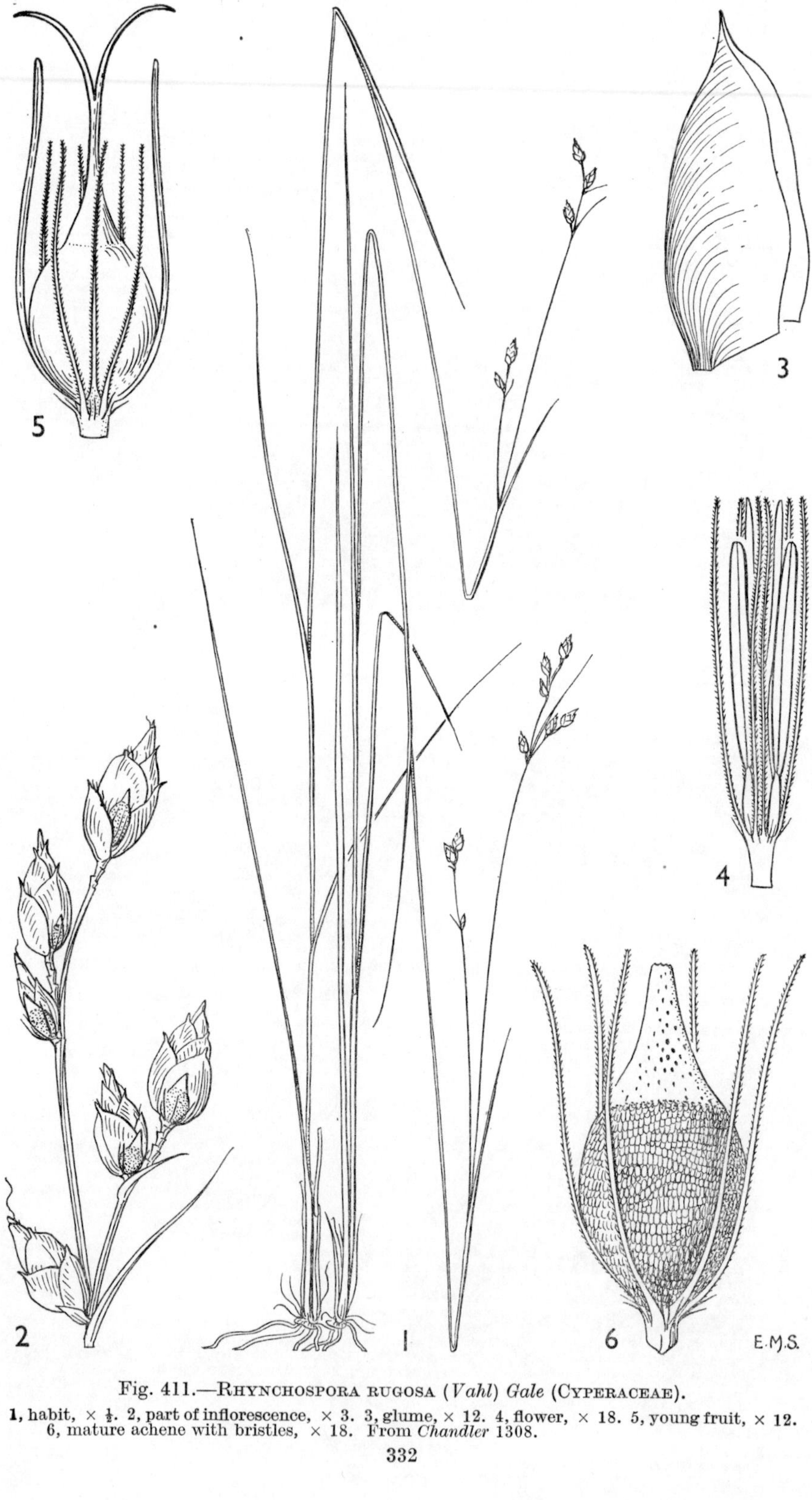

Fig. 411.—RHYNCHOSPORA RUGOSA (*Vahl*) *Gale* (CYPERACEAE).

1, habit, × ⅓. 2, part of inflorescence, × 3. 3, glume, × 12. 4, flower, × 18. 5, young fruit, × 12. 6, mature achene with bristles, × 18. From *Chandler* 1308.

8. **R. rugosa** (*Vahl*) *Gale* in Rhodora 46: 275 (1944). *Schoenus rugosus* Vahl, Ecl. Amer. 2: 5 (1798). *Rhynchospora glauca* Vahl, Enum. 2: 233 (1806); F.T.A. 8: 482. *R. brownii* Roem. & Schult., Syst. Veg. 2: 86 (1817); Robinson in Kirkia 1: 36 (1961). *R. africana* Cherm. in Arch. Bot. Caen 4, Mém. 7: 44 (1931). *R. glauca* Vahl var. *juncea* (Kunth) Cherm. of Aké Assi, Contrib. 257 (1964). Erect plant, up to 80 cm. high, with elliptical, brown spikelets about 4 mm. long; marshes or near streams.
Guin.: Labé *Adam* 14563! **Iv. C.:** Moossou (Jan.) *Aké Assi* 6881! Grand Bassam (Sept.) *de Wit* 7376! **N. Nig.:** Kakara, Mambila Dist. 4,800 ft. (Jan.) *Hepper* 1792! **S. Nig.:** Obudu, 5,000 ft. (Apr.) *Haines* 231! **W. Cam.:** Nkambe, 5,000 ft. (Feb.) *Brunt* 992! Mbiami, 6,000 ft. (June) *Brunt* 752! Pantropical.
[The pattern of variation within this aggregate species must be further studied before the West African taxa can be properly delineated. From the material cited above *Adam* 14563 stands out as having a dark achene with bristles as long as both achene and style-base together and shows other features of *R. africana* Cherm. The highland specimens from Nigeria and the W. Cameroon have 5–6 bristles shorter than the achene and may belong to the Australian taxon *R. brownii* Roem. & Schult.]

9. **R. brevirostris** *Griseb.* Cat. Pl. Cuba 246 (1866); Kük. in Engl., Bot. Jahrb. 75: 286 (1950); Robinson in Kirkia 1: 39 (1961); Berhaut, Fl. Sén. ed. 2, 360 (1967). *R. barteri* C.B. Cl. in F.T.A. 8: 482 (1902); F.W.T.A., ed. 1, 2: 470; Aké Assi, Contrib. 2: 257 (1964). Slender annual, with stems about 15 cm. high having small clusters of spikelets along their length; rice-fields and other damp places, not common.
Sen.: Niokolo-Koba *Adam* 15946! (Dec.) *J. & A. Raynal* 6884! Kanémére (Nov.) *Fotius* K692! Casamance (Oct.) *Adam* 18530! (Dec.) *Berhaut* 6701! **Mali:** (Dec.) *Schnell* 2354 *bis*! Sotuba (Nov.) *J. Raynal* 5133! **Guin.:** *Scaëtta* 3086! 3087*b*! **S. L.:** Mapotolon (Nov.) *T. S. Jones* 59! Kitchom (Jan.) *Deighton* 945! **N. Nig.:** Nupe *Barter* 1010! Gurum, Adamawa Prov. (Nov.) *Hepper* 1247! Also in E. Cameroun, C. African Rep., Middle Congo, SE. Africa and tropical America.

10. **R. perrieri** *Cherm.* in Bull. Soc. Bot. Fr. 69: 721 (1922); Kük. in Engl., Bot. Jahrb. 75: 279 (1950); Robinson in Kirkia 1: 40 (1961); Berhaut, Fl. Sén. ed. 2, 360 (1967); J. & A. Raynal in Adansonia, sér. 2: 322 (1967). *R. deightonii* Hutch. in F.W.T.A., ed. 1, 2: 468 (1936), name only. Similar to the preceding species but generally taller, up to 50 cm. and of more frequent occurrence; edge of swamps.
Sen.: Kolda, Casamance (Jan.) *J. Raynal* 7836! (Feb.) *Berhaut* 6881! Toubakouta *J. Raynal* 7932. **Guin.:** Friguiagbé (Nov.) *Chillou* 863! Hollandé Tossékré (Oct.) *Adam* 12725*b*! **S. L.:** Newton (Nov.) *Deighton* 1449! Rokupr (Nov.) *Jordan* 675! Mandu (Feb.) *Morton* SL 845! Malema (Nov.) *Deighton* 326! Kenema (Nov.) *Deighton* 5876! **Lib.:** Monrovia (Feb.) *Dinklage* 2758! Buchanan (Nov.) *Adam* 16010! Bushrod I. (Nov.) *Winne* 545! Ganta (Nov.) *Harley* 1709! Nimba Mts. (Jan.) *Adam* 20696! **Iv. C.:** Moossou (Oct.) *Aké Assi* 4407! **U. Volta:** Ouagadougou (Oct.) *Adam* 12742! **Togo Rep.:** Bafilo Sokodé (Aug.) *Mahoux* 2159! **S. Nig.:** Uke to Nobi, Onitsha Dist. (May) *Onochie* FHI 35797! Also in E. Cameroun, C. African Rep., Kenya, Pemba, Zambia, Rhodesia and Madagascar.

11. **R. tenerrima** *Nees ex Spreng.* Cur. Post. Syst. Veg. 26 (1827); F.W.T.A., ed. 1, 2: 470. In Central and tropical S. America.

11a. **R. tenerrima** subsp. **microcarpa** *J. Raynal* in J. & A. Raynal in Adansonia, sér. 2, 7: 325, fig. 5, 2 (1967). Slender tufted plant up to 20 cm. high, with filiform stems and leaves; rice-fields, marshy grassland and sometimes wet rock-faces.
Sen.: Kolda (Jan.) *J. Raynal* 7830. **Guin.:** Is. Tristao (Nov.) *Jac.-Fél* 7325! Friguiagbé *Chillou* 598*b*! Telimélé (Oct.) *Pitot*! Telimélé to Pita (Oct.) *Pitot*! Conakry (Oct.) *Adam* 12600! Dalaba (Oct.) *Scaëtta* 3087! 3090! **S. L.:** Waterloo (Aug.) *Melville & Hooker* 295! 302! Sugar Loaf Mt. (Nov.) *Deighton* 5620! Materboi (Oct.) *Glanville* 24! **Lib.:** Grand Bassa (Nov.) *Dinklage* 2304! Buchanan (Nov.) *Adam* 16057!

Excluded and doubtful species

R. senegalensis *Steud.*—F.W.T.A., ed. 1, 2: 470 = *Fuirena stricta* Steud.
R. micrantha *Vahl* Enum. 2: 231 (1806); F.T.A. 8: 481; F.W.T.A., ed. 1, 2: 470. "In montosis Guineae 1786 *Isert*". This Central American species has not otherwise been recorded from West Africa.

16. CLADIUM P.Br., Hist. Jam. 114 (1756); F.T.A. 8: 484 (1902).

By Miss S. S. Hooper

Erect robust sedge to 1·60 m. high with creeping rhizomes, rounded leafy stems and saw-edged leaves and bracts; inflorescence an interrupted compressed panicle with numerous slender rays bearing clusters of small red-brown ovoid spikelets; achene broadly ovate, 2·5 mm. long, rugulose above *mariscus* subsp. *jamaicense*

C. mariscus (*Linn.*) *Pohl* subsp. **jamaicense** (Crantz) Kük. in Peter in Fedde, Rep. Beih. 40, 1: 523 (1938), and in Fedde, Rep. 51: 189 (1942). *C. jamaicense* Crantz, Instit. 1: 362 (1766); A. Raynal in Ann. Fac. Sci. Univ. Dakar 9: 167 (1963). *C. mariscus* of Adam in Not. Syst. 16: 317 (1960); Berhaut, Fl. Sén. ed. 2, 360 (1967). Characterized by the very sharp serrated edges of the tough leaves and bracts; marshy woodland and dune slacks of the Cape Verde Peninsula.
Sen.: Dakar *Adam* 17635! 17679! 17699! M'bidjem (June) *J. & A. Raynal* 5982! Mbeubeussé (May) *Naegelé*! Also in Cape Verde Islands, Congo, Socotra, Kenya, Uganda and to S. Africa, and the Old and New World tropics; subsp. **mariscus** occurs in temperate regions.

17. MAPANIA Aubl., Pl. Guin. Franc. 1: 47 t. 17 (1775); F.T.A. 8: 489 (1902).

By Miss S. S. Hooper

Inflorescence on an axillary peduncle which bears reduced leaves only, and subtended by short scarious bracts:
- Leaf-blade oblong, 40–85 mm. broad, suddenly narrowed into a pseudopetiole below, obtuse and cuspidate above; sheath inflated, red-purple within 1. *baldwinii*
- Leaf-blade linear, 10–45 mm. broad, gradually narrowed below, acute to acuminate; sheath not inflated:
  - Stigmas 2; leaf-tip asymmetrical, acute; inflorescence straw-coloured, spikes more or less confluent .. .. 2. *linderi*
  - Stigmas 3; leaf-tip symmetrical, acuminate; inflorescence brown, spikes discrete, acute, in heads up to 2 cm. across .. .. 3. *mangenotiana*

Inflorescence on a terminal peduncle bearing normal leaves below, and subtended by leafy bracts:
- Triangular pseudopetiole present above sheath .. .. .. 4. *amplivaginata*

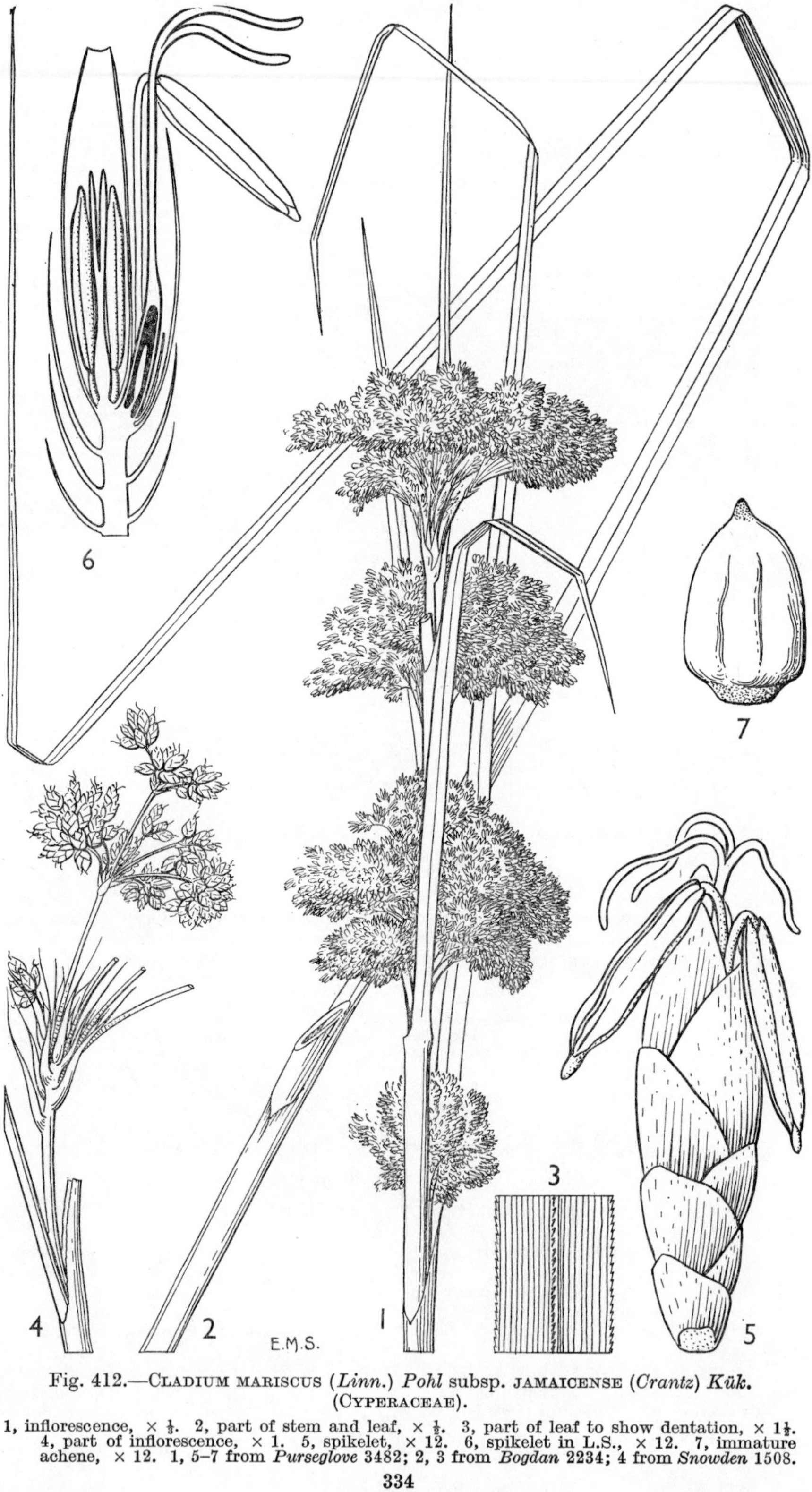

Fig. 412.—CLADIUM MARISCUS (*Linn.*) *Pohl* subsp. JAMAICENSE (*Crantz*) *Kük.* (CYPERACEAE).

1, inflorescence, × ⅓. 2, part of stem and leaf, × ½. 3, part of leaf to show dentation, × 1½. 4, part of inflorescence, × 1. 5, spikelet, × 12. 6, spikelet in L.S., × 12. 7, immature achene, × 12. 1, 5–7 from *Purseglove* 3482; 2, 3 from *Bogdan* 2234; 4 from *Snowden* 1508.

Leaf-blade narrowed above sheath but not forming a triangular pseudopetiole:
Achene with a surface reticulum of raised lines:
Spikes more or less confluent into a hemispherical inflorescence; peduncles enlarged beneath head; leaves tinged red beneath; glumes 7–10 mm. long; achene with simple longitudinal ribs .. .. 5. *coriandrum*
Spikes discrete in a spherical inflorescence; peduncle not enlarged; glumes 4–5 mm. long; achene with branched longitudinal ribs .. .. 6. *minor*
Achene with a tuberculate surface:
Leaves and bracts suddenly narrowed into a triangular tip 7. *ivorensis*
Leaves and bracts gradually acuminate:
Glumes brown; style-base not persistent .. .. .. 8. *macrantha*
Glumes white; style-base yellowish, conical and persistent 9. *rhynchocarpa*

1. **M. baldwinii** *Nelmes* in Kew Bull. 6: 420 (1951); Nelmes & Baldwin in Amer. J. Bot. 39: 369 (1952). *M. comoensis* of F.W.T.A., ed. 1, 2: 471 (1936), English descr. only. Readily distinguished from other species by a combination of pale axillary inflorescences, oblong leaves and inflated sheaths; in forest.
Lib.: Yéképa, Nimba Reserve (Feb.) *Adam* 20906! Kitoma, Sanokwele Dist. (Mar.) *Adam* 16702! Bilimu Mt., Sanokwele Dist. (Dec.) *Harley* 1994! Monroviatown, Tchien Dist. (Aug.) *Baldwin* 7097! Duoh, Sinoe Dist. (Mar.) *Baldwin* 11341! **Iv. C.:** Troya, Taï to Tabou (Mar.) *de Wilde* 3552! Taï (Jan.) *Lorougnon* 1265! 10 km. W. of Tapéguhé, Sassandra R. (June) *Leeuwenberg* 4520! Guitry to Divo (Dec.) *Boughey* GC 13564! Yapo forest, Agboville (Oct.) *Mangenot* "No. 5"! Malamalasso, Bas Comoé (Mar.) *Chev.* 17498! **Ghana:** Ankasa F.R. (July) *Hall & Enti* GC 35522! Afao Hills F.R. (Dec.) *Adams* 1959! Headwaters of Ochie R., Juaso *West-Skinn* 8! Kakum F.R. (Jan.) *Hall* 1698! Atewa Range F.R. (Jan.) *Enti & Hall* GC 36356!
2. **M. linderi** *Hutch. ex Nelmes* in Kew Bull. 6: 421 (1951); Nelmes & Baldwin in Amer. J. Bot. 39: 371 (1952). Similar to *M. baldwinii* in having several slender axillary peduncles but differing from that species in the linear leaves; in riverine or montane forest.
Guin.: Ya R., Nimba Mts. (Apr.) *Schnell* 5077! 5111! 5113! 5212! 5218! Lower Goué R. valley (Apr.) *Schnell* 5020! **S. L.:** Kambui Hills (Mar.) *Small* 506! Gola Forest (Apr.) *Small* 631! **Lib.:** Wohmen, Vongama Dist. (Oct.) *Baldwin* 10072! Ganta, Sanokwele Dist. (July) *Harley* 1233! Gbanga (Sept.) *Linder* 586! Yoma, Montserrado Dist. (Aug.) *Leeuwenberg* 4838! Dieble, Webo Dist. (July) *Baldwin* 6396! **Iv. C.:** Frontier Guin./Iv. C., Man Dist. (Sept.) *Mangenot* "No. 6"! Taï (Feb.) *Lorougnon* 1261! Taï to Tabou (Mar.) *de Wilde* 3528! Yapo Forest, Agneby Dist. (July, Dec.) *de Wilde* 140! Troya (Feb.) *Lorougnon* 1262!
3. **M. mangenotiana** *G. Lorougnon* in Bull. Jard. Bot. Brux. 34: 297, fig. 7 (1964). *M. dolichostachya* of F.W.T.A., ed. 1, 2: 471 (1936). *M. africana* Boeck. subsp. *occidentalis* J. Raynal in Adansonia, sér. 2, 5: 277 (1965). Characterized by small lateral inflorescences and leaves up to 25 mm. wide; in forest.
Lib.: Pah Mountain, Webo Dist. (June) *Baldwin* 6076! **Iv. C.:** Tabou (Feb.) *Lorougnon* 1267! Cavally basin, Grabo Dist. (July) *Chev.* 19697!
4. **M. amplivaginata** *K. Schum.* in Notizbl. Bot. Gart. Berl. 3: 105 (1901). *M. oblonga* C.B. Cl. in F.T.A. 8: 491 (1902); F.W.T.A., ed. 1, 2: 471. *Langevina monosperma* Jac.-Fél. (1947)—F.W.T.A., ed. 2, 3: 55. Distinguished from the other species with a terminal inflorescence by the oblong pseudopetiolate leaves; in forest.
S. Nig.: Kwa Falls, Calabar (Mar.) *Brenan* 9250! Oban (Mar.) *Coombe* 170! Afi F.R. (May) *Jones & Onochie* FHI 18641! **W. Cam.:** Rio del Rey *Johnston*! Banga, S. Bakundu F.R. (Mar.) *Brenan & Onochie* 9421! Kembong F.R., Mamfe Dist. (Mar.) *Onochie et al.* FHI 31182! Also in E. Cameroun, Rio Muni and Gabon.
5. **M. coriandrum** *Nelmes* in Kew Bull. 6: 419 (1951). *M. deistelii* of F.W.T.A., ed. 1, 2: 471, partly. Plant to 1 m. or more with strictly ranked leaves and large bracts ensheathing the lower part of the inflorescence; in high forest.
Guin.: Ya R., Nimba Mts. (Apr.) *Schnell* 5082! **S. L.:** York Pass, Picket Hill (Nov.) *Gledhill* 512! **Lib.:** Nimba Reserve (Aug., Dec.) *Adam* 468! 20119! Kitoma, Sanokwele Dist. (Mar.) *Adam* 16719! 16856! Medina to Bumbuna (Oct.) *Linder* 1306! **Iv. C.:** Tabou (Mar.) *Lorougnon* 1258! Source Banco R. (Jan., Mar.) *Lorougnon* 1256! 1260! Adiopodoumé (Dec.) *Aké Assi* 7164! Malamalasso, Assinie Dist. (May) *Mangenot* "No. 2"! **Ghana:** Esekawkaw F.R., Asuom (May) *Enti* GC 35393! Asin Cocoa Station, Foso *West-Skinn* 50!
6. **M. minor** (*Nelmes*) *J. Raynal* in Adansonia, sér. 2, 5: 278 (1965). *M. macrantha* (Boeck.) Pfeiff. var. *minor* Nelmes in Kew Bull. 6: 420 (1951). Plant to 1 m. or more with an inflorescence about 3 cm. across of many ovate spikes and a long, generally reflexed bract; in forest.
Lib.: Nyakake, Webo Dist. (June) *Baldwin* 6074! **Iv. C.:** Taï (Feb.) *Lorougnon* 1242! Grabo (Jan.) *Lorougnon* 1243! Tabou (Jan.) *Lorougnon* 1245! Sassandra to Gagnoa *Mangenot* "No. 1"!
7. **M. ivorensis** (*J. Raynal*) *J. Raynal* in Adansonia, sér. 2, 8: 415 (1968). *M. macrantha* (Boeck.) Pfeiff. subsp. *ivorensis* J. Raynal in Adansonia, sér. 2, 5: 278 (1965). *M. deistelii* of F.W.T.A., ed. 1, 2: 471 (1936), partly. Characterized by the spherical head in which the spikes are hardly distinguishable and the broad linear leaves; in forest.
Lib.: Bomi Hills, Montserrado Dist. (May) *Harley* 2035! **Iv. C.:** Taï (Jan.) *Lorougnon* 1253! Guitry to Divo (Dec.) *Boughey* GC 13562! Yapo (Mar., Dec.) *Lorougnon* 1247! *J. & A. Raynal* 13619! Malamalasso (May) *Mangenot* "No. 3"!
8. **M. macrantha** (*Boeck.*) *Pfeiffer* in Bot. Archiv. 12: 450 (1925). *Hypolytrum macranthum* Boeck. in Engl., Bot. Jahrb. 5: 507 (1884). *Mapania superba* C.B. Cl. in F.T.A. 8: 491 partly, illegitimate name. Robust plant with bracts about 3 cm. broad and a large brown inflorescence; in rain forest.
S. Nig.: Oban (Mar.) *Richards* 5148! Oban F.R., Calabar (Feb.) *Okafor & Daramola* FHI 36327! Also in E. Cameroun and Gabon.
9. **M. rhynchocarpa** *Lourougnon & J. Raynal* in Adansonia, sér. 2, 8: 419 (1968). Similar in habit to *M. ivorensis* but with a larger inflorescence; in forest.
S. L.: Mt. Loma (Sept.) *Jaeger* 7461. **Iv. C.:** Soubré (Apr.) *Lorougnon* 1108 & 1112.

18. HYPOLYTRUM L. C. Rich. in Pers. Syn. 1: 70 (1805); F.T.A. 8: 486 (1902); Nelmes in Kew Bull. 10: 63–82 (1955).

By Miss S. S. Hooper

Flowering stems axillary, bearing scales or reduced leaves at the base and bracts which scarcely exceed the inflorescence at the top; achene topped by a distinct inflated style-base:
Spikelets narrowly elliptic or cylindric (7–)10–15(–20) mm. long; glume uniformly brown; achene elliptic, narrowed above and below, with an enlarged style-base

somewhat smaller than the longitudinally wrinkled base of the achene; leaves 8–15(–20) mm. broad .. .. 1. *heteromorphum*
Spikelets elliptic to ovate or obovate, 4–9 mm. long:
Glume pale and red-dotted in the upper half, darker below; achene oblong or obovate; style-base pale, enlarged, conical, decurrent on the angles of the deep green irregularly reticulate-costate base of the achene; leaves (13·5–)18–25(–28) mm. broad; flowering stem often dull purple .. .. 2. *poecilolepis*
Glume rufous except for a pale margin or tip; achene obtuse with the pale enlarged style-base exceeding the dark wrinkled basal part of the achene and exserted beyond the glume at maturity:
Inflorescence contracted, not exceeding 3 cm. across; spikelets about 3 mm. wide, pale with the broad white margins of the glumes .. .. 3. *senegalense*
Inflorescence corymbose, exceeding 5 cm. across; glumes without a wide white margin .. .. 4. *africanum*
Flowering stems terminal with well-developed cauline leaves; bracts leaf-like, often exceeding the inflorescence; achene with the style-base either small or hardly distinguishable from the rest of the achene:
Achene pale, purple-spotted, smooth or slightly ridged; style-base large but at maturity not differentiated from the rest of the elliptical achene; plant rhizomatous, large, common .. .. 5. *purpurascens*
Achene brown at maturity, deeply longitudinally costate, style-base small not inflated;
Basal leaves reduced to brown-edged sheaths, cauline leaves borne towards the top of the slender stems, elliptical, somewhat glaucous; bracts leaf-like, exceeding the slender, few-spikeletted paniculate inflorescence; achene nearly spherical, dark reddish-brown, deeply longitudinally ridged and slenderly transversely lineolate with the simple style-base as a terminal apiculus .. .. 6. *costatum*
Basal and cauline leaves similar, linear-lanceolate; inflorescence corymbose with many dark spikelets; stamens 3; achene deeply longitudinally costate
7. *cacuminum*

1. **H. heteromorphum** *Nelmes* in Kew Bull. 9: 522 (1955). *H. africanum* of F.T.A. 8: 488; F.W.T.A., ed. 1, 2: 472: Nelmes & Baldwin in Amer. J. Bot. 39: 371 (1952), partly. Differs from all the following species in the flowering spikes being several times longer than broad; on wet rocks and other damp places in forest.
Guin.: Erimakuna, Sulimania (Mar.) *Sc. Elliot* 5396! Ziama (Feb., Apr.) *Adam* 3715! 4805! **S. L.:** Kamalu (May) *Thomas* 345! Bure, nr. Mange (Jan.) *Jordan* 850! Makonde (Apr.) *Sc. Elliot* 5686! Ndilajula, nr. Njala (Feb.) *Deighton* 1078! Messima Crim (Jan.) *Morton & Gledhill* SL 1698! **Lib.:** Nabenah, Cape Mount Dist. (Apr.) *Dinklage* 2584! Ganta, Sanokwele Dist. (Dec., Jan.) *Harley* 796! *Baldwin* 14050! Webo (Apr.) *Dinklage* 2653! Cavally R. (Mar.) *Bernardi* 3415! **Iv. C.:** Taï (Jan.) *Lorougnon* 1272! Troya, Taï to Tabou (Mar.) *Leeuwenberg* 3553! Nigbi II, nr. Soubré (Nov.) *de Wilde* 3275! 56 km. N. of Sassandra, E. of Beyo (Jan.) *Leeuwenberg* 2546! Agbo (Jan.) *Chev.* 16611! **Ghana:** Kakumdu (Jan.) *Hall* 2856! Agboko to Dedesa, Afram Plains (May) *Kitson* 1125! **S. Nig.:** Onitsha *Barter* 1575! Oban Rock, Oban (Mar.) *Coombe* 189! Kwa Falls, Calabar (Mar.) *Brenan et al.* 9251! Also in E. Cameroun, Gabon, Congo, Central African Rep., Uganda and Tanzania.
2. **H. poecilolepis** *Nelmes* in Kew Bull. 10: 77 (1955). Stoutly rhizomatous plant distinguished from *H. africanum* by the wide leaves and mottled appearance of the flowering spikes; damp places in high forest.
S. L.: *Lane-Poole* 143! Kenema (Jan.) *Thomas* 7566! Gorahun (Nov.) *Deighton* 450! Kambui F.R., Nongowa (Mar.) *Jordan* 2016! **Lib.:** Nimba, Yéképa (Feb.) *Adam* 20791! Gola Forest, Ba (Dec.) *Baldwin* 10720! Javajai, Boporo Dist. (Nov.) *Baldwin* 10277! Sinoe Basin *Whyte*! Kitoma, Sanokwele Dist. (Mar.) *Adam* 16660! Blazie (Nov.) *Adam* 16135! **Iv. C.:** 61 km. N. of Sassandra, W. of Niapidou (Jan.) *Leeuwenberg* 2564! Tienkula (Mar.) *Bernardi* 8366! Adiopodoumé (Dec.) *Aké Assi* 7167! Banco Forest, Abidjan (Jan., Mar.) *de Wit* 9060/9060! *J. & A. Raynal* 13551! **Ghana:** Tano Nimri F.R. (Jan.) *Ampah* FH 5946*b*! Ancobra R. (Dec.) *Johnson* 1006! Simpa (Feb.) *Vigne* 2916! Benso, Tarkwa Dist. (Dec.) *Andoh* FH 5432! Kakum F.R. (Nov.) *Hall* 2751!
3. **H. senegalense** *A. Rich.* in Pers. Syn. 1: 70 (1805); F.T.A. 8: 488; Nelmes l.c. 79. Spikes pale and congested in a head, perhaps indistinguishable from the following at the specific level; stream sides on mountains.
Guin.: Rio Nunez *Heudelot* 727! Friguiagbé (Apr.) *Chillou* 153! Télimélé (Apr.) *Pitot*! Kinkon (Apr.) *Pitot*! Sita (Apr.) *Pitot*! **S. L.:** Loma Mts. (Oct.) *Jaeger* 7853! Bintumane Peak (Nov.) *Adam* 22188!
4. **H. africanum** *Nees ex Steud.* Syn. Pl. Glum. 2: 132 (1855); Nelmes l.c. 78. *H. longiscaposum* C.B. Cl. in F.T.A. 8: 489; F.W.T.A., ed. 1, 2: 472. Distinguishable from *H. senegalense* only by the corymbose or paniculate inflorescence and less widely white-margined glumes; by streams in forest.
Guin.: Friguiagbé (Feb.) *Chillou* 2489! Kindia *Jac.-Fél.*! Mamou (Mar.) *Dalz.* 8430! Konkoué Falls (Mar.) *Pitot*! Mali (Nov.) *Chev*! **S. L.:** Regent (Dec.) *Sc. Elliot* 4014! Mt. Horton (Jan.) *T. S. Jones* 300! Makalo (Nov.) *Marmo* 79! Kitoma (Mar.) *Adam* 16757! Sankan Biriwa, 4,500 ft. (Jan.) *Cole* 161! **Lib.:** Nimba (Dec., Jan.) *Adam* 20300! 20571! 20783! Wohmen, Vonjama Dist. (Oct.) *Baldwin* 10057! Lamco Camp, Nimba Mts. (Nov.) *Adames* 763!
5. **H. purpurascens** *Cherm.* in Bull. Soc. Bot. Fr. 80: 508 (1933); Nelmes l.c. 69. *H. nemorum* of F.T.A. 8: 487, not *Schoenus nemorum* Vahl. *H. heterophyllum* of F.W.T.A. ed. 1, 2: 472. Robust plant with spreading rhizomes and spherical or hemispherical fruiting heads; stream sides and other damp places in forest.
S. L.: Konta, Kambia Dist. (Apr.) *Hepper* 2652! Hastings (Sept.) *Melville & Hooker* 445! Kantobe (Feb.) *Bakshi* 3! Njala (Nov.) *Deighton* 2561! Gbap (Mar.) *Adames* 17! **Lib.:** Marshall, Montserrado Dist. (Feb.) *Baldwin* 11047! Kakatown, Salala Dist. *Whyte*! Loffa R., nr. Kondessu, Boporo Dist. (Dec.) *Baldwin* 10659! Gbanga (Sept.) *Linder* 797! Kitoma, Sanokwele Dist. (Mar.) *Adam* 16767! **Iv. C.:** Tabou (July) *Lorougnon* 1273HL! Mt. Orumbo-Boka, nr. Toumodi (Dec.) *Boughey* 14426! Adiopodoumé (Oct., Nov., Jan.) *Lorougnon* 1274HL! *Aké Assi* 7296! *de Wilde* 1008! **Ghana:** Beyin, Cape Coast (Feb.) *Hall* 2916! Princes Town, Axim (Mar.) *Morton* A309! Essiama (Dec.) *Morton* A2500! Ancobra R., Axim (Mar.) *Morton* GC 6592! 6593! **S. Nig.:** Lagos (Feb.) *Millen* 193! Ikoyi Plains (Jan.) *Dalz.* 1304! Sapoba (Feb.) *Onochie* FHI 31939! Okomu F.R., Benin (Feb.) *Brenan* 9183! Calabar (Mar.) *Brenan et al.* 9216! **W. Cam.:** Rio del Rey *Johnston*! Lake Ejagham, Mamfe (Mar.) *Onochie* FHI 30899! **F. Po:** (Nov., Dec.) *T. Vogel* 206! *Mann* 120! Also in E. Cameroun, and through the Congo basin to Angola.
6. **H. costatum** *Nelmes* in Kew Bull. 10: 74 (1955). Plant with a horizontal rhizome and several slender leafy stems; apparently a rare montane species.
Lib.: Pah Mt., Webo Dist. (June) *Baldwin* 6072!

Fig. 413.—HYPOLYTRUM HETEROMORPHUM *Nelmes* & H. PURPURASCENS *Cherm.* (CYPERACEAE).

H. heteromorphum Nelmes—1, habit, × ½. 2, compound spikelet, × 4. 3, glume, × 12. 4, " spikelet ", × 12. 5, diagram of " spikelet ". 6, achene, × 12. *H. purpurascens* Cherm.—7, inflorescence, × ½. 1–6 from *Germain* 8181.

3

7. **H. cacuminum** *Nelmes* in Kew Bull. 10: 74 (1955). Characterized by the enlarged woody stem-bases and dense dark-spikeletted panicle; montane grassland.
Guin.: Nimba Mt. (Feb., Dec.) *Schnell* 217! 411! *Adam* 7393! Nzérékoré (Oct.) *Jac.-Fél.* 1942! **S. L.:** Loma Mts. (Aug.–Nov.) *Jaeger* 769! 1339! 1440! Sankan Biriwa, 4,500 ft., Tingi Mts. (Jan.) *Cole* 122! Neremefondi, Tingi Mts. (Dec.) *Morton & Gledhill* SL 2987! **Lib.:** (Jan.) *Adam* 20636! Mt. Wolagwisi, nr. Pandamai (Mar.) *Bequaert* 94! **Iv. C.:** Nimba Mts., 2,500–5,000 ft. (Aug.) *Boughey* GC 18055!

19. SCLERIA Berg., Vet. Acad. Handl. Stockh. 24: 142, t.4 & 5 (1765); F.T.A. 8: 493 (1902); Nelmes in Kew Bull. 10: 415 (1955), and 11: 73 (1956); E. A. Robinson in Kew Bull. 18: 487 (1966). *Hypoporum* Nees in Linnaea 9: 303 (1834).

By Miss D. M. Napper

Hypogynium enlarging at maturity into a ciliate-margined corky cupule wider than the achene; stout broad-leaved perennials with a terminal panicle:
Achene smooth:
Achene compressed, with a circular groove round the top close to the style-base, 4·5–6 mm. diam., bluish grey at maturity, shining .. .. 1. *depressa*
Achene globose to subovoid, slightly compressed, 2·5–3·25 mm. diam., ivory to honey-grey at maturity, shining .. .. 2. *vogelii*
Achene verrucose except for the smooth, shining apex, subglobose to globose, about 3 mm. diam., cream .. .. 3. *verrucosa*
Hypogynium small to very small, lobed or reduced to an undifferentiated stipe-like base to the achene, margin entire:
Fertile spikelets unisexual; hypogynium usually well developed; inflorescence of terminal and axillary panicles with leafy bracts, not composed of a series of interrupted spikes (see also Nos. 20–22):
Plant perennial with a well developed and often creeping rhizome:
Female glume 7–11 mm. long usually black except for the midrib; achene ovoid to ellipsoid, 4–6 mm. long, smooth or slightly verrucose; hypogynium much reduced .. .. 4. *melanomphala*
Female glumes 2·5–8 mm. long; achene up to 4 mm. long:
Male spikelets more or less curved, conspicuously spreading, 7–9 mm. long; achene glabrous, distinctly longitudinally ribbed except for the tuberculate apex; panicles contracted, spike-like .. .. .. 5. *spiciformis*
Male spikelet straight, 4–9 mm. long; achene lacking the above combination of characters; panicles lax:
Stems up to 10 m. long, scandent; leaves scabrid; panicles erect, branches patently spreading; achenes ovoid to cylindric-ovoid, 3–3·5 mm. diam., purple, very rarely white, smooth, hairy .. .. .. 6. *boivinii*
Stems up to 2 m. long, erect or decumbent but never scandent; leaves not scabrid:
Peduncles of the lateral panicles very short with scarcely exserted, erect panicles; male spikelet 3–4·75 mm. long; achenes 1·8–3 mm. diam., smooth, glabrous or pubescent beneath:
Tip of the female glume longer than the achene, recurved; achene globose to ovoid-globose, 1·8–2·4 mm. diam., smooth, shining, finely hairy beneath or all over, often purplish at the apex .. .. .. 7. *pterota*
Tip of the female glume long or short, but always straight; achene ovoid to globose or ellipsoid, up to 3 mm. diam., grey, honey-coloured or purplish at maturity:
Male spikelet 3–4 mm. long; fertile glume shorter than the achene; achene 2·2–2·5 mm. diam., ivory to grey at maturity .. 8. *naumanniana*
Male spikelets 4–4·75 mm. long; fertile glume with a long tip exceeding the achene; achene 2·5–3 mm. diam., the apex dark purple to blue-black at maturity .. .. 9. *iostephana*
Peduncles of the lateral panicles long, with well exserted, often pendent panicles; male spikelet 5–9 mm. long; achenes 2–2·5 mm. diam., smooth to lacunose, with tufts of fine hairs:
Lateral panicles 2–5 at each node, rarely solitary; male spikelet 5–6 mm. long; achene narrowly ovoid to subglobose, smooth or lightly striate-lacunose .. .. 10. *lagoënsis*
Lateral panicles solitary; male spikelet 7–9 mm. long; achene subglobose, faintly or distinctly lacunose .. .. 11. *achtenii*
Plants annual or at least short-lived:
Pedicel of the male spikelet equal to or longer than the spikelet; achene globose to ellipsoid, 2 mm. diam. with oblong lacunae having reddish viscid-pubescent patches on the walls .. .. 12. *mikawana*
Pedicel of the male spikelet much shorter than the spikelet:
Lateral panicles erect; peduncles rigid, usually short:
Achene globose, 2·7–3·5 mm. diam. with deep square lacunae, hairy or glabrous 13. *sphaerocarpa*

Achene ovoid or ellipsoid-cylindric, glabrous:
Achene cylindric to ellipsoid-cylindric, 1·5–2 mm. diam. obscurely or distinctly striate-lacunose, the lacunae vertically elongated .. .. 14. *tessellata*
Achene ovoid to almost globose, 2–2·5 mm. diam., smooth at the apex alveolate-lacunose below, rarely so faintly as to appear almost smooth 15. *foliosa*
Lateral panicles pendulous; peduncles long and flexuous:
Achenes smooth, glabrous:
Lateral panicles solitary; achene globose, compressed, 3·2–3·8 mm. diam. 16. *schimperana*
Lateral panicles 2 or more at least at some of the nodes; achenes oblong-cylindric, subtrigonous, 1·7–1·8 mm. diam., grey with darker longitudinal banding, shining .. .. 17. *gracillima*
Achenes lacunose:
Achene globose, 2·75–3 mm. diam., deeply lacunose, glabrous or with reddish viscid-pubescent patches on the walls; male spikelet 3–5 mm. long 18. *globonux*
Achene obovoid-globose to oblong-ellipsoid, 1·5–2·2 mm. diam., deeply and squarely lacunose, usually hairy on the walls; male spikelet 2·5–3·5 mm. long .. .. 19. *parvula*
Fertile spikelets bisexual; hypogynium little developed, usually represented by a stipitate base to the achene:
Inflorescence of terminal and lateral panicles, the latter often with foliaceous bracts; fertile spikelets androgynous or unisexual:
Leaves not more than 5 mm. broad; spikelets 4–5 mm. long; achene ovoid to ellipsoid, 2·5–3 mm. long, smooth, glabrous .. .. .. 20. *lithosperma*
Leaves 10–22 mm. broad; panicles large with simple branches; achene smooth:
Fertile spikelets 5–6 mm. long; achene ellipsoid or ovoid-ellipsoid, 2–2·5 mm. diam. .. .. 21. *lacustris*
Fertile spikelets 11 mm. long; achene subglobose, slightly compressed, 3·5–4 mm. diam. .. .. 22. *chevalieri*
Inflorescence of a terminal interrupted spike or panicle without foliaceous bracts; spikelets usually crowded into sessile or subsessile glomerules:
Inflorescence branched; branches few with few glomerules, or numerous and compound:
Achenes subglobose, up to 1·4 mm. long, tuberculate-trabeculate*; panicle loose and copiously branched:
Plants perennial with a slender rhizome; spikelets 3·5–5 mm. long, obtuse 23. *poöides*
Plants annual; spikelets 3–4 mm. long, acuminate, black 24. *robinsoniana*
Achenes 1·5–2 mm. long, smooth to lacunose-tessellate or striate or tuberculate, glabrous:
Plants annual; panicle loose and copiously branched; spikelets acute, narrowly lanceolate; achenes oblong-ellipsoid, 1·8 mm. long with a thin pericarp 25. *guineensis*
Plants perennial; panicle scanty; spikelets usually subobtuse, oblong-lanceolate; achene subglobose, shining:
Leaf-sheath more or less truncate at the mouth; achene ovoid-globose, lacunose, faintly tessellate.. .. .. 26. *striatinux*
Leaf-sheath with a long tongue opposite the blade; achene ovoid-globose to compressed-globose, smooth or tuberculate .. .. 27. *rehmannii*
Inflorescence a simple spike, very occasionally with 1 or 2 short branches bearing a single glomerule arising from the lowest glomerules:
Glomerules erect to spreading; glume glabrous to hairy but without setose bristles:
Plants rhizomatous, perennial:
Stems bulbous-based; achene obovoid to subglobose, 1·2–1·6 mm. long, usually smooth and shining but occasionally tuberculate .. .. 28. *bulbifera*
Stems not bulbous-based:
Achene globose, subtrigonous, 1·2 mm. diam., obscurely trabeculate* with a dark line down each face .. .. 29. *monticola*
Achene broadly ovoid to subglobose, 1·1–1·5 mm. long, strongly trabeculate 30. *dieterlinii*
Plants annual:
Glumes and bracteoles with few to many setose hairs, chiefly on the midrib and margins; achene oblong-ovoid to oblong-globose, 1·2–1·6 mm. long, trabeculate* .. .. 31. *hirtella*
Glumes and bracteoles glabrous; achene globose, compressed, sub-trigonous, 1·4–1·6 mm. long, trabeculate .. .. 32. *pergracilis*

*Trabeculate = transversely ridged.

Glomerules reflexed; glumes normally densely or sparsely coated with conspicuous, often black setose bristles:
Plants perennial; achene subglobose to oblong-ovoid, 1·1–1·5 mm. diam., smooth or almost so:
Rhizome stout and woody, at least 2 mm. thick, unbranched and giving rise to a uniseriate row of stems; basal leaf-sheaths distant, tight with a short or very short blade; achene subglobose, smooth or lightly tuberculate 33. *nutans*
Rhizome slender, 1–2 mm. thick, usually several per plant and the stems not conspicuously seriate; leaves tufted from the base of the plant with loose longitudinally slit sheaths and long blades; achene oblong-ovoid to oblong-globose, trabeculate* .. .. 34. *aterrima*
Plants annual: achene obovoid, compressed, 1·1–1·6 mm. diam., trabeculate* or rarely tuberculate:
Glomerules subsessile usually solitary; glumes with an awn 1·5–3 mm. long 35. *melanotricha*
Glomerules shortly pedunculate, paired, the second one on a peduncle up to 3 mm. long; glumes mucronate or with a very short awn, not over 1 mm. long .. .. 36. *grata*

1. **S. depressa** (*C.B. Cl.*) *Nelmes* in Amer. J. Bot. 39: 392 (1952), and in Kew Bull. 11: 77 (1956); Berhaut, Fl. Sén. ed. 2, 358 (1967). *S. racemosa* Poir. var. *depressa* C.B. Cl. in F.T.A. 8: 508 (1902). *S. racemosa* of F.W.T.A., ed. 1, 2: 493. *S. racemosa* subsp. *depressa* (C.B. Cl.) J. Raynal in Adansonia sér. 2, 4: 153 (1964). A stout perennial 1–2 m. high; swamps in savanna and forest.
Maur.: L. Rkiz (Sept.) *Sadio* 297! **Sen.:** *Heudelot* 368! Sangalkam (July) *de Wit* 7299! Kayar (Dec.) *Pitot*! Niokolo-Koba (Dec.) *J. & A. Raynal* 6848! Sedhiou, Casamance (Feb.) *Chev.* 2449! **Gam.:** *Ingram*! **Mali:** Gaïbala Ravine, Kita Mts. (Mar.) *Jaeger* 5014! Foot of Kita Mts. (Oct.) *Jaeger* 2! Katé (Nov.) *Duong* 183 (or 2107)! Bougouni (Jan.) *Duong*! Koutiala (Feb.) *Roberty* 3412! **Guin.:** *Farmar* 243! Kindia to Telimélé *Roberty* 10742! Fouta Djalon (Mar.) *Langdale-Brown* 2628! **S. L.:** Freetown (Dec.) *Morton* SL 218! Yonibana (Nov.) *Thomas* 5029! Gorahun (Nov.) *Deighton* 353! Kenema (Jan.) *Thomas* 7523! Kurobonla, Loma Mts. (Jan.) *Morton* SL 429! **Lib.:** Mecca, Boporo Dist. (Nov.) *Baldwin* 10407! Gondolahun, Kolahun Dist. (Nov.) *Baldwin* 10105! Zigida, Vonjama Dist. (Oct.) *Baldwin* 9990! Ganta, Sanokwele Dist. (Oct.) *Harley* 1033! Nimba Mts. (Feb.) *Adam* 21007! **Iv. C.:** *Bégué* 13! Ferkessédougou *Kerharo & Bouquet* 182! **Ghana:** Tumu to Sorabela (Mar.) *Morton* GC 8877! Banda (Dec.) *Morton* GC 25110! Ancobra (Mar.) *Morton*! Sese to Dixcove (Mar.) *Morton* A469! Pokoase (Apr.) *Irvine* 1585! **N. Nig.:** Nupe *Barter* 921! Zaria F.R. (Jan.) *Kershaw* 900024! Kontagora (Nov.) *Dalz.* 237! Naraguta, Jos Plateau (Oct.) *Hepper* 1167! Gangumi, Adamawa Prov. (Dec.) *Latilo & Daramola* FHI 28800! **S. Nig.:** Lagos (Mar.) *Millen* 96! Olokemeji (Nov.) *Onochie* FHI 8130! Ogbomosho (Dec.) *Haines* 75! Ekowe, Nun R. (Oct.) *Anderson* 1408! **W. Cam.:** Bamali, Ndop Plain (Mar.) *Brunt* 158!

2. **S. vogelii** *C.B. Cl.* in F.T.A. 8: 508 (1902); Nelmes in Kew Bull. 11: 78 (1956). Stout perennial to 2·5 m. high; swampy places in forest and savanna.
Lib.: Grand Bassa *T. Vogel* 59! (Aug.) *Dinklage* 2001! **Iv. C.:** L. Irobo (Aug.) *Monod*! Audouin Forest (Aug.) *de Wit* 7309! Dabou to Abidjan (July) *de Wilde* 154! Grand Bassam (Sept.) *de Wit* 7394! Ebrié Lagoon (July) *de Wilde* 609! **Ghana:** *Boughey* GC 14258! Awiebo (Aug.) *Hall* 3395! **S. Nig.:** Lagos (Feb., June) *Millen* 154! *Dalz.* 1301! Abeokuta *Irving*! Oyeregbene, Lower Nun R. (July) *Anderson* 1348! **W. Cam.:** Buea (Mar.) *Maitland* 553! Also in E. Cameroun and Gabon.

3. **S. verrucosa** *Willd.* Sp. Pl. 4: 313 (1805); F.T.A. 8: 509; Nelmes in Kew Bull. 11: 79 (1956). Stout perennial up to 2·5 m. high with a narrow panicle; on riverbanks and in marshes.
Sen.: Saré Doma, nr. Kolda (Jan.) *Raynal* 7816! **S. L.:** *Thomas* 8692! Moria to Fintonia (Jan.) *Morton & Gledhill* SL 488! Samu (Dec.) *Sc. Elliot* 4218! Njala (Nov.) *Deighton* 1777! Senihun to Njala (Sept.) *Deighton* 3049! **Lib.:** Wohmen, Vonjama Dist. (Oct.) *Baldwin* 12045! Javajai, Boporo Dist. (Nov.) *Baldwin* 10268! Jabrocca, Grand Cape Mount Dist. (Dec.) *Baldwin* 10848! Peahtah, Salala Dist. (Oct.) *Linder* 1080! Dieble, Webo Dist. (July) *Baldwin* 6278! **Ghana:** Ankaful (Nov.) *Hall* 1666! Aburi (Oct.) *Johnson* 232! 474! Abetifi, Kwahu (Jan.) *Irvine* 1829! Akroso (Aug.) *Howes* 953! **N. Nig.:** Jos Plateau (Oct.) *Lely* P799! **S. Nig.:** Lagos *W. MacGregor* 101! (Jan.) *Haines* 89! Ibadan (Nov.) *Newberry & Etim* 153! Ebute Metta (Jan.) *Millen* 94! Okomu F.R., Benin (Feb.) *Onochie* 9077! **W. Cam.:** Victoria (Jan.) *Maitland* 901! Nkom to Wum, Bamenda (July) *Ujor* FHI 29280! Eastwards to the Congo Republic and the western parts of Uganda and Tanzania.

4. **S. melanomphala** *Kunth* Enum. Pl. 2: 345 (1837); F.T.A. 8: 506; Nelmes in Kew Bull. 11: 88 (1956); Aké Assi, Contrib. 2: 274 (1964); E. A. Robinson in Kew Bull. 18: 546 (1966). Stout perennial up to 2 m. high with broad leaves and dense panicles, the lateral ones pendulous; swamps and very wet places in grassland and forest.
Guin.: Fetoré Plain, Fouta Djalon (Sept.) *Adames* 372! 373! **S. L.:** Franziga, Tambakka (Feb.) *Sc. Elliot* 5046! Kabala (Sept.) *Thomas* 2258! Mabonto (Aug.) *Deighton* 1398! Jaima to Baima (July, Aug.) *Dawe* 536! Pujehun (Apr.) *Deighton* 1656! **Lib.:** Gondolahun, Kolahun Dist. (Nov.) *Baldwin* 10104! Gbanga (Sept.) *Linder* 513! Bushrod Is., Montserrado Dist. (Feb.) *Baldwin* 11067! Yéképa, Nimba Mts. (Jan.) *Adam* 20518! Jabroke, Webo Dist. (July) *Baldwin* 6508! **Iv. C.:** Toulépleu (Oct.) *Aké Assi* 4841! Gan (Nov.) *Aké Assi* 8293! **Ghana:** Axim (Dec.–Feb.) *Johnson* 1001! *Akpabla* 823! *Morton* GC 8441! **N. Nig.:** (Aug.) *Lely* P486! Bomo Fadama, Zaria Prov. (Aug.) *Thatcher* S.482! Vom, Bauchi Plateau (Feb.) *Saunders* 26! Sha, Plateau Prov. (Sept.) *Lawlor & Hall* 729! Bellel, Mambila Dist. (Jan.) *Hepper* 1697! **S. Nig.:** Ilesha Rd., Shaki Dist. (Aug.) *Latilo & Taylor* FHI 43399! **W. Cam.:** Mencham R. (Feb.) *Brunt* 954! Babeni, Ndop Plain (Apr.) *Brunt* 373! Madele, Wum Dist. (June) *Daramola* FHI 41097! Throughout tropical and S. Africa, Madagascar and S. America.

5. **S. spiciformis** *Benth.* in Fl. Nigrit. 556 (1849); F.T.A. 8: 506; Nelmes in Kew Bull. 11: 100 (1956); Aké Assi, Contrib. 2: 276 (1964). Tufted perennial 30–75 cm. high with dense, spikelike panicles, the lateral ones solitary and erect. Seepage areas on rocky outcrops and seasonally swampy grassland.
Guin.: Conakry (Oct.) *Adam* 12611! Friguiagbé (Sept.) *Chillou* 706! Macenta (Oct.) *Baldwin* 9813! Nimba Mt. (Nov.) *de Wilde* 885! **S. L.:** York (Nov.) *Morton* SL 1452! Sugarloaf Mt. (Dec.) *Sc. Elliot* 3952! Bonthe Is. (Nov.) *Deighton* 2292! Gbengbe Hills, Bumban (Nov.) *Morton* SL 2910! Bintumane, Loma Mts. (Nov.) *Jaeger* 784! **Lib.:** Soplima, Vonjama Dist. (Nov.) *Baldwin* 10056! Tawata, Boporo Dist. (Nov.) *Baldwin* 10350! Duport (Nov.) *Linder* 1456! Grand Bassa (July) *T. Vogel* 107! Bele Mt., Nimba Mts. (Oct.) *Adames* 706! **Iv. C.:** Tonkoui Mt. (Sept.) *Aké Assi* 3330! Also in Congo Republic.

6. **S. boivinii** *Steud.* Syn. Pl. Glum. 2: 173 (1855); Napper in Kew Bull. 25: 441 (1971). *S. barteri* Boeck. in Linnaea 38: 504 (1874); F.T.A. 8: 507; F.W.T.A., ed. 1, 2: 493; Nelmes in Kew Bull. 11: 92 (1956). Scandent perennial with leaves up to 6 mm. wide; climbing in scrub, bush and forest.
Sen.: *Naegelé*! **Guin.:** Kakoulima (Feb.) *Schnell* 2495! Nion, Nimba Mts. (Mar.) *Schnell* 683! **S. L.:** (Sept.) *Deighton* 2112! Kambia (Jan.) *Sc. Elliot* 4388! Mabanda (Aug.) *Thomas* 1589! Rowalla (July)

*Trabeculate = transversely ridged.

Fig. 414.—Scleria depressa (*C.B.Cl.*) *Nelmes* (Cyperaceae).

1, inflorescence, × 1. 2, male and female spikelets, × 4. 3, male spikelet, × 6. 4, stamen, × 14. 5, female spikelet with fruit, × 3.

Thomas 1195! Central Prov. (July–Aug.) *Dawe* 545! **Lib.:** Du R. (Aug.) *Linder* 269! Grand Bassa (Aug. *Dinklage* 2010! Suacoco, Gbanga Dist. (Oct.) *Daniel* 26! Ganta, Sanokwele Dist. (Sept.) *Harley* 98! Mnanulu, Webo Dist. (June) *Baldwin* 6053! **Iv. C.:** Taï (Aug.) *Schnell* 1589! Guiglo *Kerharo & Bouquet* 575! Ayamé to Bianoua (Nov.) *Aké Assi* 9759! Cosrou (Aug.) *Monod*! Abidjan (Dec.) *Pitot*! **Ghana:** Anaji (Oct.) *Howes* 988! Assin Yan Kumasi *Cummins* 238! Tarkwa (Dec.) *Johnson* 1000! Kumasi (Aug.) *Vigne* 2407! Alavanyo, Hohoe Dist. (Sept.) *St. Cl-Thompson* 1472! **N. Nig.:** Sanga F.R., Jemaa Dist. (Dec.) *Olorunfemi* FHI 55663! **S. Nig.:** Ilaro (Jan.) *Millen* 97! Lagos (Dec.) *Hagerup* 804! Onitsha *Barter* 1786! Oban *Talbot* 855! Onicha Olona (Oct.) *Thomas* 1854! **W. Cam.:** Barombi F.R., Kumba (Feb.) *Dundas* FHI 13985! **F. Po:** *T. Vogel* 87! (Dec.) *Mann* 113! Tropical Africa to Uganda and Angola, also Pemba Island and Madagascar.

7. **S. pterota** *Presl* in Isis 21: 268 (1826); Nelmes in Kew Bull. 11: 91 (1956); Aké Assi, Contrib. 2: 276 (1964); Berhaut, Fl. Sén. ed. 2, 358 (1967). *S. naumanniana* of F.W.T.A., ed. 1, 2: 493, partly. Perennial up to 120 cm. high with leaves 5–11 mm. wide and loose panicles; in damp shady places and river banks.
Sen.: Kouloutntou (Apr.) *Adam* 17989! Niokolo-Koba (Dec.) *J. & A. Raynal* 6847! Sédhiou, Casamance (Nov.) *Berhaut* 6384! **Mali:** Kita (Oct.) *Jaeger* 3391! **S. L.:** *Thomas*! Rokupr (Nov.) *Jordan* 677! Njala (May, Sept.) *Deighton* 1185! 2150! Mange (Dec.) *Morton & Gledhill* SL 97! **Lib.:** Gbanga (Sept.) *Linder* 516! Nyaake, Webo Dist. (June) *Baldwin* 6119! Sarbo, Webo Dist. (July) *Baldwin* 6385! **Iv. C.:** Dakpadou Forest (Jan.) *Aké Assi* 7308! **Ghana:** Elmina (June) *Hall* 1473! Princes Town (Feb.) *Morton* GC 8461! Domenasi Bridge, Ankobra R. (Aug.) *Hall* 3338! Afram Mankrong F.R. (Dec.) *Adams* 4968! Aframso, nr. Mampong *Hall* 3031! **N. Nig.:** Bonu Kurmi, Niger Prov. (June) *E. W. Jones* 266! **S. Nig.:** Akpaka F.R., Onitsha Dist. (Nov.) *Okeke* FHI 38444! Throughout tropical Africa, Madagascar and south tropical America.

8. **S. naumanniana** *Boeck.* in Engl., Bot. Jahrb. 5: 94 (1883); Nelmes in Kew Bull. 11: 96 (1956); Berhaut, Fl. Sén. ed. 2, 358 (1967). *S. ovuligera* Boeck. in Linnaea 38: 497 (1874); F.T.A. 8: 507; not of Steud. Perennial 60–120 cm. high with leaves 5–10 mm. wide and loose, lanceolate to pyramidal inflorescences; in damp places in forest or bush, marshes, among sand dunes and near mangrove swamps.
Sen.: Diantême (Oct.) *Adam & Kerharo* 1547! Sangalkam (Oct.) *J. & A. Raynal* 6405! Oussouye (Apr., Sept.) *Broadbent* 123! *Berhaut* 5813! **Mali:** Niger Delta *Duong*! **Guin.:** Kanéa (Oct.) *Adam* 12631! Macenta (Oct.) *Baldwin* 9827! Kindia Region (Oct.) *Kent* 46! **S. L.:** *Deighton* 2028! Bathurst (Jan.) *Dalz.* 8251! Binkolo (Aug.) *Thomas* 1685! Njala (Sept.) *Deighton* 1994! Loma Mts. (Sept.) *Jaeger* 7451! **Lib.:** Jabrocca, Grand Cape Mt. Dist. (Dec.) *Baldwin* 10873! Grand Bassa (July) *T. Vogel* 60! Sanokwele (Sept.) *Baldwin* 9549! Dieblo, Webo Dist. (July) *Baldwin* 6306! Cavally, Maryland Dist. (June) *Baldwin* 6014! **Iv. C.:** Danané (Oct.) *Aké Assi* 9208! Issia Rock (Aug.) *Aké Assi* 4832! Cosrou (Aug.) *Monod*! Banco Forest (Nov.) *Bégué* 97! Adiopodoumé (Oct.) *Leeuwenberg* 1783! **Ghana:** Atwabo *Fishlock* 6 of 1931! Essiama (Oct.) *Hall* 2671! Axim (Mar.) *Morton* A372! Anaje, nr. Sekondi (Oct.) *Howes* 989! Maliato, Togo Plateau (Nov.) *Morton* A3531! **Dah.:** Sémé, nr. Cotonou (Feb.) *J. & A. Raynal* 13511! **S. Nig.:** Lagos (Dec.) *Dalz.* 1302! Epe (Aug.) *Gillett* 15272! Ajagbodudu, Benin Dist. (Aug.) *Wright* 57a! Mamu River F.R., Awku Dist. (Aug.) *Olorunfemi* FHI 34184! Oban *Talbot* 854! Also from the Congo to Angola.

9. **S. iostephana** *Nelmes* in Kew Bull. 11: 94 (1956); Aké Assi, Contrib. 2: 274 (1964). Stout erect perennial up to 2 m. high with spreading panicles and large buff-coloured glumes streaked with red; stream-sides and damp places in rain-forest.
Iv. C.: Touba *Aké Assi* 4786! Mont Tonkoui *Aké Assi* 5583! **Ghana:** Shiare, Togo Hills (Dec.) *Jenik & Hall* CC 1134! **N. Nig.:** Kachia to Kaduna (Dec.) *Meikle* 778! Also in the Congo, Uganda and southwards to Zambia.

10. **S. lagoënsis** *Boeck.* in Vidensk. Medd. Dansk Naturrh. Foren. 1869: 151 (1869); E. A. Robinson in Kew Bull. 18: 538 (1966). *S. canaliculato-triquetra* Boeck. in Flora 62: 573 (1879); F.T.A. 8: 505; F.W.T.A., ed. 1, 2: 493; Nelmes in Kew Bull. 11: 84 (1956); Aké Assi, Contrib. 2: 273 (1964); Berhaut, Fl. Sén. ed. 2, 358 (1967). Tufted perennial 60–150 cm. high with swollen stem-bases, and 5–12 mm. wide leaves; lateral panicles pendulous on slender exserted peduncles; in damp grasslands, usually seasonally water-logged.
Sen.: Casamance (Oct.) *Adam* 18486! Ziguinchor (Sept.) *Broadbent* 83! **Guin.:** Kouroussa (July, Aug.) *Pobéguin* 401! 421! Nimba Mts. (Sept.) *Schnell* 3593! **S. L.:** (Oct.) *Deighton* 2150! Kasanko (Oct.) *Jordan* 642! Makump (Aug.) *Deighton* 1357! Tabe (Sept.) *Deighton* 3040! Falaba (Oct.) *Small* 449! **Iv. C.:** Dabakala to Comoë **R.** (Nov.) *Aké Assi* 9241! Toumodi (Aug.) *de Wit* 7343! Dimbokro (Aug.) *Roberty* 3081! Dabou (Aug.) *Aké Assi* 4386! Loupou (July) *de Wilde* 372! **Ghana:** Elmina (May) *Morton* A904! Ejura Scarp (Aug.) *Hall* CC 338! Banda Hills (Nov.) *Harris*! Mankesim to Accra (May) *Morton* A3363! Amedzofe (Nov.) *Morton*! **N. Nig.:** Katsina Allah (June) *Dalz.* 815! Fadama, Zaria Prov. (Aug.) *Kershaw* 900715! Richa, Jos Plateau (Sept.) *Lawlor & Hall* 664a! Panshanu, Bauchi Prov. (Aug.) *Lawlor & Hall* 359! Vogel Peak (Nov.) *Hepper* 1417! **S. Nig.:** Igboora (Oct.) *Haines* 227! Abeokuta *Irving*! Ifon F.R., Owo Dist. (Oct.) *Adebusuyi* FHI 43569! Mamu River F.R., Awka Dist. (Sept.) *Onochie* FHI 34076! Throughout tropical Africa and south tropical America.

11. **S. achtenii** *De Wild.* in Rev. Zool. Afr., Suppl. Bot. 14, 2: 16 (1926); E. A. Robinson in Kew Bull. 18: 534 (1966); Berhaut, Fl. Sén. ed. 2, 358 (1967). *S. nyasensis* of Nelmes in Kew Bull. 11: 86 (1956), partly; Aké Assi, Contrib. 2: 275 (1964); not of C.B. Cl. Perennial up to 2 m. high with swollen stem-bases and 2–5 mm. wide leaves; lateral panicles pendulous on long slender peduncles; by stream-sides and in marshy grasslands.
Sen.: Sangalkam (June) *Berhaut* 1736! Oussouye (July) *Berhaut* 6214! **Guin.:** Friguiagbé (May) *Chillou* 367! Madina Tossekré (Oct.) *Adam* 12540! Koba (Nov.) *Jac.-Fél.* 7267! **S. L.:** Rokon, Samu (July) *Jordan* 482! Waterloo (Aug.) *Melville & Hooker* 307! Gbinti (Jan.) *Deighton* 982! Masingbi (Feb.) *Deighton* 4076! Mano Salija (Nov.) *Deighton* 345! **Lib.:** Monrovia (July) *Künkel* 78! Mesurado Swamp, nr. Monrovia (Sept.) *Künkel* 63! Duport (Nov.) *Linder* 1459! Buchanan (Dec.) *Adam* 16424! Bele Mt. (Oct.) *Adames* 705! **Iv. C.:** Moossou *Aké Assi* 4783! *Bodard* 44! Toupa *Aké Assi* 4771! Grand Lahou *Aké Assi* 4773! **Ghana:** Mpeasem, Cape Coast (July) *Hall* 3235! Sakogu (Aug.) *Hall* CC 451! Kpandu (Jan.) *Hall* GC 40068! **Dah.:** Porto Novo (Mar.) *Adjanohoun* 352! **N. Nig.:** Bida to Abuja (Dec.) *Haines* 77! **S. Nig.:** Topo, nr. Badagri (Feb.) *Haines* 209! Lagos (Dec.) *Hagerup* 782! Throughout tropical Africa and south to Natal.

12. **S. mikawana** *Makino* in Bot. Mag. Tokyo 27: 57 (1913); Nelmes in Kew Bull. 11: 107 (1956); E. A. Robinson in Kew Bull. 18: 525 (1966); Berhaut, Fl. Sén. ed. 2, 359 (1967). *S. glandiformis* of F.T.A. 8: 504; F.W.T.A., ed. 1, 2: 493, partly. Tall slender annual up to 120 cm. high with leaves 3–7 mm. wide; lateral panicles solitary, erect, scarcely exserted; swampy places.
Sen.: *Heudelot* 662! Niokolo-Koba (Dec.) *J. & A. Raynal* 6894! **S. L.:** Robumbe, Magbema (Oct.) *Jordan* 803! Rokupr, Magbema (Nov.) *Jordan* 694! Rokon, Samu (Oct.) *Jordan* 572! Nganyahun, Senihun to Njala (Sept.) *Deighton* 3048! Njala (Dec.) *Katta* 6! **Iv. C.:** Abouabou (Mar.) *Aké Assi* 10010! Across tropical Africa and widespread in Asia.

13. **S. sphaerocarpa** (*E. A. Robinson*) *Napper* in Kew Bull. 25: 441 (1971). *S. tessellata* Willd. var. *sphaerocarpa* E. A. Robinson in Kew Bull. 18: 526 (1966). *S. glandiformis* of F.W.T.A., ed. 1, 2: 493, partly. *S. globonux* of Nelmes in Kew Bull. 11: 105 (1956), partly. Densely tufted annual up to 1 m. high with leaves 2–7 mm. wide; lateral panicles solitary, erect, scarcely exserted; in swampy grassland.
Sen.: Ziguinchor (Dec.) *Berhaut* 6676! Gambia R., E. Senegal (Sept.) *Fotius* K438! **S. L.:** Mateboi, Sanda Tenraran (Nov.) *Jordan* 841! **Lib.:** Gbanga (Sept.) *Linder* 515! **Iv. C.:** Kodiokoffi (Aug.) *Chev.* 22368! Dabakala to Toupé (Nov.) *Aké Assi* 9244! Bouaké (Sept.) *de Wilde* 951! **Ghana:** Ankaful, nr.

Cape Coast (Sept.) *Hall* 2357! Achimota (Aug.) *Adams* 4901! Dabala (July) *Hall* 3241! Lumbaga (Dec.) *Morton* GC 6244! Tamale (Dec.) *Adams & Akpabla* 4143! **N. Nig.:** Gurum, Adamawa Dist. (Nov.) *Hepper* 1246! Gotel Mts. *Tuley* 1728! **S. Nig.:** Igboora (Oct.) *Haines* 226! Also from the Sudan southwards to Zambia and Angola.

14. **S. tessellata** *Willd.* Sp. Pl. 4: 315 (1805); Nelmes in Kew Bull. 11: 108 (1956); E. A. Robinson in Kew Bull. 18: 526 (1966); Berhaut, Fl. Sén. ed. 2, 359 (1967). *S. glandiformis* Boeck. in Linnaea 38: 458 (1874); F.T.A. 8: 503; F.W.T.A., ed. 1, 2: 493, partly. Densely tufted annual up to 1 m. high, with leaves 2–7 mm. wide; lateral panicles solitary, erect, scarcely exserted; seasonally swampy grasslands.
Sen.: Gassane, Mbacké to Thiel (Oct.) *J. & A. Raynal* 7695! Fatick (Dec.) *Maymard*! *J. & A. Raynal* 6714! Niokolo-Koba (Oct.) *Adam* 15713! Velingara, Casamance (Oct.) *Adam* 18598! **Mali:** Kita Mts. (Sept.) *Jaeger* 1! Bamako *Adam* 15355! Sotuba (Oct.) *Raynal* 5017! Koulikoro (Oct.) *Chev.* 2468! Bokywere (Oct.) *Davey* 366! **Guin.:** Kiffaya to Seriba (Oct.) *Pitot*! Thianguel-Bori (Oct.) *Pitot*! Sambaïlo (Oct.) *Adam* 12716! **Iv. C.:** *Bégué* 91! Tiefinzo, Odienné Region (Oct.) *Aké Assi* 8264! Séguélo (Oct.) *Aké Assi* 9705! **Ghana:** Lawra to Wa (Oct.) *Ankrah* GC 20449! Bolgatanga to Bawku (Oct.) *Adei* SLUS/C98! Zowse, nr. Bawku (Nov.) *Hall & Enti* GC 35989! **N. Nig.:** Nupe *Barter* 1042! Minna (Dec.) *Meikle* 723! Zaria (Dec.) *C. B. Taylor* 10! Kufena Hill, Zaria (Sept.) *Clayton* 1314! Biu (Nov.) *De Leeuw* 1327*c*! Throughout tropical Africa, Madagascar and also in India.

15. **S. foliosa** *Hochst. ex A. Rich.* Tent. Fl. Abyss. 2: 509 (1851); F.T.A. 8: 503; Nelmes in Kew Bull. 11: 102 (1956); E. A. Robinson in Kew Bull. 18: 525 (1966); Berhaut, Fl. Sén. ed. 2, 359 (1967). *S. glandiformis* of F.W.T.A., ed. 1, 2: 493, partly. Tufted annual 60–100 cm. high, rarely more, with leaves 2–8 mm. wide; lateral panicles solitary, erect, scarcely exserted; in grassy, seasonally wet places.
Sen.: Ndiar (Oct.) *J. & A. Raynal* 7664! Niassia, Casamance (Nov.) *Berhaut* 6656! **Iv. C.:** Dinderesso (Nov.) *Néa* 145! **Ghana:** Bole (Nov.) *Harris*! **N. Nig.:** (Sept.) *Lely* P775! Kufena Hill, Zaria (Sept.) *Clayton* 1316! Bauchi (Dec.) *Haines* 76! Biu, Bornu (Nov.) *Magaji & Oche* MG79! Maiduguri, Bornu (Oct.) *H. B. Johnston* N.36*a*! Widespread in tropical Africa, also in Madagascar.
[A few plants have been collected with characters reminiscent of both *S. foliosa* and *S. schimperana*. These represent such a wide range of variation in character association that it seems possible that they may be samples from hybrid populations. Further detailed examination in the field is essential.]

16. **S. schimperana** *Boeck.* in Linnaea 38: 466 (1874); F.T.A. 8: 504; Nelmes in Kew Bull. 11: 104 (1956), as *schimperiana*. Tufted annual up to 75 cm. high with leaves up to 8 mm. wide; lateral panicles erect, on long or short villous-hispidulous peduncles; in marshy grassland.
N. Nig.: Biu (Nov.) *De Leeuw* 1327*a*! Across north-eastern tropical Africa and south to Uganda.

17. **S. gracillima** *Boeck.* in Flora 62: 570 (1879); F.T.A. 8: 505; Nelmes in Kew Bull. 11: 110 (1956); E. A. Robinson in Kew Bull. 18: 534 (1966); Berhaut, Fl. Sén. ed. 2, 358, 360. Slender annual (or perennial?) up to 75 cm. high with leaves 1–2 mm. wide; panicles small and delicate, almost spike-like, the lateral ones pendulous on slender or filiform peduncles thickening upwards, usually paired; in seasonally swampy grasslands.
Sen.: Niokolo-Koba (Nov., Dec.) *Adam* 17104! 17150! *J. & A. Raynal* 6892! **Iv. C.:** Séguéla (Nov.) *Aké Assi* 9720! **Ghana:** Ejura Scarp, Ashanti (Dec.) *Morton* GC 9630! 9707! S. of Yendi (Nov.) *Hall* GC 42171! Kpandai (Nov.) *Hall* GC 42202! **S. Nig.:** Igbete to Kishi (Mar.) *Gledhill* 732! Throughout tropical Africa, also in Brazil.

18. **S. globonux** *C.B. Cl.* in F.T.A. 8: 504 (1902); Nelmes in Kew Bull. 11: 104 (1956); Berhaut, Fl. Sén. ed. 2, 359 (1967). *S. glandiformis* of F.W.T.A., ed. 1, 2: 493, partly. Sparingly villous annual up to 1 m. high with leaves 3–9 mm. wide; lateral panicles solitary pendulous on long flexuous villous hispidulous peduncles; seasonal or permanently swampy areas in grassland.
Sen.: Niokolo-Koba (Nov.) *Adam* 15920! 17097! **Port. G.:** Bafata (Nov.) *Esp. Santo* 2825! **S. L.:** Jokibu (Nov.) *Deighton* 3776! **Ghana:** Burufo (Dec.) *Adams & Akpabla* 4386! 4392*a*! **N. Nig.:** Kontagora (Nov.) *Dalz.* 238! Across tropical Africa to the Sudan and Uganda.

19. **S. parvula** *Steud.* Syn. Pl. Glum. 2: 174 (1855); Nelmes in Kew Bull. 11: 105 (1956), partly; E. A. Robinson in Kew Bull. 18: 532 (1966); Napper in Kew Bull. 25: 442 (1971). *S. bambariensis* Cherm. var. *b* of E. A. Robinson l.c. 527 (1966), partly. Annual with leaves up to 6 mm. wide, lateral panicles 1–3 together, pendulous, on long slender peduncles; in seasonally swampy grassland and stream-sides.
Sen.: *Adam* 12491! Niokolo-Koba (Dec.) *J. & A. Raynal* 6802! **S. L.:** Picket Hill (Nov.) *T. S. Jones* 204! **Iv. C.:** Momi Mt. (Oct.) *Aké Assi* 9212*b*! 9214! 9215! Widespread in tropical Africa.

20. **S. lithosperma** (*Linn.*) *Swartz* Prodr. Veg. Ind. Occ. 18 (1788); F.T.A. 8: 502; Nelmes in Kew Bull. 10: 421 (1955); E. A. Robinson l.c. 517 (1966). *Scirpus lithospermus* Linn., Sp. Pl. 51 (1753), partly. Slender rhizomatous perennial 30–100 cm. high with narrow leaves and both terminal and lateral panicles; usually in dense shade of forest or woodland.
Iv. C.: Oroumba-Boka (May, Aug.) *Boughey* GC 18588! *Aké Assi* 8863! Lamto (Dec.) *Aké Assi* 9855! Téhini, Bouna Reserve (Aug.) *de Wilde* 759! **Ghana:** Abontsin, Mankessim to Winneba (Nov.) *Adams* 4916! Odumase (June) *Irvine* 1644! Shai Hills (Jan.) *Morton* 8265! Volta River F.R. (Nov.) *Morton* GC 6066! Ajena, Volta River (Nov.) *Morton* GC 8031! **S. Nig.:** Olokemeji F.R., Abeokuta Prov. (Nov.) *Jones, Keay & Onochie* FHI 14189! Pantropical.

21. **S. lacustris** *Wright ex Sauvelle* in Anal. Acad. Cienc. Habana 8: 152 (1871); Nelmes in Kew Bull. 10: 422 (1955); E. A. Robinson in Kew Bull. 18: 517 (1966); Berhaut, Fl. Sén. ed. 2, 358 (1967). Stout perennial 60–150 cm. high often rooting from the lower nodes, stems and leaves spinulose on the angles and veins; in shallow water at edge of swamp etc.
Sen.: Mbao (Sept.) *J. & A. Raynal* 6347! Malika *Adam* 2351! L. Youi (Nov.) *Pitot*! Sangalkam (Oct.) *J. & A. Raynal* 7672! Niokolo-Koba (Dec.) *J. & A. Raynal* 6893! **S. L.:** Kasewe Hills (Nov.) *Morton* SL 1526! Moyamba (Sept.) *Morton & Jarr* SL 2276! Rosino, Samu (Oct.) *Jordan* 797! Nganyahun, Senihun to Njala (Sept.) *Deighton* 3074! Njala (Feb.) *Deighton* 2597! **Ghana:** Beyin (July) *Hall* GC 35542! Also in the Congo and tropical South America.

22. **S. chevalieri** *J. Raynal* in Adansonia, sér. 2, 4: 150 (1964); Berhaut, Fl. Sén. ed. 2, 358. Tall stout perennial differing from the above in the broader leaves and in the panicle being larger and stiffer in all its parts, also the fertile spikelets usually unisexual; swamps.
Sen.: Dakar to Rufisque (Nov.) *Chev.* 33902!

23. **S. poöides** *Ridley* in Trans. Linn. Soc., ser. 2, 2: 170 (1884); F.T.A. 8: 502 (1902); Nelmes in Kew Bull. 10: 433 (1955), as *poaeoides*; E. A. Robinson in Kew Bull. 18: 512 (1966). *S. multispiculata* Boeck., Cyper. Nov. 1: 36 (1888); F.T.A. 8: 501; F.W.T.A., ed. 1, 2: 493. Very slender perennial up to 75 cm. high with leaves not more than 2 mm. wide; panicle loose and copiously branched; locally dominant in swampy grassland.
N. Nig.: Nupe *Barter* 1349! Vodni, Pankshin *Saunders* 48! Naraguta (Aug.) *Lely* P446! Widespread throughout tropical Africa and Madagascar.

24. **S. robinsoniana** *J. Raynal* in Adansonia, sér. 2, 7: 239 (1967). Slender annual up to 45 cm. high with a loose copiously branched panicle very similar to the above; seepage areas in rocky places.
Guin: Kindia (Oct.) *Jac.-Fél.* 2130! Dalaba (Sept.) *Adames* 336! Hollandé Tossékré (Oct.) *Adam* 12726! Ditinn (Oct.) *Pitot*! Mali (Dec.) *Schnell* 2354 (partly)! **S. L.:** Kanya (Oct.) *Thomas* 2967!

25. **S. guineensis** *J. Raynal* in Adansonia, sér. 2, 4: 148 (1964). Slender annual up to 45 cm. high with a loose, copiously branched panicle differing from *S. robinsoniana* in little except the achene; damp places in grassland.
Guin.: Friguiagbé to Toumou (Sept.) *Boismaré* 387 in *Hb. Chillou* 3905!

26. **S. striatinux** *De Wild.* in Rev. Zool. Afr. 14, Suppl. Bot.: 22, fig. 5 (1926), as *striatonux*; Nelmes in Kew Bull. 10: 429 (1955); Napper in Kew Bull. 25: 442 (1971). *S. lelyi* Hutch. & Dalz., F.W.T.A., ed. 1, 2:

493, name only. *S. rehmannii* C.B. Cl. var. *ornata* Cherm. in Bull. Jard. Bot. Brux. 13: 283 (1935). Rhizomatous perennial 45–60 cm. high with leaves 1–3 mm. wide; panicle 3·5–10 cm. long with 6–9 sessile glomerules on the axis and 2–4 simple branches from basal glomerules; in seasonally damp or swampy grassland.

Ghana: Wuropong (May) *Morton* A3928! Amedzofe (May, June) *Morton* A3426! GC 7213! **N. Nig.:** Maska, Katsina Prov. (June) *Keay* FHI 25900! Anara F.R. (May) *Keay* FHI 22936! Kurni, Kaduna to Keffi (June) *Kershaw* 900612! Jos Plateau (May) *Lely* P292! Naraguta (June) *Lely* P299! **S. Nig.:** Obudu Plateau, 5,500 ft. *Tuley* 970! **W. Cam.:** Mbaw Plain (Apr.) *Brunt* 387! Bali (Apr.) *Brunt* 1102! Bambulue, Bamenda (May) *Daramola* FHI 41058! Bali-Ngemba F.R. (May) *Ujor* FHI 30321! Bum, Bamenda (May) *Maitland* 1527! Throughout tropical Africa.

27. **S. rehmannii** *C.B. Cl.* in Fl. Capensis 7: 295 (1898); F.T.A. 8: 501; Nelmes in Kew Bull. 10: 425 (1955); E. A. Robinson in Kew Bull. 18: 507 (1966). *S. rehmannii* var. *ornata* of Berhaut, Fl. Sén. ed. 2, 358 (1967); not of Cherm. Slender perennial with a long rhizome bearing well-spaced stems up to 1·5 m. high with leaves 1·5–3·5 mm. wide; panicle 3–17 cm. long with 4–12 sessile glomerules on the axis and 1–7 simple branches from the lower glomerules; in seasonal and permanent swamps.

Sen.: Niokolo-Koba *Adam* 15913! 17088! **Guin.:** Sambaïlo (Oct.) *Pitot*! Sambaïlo stream (Oct.) *Pitot*! Widespread in tropical and South Africa.

28. **S. bulbifera** *A. Rich.* Tent. Fl. Abyss. 2: 510 (1851); F.T.A. 8: 500; Nelmes in Kew Bull. 10: 438 (1955); E. A. Robinson l.c. 503 (1966). *S. schweinfurthiana* Boeck. in Flora 62: 570 (1879); F.T.A. 8: 500; F.W.T.A., ed. 1, 2: 491. Polymorphic rhizomatous perennial with swollen culm-bases and leaves 1–5 mm. wide, the lower ones reduced almost to bladeless sheaths; spike up to 20 cm. long with 3–17 glomerules; damp places in woodland and savanna.

Sen.: Kanouméri, Kédougou to Bandiafassi (July) *Fotius* K215*a*! **Mali:** Kita (July) *Duong* 438! **Iv. C.:** Gouékouma, Upper Sassandra R. (May) *Chev.* 21668! Niangbo, nr. Tafiré (June) *Leeuwenberg* 4446! Téhini, Bouna Reserve (Aug.) *Oldeman* 315! **U. Volta:** Bandakoro to Sourkotomo (July) *Scholz* 301! **Ghana:** Bangmobangmo F.R., Nandom (Sept.) *Hall* CC 785! Damongo (Sept.) *Hall* CC 813! Banda (Apr.) *Morton* A3285! Yeji (July) *Vigne* 3354! Shiare to Chilinga, Nkwanta Hills (May) *Morton* A3991! **N. Nig.:** Patti Reserve, Lokoja (June) *Daramola* FHI 36933! Kaduna, Zaria Prov. (May) *Thatcher* S358! Kilba, bush country (July) *Dalz.* 266! Jos Plateau (May) *Lely* P293! Naraguta (June) *Lely* 300! **W. Cam.:** Mbiami to Mbaw (June) *Brunt* 734! Throughout tropical Africa and Madagascar.

29. **S. monticola** *Napper* in Kew Bull. 25: 444 (1971). Perennial with slender rhizome and slender stems 30–60 cm. high, leaves 1–2 mm. wide; spike 3–5 cm. long with 3–7 glomerules; grassland in rocky places.

S. L.: Loma Mts., 5,750 ft. (Nov.) *Morton* SL 2661! Bintumane Peak (Nov.) *Jaeger* 321! 472!

30. **S. dieterlinii** *Turrill* in Kew Bull. 1914: 20; Napper in Kew Bull. 25: 443 (1971). *S. schweinfurthiana* of F.W.T.A., ed. 1, 2: 491, partly. *S. flexuosa* of E. A. Robinson in Kew Bull. 18: 505 (1966), partly. Perennial up to 60 cm. high having long slender rhizomes with a terminal fleshy tuber; leaves 1–2 mm. wide; inflorescence a simple spike up to 10 cm. long with 7–10 glomerules; grassland in rocky places.

Guin.: Macenta (Oct.) *Baldwin* 9761! **S. L.:** Bumban *Deighton* 1248! Loma Mts. (Sept.) *Jaeger* 1045! 1763! 7712! *T. S. Jones* 74! **Iv. C.:** Nimba Mts. *Schnell*! Tonkoui Mt. (Sept.) *Schnell* 1732! 1738! Little Tonkoui (Sept.) *Schnell*! Western tropical Africa to Angola.

31. **S. hirtella** *Swartz* Prodr. Veg. Ind. Occ. 19 (1788); E. A. Robinson in Kew Bull. 18: 499 (1966). *S. melanotricha* of F.W.T.A., ed. 1, 2: 491, partly. *S. tricholepis* Nelmes in Kew Bull. 10: 447 (1955). *S. melanotricha* A. Rich. var. *glabrior* C.B. Cl. in F.T.A. 8: 496 (1902). Glabrous or villous slender annual up to 60 cm. high with leaves 1–3 mm. wide; inflorescence up to 13 cm. long with 4–15 glomerules, more widely spaced than in *S. pergracilis*; seasonally swampy or damp places.

Sen.: Niokolo-Koba (Dec.) *J. & A. Raynal* 6803! Gouloumbo to Tambacounda (Oct.) *Pitot*! Niassia (Nov.) *Berhaut* 6654! **Guin.:** Friguiagbé *Chillou* 598! Timbo (Oct.) *Pobéguin* 1790*a*! Dalaba (Sept.) *Adames* 335! Conakry (Oct.) *Adams* 12604! Macenta (Oct.) *Baldwin* 9791! **S. L.:** *Morson*! Regent (Dec.) *Sc. Elliot* 4187! Picket Hill (Nov.) *T. S. Jones* 194! Brookfields (Oct.) *Deighton* 2141! Kambia (Nov.) *Morton & Gledhill* SL 71! **Lib.:** Grand Bassa (Nov.) *Dinklage* 2300! Monrovia (Aug.) *Baldwin* 9204! Genne-Loffa, Kolahun Dist. (Nov.) *Baldwin* 10101! Palilah, Gbanga Dist. (Aug.) *Baldwin* 9152! Nimba Mts. (Dec.) *Adam* 16465! **Iv. C.:** Momi Mt. (Oct.) *Aké Assi* 9213! **Ghana:** Nyinahin Hills (Nov.) *Morton* A2790! **N. Nig.:** Idon, Kaduna to Keffi (Aug.) *Kershaw* 900712! Kufena Hill, Zaria Dist. (Sept.) *Clayton* 1315! Farin Rua, Plateau Prov. (Aug.) *J. Hall* 658! Panshanu, Bauchi Prov. (Aug.) *Lawlor & Hall* 621! **W. Cam.:** Mamfe (Mar.) *Richards* 5246! Widespread in tropical Africa and tropical America.

32. **S. pergracilis** (*Nees*) *Kunth* Enum. Pl. 2: 354 (1837); F.T.A. 8: 495; Nelmes in Kew Bull. 10: 445 (1955); E. A. Robinson in Kew Bull. 18: 494 (1966); Berhaut, Fl. Sén. ed. 2, 356 (1967). *Hypoporum pergracile* Nees in Edinb. New Phil. Journ. 17: 267 (1834). Slender annual 45 cm. high with leaves 1–2 mm. wide; inflorescence a simple 5–13 cm. long spike of 5–30 glomerules; seasonally wet and swampy places.

Sen.: Niokolo-Koba (Oct., Nov.) *Adam* 15728! 15839! 17021! Bignona (Nov.) *Berhaut* 6441! nr. Gambia R., E. Senegal (Sept.) *Fotius* K437! **Mali:** Kita Mts. (Oct.) *Jaeger* 53! 5768! Niger Delta (Dec.) *Duong*! Bamako *Adam* 15343! 15356! **Guin.:** Kaiserville to Youkounkoun (Dec.) *Pitot*! Youkounkoun (Oct.) *Pitot*! **Iv. C.:** Bouaké to Katiola (Sept.) *de Wilde* 995! Sanlo to Kakpin (Nov.) *Aké Assi* 9284! **Ghana:** Yagha, Wa to Lawra (Sept.) *Hall* CC 631! Gbinderi, Yendi to Gambaga (Oct.) *Rose Innes* GC 30490! Damongo (Dec.) *Morton* GC 9966! Bimbila (Jan.) *Akpabla* 1840! Ho to Adidome (Sept.) *Hall* GC 37060! **Togo Rep.:** Sokode to Basari *Schroeder* 127! **N. Nig.:** Nupe *Barter* 1006! Kufena Hill, Zaria Dist. (Sept.) *Clayton* 1306! Funtua to Sokoto (Sept.) *De Leeuw* 1547! Biu (Oct.) *De Leeuw* 1303! Gurum, Adamawa Prov. (Nov.) *Hepper* 1241! **S. Nig.:** Lagos *W. Macgregor* 243! Igboora (Oct.) *Haines* 217! Widespread in tropical Africa, also in Asia and New Guinea.

33. **S. nutans** *Kunth* Enum. Pl. 2: 351 (1837); E. A. Robinson l.c. 502 (1966). *S. hirtella* of F.T.A. 8: 497. Rhizomatous perennial 30–100 cm. high with solitary distant subseriate narrow stems bearing glabrous or villous leaves 1–3 (–5) mm. wide, the basal ones with much reduced blades; spike up to 18 cm. long with 4–15 glomerules; achene subglobose, 1·4–1·8 mm. long, 1–1·5 mm. diam., smooth or lightly tuberculate; swamps.

N. Nig.: Shere Mts., Bauchi Prov. (Aug.) *Hall* 555! Farin Rua, Plateau Prov. (Aug.) *J. Hall* 657! **S. Nig.:** Lagos, *W. MacGregor* 245! **W. Cam.:** Mamfe (Dec.) *Migeod* 280! Widespread in tropical and southern Africa, Madagascar and tropical America.

34. **S. aterrima** (*Ridley*) *Napper* in Kew Bull. 25: 445 (1971). *S. hirtella* Sw. var. *aterrima* Ridley in Trans. Linn. Soc., Ser. 2, 2: 166 (1884). *S. catophylla* C.B. Cl. in Dur. & Schinz, Consp. Fl. Afr. 5: 670 (1895); F.T.A. 8: 498; E. A. Robinson l.c. 501 (1966). *S. hirtella* of Boeck. in Linnaea 38: 439 (1874), partly; F.W.T.A., ed. 1, 2: 491; Berhaut, Fl. Sén. ed. 2, 356 (1967). Tufted rhizomatous perennial up to 1 m. high, with densely villous leaves crowded at the base of the stem; spike 3–13 cm. long with 4–15 glomerules; in seasonally swampy places.

Sen.: Niokolo-Koba (Dec., Feb.) *J. & A. Raynal* 6919! *Adam* 17505! Oussouye (Nov.) *Berhaut* 6492! Bayotte, Casamance (Nov.) *Berhaut* 6460! **Guin.:** Youkounkoun (Oct.) *Pitot*! Friguiagbé *Boismaré* 354! Kinkon to Pita (Oct.) *Pitot*! Madina Tossekré (Oct.) *Adam* 12567! Fetoré plain, Fouta Djalon (Sept.) *Adames* 354! **S. L.:** Dimbaia (Sept.) *Adames* 250! Kasokora to Bumban (Aug.) *Deighton* 1247! Mateboi (Oct.) *Glanville* 88! Loma Mt. (Sept.) *Jaeger* 7606! Mano Salija (Nov.) *Deighton* 342! **Lib.:** Monrovia (Aug.–Dec.) *Baldwin* 9179! *Kunkel* 60! *Dinklage* 3013! Paynesville (Sept.) *Van Dillewijn* 83! **Iv. C.:** N'Douci, N'zi R. (Aug.) *de Wit* 7391! Dimbokro (Aug.) *Roberty* 3077 (88)! Moossou, Grand Bassam (Aug.) *J. J. F. E. de Wilde* 221! **Ghana:** Banda (Aug.) *Hall* 1983! Kintampo (Sept.) *Hall* CC 904! Ejura (Dec.) *Morton* GC 9708! Tiasi, Afram Plains (Aug.) *Hall* CC 154! Essiama (Oct.) *Hall* 2662! **Togo Rep.:** Sokodé to Basari *Schroeder* 126! **Dah.:** Porto Novo (Sept.) *Adjanohoun* 283! **N. Nig.:** Nupe

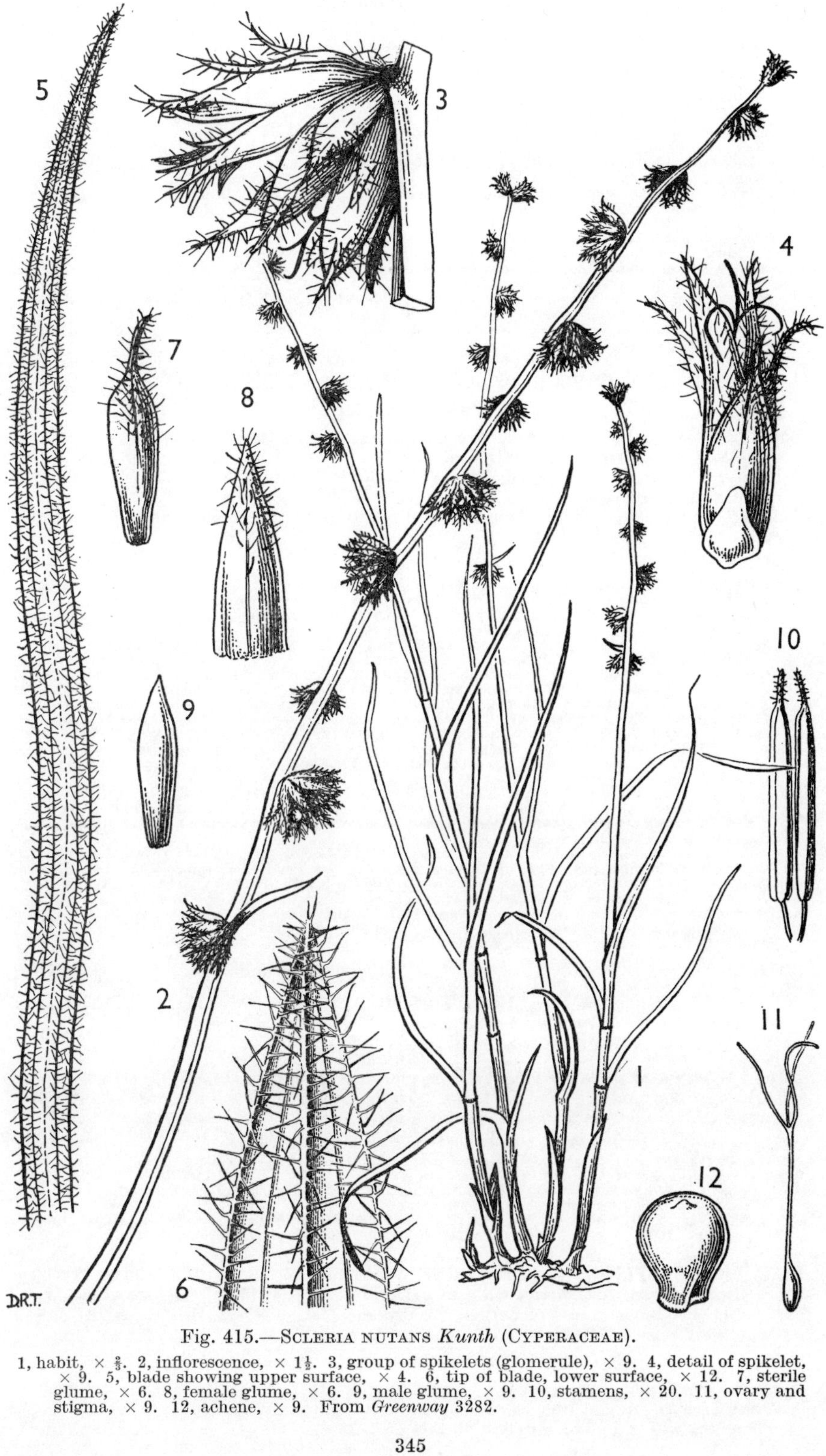

Fig. 415.—SCLERIA NUTANS *Kunth* (CYPERACEAE).

1, habit, × ⅔. 2, inflorescence, × 1⅓. 3, group of spikelets (glomerule), × 9. 4, detail of spikelet, × 9. 5, blade showing upper surface, × 4. 6, tip of blade, lower surface, × 12. 7, sterile glume, × 6. 8, female glume, × 6. 9, male glume, × 9. 10, stamens, × 20. 11, ovary and stigma, × 9. 12, achene, × 9. From *Greenway* 3282.

Barter 1561! Abinsi (Oct.) *Dalz.* 821! **S. Nig.:** Badagry (Aug.) *Onochie* FHI 33474! Enugu to Agbani Udi Dist. (May) *Onochie* FHI 35815! Widespread in tropical Africa and America.

35. **S. melanotricha** *Hochst. ex A. Rich.* Tent. Fl. Abyss. 2: 511 (1851); F.T.A. 8: 495; F.W.T.A., ed. 1, 2: 491, partly; E. A. Robinson l.c. 501 (1966); Berhaut, Fl. Sén. ed. 2, 356 (1967). Slender annual up to 45 cm. high with linear leaves; spike 5–20 cm. long with 5–17 sessile reflexed glomerules; achenes obovoid, compressed, 1·25–1·75 mm. long, 1·1–1·6 mm. diam., lacunose, often with three smooth longitudinal panels; in seasonally swampy places in shallow soils.
Sen.: E. Senegal (Sept.) *Fotius* K393! **Mali:** Kita Mts. (Sept.) *Jaeger* 76! 2672! 5769! Koulikoro (Oct.) *Chev.* 2467! **Guin.:** Conakry (Oct.) *Adam* 12602! Kinkon to Pita (Oct.) *Pitot*! Thianguel-Bori (Oct.) *Pitot*! **S. L.:** Konta to Kukuna (Nov.) *Morton & Gledhill* SL 51! Falaba (Oct.) *Small* 430! **Ghana:** Nakpanduri (Aug.) *Hall* CC 520! Bangmobangmo F.R., Nandom (Sept.) *Hall* CC 789! **N. Nig.:** (Aug.) *Lely* P489! Sangi F.R., Jemaa Div. (Sept.) *Killick* 72! Panshanu, Bauchi Prov. (Aug.) *Lawlor & Hall* 396! Gotel Mts. *Tuley* 1714! Widespread in tropical Africa.

36. **S. grata** *Nelmes* in Kew Bull. 10: 453 (1955). *S. melanotricha* of F.W.T.A., ed. 1, 2: 491, partly. Slender glabrescent or villous annual up to 22 cm. high with leaves not exceeding 2 mm. wide; spike 3–7 cm. long with 4–7 glomerules of sessile and shortly pedunculate clusters, otherwise very like *S. melanotricha*; in seasonally swampy places on shallow soils.
S. L.: Tingi Mts., Sankan Biriwa (Dec.) *Morton & Gledhill* SL 3109! **Iv. C.:** Momi Mt. (Oct.) *Aké Assi* 9212*a*! **N. Nig.:** Jos Plateau *Batten-Poole* 143! Vom, Jos Plateau (Feb.) *Saunders* 59! **W. Cam.:** Bambili, Bamenda, 6,000 ft. (Nov.) *Bauer* 122! Infrequent, but occurs over a wide area of tropical Africa.

20. DIPLACRUM R. Br., Prodr. 240 (1810); F.T.A. 8: 510 (1902).

By Miss D. M. Napper

Small herb up to 12 cm. high; leaves 3·5–5 cm. long, 2 mm. broad; inflorescence of subsessile or shortly pedunculate clusters of spikelets in the leaf-axils, 3–4 mm. broad; bracts subulate; spikelets 1–2 mm. long, glabrous; achene subglobose, longitudinally ribbed .. .. 1. *africanum*

Tall herb up to 100 cm. high; leaves long, 1 cm. broad; inflorescence of shortly pedunculate dense axillary clusters 1 cm. wide; spikelets 2–3 mm. long; achene subglobose, smooth .. .. 2. *longifolium*

1. **D. africanum** *C.B. Cl.* in Dur. & Schinz, Consp. Fl. Afr. 5: 668 (1895); F.T.A. 8: 510; Berhaut, Fl. Sén. ed. 2, 358 (1967); Napper in Kew Bull. 25: 445 (1971). *D. pygmaeum* Boeck. in Linnaea 38: 434 (1874), partly, not of (R. Br.) Nees. Annual, 2·5–12 cm. high, with green clusters of spikelets; in moist grassland and swamps, frequently associated with rice fields.
Sen.: Niokolo-Koba (Nov.) *Adam* 15962! 16996! Oussouyé, Casamance (Nov.) *Berhaut* 6639! **Port. G.:** Teixeira Pinto, Cacheu (Dec.) *Raimundo & Guerra* 548! Bissau, Prabis (Nov.) *Raimundo & Guerra* 171! **Guin.:** Is. Tristao *Jac.-Fél.* 7328! Senegambia *Heudelot* 675! **S. L.:** Morea to Fintonia, Kuru Hills *Morton & Gledhill* SL 597! Kitchom (Jan.) *Deighton* 946! Njala (Oct.) *Deighton* 1350! Magbema (Nov.) *Jordan* 699! Mano Salija (Nov.) *Deighton* 323! **Lib.:** Duport (Nov.) *Linder* 1446! **Iv. C.:** Singrobo (Dec.) *Aké Assi* 7238! **Ghana:** Lumbunga, Tamale Dist. (Dec.) *Morton* 6252! Banda to Menji, Wenchi Dist. (Dec.) *Morton* GC 25240! Ejura Scarp (Dec.) *Morton* GC 9631! Ankobra (Feb.) *Morton* A1629! **N. Nig.:** Nupe *Barter* 1041! Minna (Dec.) *Meikle* 727! Kainji (Nov.) *Hepper* 3897! Gurum, Sardauna Prov. (Nov.) *Hepper* 1239! **S. Nig.:** Olokemeji to Eruwa, Abeokuta Dist. (Nov.) *Daramola* FHI 43845! Sporadically throughout tropical Africa.
2. **D. longifolium** (*Griseb.*) *C.B. Cl.* in Dur. & Schinz, Consp. Fl. Afr. 5: 669 (1895); F.T.A. 8: 511. *Pteroscleria longifolia* Griseb., Fl. Brit. W. Ind. 579 (1864). *Scleria longifolia* (Griseb.) Roberty in Bull. I.F.A.N. 17, Sér. A: 37 (1955), not of Boeck. *Scleria amphigaea* Raymond in Nat. Canad. 91: 132 (1964). Stout perennial in swamps and streams.
S. L.: Freetown (Dec.) *Gledhill* 524! Kamakwie (Sept.) *Jordan* 551! Malema (Nov.) *Deighton* 333! Gegbwema to Faiama (Oct.) *Pyne* 27! Baoma (Dec.) *Deighton* 3805! **Lib.:** Gondolahun, Kolahun Dist. (Nov.) *Baldwin* 10103! Zuie, Boporo Dist. (Nov.) *Baldwin* 12049! Fishtown, Grand Bassa Dist. (Oct.) *Dinklage* 1867! Tappita (Aug.) *Baldwin* 9033! Ganta, Sanokwele Dist. (Oct.) *Harley* 1017! **Iv. C.:** Zétouo forest, Toulépleu Dist. (Oct.) *Aké Assi* 4842! **Ghana:** Domenasi Bridge to Nkroful (Aug.) *Hall* 3369! Tarkwa (Feb.) *Hall* 2484! **S. Nig.:** Lagos *Batten-Poole* 66! (Nov., Dec.) *Dalz.* 1294! *Hagerup* 760! Also in tropical S. America and the W. Indies.

21. COLEOCHLOA Gilly in Brittonia 5: 12 (1943); Nelmes in Kew Bull. 8: 374 (1953). *Eriospora* of A. Rich., Tent. Fl. Abyss. 2: 508 (1851); F.T.A. 8: 511 (1902); not of Berkeley & Broome.

By Miss D. M. Napper

Tufted perennial with flat or folded leaves, glabrous except for the shortly strigose upper surface of the midrib; panicle loose, terminal, with spikelets in dense clusters 5–6 mm. long; glumes dark chestnut brown, acuminate .. *abyssinica* var. *castanea*

Coleochloa abyssinica (*A. Rich.*) *Gilly* in Brittonia 5: 14 (1943). *Eriospora abyssinica* A. Rich., Tent. Fl. Abyss. 2: 508 (1851); F.T.A. 8: 512. Var. *abyssinica* occurs in Ethiopia and Eritrea.

C. abyssinica var. castanea (*C.B. Cl.*) *Pichi-Sermolli* in Webbia 7: 347 (1950); Nelmes in Kew Bull. 8: 377. *Eriospora abyssinica* var. *castanea* C.B. Cl. in F.T.A. 8: 513 (1902). Tufted perennial 60–100 cm. high; in crevices of rock pavements and outcrops.
N. Nig.: Vogel Peak, Sardauna Prov. (May) *Daramola* FHI 62913! **W. Cam.:** 4 miles S.W. of Bali, 4,000 ft. (Apr.) *Brunt* 1086! Also in Ethiopia, the Sudan, Uganda and Tanzania.

22. AFROTRILEPIS (Gilly) J. Raynal in Adansonia, sér. 2, 3: 258 (1963). *Trilepis* Nees subgen. *Afrotrilepis* Gilly in Brittonia 5: 18 (1943). *Eriospora* of F.T.A. 8: 511 (1902) partly, not of Berkeley & Broome. *Catagyna* of F.W.T.A., ed. 1, 2: 490 (1936), not of P. Beauv. ex Lestib.

By Miss D. M. Napper

A stout tufted plant often with an ascending or erect simple or branched rhizome; leaves at least 10 cm. long, glabrous, pilose or canescent on the back; beak of the achene scabrid or ciliate, rarely glabrous .. .. 1. *pilosa*

A procumbent cushion plant; leaves not over 3 cm. long; beak of the achene smooth 2. *jaegeri*

1. **A. pilosa** (*Boeck.*) *J. Raynal* l.c. 258 (1963). *Trilepis pilosa* Boeck. in Linnaea 39: 10 (1875). *Eriospora pilosa* (Boeck.) Benth. in Hook. Ic. Pl. 14: 30, t. 1342 (1881); F.T.A. 8: 511. *E. pilosa* var. *longipes* C.B. Cl. in F.T.A. 8: 512 (1902). *Catagyna pilosa* (Boeck.) Hutch. in F.W.T.A., ed. 1, 2: 490. *Afrotrilepis pilosa* (Boeck.) J. Raynal var. *trichocarpa* J. Raynal l.c. 258 (1963). Stout perennial often with an erect or ascending trunk-like rhizome covered with the remains of old leaves, or with roots, and golden-green to brown and chestnut spikelets in clusters on long or short peduncles; on granite outcrops rooting among rocks, or as a mat on the rock-slab surface.
Mali: *Duong* 1745! Sibi, 50 km. S. of Bamako (May) *Duong* 1594! **Guin.:** Friguiagbé (May) *Chillou* 524! Kindia *Pobéguin* 1363! Mamou (May) *Pitot*! Macenta (Oct.) *Baldwin* 9759! Nzérékoré (Apr.) *Pitot*! **S. L.:** Binkolo (Aug.) *Thomas* 1741! Kondembaia to Dalakuru (Apr.) *Deighton* 5072! Koinadugu Dist. (Aug.) *Haswell* 230! Bintimani, Loma Mts. (June) *Morton* SL 3574! Sefadu (Apr.) *Deighton* 3221! Mandu (July) *Deighton* 1966! **Lib.:** Yoma, Gola Forest (Apr.) *de Wilde* 3856! Palilah, Gbanga Dist. (Aug.) *Baldwin* 9147! Zwedru, Tchien Dist. (Aug.) *Baldwin* 7056! Mt. Bélé, Nimba (June) *Adam* 21385! Gio Forest (May) *Voorhoeve* 278! **Iv. C.:** Nimba Mts., 2,500–5,000 ft. (Aug.) *Boughey* GC 18086! Séguéla (Apr.) *Leeuwenberg* 3265! Mankono (July) *Aké Assi* 4806! Toulépleu *Adam* 6341! Issia rock (May) *Leeuwenberg* 4129! **U. Volta:** Sindou, 50 km. W. of Banfora (June) *Leeuwenberg* 4327! Toussiana Waterfall, Bobo-Dioulasso (June) *Scholz* 305! **Ghana:** Wa to Tumu (Apr.) *Addei* SLUS 20! 20 miles N. of Sawla (Apr.) *Morton* A3308! Kwahu Tafo to Mankrong (Apr.) *Morton* A686! Amedzofe (June) *Boughey* 13066! Afram Plains (Mar.) *W. H. Johnson* 708! **Dah.:** Agouagon (Apr.) *Chev.* 23528! **N. Nig.:** Nupe *Barter* 1560! Kufena, Zaria (May) *Kershaw* 900340! Anara F.R., Zaria (May) *Keay* FHI 22877! Lapai to Gawu, Niger Prov. (July) *Onochie* FHI 35392! Wana, Plateau Prov. (June) *Hepburn* 116! **S. Nig.:** Lagos *Dawodu* 59! *Rowland*! Ado Rock, Iseyin (Apr.) *Onochie* FHI 34689! Mt. Orosun, Ondo (Aug.) *Jones* 20708! Owo F.R. (July) *Onochie* FHI 33250! **W. Cam.:** Mamfe (June) *Gregory* 128! Bum, Bamenda, 4,000 ft. (June) *Maitland* 1603! Also in Gabon.
[*A. pilosa* is a very polymorphic species, and as used here the name covers a wide range of forms for which there seems no adequate means of characterization except perhaps in the case of var. *trichocarpa* J. Raynal from the Ivory Coast. Rather than describe a number of new varieties of doubtful validity it seems preferable to avoid categorizing the infra-specific variation for the time being.]

2. **A. jaegeri** *J. Raynal* l.c. 259 (1963). A low growing perennial herb with branching leafy woody stem and few very dark spikelets.
S. L.: Da Oulen, Loma Mts. (Aug., Oct.) *Jaeger* 1331! 7822! 7869! Dawuli Peak, Loma Mts. (June) *Morton* SL 3611! Sankan Biriwa, 5,000 ft., Tingi Mts. (Apr.) *Morton & Gledhill* SL 1895!

23. MICRODRACOIDES Hua in Bull. Mus. Hist. Nat. Paris 12: 421 (1906). *Schoenodendron* Engl., Bot. Jahrb. 44, Beibl. 101: 34 (1910).

By Miss S. S. Hooper

Dioecious; forming branched 'trunks' up to 100 cm. high and several cm. across; leaves in a close-set spiral forming a terminal tassel, resembling pine-needles, their bases clothing the stem and producing the 'trunks'; inflorescence elongate, branched, erect .. .. *squamosus*

M. squamosus *Hua* in Bull. Mus. Hist. Nat. Paris 12: 422 (1906); Chev. in La Terre et la Vie 3: 131 (1933); Cherm. in Bull. Soc. Bot. Fr. 80: 90 (1933). *Vellozia*-like in habit and unlike any other W. African sedge; on wet granitic outcrops.
Guin.: Friguiagbé (July) *Chillou* 3295! Kindia *Pobéguin*! Mt. Gangan (Nov.) *Schnell* 2187! Grandes Chutes (Dec.) *Chev.* 20222! Kinkon (Apr.) *Pitot*! **S. L.:** Gbengbe Hills, Bumban (Aug., Nov.) *Deighton* 1309! *Morton* SL 2917! **S. Nig.:** Boki Hills, 3,000 ft. *Catterall* 14/1934! **W. Cam.:** Cross R. (Dec.) *Schinz*! Mamfe (June, Dec.) *Gregory* 127! *Morton* K685! *Migeod* 259! Fongom, Bamenda, 3,000 ft. *Maitland*! Also in E. Cameroun.

24. CAREX Linn., Sp. Pl. 972 (1753), and Gen. Pl. ed. 5, 420 (1754); F.T.A. 8: 514 (1902); Kükenthal in Engl., Pflanzenr. Cyperaceae—Caricoideae: 67–767 (1908).

By Miss D. M. Napper

Inflorescence with 5–10 dense cylindric spikes 4–6 cm. long on long solitary peduncles:
Terminal spike male, usually with some utricles in the upper part; spikes 3–6; perennial with pale brown or straw-coloured sheaths to the lower leaves 1. *preussii*
Terminal spike male throughout, lateral spikes female or the upper ones with some male flowers at the apex; spikes more numerous; perennial with rich purplish-red sheaths to the lower leaves .. .. 2. *mannii*
Inflorescence a spreading or very dense panicle with numerous sessile short spikes up to 1·5 mm. long; perennials with pale brown or straw-coloured sheaths to the lower leaves:
Panicle lax:
Glumes broadly acute with a 1–2 mm. long scabrid awn; panicle scanty; utricle sparingly ciliolate above .. .. 3. *neo-chevalieri*
Glumes acuminate or with a short smooth or scabrid awn not exceeding 1 mm. long; panicle much branched; utricle glabrous:
Utricle 3 mm. long including the stout 0·75 mm. long beak .. 4. *echinochloë*
Utricle 4·5–5 mm. long including the slender 1·5–2 mm. long beak 5. *chlorosaccus*
Panicle very dense, almost spike-like.. .. .. 6. *lycurus*

1. **C. preussii** *K. Schum.* in Engl., Bot. Jahrb. 24: 340 (1897). *C. simensis* of F.T.A. 8: 522, partly; F.W.T.A., ed. 1, 2: 495. *C. preussii* var. *camerunensis* Nelmes in Kew Bull. 1938: 247. *C. longipedunculata* K. Schum. var. *preussii* (K. Schum.) Kük. 1. c. 652 (1909). Tufted perennial 30–100 cm. high with scanty inflorescence; spikelets usually rust and green, but sometimes redder; in upland forest.
N. Nig.: Gotel Mts., Mambila Plateau *Tuley* 2033! **W. Cam.:** Cam. Mt. (Dec.) *Mann* 2099! Ukile, Mann's

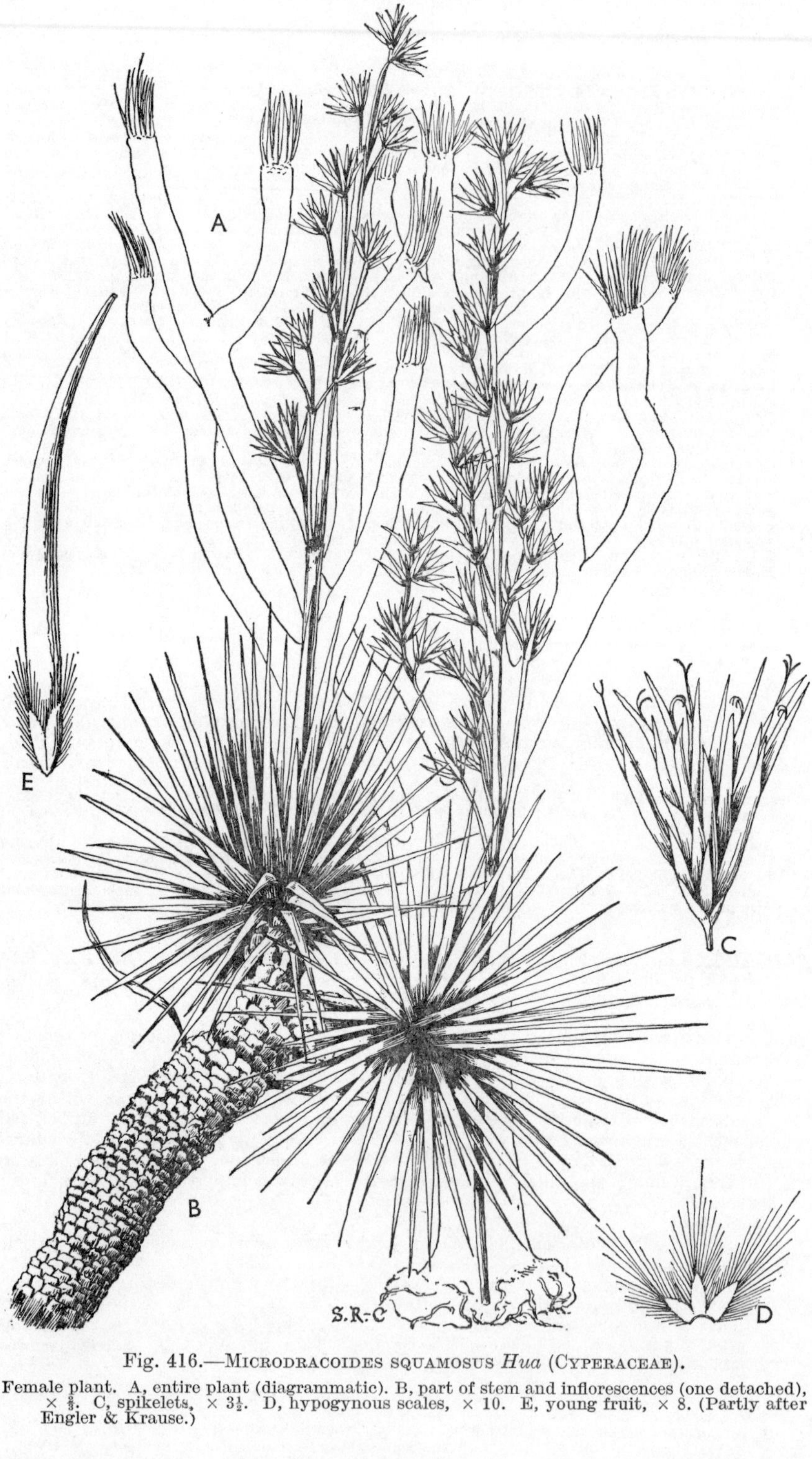

Fig. 416.—Microdracoides squamosus *Hua* (Cyperaceae).

Female plant. A, entire plant (diagrammatic). B, part of stem and inflorescences (one detached), × ⅔. C, spikelets, × 3½. D, hypogynous scales, × 10. E, young fruit, × 8. (Partly after Engler & Krause.)

Spring (Mar.) *Brenan & Onochie* 9538! Bafut-Ngemba F.R., Bamenda (Feb.) *Hepper* 2085! Mba Kokeka, Bamenda (Mar.) *Coombe* 229! L. Oku, Bamenda (Jan.) *Keay & Lightbody* FHI 28465!

2. **C. mannii** *E. A. Bruce* in Kew Bull. 1933: 150. *C. boryana* of F.T.A. 8: 523, not of Schkuhr. *C. boryana* Schkuhr var. *simplicissima* Kük. l.c. 651 (1909). Tufted perennial up to 140 cm. high with 6–10 solitary dark reddish brown spikes; in forest.
W. Cam.: Ukele Camp, Cam. Mt., 6,000 ft. (Feb.) *Maitland* 1341! **F. Po:** (Dec.) *Boughey* GC 10794! St. Isabel Peak *Guinea* 2683! Clarence Peak, 7,500–8,500 ft. (June–Dec.) *Mann* 661! 1478! Also in eastern Africa.

3. **C. neo-chevalieri** *Kük.* in Bull. Mus. Nat. Hist. Paris sér. 2, 3: 467 (1931). *C. echinochloe* of F.W.T.A., ed. 1, 2: 495, partly. Tufted perennial 60–100 cm. high with scanty panicles of light brown spikes with glabrous or shortly hairy glumes; upland forests, rarely below 4,000 ft.
Guin.: La Sala Falls (Oct.) *Pitot*! **Mali:** (Nov.) *Chev.* 37803! **S. L.:** Séréleu Massif (Sept.) *Jaeger* 7665! Bintimani Mts., 4,600 ft. (July) *Bakshi* 283! **Ghana:** Shiare, Togo Hills (Nov.–Dec.) *Morton* A4078! *Jenik & Hall* CC 1128! **N. Nig.:** Richa, Jos Plateau, 4,000 ft. (Sept.) *Lawlor & Hall* 664*b*! Also from E. Cameroun and C. African Republic to Burundi.

4. **C. echinochloë** *G. Kunze* Suppl. Schkuhr's Riedgr. 47, t. 12 (1841); F.T.A. 8: 518; Kük. l.c. 270 (1909); F.W.T.A., ed. 1, 2: 495, partly, excl. synonymy. Tufted perennial about 1 m. high with simple or compound panicles of short greenish spikes; utricles ciliolate; forests and shade, 5,500–8,000 ft.
W. Cam.: Cam. Mt., 7,000 ft. (Jan., Nov.) *Mann* 1359! 2106! Mann's Spring, Cam. Mt. (June, Dec.) *Mildbraed*! *Boughey* GC 12609! Bafut-Ngemba F.R., Bamenda, 7,300 ft. (Feb.) *Hepper* 2191! L. Bambulue, 5,500 ft. (July) *Bumpus* 46/Bam/29! Also on mountains in eastern and north-eastern Africa.

5. **C. chlorosaccus** *C.B. Cl.* in J. Linn. Soc. 34: 298 (1899); F.T.A. 8: 519. *C. echinochloë* G. Kunze var. *chlorosaccus* (C.B. Cl.) Kük. l.c. 271 (1909). Tufted perennial sedge 60–100 cm. high with pale brown or greenish basal leaf-sheaths and much branched panicles with short green spikes; montane forest 6,000–8,000 ft.
F. Po: S.W. of L. Biaó (Sept.) *Wrigley & Melville* 663! Clarence Peak, 6,000–8,000 ft. (Dec.) *Boughey* 126! *Mann* 653! Also in eastern and northern tropical Africa.

6. **C. lycurus** *K. Schum.* in Engl., Pflanzenw. Ost-Afrika C: 129 (1895); F.T.A. 8: 517; Kük. l.c. 127. Perennial 30–60 cm. high with a long creeping rhizome, pale brown basal leaf-sheaths and a dense, light yellowish brown inflorescence; montane forests 7,000–8,000 ft.
W. Cam.: Bambili Lakes, Mezam Division (May) *Bauer* 67! Also east and north-east tropical Africa.
[The inflorescence of this specimen is very immature but apart from a slightly smaller and more slender habit than is usually met with on the mountains of East Africa there are no significant differences.]

202. GRAMINEAE*

By W. D. Clayton

Annual or perennial herbs, rarely shrubs or trees, sometimes with rhizomes or stolons; stems erect, ascending or creeping, usually branched at the base, in perennials with sterile shoots and flowering stems (*culms*) mixed, in annuals only the latter present; culms cylindrical, rarely flattened, jointed, usually hollow in the internodes, closed at the nodes; branches subtended by a leaf, and with a 2-keeled hyaline leaflet (*prophyll*) at the base. Leaves solitary at the nodes, sometimes crowded at the base of the stem, alternate and 2-rowed, consisting of sheath, ligule and blade; sheaths encircling the culm, with the margins free and overlapping or ± connate, frequently swollen at the base, the shoulders sometimes extended upwards into triangular auricles; ligule adaxial, placed at the junction of sheath and blade, membranous or reduced to a fringe of hairs, rarely absent (very rarely with a similar abaxial structure—the external ligule); blades usually long and narrow, rarely broad, flat or sometimes rolled or terete, parallel-nerved, rarely with transverse connections, usually passing gradually into the sheath, sometimes amplexicaul or with falcate auricles, rarely narrowed into a false petiole or articulated with the sheath. Inflorescence made up of spikelets arranged in a panicle, or in spikes or racemes, these either solitary, digitate, or disposed along a central axis; usually terminal, sometimes (especially in *Andropogoneae*) numerous, each inflorescence being subtended by a bladeless sheath (*spatheole*) and the whole flowering branch system condensed into a leafy false panicle. Spikelets consisting of bracts distichously arranged along a slender axis (*rhachilla*); the two lower bracts (*glumes*) empty; the succeeding 1 to many bracts (*lemmas*) each enclosing a flower and opposed by a hyaline scale (*palea*), the whole (lemma, palea and flower) termed a *floret*; base of spikelet or floret sometimes with a horny prolongation downwards (*callus*); glumes or lemmas often bearing 1 or more stiff bristles (*awns*); this basic pattern of spikelet structure consistent throughout the family, though often much modified by reduction, suppression or elaboration of parts. Flowers usually ⚥, sometimes unisexual, small and inconspicuous; perianth represented by 2, rarely 3, minute hyaline or

* The traditional name *Gramineae* is permitted by the International Code of Botanical Nomenclature, which also provides for the use of *Poaceae*, based upon the type genus *Poa* L., as an alternative.

fleshy scales (*lodicules*); stamens hypogynous, 1–6, rarely more, usually 3, with delicate filaments and 2-thecous anthers opening by a longitudinal slit or rarely a terminal pore; ovary 1-locular, with 1 anatropous ovule often adnate to the adaxial side of the carpel; styles usually 2, rarely 1 or 3, generally with plumose stigmas. Fruit mostly a caryopsis with thin pericarp adnate to the seed, rarely with free seed, still more rarely a nut or berry; caryopsis commonly combined with various parts of the spikelet, or less often the inflorescence, to form a false fruit; seed with starchy endosperm, an embryo at the base of the abaxial face, and a point or line (*hilum*) on the base or adaxial face marking the connection between pericarp and seed.

About 660 genera and 9,000 species, in 50–60 tribes; throughout the world.

Key to the tribes

Spikelets 1- to many-flowered, breaking up at maturity above the more or less persistent glumes, or if falling entire then *not* 2-flowered with the lower floret male or barren and the upper female or hermaphrodite; spikelets usually laterally compressed or terete (to p. 351):
 Spikelets unisexual, dissimilar, the sexes mixed or in different parts of the same inflorescence:
  Leaf-blades with the lateral nerves parallel to the midrib; ♀ lemma indurated, shorter than the glumes .. .. II. OLYREAE (p. 352)
  Leaf-blades with slanting lateral nerves running obliquely from the midrib to the margin; ♀ lemma papery, much longer than the glumes III. PHAREAE (p. 352)
 Spikelets bisexual, similar:
  Tall woody bamboos, less often perennial herbs; leaf-blades flat, lanceolate to ovate, many-nerved, often with transverse veins, usually with a petiole-like base and articulated with the sheath; lemmas several, 5- to many-nerved, awnless; lodicules usually 3 .. .. I. BAMBUSEAE (p. 351)
  Perennial or annual herbs with herbaceous culms; leaf-blades usually sessile and not articulated with the sheath, or if with a petiole-like base then not with the other characters given above:
   Spikelets borne on opposite sides of the rhachis of solitary spikes or racemes, the spikelets placed broadside on to the rhachis:
   Spikelets borne upon a pedicel 1·5 mm. long, several-flowered
   VIII. BRACHYPODIEAE (p. 352)
   Spikelets quite sessile, or single-flowered .. .. IX. TRITICEAE (p. 352)
   Spikelets in panicles or 1-sided spikes or racemes; very rarely on opposite sides of the rhachis of solitary spikes, but then edgeways on with the back of the lemmas pressing against the rhachis:
    Leaf-blades transversely veined between the main nerves, narrowly lanceolate to ovate; spikelets several-flowered:
     Spikelets in a 1-sided raceme; lemmas awned IV. STREPTOGYNEAE (p. 352)
     Spikelets in a panicle; lemmas awnless .. XVI. CENTOTHECEAE (p. 353)
    Leaf-blades without transverse veins, linear:
     Lemmas deeply cleft into 9 lobes or awns XVII. PAPPOPHOREAE (p. 353)
     Lemmas entire or bilobed, rarely 3-awned:
      *Spikelets containing 2 or more fertile florets (except *Leptochloa uniflora*):
       Rhachilla internodes bearing long silky hairs which envelop the lemmas, the latter glabrous (in only W. African genus); tall grasses with large plume-like panicles.. .. .. XIII. ARUNDINEAE (p. 352)
       Rhachilla internodes glabrous or shortly hairy, in the latter case with the hairs not enveloping the lemmas:
        Florets 2 per spikelet, one or both of the florets hardened or leathery; spikelets awnless .. .. XXIV. ISACHNEAE (p. 354)
        Florets nearly always 3 or more per spikelet; lemmas membranous or awned:
         Inflorescence made up of racemes either solitary, digitate or scattered along an axis .. .. XIX. ERAGROSTIDEAE (p. 353)
         Inflorescence an open or contracted panicle, or globose cluster:
          Glumes shorter than the lowest floret with the upper florets distinctly exserted; lemmas awnless or with straight awns from the entire, or lobed tip:
           Lemmas 1–3-nerved:
            Spikelets awnless.. .. .. XIX. ERAGROSTIDEAE (p. 353)
            Spikelets awned .. .. .. XIV. DANTHONIEAE (p. 352)
           Lemmas 5 to many-nerved:
            Lemmas bearing long slender awns, closely coiled and entangled towards the tips (in the only West African genus) .. X. MELICEAE (p. 352)

Lemmas awnless or with stiff straight awns:
Spikelets densely imbricate in short globose or oblong heads XVIII. AELUROPODEAE (p. 353)
Spikelets arranged in open or contracted panicles:
Ovary glabrous or hairy, the styles arising from its tip; lemmas awnless or awned from the tip (except *Festuca mekiste* which has glabrous leaf-sheaths and scabrid palea keels) .. VI. POEAE (p. 352)
Ovary crowned by a fleshy hairy cap, the styles arising from beneath the latter; lemmas awned from just below the tip (leaf-sheath hairy and palea keels ciliolate in W. African species) VII. BROMEAE (p. 352)
Glumes longer than the lowest floret, usually as long as the spikelet and enclosing the florets, rarely shorter but then the lemmas with geniculate awns; lemmas 5- to many-nerved:
Lemmas with a geniculate dorsal awn, rarely awnless (and then with pubescent inflorescence axis, otherwise see *Festuca*); ligule membranous XI. AVENEAE (p. 352)
Lemmas with a straight or geniculate awn from the sinus of the prominently 2-lobed tip; ligule a ciliate fringe XIV. DANTHONIEAE (p. 352)
*Spikelets with 1 fertile floret (very rarely more, *Tetrapogon*, and then with long-ciliate lemmas), with or without 1 or 2 male or barren florets below it, and 1 or more above:
Glumes very minute or suppressed (in *Oryza* 2 sterile lemmas at the base of the floret simulate glumes, the true glumes being barely discernable as minute collars at the tip of the pedicel); palea 3-nerved, similar to the lemma in texture; stamens often 6; spikelets laterally compressed V. ORYZEAE (p. 352)
Glumes usually well-developed, or at least the upper; palea usually 2-nerved and hyaline; stamens 3 or less; spikelets usually not flattened:
Spikelets falling entire at maturity, either singly or in clusters, from the persistent axis of slender spike-like panicles or racemes XXII. ZOYSIEAE (p. 354)
Spikelets breaking up at maturity above the persistent glumes (rarely embedded in a segment of the fragile rhachis and falling with it):
Inflorescence made up of racemes or spikes (solitary or digitate in the West African species) .. .. XX. CHLORIDEAE (p. 353)
Inflorescence a panicle, either open or contracted and spiciform:
Florets 2, the lower barren or male, the upper hermaphrodite XXIII. ARUNDINELLEAE (p. 354)
Floret 1:
Spikelets awnless .. .. .. XXI. SPOROBOLEAE (p. 354)
Spikelets awned (in the West African species):
Awn divided into 3 branches (except *Aristida diminuta*); lemma terete and rigid at maturity; ligule ciliolate XV. ARISTIDEAE (p. 353)
Awn entire; ligule membranous.. XII. AGROSTIDEAE (p. 352)
Spikelets 2-flowered, falling entire at maturity, with the upper floret (of fertile spikelets) female or hermaphrodite and the lower male or barren and in the latter case often much reduced; spikelets usually dorsally compressed:
Spikelets all hermaphrodite, or with male or barren and hermaphrodite spikelets mixed in the same inflorescence:
Spikelets solitary, rarely paired with the spikelets all more or less similar; glumes usually membranous, the lower usually smaller or sometimes suppressed; upper lemma papery to polished and stony, usually awnless XXV. PANICEAE (p. 354)
Spikelets typically paired, with 1 sessile and the other pedicelled, those of each pair usually dissimilar in shape and sex (the pedicelled sometimes much reduced), rarely with the spikelets all alike; glumes as long as the spikelets and enclosing the florets, more or less rigid and firmer than the hyaline or membranous lemmas; upper lemma often with a geniculate awn XXVI. ANDROPOGONEAE (p. 356)
Spikelets unisexual, the male and female spikelets in separate inflorescences or in different parts of the same inflorescence .. .. XXVII. MAYDEAE (p. 358)

Key to the genera

Tribe I—Bambuseae

Tall woody bamboos with culms up to 12 m. high:
Stamens 3, filaments free; leaf-blades attenuate to a filiform tip about 10 mm. long 1. **Arundinaria**
Stamens 6, leaf-blades acuminate to a pungent tip about 5 mm. long:
Filaments free; mouth of leaf-sheath auriculate, glabrous .. 2. **Bambusa**
Filaments united; mouth of leaf-sheath without auricles, long-ciliate 3. **Oxytenanthera**

Herbaceous plants with culms up to about 1 m. high:
  Spikelets linear, with several to many bisexual florets, the rhachilla distinct; filaments of stamens free .. .. 4. **Guaduella**
  Spikelets narrowly ovate, with 3–5 male and a terminal female floret, the rhachilla represented by a rounded callus at the base of each floret; filaments of stamens united .. .. 5. **Puelia**

Tribe II—Olyreae

Only West African genus .. .. 6. **Olyra**

Tribe III—Phareae

Only West African genus .. .. 7. **Leptaspis**

Tribe IV—Streptogyneae

Only genus .. .. 8. **Streptogyna**

Tribe V—Oryzeae

Spikelets with 2 reduced sterile lemmas at the base of the fertile floret; fertile lemma coriaceous .. .. 9. **Oryza**
Spikelets without sterile lemmas; fertile lemma papery .. .. 10. **Leersia**

Tribe VI—Poeae

Lemmas awned, rounded on the back or only keeled at the tips:
  Perennial; lemmas lanceolate to oblong .. .. 11. **Festuca**
  Annual; lemmas linear-lanceolate .. .. 12. **Vulpia**
Lemmas awnless, compressed and keeled .. .. 13. **Poa**

Tribe VII—Bromeae

Only West African genus .. .. 14. **Bromus**

Tribe VIII—Brachypodieae

Only West African genus .. ..15. **Brachypodium**

Tribe IX—Triticeae

Spikelets solitary at each node of the rhachis, several-flowered .. 16. **Triticum**
Spikelets in clusters of three at each node, one-flowered .. .. 17. **Hordeum**

Tribe X—Meliceae

Only West African genus .. .. 18. **Streblochaete**

Tribe XI—Aveneae

Lemmas awnless or mucronate .. .. 19. **Koeleria**
Lemmas long awned:
  Annual; spikelets 2-flowered; rhachilla glabrous; lemmas 2-toothed at apex 20. **Aira**
  Perennials; spikelets 2–6-flowered; rhachilla more or less hairy:
    Lemma bifid at the tip, awned from the middle or above the middle of the back 21. **Helictotrichon**
    Lemma 4-toothed at the tip, awned from near the base of the back 22. **Deschampsia**

Tribe XII—Agrostideae

Glumes 1–3-nerved; lemma hyaline at maturity, truncate or denticulate truncate with percurrent nerves .. .. 23. **Agrostis**
Glumes 5-nerved; lemma indurated and rigid at maturity, 2-toothed at apex 24. **Hypseochloa**

Tribe XIII—Arundineae

Only West African genus .. .. 25. **Phragmites**

Tribe XIV—Danthonieae

Spikelets in open or contracted panicles:
  Lemmas 5–many-nerved; glumes as long as the spikelet:
    Lobes of the lemma acute, transversely bearded at their base; glumes 5–11-nerved; spikelets 2–6-flowered .. .. 26. **Asthenatherum**

Lobes of the lemma produced into a fine bristle, not bearded; glumes 1–3-nerved; spikelets 2-flowered .. .. 27. **Pentaschistis**
Lemmas 3-nerved and 3-awned; glumes shorter than the spikelet .. 28. **Triraphis**
Spikelets in globose clusters, these sometimes confluent into a cylindrical spike 29. **Elytrophorus**

Tribe XV—Aristideae

Awns, at least the central one, plumose; column more or less developed, articulated at its base .. .. 30. **Stipagrostis**
Awns never plumose .. .. 31. **Aristida**

Tribe XVI—Centotheceae

Spikelets mostly 3-flowered; upper lemmas with reflexed bristles .. 32. **Centotheca**
Spikelets mostly 10–20-flowered; lemmas glabrous .. .. .. 33. **Megastachya**

Tribe XVII—Pappophoreae

Lemmas divided into 9 awns; fertile floret 1 .. .. 34. **Enneapogon**
Lemmas divided into 4 membranous lobes alternating with 5 awns; fertile florets usually 4 or more .. .. 35. **Schmidtia**

Tribe XVIII—Aeluropodeae

Only genus .. .. 36. **Aeluropus**

Tribe XIX—Eragrostideae

Spikelets pedicelled in open or contracted panicles .. .. .. 37. **Eragrostis**
Spikelets sessile or very short-pedicelled in secund 2-rowed spikes:
Spikes solitary and terminal; ligule ciliate .. .. 38. **Tripogon**
Spikes digitate or scattered along an axis:
Lemmas conspicuously ciliate on the margins .. .. .. 39. **Trichoneura**
Lemmas glabrous or apparently so (pubescent at the base beside the keel in *Bewsia*):
Racemes deciduous at maturity with the spikelets attached .. 40. **Dinebra**
Racemes not deciduous:
Spikes digitate or subdigitate:
Spikes terminating in a sharp point .. .. .. 41. **Dactyloctenium**
Spikes terminating in a spikelet:
Lemmas drawn out into a short awn-point; racemes whorled on a short common axis .. .. 42. **Acrachne**
Lemmas at most acute:
Glumes strongly keeled, the spikelets laterally compressed .. 43. **Eleusine**
Glumes rounded on the back, the spikelets plump .. .. 44. **Coelachyrum**
Spikes scattered on a long central axis:
Awn of the lemma dorsal; callus bearded .. .. .. 45. **Bewsia**
Awn of the lemma (if any) apical or from between 2 teeth:
Ligule reduced to a line of hairs .. .. 46. **Pogonarthria**
Ligule hyaline:
Lemmas acuminate to a short awn-point; spikelets laterally compressed 42. **Acrachne**
Lemmas obtuse to acute, sometimes mucronate but then not compressed:
Spikelets 2–3 mm. long, 1–6-flowered, laterally compressed .. 47. **Leptochloa**
Spikelets 5–9 mm. long, 6–10-flowered, more or less terete .. 48. **Diplachne**

Tribe XX—Chlorideae

Spikelets with 1 (except *Tetrapogon*) fertile and 1 or more imperfect florets:
Fertile floret with 2 imperfect florets below and one above it; upper glume with an oblique dorsal awn .. .. 49. **Ctenium**
Fertile floret with imperfect florets only above it; upper glume without a dorsal awn:
Fertile florets 2–4; lemma villous, broadly ovate; raceme usually solitary 50. **Tetrapogon**
Fertile floret 1:
Spikes usually 2 or more:
Glumes unequal, the lower shorter than the florets .. .. 51. **Chloris**
Glumes equal or nearly so, as long as or slightly longer than the florets 52. **Chrysochloa**
Spikes solitary; lemma narrow, glabrous or scaberulous .. .. 53. **Enteropogon**
Spikelets with 1 fertile floret and no imperfect florets:
Spikelets with flexuous awns about 3 cm. long .. .. .. 54. **Schoenefeldia**

Spikelets awnless or with a short stiff awn up to 0·5 cm. long:
Upper and lower glumes much compressed laterally and both keeled:
Glumes longer than the floret and enclosing it; spike solitary .. 55. **Brachyachne**
Glumes shorter than the floret; spikes digitate .. .. .. 56. **Cynodon**
Upper glume with a flat back, the lower boat-shaped or much reduced; spikes solitary:
Lower (adaxial) glume present; spikelets arranged on one side of the slender rhachis, not sunken, awnless.. .. .. 57. **Microchloa**
Lower glume absent (except for the terminal spikelet); spikelets sunk in hollows on opposite sides of the rhachis, awned or awnless.. .. .. 58. **Oropetium**

Tribe XXI—Sporoboleae

Inflorescence a panicle, rarely spike-like .. .. 59. **Sporobolus**
Inflorescence an ovoid head .. .. 60. **Crypsis**

Tribe XXII—Zoysieae

Glumes long and finely awned; spikelets solitary .. .. 61. **Perotis**
Glumes awnless:
Spikelets clustered on short branches, these densely crowded into a narrow cylindrical inflorescence; lower glume very small or suppressed, the upper armed with hooked spines on the back .. .. 62. **Tragus**
Spikelets paired or solitary on flattened cuneiform branches 2–5 mm. long, these distant along a central axis; both glumes prominent, smooth on the back or the upper tuberculate, their margins pectinate .. .. 63. **Leptothrium**

Tribe XXIII—Arundinelleae

Callus of upper floret pungent, 3–3·5 mm. long; spikelets 26–28 mm. long in groups of three; lower lemma 5–7-nerved .. .. 64. **Tristachya**
Callus of upper floret obtuse, truncate, 2-toothed or obliquely 1-toothed:
Lemma of upper floret scabrid or scaberulous, the callus very short and broadly rounded; spikelets paired or solitary .. .. 65. **Arundinella**
Lemma of upper floret usually hairy, or if glabrous then quite smooth:
Spikelets in close clusters of 3, the pedicels distinct and 1–3 mm. long; rarely in groups of 2–3 on longer pedicels, but then the lobes of the upper lemma awned and the callus linear (0·7–1 mm. long) .. .. 66. **Loudetiopsis**
Spikelets paired or solitary:
Upper lemma evenly pubescent or pilose, rarely glabrous; callus oblong to linear, 0·5–1 mm. long, truncate, 2-toothed, or obliquely 1-toothed; lower lemma 3-nerved (except *L. togoensis*) .. .. 67. **Loudetia**
Upper lemma transversely bearded or with a tuft of hairs near each margin; callus short and obtuse:
Lower lemma 5–9-nerved; upper lemma lobes acute .. 68. **Danthoniopsis**
Lower lemma 3-nerved; upper lemma awned from the lateral lobes, with a tuft of hair near each margin .. .. 69. **Trichopteryx**

Tribe XXIV—Isachneae

Spikelets arranged in panicles .. .. 70. **Isachne**
Spikelets arranged in one-sided spikes .. .. 71. **Heteranthoecia**

Tribe XXV—Paniceae

Spikelets falling singly, not subtended by bristles, or if so then the bristles persisting on the axis after the spikelets have fallen (to p. 356):
Inflorescence an open or spike-like panicle (rarely the branchlets contracted about the primary branches—*Panicum laxum*):
Spikelets, or some of them, subtended by one to many long bristles 72. **Setaria**
Spikelets not subtended by bristles:
Spikelets arranged in cylindrical spike-like panicles .. .. 73. **Sacciolepis**
Spikelets arranged in open or contracted panicles:
Lower lemma entire at the tip, awnless; upper lemma crustaceous:
Spikelets distinctly gibbous and laterally compressed .. .. 74. **Cyrtococcum**
Spikelets not gibbous (sometimes slightly so, but then not laterally compressed):
Upper floret without lateral basal wings or scars .. .. 75. **Panicum**
Upper floret with lateral basal wings or scars .. .. .. 80. **Ichnanthus**
Lower lemma bilobed at the tip; upper lemma leathery or chartaceous:
Spikelets clothed in silky hairs extending beyond the tip of the spikelet (in West African species); upper glume gibbous or straight on the back; lower lemma shortly awned .. .. 95. **Rhynchelytrum**

Spikelets glabrous or at most pubescent; upper glume more or less straight on the back:
Upper glume narrower than the lower lemma, the latter awnless; upper lemma leathery .. .. 96. **Tricholaena**
Upper glume wider than the lower lemma, the latter usually awned; upper lemma thinner, chartaceous .. .. 97. **Melinis**
Inflorescence consisting of one-sided spikes or racemes, these either digitate or scattered along a central axis, rarely solitary; the racemes sometimes with short secondary racemelets (especially *Echinochloa crus-pavonis* and *Entolasia olivacea*) or with the spikelets long-pedicelled and distant (*Brachiaria deflexa*):
Racemes very short, bearing 1–5 spikelets, and sunken in hollows in a thickened corky rhachis, the latter disarticulating at maturity with the spikelets attached; lower glume abaxial; both lemmas coriaceous .. .. 76. **Stenotaphrum**
Racemes free, usually bearing numerous spikelets:
Upper lemma hard, crustaceous or coriaceous, usually with narrow inrolled margins and exposing much of the palea:
Glumes and lemmas with laterally compressed thickened apices; lower glume abaxial .. .. 77. **Acroceras**
Glumes and lower lemma obtuse to acuminate or awned, but not thickened and laterally compressed at the tip:
Leaf-blades broadly lanceolate to ovate-elliptic, with distinct transverse veins below; spikelets glabrous, awnless; back of upper lemma glabrous:
Upper glume half as long as the spikelet .. .. 78. **Microcalamus**
Upper glume as long as the spikelet:
Base of upper lemma rounded, without lateral wings 79. **Commelinidium**
Base of upper lemma extended into a short stipe flanked by hyaline wings 80. **Ichnanthus**
Leaf-blades linear to lanceolate, transverse veins absent (or if present then spikelets not as above); upper lemma without basal wings:
Spikelets laterally compressed:
Spikelets lanceolate, hairy with stiff bristles based in yellowish tubercles 81. **Chloachne**
Spikelets obliquely ovate, the upper glume gibbous towards the base and usually clothed in hooked hairs when mature .. .. 82. **Pseudechinolaena**
Spikelets dorsally compressed:
Spikelet with a bead-like swelling at its base, formed from the swollen lowest rhachilla-internode covered by the thin lower glume; upper lemma obtuse, mucronate (if not mucronate see *Brachiaria callopus*) .. 83. **Eriochloa**
Spikelet passing smoothly into the pedicel without a bead-like swelling:
Lower glume present:
Glumes acuminate or awned, or if not then the spikelets subsessile in 4 rows; upper lemma not obtuse and mucronate:
Racemes 1–2-rowed; culms creeping or ascending; leaf-blades lanceolate 84. **Oplismenus**
Racemes 4 or more rowed; culms erect or suberect, sometimes floating; leaf-blades linear .. .. 85. **Echinochloa**
Glumes obtuse to acute, if acuminate then with the upper lemma mucronate; spikelets usually in 1–2 rows, sometimes irregularly arranged with some of the spikelets pedicelled:
Lower glume abaxial (facing away from the rhachis); spikelets solitary, sessile (in West African spp.):
Upper lemma acute; spikelets glabrous .. .. .. 86. **Paspalidium**
Upper lemma obtuse, mucronate; spikelets hairy .. 87. **Urochloa**
Lower glume adaxial (facing towards the rhachis), or with some of the spikelets pedicelled:
Upper lemma glabrous, usually shiny or rugose.. .. 88. **Brachiaria**
Upper lemma hairy.. .. .. 89. **Entolasia**
Lower glume absent:
Back of upper lemma facing axis; spikelets plano-convex, ovate to orbicular 90. **Paspalum**
Back of upper lemma turned away from axis; spikelets lanceolate 91. **Axonopus**
Upper lemma thinly cartilaginous, usually with flat thin margins covering much of the palea:
Spikelets awned:
Upper glume and lower lemma with fine entangled awns 15 mm. long; upper lemma awnless; lower glume adaxial .. .. 92. **Acritochaete**
Upper glume and lower lemma awnless, or with an awn point up to 1 mm. long; upper lemma awned; lower glume abaxial; spikelets in untidy groups of 2–3 93. **Alloteropsis**

Spikelets awnless, very rarely with an awnlet up to 1·5 mm. long; lower glume (if present) abaxial, very small .. .. 94. **Digitaria**

Spikelets falling singly or in clusters, surrounded by 1 or more bristles which fall with the spikelets; if without bristles, then falling in clusters:

Spikelets in clusters of 3–9, not surrounded or subtended by bristles; lower glume indurated .. .. 98. **Anthephora**

Spikelets surrounded or subtended by 1 to many bristles; lower glume not indurated:

Bristles solitary below each spikelet:

Racemes loose; spikelets 6–12 mm. long; peduncles 2–15 mm. long and falling with the spikelet .. .. 99. **Paratheria**

Racemes dense; spikelets about 3 mm. long; peduncles up to 1 mm. long and remaining on the axis after the spikelet has fallen .. .. 100. **Beckeropsis**

Bristles few to many, forming an involucre around the solitary or clustered spikelets:

Bristles free to the base, fine to very fine, more or less filiform .. 101. **Pennisetum**

Bristles connate at base or to beyond the middle, rigid or spinous, at least the longest bristles flattened at the base .. .. 102. **Cenchrus**

Tribe XXVI—Andropogoneae

Internodes of rhachis and pedicels slender, filiform or linear, rarely thickened upwards and then with the upper lemma awned:

Racemes arranged in panicles with a common axis longer than the lowest raceme, not supported by spathes; pedicels solid:

Spikelets of each pair alike, enveloped or subtended by long hairs from the callus:

Spikelets all pedicelled; panicle spiciform, the rhachis continuous and tough 103. **Imperata**

Spikelets paired, with 1 sessile and the other pedicelled; rhachis fragile:

Panicle large and broad, silvery, the racemes long, flexuous and densely overlapping 104. **Saccharum**

Panicle narrow, fulvous, of several short racemes scattered along a central axis 105. **Eriochrysis**

Spikelets of each pair different, sometimes the pedicelled modified or suppressed:

Spikelets dorsally compressed; lower glume more or less flattened on the back:

Pedicelled spikelets present .. .. 106. **Sorghum**

Pedicelled spikelets suppressed, but pedicels present .. .. 107. **Sorghastrum**

Spikelets laterally compressed or terete; lower glume rounded on the back:

Racemes reduced to 1 sessile and 2 pedicelled spikelets .. .. 108. **Chrysopogon**

Racemes composed of many pairs of spikelets .. .. .. 109. **Vetiveria**

Racemes solitary, paired or sub-digitate, and usually supported by spathes (very rarely the lowest racemes overtopped by the common axis, but then the pedicels with a translucent middle line):

Fertile lemma awned from low down on the back; leaf-blades lanceolate, cordate-amplexicaul at the base .. .. 110. **Arthraxon**

Fertile lemma awned from the tip, or the sinus of the 2-toothed tip, rarely awnless:

Awn from the tip of the narrow fertile lemma:

Racemes digitate, or racemosely arranged on a common axis; lower glume of sessile spikelet flat on the back, the margins sharply involute; awn glabrous or puberulous; callus obtuse:

Sessile spikelets of all the pairs hermaphrodite and awned; internodes and pedicels with a translucent middle line; lower glume of sessile spikelets pitted (in West African species) .. .. 111. **Bothriochloa**

Sessile spikelets of the lowest 1–3 pairs male or barren and awnless; lower glume of sessile spikelets not pitted:

Internodes and pedicels solid; racemes ascending or divergent 112. **Dichanthium**

Internodes and pedicels with a translucent middle line; racemes nodding 113. **Euclasta**

Racemes solitary or paired:

Lower glume of sessile spikelet with a deep circular pit on the back, sharply inflexed on the margins; awn glabrous; callus obtuse 114. **Eremopogon**

Lower glume of sessile spikelet without a pit, rounded on the back; awn pubescent to pilose; callus pungent:

Racemes composed of a single awned sessile spikelet and 2 pedicelled spikelets, enclosed by an involucre of 4 sterile spikelets .. .. 115. **Themeda**

Racemes composed of more than 3 pairs:

Upper spikelet of each pair awned and coriaceous; only the lowermost pair of spikelets homogamous; rhachis of racemes tough, flexuous 116. **Trachypogon**

Upper spikelet of each pair awnless and herbaceous, the other awned and sessile;

lower 1–10 pairs homogamous, the rhachis fragile above them; racemes solitary; callus pungent .. .. 117. **Heteropogon**

Awn from the sinus of the 2-toothed fertile lemma, rarely entire (*Schizachyrium pulchellum*, *S. scintillans*), or awnless:

Lower glume of sessile spikelets transversely rugose; pedicelled spikelets reduced to the narrow curved pedicel; lower floret male .. 118. **Thelepogon**

Lower glume of sessile spikelets smooth, rarely rugose and then the pedicelled spikelet well developed:

Callus of sessile spikelet inserted into the crateriform or cupuliform tip of the internode, at least the rim of the internode lapping over and concealing the tip of the callus; lower glume of the sessile spikelets 2-keeled or with the margins sharply inflexed, and usually depressed between the keels, rarely the keels rounded but then deeply grooved between them:

Lower floret of sessile spikelets male, with a palea; internodes of rhachis and pedicels more or less stout; callus of sessile spikelets obtuse:

Racemes paired or digitate; upper glume of sessile spikelet awnless; awn of upper lemma glabrous; internodes of rhachis and pedicels 3-angled 119. **Ischaemum**

Racemes solitary; upper glume of sessile spikelets awned; awn of upper lemma ciliate along the coils; internodes of rhachis and pedicels linear, flat 120. **Sehima**

Lower floret of sessile spikelet barren and reduced to the lemma:

Spikelets of each pair alike in shape and sex .. .. 121. **Microstegium**

Spikelets of each pair conspicuously different:

Callus of sessile spikelets obtuse, usually very short; awn glabrous or scaberulous:

Racemes solitary; lower glume of sessile spikelet convex on the back, the keels lateral or frontal with several intercarinal nerves; internodes of the the rhachis and pedicels clavate to linear, the apex deeply hollowed into a cup with a fimbriate rim .. .. 122. **Schizachyrium**

Racemes paired or digitate, rarely solitary and then with the lower glume of the sessile spikelet concave and nerveless between the keels:

Racemes deflexed at maturity, borne upon subequal flattened raceme-bases, seldom exceeding the spatheole in length; internodes of the rhachis and pedicels linear; leaves aromatic; panicle dense, decompound 123. **Cymbopogon**

Racemes not deflexed, borne upon unequal, more or less terete, raceme-bases; leaves not aromatic .. .. 124. **Andropogon**

Callus of sessile spikelets acute to pungent, 1–5 mm. long; awn puberulous to hirsute; racemes paired .. .. 125. **Diheteropogon**

Callus of sessile spikelet applied obliquely to the apex of the internode with its tip free, usually acute to pungent; lower glume of sessile spikelet convexly rounded on the back without keels (rarely with a median groove); internodes and pedicels linear:

Pedicelled spikelet without an appreciable callus; upper glume of sessile spikelet usually awnless; fertile lemma usually minutely bidentate, its awn pubescent to hirtellous; racemes paired:

Lower glume of sessile spikelet rounded on the back, or rarely with 2 shallow striations; upper glume obtuse to acute or mucronate 126. **Hyparrhenia**

Lower glume of sessile spikelet with a median longitudinal groove:

Raceme-bases produced at the tip into a long scarious appendage; racemes 2-awned per pair; upper glume acute to apiculate .. 127. **Hyperthelia**

Raceme-bases without distinct appendages; racemes 9–14-awned per pair; upper glume muticous or shortly awned .. .. 128. **Parahyparrhenia**

Pedicelled spikelet prolonged at the base into a callus 0·5–3 mm. long; upper glume of sessile spikelet nearly always awned; fertile lemma usually bifid for $\frac{1}{4}$–$\frac{1}{2}$ its length, its awn shortly pubescent to glabrescent:

Racemes paired (except *E. gossweileri*), with 1–10 homogamous pairs at the base, exserted from the narrow spatheole .. .. .. 129. **Elymandra**

Racemes solitary, without homogamous pairs:

Spatheoles cymbiform, coloured and enclosing the racemes; racemes dense with numerous spikelets concealing the short internodes; pedicelled spikelets broadly lanceolate, subacute, hairy or merely spinose-ciliate on the margins; callus of sessile spikelets obtuse .. .. .. 130. **Monocymbium**

Spatheoles linear to narrowly lanceolate, usually greenish, embracing the racemes or not; racemes loose, with few spikelets and long internodes visible between them; pedicelled spikelets linear-lanceolate, acuminate, nearly always glabrous; callus of sessile spikelets usually acute to pungent 131. **Anadelphia**

Internodes of the rhachis and pedicels stout, 3-angled, rounded, or flattened and thickening upwards; upper lemma awnless:
Lower glume produced into a long flattened tail; spikelets similar 132. **Vossia**
Lower glume without a herbaceous tail:
Pedicelled spikelets long-awned from the lower glume (rarely awnless in *U. giganteum*); lower glume of sessile spikelet entire; callus short, rounded, at most pubescent (if top of internode ciliate from bearded callus see *Loxodera*) .. 133. **Urelytrum**
Pedicelled spikelets awnless, or sometimes awned but then the lower glume of the sessile spikelets bifid or aristulate at the tip:
Racemes numerous on a short common axis .. .. .. 134. **Jardinea**
Racemes solitary:
Racemes more or less villous:
Articulation of the rhachis transverse, the callus very short; lower glume smooth, hirsute; lower floret male with a well-developed palea .. 135. **Lasiurus**
Articulation of the rhachis oblique, the callus up to 2 mm. long; spikelets paired:
Lower glume of the sessile spikelet without ridges, bimucronate or biaristate at the tip, hirsute or with a ciliate fringe; lower floret reduced to a barren lemma 136. **Elionurus**
Lower glume of the sessile spikelet longitudinally ridged, entire at the tip, spinulose on the nerves but otherwise glabrous; top of internode fringed with hairs from the bearded callus; lower floret male with a palea 137. **Loxodera**
Racemes glabrous:
Sessile spikelet globose, the lower glume muricate .. 138. **Hackelochloa**
Sessile spikelet not globose:
Spikelets solitary, the pedicelled and its pedicel completely suppressed 139. **Oxyrhachis**
Spikelets paired, or at least the pedicel present; articulation of the rhachis transverse, or the raceme tough:
Pedicels wholly or partially fused to the internodes of the rhachis:
Raceme tough, its flattened rhachis formed from the fused pedicels and internodes; spikelets almost alike; callus oblique .. .. 140. **Hemarthria**
Raceme fragile; pedicelled and sessile spikelets different:
Spikelets sunken in a cylindrical segment formed from the fused pedicel and internode; callus circular with a knob-like projection from its centre 141. **Rottboellia**
Spikelets not sunken, the flattened or triquetrous pedicel and internode usually fused only in the lower half, the keeled upper glume of the sessile spikelet thrust between them above; callus elliptic with a low central ridge .. .. 142. **Robynsiochloa**
Pedicels quite free from the internodes of the rhachis; racemes fragile:
Racemes borne in spathate panicles; lower glume of the sessile spikelet winged; spikelets awnless:
Sessile spikelets 6–7 mm. long; internode and pedicel gaping, the lower glume of the sessile spikelet thrust between; panicle scanty 143. **Chasmopodium**
Sessile spikelets small (3–4 mm. long in our species), imbricate; internode and pedicel approximate; racemes fasciculate .. 144. **Coelorhachis**
Racemes long-exserted, terminal, the culms usually simple but sometimes with a few side branches; internode and pedicel appressed to approximate; pedicelled spikelet usually reduced.. .. .. .. 145. **Rhytachne**

Tribe XXVII—Maydeae

Male spikelets in large terminal panicles; female spikelets in axillary sheathed " ears " 146. **Zea**
Male spikelets in a short raceme projecting from an ovoid ivory-like sheath containing a solitary female spikelet .. .. 147. **Coix**

1. ARUNDINARIA Michx., Fl. Bor. Am. 1: 73 (1803).

Leaf-blades lanceolate, 6–10 cm. long and 0·5–1 cm. broad, narrowed at the base into a very short false petiole, long-attenuate at the apex to a flexuous filiform tip; leaf-sheath glabrous, with small deciduous auricles bearing long setae; culm sheath without auricles, the blade linear-lanceolate, 1·5–7 cm. long; spikelets 2–4 cm. long, in a panicle; glumes 2; florets usually 5–10, distant, the rhachilla visible; lodicules 3; stigmas 2, rarely 3 .. .. *alpina*

A. alpina *K. Schum.* in Engl., Pflanzenw. Ost-Afr. C: 116 (1895); Camus, Bambus. 48; Hubbard in Hook., Ic. Pl. 6: t. 3594. Culms up to about 6 m. high and 3–5 cm. diam., hollow, with thick walls; flowers infrequently; occurs in gregarious patches.
W. Cam.: *Johnstone*! Mba Kokeka, Bamenda, 7,500 ft. *Richards* 5332! *Keay* FHI 28400! *Brunt* 795! Also in the highland regions of eastern Africa.

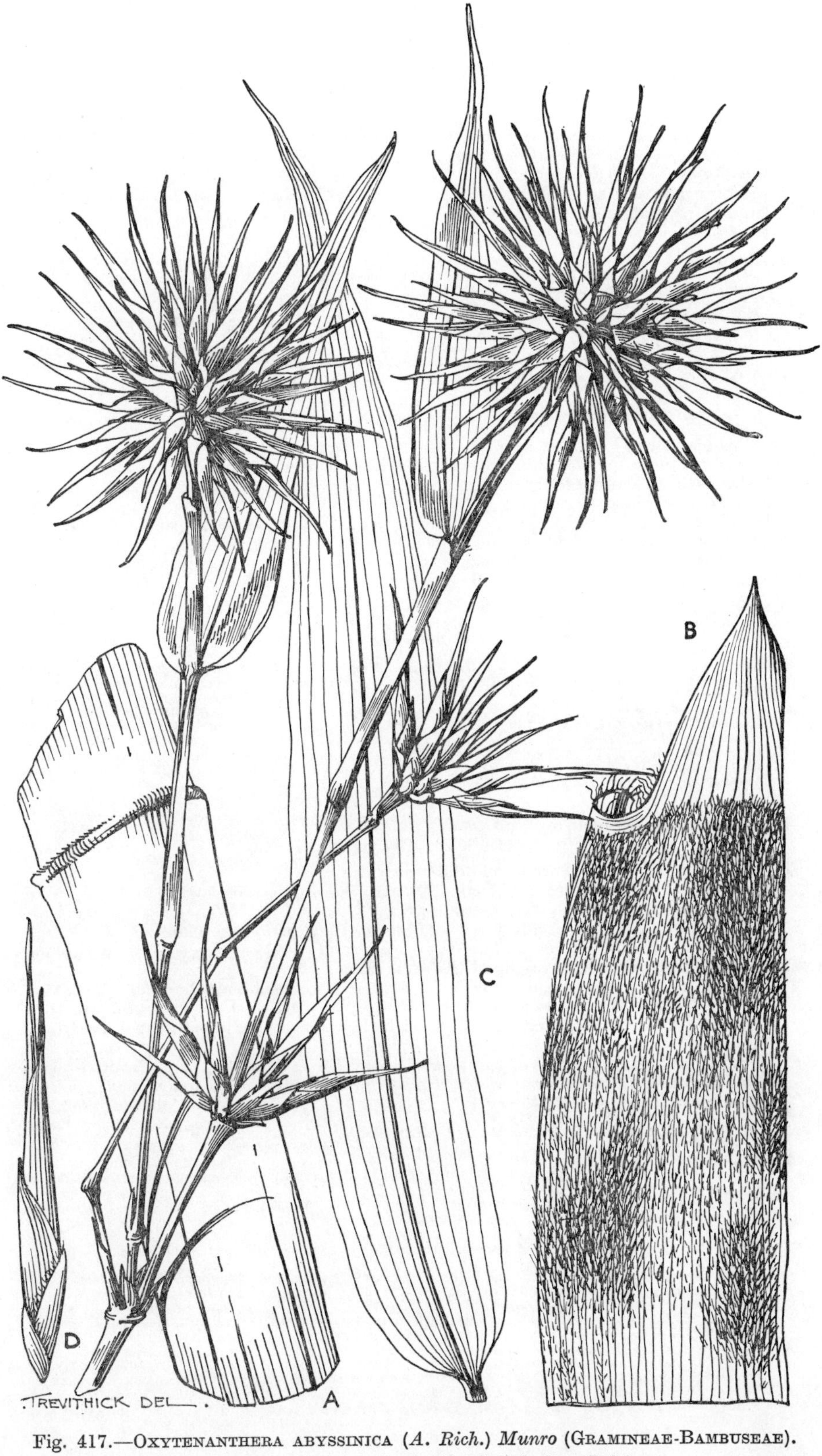

Fig. 417.—OXYTENANTHERA ABYSSINICA (*A. Rich.*) *Munro* (GRAMINEAE-BAMBUSEAE).
A, part of culm. B, culm-sheath. C, leaf. D, spikelet.

2. BAMBUSA Retz. corr. Schreb., Gen. Plant. 236 (1789), *nom. cons.*

Leaf-blades oblong-lanceolate, mostly 10–25 cm. long and 1–4 cm. broad, broadly rounded at the base and abruptly narrowed to a short false petiole, acuminate to a pungent tip; shoulders of leaf-sheath usually produced into tiny auricles, glabrous or with a few very short bristles; culm sheath bearing auricles up to 1 cm. long, the blade broadly ovate, 3–5 cm. long; spikelets 1–2 cm. long, in clusters of 1–10 or more along the branches of a spreading panicle; florets 4–12, distichous, laterally compressed; lemmas merging into glumes below, the lowermost of which bear spikelet buds in their axils; style 1, divided at the apex into 2–3 stigmas; lodicules 3 .. *vulgaris*

B. vulgaris *Schrad. ex Wendel.* Collect. Pl. 2: 26, t. 47 (1810); Munro in Trans. Linn. Soc. 26: 106; Camus, Bambus. 122; Chev. Bot. 752, and in Rev. Bot. Appliq. 14: 136. Culms up to about 13 m. high and 10 cm. diam., hollow; flowers infrequently; moist soils near streams.
S.L.: Njala *Deighton* 5517! Loma Mts. *Jaeger* 8951! **Lib.:** *Johnston*! Monrovia *Baldwin* 13074! Nyaake, Webo *Baldwin* 6083! **Iv. C.:** Kéeta, Cavally Basin *Chev.* 19354! **U. Volta:** Bobo-Dioulasso *Scholz* 96! **Ghana:** Accra *T. Vogel*! Kumasi *Taylor* 5804! Aiyinabrem (Mar. 1925) *Fishlock* 85! **Dah.:** Pobé *Adjanohoun* 96! **S. Nig.:** Ibadan *Van Eynatten* 45! *Onochie* FHI 8215! Although poorly represented by herbarium specimens, this bamboo appears to have become naturalized throughout the forest zone of West Africa. An Asiatic species, introduced into most tropical countries.

3. OXYTENANTHERA Munro in Trans. Linn. Soc. 26: 126 (1868).

Leaf-blades oblong-lanceolate, mostly 8–20 cm. long and 1–3 cm. broad, rounded at the base and contracted to a short false petiole, acuminate to a pungent tip; shoulders of leaf-sheath bearing setae 2–5 mm. long, without auricles; culm sheath without auricles, the blade linear and 1–3 cm. long; spikelets 15–40 mm. long, narrowly lanceolate, acuminate, in globose clusters 5–6 cm. diam., which are sometimes aggregated into an interrupted false spike; fertile florets 1–2, the lemmas sharply mucronate, convolute, merging into glumes below, the lowest of these with spikelet buds in their axils; stigmas 3; lodicules absent .. .. *abyssinica*

O. abyssinica (*A. Rich.*) *Munro* in Trans. Linn. Soc. 26: 127 (1868); Camus, Bambus. 144; Chev. Bot. 752, and in Rev. Bot. Appliq. 14: 136; Berhaut, Fl. Sén. ed. 2, 388. *Bambusa abyssinica* A. Rich., Tent. Fl. Abyss. 2: 439 (1851). Culms up to about 6 m. high and 5 cm. diam.; stems thick-walled, almost solid; in savanna.
Sen.: Tambacounda (Jan.) *Roberty* 16616! **Gam.:** Berief *Unknown Coll.*! Gambisara *Unknown Coll.*! Bakadaji (July) *Deighton* 5581! **Port. G.:** Teixeira Pinto (Dec.) *Raimundo & Guerra* 512! **Guin.:** Kouroussa *Pobéguin* 1093! Timbo (Apr.) *Chev.* 13274! Farana *Chev.* 20407! **S.L.:** Falaba *Sc. Elliot* 5205! Musaia (Apr.) *Deighton* 5479! Kalia to Kamba (Nov.) *Glanville* 314! Mt. Babaana *Small* 440! Bintumane Peak, 3,000 ft. (July) *Bakshi* 225! **Iv. C.:** Séguélo *Aké Assi* 6581! **Ghana:** Gambaga (Oct.) *Vigne* FH 4638! *Morton* GC 7411! Gambaga to Bawku *Rose Innes* GC 30494! **Togo Rep.:** Pega *Kersting* A498! **Dah.:** Atacora Mt. *Chev.* 23983! **N. Nig.:** Nupe *Barter* 805! Minna *Meikle* 734! Panshanu (Aug.) *Lawlor & Hall* 374! Mongu F.R. (July) *Keay & King* FHI 37086! Anara F.R. (May) *Keay* FHI 25792! Throughout tropical Africa. (See Useful Plants.)

4. GUADUELLA Franch. in Bull. Soc. Linn. Paris 1: 676 (1887); Clayton in Kew Bull. 16: 247 (1962).

Leaves and inflorescence borne upon different culms:
- Culms up to 14 cm. high; leaf-blades 2–4 cm. long, pilose above and below; palea conspicuously exceeding the lemma .. .. 1. *humilis*
- Culms 40–90 cm. high, rarely less; leaf-blades 10–20 cm. long, glabrous above and glabrous or pubescent below; lemma and palea subequal .. 2. *densiflora*

Leaves and inflorescence borne upon the same culm:
- Lemmas not or scarcely imbricate, appressed to the rhachilla internode, the latter half as long as the lemma; leaf-blades abruptly narrowed to an acuminate tip; lemmas glabrous except for the pubescent base; paleas usually with 1–2 intercarinal nerves .. .. 3. *oblonga*
- Lemmas imbricate, the keel of the lemma above covered for $\frac{1}{4}$–$\frac{3}{4}$ of its length; leaf-blades gradually acuminate at the tip; lemmas more or less pubescent; paleas usually with 4 intercarinal nerves .. .. 4. *macrostachys*

1. **G. humilis** *W. D. Clayton* in Kew Bull. 16: 248 (1962). Creeping plant of forest shade.
S. Nig.: Oban (Mar.) *Coombe* 173!

2. **G. densiflora** *Pilger* in Engl., Bot. Jahrb. 30: 123 (1902). *G. ledermannii* Pilger in Engl., Bot. Jahrb. 43: 387 (1909); F.W.T.A., ed. 1, 2: 503. *G. foliosa* Pilger l.c. 45: 211 (1910). Rhizomatous perennial of forest undergrowth.
S. Nig.: Cross River North F.R. (Dec.) *Keay* FHI 28156! Oban *Talbot* 956! 958! **W. Cam.:** S. Bakundu F.R. (Mar.) *Brenan* 9441! 9441*a*! Also in Gabon and Cabinda.

3. **G. oblonga** *Hutch. ex W. D. Clayton* in Kew Bull. 16: 247 (1962). *Puelia ciliata* of Chev. Bot. 753, and in Rev. Bot. Appliq. 14: 137, not of Franch. An *Aframomum*-like rhizomatous perennial 0·30–1 m. high; in forest shade.
Guin.: Dyeke (Oct.) *Baldwin* 9653! **S.L.:** Yonibana (Nov.) *Thomas* 4813! Bumbuna (Oct.) *Thomas* 3888! Mamaba (Nov.) *Thomas* 4405! Makump (Oct.) *Glanville* 68*a*! Kailahun (Sept.) *Jordan* 525! **Lib.:** Gbanga (Sept.) *Linder* 564! Tawata, Boporo (Nov.) *Baldwin* 12037! Genna Tanyehun, Grand Cape Mt. (Dec.) *Baldwin* 10727! Wohmen, Vonjama (Oct.) *Baldwin* 12028! Bushrod I. (Aug.) *Baldwin* 13094! **Iv. C.:** Grabo to Taté (Aug.) *Chev.* 19758! Taï (Oct.) *Adjanohoun* 174*a*!

4. **G. macrostachys** (*K. Schum.*) *Pilger* in Mildbraed, Wiss. Ergebn. Deutsch. Zentr.-Afr.-Exped. 1910–11, 2: 93 (1922). *Microbambus macrostachys* K. Schum. in Engl., Bot. Jahrb. 24: 336, t. 4 (1897). *Guaduella zenkeri* Pilger in Engl., Bot. Jahrb. 30: 123 (1902). Rhizomatous perennial 30–60 cm. high; in forest shade.
Ghana: Obepehia forest (Oct.) *Ankrah* SLUS 305! Subiri F.R., Tarkwa (Sept.) *Andoh* FH 5564! Ankasa F.R. (June) *Hall* GC 35523! **S. Nig.:** Oban *Talbot* 957! **W. Cam.:** Barombi *Preuss*. Also in E. Cameroun.

Fig. 418.—GUADUELLA OBLONGA *Hutch. ex W.D. Clayton* (GRAMINEAE-BAMBUSEAE).

A, rhizome. B, flowering culm. C, portion of spikelet. D, lower glume. E, upper glume. F, lemma and palea. G, lemma. H and I, palea. J, flower. K, lodicule and two stamens.

5. PUELIA Franch. in Bull. Soc. Linn. Paris 1: 674 (1887); Clayton in Hook., Ic. Pl. 37: t. 3642 (1967).

Spikelets pedicelled in a lax panicle 9–15 cm. long; leaf-sheaths glabrous 1. *olyriformis*
Spikelets subsessile in a dense 1-sided panicle 1·5–3·5 cm. long; leaf-sheaths puberulous 2. *ciliata*

1. **P. olyriformis** (*Franch.*) *W. D. Clayton* in Kew Bull. 20: 271 (1966). *Atractocarpa olyriformis* Franch. in Bull. Soc. Linn. Paris 1: 675 (1887). Perennial, about 60 cm. high.
S.L: York Pass (Feb., Sept.–Dec.) *Melville* 481*b*! *Deighton* 3319! *Morton* SL 3365! *Morton & Gledhill* SL 786! **Lib.**: Vahun (Nov.) *Baldwin* 10217! Also in the Congos and Tanzania.
2. **P. ciliata** *Franch.* in Bull. Soc. Linn. Paris 1: 674 (1887). *P. acuminata* Pilger in Engl., Bot. Jahrb. 30: 125 (1901); F.W.T.A., ed. 1, 2: 503. *P. subsessilis* Pilger l.c. 124 (1901).
W. Cam.: Barombi (Apr.) *Preuss* 277. Also in E. Cameroun, Gabon and Middle Congo.

6. OLYRA Linn., Syst. Nat. ed. 10: 1261 (1759).

Erect or scandent perennial; leaf-sheaths closely nerved, joined to the blade by a short false petiole; blades ovate-elliptic or ovate-oblong, tailed-acuminate, asymmetrically rounded at the base, 10–20 cm. long, 3–6 cm. broad, with about 8 primary nerves on each side of the midrib and fainter ones between, and with fairly distinct transverse nerves; panicle narrow to sub-pyramidal, the branches with female spikelets above and male below; male spikelets on filiform pedicels; glumes rudimentary; lemma lanceolate, 4 mm. long, awned; female spikelets on clavate pedicels, ovoid, long-awned from the lower 7–9-nerved glume; grain tightly enclosed by the hardened shining white lemma .. .. *latifolia*

O. latifolia *Linn.* Syst. Nat. ed. 10: 1261 (1759); Chev. Bot. 737, and in Rev. Bot. Appliq. 14: 135; Berhaut, Fl. Sén. ed. 2, 410, 412. *O. brevifolia* Schumach. (1827)—Chev. Bot. 737. *O. guineensis* Steud. (1854). A tall cane-like grass up to 3 m. high; fascicles of lateral shoots bearing very small leaves occasionally occur; in forests and forest margins.
Sen.: Forêt de Boudié, Casamance (Nov.) *Berhaut* 6380! **Guin.**: Bambaya (Jan.) *Jaeger* 3874! Baffing R. (Nov.) *Pobéguin* 1771! **S.L.**: Njala (Sept.) *Deighton* 117! Jigaya (Sept.) *Thomas* 2539! Mt. Aureol, Freetown (Sept.) *Melville & Hooker* 447! Gberia Fotombu (Oct.) *Small* 408! Bintumane Peak, 3,200 ft. (Jan.) *T. S. Jones* 121! **Lib.**: Robertsport (Dec.) *Baldwin* 10925! Vonjama (Oct.) *Baldwin* 9898! Gbanga (Sept.) *Linder* 600! Ganta (Sept.) *Baldwin* 12018! Webo (June) *Baldwin* 6086! **Iv. C.**: Adiopodoumé (Nov.) *Leeuwenberg* 1929! Soubré (Nov.) *de Wilde* 3267! Voguié (Jan.) *Chev.* 17102! Yapo Nord (Mar.) *Bernardi* 8717! Mbasso (July) *de Wilde* 549! **Ghana**: Aburi Hills (Nov.) *Johnson* 287! Adankrono (Aug.) *Chipp* 558! Ejura (Nov.) *Vigne* FH 3454! Kumasi *Cummins* 1965! Kumasi to Sunyani (Sept.) *Rose Innes* GC 30141! **N. Nig.**: Kabba (Oct.) *Parsons* 30! Abugi (Dec.) *Clayton* 623! Jemaa (Apr., Oct., Dec.) *E. W. Jones* 45! *Gambles* 63! *Olorunfemi* FHI 55148! **S. Nig.**: Olokemeji (Nov.) *Nchami* FHI 14327! Iseyin *W. MacGregor* 237! Onitsha *Barter* 1778! Kukuruku (June) *Dundas* FHI 21479! Oban *Talbot*! **W. Cam.**: Buea (Mar.) *Maitland* 567! Gangumi, Gashaka (Dec.) *Latilo & Daramola* FHI 28888! Bambui (Aug.) *Brunt* 1207! **F. Po**: (Dec.) *T. Vogel* 263! *Barter*! *Mann* 117! Throughout tropical Africa and tropical America. (See Useful Plants.)

7. LEPTASPIS R. Br., Prod. 211 (1810).

Perennial; culms ascending, rooting at the lower nodes; leaf-sheaths closely nerved, joined to the blade by a false petiole about 1 cm. long; blade oblong or oblong-oblanceolate, abruptly narrowed at tip, asymmetrically subacute at base, 10–25 cm. long, 2·5–3 cm. broad, subpinnately nerved with about 4 pairs of principal lateral nerves, and distinct transverse nerves; panicle terminal, branches 2–3 in a whorl, bearing short side branches with 1 male and 1 female spikelet, the male above; male spikelets 1-flowered, the lemma 4 mm. long, the glumes half as long; stamens 6; female spikelets 1-flowered, 6 mm. long; lemma inflated conchiform, closed except for a small hole near the apex, covered in hooked hairs .. .. .. *cochleata*

L. cochleata *Thwaites* Enum. Pl. Zeyl. 357 (1864). *L. conchifera* Hack. (1887)—A. Chev. in Rev. Bot. Appliq. 14: 135. Up to 1 m. high; the false petiole is twisted bringing the abaxial leaf surface uppermost; forest undergrowth.
Guin.: Dioromandou to Koundian (Feb.) *Chev.* 20738! Mt. Boola (Mar.) *Chev.* 20912! **S.L.**: Zimi (Apr.) *Deighton* 3648! Foya (Feb.) *Marmo* 188! Kasewe Hills F.R. (June) *Deighton* 5830! Bafodeya (Apr.) *Sc. Elliot* 1892! Makute (June) *Thomas* 461! **Lib.**: Gbanga (Sept.) *Linder* 803! Bobei Mt. (Sept.) *Baldwin* 9565! Bilimu Mt. (Jan.) *Baldwin* 1323! Ganta (Feb.) *Harley* 328! Nikabuyu to Zigida (Mar.) *Bequaert* 124! **Iv. C.**: Agboville (Sept.) *Adjanohoun* 392*a*! Mbasso (Mar.) *Chev.* 17598! Sassandra (Feb.) *Leeuwenberg* 2758! Guéyo (Feb.–Mar.) *Leeuwenberg* 3799! *Chev.* 17044! **Ghana**: Ademsu (Feb.) *Thomas* D154! Bompata (Jan.) *Vigne* FH 2734! Aburi (Aug.) *Irvine* 98! Fumso (Mar.) *Darko* 549! Kade (Mar.) *Ankrah* GC 20191! **S. Nig.**: Abeokuta *Barter* 3408! Ikorodu (Mar.) *Schlechter* 12998! Okomu F.R., Benin (Dec.) *Brenan* 8504! Sapoba *Kennedy* 2395! Oban *Talbot*! **W. Cam.**: Cam. Mt., 3,500 ft. (Feb.) *Mann* 1361! Likombe (Apr.) *Maitland* 1125! Bamenda, 4,000 ft. *Maitland*! **F. Po**: *T. Vogel*! *Mann* 245! *Guinea* 995! Tropical Africa generally, and in Madagascar, Ceylon, Malaysia and the Solomon Is.

8. STREPTOGYNA P. Beauv., Ess. Agrost. 80, t. 16, fig. 8 (1812).

Leaf-blades narrowly lanceolate, tessellately nerved, 15–40 cm. long and 10–36 mm. broad, acuminate at the tip, narrowed at the base into a short false petiole; spikelets linear, 22–30 mm. long, thinly coriaceous, dark green or brownish, borne upon very short pedicels in a narrow erect raceme 11–30 cm. long; lemmas 5–7, awned, the lower fertile and appressed-pubescent below, the upper sterile; stigmas 2, retrorsely spinulose, 1–2 cm. long, long-exserted and tangled; lodicules 3 .. .. *crinita*

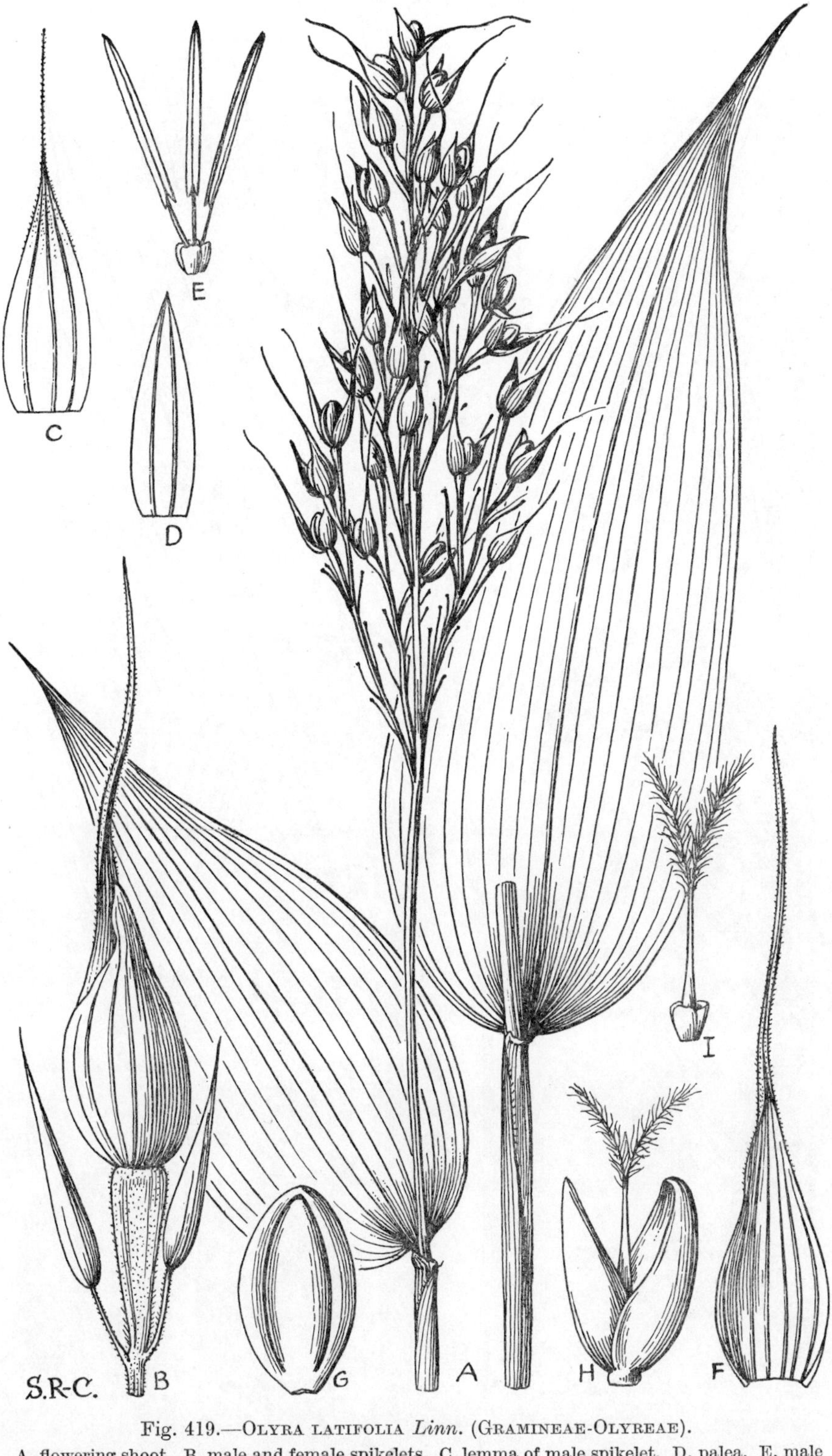

Fig. 419.—OLYRA LATIFOLIA *Linn.* (GRAMINEAE-OLYREAE).

A, flowering shoot. B, male and female spikelets. C, lemma of male spikelet. D, palea. E, male flower. F, lower glume of female spikelet. G and H, female floret. I, female flower.

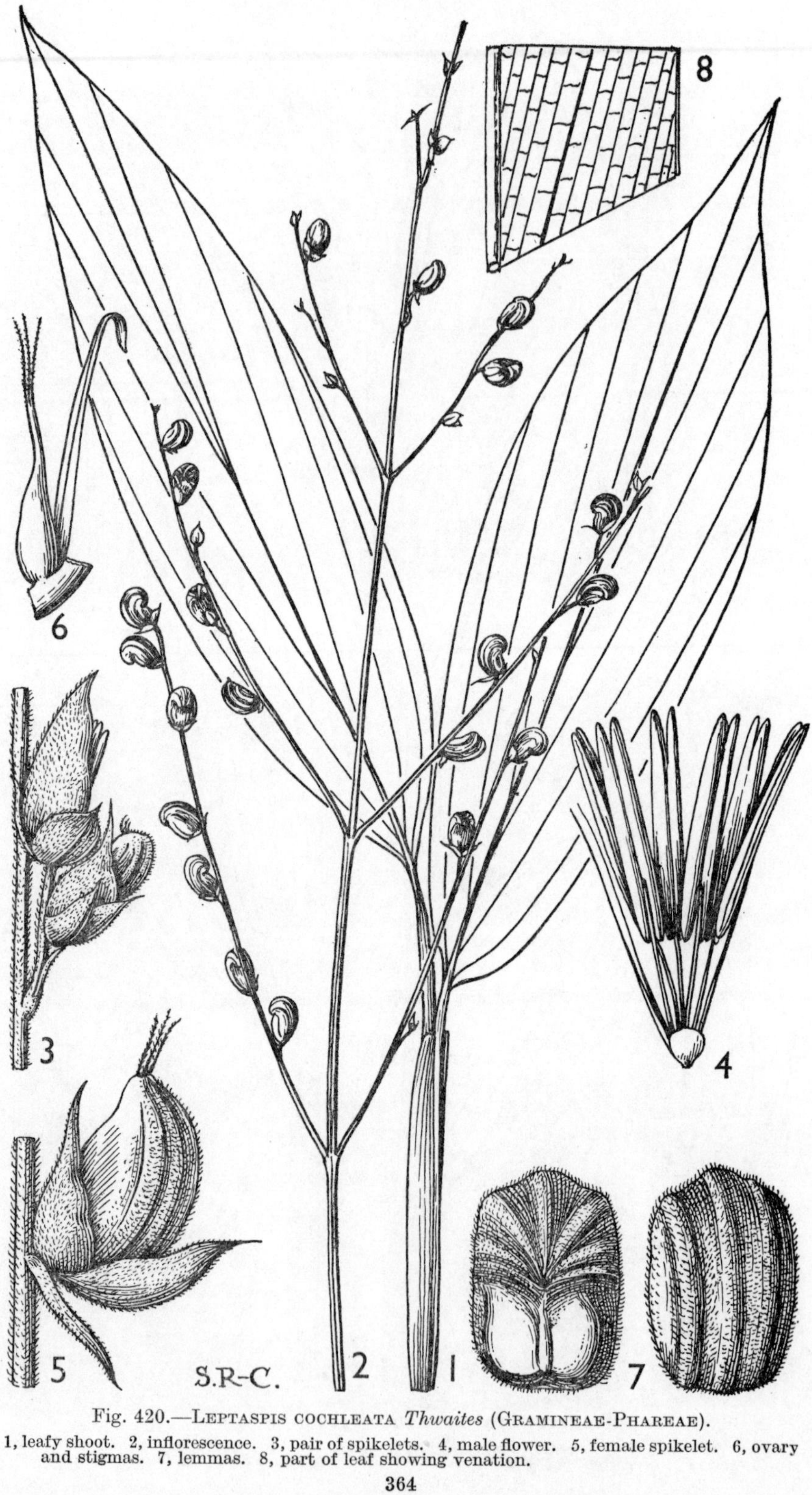

Fig. 420.—LEPTASPIS COCHLEATA *Thwaites* (GRAMINEAE-PHAREAE).

1, leafy shoot. 2, inflorescence. 3, pair of spikelets. 4, male flower. 5, female spikelet. 6, ovary and stigmas. 7, lemmas. 8, part of leaf showing venation.

S. crinita *P. Beauv.* Ess. Agrost. 80, t. 16, fig. 8 (1812); C. E. Hubbard in Hook., Ic. Pl. 36: t. 3572 (1956); Chev. Bot. 751; Berhaut, Fl. Sén. ed. 2, 392. *S. gerontogaea* Hook. f. in Trimen, Fl. Ceyl. 5: 301 (1900); F.W.T.A. ed. 1, 2: 505; A. Chev. in Rev. Bot. Appliq. 14: 122. A broad-leaved perennial of forest shade, 0·3–1·7 m. high, spreading by scaly rhizomes. The rhachilla segment adjacent to the base of each lemma forms a trap for catching hairs and thus effecting dispersal.

Sen.: Forêt de Boudié, Casamance (Nov.) *Berhaut* 6385! **Port. G.:** Cadique (Jan.) *Pereira & Correia* 2621! Bedanda, Cantanhex (Jan., May) *Pereira & Correia* 2798! *Pereira* 1896! Catio (Jan.) *Raimundo & Guerra* 811! **S.L.:** Kuntaia (June) *Thomas* 441! Njala (Jan.) *Dalz.* 8420! Zimi (Jan.) *Fisher* 1! Kambia (Jan.) *Sc. Elliot* 4389! Kenema (Nov.) *Deighton* 397! **Lib.:** Monrovia (Jan.) *Dinklage* 3375! Gretown (July) *Baldwin* 6924! Mnanulu (June) *Baldwin* 6057! Wohmen (Oct.) *Baldwin* 12027! Peahtah (Oct.) *Linder* 1012! **Iv. C.:** Bingerville (Dec.) *Chev.* 16036! Sassandra (Dec.) *Leeuwenberg* 2261! N'douci (Dec.) *Oldeman* 701! Abidjan (Dec.) *de Wit* 7022! Daloa to Abidjan (Dec.) *Boughey* GC 13609! **Ghana:** Amuni (Dec.) *Chipp* 53! Atuna (Dec.) *Vigne* FH 3522! Anyaboni (Dec.) *Morton* 6095! Krobo hill, Mampong (Oct.) *Rose Innes* GC 31163! Kade (Mar.) *Ankrah* GC 20190! **N. Nig.:** Acharane F.R., Kabba Prov. (Nov.) *Latilo* FHI 34018! Jemaa (Nov., Dec.) *Clayton* 1425! *Olorunfemi* FHI 55140! Bonu F.R., Niger Prov. (June) *Onochie & E. W. Jones* FHI 38472! Kwagiri, Gashaka (Nov.) *Latilo & Daramola* FHI 28770! **S. Nig.:** Olokemeji (Nov.) *Onochie* FHI 8136! Okelifi, Ondo (Nov.) *Onochie* FHI 34228! Idanre (Mar.) *Coombe* 159! Onitsha *Barter* 1814! Itu (Jan.) *Jones* FHI 6870! **F. Po:** *Mann* 108! Extends south to the Congo, and east to Uganda and the Sudan; also in S. India and Ceylon. (See Useful Plants.)

[The name *S. crinita* has been commonly misapplied to the American species *S. americana* C. E. Hubbard.]

9. ORYZA Linn., Sp. Pl. ed. 1, 333 (1753), and Gen. Pl. ed. 5, 155 (1754); Tateoka in Bot. Mag. Tokyo 76: 165 (1963).

Ligule of the lower leaves very long, 15–45 mm., acute:
- Annual; spikelets persistent, usually awnless; cultivated rice .. .. 1. *sativa*
- Rhizomatous perennial; spikelets deciduous, awned; wild .. 2. *longistaminata*

Ligule of the lower leaves short, rarely over 6 mm. long, rounded or truncate at tip:
- Spikelets broadly linear, 1–1·5 mm. broad and 6–8 mm. long; sterile lemmas subulate; awns 6–17 cm. long .. .. 3. *brachyantha*
- Spikelets oblong or oblong-elliptic; sterile lemmas flat, membranous, linear:
  - Spikelets 4·5–6 mm. long, hispid, awned:
    - Panicle contracted with ascending branches; ligule less than 3·5 mm. long; spikelets narrowly elliptic, less than 2 mm. broad, the length about 3 times the breadth 4. *eichingeri*
    - Panicle loose with spreading branches; ligule 3–10 mm. long; spikelets elliptic, more than 2 mm. broad, the length about 2·5 times the breadth .. 5. *punctata*
  - Spikelets over 7 mm. long (shorter in some forms of *O. glaberrima* but then the spikelets glabrous):
    - Fertile lemma and palea hispid; spikelets deciduous, awned; awns hispidulous; wild .. .. 6. *barthii*
    - Fertile lemma and palea glabrous or sometimes hispid; spikelets persistent, usually awnless; awns, when present, smooth or at most scaberulous; cultivated rice 7. *glaberrima*

1. **O. sativa** *Linn.* Sp. Pl. 333 (1753); Chev. Bot. 739; Berhaut, Fl. Sén. ed. 2, 416; Portères in J. Agric. Trop. 3: 341, 541, 627. Many cultivars; Portères, l.c. discusses infra-specific taxa in detail, with keys and synonymy. Widely cultivated on river flood-plains in W. Africa, and in a few localities grown as an upland crop. Throughout the tropics, extending into warm-temperate regions. (See Useful Plants.)

2. **O. longistaminata** *A. Chev. & Roehr.* in Compt. Rend. Acad. Sci. 159: 561 (1914); Chev. Bot. 739. *O. barthii* of F.W.T.A. ed. 1, 2: 537, not of A. Chev; Berhaut, Fl. Sén. ed. 2, 416; Tateoka in Bot. Mag. Tokyo 75: 455; (see Clayton in Kew Bull. 21: 488). *O. silvestris* Stapf ex A. Chev. (1911), name only. *O. silvestris* Stapf var *punctata* Stapf f. *longiligulata* A. Chev., Et. Fl. Afr Centr. Fr. 1: 369 (1913), name only. *O. dewildemanii* Vanderyst (1920), in syn. Swamps, riversides and as a weed in rice fields.

Sen.: (Aug.) *Roger*! N'tiago (Jan.) *Henry*! Richard Toll (Jan.) *Henry*! Dakar (Oct.) *Chev.* 33841! **Gam.:** *Saunders* 71! Wallikunda (Sept.) *Devel. Corp.*! Ka-uur (Jan.) *Sampson* 48! **Mali:** Teheri, Kansu to Degué (Jan.) *Davey* 1! Bandiagara (May) *Rogeon & Leclercq* 253! Mopti (Sept.) *Chev.* 24924! Gao (Jan.) *Chev.* 43058! **S.L.:** Gbinti, L. Scarcies (Apr.) *Glanville* 223! Njala (Oct.) *Deighton* 1416! Tombo (Jan.) *Deighton* 1544! Bap to Pazehun (Dec.) *Sampson* 52! Rokupr, Magbema (July) *Jordan* 460! **Iv. C.:** Béréby (Nov.) *Oldeman* 576! Ferkessédougou (Nov.) *Leeuwenberg* 2027! Vavoua to Séguéla (Oct.) *Adjanohoun* 294*a*! **U. Volta:** Leo to Po (Oct.) *Rose Innes* 31088! Samendeni (Oct.) *Kmoch* 155! Tondoura (Oct.) *Scholz* 79! **Ghana:** Salaga to Bimbila (Oct.) *Rose Innes* GC 30959! Damongo (Mar.) *Hepper & Morton* A3178! Lawra to Wa (Oct.) *Ankrah* GC 20448! Navrongo (Oct.) *Vigne* FH 4639! Bawku (Oct.) *Rose Innes* GC 30255! **Dah.:** Ouidah (Aug.) *Adjanohoun* 203! **Niger:** Niamey (Oct.) *Hagerup* 526! S. Dallol (Jan.) *Virgo* 71! **N. Nig.:** Badeggi (Sept.) *Clayton* 312! Abinsi (Nov.) *Dalz.* 909! Kwarre, Sokoto (Sept.) *Palmer* 18! Katsina (Oct.) *Clayton* 1365! Damaturu Kabau (Oct.) *Daggash* FHI 22016! Gurum, near Vogel Peak (Nov.) *Hepper* 1256! Throughout tropical and S. Africa. (See Useful Plants.)

[The name *O. perennis* Moench (1794) has sometimes been used for our species, but the correct application of this name is quite uncertain and it is best discarded.]

3. **O. brachyantha** *A. Chev. & Roehr.* in Compt. Rend. Acad. Sci. 159: 561 (1914); Chev. Bot. 738, and in Rev. Bot. Appliq. 12: 1022 & 14: 133; Berhaut, Fl. Sén. ed. 2, 416; Jacques-Félix in J. Agric. Trop. 6: 586. *O. brachyantha* var. *guineensis* A. Chev. in Rev. Bot. Appliq. 12: 1022 (1932). *O. guineensis* A. Chev. l.c., in syn. Culms 30–45 cm. high; normally behaving as an annual, but Jacques-Félix (l.c.) states that it will perennate if watered during the dry season; in shallow pools on ironstone.

Mali: Ségou (Sept.) *Chev.* 24977! **Guin.:** Dubréka to Conakry (Nov.) *Chev.* 34616 *bis*! Kiffaya (Oct.) *Adam* 12701! Baffing R. (Mar.) *Pobéguin* 1818! **S.L.:** Lungi (Nov.) *Glanville* 102! Kambia (Sept.) *Jordan* 305! Matanka to Binkolo (Sept.) *Glanville* 424! Wonkifu (Sept.) *Jordan* 347! Kambia (Aug.) *Hooper* 100! Also in the Congo and Sudan.

4. **O. eichingeri** *A. Peter* in Fedde Rep., Beih. 40, 1, Anhang: 74 (1930); Tateoka in Bot. Mag. Tokyo 78: 156, 198; Aké Assi, Contrib. 2: 281. *O. latifolia* Hook. f. var. *collina* (Trimen) Hook. f. (1896). *O. sativa* L. var. *collina* Trimen (1889). *O. glauca* Robyns (1936), without description. Culms slender, firm; damp places in forest undergrowth.

Iv. C.: Ouango–Fitini (July) *Aké Assi* 5828! Also in the Congo, Uganda, Tanzania and Ceylon.

5. **O. punctata** *Kotschy ex Steud.* Syn. Pl. Glum. 1: 3 (1854); Tateoka in Bot. Mag. Tokyo 78: 156. *O. sativa* var. *punctata* (Kotschy ex Steud.) Schweinf. in Ber. Deutsch. Bot. Gesell. 44: 165 (1926). *O. schweinfurthiana* Prod. in Bot. Archiv. 1: 231 (1922). Culms thick, spongy; swampy streamsides.

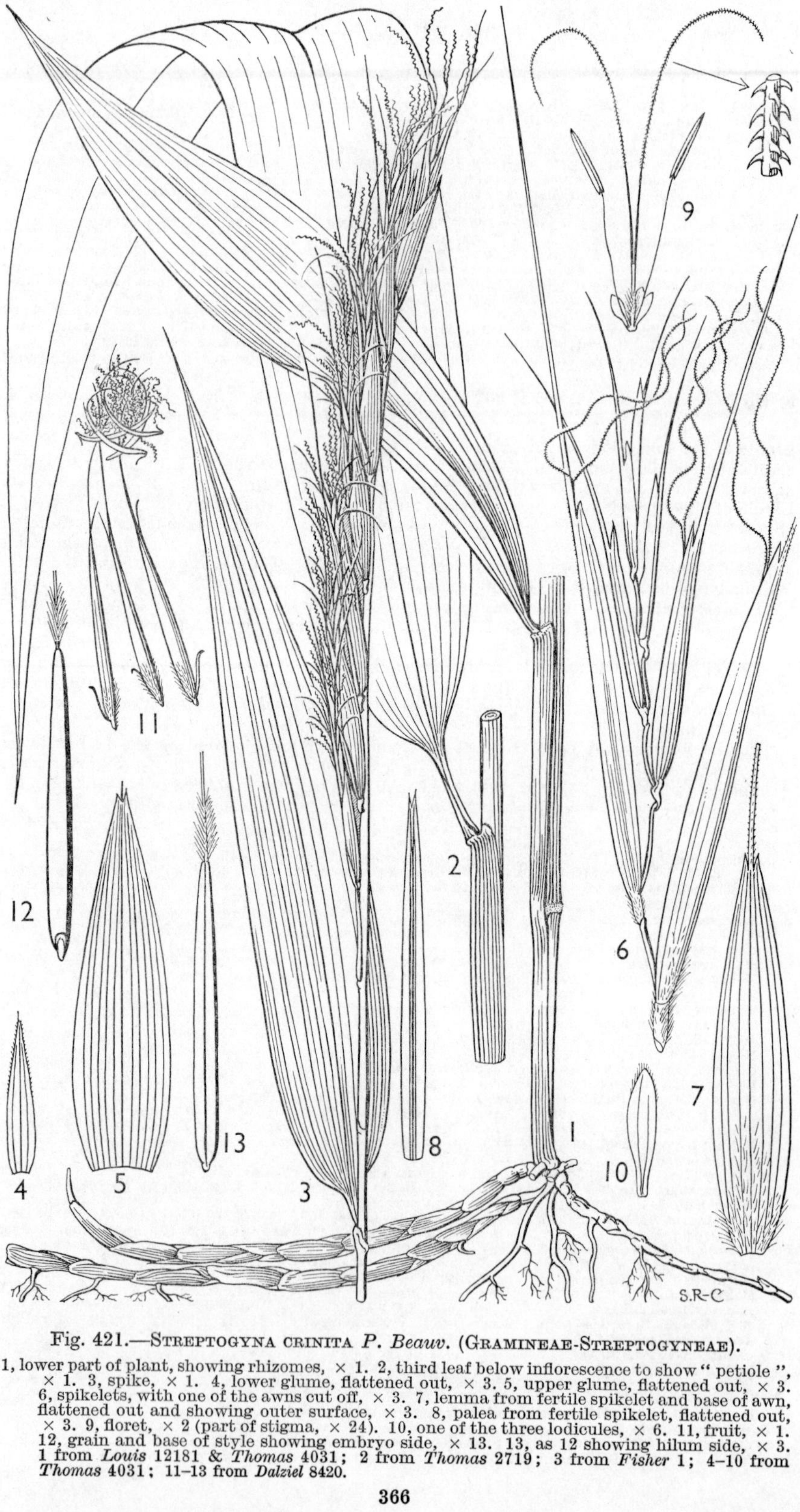

Fig. 421.—STREPTOGYNA CRINITA *P. Beauv.* (GRAMINEAE-STREPTOGYNEAE).

1, lower part of plant, showing rhizomes, × 1. 2, third leaf below inflorescence to show " petiole ", × 1. 3, spike, × 1. 4, lower glume, flattened out, × 3. 5, upper glume, flattened out, × 3. 6, spikelets, with one of the awns cut off, × 3. 7, lemma from fertile spikelet and base of awn, flattened out and showing outer surface, × 3. 8, palea from fertile spikelet, flattened out, × 3. 9, floret, × 2 (part of stigma, × 24). 10, one of the three lodicules, × 6. 11, fruit, × 1. 12, grain and base of style showing embryo side, × 13. 13, as 12 showing hilum side, × 3. 1 from *Louis* 12181 & *Thomas* 4031; 2 from *Thomas* 2719; 3 from *Fisher* 1; 4–10 from *Thomas* 4031; 11–13 from *Dalziel* 8420.

Iv. C.: Mankono to Boundiali (Dec.) *Adjanohoun* 202*a*! **Ghana**: Amelorkope, Trans Volta (Nov.) *Ankrah* GC 20512! Kpandu to Accra (Sept.) *Ankrah* GC 20216! Hacho, Legon (Nov.) *Hall* GC 37171! Cape Coast (July) *Hall* 3225! Folifoli (Aug.) *Hall* CC 192! **S. Nig.**: Ibadan (Jan.) *Clayton* 414! Also in Sudan, E. Africa, Congo and Angola, Madagascar and Thailand.

6. **O. barthii** *A. Chev.* in Bul!. Mus. Hist. Nat. Paris 16: 405 (1911), not of F.W.T.A. ed. 1, 2: 537. *O. breviligulata* A. Chev. & Roehr. in Compt. Rend. Acad. Sci. 159: 560 (1914); F.W.T.A. ed. 1, 2: 537; Chev. Bot. 738, and in Rev. Bot. Appliq. 12: 1018 & 14: 133; Berhaut, Fl. Sén. ed. 2, 416; Tateoka in Bot. Mag. Tokyo 75: 457; Aké Assi, Contrib. 2: 281. *O. stapfii* Roshev. in Bull. Appl. Bot. Leningrad 27, 4: 51 (1931); F.W.T.A., ed. 1, 2: 537. *O. mezii* Prod. in Bot. Archiv. 1: 223 (1922). *O. silvestris* Stapf ex A. Chev. var. *barthii* Stapf ex A. Chev. in Bull. Mus. Nat. Hist. Paris 16: 405 (1911), name only. *O. perennis* Moench. subsp. *barthii* (A. Chev.) A. Chev. in Rev. Bot. Appliq. 12: 1028 (1932) and 14: 132. Growing in water, often as a weed in rice fields.
Maur.: N. of Nioro *Rossetti* 61/302! **Sen.**: Kaolack to Nioro (Oct.) *Chev.* 33934! Oussouye, Casamance (Oct.) *Adam* 18414! **Gam.**: Georgetown (Sept.) *Pirie* 46/33! Bansang, Macarthy Div. *Duke* 12! **Mali**: Tanfola (Mar.) *Chev.* 642! Faguibine to Bomma (Aug.) *Chev.* 2208! Nafadie (Oct.) *Chev.* 2206! 2207! Ségou (Sept.) *Chev.* 24981! **Guin.**: Sébori (Oct.) *Adam* 12679! Coniré R. (Sept.) *Adam* 15249! **S.L.**: Kasewe, Yoni (Feb.) *Jordan* 1052! Njala (Dec.) *Deighton* 2824! Makump (Nov.) *Glanville* 428! Bumban (Dec.) *Deighton* 2938! Kasanko, Maforki (Oct.) *Jordan* 643! **Iv. C.**: Bassawa (Oct., Nov.) *Adjanohoun* 279*a*! *Aké Assi* 9261! **U. Volta**: Ouagadougou to Sikasso (Sept.) *Adam* 15424! Zabré (Oct.) *Scholz* 156! Tougan (Sept.) *Aké Assi* 6428! **Ghana**: Berekum to Sampa (Dec.) *Adams* 5285! Bolgatanga (Aug., Oct.) *Rose Innes* GC 30263! *Hall* CC 564! Zuarungu (Sept.) *Thorold* CB 200! **Niger**: Toukounous *Bartha* 47! Niamey to Zinder *Hagerup* 569! **N. Nig.**: Damaturu (Aug.) *De Leeuw* 1113! Katagum Dist. *Dalz.* 268! Maiduguri (Oct.) *Johnston* N 35! Gajibo–Fadama, Dikwa Dist. (Nov.) *Davey* FHI 27106! **S. Nig.**: Okomu F.R., Benin (Dec.) *Brenan & Jones* 8593! Also in C. African Rep., Sudan, Tanzania and Zambia.

7. **O. glaberrima** *Steud.* Syn. Pl. Glum. 1: 3 (1854); Chev. Bot. 738; Berhaut, Fl. Sén. ed. 2, 416; Portères in J. Agric. Trop. 3: 833. Many cultivars; Portères (l.c.) gives a key to varieties and forms, and discusses synonymy.
An indigenous cultivated rice, grown on flood-plains from Senegal to L. Chad. (See Useful Plants.)

10. LEERSIA Soland. ex Sw., Prod. Veg. Ind. Occ. 21 (1788), *nom. cons.*; Launert in Senck. Biol. 46: 129 (1965).

Lemma with an awn about 5 mm. long; spikelets 4–5 mm. long, up to 1 mm. broad 1. *tisserantii*

Lemma awnless:
- Spikelets glabrous or puberulous, 1·7–2 mm. long; stamens 3; culms slender, trailing 2. *triandra*
- Spikelets more or less pubescent on the sides, and prominently ciliate on the keel with curving glassy hairs; stamens 6:
  - Hairs on lemma and palea curved almost to a semicircle; spikelets 2–2.5 mm. long; panicle open, the branches spreading; culms erect .. .. 3. *drepanothrix*
  - Hairs on lemma and palea more or less straight, the keels pectinate-ciliate; spikelets 3·5–4 mm. long; panicle narrow, the branches ascending; culms weak, usually long decumbent .. .. 4. *hexandra*

1. **L. tisserantii** (*A. Chev.*) *Launert* in Senck. Biol. 46: 137 (1965). *Oryza tisserantii* A. Chev. in Rev. Bot. Appliq. 12: 1024 (1932); Jacques-Félix in J. Agric. Trop. 6: 582. Slender grass growing in streams.
Guin.: Fouta Djalon *Jac.-Fél.* 7419. Also in C. African Rep., Congo, Kenya and Zambia.

2. **L. triandra** *C. E. Hubbard* in Kew Bull. 1936: 313. A straggling grass of marshy places.
S.L.: Hangha (Sept.) *Deighton* 4031! Konta (Aug.) *Deighton* 1241! **Lib.**: Karmahun, Kolahun Dist. (Nov.) *Baldwin* 10227! Dubo, Sanokwele Dist. (Sept.) *Baldwin* 9485!

3. **L. drepanothrix** *Stapf* in Journ. de Bot. 19: 107 (1905); A. Chev. in Rev. Bot. Appliq. 14: 134; Berhaut, Fl. Sén. ed. 2, 406. Up to 1 m. high, erect from a short rhizome; marshy places.
Sen.: Casamance: Djembering (Oct.) *Adam* 18346! Bougnadou (Oct.) *Adam* 18481! Ouassadou (Dec.) *Berhaut* 2889. Sangalkam (June) *Berhaut* 927. **Guin.**: Kindia (Nov.) *Jac.-Fél.* 216! Baffing R. (Mar.) *Pobéguin* 1829! Kouroussa *Pobéguin* 495! **U. Volta**: Tindangu to Porga (Aug.) *Scholz* 140! **Ghana**: Sunyani to Wenchi (Dec.) *Adams* 5232! Kintampo to Yapei ferry (Nov.) *Rose Innes* GC 30846! Tamale (Sept.) *Thorold* 211! Pong Tamale to Savelugu (Sept.) *Oppong* 154! Nakpanduri to Yendi (Sept.) *Rose Innes* GC 32151! **N. Nig.**: Demboa F.R. (Oct.) *De Leeuw* 1853*a*! Also in Uganda and the Sudan.

4. **L. hexandra** *Sw.* Prod. Veg. Ind. Occ. 21 (1788); A. Chev. in Rev. Bot. Appliq. 14: 134; Berhaut, Fl. Sén. ed. 2, 403. *L. mauritanica* Salzm. ex Trin. (1840). *L. parviflora* Desv. (1831). *L. abyssinica* Hochst. ex A. Rich. (1851). *L. capensis* C. Muell. (1856). *L. aegyptiaca* Fig. & De Not. (1853). *L. ferox* Fig. & De Not. (1853). *Asprella hexandra* (Sw.) Roem. & Schult. (1817). *Oryza hexandra* (Sw.) Doell. (1871). *Homalocenchrus hexandrus* (Sw.) Kuntze (1891). Culms ascending from a rhizomatous base, forming matted carpets 30–90 cm. high in shallow water; the retrorsely spinulose midrib of the leaf can inflict most painful lacerations.
Gam.: *Ruxton* 72! 73! 104! **Mali**: Klela (Oct.) *Farrow* 67! **Guin.**: Timbo (Mar.) *Pobéguin* 1714! **S.L.**: Yonibana (Jan.) *Glanville* 154! **Iv. C.**: Zuénoula (Oct.) *Adjanohoun* 332*a*! Mankono (July) *Chev.* 21969! **U. Volta**: Samandéni (Sept.) *Kmoch* 102! **Ghana**: Awuabo to Krisin (Mar.) *Rose Innes* GC 30879! Pokoase (Sept.) *Ankrah* GC 20076! Nungua (June) *Ankrah* GC 20178! Yapei ferry to Kintampo (Nov.) *Rose Innes* GC 30854! Lawra (Oct.) *Ankrah* GC 20459! **Togo Rep.**: Palimé *Stage* 71! **N. Nig.**: Nupe *Barter* 783! Badeggi (Sept.) *Clayton* 313! Samaru, Zaria (Aug.) *Thatcher* S490! Gindiri, Jos Plateau (Oct.) *Hepper* 1107! Abinsi (Oct.) *Dalz* 884! L. Chad *Golding* 84! Gembu, Mambila Plateau (Jan.) *Hepper* 1840! **S. Nig.**: Lagos (Feb., May, Nov.) *Dawodu* 21! *Dalz.* 1312! *Haines* 356! Badagry (Aug.) *Onochie* FHI 33466! Obu Dist. *Thomas* 430! **W. Cam.**: Bafut-Ngemba F.R., Bamenda (June) *Daramola* FHI 41558! Jakiri to Kumbo (Feb.) *Charter* FHI 38006! Jakiri (June) *Brunt* 532! Babungo, Ndop (Feb.) *Brunt* 963! **F. Po**: Lago Biao (Sept.) *Wrigley & Melville* 602! Throughout the tropics. (See Useful Plants.)

Lachnochloa pilosa *Steud.* Syn. Pl. Glum. 1: 5 (1854). Senegambia. Although included in *Oryzeae* by Steudel, the description will fit no West African grass. The type, which cannot now be found, was incomplete and has obviously been misinterpreted.

11. FESTUCA Linn., Sp. Pl. 73 (1753), and Gen. Pl. ed. 5, 33 (1754); St.-Yves in Candollea 4: 65 (1929).

Awns up to 3 (rarely 5) mm. long; leaf-blades narrowly linear, convolute; panicle narrow and subspiciform, dense, usually 10–15 cm. long, the branches scabrid;

Fig. 422.—LEERSIA HEXANDRA *Swartz* (GRAMINEAE-ORYZEAE).

1, ligule. 2, spikelet showing lemma (left) and palea (right). 3, side view of same. 4, palea showing stamens. 5, flower.

spikelets 6·5–12 mm. long; glumes broad with hyaline margins, 5·5–10 mm. long, about ¾ the length of the spikelet, the lower often 3-nerved; lemmas 6·5–9·5 mm. long, with a terminal awn-point .. .. 1. *abyssinica*
Awns 8–15 mm. long:
Leaf-blades 1–2 mm. wide; lemmas smooth; awn terminal; ovary hairy on top 2b. *camusiana* subsp. *chodatiana*
Leaf-blades 7–17 mm. wide; lemmas scabrid; awn subterminal; ovary glabrous 3. *mekiste*

1. **F. abyssinica** *Hochst. ex A. Rich.* Tent. Fl. Abyss. 2: 432 (1851); St.-Yves l.c. 80; Hedberg in Sym. Bot. Uppsal. 15: 37 (1957). *F. schimperana* A. Rich. l.c. 433 (1851); F.W.T.A. ed. 1, 2: 508. *F. rigidula* Steud., Syn. Pl. Glum. 1: 314 (1854). *F. gelida* Chiov. in Ann. Bot., Roma 6: 147 (1907). *Koeleria afromontana* Jac.-Fél., Gram. Afr. Trop. 186 (1962), name only. Densely caespitose perennial, 15–60 cm. high.
W. Cam.: Cam. Mt., 10,000–13,300 ft. (Dec., Jan.) *Mann* 1349! *Mildbr.* 10909! *Boughey* GC 12627! *Hinds* C13! *Maitland* 1247! **F. Po:** Clarence Peak, 8,500 ft. (Mar.) *Mann* 656! 1465! 1477! On mountain peaks throughout tropical Africa.
2a. **F. camusiana** *St.-Yves* in Bull. Soc. Bot. Genève, sér. 2, 18: 156 (1926).
2b. **F. camusiana** subsp. **chodatiana** *St. Yves* l.c. 158 (1926). *F. simensis* of F.W.T.A. ed. 1, 2: 508, not of Hochst. ex A. Rich. (1851). Loosely tufted perennial 60–90 cm. high. Dimensions of spikelets and their parts are slightly smaller in E. Africa.
W. Cam.: Cam. Mt., 7,000 ft. (Apr., Dec.) *Mann* 2069! *Boughey* 6383! 6775! Above Bafut-Ngemba F.R., 7,050 ft. (July) *Brunt* 775! **F. Po:** 8,500 ft. (Mar.) *Mann* 1473! Also on mountain peaks in Kenya, Uganda, Tanzania, southern Sudan and eastern Congo.
3. **F. mekiste** *W. D. Clayton* in Kew Bull. 23: 293 (1969). *F. gigantea* of F.W.T.A. ed. 1, 2: 508, not of (Linn.) Vill. Perennial 0·6–2 m. high.
W. Cam.: Cam Mt., 7,000–7,600 ft. (Dec.) *Mann* 2078! *Maitland* 1338! 1338*a*! 8,000 ft. (Feb.) *Maitland* 1038! Ukele, 8,000 ft. (Feb.) *Maitland* 1336! **F. Po:** 8,500 ft. (Mar.) *Mann* 1462! 1468! Also in Kenya. (See Useful Plants.)

12. VULPIA Gmel., Fl. Bad. 1: 8 (1806).

Leaf-sheaths glabrous; blades linear, 1–2 mm. broad; panicle erect and narrow, more or less secund 2·5–7 cm. long; spikelets 7–14 mm. long without the awns, 5–10-flowered; glumes subulate-lanceolate, the lower 1-nerved and ½–¾ as long as the spikelet, the upper 3-nerved; lemmas 5–9 mm. long, linear-lanceolate, faintly 5-nerved, narrowed into a slender awn up to 13 mm. long .. .. *bromoides*

V. bromoides (*Linn.*) *S. F. Gray* Nat. Arr. Brit. Pl. 2: 124 (1821). *Festuca bromoides* Linn., Sp. Pl. 75 (1753). A very slender annual, up to about 23 cm. high in our area.
W. Cam.: Cam. Mt., 7,000–10,000 ft. (Dec.) *Mann* 2091! *Boughey* GC 12537! A species from Europe and the Mediterranean, extending to mountain peaks in tropical Africa; introduced elsewhere.

13. POA Linn., Sp. Pl. 67 (1753), and Gen. Pl. ed. 5, 31 (1754).

Panicle contracted, almost linear, the spikelets densely clustered 1. *leptoclada*
Panicle open, with spreading branches:
Annual; lemmas obtuse .. .. 2. *annua*
Perennial; lemmas acute .. .. 3. *schimperana*

1. **P. leptoclada** *Hochst. ex A. Rich.* Tent. Fl. Abyss. 2: 422 (1851); Hedberg in Sym. Bot. Uppsal. 15: 39 (1957). Straggling or tufted perennial 15–60 cm. high.
W. Cam.: Cam. Mt., 6,000–12,000 ft. (Dec.–Feb.) *Mann* 2071! *Dalz.* 8351! *Maitland* 1273! *Hinds* C83! *Morton* GC 6901! **F. Po:** 7,500 ft. (Apr.) *Mann* 1480! Mountain peaks throughout tropical Africa.
2. **P. annua** *Linn.* Sp. Pl. 68 (1753). Loosely tufted annual up to about 15 cm. high.
W. Cam.: Buea, 3,000 ft. (Apr., Dec.) *Brenan & E. W. Jones* 9583! *Hinds* C5! Cam. Mt., 5,600–7,500 ft. (Nov.–Jan.) *Maitland* 1262! *Hinds* C57! *Migeod* 218! **F. Po:** Moka, 4,000 ft. (Dec.) *Boughey* 58! A very common weed of temperate regions, extending to highland areas in the tropics.
3. **P. schimperana** *Hochst. ex A. Rich.* Tent. Fl. Abyss. 2: 423 (1851); Hedberg in Sym. Bot. Upsal. 15: 41 (1957). *P. binata* of F.W.T.A., ed. 1, 2: 509, not of Nees. Tufted perennial about 60 cm. high.
N. Nig.: Chappal Waddi (Nov.) *Jackson, Magaji & Tuley* 2036! **W. Cam.:** Cam. Mt., 9,000 ft. (Nov., Dec.) *Migeod* 173! *Mann* 2083! Above Bamenda, 7,500 ft. (July) *Brunt* 781*b*! Mountain peaks in tropical Africa.

14. BROMUS Linn., Sp. Pl. 76 (1753), and Gen. Pl. ed. 5: 33 (1754).

Leaf-blades flat, 5–10 mm. broad, usually with falcate auricles at the base; leaf-sheaths laxly pilose with deflexed hairs, retrorsely scabrid on the nerves; panicle very lax with slender branches; spikelets 1·5–3 cm. long without the awns, laterally compressed; lower glume 3–6 mm. long, 1-nerved, the upper 5–12 mm. long, 3-nerved; lemmas 7–14 mm. long, pubescent on the nerves, glabrous or appressed-pubescent between, with an awn 2–12 mm. long inserted just below the obtuse tip; palea finely ciliolate on the keels .. .. *leptoclados*

B. leptoclados *Nees* Fl. Afr. Austr. 453 (1841). *B. cognatus* Steud., Syn. Pl. Glum. 1: 321 (1854). *B. scabridus* Hook. f. in J. Linn. Soc. 7: 231 (1864); F.W.T.A., ed. 1, 2: 508. A handsome perennial up to 2 m. high. May be confused with *Festuca mekiste*, in which the lemma and keels of the palea are merely scabrid.
W. Cam.: Cam. Mt., 9,000 ft. (Nov.–Mar.) *Mann* 2085! *Dalz.* 8352! *Migeod* 175! *Maitland* 1261! *Hinds* C47! On highlands throughout tropical and S. Africa.

Fig. 423.—VULPIA BROMOIDES (*Linn.*) *S. F. Gray* (GRAMINEAE-POEAE).

1, habit, × ⅔. 2, spikelet, × 3. 3, first glume, × 6. 4, second glume, × 6. 5, floret, × 6. 6, lemma, × 6. 7, palea, × 6. 8, flower, × 6. 9, grain, × 6. 10, ligule, × 4. From *I. J. Blair* 292.

15. BRACHYPODIUM P. Beauv., Ess. Agrost. 100 (1812).

Leaf-sheaths nearly always scabrid; racemes 6–12 cm. long, usually with 5–8 spikelets borne upon pedicels 1–2 mm. long; spikelets narrowly lanceolate, 1·5–3 cm. long, 6–12-flowered; glumes subequal, 4–10 mm. long; lemmas 7–11 mm. long, 7-nerved, tipped with a straight awn 4–8 mm. long. .. .. *flexum*

B. flexum *Nees* Fl. Afr. Austr. 1: 456 (1841). Weak-stemmed perennial 30–90 cm. high.
S.L.: Tingi Hills (May) *Gledhill* 222! **N. Nig.:** Chappal Waddi, Sardauna Prov. (Nov.) *Jackson, Magaji & Tuley* 2035! **F. Po:** Clarence Peak, 7,000 ft. *Mann* 321! Also in Sudan and Ethiopia to South Africa and Madagascar.
[*B. flexum* is very closely related to the European *B. sylvaticum* (Huds.) P. Beauv., which has smooth leaf-sheaths. The Sierra Leone specimen cited above has smooth sheaths, the Nigerian scabrid, and the one from F. Po is very poorly preserved and smothered in glue. The identity of the West African plants must remain uncertain pending further study based on better material.—W.D.C.]

16. TRITICUM Linn., Sp. Pl. 85 (1753), and Gen. Pl. ed. 5, 37 (1754).

Spikelets 10–15 mm. long and 9–18 mm. across the face, 5–9-flowered, imbricate in a dense spike which is square in section or more commonly broadest across the face of the spikelets; glumes oblong, shorter than the spikelet, keeled above, the central nerve prolonged into a tooth or short awn; lemmas glabrous or pubescent, awned or awnless .. .. *aestivum*

T. aestivum *Linn.* Sp. Pl. 85 (1753). *T. vulgare* Vill., Hist. Pl. Dauph. 2: 153 (1787); F.W.T.A., ed. 1, 2: 509; De Miré & Gillet in J. Agric. Trop. 3: 740 (1956). *T. sativum* Lam. (1778).
Mali: Bamako (Jan.) *Chev.* 217! **N. Nig.:** Katsina *Rae*! *Thornton*! Kari *Foster* 21! Kalkala (Aug.) *Gwynn*! Bornu *Thornton*! The common bread wheat of temperate regions. (See Useful Plants.)

17. HORDEUM Linn., Sp. Pl. 84 (1753), and Gen. Pl. ed. 5, 37 (1754).

Spikelets of each triad alike and fertile, sessile, arranged in six rows along the spike; glumes very narrow, about 1 cm. long, side by side in front of the floret; lemma as long as the glumes, prolonged in a straight, stiff, scabrid bristle 5–10 cm. long *vulgare*

H. vulgare *Linn.* Sp. Pl. 84 (1753); Chev. Bot. 752.
N. Nig.: Bornu *Thornton*! Mongonu *Dept. of Agric.*! Mobber *Dept. of Agric.*! The common barley of temperate regions. (See Useful Plants.)

18. STREBLOCHAETE Hochst. ex Pilger in Engl., Bot. Jahrb. 37, Beibl. 85: 81 (1906); F.T.A. 10: 101; Tateoka in Bot. Mag. Tokyo 78: 289 (1965).

Leaf-blades narrowly lanceolate-linear, up to 1 cm. broad; leaf-sheath entire, the margins united; panicle slender, narrow; spikelets narrowly lanceolate, 16–28 mm. long, loosely 2–6-flowered; glumes membranous with hyaline margins, the upper up to 12 mm. long; lemmas up to 15 mm. long, 7-nerved, entire or shortly bilobed with the lobes up to 2 mm. long, awned from the back just below the sinus of the lobes; awn flexuous, up to 4 cm. long .. .. *longiarista*

S. longiarista (*A. Rich.*) *Pilger* in Notizbl. Bot. Gart. Berl. 9: 516 (1926); F.T.A. 10: 102. *Trisetum longiaristum* A. Rich., Tent. Fl. Abyss. 2: 417 (1851). A loosely tufted perennial up to 1 m. high; the awns become entangled so that the florets are detached from the inflorescence in bunches; margins of montane forests.
W. Cam.: Cam. Mt., 8,000 ft. (Dec.) *Mann* 2077! Jonjo, 7,600 ft. (Feb.) *Maitland* 1335! Mann's Spring, 7,000 ft. (Mar.) *Brenan* 9512! Summit Peak, S.E. Bamenda, 8,200 ft. (Dec.) *Boughey* GC 11031! Bambui, 7,400 ft. (Dec.) *Boughey* GC 10865! **F. Po:** (Dec.) *Boughey* GC 10817! Mountain peaks throughout Africa; also in Réunion, the Malayan Islands and the Philippines.

19. KOELERIA Pers., Syn. Pl. 1: 97 (1805); Domin in Biblio. Bot. 14 (65) (1907); F.T.A. 10: 94 (1937).

Leaf-blades strongly nerved, convolute; panicles spiciform, dense, 4–9 cm. long, straw-coloured or purplish, the axis pubescent or tomentose; spikelets 3·5–7 mm. long, 1–3-flowered; glumes slightly shorter than the spikelet, the upper 1–3-nerved; lemma 3-nerved .. .. *capensis*

K. capensis (*Steud.*) *Nees* in Linnaea 7: 321 (1832). *Aira capensis* Steud. in Flora 12: 469 (1829). *Koeleria cristata* (Linn.) Pers. var. *convoluta* (Hochst. ex Steud.) C. E. Hubbard in Kew Bull. 1936: 500; F.T.A. 10: 95. *K. convoluta* Hochst. ex Steud., Syn. Pl. Glum. 1: 293 (1854); F.W.T.A., ed. 1, 2: 527; Domin in Biblio. Bot. 14 (65): 109. *K. gracilis* Pers. var. *convoluta* (Hochst. ex Steud.) Hedberg in Symb. Bot. Uppsal. 15: 42. Tufted perennial, 23–45 cm. high.
W. Cam.: Cam. Mt., 10,000–12,000 ft. (Dec.–Apr.) *Mann* 1357! *Maitland* 1259! *Boughey* GC 12644! *Keay* FHI 28617! *Thresh* 6! E. Cameroun and Ethiopia to the Cape.

20. AIRA Linn., Sp. Pl. 63 (1753), and Gen. Pl. ed. 5: 31 (1754); F.T.A. 10: 87 (1937).

Panicle delicate, with spreading filiform branches; spikelets 2·7–3·2 mm. long (down to 2·2 mm. long in Europe); glumes acute, membranous, shining, enclosing 2 florets; lemma about 2 mm. long, glabrous, with a geniculate awn 3–4 mm. long arising from about the middle of the back; callus of the florets normally bearded with white hairs, but glabrous in our area .. .. *caryophyllea*

A. caryophyllea *Linn.* Sp. Pl. 66 (1753); F.T.A. 10: 87. Slender annual 5–30 cm. high.
N. Nig.: Chappal Waddi (Apr., Nov.) *Jackson, Magaji & Tuley* 2024! 2055! *Gbile & Daramola* FHI 63274! **W. Cam.:** Cam. Mt., 8,000–12,000 ft. (Dec.–Jan.) *Mann* 1356! *Steele* 60! *Maitland* 868! 1274! *Mildbr.* 10908! Europe and western Asia, extending to tropical African mountains and to South Africa; introduced into most other temperate regions.
[The glabrous callus is found in *A. cupaniana* Guss., but the size and shape of the glumes suggest that our species is best treated as a form of *A. caryophyllea.*]

Imperfectly known species.

A. paradoxa *Steud.* Syn. Pl. Glum. 1: 223 (1854). " Guinea " *Lenormand.* The plant cannot be identified from the description.

21. HELICTOTRICHON Bess. ex Roem. & Schult., Syst. Veg. 2, Addit. 52 (1827); F.T.A. 10: 103 (1937).

Lower glume 3·5–5 mm. long, up to half the length of the spikelet; spikelets 1·4–1·9 cm. long, arranged in a narrow panicle; florets 3–4, long exserted from the glumes; lemmas with a geniculate awn from the upper third of the back; leaf-blades up to 10 mm. broad .. .. 1. *mannii*
Lower glume 5–12 mm. long, amost as long as the spikelet; spikelets 0·8–1·6 cm. long, arranged in a narrow panicle; florets 2–3; lemmas with a geniculate awn from about the middle of the back; leaf-blades up to about 5 mm. broad .. 2. *elongatum*

1. **H. mannii** (*Pilger*) *C. E. Hubbard* in Kew Bull. 1936: 334; F.T.A. 10: 110. *Avenastrum mannii* Pilger in Notizbl. Bot. Gart. Berl. 9: 520 (1926). *Arrhenatherum mannii* (Pilger) Potztal in Engl., Bot. Jahrb. 75: 329 (1951). Perennial up to 1·2 m. high.
W. Cam.: Cam. Mt., 6,000–8,000 ft. (Dec.–Feb.) *Mann* 2089! *Maitland* 855! *Hinds* C58! *Dunlap* 43! *Breteler* 132! **F. Po:** Clarence Peak (Pico de S. Isabel) 8,500 ft. (Mar.) *Guinea* 2658! *Mann* 657! 1476!
2. **H. elongatum** (*Hochst. ex A. Rich.*) *C. E. Hubbard* in Kew Bull. 1936: 335. *Danthonia elongata* Hochst. ex A. Rich., Tent. Fl. Abyss. 2: 419 (1851). *Helictotrichon maitlandii* C. E. Hubbard in Kew Bull. 1936: 332; F.T.A. 10: 112; F.W.T.A., ed. 1, 2: 528. *H. rigidulum* (Pilger) C. E. Hubbard in Kew Bull. 1936: 335; F.T.A. 10: 113; F.W.T.A. ed. 1, 2: 528. *Avenastrum rigidulum* Pilger in Notizbl. Bot. Gart. Berl. 9: 519 (1926). *A. elongatum* (Hochst. ex A. Rich.) Pilger, l.c. (1926), incl. var. *preussii* Pilger. *Arrhenatherum rigidulum* (Pilger) Potztal in Engl., Bot. Jahrb. 75: 329 (1951). Perennial up to 1·5 m. high.
N. Nig.: Chappal Waddi (Nov.) *Jackson, Magaji & Tuley* 2001! **W. Cam.:** Cam. Mt., 7,000–8,000 ft. (Dec.–Feb.) *Mann* 2068! *Maitland* 965! 1039! 1222! Bafut-Ngemba F.R., Bamenda (Mar.) *Hepper* 2243! Lakom, Bamenda (May) *Maitland* 9a! Kishong, Bamenda (Mar.) *Brunt* 1022! Sudan and Ethiopia southwards to Rhodesia, and in Madagascar.

22. DESCHAMPSIA P. Beauv., Ess. Agrost. 91 (1812); F.T.A. 10: 89 (1937).

Leaf-blades flat, up to about 5 mm. broad, scabrid upon the prominent ridges above; awns straight, not projecting beyond the glumes; spikelets 4–5 mm. long 1. *caespitosa*
Leaf-blades tightly rolled, up to 1 mm. diam.; awns slightly geniculate, projecting about 2 mm. beyond the glumes; spikelets 4–8 mm. long .. .. 2. *mildbraedii*

1. **D. caespitosa** (*Linn.*) *P. Beauv.* Ess. Agrost. 91: 160 (1812), incl. var. *mannii* C. E. Hubbard in Kew Bull. 1935: 311; F.T.A. 10: 91. *Aira caespitosa* Linn., Sp. Pl. 64 (1753). A harsh-leaved perennial up to 1 m. high.
F. Po: Clarence Peak (Pico de S. Isabel), 9,000–10,000 ft. (Mar.) *Mann* 322! 1464! *Guinea* 2793! Temperate regions of both hemispheres, and on the mountains of East Africa.
2. **D. mildbraedii** *Pilger* in Notizbl. Bot. Gart. Berl. 10: 769 (1929); F.T.A. 10: 92. Caespitose perennial up to 1·2 m. high, with stiff rush-like leaves.
W. Cam.: Cam. Mt., 10,000–12,000 ft. (Dec.–Feb.) *Maitland* 1246! *Hinds* C31! *Boughey* GC 12661! *Keay* FHI 28616! *Dalz.* 8350!

23. AGROSTIS Linn., Sp. Pl. 61 (1753), and Gen. Pl. ed. 5, 30 (1754); F.T.A. 10: 166 (1937).

Panicle loose, open, with flexuous filiform branches; spikelets 4–4·5 mm. long; lemmas pilose; awn basal, about 5 mm. long; palea about as long as the lemma 1. *mannii*
Panicle dense and spike-like; spikelets 3·5–4·5 mm. long; lemma glabrous or pubescent; awn basal, 3·5–5 mm. long; palea two thirds as long as the lemma 2. *quinqueseta*

1. **A. mannii** (*Hook. f.*) *Stapf* in Fl. Capensis 7: 549 (1899); F.T.A. 10: 184. *Deyeuxia mannii* Hook. f. in J. Linn. Soc. 7: 228 (1864). *Calamagrostis mannii* (Hook. f.) Engl. (1892). Tufted perennial up to 1 m. high; the spikelets sometimes with a rhachilla extension up to 3 mm. long. Very closely related to *A. kilimandscharica* Mez (1922) from East Africa.
W. Cam.: Cam. Mt., 6,700–12,000 ft. (Dec.–Apr.) *Mann* 2096! *Maitland* 1033! *Hinds* C87! *Boughey* GC 6956! *Morton* GC 6887! **F. Po:** Clarence Peak (Pico de S. Isabel), 8,500 ft. (Mar.–Dec.) *Mann* 1469! *Boughey* GC 10801! *Guinea* 2791! 2795! Also in E. Cameroun and Ethiopia.
2. **A. quinqueseta** (*Steud.*) *Hochst.* in Flora 38: 285 (1855); F.T.A. 10: 182 (1937). *Anomalotis quinqueseta* Steud., Syn. Pl. Glum. 1: 198 (1854). *Agrostis mildbraedii* Pilger in Wiss. Ergebn. deutsch. Zentr.-Afr. Exped. 1907–1908, 2: 47 (1910); F.T.A. 10: 181. *A. congesta* C. E. Hubbard in Kew Bull. 1936: 301; F.T.A. 10: 182; F.W.T.A. ed. 1, 2: 530. Perennial up to 1 m. high.
W. Cam.: Cam. Mt., 8,000–10,000 ft. (Dec.–Feb.) *Mann* 2086! *Maitland* 930! 1034! Also in E. Cameroun, Congo, Ethiopia, Uganda and Kenya.

24. HYPSEOCHLOA C. E. Hubbard in Kew Bull. 1936: 300; F.T.A. 10: 165 (1937).

Inflorescence a delicate panicle, 5–10 cm. long; spikelets cleistogamous, 3 mm. long; glumes equal, acutely acuminate, as long as the spikelet; lemma indurated, with a geniculate awn 4–6 mm. long from the middle of its back.. .. *cameroonensis*

Fig. 424.—DESCHAMPSIA MILDBRAEDII *Pilger* (GRAMINEAE-AVENEAE).
A, plant. B, section of leaf. C, spikelet. D, rhachilla. E, floret. F, lemma. G, palea. H, lodicules. I, flower. J, ovary and stigmas.

H. cameroonensis *C. E. Hubbard* l.c. (1936); F.T.A. 10: 165. A delicate slender annual up to 45 cm. high; closely resembles *Aira caryophyllea*, with which it is often found mixed.
W. Cam.: Cam. Mt., 7,000–12,000 ft. (Dec.) *Mildbr.* 10881! *Maitland* 874! 960! 1228*a*! 1274*a*! *Hinds* C19*a*!

Additional genus.

Polypogon monspeliensis (*Linn.*) *Desf.* Fl. Atlant. 1: 67 (1798). *Alopecurus monspeliensis* Linn., Sp. Pl. 61 (1753). Differs from *Agrostis* in the long-awned glumes and dense panicle. A cosmopolitan species of warm-temperate regions, recorded from the northern edge of our area at Aïr, Niger Republic by De Miré & Gillet in J. Agric. Trop. 3: 735 (1956).

25. PHRAGMITES Adans., Fam. Pl. 2: 34, 559 (1763); F.T.A. 10: 152 (1937); Clayton in Kew Bull. 21: 113 (1967).

Leaf-blades usually smooth beneath, with filiform flexuous tips; rhachilla hairs 5–11 mm. long, copious, silky; upper glume 5–9 mm. long, obtuse to tridenticulate
1b. *australis* subsp. *altissimus*
Leaf-blades usually scabrid beneath (at least in the upper half), with stiff attenuate tips; rhachilla hairs 5–7 mm. long, rather sparse; upper glume 4–7 mm. long, acute to subacute .. .. 2. *karka*

1a. **P. australis** (*Cav.*) *Trin. ex Steud.* Nom. Bot. ed. 2, 2: 324 (1841). *Arundo australis* Cav. in Ann. Hist. Nat. 1: 100 (1799). *A. phragmites* Linn. (1753). *A. vulgaris* Lam. (1778), illeg. name. *Phragmites communis* Trin., Fund. Agrost. 134 (1820); F.T.A. 10: 153; Chev. Bot. 748; De Miré & Gillet in J. Agric. Trop. 3: 735. *P. vulgaris* Crép. (1866)—F.W.T.A., ed. 1, 2: 510, partly; Berhaut, Fl. Sén. ed. 2, 419.

1b. **P. australis** subsp. **altissimus** (*Benth.*) *W. D. Clayton* in Taxon 17: 158 (1968). *Arundo altissima* Benth., Cat. Pl. Pyr. 62 (1826). *A. maxima* Forsk., Fl. Aegypt. Arab. 24 (1792), name of doubtful application. *A. isiaca* Del. (1813), illeg. name. *P. communis* subsp. *maximus* (Forsk.) W. D. Clayton in Kew Bull. 21: 116 (1967), doubtful name. A rhizomatous perennial up to 3·5 m. or more high, with handsome plumose panicles; found near water. Subsp. *australis*, a cosmopolitan grass of temperate regions, is of smaller stature and has an acute to apiculate upper glume.
Sen. *Adanson*! *Richard*! (Nov.) *Chev.* 33980! Senegal R. *Brunner* 15! **Gam.:** Karinikunda (Feb.) *Frith* 169! Massembe (July) *Deighton* 5601! **Niger:** Niamey (Jan.) *Chev.* 43197! Zinder (Feb.) *Chev.* 43664! **N. Nig.:** Chad (Oct.–Dec.) *Elliot* 159! *Golding* 36! *Davey* FHI 27114! *Johnston* N 17! Shores of the Mediterranean, extending eastwards to Iran, Arabia and northern Kenya. (See Useful Plants.)

2. **P. karka** (*Retz.*) *Trin. ex Steud.* Nom. Bot. ed. 2, 2: 324 (1841). *Arundo karka* Retz., Obs. Bot. 4: 21 (1786). *P. laxiflorus* Steud. (1854). *P. mauritianus* of F.T.A. 10: 155, partly. *P. vulgaris* of F.W.T.A., ed. 1, 2: 510, partly. A rhizomatous perennial up to about 7 m. high, with large plumose panicles; found near water. Very closely related to *P. mauritianus* Kunth, which is distinguished by its broader acuminate subequal glumes.
Sen.: *Roger*! **Mali:** Kayes (Nov.) *Jaeger* 5667! **Port. G.:** Bissau (Feb.) *Esp. Santo* 1798! **S.L.:** Freetown (Dec.) *Hagerup* 762! **Ghana:** Kpong (June, Jan.) *Ankrah* GC 20054! *Irvine* 1918! Aburi (Nov.) *Johnson* 1039! Kete Krachi (Dec.) *Adams* 4558! Sawla to Tamale (Oct.) *Rose Innes* GC 30672! **Dah.:** Porto Novo (Mar., Nov.) *Adams* K450! *De Kimpe* 113! **Niger:** Bengou (Feb.) *Virgo* 4! **N. Nig.:** Nupe *Barter* 1031! Vom (Dec.) *Haines* 120! Fodama (Dec.) *Moisier* 236! Katagum *Dalz.* 285! Yola (Dec.) *Hepper* 1616! **S. Nig.:** Old Oyo (Feb.) *Keay* FHI 16260! Olokemeji (Nov.) *Jones, Keay & Onochie* FHI 14233! Mbiakpaba (Mar.) *Jackson* 514! *Daramola* FHI 46341! Also in E. Cameroun, Uganda, Ethiopia and Sudan, extending eastwards through India and Malesia to N. Australia.

Additional genus.

Gynerium sagittatum (*Aubl.*) *P. Beauv.* Ess. Agrost. 138 (1812). *Saccharum sagittatum* Aubl., Pl. Gui. 1: 50 (1775). A huge plant up to 8 m. high; leaf-blades up to 7 cm. wide, spinulose on the margins; plants dioecious, the ♀ panicle plumose, the ♂ glabrous; spikelets 3·5–5·5 mm. long. A species from tropical America introduced to Ghana.
Ghana: Aburi (♂) *Howes* 1200! Yawkoko (♂) (Jan.) *Morton* GC 25335!

26. ASTHENATHERUM Nevski in Act. Univ. As. Med. sér. 8b, Bot. Fasc. 17: 8 (1934); Conert in Senck. Biol. 43: 239 (1962).

Panicle dense, embraced below by the broad uppermost leaf-sheath; glumes subequal, 7–9 mm. long, prominently 7–9-nerved, enclosing 2 or 3 florets; lemmas about 3 mm. long (entire part), 9-nerved, the lateral lobes 2 mm. long, acuminate, the awn straight or slightly geniculate .. .. *forskalii*

A. forskalii (*Vahl*) *Nevski* in Act. Univ. As. Med. sér. 8b, Bot. Fasc. 17: 8 (1934). *Avena forskalii* Vahl, Symb. Bot. 2: 25 (1791). *Danthonia forskalii* (Vahl) R. Br. in Denham & Clapperton, Narr. Trav. North & Centr. Afr., Append. 244 (1826); F.T.A. 10: 140; F.W.T.A., ed. 1, 2: 527; Chev. Bot. 744, and in Rev. Bot. Appliq. 14: 44; De Miré & Gillet in J. Agric. Trop. 3: 730. Loosely tufted perennial with pungent glaucous leaves and coarse stringy roots; a desert grass.
Maur.: Oualata *Rossetti* 61/161! Oualata to Nèma *Jumelle* 1! N. Africa and the Middle East, extending southward to Sudan and N. Kenya; also in Angola.

27. PENTASCHISTIS Stapf in Fl. Capensis 7: 315 (1898); F.T.A. 10: 123 (1937).

Leaf-blades tightly involute, stiff and rigid; panicle dense, contracted; spikelets 5·5–6.5 mm. long, strictly 2-flowered, enclosed by the thin 1-nerved glumes; lemmas 2–2·5 mm. long (entire part), obscurely 7-nerved, glabrous or sparsely pubescent below, the lateral lobes drawn out into fine bristles 2 mm. long, the awn geniculate with a twisted column .. .. *pictigluma*

P. pictigluma (Steud.) Pilger in Not. Bot. Gart. Berlin 9: 517 (1926); F.T.A. 10: 133; Wickens in Kew Bull. 26: 41 (1971). *Aira pictigluma* Steud., Syn Pl. Glum. 1: 221 (1854). *Pentaschistis mannii* Stapf ex C. E. Hubbard in Kew Bull. 1936: 501; F.T.A. 10: 134; F.W.T.A., ed. 1, 2: 528. Tufted perennial up to 30 cm. high.
W. Cam.: Cam. Mt., 10,000ft.-summit (Nov.–Jan.) *Mildbr.* 10894! *Hinds* C37! *Maitland* 1244! *Migeod* 197! *Keay* FHI 28625! Also in Sudan, Ethiopia, Uganda and Tanzania.

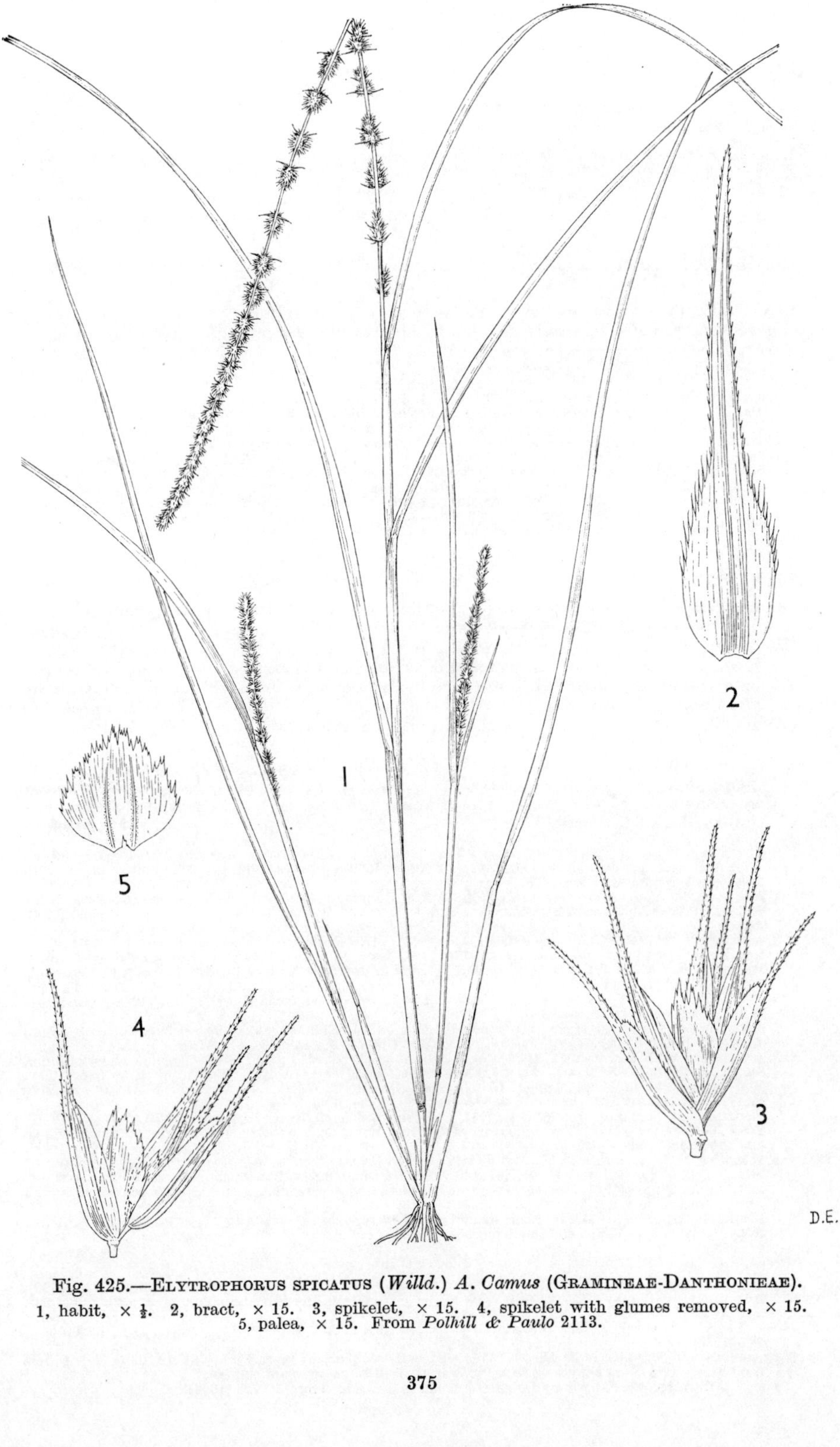

Fig. 425.—ELYTROPHORUS SPICATUS (*Willd.*) *A. Camus* (GRAMINEAE-DANTHONIEAE).
1, habit, × ½. 2, bract, × 15. 3, spikelet, × 15. 4, spikelet with glumes removed, × 15.
5, palea, × 15. From *Polhill & Paulo* 2113.

28. **TRIRAPHIS** R. Br., Prod. 185 (1810).

Panicle very dense, ovoid, 1–3 cm. long; spikelets about 3 mm. long and 3–11-flowered; lemmas 3-nerved, pilose, with a central awn 2–3 mm. long from the sinus of the briefly bilobed tip, and with the side nerves excurrent into shorter lateral awns; palea wingless; grain fusiform, 1·5 mm. long .. .. *pumilio*

T. pumilio *R. Br.* in Denham & Clapperton, Narr. Trav. North & Centr. Afr., Append. 245 (1826); A Chev. in Rev. Bot. Appliq. 14: 131; De Miré & Gillett in J. Agric. Trop. 3: 701. *T. glomerata* A. Camus (1931). *T. nana* (Nees) Hack. (1890). Annual, 2–25 cm. high; dry soils.
Maur.: Mantounsi *Rossetti* 61/393! **Niger:** Teneré (Sept.) *P. de Fabrègues* 922. Tazolé *Chev.* 42979. Aïr (*fide* De Miré & Gillett *l.c.*). N. Africa from Mauritania to Arabia, and in S. Africa. (See Useful Plants.)

29. **ELYTROPHORUS** P. Beauv., Ess. Agrost. 67 (1812); Schweickerdt in Ann. Nat. Mus. 10: 191; Jacques-Félix in J. Agric. Trop 5: 304.

Leaf-blades finely ridged above and below; ligule membranous, up to 1 mm. long; glumes lanceolate, aristately acuminate; lemma 0·6–1·5 mm. long, terminating in an awn 0·5–1·8 mm. long; keels of palea broadly winged or sometimes wingless; anthers 1–2 per floret .. .. *spicatus*

E. spicatus (*Willd.*) *A. Camus* in Lecomte, Fl. Gen. Indo-Chine 7: 547 (1923); Berhaut, Fl. Sén. ed. 2, 396. *Dactylis spicata* Willd. in Ges. Naturf. Freunde Berlin, Neue Schrift 3: 416 (1801). *Elytrophorus articulatus* P. Beauv. (1812),—Chev. Bot. 748, and in Rev. Bot. Appliq. 14: 134. *Phleum glomeruliflorum* Steud., Syn. Pl. Glum. 1: 150 (1854). Annual, 15–60 cm. high; swamps and old rice fields.
Sen.: Richard Toll (Jan.) *Roger* 30! Tamboukané (Oct.) *Chev.* 2383! **Mali:** Dogo *Demange* 21/1957! Sole (Apr.) *Davey* 604! Mema (Oct.) *Davey* 497! Ansongo (Sept.) *Hagerup* 407! Macina (Sept.) *Chev.* 24892! **Iv. C.:** Ouango-Fitini (Dec.) *Aké-Assi* 7526! **Ghana:** Bolgatanga to Tamale (Oct.) *Ankrah* GC 20299! *Rose Innes* GC 31114! Pong Tamale (Nov.) *Addei* SLUS/296! Kamba (Jan.) *Harris*! Yendi (Dec.) *Adams & Akpabla* 4130! **N. Nig.:** *Ward* 89! Shendam (Nov.) *Clayton* 1477! Dikwa (Nov.) *Johnston* N. 93! Bama (Dec.) *McClintock* 86! Yola (Nov.) *Hepper* 1203! Throughout the Old World tropics.

30. **STIPAGROSTIS** Nees in Linnaea 7: 290 (1832); de Winter in Bothalia 8: 307 (1965).

Glumes pubescent; lemma strongly papillose; awn 5–6 cm. long, the column hairy below the insertion of the lateral awns .. .. 1. *hirtigluma*
Glumes glabrous; lemma smooth or weakly papillose:
Awn with a dense tuft of hairs at the insertion of the lateral awns; column 5–6 mm. long, the central awn 20–25 mm.; lemma about 2 mm. long with a callus about 0·5 mm. long; sheaths and internodes glabrous .. .. .. 2. *uniplumis*
Awn glabrous at the insertion of the lateral awns; lemma 3–4 mm. long, with a callus 1–1·5 mm.:
Central awn 35–45 mm. long, the column 4–10 mm. long; basal sheaths and internodes densely woolly tomentose .. .. 3. *plumosa*
Central awn 10–15 mm. long, the column 1 mm. long; basal sheaths glabrous, the internodes silky hairy .. .. 4. *acutiflora*

1. **S. hirtigluma** (*Steud. ex Trin. & Rupr.*) *de Winter* in Kirkia 3: 134 (1963). *Aristida hirtigluma* Steud. ex Trin. & Rupr., Sp. Gram. Stip. 171 (1842); A. Chev. in Rev. Bot. Appliq. 14: 47. Tufted annual with convolute leaf-blades, 45 cm. high.
Mali: Région nord *Leclercq* 42686! **N. Nig.:** Arege, Bornu (Sept.) *Goss* 5! Extends eastward through the Sudan, Ethiopia and Kenya to Arabia, the Middle East and India; also in Angola, Botswana and South Africa.
2. **S. uniplumis** (*Licht.*) *de Winter* in Kirkia 3: 136 (1963). *Aristida uniplumis* Licht. in Roem. & Schult., Syst. Veg. 2: 401 (1817). *A. papposa* Trin. & Rupr., Sp. Gram. Stip. 173 (1842), incl. var. *senegalensis* Trin. & Rupr.; F.W.T.A., ed. 1, 2: 534; Chev. Bot. 741, and in Rev. Bot. Appliq. 14: 48; Berhaut, Fl. Sén. ed. 2, 406; De Miré & Gillet in J. Agric. Trop. 3: 728. *A. concinna* Sander in J. A. Schmidt, Beitr. Fl. Cap. Verd. Ins. 140 (1852). *Stipagrostis papposa* (Trin. & Rupr.) de Winter in Kirkia 3: 135 (1963). Caespitose perennial up to about 60 cm. high.
Sen.: *Jardin*! *Leprieur*! **Mali:** Région nord *Leclercq* 42740! Bamba (Sept.) *Hagerup* 324! Gidigada (Aug.) *Lean* 86! Kidal (Sept.) *Rossetti* 59/154! Throughout much of tropical and S. Africa.
3. **S. plumosa** (*Linn.*) *Munro ex T. Anders.* in J. Linn. Soc. 5, Suppl. 1: 40 (1860). *Aristida plumosa* Linn., Sp. Pl. ed. 2, 1666 (1763), incl. var. *floccosa* (Coss. & Dur.) Henr. (1926); F.W.T.A., ed. 1, 2: 534; Chev. Bot. 741, and in Rev. Bot. Appliq. 14: 48, 109; De Miré & Gillet in J. Agric. Trop. 3: 728. Densely caespitose perennial with convolute leaf-blades; 30–45 cm. high.
Maur.: Twisti Legehatin, Aouker *Rossetti* 61/352! **Sen.:** *Leliévre*! **Niger:** In Abangarit (Dec.) *Proudlock* 5! Tasolé (Feb.) *Chev.* 44195! Throughout N. Africa from Mauritania to the Sudan, extending through Arabia and the Middle East to India.
4. **S. acutiflora** (*Trin. & Rupr.*) *de Winter* in Kirkia 3: 133 (1963). *Aristida acutiflora* Trin. & Rupr., Sp. Gram. Stip. 167 (1842); F.W.T.A., ed. 1, 2: 534; Chev. Bot. 740, and in Rev. Bot. Appliq. 14: 45; De Miré & Gillet in J. Agric. Trop. 3: 726. Slender caespitose or stoloniferous perennial, with narrowly convolute curved leaf-blades; up to about 45 cm. high.
Maur.: Oualata *Rossetti* 61/157! **Mali:** Région nord *Leclercq* 42589! Oualata to Neissa *Jumelle* 4! N. Africa from Mauritania to Egypt and the Sudan.

Additional species.

S. ciliata (*Desf.*) *de Winter* in Kirkia 3: 133 (1963). *Aristida ciliata* Desf. in Schrad., Neues J. Bot. 3: 225 (1809). Awn disarticulating by a break in the body of the lemma itself, the central limb plumose in the upper half and up to 45 mm. long.
A species from the Mediterranean and Orient, recorded at Aïr by De Miré & Gillet in J. Agric. Trop. 3: 727 (1956).

S. pungens (*Desf.*) *de Winter* in Kirkia 3: 135 (1963). *Aristida pungens* Desf., Fl. Atlant. 1: 109 (1798). Differs from other W. African species of *Stipagrostis* in that all three awns are plumose.
N. Africa, recorded at Aïr by De Miré & Gillet in J. Agric. Trop. 3: 728 (1956).

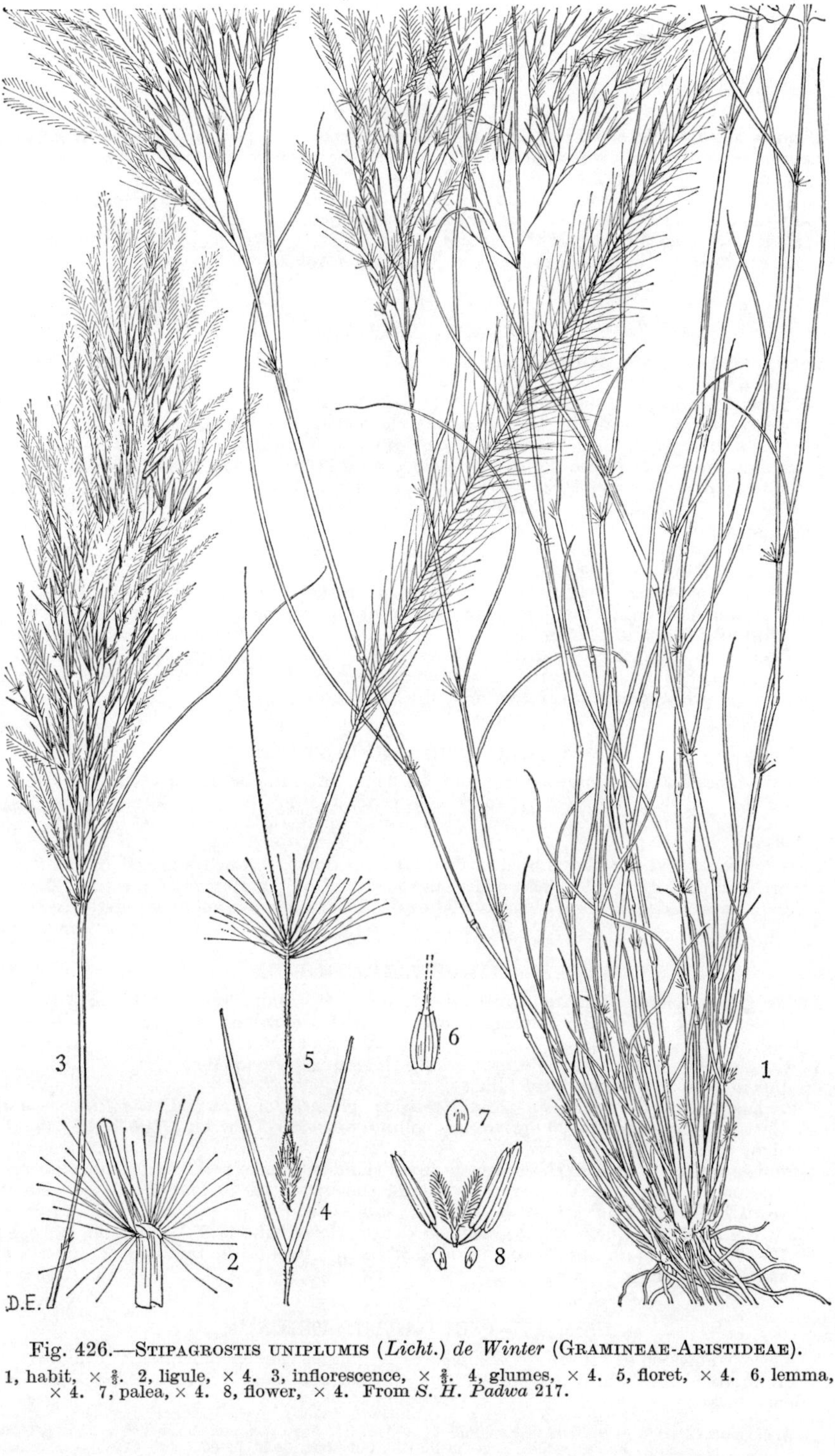

Fig. 426.—STIPAGROSTIS UNIPLUMIS (*Licht.*) *de Winter* (GRAMINEAE-ARISTIDEAE).
1, habit, × ⅔. 2, ligule, × 4. 3, inflorescence, × ⅔. 4, glumes, × 4. 5, floret, × 4. 6, lemma, × 4. 7, palea, × 4. 8, flower, × 4. From *S. H. Padwa* 217.

31. ARISTIDA Linn., Sp. Pl. 82 (1753), and Gen. Pl. ed. 5, 35 (1754); Henrard in Mededeel. Rijks Herb. Leiden 54 (1926), 54A (1927), 54B (1928), 54C (1933), 58 (1929) and 58A (1932).

KEY TO THE SECTIONS

Lemma not articulated at its junction with the awns.. .. .. Sect. I. **Aristida**
Lemma or column of awns articulated, and breaking cleanly across when mature:
Lemma not produced into a column, the body of the lemma immediately passing into the three awns .. .. Sect. II. **Pseudochaetaria**
Lemma prolonged into a more or less twisted column or beak:
Articulation at the base of the column .. .. .. Sect. III. **Arthratherum**
Articulation at the top of the column, just below the awns Sect. IV. **Pseudarthratherum**

Sect. I—ARISTIDA

Lateral awns absent; delicate annual .. .. 1. *diminuta*
Lateral awns present:
Plants perennial; densely tufted:
Lemma at most beaked; base of plant clothed in the fibrous remains of old leaf-sheaths; upper glume 5–7.5 mm. long; awns 6–8 mm. long .. .. 3. *recta*
Lemma drawn out into a slightly twisted column 1–3 mm. long; base of plant lacking fibres, often shortly rhizomatous; upper glume 6–9 mm. long; awns 8–20 mm. long .. .. 4. *junciformis*
Plants annual:
Spikelets 2–2·5 mm. long (excluding awns) .. .. 2. *cumingiana*
Spikelets 5 mm. long or more:
Glumes subequal, 9–12 mm. long; panicle rather stiff, the branches ascending; lemma 9–11 mm. long, bearing longitudinal rows of spiny hooked hairs; awn usually 15–25 mm. long .. .. 5. *rhiniochloa*
Glumes unequal, the lower about $\frac{2}{3}$ as long as the upper which is 6–10 mm. long; panicle flexuous, the branches secund; lemma 5–15 mm. long, glabrous or scabrid, very rarely tuberculate; awn usually 10–20 mm. long.. .. 6. *adscensionis*

Sect. II—PSEUDOCHAETARIA

Panicle dense and continuous, spike-like, 4–12 cm. long, the branches always very short and the spikelets fascicled; glumes with awns 2–5 mm. long; internodes and leaf-sheaths densely but very shortly pubescent; blades glaucous, flat, up to 6 mm. broad .. .. 7. *hordeacea*
Panicle loose, with slender branches up to 10 cm. long; branches naked in the lower part, and densely clothed with spikelets towards the tip; glumes with awns 1–2 mm. long; internodes and leaf-sheaths glabrous; blades narrow, more or less convolute 8. *kunthiana*

Sect. III—ARTHRATHERUM

Lower glume longer than the upper, the former 20–30 mm. long, the latter 3–6 mm. shorter; callus conical; column of awn up to 4·5 cm. long, the limbs 3·5–4·5 cm. long .. .. 9. *funiculata*
Lower glume shorter than the upper, rarely the glumes subequal:
Callus at base of mature floret bifid:
Panicle linear, the spikelets almost sessile; glumes long awned, the lower about 20 mm., the upper 25–35 mm. long; column of awn 5–7 cm. long, the limbs 6–8 cm. long .. .. 10. *kerstingii*
Panicle diffuse, delicate, the spikelets upon slender filiform pedicels; glumes obtuse, at most mucronate, the lower 5–7 mm., the upper 14–20 mm. long; column of awn 1·5–3 cm. long .. .. 11. *stipoides*
Callus conical, obliquely acute; panicle dense; lower glume 9–12 mm. long, upper 12–18 mm. with an awn 5–10 mm. long from the sinus of the bifid apex; column of awn 1–3 cm. long .. .. 12. *sieberana*

Sect. IV—PSEUDARTHRATHERUM

Spikelets congested at the tips of the branches; glumes usually subequal, 4–7 mm. long; lemma 4–5 mm. long, surmounted by a column 2–5 mm. long; leaf-blades narrow, convolute .. .. 13. *mutabilis*

1. **A. diminuta** (*Mez*) *C. E. Hubbard* in Kew Bull. 4: 480 (1949). *Stipa diminuta* Mez in Fedde, Rep. 17: 208 (1921). *A. cumingiana* var. *reducta* Pilger in Notizbl. Bot. Gart. Berl. 11: 805 (March 1933), var. *uniseta*

Stent & Rattray (May 1933), var. *diminuta* (Mez) Jac.-Fél. in J. Agric. Trop. 13: 51 (1966). Delicate annual 15–30 cm. high.
Mali: Sotuba to Bamako (Dec.) *Adam* 11427*b*! Tanzania to Rhodesia.

2. **A. cumingiana** *Trin. & Rupr.* Sp. Gram. Stip. 141 (1842); Henrard in Med. Rijks Herb. 58: 159; A. Chev. in Rev. Bot. Appliq. 14: 46; Lebrun in Bull. Soc. Bot. Fr. 116: 257 (1969). *A. delicatula* Hochst. ex A. Rich. (1851). Annual 15–30 cm. high; in damp soils.
Sen.: Teyel (Nov.) *Boudet* 4620. Kanéméré (Oct.) *Fotius* K570. Salikénié (Jan.) *Trochain* 1326. Kolda (Jan.) *J. Raynal* 7849. **Mali:** Bamako (Dec.) *Adam* 11427! **Guin.:** Kouroussa (Dec.) *Pobéguin* 542! **S.L.:** Benekoro (Nov.) *Glanville* 322! **N. Nig.:** Borgu *Barter* 1394! Jos (Oct.) *Hepper* 1175! Naraguta (Dec.) *Kennedy* 28! Anara F.R., Zaria (Oct.) *Hepper* 996! Throughout tropical Africa, extending eastwards through India to the Phillipines.

3. **A. recta** *Franch.* in Bull. Soc. Hist. Nat. Autun 8: 365 (1896); Henrard l.c. 261; Aké Assi, Contrib. 2: 280. *A. atroviolacea* Hack. in Bull. Herb. Boiss. sér. 2, 6: 707 (1906). *A. elliotii* A. Chev. in Rev. Bot. Appliq. 14: 46 (1934); Chev. Bot. 740. Perennial 20–45 cm. high; in damp soils.
Mali: Balandougou (Feb.) *Chev.* 324! **Guin.:** Farana *Sc. Elliot* 5332! **Iv. C.:** Korogo (July) *Adjanohoun* 401*a*! Zuenoula to Vavoua (Jan.) *Adjanohoun* 185*a*! **Ghana:** Kete Krachi to Attebebu (Oct.) *Rose Innes* GC 30609! Ejura scarp (Mar.) *Hepper & Morton* A3203! Yendi to Kete Krachi (Apr.) *Morton* GC 9138! **N. Nig.:** Abinsi (Mar.) *Dalz.* 875! Keffi (Mar.) *Swire* 6! **S. Nig.:** Ago-Are F.R., Oyo (Feb.) *Keay* FHI 37739! Igbetti (July) *J. Hall* 1322! **W. Cam.:** Bamenda (Jan.) *Daramola* FHI 40600! Southward to the Transvaal.

4. **A. junciformis** *Trin. & Rupr.* Sp. Gram. Stip. 143 (1842). Tufted perennial 30–60 cm. high; poor stony soils.
N. Nig.: Chappal Waddi (Apr., Nov.) *Jackson, Magaji & Tuley* 2030! *Gbile & Daramola* FHI 63287! Kenya to South Africa.

5. **A. rhiniochloa** *Hochst.* in Flora 38: 200 (1855). *A. rigidiseta* Pilger (1914). Annual with flat leaf-blades 2–4 mm. wide; up to 60 cm. high.
Maur.: (Aug.) *Rossetti* 61/170! **Mali:** Boré (Oct.) *Davey* 3/1957! **Niger:** Kao-Kiloum (Sept., Dec.) *P. de Fabrégues* 1337, 2072 (fide Lebrun in Adansonia 7: 395 (1967)). Also in Ethiopia, Sudan and Eritrea; and from Tanzania south to the Transvaal.

6. **A. adscensionis** *Linn.* Sp. Pl. 82 (1753); Henrard l.c. 322; Chev. Bot. 740, and in Rev. Bot. Appliq. 14: 45; Berhaut, Fl. Sén. ed. 2, 407; De Miré & Gillet in J. Agric. Trop. 3: 726; incl. subsp. *guineensis* (Trin. & Rupr.) Henr. (1926), var. *festucoides* (Poir.) Henr. (1926), and var. *senegalensis* (Trin. & Rupr.) Dur. & Schinz (1894). *A. submucronata* Schumach., Beskr. Guin. Pl. 67 (1827). *A. festucoides* Poir., Encycl. Meth. Bot., Suppl. 1: 453 (1810). *A. guineensis* Trin. & Rupr., Sp. Gram. Stip. 137 (1842). *A. vulgaris* var. *senegalensis* Trin. & Rupr. l.c. 135. *A. thonningii* Trin. & Rupr. l.c. 137. Annual, up to about 1 m. high, leaf-blades often becoming involute; common on dry, shallow or over-farmed soils. Very variable in shape of inflorescence and length of glumes and lemmas; a number of varieties have been proposed (see Henrard l.c.), but they intergrade one with another and their significance is doubtful.
Sen.: *Leprieur*! *Roger*! *Heudelot* 539! Longa (Oct.) *Berhaut* 2807! Malika (Aug.) *Broadbent* 27! **Mali:** *Leclercq* 42706! Sompi (Aug.) *Chev.* 2381! Gao (Sept.) *Hagerup* 357! Timbuktu *Hagerup* 158! L. Horo, nr. Gidijada (Aug.) *Lean* 85! **Iv. C.:** Bouaké (May) *de Wilde* 82! **U. Volta:** Ouassa to Diebougou (Oct.) *Rose Innes* GC 31512! Ouagadougou (May, July) *Scholz* 92! 92*a*! **Ghana:** Christiansborg (Apr.) *Johnson* 1023! Cape St. Paul (Apr.) *Rose Innes* GC 31223! Zuarungu (June) *Williams* 514! Navrongo (June) *Vigne* FH 4542! Bawku (Oct.) *Rose Innes* GC 30501! **Dah.:** Cotonou (Nov.) *Clayton* 396! Kraké (Oct.) *Risopoulos* 1233! **Niger:** Tigidit, Agades (Aug.) *Davies, Baker & Dean* 17*c*! **N. Nig.:** Ilorin (July) *Ward* 84! Fadan Karshi, Wamba (Oct.) *Kennedy* FHI 8036! Samaru, Zaria (Sept.) *Freeman* S129! Dikwa (Nov.) *Johnston* W81! Arege, Chad (Oct.) *Golding* 32! **S. Nig.:** *Dodd* 386! Lagos *W. MacGregor* 252! Throughout the tropics. (See Useful Plants.)

7. **A. hordeacea** *Kunth* Rev. Gram. 2: 517, t. 173 (1831); Henrard l.c. 140, incl. var. *longiaristata* Henr.; Berhaut, Fl. Sén. ed. 2, 393; De Miré & Gillet in J. Agric. Trop. 3: 727. *A. steudeliana* Trin. & Rupr., Sp. Gram. Stip. 155 (1842); Chev. Bot. 741, and in Rev. Bot. Appliq. 14: 109. *A. densispica* Steud., Syn. Pl. Glum. 1: 139 (1854). Annual, 10–45 cm. high.
Sen.: Cap Vert *Leprieur*! Dakar (Jan.) *Hagerup* 783! Moao (Oct.) *Berhaut* 1188! **Mali:** Dioura (Jan., Aug.) *Davey* 19! 221! Kidal (Sept.) *Rossetti* 59/146! Tilemsi (Sept.) *Popov* 113! **U. Volta:** Bousséra (Dec.) *Aké Assi* 7546! Soudougui (Sept.) *Scholz* 51*a*! **Ghana:** Mirigu (Oct.) *Vigne* FH 4646! Dedoro-Tankara (Dec.) *Adams* 4350! Navrongo (Nov.) *Akpabla* 386! Bolgatanga to Navrongo (Oct.) *Rose Innes* GC 31096! Bawku (Oct.) *Ankrah* GC 20288! **Togo Rep.:** Kande to Sansanne Mango (Oct.) *Rose Innes* GC 31412! **Niger:** Agades (Feb., Mar.) *Bradley* 23*b*! 24*b*! **N. Nig.:** Samaru, Zaria (Oct.) *Freeman* S147! Kano (Oct.) *Keay* FHI 21137! Yola (Nov.) *Hepper* 1194! Katsina (Oct.) *Clayton* 1367! Ngala (Nov.) *Johnston* W77! Throughout tropical Africa.

8. **A. kunthiana** *Trin. & Rupr.* Sp. Gram. Stip. 151 (1842); Henrard l.c. 142. Annual, 45 cm. high, closely resembling *A. mutabilis*.
Sen.: *Roger*! Richard Tol *Leliévre*! **Mali:** Dioura (Sept.) *Farrow* 81!

9. **A. funiculata** *Trin. & Rupr.* Sp. Gram. Stip. 159 (1842); Henrard l.c. 90; Chev. Bot. 740, and in Rev. Bot Appliq. 14: 47; Berhaut, Fl. Sén. ed. 2, 407; De Miré & Gillet in J. Agric. Trop. 3: 727. *A. funicularis* Trin ex Steud. (1841), without description. Annual, up to 45 cm. high, with pale brown glumes.
Sen.: *Leprieur*! Saint-Louis (Oct.) *Berhaut* 1742! **Mali:** Région nord *Leclercq* 42655! Goundam (Aug.) *Chev.* 2371! Gidigada (Aug.) *Lean* 84! Bamba (Sept.) *Hagerup* 325! Macina (Aug.) *Davey* 42! **Niger:** Toukounous *Bartha* 71! **N. Nig.:** Galarige, Chad (Oct.) *Golding* 31! Arege, Bornu (Oct.) *De Leeuw* 2074! Extends eastward through Sudan, Somalia and Arabia to India. (See Useful Plants.)

10. **A. kerstingii** *Pilger* in Engl., Bot. Jahrb. 34: 127 (1904); Henrard l.c. 91; A. Chev. in Rev. Bot. Appliq. 14: 47; Berhaut, Fl. Sén. ed. 2, 393, 407. *A. plicapolonica* Mez in Fedde Rep. 17: 146 (1921). Annual or short-lived perennial 30–90 cm. high, with filiform leaf-blades, and large handsome awns.
Sen.: *Heudelot* 306! Koungheul (Jan.) *Berhaut* 2811! Vélingara (Oct.) *Adam* 18571! Ndioridi (Oct.) *Raynal* 7724, 7727. **Guin.:** Gali (July) *Adames* 312! Dabola (Sept.) *Jaeger* 4900! **Iv. C.:** Bouna to Batie (Oct.) *Rose Innes* GC 31544! Bouna to Bondoukou (Oct.) *Rose Innes* GC 31548! Bouna (Aug., Sept.) *Aké Assi* 6517! *de Wilde* 785! *Oldeman* 288! **U. Volta:** Leo to Ouassa (Oct.) *Rose Innes* GC 31494! Doutié (Sept.) *Scholz* 38*b*! Kouéndi (July) *Scholz* 38*a*! **Ghana:** Kete Krachi to Yendi (Sept.) *Rose Innes* GC 30446! Bimbila (Oct.) *Ankrah* GC 20413! Tamale (Oct.) *Baldwin* 13558! Bole to Bamboi (Oct.) *Rose Innes* GC 30652! Bawku to Bolgatanga (Oct.) *Rose Innes* GC 31451! **Togo Rep.:** Sansanné Mango to Dapango (Oct.) *Rose Innes* GC 31430! Lama Kara to Sokode (Oct.) *Rose Innes* GC 31399! **Dah.:** Boukombe (Nov.) *Risopoulos* 1254! **N. Nig.:** Jebba *Barter* 753*a*! Yola (Sept.) *Kennedy* FHI 7285! Jos (Oct.) *Hepper* 1056! Anara F.R. (Oct.) *Keay* FHI 5493! Maiduguri (Oct.) *Johnston* N41!

11. **A. stipoides** *Lam.* Tab. Encycl. 1: 157 (1791); Henrard l.c. 93; Chev. Bot. 741, and in Rev. Bot. Appliq. 14: 110; Berhaut, Fl. Sén. ed. 2, 407. *A. stipiformis* Poir., Encycl. Suppl. 1: 452 (1810), superfluous name. *A. amplissima* Trin. & Rupr., Sp. Gram. Stip. 155 (1842), superfluous name. *A. lamarckii* Steud., Nom. Bot. ed. 1, 69 (1821), superfluous name. *Chaetaria lamarckii* Roem. & Schult., Syst. 2: 393 (1817), superfluous name. Robust annual 90–120 cm. high, with convolute leaf-blades and purplish glumes.
Maur.: Goberni *Rossetti* 61/335! **Sen.:** *Roger*! *Heudelot* 307! *Farmar* 17*a*! Dakar *Huard*! **Mali:** *Davey*! Sansanding to Ségou (Sept.) *Chev.* 2375! Gao (Sept.) *Rossetti* 59/142! Tondidarou (Oct.) *Farrow* 78! **Guin.:** Mamou (Feb.) *Roberty* 10500! **Niger:** Niamey (Oct.) *Vaillant* 2586! Toukounous *Bartha* 44! Wadi Dollol *White* 20! 39! Matankari (Mar.) *Virgo* 16! **N. Nig.:** Sokoto (Oct.) *Moisier* 146! Katsina (Oct.) *Clayton* 1380! Damaturu (Oct.) *Daggash* FHI 22002! Katagum *Dalz.* 292! Kalkala (Oct.) *Gwynn* 106!

Fig. 427.—Aristida adscensionis *Linn.* (Gramineae-Aristideae).
1, plant. 2, spikelet. 3, lower glume. 4, upper glume. 5, floret. 6, flower.

Extends eastward to the Sudan, with sporadic occurrences southwards to S.W. Africa. A closely related perennial species *A. meridionalis* Henr. occurs south of the Equator.

12. **A. sieberana** *Trin.* in Spreng., Neue Entd. 2: 61 (1821); Henrard l.c. 564; Chev. Bot. 741, and in Rev. Bot. Appliq. 14: 109. *A. longiflora* Schumach., Beskr. Guin. Pl. 68 (1827); F.W.T.A., ed. 1, 2: 534; Henrard l.c. 114; Berhaut, Fl. Sén. ed. 2, 407; De Miré & Gillet in J. Agric. Trop. 3: 727. *A. pallida* Steud., Syn. Pl. Glum. 1: 143 (1854); F.W.T.A., ed. 1, 2: 534; Henrard l.c. 412; A. Chev. in Rev. Bot. Appliq. 14: 48; Berhaut, Fl. Sén. ed. 2, 407; De Miré & Gillet in J. Agric. Trop. 3: 728. *A. aristidis* Coss. (1895); Chev. Bot. 740. *A. schebehlensis* Henr. in Med. Rijks Herb. Leiden 54B: 537 (1928). *A. leiocalycina* Trin. & Rupr. (1842). Loosely tufted perennial, with tough wiry culms and glaucous involute leaf-blades; on coastal sands, and on dry sandy soils inland. Very similar to *A. stipitata* Hack., a species from southern Africa with the glumes awnless or at most mucronate.
Maur.: Aouker *Rossetti* 61/341! **Sen.:** *Adanson*! *Spire*! Dakar (May) *Baldwin* 5731! Saint Louis (Feb.) *Berhaut* 2812! **Mali:** Timbuktu (June, July) *Chev.* 1228! *Hagerup* 106! Saraféré (July) *Lean* 59! Diré to Bandiagara (May) *Rogeon & Leclercq* 227! Kersani (Aug.) *Demange* 23/1957! **Guin.:** *Pobéguin* 550! **Iv. C.:** Ferkessédougou (July) *Adjanohoun* 433a! Bouna (July) *Adjanohoun* 434a! **U. Volta:** Sindou (June) *Leeuwenberg* 4316! Banfora (June) *Scholz* 7! **Ghana:** Cape St. Paul (Apr.) *Rose Innes* GC 31221! Ada (Jan.) *De Wit & Morton* A2978! Old Ningo (May) *Rose Innes* GC 31265! Tamale (June) *Andoh* FH 5915! Navrongo to Lawra (Oct.) *Rose Innes* GC 30276! **Togo Rep.:** Lomé *Warnecke* 168! **Dah.:** Cotonou (Nov.) *Clayton* 395! *Debeaux* 164! *Chev.* 4459! Ajuda (Oct.) *Newton* 15! **Niger:** Niamey (Jan., Oct.) *Chev.* 43172! *Hagerup* 491! *Vaillant* 2587! Dakoo (Dec.) *Vaillant* 2589! Toukounous *Bartha* 10! **N. Nig.:** Kontagora (Dec.) *Hubbard* S584! Kano (Dec.) *Meikle* 807! Katsina *Imam* FHI 27804! Hadejia (Dec.) *Clayton* 468! Mongonu (Oct.) *Golding* 29! **S. Nig.:** Lagos *Dalz.* 1126! Extends eastward to Somali Rep., and north to Tunisia and Israel.

13. **A. mutabilis** *Trin. & Rupr.* Sp. Gram. Stip. 150 (1842), incl. var. *senegalensis* Trin. & Rupr. l.c. 151 (1842), and var. *nigritiana* (Hack.) Bourreil in Taxon 18: 518 (1969); Henrard l.c. 186; A. Chev. in Rev. Bot. Appliq. 14: 47; Berhaut, Fl. Sén. ed. 2, 407; De Miré & Gillet in J. Agric. Trop. 3: 727. *A. nigritiana* Hack. in Fedde Rep. 10: 166 (1911); F.W.T.A., ed. 1, 2: 534. *A. longeradiata* Steud., Syn. Pl. Glum. 1: 140 (1854); F.W.T.A., ed. 1, 2: 534. *A. tenuiflora* Steud. l.c. 138 (1854). *A. hoggariensis* Batt & Trab. (1906). *A. lauriolii* Maire (1934). *A. meccana* Hochst. ex Trin. & Rupr., Sp. Gram. Stip. 152 (1842); F.W.T.A., ed. 1, 2: 534; A. Chev. in Rev. Bot. Appliq. 14: 47; De Miré & Gillet in J. Agric. Trop. 3: 727. Annual, 10–60 cm. high; rather variable, and several varieties have been described (see Henrard l.c.).
Maur.: Mahmoude *Rossetti* 61/281! Nima (Oct.) *Hepper* 3719! Aouker *Rossetti* 61/336! **Sen.:** Dagana (Oct.) *Hepper* 3644! Wallo *Leprieur*! **Mali:** Région Nord *Leclercq* 42658! Timbuktu (Aug.) *Lean* 79! *Hagerup* 260! Korienza (Nov.) *Davey* 206! Dioura (Nov.) *Davey* 147! **Ghana:** Tamale to Bolgatanga (Sept.) *Addei* SLUS 761! Navrongo (Aug.) *Rose Innes* GC 31672! GC 31898! **Niger:** *Vaillant* 2590! D'in Abbangarit (Nov.) *Money-Kyrle* Sah 41! Agadez (Nov.) *Money-Kyrle* Fe 7! Zinder (Nov.) *Money-Kyrle* Fe 16! 80 miles S. of In Guezzam (Oct.) *Everett* 6! **N. Nig.:** Katagum *Dalz.* 290! Maiduguri (Oct.) *Johnston* W42! Kauwa (Oct.) *Gwynn* 126! Mongonu (Oct.) *Golding* 15! Extends eastwards through the Sudan, Kenya and Arabia to India, southwards into Tanzania. (See Useful Plants.)
A. mutabilis may be separated from *A. meccana* by its subequal glumes (greatest difference in glume length 0·2–2·2 mm. in the former, 2·2–4·2 mm. in the latter). African populations clearly show modal values corresponding to one or both of these ranges, but the distinction is not sharp; moreover Indian plants display an intermediate range. These entities seem to be geographical races rather than independent species.

Additional species.

A. caerulescens *Desf.* Fl. Atlant. 1: 109 (1798). Distinguished from *A. adscensionis* by its perennial habit. Mediterranean and Orient; recorded on the northern limit of our area at Aïr by De Miré & Gillet in J. Agric. Trop. 3: 727 (1956).

32. CENTOTHECA Desv. in Nouv. Bull. Soc. Philom. 2: 189 (1810).

Leaf-blades lanceolate, 1·5–3 cm. broad; spikelets lanceolate, up to 8 mm. long, terete or slightly compressed; lemmas emarginate and mucronate at the tip, the lowest glabrous, the upper with reflexed tubercle-based bristles .. .. .. *lappacea*

C. lappacea (*Linn.*) *Desv.* in Nouv. Bull. Soc. Philom. 2: 189 (1810); Chev. Bot. 751, and in Rev. Bot. Appliq. 14: 135; Berhaut, Fl. Sén., ed. 2, 408. *Cenchrus lappaceus* Linn., Sp. Pl. ed. 2, 1488 (1763). *Uniola lappacea* (Linn.) Trin. (1830). *Holcus latifolius* Osb. (1757). *Centotheca latifolia* Trin. (1820), illegitimate name. Annual, 30–90 cm. high; in forest shade.
Sen.: Forêt de Boudié, Casamance (Nov.) *Berhaut* 6388! **Guin.:** Macenta (Oct.) *Baldwin* 9816! **S.L.:** *Barter*! Freetown (Oct.) *Deighton* 2184! Ronietta (Nov.) *Thomas* 5474! Njala (Jan.) *Dalz.* 8421! Rokupr (Nov.) *Jordan* 680! **Lib.:** Kle (Dec.) *Baldwin* 10569! Soplima (Nov.) *Baldwin* 10115! Ganta (Dec.) *Harley* 1729! Mecca (Dec.) *Baldwin* 10818! Gretown (July) *Baldwin* 6800! **Iv. C.:** Abidjan (Dec.) *Leeuwenberg* 2283! Guéyo (Mar.) *Leeuwenberg* 3809! Yapo (Sept.) *Roberty* 12035! Adiopodoumé (Oct.) *Roberty* 15321! Guitry (Dec.) *Boughey* GC 13521! **Ghana:** Aburi (Apr.) *Brown* 323! Oda (Aug.) *Howes* 967! Tafo (Feb.) *Thorold* 313! Vane (Dec.) *Rose Innes* GC 31186! Kade (Oct.) *Ankrah* GC 20322! **N. Nig.:** Jemaa (Oct., Nov.) *Clayton* 1427! *Gambles* 61! 65! Donga Plain (Feb.) *Brunt* 1011! **S. Nig.:** Lagos *W. MacGregor* 300! Ibadan (Nov.) *Ogua* FHI 7946! Okomu F.R., Benin (Dec.) *Brenan* 8395! Sapoba (Nov.) *Keay & Onochie* FHI 21613! Oban *Talbot* 842! **W. Cam.:** Victoria (July) *Maitland* 12! Banga (Nov.) *Dundas* FHI 13901! **F. Po:** (Nov., Dec.) *T. Vogel* 230! 260! *Mann* 109 (partly)! Monte Balea (Dec.) *Guinea* 351! Old World tropics. (See Useful Plants.)

33. MEGASTACHYA P. Beauv., Ess. Agrost. 74 (1812).

Leaf-blades lanceolate to ovate, 1·5–3 cm. broad; spikelets narrowly oblong, laterally compressed, on slender pedicels; lemmas truncate and mucronate at tip *mucronata*

M. mucronata (*Poir.*) *P. Beauv.* Ess. Agrost. 74, 167 (1812); Aké Assi, Contrib. 2: 277. *Poa mucronata* Poir., Encyl. 5: 91 (1804). *Centotheca mucronata* (Poir.) O. Ktze., Rev. Gen. Pl. 765 (1891); F.W.T.A., ed. 1, 2: 505. *C. owariensis* (P. Beauv.) Hack. ex C.B. Cl. (1896). *Megastachya owariensis* P. Beauv., Ess. Agrost. t. 15, fig. 5 (1812). *Eragrostis owariensis* (P. Beauv.) Steud. (1854). *E. beninensis* Steud. (1840). *E. jardinii* Steud. (1854). *E. palisotii* Dur. & Schinz, Consp. Fl. Afr. 5: 887 (1894). *Leersia disticha* Benth. (1849). *Centotheca urekana* Guinea in An. Agric. Terr. Esp. Golfo Guin. 1947: 142 (1948). Tufted annual 60–90 cm. high, in forest shade.
S.L.: Mesima (Apr.) *Deighton* 3696! **Iv. C.:** Abidjan (June, Dec.) *Adjanohoun* 407a! *Pitot*! (Sept.) *de Wilde* 887! **Ghana:** Nzima (Jan.) *Thorold* 296! Atwabo *Fishlock* 32! **S. Nig.:** Lagos *Dalz.* 1136! Okomu F.R., Benin (Jan.) *Brenan* 8868! Aboh *Barter* 12! Mamu F.R., Onitsha (Nov.) *Keay* FHI 22293! Opobo (Apr.) *Jeffreys* 32! **W. Cam.:** Mamfe (Dec.) *Morton* K612! **F. Po:** Ureka (Nov.) *Guinea* 2341! Tropical Africa generally.

Fig. 428.—CENTOTHECA LAPPACEA (*Linn.*) *Desv.* (GRAMINEAE-CENTOTHECEAE).

1, whole plant. 2, spikelet. 3, lower glume. 4, upper glume. 5 and 9 lemmas. 6, palea. 7, flower and palea. 8, ovary and stigmas.

34. ENNEAPOGON Desv. ex P. Beauv., Ess. Agrost. 81 (1812); Renvoize in Kew Bull. 22: 393 (1968).

Uppermost (3rd) floret 0·5–2 mm. long (incl. awns); anthers 0·3–0·5 mm. long; basal leaf-blades filiform .. .. 1. *desvauxii*
Uppermost (3rd) floret 4–7 mm. long (incl. awns); anthers 0·5–1·3 mm. long; leaf-blades usually flat, 2–3 mm. broad .. .. 2. *schimperanus*

1. **E. desvauxii** *P. Beauv.* Ess. Agrost. 82, 161, f. 16/11 (1812). ***E. brachystachyus*** (Jaub. & Spach) Stapf in Fl. Capensis 7: 654 (1900); A. Chev. in Rev. Bot. Appliq. 14: 132; De Miré & Gillet in J. Agric. Trop. 3: 731. *Pappophorum brachystachyum* Jaub. & Spach in Ann. Sci. Nat. Bot. sér. 3, 14: 365 (1850). *P. vincentianum* Schmidt, Beitr. Fl. Cap. Verd. Ins. 144 (1852). *P. nanum* Steud., Syn. Pl. Glum. 1: 200 (1854). Tufted perennial up to 23 cm. high.
Niger: Aïr (*fide* De Miré & Gillet). From the Cape Verde Is. and Algeria eastwards through Ethiopia and Kenya to Arabia; southwards to S. Africa; central & S. America.

2. **E. schimperanus** (*Hochst. ex A. Rich.*) *Renvoize* in Kew Bull. 22: 400 (1968). *Pappophorum schimperanum* Hochst. ex A. Rich., Tent. Fl. Abyss. 2: 403 (1851). *P. elegans* Nees ex Steud., Syn. Pl. Glum. 1: 199 (1854). *P. glumosum* Hochst. in Flora 38: 203 (1855). ***Enneapogon glumosus*** (Hochst.) Maire & Weiller, Fl. Afr. Nord. 2: 193 (1953); De Miré & Gillet in J. Agric. Trop. 3: 732. ***E. elegans*** (Nees ex Steud.) Stapf in Kew Bull. 1907: 224; A. Chev. in Rev. Bot. Appliq. 14: 132. Tufted perennial up to 1 m. high.
Mali: Ansongo (Sept.) *Hagerup* 397! **Niger:** Aïr (*fide* De Miré & Gillet). Extends eastwards to Sudan, Ethiopia, and Arabia.
[Note: *Pappophorum senegalense* Steud (1854) (= *Enneapogon cenchroides* (Roem. & Schult.) C. E. Hubbard) was actually collected in Gabon, not Senegal.]

35. SCHMIDTIA Steud. ex J. A. Schmidt, Beitr. Fl. Cap. Verd. Ins. 144 (1852), *nom. cons.*; Launert in Bol. Soc. Brot. 39: 303 (1965).

Panicle narrow, 6–12 cm. long, the branches and spikelets softly pubescent all over; spikelets about 1 cm. long; glumes membranous, ovate-lanceolate, acute, 4·5–9 mm. long, 9-nerved; lemmas broadly obovate, with 5 bristles 4·5–9 mm. long, alternating with 6 hyaline lobes 2 mm. long .. .. *pappophoroides*

S. pappophoroides *Steud. ex J. A. Schmidt* Beitr. Fl. Cap. Verd. Ins. 145 (1852); A. Chev. in Rev. Bot. Appliq 14: 132; Berhaut, Fl. Sén. ed 2, 408. *S. bulbosa* Stapf in Fl. Capensis 7: 658 (1900). Caespitose or almost suffrutescent perennial, softly pilose all over.
Maur.: Néma *Rossetti* 61/137! **Sen.:** Walo *Heudelot* 516! Extends eastward to the Sudan and Kenya, and south to Angola.

36. AELUROPUS Trin., Fund. Agrost. 143 (1820).

Culms procumbent, stoloniferous, wiry and woody; leaf-blades subulate-lanceolate, pungent, glabrous, 1·5–3 cm. long; spikelets ovate, 2–5 mm. long, softly villous; lemmas 5–9-nerved, mucronate .. .. *lagopoides*

A. lagopoides (*Linn.*) *Trin. ex Thw.* Enum. Pl. Zeyl. 374 (1864); Bor in Webbia 24: 401 (1969). *Dactylis lagopoides* Linn., Mant. 33 (1767). *D. brevifolia* Koenig ex Willd., Sp. Pl. 1: 410 (1797). *Aeluropus laevis* Trin., Fund. Agrost. 143 (1820). *A. repens* (Desf.) Parl. (1848)—A. Chev. in Rev. Bot. Appliq. 14: 134. Perennial; seashores, and saline soils inland.
Maur.: Keur Masséne *Rossetti* 59/8! Nouakchott to Rosso (Nov.) *Adam* 19584! From the Mediterranean to the Punjab, reaching southwards into the Sudan, Ethiopia and Somaliland.

37. ERAGROSTIS Wolf, Gen. Pl. Vocab. Char. Def. 23 (1776).

Keels of paleas ciliate, the hairs spreading beyond the margins of the lemmas; spikelets 1·5–4·5 mm. long and 1–2·5 mm. wide, breaking up from the apex downwards; lemmas obtuse:
- Lemmas, or at least the upper, with a few short stiff hairs on the keel; panicle spike-like, woolly, more or less lobed or interrupted, typically 4–12 cm. long and 5–15 mm. wide, the spikelets densely clustered; spikelets 2–4·5 mm. long and 1·5–2 mm. wide; upper glume 1–1·2 mm. long; lemma narrowly elliptic in side view, 0·9–1·5 mm. long; palea hairs much longer than the width of the floret; anthers 2; grain 0·3–0·5 mm. long .. .. 1. *ciliaris*
- Lemmas smooth or scaberulous on the keel; anthers 3:
  - Inflorescence sticky from glands on branchlets and glumes, to which grains of sand adhere .. .. 2. *viscosa*
  - Inflorescence with or without glands, but not sticky:
    - Panicle dense, linear-elliptic, 4–22 cm. long and 5–25 mm. wide, not lobed or woolly, nor with the spikelets clustered; spikelets 2–4·5 mm. long and 1·5–2·5 mm. wide; upper glume 1–1·5 mm. long; lemma narrowly elliptic in side view, 1–1·5 mm. long, glabrous on the keel; palea hairs about as long as the width of the floret; grain 0·5 mm. long .. .. 3. *arenicola*
    - Panicle open, pyramidal to elliptic, 2–14 cm. long and 10–30 mm. wide, the spikelets scattered; spikelets 1·5–2·5 mm. long and 1–1·5 mm. wide; upper glume 0·7–1 mm. long, lemma elliptic in side view, 0·5–1 mm. long, scaberulous on the keel; palea hairs usually shorter than the width of the floret; grain 0·4–0·6 mm. long 4. *tenella*

Keels of paleas glabrous or minutely ciliolate (rarely ciliate, but then concealed within the lemma—*E. plurigluma*):

Spikelets breaking up from the apex downwards, the rhachilla disarticulating between the florets so that lemma and palea fall together enclosing the grain; spikelets up to 1·5 mm. wide; lemmas obtuse:

Panicle ovate-elliptic, 20–35 cm. long and 7–15 cm. wide, loose, the spikelets on long fine pedicels; spikelets 4–10 mm. long, 6–22-flowered, 1–1·5 mm. wide; ligule reduced to a ciliate rim; lemma 1·2–1·7 mm. long; keels of palea scabrid; anthers 3; grain subglobose .. .. 5. *aspera*

Panicle linear to lanceolate, 10–60 cm. long and 1–5 cm. wide, dense, the spikelets short pedicelled; spikelets 1·3–2·2 mm. long, 4–10-flowered, 0·8–1 mm. wide; ligule membranous, 0·5 mm. long; lemma 0·7 mm. long; anthers 2; grain ellipsoid 6. *namaquensis*

Keels of palea smooth.. .. .. 6a. var. *namaquensis*

Keels of palea scaberulous .. .. 6b. var. *diplachnoides*

Spikelets breaking up from the base upwards, the rhachilla persistent (except *E. egregia*), and the lemma and palea falling separately:

Glumes pubescent to pilose; lemmas hairy near the margins or only with a few short hairs at the base; spikelets ovate to oblong; paleas deciduous; anthers 3:

Perennial; spikelets 4–6·5 mm. wide; lemmas 3–3·5 mm. long, acute, glandular on keel and nerves, hairy near the margins; panicle glandular on branches and pedicels; leaf-blades flat, 3–9 mm. wide and 10–30 cm. long; grain broadly oblong, flattened, 1–6 mm. long .. .. 7. *blepharostachya*

Annuals; spikelets 2·5–4·5 mm. wide; lemmas 1·8–2·5 mm. long:

Lemmas thinly chartaceous, with a line of white tubercle based hairs near the margins and a few glands on the nerves, obtuse to subacute; panicle glandular on branches and pedicels; spikelets 3–8 mm. long and 2·5–3·5 mm. wide; leaf-blades flat or rolled, 2–18 cm. long and 2–7 mm. wide; grain broadly oblong, plump, 0·7 mm. long .. .. 8. *scotelliana*

Lemmas coriaceous, glabrous except for a few short rigid hairs at the base, closely packed, broadly rounded at the apex; panicle branches and pedicels usually without glands; spikelets 3–20 mm. long and 3–4 mm. wide; leaf-blades flat, 4–18 cm. long and 2–8 mm. wide; grain broadly elliptic, plano-convex, 1–1.5 mm. long .. .. 9. *turgida*

Glumes and lemmas glabrous (or the latter with a few scurfy hairs when young in *E. pobeguinii*):

Lower 2–5 florets barren, smaller in size, the palea absent or reduced; anthers 3:

Spikelets linear, 9–15 mm. long and 2·5–4 mm. wide, breaking above the sterile lemmas and the upper part falling entire at maturity; lower 3–4 lemmas reduced, the lowest without a palea and glume-like; fertile lemmas imbricate, acute, 2·5–3 mm. long; anthers 2–3, 0·7–1·3 mm. long .. .. 10. *egregia*

Spikelets ovate to broadly elliptic, 4–6 mm. long and 2–4 mm. wide, the florets falling separately; paleas deciduous:

Fertile lemmas obtuse at the tip, chartaceous, imbricate; paleas ciliate on the keels, the cilia 0·2–0·4 mm. long; barren lemmas 3–5, without paleas 11. *plurigluma*

Fertile lemmas acute at the tip, firmly membranous, contiguous; paleas shortly ciliolate on the keels, the cilia 0·1–0·2 mm. long; barren lemmas 2, without paleas .. .. 12. *cenolepis*

Lower florets fertile, as large as or larger than the upper:

Keels of paleas broadly winged, the wings up to 0·5 mm. wide and folding over the sides of adjacent lemmas; spikelets ovate to oblong, 6–13 mm. long and 4–5 mm. wide, light olive green; lemmas lanceolate, acute, 3–4 mm. long; paleas deciduous; leaf-blades up to 5 mm. wide, but usually narrower and tightly convolute; anthers 3 .. .. 13. *invalida*

Keels of paleas not or only very narrowly winged:

Paleas falling at about the same time as the lemmas:

Leaf-blades spreading horizontally or becoming reflexed, 0·5–6 cm. long and 0·5–2 mm. wide, flat or rolled; spikelets ovate-oblong to oblong, 4–7 mm. long and 2·5–4 mm. wide, dark olive green; lemmas broadly ovate-oblong, 1·3–2·5 mm. long, loosely contiguous, obtuse to subacute, firmly chartaceous, dully shining; keels of the palea narrowly winged, ciliolate; anthers 3 14. *volkensii*

Leaf-blades mostly erect or sometimes obliquely spreading:

Annuals; anthers 0·2 mm. long:

Spikelets ovate to ovate-oblong, 2–3·5 mm. wide; anthers 2:

Lemmas suborbicular in side view, obtuse at the tip when flattened, firmly chartaceous, olive green, dully shining; panicle stiff, 5–12 cm. long and 3–7 cm. wide, the lateral pedicels 1–2 mm. long; spikelets 3–5·5 mm. long and 2–2·5 mm. wide .. .. 15. *mokensis*

Lemmas narrowly ovate in side view, acute at the tip when flattened, membranous, tinged with pink or purple; panicle delicate, 5–12 cm. long and 2–6 cm. wide, the lateral pedicels 2–10 mm. long; spikelets 4–12 mm. long and 2–3·4 mm. wide .. .. 16. *unioloides*

Spikelets linear, 1–1·5 mm. wide:

Grain subglobose, 0·4–0·6 mm. long; lowest branches of the panicle not whorled, usually solitary, glabrous in the axil; lemmas narrowly ovate, acute to subobtuse, 1–1·5 mm. long, diverging at about 45° from the rhachilla and 3–4 times as long as its internodes; spikelets 4–10 mm. long; glumes lanceolate, about ½ as long as the adjacent lemmas; leaf-blades up to 15 cm. or more long, rolled or flat, 1–3 mm. wide; anthers 2 17. *gangetica*

Grain ellipsoid, 0·6–1 mm. long, flattened on one side; lowest branches of the panicle whorled (except in the smallest panicles), usually with a few long white hairs in the axil; lemmas lanceolate-oblong, subacute or obtuse, 1·4–2 mm. long, more or less appressed to the rhachilla and 2–2½ times as long as its internodes; spikelets 3–7 mm. long, 4–14-flowered; leaf-blades up to 20 cm. long, flat or rolled, 1–4 mm. wide; anthers 3 .. 18. *pilosa*

Perennials; spikelets 1·5–3·5 mm. wide; grain ellipsoid, 0·7–0·9 mm. long; anthers 3:

Spikelets ovate to ovate oblong, 5–15 mm. long and 3–3·5 mm. wide, long-pedicelled in large open loose panicles up to 35 cm. long and 25 cm. wide; culms herbaceous; lemmas obliquely ovate in side view, 2–3 mm. long, grey, firmly chartaceous; anthers 1 mm. long .. .. .. 19. *chalarothyrsos*

Spikelets linear:

Flowering culms woody, branched and geniculate, rooting at the nodes, and eventually functioning as stolons; lemmas 2·2–3·2 mm. long, lanceolate in side view, acute at the tip when flattened, thinly chartaceous; anthers 0·9–1·3 mm. long; spikelets 2–3 mm. wide and 8–16 mm. long, pallid, loose, the rhachilla clearly visible; panicle lax.. .. .. .. 20. *barteri*

Flowering culms herbaceous, erect; lemmas 1·4–2 mm. long, narrowly ovate in side view, firmly chartaceous; anthers 3, 0·4–1 mm. long; spikelets 1·5–2·5 mm. wide and 5–20 mm. long, dense, the rhachilla concealed by the florets:

Leaf-blades 15–30 cm. long and 2–4 mm. wide, flat or rolled; culms 45–100 cm. high and 1·5–3 mm. diam. at base, leafy; panicle ovate or oblong, spikelets more or less clustered about the obliquely ascending primary branches; spikelets pallid to grey or purple; lemmas 1·4–2 mm. long, acute to subacute at the tip when flattened .. .. .. 21. *atrovirens*

Leaf-blades 3–7 cm. long and 1–2 mm. wide, usually rolled; culms about 30 cm. high and 1 mm. diam. at base, wiry, leafless, the leaves mostly basal; panicle linear, the spikelets clustered about the main axis; spikelets dark grey; lemmas 1·4–1·7 mm. long, obtuse at the tip when flattened 22. *camerunensis*

Paleas not deciduous, persisting long after the lemmas have fallen:

Margins of leaf-blades bearing warty glands, the blades flat, 3–11 mm. wide, glabrous; panicle ovate to oblong, open or contracted, loose or dense, 3–25 cm. long and 2–12 cm. wide, usually bearing glands on branches and pedicels; spikelets narrowly oblong to oblong or ovate-oblong, pallid yellowish green to leaden grey; lemmas ovate to ovate-oblong, obtuse; grain subglobose, up to 0.6 mm. long:

Spikelets 2–4 mm. wide; lemmas 2–2·8 mm. long .. .. 23. *cilianensis*

Spikelets 1·3–2 mm. wide; lemmas 1·5–2 mm. long.. .. .. 24. *minor*

Margins of leaf-blades not glandular (rarely with tubercle-based hairs in *E. pobeguinii*); spikelets linear to narrowly oblong:

Perennials:

Panicle scantily branched, bearing up to 15 spikelets, linear, 3–12 cm. long and 0·5–2 cm. wide; spikelets 3–5 mm. wide and 7–18 mm. long, narrowly oblong, pallid to olive-grey; basal sheaths bulbously swollen and hardened below; leaf-blades up to 2 mm. wide, usually rolled and setaceous, pilose with tubercle-based hairs; lemmas 3–4 mm. long, firmly chartaceous, acute, with or without a few deciduous scurfy hairs near the base 25. *pobeguinii*

Panicle bearing numerous spikelets; spikelets up to 2·5 mm. wide; basal sheaths not swollen:

Lemmas closely imbricate, their sides pressed flat against adjacent lemmas for almost their entire length; spikelets 5–25 mm. long and 2–2·5 mm. wide, purplish grey; leaf-blades stiff, up to 6 mm. wide, but usually much narrower and rolled, densely scabrid above; panicle elliptic, stiff, open, the spikelets short-pedicelled and loosely appressed to the branches; anthers 2 .. .. 26. *squamata*

Lemmas contiguous, free from adjacent lemmas for much of their length; pedicels long and slender; anthers 3:

Lower glume about ⅔ as long as the upper, the latter 0·8–1·4 mm. long and barely reaching the base of the adjacent lemma; spikelets dark green, the margins conspicuously saw-toothed; panicle elliptic, open, 5–20 cm. long 27. *tenuifolia*

Lower glume about as long as the upper, the latter 1·5–2 mm. long and reaching at least to the middle of the adjacent lemma; spikelets pale brownish-green; panicle linear to lanceolate, 10–30 cm. long, the slender flexuous branches ascending close to the axis .. .. 28. *domingensis*

Annuals:

Florets appressed to the rhachilla; spikelets 0·8–1·5 mm. wide; anthers 3; grain ellipsoid, 0·5–0·7 mm. long:

Lower glume as long as lower lemma; panicle narrowly ovate, the branches in successive whorls; spikelets grey, 3–8 mm. long; lemmas 1·5–1·7 mm. long, obtuse; anthers 0·8–1 mm. long .. .. .. 29. *cylindriflora*

Lower glume half as long as lower lemma; panicle at first linear and embraced by the uppermost sheath, but the branches spreading at maturity; spikelets pallid, 3·5–15 mm. long; lemmas 1·2–1·7 mm. long, subacute; anthers 0·2 mm. long .. .. 30. *aegyptiaca*

Florets diverging at 45°; panicle not whorled (sometimes the lowest branches irregularly whorled in *E. macilenta*):

Spikelets very dark green to almost black, oblong, 3–6 mm. long and 1–2 mm. wide, loosely 6–14-flowered; glumes acuminate, 1–1·5 mm. long; lemmas 1·3–1·6 mm. long, acute; anthers 3, 0·5 mm. long; grain broadly oblong, 0·5 mm. long; panicle ovate-elliptic, up to 35 cm. long, diffuse 31. *macilenta*

Spikelets pallid to purplish, 1·5–2·5 mm. wide:

Lemmas obtuse at the tip when flattened, 1·5–2 mm. long; anthers 2, 0·1–0·3 mm. long; grain subglobose, 0·5 mm. long:

Base of the florets interleaved with those of the opposite row; panicle ovate to elliptic, 5–50 cm. long, very loose, the spikelets trembling on slender pedicels; spikelets 8–50 mm. long, 10–100-flowered 32. *tremula*

Base of the florets free from those of the opposite row; panicle ovate and open to linear and contracted, 4–15 cm. long; spikelets 8–25 mm. long, 14–50-flowered .. .. 33. *elegantissima*

Lemmas acute at the tip when flattened:

Panicle linear to narrowly elliptic, dense, the spikelets closely clustered about the primary branches; spikelets 7–35 mm. long and 2–2·5 mm. wide, 10–100-flowered, pallid or pinkish brown; lemmas 2·5–2·7 mm. long, imbricate; anthers 2, 0·1–0·2 mm. long; grain subglobose, 0·3–0·4 mm. long; leaf-blades flat, up to 4 mm. wide .. .. 34. *lingulata*

Panicle elliptic, open, the spikelets uniformly distributed; spikelets 3–9 mm. long and 1·5–2 mm. wide, 7–30-flowered, purplish; glumes 1–2 mm. long, lanceolate, covering ¾ of the adjacent lemma or more; lemmas 1·5–2 mm. long, contiguous; anthers 3, 0·3–0·4 mm. long; grain subglobose, 0·4 mm. long .. .. 35. *welwitschii*

1. **E. ciliaris** (*Linn.*) *R. Br.* in Tuckey, Narr. Exp. Congo, Append. 478 (1818); Chev. Bot. 749, and in Rev. Bot. Appliq. 14: 117; Berhaut, Fl. Sén. ed. 2, 396; De Miré & Gillet in J. Agric. Trop. 3: 732. *Poa ciliaris* Linn., Syst. Nat. ed. 10, 2: 875 (1759). *Eragrostis pulchella* Parl. in Atti Riun. Sci. Ital. 8: 586 (1847). Annual 15–60 cm. high; a common weed of roadsides, cultivated land and waste places.
Sen.: *Farmar* 124! *Roger* 12! 41! Casamance (Oct.) *Adam* 18276! **Gam.:** *Saunders* 119! Genieri (July) *Fox* 128! Sankuli-Kunda (Sept.) *Pirie* 10/33! Yundum, W. Div. (Dec.) *Austin* 16! *Ashrif* 14! **Mali:** Labézenga (Sept.) *Hagerup* 452! Kersani (Aug.) *Farrow* 22! Dioura (Sept.) *Farrow* 139! Korienza (Nov.) *Davey* 200! **Guin.:** Baffing R. (Nov.) *Pobéguin* 1819! **S.L.:** *Morson*! Freetown (Oct.) *Deighton* 2155! Lumley (Jan.) *Gledhill* 145! Bonthe (Mar.) *Deighton* 2487! Musaia (Dec.) *Deighton* 4458! **Lib.:** Buchanan (Mar.) *Baldwin* 11206! Timbo (Mar.) *Baldwin* 11222! Sangwin (Mar.) *Baldwin* 11326! Harper Common (June) *Baldwin* 5972! **Iv. C.:** Diahbo to Bouaké (July) *Chev.* 22056! Adiopodoumé (Dec.) *Leeuwenberg* 2129! Abidjan to Dabou (Dec.) *Boughey* GC 14453! Cocody, Abidjan (June) *Adjanohoun* 425a! Abouabou (May) *Leeuwenberg* 4239! **U. Volta:** Ouagadougou (June) *Scholz* 110! Nadjogo to Nadiokan (June) *Scholz* 118! Samandeni (Oct.) *Kmoch* 170! **Ghana:** Cape St. Paul (Feb.) *Thorold* 4! Nungua (Aug.) *Rose Innes* GC 30057! Aburi (Sept.) *Johnson* 840! Kintampo (Dec.) *Vigne* FH 3196! Tamale (Dec.) *Lloyd Williams* 460! **Togo Rep.:** Lomé *Warnecke* 170! Palime *Stage* 80! **Dah.:** *Spire* 1918! Agouagon (May) *Chev.* 23494! Ina Farm No. 24 *Serv. Agr.*! **N. Nig.:** Sokoto (Nov.) *Moiser* 148! Nupe *Barter* 864! Abinsi *Dalz.*! Katagum *Dalz.* 248! Yola (Sept.) *Kennedy* FHI 7284! **S. Nig.:** Ibadan (Jan.) *Meikle* 946! Lagos (Feb.) *W. MacGregor* 17! Owo (July) *Jones* FHI 4375! Agolo, Awka Dist. *Thomas*! Opobo (Apr.) *Jeffreys* 2! **W. Cam.:** Victoria (Nov.) *Maitland* 83! Throughout the tropics. (See Useful Plants.)

2. **E. viscosa** (*Retz.*) *Trin.* in Mém. Acad. Sci. Petersb. sér. 6, 1: 397 (1830). *Poa viscosa* Retz., Obs. Bot. 4: 20 (1786). *Eragrostis retinorrhea* Steud. (1854). Annual, 20–45 cm. high; culms, leaf-sheaths and base of blade sticky; lower glume with a sticky yellow mid-nerve; dry sandy soils.
N. Nig.: Yola to Biu (Nov.) *Magaji & Tuley* 1847! Eastwards to the Philippine Islands, south to South Africa; a few records in tropical America.

3. **E. arenicola** *C. E. Hubbard* in Kew Bull. 4: 345 (1949). *E. tenella* var. *compacta* Rendle, Cat. Welw. 2: 233 (1899). Tufted annual 15–45 cm. high; old fields and disturbed land.
N. Nig.: Zaria Dist. (Dec.) *Taylor* 22! Samaru (Nov.) *Thatcher* S552! Gindiri (Oct.) *Hepper* 1132! Yola to Biu (Nov.) *Magaji & Tuley* 1847! **W. Cam.:** Gurum (Nov.) *Hepper* 1281! Tropical Africa generally, but more frequent towards the south.

4. **E. tenella** (*Linn.*) *P. Beauv. ex Roem. & Schult.* Syst. Veg. 2: 576 (1817); Berhaut, Fl. Sén. ed. 2, 408. *Poa tenella* Linn., Sp. Pl. 69 (1753). *P. amabilis* Linn. l.c. 68 (1753). *P. plumosa* Retz. (1786).

P. lepida Hochst. ex A. Rich. (1851). *Eragrostis amabilis* (Linn.) Wight & Arn. ex Hook. & Arn. (1838). *E. plumosa* (Linn.) Link (1827)—Chev. Bot. 750, and in Rev. Bot. Appliq. 14: 119. Delicate annual 7–30 cm. high; cultivated land, roadsides and waste places.
Sen.: Dakar (Oct.) *Berhaut* 2821! *Adam* 16907! Marc Tally (Oct.) *Adam* 9873! Kayes (June) *Chev.* 26022! Kaolak (Sept.) *Berhaut* 307! **Gam.:** Yundum *Ashrif* 32! **Mali:** Bamako (Aug.) *Waterlot* 1227! Kara (Dec.) *Davey* 186! **Port. G.:** Bubaque (Feb.) *Pereira* 3720! **Guin.:** Dyeke (Oct.) *Baldwin* 9664! **S.L.:** Rokupr (Apr.) *Jordan* 438! Freetown (July) *Deighton* 2025! Njala (Nov.) *Deighton* 1503! Bonthe (Nov.) *Deighton* 2269! Mesima (Feb.) *T. S. Jones* 392! **Lib.:** Sinkor (July) *Barker* 1380! Monrovia (Nov.) *Dinklage* 3260! R. Cess (Mar.) *Baldwin* 11256! Nyaake (June) *Baldwin* 6070! Harper (June) *Baldwin* 5994! **Iv. C.:** Téhini (Aug.) *de Wilde* 773! Adiopodoumé (Dec.) *Leeuwenberg* 2127! **U. Volta:** Diebougou (Oct.) *Rose Innes* GC 31515! **Ghana:** Bawku (Aug.) *Irvine* 5165! Tamale (Oct.) *Rose Innes* GC 30677! Wenchi (Oct.) *Rose Innes* GC 30520! Cape St. Paul (Apr.) *Rose Innes* GC 31227! Achimota (May) *Ankrah* GC 20143! **Niger:** Toukounous *Bartha* 77! **N. Nig.:** Biu (Nov.) *Magaji & Oche* MG65! Naraguta (July) *Lawlor & Hall* FHI 45570! Jebba (Apr.) *Meikle* 1388! Ilorin (Oct.) *Ward* 76! Badeggi, Bida (June) *Vaillant*! **S. Nig.:** Ibadan (Dec.) *Jones* FHI 4946! Lagos (Feb.) *W. MacGregor* 338! Benin (Dec.) *Brenan* 8629! Onitsha (Nov.) *Baldwin* 13733*a*! Port Harcourt *Maitland*! **W. Cam.:** Bota, Victoria *Brunt* 1043! Victoria (July) *Maitland* 14! Throughout the tropics.

5. **E. aspera** (*Jacq.*) *Nees* Fl. Afr. Austr. 408 (1841); A. Chev. in Rev. Bot. Appliq. 14: 117; Berhaut, Fl. Sén ed. 2, 408; Aké Assi, Contrib. 2: 277. *Poa aspera* Jacq., Hort. Vindob. 3: 32 (1776). *P. hippuris* Schumach., Beskr. Guin. Pl. 69 (1827). Annual, 30–90 cm. high, with large feathery panicles; a weed of old farmland.
Gam.: Sankuli Kunda (Sept.) *Pirie* 50/33! **Mali:** Ségou *Rogeon* 285! Dinderesso (Nov.) *Néa* 89! **Guin.:** Baffing R. (Nov.) *Pobéguin* 1799! Tinkisso Valley (Mar.) *Pobéguin* 1755! Kouroussa (Nov.) *Pobéguin* 528! 533! **S.L.:** Musaia (Dec.) *Deighton* 4457! **Iv. C.:** Nambonkaha (Nov.) *Leeuwenberg* 2064! **U. Volta:** Linglìékoro (Nov.) *Scholz* 87! Nouna to Dedougou (Oct.) *Scholz* 56! **Ghana:** Gambaga (Nov.) *Vigne* FH 4680! Zongoiri to Binaba & Tili (Oct.) *Rose Innes* GC 31102! Tamale (Dec.) *Lloyd Williams* 463! Samu, near Yendi (Sept.) *Thorold* 204! Salaga (Nov.) *Akpabla* 362! **Niger:** Agades (July) *Chopard & Villiers*! **N. Nig.:** Samaru (Oct.) *Freeman* S136! Katagum Dist. *Dalz.* 272! Nupe *Barter* 859! Makurdi, Benue Prov. (Nov.) *Keay* FHI 22273! Ilorin (Oct.) *Ward* 77! **S. Nig.:** (Jan.) *Dodd* 384! Awka *Thomas*! Aguku *Thomas* 1379! **W. Cam.:** Gurum (Nov.) *Hepper* 1268! In tropical and S. Africa, extending through Egypt and Arabia to India.

6. **E. namaquensis** *Nees* in Linnaea 12: 452 (1838); A. Chev. in Rev. Bot. Appliq. 14: 119. *E. maritima* A. Chev. in Bull. Mus. Hist. Nat., sér. 2, 20: 473 (1948), and in Rev. Bot. Appliq. 29: 134. *Diandrochloa namaquensis* (Nees) de Winter in Bothalia 7: 388 (1960). *Poa sporoboloides* A. Rich., Tent. Fl. Abyss. 2: 426 (1851). *Sporobolus confertiflorus* A. Rich., Tent. Fl. Abyss. 2: 397 (1851). Tufted annual 15–120 cm. high; sandy stream sides.

6a. **E. namaquensis** Nees var. **namaquensis**
Sen.: *Chev.*! M'Bidjem (Dec.) *Chev.* 2232! Fadiout (Dec.) *Ezanno* 29! **Mali:** Ségou (Oct.) *Roberty* 3156! Tondigame (Oct.) *Farrow* 26! Dioura (Nov.) *Farrow* 113! Keribane (Oct.) *Davey* 477! **Ghana:** Navrongo to Tumu (Dec.) *Adams* 4360! **Togo Rep.:** Sokodé *Schroeder* 124! **Niger:** Zinder (Nov.) *Hagerup* 613! **N. Nig.:** Naraguta F.R. (Dec.) *Kennedy* 35! Jos (Dec.) *Haines* 150! Nupe *Barter* 993! Vogel Peak (Nov.) *Hepper* 1328! 1452! Tropical and S. Africa.

6b. **E. namaquensis** var. **diplachnoides** (*Steud.*) *W. D. Clayton* in Kew Bull. 25: 251 (1971). *E. diplachnoides* Steud., Syn. Pl. Glum. 1: 268 (1854); A. Chev. in Rev. Bot. Appliq. 14: 117; Berhaut, Fl. Sén. ed 2,. 408; De Miré & Gillet in J. Agric. Trop. 3: 732; F.W.T.A. ed. 1, 2: 512. *E. leprieurii* Steud. l.c. 269 (1854). *E. interrupta* of A. Chev. in Rev. Bot. Appliq. 14: 118, not of (R. Br.) P. Beauv. *E. interrupta* var. *diplachnoides* (Steud.) Hook. f., Fl. Brit. Ind. 7: 316 (1896); Chev. Bot. 749. Tufted annual 30–90 cm. high; sandy, cultivated soils.
Sen.: *Leprieur*! (Nov.) *Roberty* 10159! Niokolo-koba (Nov.) *Adam* 17189! Nyazobile *Chev.* 33916! **Gam.:** *Saunders* 110! Wali (Oct.) *Pirie* 58/33! Georgetown (Jan.) *Fairchild* 484! **Mali:** Tamboukané (Dec.) *Chev.* 2227! **Port. G.:** Boruntuma (Dec.) *Pereira & Correia* 2222! Bafata (Dec.) *Pereira & Correia* 2502! **Guin.:** Baffing R. (Nov.) *Pobéguin* 1820! Kouroussa (Dec., Jan.) *Pobéguin* 539! 624! **S.L.:** Wania (Jan.) *Morton & Gledhill* SL575! **Iv. C.:** Dabou (Dec.) *Aké Assi* 7157! **Ghana:** *Beal* 39! Gushiego to Pigu (Oct.) *Rose Innes* GC 30705! Yapei Ferry, White Volta R. (Oct.) *Rose Innes* GC 31117! Tamale (Dec.) *Williams* 464! Saboba (Mar.) *Hepper & Morton* A3123! **Niger:** Casolé (Feb.) *Chev.* 42967! **N. Nig.:** Marte (Nov.) *Golding* 39! Dikwa (Nov.) *Johnston* N97! Abinsi (June) *Dalz.* 908! Nupe *Barter* 1361! Lokoja *Ansell*! **S. Nig.:** Onitsha (Feb.) *Jones* FHI 6890! Tropical Africa, extending eastwards through Egypt, Ethiopia and the Sudan to Iraq, India, Ceylon and Thailand.

7. **E. blepharostachya** *K. Schum.* in Engl., Bot. Jahrb. 24: 336 (1897). Coarse tufted perennial 45–100 cm. high, with a stiff open panicle of straw-coloured spikelets; roadside gravels and disturbed, often damp, soils.
Iv. C.: Oroumba Boka (Aug.) *Boughey* GC 18537! Lamto (Dec.) *Aké Assi* 9803! Bouaké (Sept.) *de Wilde* 980! **Ghana:** Banda (Nov.) *Harris*! Kintampo (Dec.) *Vigne* FH 3207! Dafo to Winnta (Sept.) *Ankrah* GC 20353! Amisano, Cape Coast (Mar.) *Hall* 1288! Kpedsu (Dec.) *Howes* 1063! **Togo Rep.:** Misahöhe *Kling* 79. Bodome (July) *Baumann* 247! Bismarckburg (Aug.) *Büttner* 100! 120! Palime (Mar.) *Stage* 14! **S. Nig.:** Iwo (May) *Clayton* 534! Ife (Mar.) *Coombe* 149!

8. **E. scotelliana** *Rendle* in Elliott in J. Linn. Soc. 30: 99 (1894); A. Chev. in Rev. Bot. Appliq. 14: 120; Aké Assi, Contrib. 2: 279. Loosely tufted annual 15–60 cm. high, with a delicate spreading panicle of purplish spikelets; shallow soils over iron pan or rock.
Guin.: Mali (Apr.) *Pitot* 1211! Séguéa (Oct.) *Adam* 12639! Dalaba (Oct., Nov.) *Adames* 389! *Chev.* 34594*a*! Timbo (Oct.) *Pobéguin* 1748! **S.L.:** *Morson*! Regent (Dec.) *Sc. Elliot* 4114! York (Dec.) *Deighton* 3291! Tambiama (Feb.) *Glanville* 157! Kissy (Nov.) *Deighton* 2923! **Lib.:** Genne Loffa, Kolahun Dist. (Nov.) *Baldwin* 10092! **Iv. C.:** Singrobo (Nov.) *Adjanohoun* 370*a*! Séguéla (June) *Adjanohoun* 234*a*! **Ghana:** Berekum to Sampa (Oct.) *Rose Innes* GC 31643! Mampong (Oct.) *Rose Innes* GC 31157! Kumasi (Dec.) *Adams* 4445! Tonogo (Nov.) *Morton* A3460! Mpraeso to Adawso (Nov.) *Rose Innes* GC 31658! **N. Nig.:** Sha (Sept.) *J. Hall* K41! Omu-Aran (Nov.) *Latilo* FHI 62285! **S. Nig.:** Bori to Ogoni (July) *Nwanzo* 780! **W. Cam.:** Bamenda (Dec.) *Baldwin* 13837! Extends southward to the Congo.

9. **E. turgida** (*Schumach.*) *De Wild.* Compagnie Kasai 250 (1910); Berhaut, Fl. Sén. ed. 2, 409; incl. var. *ivorensis* A. Chev. in Bull. Mus. Hist. Nat., sér. 2, 20; 473 (1948). *Poa turgida* Schumach., Beskr. Guin. Pl. 86 (1827). *P. rubiginosa* (Trin.) Kunth, Enum. Pl. 1: 339 (1833). *Eragrostis rubiginosa* Trin. in Mém. Acad. Sci. Petersb. sér. 6, 1: 401 (1830); Chev. Bot. 750, and in Rev. Bot. Appliq. 14: 119. *E. ledermannii* Pilger in Engl., Bot. Jahrb. 45: 210 (1910). *Briza rubella* Steud., Syn. Pl. Glum. 1: 282 (1854). About 15–60 cm. high, with pink-tinged spikelets subsessile on the branches of a stiff panicle; roadsides and old farmland.
Sen.: *Leprieur*! Diaroumé, Casamance (Oct.) *Adam* 18410! **Mali:** Fada to Koupéla (Sept.) *Chev.* 24550! Guelia (Nov.) *Chev.* 315! San (Sept.) *Chev.* 2423! Dari (Sept.) *Farrow* 19! **Port. G.:** Piche (Sept.) *Pereira* 3196! 3245! **Guin.:** *Pobéguin* 574! Baffing Valley (Sept.) *Pobéguin* 1752! Timbo (Oct.) *Jac.-Fél.* 1887! Kouroussa (Aug.) *Pobéguin* 1094! **Iv. C.:** N'Zi Plain, Bocanda (June) *Adjanohoun* 235*a*! Bouaké (May) *de Wilde* 56! Confluence of Sassandra & Baffing R. *Chev.* 21772! **U. Volta:** Konpienga (June) *Scholz* 52*a*! **Ghana:** Navrongo (June) *Vigne* FH 4620! Saboba (Sept.) *Rose Innes* GC 30468! Tamale (Oct.) *Baldwin* 13572*a*! Kete Krachi to Kpetchu (Oct.) *Ankrah* GC 20375! Accra *Don*! **N. Nig.:** Katagum *Dalz.* 251! Naraguta (July) *Lely* P403! Samaru (July) *Clayton* 1242! Nupe *Barter* 401! Kankiya (Aug.) *Onwudinjoh* FHI 24009! **S. Nig.:** Oyo (Oct.) *Onochie* FHI 34922! Ibadan (Aug.) *Tamajong* FHI 19514! Iseyin (Aug.)

Fig. 429.—ERAGROSTIS ASPERA (*Jacq.*) *Nees* (GRAMINEAE-ERAGROSTIDEAE).
1, ligule. 2, spikelet. 3, portion of spikelet showing axis. 4, lower glume. 5, upper glume. 6, lemma. 7, side view of palea. 8, grain.

J. B. Gillett 15401! Lagos (Feb.) *MacGregor* 175! Enugu (Sept.) *Jones* 6743! Extends eastwards through Uganda to the Sudan, and south to the Congo.

10. **E. egregia** *W. D. Clayton* in Kew Bull. 20: 271 (1966); Berhaut, Fl. Sén. ed. 2, 410. Caespitose perennial 0·6–1·5 m. high; shallow or disturbed sandy soils.
Sen.: Betenty (Dec.) *Adam* 18931! **Iv. C.:** Bouaké (Sept.) *de Wilde* 994! Toumodi to Dimbokro (Aug.) *Chev.* 22261! Toumodi Nord (Oct.) *Adjanohoun* 347*a*! **Ghana:** Mishuo (Oct.) *Rose Innes* GC 30767! Tumu to Walembele (Oct.) *Rose Innes* GC 30791! Banda to Bui (Oct.) *Rose Innes* GC 31595! Salaga (Oct.) *Ankrah* GC 20302! Legon (Nov.) *Adams* 3551!

11. **E. plurigluma** *C. E. Hubbard* in Kew Bull. 1934: 116; Berhaut, Fl. Sén. ed. 2, 409. *E. trepidula* C. E. Hubbard in Kew Bull. 12: 58 (1957). Tufted perennial about 1 m. high with rolled leaf-blades and delicate open panicle; moist soils.
Sen.: Ziguinchor (Jan.) *Raynal* 7936. **Mali:** Ségou (Nov.) *Adam* 19932! **Guin:** *Farmar* 171! Conakry (Oct.) *Adam* 12616! Kindia (Dec.) *Pobéguin* 1313! Friguiagbé (Nov.) *Chillou* 840! **S.L.:** Mateboi (Oct.) *Glanville* 79! Yonibana (Oct.) *Jordan* 583! Also in Congo, Zambia & Angola.

12. **E. cenolepis** *W. D. Clayton* in Kew Bull. 20: 266 (1966). Tufted perennial, 45–90 cm. high; with or without long white hairs from the axils of the panicle; moist soils.
Guin.: Mamou (Oct.) *Adam* 12754! Timbi Madina (Jan.) *Langdale-Brown* 2552! **S.L.:** Tingi Mts. (Dec.) *Morton & Gledhill* SL3049! SL3139! Loma Mts. (Nov.) *Morton* SL2784! *Jaeger* 8376! 9458! **N. Nig.:** Naraguta *Kennedy* 37! Jos (Dec.) *Monod* 9724*bis*! Samaru, Zaria (Oct.) *Freeman* S164! Anara F.R. (Nov.) *Keay* FHI 28120! Kaduna (Nov.) *Keay* FHI 28111! Also in E. Cameroun.

13. **E. invalida** *Pilger* in Engl., Bot. Jahrb. 34: 129 (1904). *E. glanvillei* C. E. Hubbard in Kew Bull. 1936: 312; F.W.T.A., ed. 1, 2: 516; Aké Assi, Contrib. 2: 279. A densely tufted perennial 15–90 cm. high, the panicle usually loose and open; inselbergs and rock outcrops.
Guin.: Macenta (Oct.) *Baldwin* 9754! **S.L.:** Kayihi (Nov.) *Glanville* 426! Bintumane Peak, 5,400 ft. (Jan.) *T. S. Jones* 86! 135! Bumban (Oct.) *Deighton* 3275! Mt. Loma (Sept.) *Jaeger* 7717! **Iv.C.:** Tonkoui Mt., 4,800 ft. (Oct.) *Aké Assi* 5717! **N. Nig.:** Naraguta (Aug.) *Lely* P450! Jos Plateau *Batten-Poole* 129! **S. Nig.:** Idanre Hills (Oct.) *Keay* FHI 22596! 22679! Also in Gabon.

14. **E. volkensii** *Pilger* in Engl., Bot. Jahrb. 43: 95 (1909). *E. seticaulis* Chiov. (1919). Perennial with a small stiff panicle, and slender wiry many-noded culms 30–90 cm. high, forming a straggly growth from a dense tussock.
N. Nig.: Chappal Waddi (Nov.) *Jackson, Magaji & Tuley* 2032! **W. Cam.:** Above Lake Oku, 7,900 ft. (Jan.) *Keay & Lightbody* FHI 28500! Bagango to Bafut-Ngemba (July) *Daramola* FHI 43205! Bafut-Ngemba F.R., 7,000 ft. (Feb.) *Hepper* 2205! Bamenda, 7,500 ft. (Jan., Dec.) *Migeod* 389! *Akpabla* GC 10579! On the highlands of East and Central Africa, and in Transvaal.

15. **E. mokensis** *Pilger* in Engl., Bot. Jahrb. 51: 419 (1914). Slender annual up to 30 cm. high, much branched from the base.
N. Nig.: Riyom (Nov.) *Clayton* 1407! Wana (Nov.) *Hepburn* 158! Tiba Plateau (Dec.) *Peter & Tuley* 48! Mandara, Mambila Plateau (Jan.) *Latilo & Daramola* FHI 28996! Njawai (Nov.) *Jackson, Magaji & Tuley* 1893! **S. Nig.:** Bebi, near Obudu (Dec.) *Tuley* 64! Obudu Plateau (Apr.) *Tuley* 967! *Haines* 358! **W. Cam.:** Kishong, 6,700 ft. (Jan.) *Keay & Russell* FHI 28445! **F. Po:** Moka, 3,900–5,800 ft. *Mildbr.* 7102! (Dec.) *Boughey* 154!

16. **E. unioloides** (*Retz.*) *Nees ex Steud.* Syn. Pl. Glum. 1: 264 (1854). *Poa unioloides* Retz., Obs. Bot. 5: 19 (1789). Annual, 5–45 cm. high, the culms mostly simple; there is considerable variation in the size of spikelet and density of panicle; a wayside weed.
Guin.: Conakry (Oct.) *Adam* 12615! Maneah (Dec.) *Chillou* 2312! **S.L.:** Port Loko (Oct.) *Jordan* 805! Brookfields (Aug.) *Jordan* 787! Kissy (Nov.) *Deighton* 2922! Hill Station (Aug.) *Deighton* 2031! **Lib.:** Monrovia (Aug.) *Baldwin* 13041! Kakota (July) *Harley* 1080! Harbel (Sept.) *Baldwin* 13387! Buchanan (Nov.) *Adam* 16059! Bahtown (Aug.) *Baldwin* 8040! Widespread in tropical Asia; now established in W. Africa and Guiana, and possibly in other localities in the tropics.

17. **E. gangetica** (*Roxb.*) *Steud.* Syn. Pl. Glum. 1: 266 (1854); Berhaut, Fl. Sén., ed. 2, 409. *Poa gangetica* Roxb., Fl. Ind. 1: 340 (1820). *P. cambessediana* Kunth, Rev. Gram. 2: 469 (1831). *P. ovina* A. Rich. (1851). *Eragrostis cambessediana* (Kunth) Steud., Syn. Pl. Glum. 1: 269 (1854); F.W.T.A. ed. 1, 2: 515; Chev. Bot. 748. *E. stenophylla* Hochst. ex Miq. (1851)—Chev. Bot. 750. *E. ovina* (A. Rich.) Steud. (1854)—A. Chev. in Rev. Bot. Appliq. 14: 121. *E. flamignii* De Wild. (1919). *E. vinicolor* A. Chev. in Bull. Mus. Hist. Nat. sér. 2, 20: 472 (1948), incl. var. *pallida* A. Chev. l.c. (1948); A. Chev. in Rev. Bot. Appliq. 29: 130. *E. dumasiana* A. Chev. in Bull. Mus. Hist. Nat., sér. 2, 20: 472 (1948), and in Rev. Bot. Appliq. 29: 132. *E. dakarensis* A. Chev. in Bull. Mus. Hist. Nat., sér. 2, 20: 470 (1948), and in Rev. Bot. Appliq. 29: 128. *E. pumila* A. Chev. in Bull. Mus. Hist. Nat., sér. 2, 20: 472 (1948), and in Rev. Bot. Appliq. 29: 130. Loosely tufted annual, with slender geniculate culms 15–45 cm. high; in weedy places.
Maur.: Lake Rkiz (Sept.) *Sadio* 104! Podor to Boghé (Oct.) *Hepper* 3673! **Sen.:** Saint-Louis (Feb.) *Berhaut* 2838! Sangomar (Dec.) *Adam* 19950! Wallo *Leprieur*! Tambacounda (Nov.) *Berhaut* 2840! Diouloulou (Sept.) *Adam* 18123! **Gam.:** *Saunders* 86! **Mali:** Timbuktu *Hagerup* 235! Kabarah (Aug.) *Chev.* 1326! San (Jan.) *Demange* 1013! Bamako (Oct.) *Adam* 15360! Togueré of Saredina (May) *Davey* 88! **Port. G.:** Nova Lamego (Oct.) *Pereira* 3354! Fa (Oct.) *Guerra* 426! Bedanda (Jan.) *Pereira & Correia* 2777! **Guin.:** Sériba (Oct.) *Adam* 19708! Dalaba (Nov.) *Chev.* 34599! Mamou (Nov.) *Chev.* 34896! Timbo (Oct.) *Pobéguin* 1794! Kouroussa (July) *Pobéguin* 489! **S.L.:** Musaia (Dec.) *Deighton* 4454! Kaballa (Sept.) *Thomas* 2180! Makump (Oct.) *Glanville* 67! Freetown (Aug.) *Deighton* 2051! Brookfields (Aug.) *Jordan* 786! **Lib.:** Kailahun (Nov.) *Baldwin* 10125! Vonjama (Oct.) *Baldwin* 9862! Ganta (Jan.) *Harley* 1742! Robertsfield (July) *Baldwin* 6658! Nimba (Jan.) *Adam* 20490! **Iv. C.:** Bouaké (Dec.) *Pitot*! Divo F.R., Gagnoa (Aug.) *Boughey* GC 14650! Dabou (Feb.) *Chev.*! Bériby (Nov.) *Oldeman* 636! **U. Volta:** Bobo-Dioulasso to Bonfosa (Jan.) *Chev.*! Bobo-Dioulasso (May) *Chev.* 907! Samandeni (Oct.) *Kmoch* 160! Farakoba (June) *Scholz* 97! **Ghana:** Bolgatanga to Navrongo (Oct.) *Rose Innes* GC 31095! Bawku to Bolgatanga (Oct.) *Rose Innes* GC 30502! Sunyani to Wenchi (Sept.) *Rose Innes* GC 30146! Kumasi (June) *La-Danso* 77! Accra to Weija (Mar.) *Rose Innes* GC 30559! **Togo Rep.:** Badou to Atakpame (Sept.) *Rose Innes* GC 31353! **N. Nig.:** Katsina (Oct.) *Clayton* 1370! Maiduguri (Oct.) *Johnston* N39! Zaria Dist. *C. B. Taylor* 34! Shendam (Nov.) *Clayton* 1456! Minna to Zungeru (Jan.) *Keay* FHI 37319! **S. Nig.:** Ibadan (Feb.) *Brenan & Keay* 8950! Idanre (Jan.) *Brenan* 8744! Lagos (Feb.) *W. MacGregor* 176! Awka (June) *Jones* 1787! Opobo *Jeffreys* 19! **W. Cam.:** Mamfe (Dec.) *Migeod* 262! Bambui (Feb.) *Cowell* G20! Victoria (July) *Maitland* 13! Throughout tropical Africa, and in India. (See Useful Plants.)

A rather variable species; panicle typically condensed about the primary branches, but sometimes diffuse with capillary branchlets; in the northern part of our area the spikelets tend to be longer than elsewhere, and the paleas may be only tardily deciduous; a form found on ironstone pans (*E. dumasiana* see Lebrun in Bull. Soc. Bot. Fr. 116: 260 (1969)) appears to be a short-lived perennial, but is otherwise identical with plants which are certainly annual; it may be confused with the very similar perennial species *E. atrovirens*, which has 3 stamens.

18. **E. pilosa** (*Linn.*) *P. Beauv.* Ess. Agrost. 162, 175 (1812); Berhaut, Fl. Sén. ed. 2, 408; De Miré & Gillet in J. Agric. Trop. 3: 733. *Poa pilosa* Linn., Sp. Pl. 68 (1753). *P. senegalensis* Desv., Opusc. 100 (1831). *P. linkii* Kunth (1829). *Eragrostis linkii* (Kunth) Steud. (1854). *E. filiformis* Link (1827). *E. tenuiflora* Rupr. ex Steud., Syn. Pl. Glum. 1: 268 (1854); A. Chev. in Rev. Bot. Appliq. 14: 120. *E. baguirmensis* A. Chev. in Rev. Bot. Appliq. 29: 135 (1949). A slender ascending annual 30–60 cm. high, with flexuous panicle branches; path sides and fallow land.
Maur.: Gujaf *Rossetti* 61/211! 61/237! **Sen.:** *Heudelot* 2837! Dakar (Mar.) *Sadio* 18! Djoudj (Oct.) *Adam* 21732! Thiés (Aug.) *Adam* 14936! Saint-Louis (Feb.) *Berhaut* 2822! **Gam.:** *Ruxton* 60! *Saunders*

113! **Mali:** Kidal to Bousena (Sept.) *Rossetti* 59/173! Timbuktu (Aug.) *Lean* 78! Sahel, N. Region *Leclercq* 42685! Dioura (Oct.) *Davey* 45! Gao (Aug.) *Wailly* 4724! **S.L.:** Newton (June) *Deighton* 5552! **Lib.:** Nimba (Dec.) *Adam* 20283! **U. Volta:** Ouagadougou (Aug.) *Chev.* 24748! Garango *Prost*! **Ghana:** Kintampo (Dec.) *Vigne* FH 3197! Kumasi (Mar.) *La-Danso* 24! Cape Coast (May) *Morton* A3672! Achimota (Sept.) *Milne-Redh.* 5134! Yendi (Sept.) *Rose Innes* GC 32166! **Dah.:** Cana (Apr.) *Newton* 18! **Niger:** Agades (Feb.) *Chev.* 43452! Toukounous *Bartha* 14! Maradi (Aug.) *Hall* 394! **N. Nig.:** Sokoto Prov. (Aug.) *Dalz.* 492! Gajibo (Nov.) *Johnston* N63! Kano (Dec.) *Hagerup* 646a! Samaru (July) *Clayton* 1236! Naraguta (June) *Lely* 294! **S. Nig.:** Aboh *Barter* 400! Lagos (Feb.) *W. MacGregor* 197! Igboora (Oct.) *Haines* 351! Ipapo (May) *Denton* 28! Idanre (May) *Brenan* 8744a! Widespread in tropical and warm temperate regions. (See Useful Plants.)

19. **E. chalarothyrsos** *C. E. Hubbard* in Kew Bull. 1936: 310. Perennial with stout culms up to 1·2 m. high and large loose panicles; swampy soils.
S.L.: Mange (July) *Glanville* 240! **Ghana:** Wa to Sawla (Oct.) *Rose Innes* GC 31026! Ejura (Aug.) *Hall* CC322! Also in Uganda, Kenya and Tanzania.

20. **E. barteri** *C. E. Hubbard* in Kew Bull. 1936: 311; Lebrun in Bull. Soc. Bot. Fr. 116: 260 (1969). *E. fluviatilis* A. Chev. in Bull. Mus. Hist. Nat. Paris sér. 2, 20: 472 (1948), and in Rev. Bot. Appliq. 29: 132; C. E. Hubbard in Kew Bull. 4: 343. Caespitose grass about 1 m. high; the woody culms, left lying obliquely by the subsiding flood, produce fascicles of short flowering shoots from the nodes during the dry season; rocks and sandbanks in river beds.
Sen.: Tiquet (Oct.) *Audru* 2932! **Mali:** Bamako to Quignola (Jan.) *Chev.* 254. Niger "delta" *Duong*! **S.L.:** Bure (Jan.) *Jordan* 856! Mafindo Falls, Kailahun (Feb.) *Deighton* 4005! **Iv. C.:** Tiassalé (Feb.) *Chev.*! Soubré (Nov.) *de Wilde* 3320! **Ghana:** Senchi (Jan.) *Irvine* 2627! Kete Krachi (Dec., Apr.) *Adams* 4556! *Hall* 1288! **N. Nig.:** Hadejia (Dec.) *Clayton* 471! Bonu (Dec.) *Keay* FHI 37312! Kontagora (Dec.) *Hubbard* S582! Nupe *Barter* 877! Gurara falls, Abuja (Mar.) *Vaillant* 2772! **S. Nig.:** Oyo to Iseyin (Feb.) *Brenan* 8958! Olokemeji F.R. (Jan.) *Keay* FHI 21155! Abeokuta (Dec.) *Meikle* 924! **W. Cam.:** Bamenda (May) *Daramola* FHI 41194! Also in the Congo.
[Closely allied to *E. atrovirens*, with which it intergrades.]

21. **E. atrovirens** (*Desf.*) *Trin. ex Steud.* Nom. Bot. ed. 2, 1: 562 (1840); Berhaut, Fl. Sen. ed. 2, 410. *Poa atrovirens* Desf., Fl. Atlant. 1: 73 (1798). *P. biformis* Kunth, Rev. Gram. 2: 471 (1831). *Eragrostis biformis* (Kunth) Benth. in Fl. Nigrit. 568 (1849). *E. multiflora* var. *biformis* (Kunth) A. Chev. in Rev. Bot. Appliq. 15: 1042 (1935). *E. bromoides* Jedw. in Bot. Arch. 5: 190 (1924); A. Chev. in Rev. Bot. Appliq. 14: 121; not of Steud. (1854). *E. atroviridis* Maire (1937). *E. sudanica* A. Chev. in Bull. Mus. Hist. Nat., sér, 2, 20: 471 (1948), and in Rev. Bot. Appliq. 29: 128. *E. gangetica* of F.W.T.A., ed. 1, 2: 516; of Chev. Bot. 749, and in Rev. Bot. Appliq. 14: 118; not of (Roxb.) Steud. A widespread and very variable species; moist or swampy soils.
Maur.: Lake Rkiz (Sept.) *Sadio* 27! **Sen.:** Djoudj (Oct.) *Adam* 21742! Kaëdi (Nov.) *Roberty* 10136! Niokolo-Koba (Oct.) *Adam* 15641! Oussouye (July) *Berhaut* 6211 *bis*! Casamance (Oct.) *Adam* 18366! **Gam.:** Sankuli-Kunda (Sept.) *Pirie* 4/33! **Mali:** Dogo (Apr.) *Davey* 583! Gao (Apr., Sept.) *Hagerup* 329! *De Wailly* 5025! Bandiagara to Mopti (Sept.) *Chev.* 24938! Mopti (Sept.) *Lean* 92! **Port. G.:** Bafata (July, Aug.) *Pereira* 3070! 3096! 3097! Bambadinca (July) *Pereira* 3037! **Guin.:** Conakry (May) *Baldwin* 5793! Ditinn (Apr.) *Pitot*! Dalaba *Monod*! Dabola (Apr.) *Pitot*! Faranah (Apr.) *Pitot*! **S.L.:** Port Loko to Mange (May) *Jordan* 870! Mayoso (Aug.) *Thomas* 1495! Sondugu (June) *Thomas* 689! Kichom (July) *Deighton* 3757! Toma, Bonthe Is. (Nov.) *Deighton* 2279! **Lib.:** Robertsport (Dec.) *Baldwin* 10907! Monrovia (Dec.) *Baldwin* 10492! Buchanan (Mar., Dec.) *Adam* 16425! *Baldwin* 11194! Nimba (Nov.) *Adames* 738! **Iv. C.:** Bassam (Jan.) *Hédin*! Abidjan (Oct.) *de Wilde* 1104! Toussiana (June) *Scholz* 93! **U. Volta:** Banfora (June) *Leeuwenberg* 4354! Ouagadougou (Oct.) *Stolz* 42! 109! **Ghana:** Bawku (Oct.) *Rose Innes* GC 30259! Zuarungu (June) *Williams* 507! Burufo Plateau, nr. Lawra (Apr.) *Adams* 4034! Kumasi (June) *La-Danso* 75! Nungua (Mar.) *Ankrah* GC 20120! **Dah.:** Boukombe (Nov.) *Risopoulos* 1253! **Niger:** Wadi Dollol *White* 60! Maradi (Aug.) *Hall* 395! **N. Nig.:** Sokoto (July) *Dalz.* 494! Baga Seyoram, Kukuwa (Feb.) *Davey* FHI 27170! Rafin Bauna F.R., Jos (Oct.) *Hepper* 1173! Nupe *Barter* 622! Abinsi *Dalz.* 877! **S. Nig.:** Onitsha (June) *Onochie* FHI 35860! Ogurude (Jan.) *Holland* 273! 281! Obu *Thomas* 431! Port Harcourt (Dec.) *Maitland* 713! Awgu to Okigwi (July) *Tuley* 821! **W. Cam.:** Nkambe, 5,800 ft. (Feb.) *Hepper & Charter* 1935! Bamenda, 5,000–6,000 ft. (Jan.) *Migeod* 405! *Brunt* 700! 960! 1023! N. Africa to Zambia and Angola, eastwards to the Philippines.
[The name *E. chariis* (Schult.) Hitchc. has sometimes been applied to our species, but it is a synonym of *E. nutans* (Retz.) Nees ex Steud., a different species from India.]

22. **E. camerunensis** *W. D. Clayton* in Kew Bull. 20: 265 (1966). Densely tufted perennial, 30 cm. high.
S. Nig.: Obudu Plateau (Dec., Feb., Apr.) *Tuley* 44! 480! *Haines* 142! **W. Cam.:** Bafut-Ngemba F.R. (Mar.) *Coombe* 220! Nchan *Maitland* 10a! Bambui *Pedder* 23! Kishong *Brunt* 1027! Ndu (Jan.) *Keay & Russell* FHI 28433!

23. **E. cilianensis** (*All.*) *Lut.* in Malpighia 18: 386 (1904); Berhaut, Fl. Sén. ed. 2, 409; De Miré & Gillet in J. Agric. Trop. 3: 732; Aké Assi, Contrib. 2: 278. *Poa cilianensis* All., Fl. Pedem. 2: 246 (1785). *P. megastachya* Koel., Descr. Gram. 181 (1802), superfluous name. *P. cachetica* Schumach., Beskr. Guin. Pl. 66 (1827). *Briza eragrostis* Linn., Sp. Pl. ed. 1, 70 (1753). *Eragrostis major* Host, Gram. Austr. 4: 14 (1809); Sprague & Hubbard in Kew Bull. 1933: 15; Chev. Bot. 749, and in Rev. Bot. Appliq. 14: 118. *E. megastachya* Link, Hort. Berol. 1: 187 (1827); Henrard in Blumea 3: 420. *E. articulata* De Wild. in Bull. Jard. Bot. Brux. 6: 58 (1919). *E. schweinfurthiana* Jedw. in Bot. Arch. 5: 194 (1924). Loosely tufted annual 15–60 cm. high, usually with pallid panicles; roadsides and fallow land.
Sen.: *Perrottet* 324! 959! (Jan.) *Roger* 4! *Heudelot* 519! Dakar (Sept.) *Berhaut* 1148! Mamelles (Dec.) *I.F.A.N.*! **Mali:** Kidal (Sept.) *Rossetti* 59/155! Eloualadji (Mar.) *Chev.* 43914! Timbuktu (Aug.) *Hagerup* 272! Sahel zone *Leclercq* 42582! Dioura (Sept.) *Davey* 41! **Iv. C.:** Gaoua (Oct.) *Rose Innes* GC 31661! Tehini (Aug.) *de Wilde* 743! **U. Volta:** Ouahigouya, Mossi (Aug.) *Chev.* 24775! **Ghana:** Navrongo (Aug.) *Vigne* FH 4623! Gambaga Scarp (Oct.) *Rose Innes* GC 30495! Wenchi (Oct.) *Rose Innes* GC 30521! Tong Hills (July) *Williams* 522! Nungua (Oct.) *Ankrah* GC 20096! **Niger:** Agades (Feb.) *Bradley* 19a! **N. Nig.:** Kworre, Sokoto (Sept.) *Palmer* 33! Mongonu (Oct.) *Golding* 2! Naraguta (July) *Lely* P428! Nupe (Jan.) *Baikie*! Abinsi *Dalz.* 874! **S. Nig.:** (Jan.) *Dodd* 383! Ibadan *Millen* 104! Lagos (Feb.) *W. MacGregor* 5! Aboh *Barter* 336! Widespread in tropical and warm temperate regions. (See Useful Plants.)

24. **E. minor** *Host* Gram. Austr. 4: 15 (1809). *Poa eragrostis* Linn., Sp. Pl. 68 (1753). *Eragrostis poaeoides* P. Beauv., Ess. Agrost. 162 (1812). *E. poaeiformis* Link (1827). Loosely tufted annual, 15–45 cm. high.
Maur.: Dahr of Néma *Rossetti* 61/248! **Sen.:** Hann *Adam* 16909! Also in Ethiopia, Sudan, Kenya and Tanzania; widespread in temperate and subtropical regions.
Closely allied to *E. cilianensis*, differing in little but the size of the spikelet, and the broadly oblong grain, but the species intergrade and in our area the grain is similar to that of *E. cilianensis*.

25. **E. pobeguinii** *C. E. Hubbard* in Kew Bull. 1936: 312; Berhaut, Fl. Sén, ed. 2, 409. *E. fleuryi* A. Chev. in Bull. Mus. Hist. Nat. Paris, sér. 2, 20: 47 (1948), and in Rev. Bot. Appliq. 29: 128. Densely caespitose, about 30 cm. high; shallow soils over rock and ironstone.
Sen.: Niokolo-Koba (June) *Adam* 14382! Niokolo-Koba to Kedougou (June) *Adam* 14420! **Guin.:** Timbo (June) *Pobéguin* 1092! **Ghana:** *Agric. Dept.*! **W. Cam.:** Bali-Ngemba F.R., Bamenda (May) *Ujor* FHI 30323! Also in E. Cameroun.
[A disconcertingly variable species, possibly conspecific with *E. longifolia* Hochst. ex Steud. from Ethiopia, but neither is well known. They are both related to the common African species *E. racemosa* (Thunb.) Steud.]

26. **E. squamata** (*Lam.*) *Steud.* Syn. Pl. Glum. 1: 274 (1854); Berhaut, Fl. Sén. ed. 2, 410; Aké Assi, Contrib. 2: 279. *Poa squamata* Lam., Tab. Encycl. 1: 185 (1791). *Eragrostis halophila* A. Chev. in Bull. Mus. Hist. Nat., sér. 2, 20: 473 (1948), and in Rev. Bot. Appliq. 29: 134. Caespitose perennial 30–120 cm. high, often with glaucous foliage, and sometimes pilose on the sheath; roadsides and waste places, particularly on sandy soils; also on sandy beaches.
Maur.: L. Rkiz (Sept.) *Sadio* 23! **Sen.:** Kaolak *Chev.* 1945! Casamance (Oct.) *Adam* 18296! 18365! Kolda, Casamance (Oct.) *Adam* 18558! Sangalkam (Sept.) *Berhaut* 2846! **Mali:** Sotuba, Bamako *Adam* 11251! 11311! **Port. G.:** Bissau (Dec.) *Pereira & Correia* 2145! Bedanda (Jan.) *Pereira & Correia* 2774! 2775! Boruntuma (Dec.) *Pereira & Correia* 2320! **Guin.:** Fouta Djalon, 3,300 ft. (Jan.) *Langdale-Brown* 2563! Baffing Valley (Oct.) *Pobéguin* 1750*bis*! Timbo (Oct.) *Pobéguin* 1750! Filicomedji (Oct.) *Adam* 12591! Nzérékoré (Oct.) *Baldwin* 9699! **S.L.:** Aberdeen (Nov.) *Small* 786! Njala (Nov.) *Deighton* 1778! Bubuya (Sept.) *Jordan* 319! Mambolo (Dec.) *Jordan* 738! Samaia (May) *Thomas* 236! **Lib.:** Bassa Cove *Ansell*! Monrovia (May) *Baldwin* 5811! Ganta (Nov.) *Harley* 1716! Sangwin (Mar.) *Baldwin* 11324! Harper (June) *Baldwin* 5975! **Iv. C.:** Adiopodoumé (Dec.) *Pitot*! Dabou (Nov.) *Leeuwenberg* 1944! Bériby (Nov.) *Oldeman* 661! **Ghana:** Mimima to Nasia (Oct.) *Rose Innes* GC 30765! Dedoro-Tankara (Dec.) *Adams & Akpabla* 4346! Awiebo to Krisin (Dec.) *Rose Innes* GC 30536! Atwabo *Fishlock* 71 of 1931! Nzima (Jan.) *Thorold* 295! **N. Nig.:** Okene (Mar.) *Maggs* OK53! Nupe *Barter* 729! Gombe (Oct.) *De Leeuw & Magaii* 1978! **S. Nig.:** Oyo (Feb.) *Keay* FHI 16257! Lagos (June) *Dalz.* 1308! Awka (May) *Jones* 1740! Mbieri, near Owerri (Mar.) *Brenan* 9205! Port Harcourt *Maitland*! Also in E. Cameroun and the Congos.
[Very similar to the Mexican species *E. plumbea* Scribn.]

27. **E. tenuifolia** (*A. Rich.*) *Hochst. ex Steud.* Syn. Pl. Glum. 1: 268 (1854). *Poa tenuifolia* A. Rich., Tent. Fl. Abyss. 2: 425 (1851). *Eragrostis parviglumis* Hochst. ex Steud. (1854). *E. collocarpa* K. Schum. (1895). Slender caespitose grass 30–60 cm. high, the leaves mostly from the tussocky base; highlands.
N. Nig.: Naraguta (July) *Lawlor & Hall* FHI 45576! Sha (Sept.) *Hall* K 40! Riyom, 4,000 ft. (Nov.) *Clayton* 1406! Vom, 4,000 ft. (Oct.–Dec.) *Gambles* 9! 21! *Haines* 127! **S. Nig.:** Obudu Plateau, 5,000 ft. (Dec., Feb., Apr.) *Tuley* 43! 481! *Haines* 361! **W. Cam.:** *Cowell* G 25! Bambui, 5,000 ft. (June) *Brunt* 508! Bamenda *Dept. of Agric.* 5! Nsongwa, 4,000 ft. (Apr.) *Brunt* 1052! Bambili (Feb.) *Bauer* 7! Throughout tropical Africa; also in India, Australia and S. America.

28. **E. domingensis** (*Pers.*) *Steud.* Syn. Pl. Glum. 1: 278 (1854). *Poa domingensis* Pers., Syn. Pl. 1: 88 (1805). *P. linearis* Schumach., Beskr. Guin. Pl. 67 (1827). *P. subulata* Desv., Opusc. 102 (1831). *Eragrostis guineensis* Trin. in Mém. Acad. Sci. Petersb., sér. 6, 1: 404 (1830). *E. linearis* (Schumach.) Benth. in Fl. Nigrit. 567 (1849); F.W.T.A., ed. 1, 2: 515; Berhaut, Fl. Sén. ed. 2, 409. *E. albescens* Steud., Syn. Pl. Glum. 1: 269 (1854). *E. pallescens* Hitchc. in J. Wash. Acad. Sci. 19: 303 (1929); A. Chev. in Rev. Bot. Appliq. 14: 119. *E. hagerupii* Hitchc. in Proc. Biol. Soc. Wash. 43: 92 (1930). Coarse tufted perennial with hard rigid culms, 90–120 cm. high; leaves mostly cauline, somewhat glaucous; roots clothed in a dense sheath of hairs; coastal and inland sands.
Maur.: L. Rkiz (Oct.) *Adam* 21827! **Sen.:** *Heudelot* 394! *Adanson*! Richard-Toll (May, Oct.) *Döllinger*! *Berhaut* 2832! Dakar (Aug.) *Berhaut* 1043! Hann (May) *Adam* 9200! **Gam.:** *Boteler*! **Mali:** Timbuktu (June, Aug.) *Hagerup* 105! *Lean* 73! Kabara (Aug.) *Lean* 81! Kadigue to Banguita (Jan.) *Davey* 9! Banguita (Sept.) *Farrow* 13! **Iv. C.:** Abidjan to Grand Bassam (Jan.) *Leeuwenberg* 2341! **U. Volta:** Gourma (July) *Rogeon* 481! **Ghana:** Labadi to Teshi (Apr.) *Irvine* 2859! Tema (June) *Rose Innes* GC 30359! Cape Coast (Apr.) *Hall* 1905! Cape St. Paul (Apr.) *Rose Innes* GC 31219! Dawa, Ada Road (May) *Adams* 4314! **Togo Rep.:** Porto Segoro (July) *Rose Innes* GC 30085! **Dah.:** Cotonou (July, Nov.) *Clayton* 405! *Chev.* 4456! 4460! 4467! Cotonou to Allada (Oct.) *Risopoulos* 1222! **N. Nig.:** Niger R. *T. Vogel*! **S. Nig.:** Ikorodu (Dec.) *Onochie* FHI 32032! Lagos (Sept.) *Stubbings* 27! *Dalz.* 1118! Yaba (Apr.) *Haines* 163! Ogun River F.R. (Mar.) *Killick* 216! Also in Sao Tome and tropical America.
[*E. domingensis* is closely allied to *E. prolifera* (Sw.) Steud. (= *E. fascicularis* Trin.), a species from the Congo Republic, Angola and tropical America with obtuse lemmas, fascicled leaves and a narrowly oblong panicle with short divergent branches. In America, however, the two species are not so clearly separated and tend to intergrade.]

29. **E. cylindriflora** *Hochst.* in Flora 38: 224 (1855). *E. horizontalis* Peter in Fedde, Rep., Beih. 40: Anh. 107 (1930). *E. adenocoleos* Pilger in Not. Bot. Gart. Berl. 15: 705 (1942). Loosely tufted annual 30–76 cm. high.
Niger: Tasker to Boultoum (Sept.) *P. de Fabrègues* 949! **Ghana:** Accra to Ada (Aug.) *Ankrah* GC 20154! Nungua (May, Sept.) *Rose Innes* GC 30045! *Ankrah* GC 20073! Accra (Nov.) *Thollon* 670! Prampram (July) *Irvine* 3061! Tropical and South Africa.

30. **E. aegyptiaca** (*Willd.*) *Link* Hort. Berol. 1: 191 (1827); A. Chev. in Rev. Bot. Appliq. 14: 116; Berhaut, Fl. Sén. ed. 2, 409. *Poa aegyptiaca* Willd., Hort. Berol. 107 (1809). *P. antennata* Del. ex Poir., Encycl. Suppl. 4: 330 (1816), in synon. *P. pilosa* var. *aegyptiaca* (Willd.) Fedtsch. in Bull. Jard. Bot. Pierre Grand 14 (Suppl. 2): 69 (1915). *Eragrostis senegalensis* Nees in Linnaea 10: Litt. 116 (1836); A. Chev. in Rev. Bot. Appliq. 14: 120. *E. minima* Jedw. in Bot. Arch. 5: 188 (1924). *E. albida* Hitchc. in J. Wash. Acad. Sci. 19: 304 (1929); F.W.T.A., ed. 1, 2: 515; A. Chev. in Rev. Bot. Appliq. 14: 116. *E. nigerica* A. Chev. in Bull. Mus. Hist. Nat. sér. 2, 20: 470 (1948), and in Rev. Bot. Appliq. 20: 126. A tufted annual up to about 30 cm. high, with slender geniculate culms and pale spikelets.
Sen.: St. Louis *Leprieur*! Matam (Feb.) *Berhaut* 1441! **Mali:** Timbuktu (July) *Hagerup* 160! Gourma (Aug.) *Rossetti* 59/105! Laougna (Apr.) *Davey* 585! Macina (May) *Lean* 10! San (July) *Chev.* 1093! **N. Nig:** Kalkala *Gwynn* 132! Nguru (Apr.) *Tuley* 1137! Nupe *Barter* 321! Also in Egypt and the Sudan.

31. **E. macilenta** (*A. Rich.*) *Steud.* Syn. Pl. Glum. 1: 268 (1854). *Poa macilenta* A. Rich., Tent. Fl. Abyss. 2: 428 (1851). Slender annual 30–60 cm. high.
Iv. C.: Samou to Ouodé (May) *Chev.* 21620! **N. Nig.:** Naraguta F.R. (July) *Lawlor & Hall* FHI 45567! **W. Cam.:** Bamenda *Maitland* 1448! Nchan, 5,000 ft. (May *Maitland* 10*a*! 10*d*! Santa & Bambui Farms (May, June) *Pedder* 15! Bamali to Bali Kunbat (May) *Brunt* 441! In the Congo, and in E. Africa from the Sudan to Zambia; recorded once from India.

32. **E. tremula** *Hochst. ex Steud.* Syn. Pl. Glum. 1: 269 (1854); Chev. Bot. 750, and in Rev. Bot. Appliq. 14: 120; Berhaut, Fl. Sén. ed 2, 409; De Miré & Gillet in J. Agric. Trop. 3: 733. *E. lamarckii* Steud. l.c. (1854). *E. multiflora* Trin. in Mém. Acad. Sci. Petersb. sér. 6, 1: 401 (1830); A. Chev. in Rev. Bot. Appliq. 14: 119. *Poa tremula* Lam., Tab. Encycl. Meth. Bot. 1: 185 (1791). *P. multiflora* Roxb. (1820), not of Forsk. (1775). Loosely tufted annual 60–90 cm. high, with pretty trembling panicles, and pallid spikelets tinged with light brown or purple; roadsides and waste places.
Maur.: Dahr of Néma *Rossetti* 61/203! Gaberni *Rossetti* 61/315! **Sen.:** *Roger*! Dahra to Linguère *Adam* 11088*bis*! Dakar (Oct.) *Berhaut* 2833! Bouclau *Heudelot* 166! M'Bambey (Nov.) *Chev.* 33951! **Gam.:** *Saunders* 52! Karawan (Jan.) *Dalz.* 8416! Kudang (Jan.) *Dalz.* 8417! Yundum *Austin* 14! Sankuli-Kunda (Sept.) *Pirie* 9/33! **Mali:** Timbuktu (Aug.) *Hagerup* 275! Niafounké to Diré *Rogeon* 129! Kami, Mopti (Oct.) *Davey* 196! Dioura, Macina (Nov.) *Davey* 151! Bamako (Oct.) *Adam* 15307! **Port. G.:** Nhacra (Nov.) *Raimundo & Guerra* 27! Bafata (Dec.) *Pereira & Correia* 2482! **Guin.:** Kindia *Jac.-Fél.* 198! Baffing R. (Nov.) *Pobéguin* 1830! Timbo *Pobéguin* 1830*bis*! Kouroussa (Nov.) *Pobéguin* 529! Sambaïlo (Sept.) *Pitot*! **S.L.:** Yongaro (Oct.) *Glanville* 89! Lumley (Aug.) *Melville & Hooker* 242! Waterloo to Newton (Nov.) *Deighton* 5625! **Iv. C.:** *Bégué* 76! Bemouni (Jan.) *De Wit* 7531! Bouaké to Katiola (Apr.) *Leeuwenberg* 3314! Abidjan to Dabou (Dec.) *Boughey* GC 14456! Tehini (Aug.) *de Wilde* 816! Bouaké (May) *de Wilde* 55! **U. Volta:** Ouagadougou (Oct.) *Scholz* 3*a*! **Ghana:** Navrongo (Sept.) *Vigne* FH 4631! Bawku (Aug.) *Irvine* 4615! Tamale (Dec.) *Williams* 450! Ejura (June) *Thorold* 84! Kpandu to Accra (Sept.) *Ankrah* GC 20198! **Togo Rep.:** Atakpame to Sokode (Oct.) *Rose Innes* GC 31375! Lomé

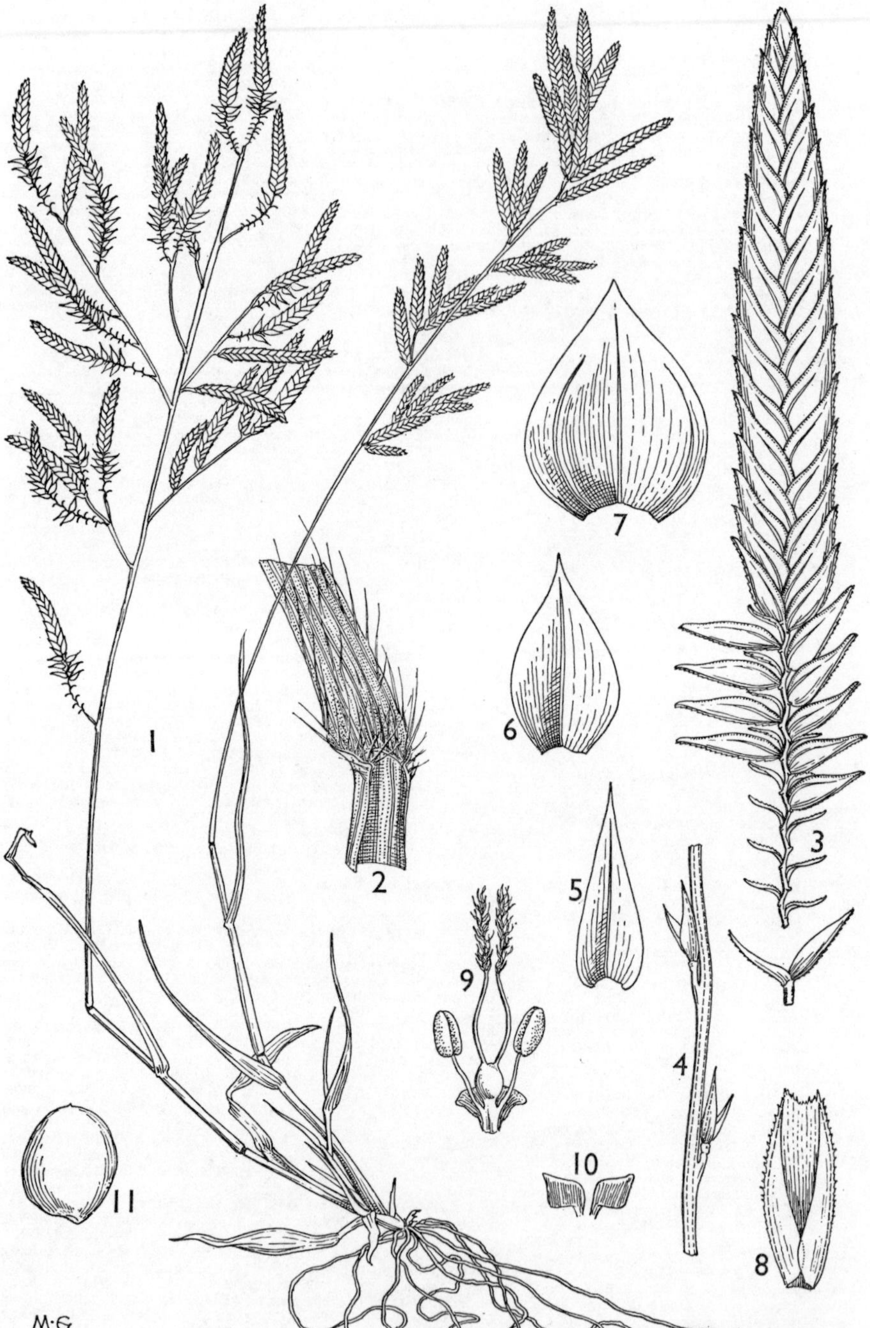

Fig. 430.—Eragrostis lingulata *W. D. Clayton* (Gramineae-Eragrostideae).

1, habit, × ⅗. 2, ligule, × 4. 3, spikelet showing disarticulation, × 6. 4, branchlet after spikelets fallen, × 6. 5, lower glume, × 14. 6, upper glume, × 14. 7, lemma, × 14. 8, palea, × 20. 9, flower, × 20. 10, lodicules, × 20. 11, grain, × 20. From *Pitot* s.n.

Warnecke 164! **Dah.:** Gebé *Newton* 21! Boukombe (Nov.) *Risopoulos* 1267! **Niger:** Zinder (Nov.) *Money-Kyrle* Fe 6! Fe 17! Niamey (Oct.) *Hagerup* 464! Toukounous *Bartha* 9! Dasso *White* 16! **N. Nig.:** Sokoto (July) *Dalz.* 493! Katsina (Oct.) *Clayton* 1373! Dikwa (Dec.) *Johnston* N100! Gurum (Nov.) *Hepper* 1312! Lokoja (Sept.) *T. Vogel* 156! Idoma (Oct.) *Kennedy* FHI 8025! **S. Nig.:** Fashola *Thorold*! Ibadan (Nov.) *Meikle* 667! Lagos (June) *Dalz.* 1320! Awka Div. (July) *Jones* 6218! Aboh *Barter* 351! Throughout tropical Africa and in India. (See Useful Plants.)

33. **E. elegantissima** *Chiov.* in Ann. Ist. Bot. Roma 8: 367 (1908). Annual, about 30 cm. high.
U. Volta: Dedougou (Nov.) *Bille* 3005! **Niger:** Takieta (Sept.) *P. de Fabrègues* 2394! Chad and Eritrea.

34. **E. lingulata** *W. D. Clayton* in Kew Bull. 20: 269 (1966); Lebrun in Bull. Soc. Bot. Fr. 116: 260 (1969). *E. perbella* of Berhaut, Fl. Sén. ed. 2, 410, not of K. Schum. Loosely tufted annual about 30 cm. high; weedy places.
Sen.: Vélingara (Oct.) *Adam* 18569! Kaffrine (Nov.) *Berhaut* 662! Tambacounda (Apr.) *Berhaut* 2842! Richard-Toll (Oct.) *Berhaut* 2843! **Mali:** Boré (Sept.) *Davey* 7/1957! **Port. G.:** Pirada (Nov.) *Esp. Santo* 3157! **Guin.:** Sériba (Oct.) *Pitot*!

35. **E. welwitschii** *Rendle* Cat. Welw. 2: 234 (1899). *E. densa* De Wild. in Bull. Jard. Bot. Brux. 6: 61 (1919); F.W.T.A., ed 1, 2: 515. Loosely tufted annual 15–45 cm. high; roadsides and old farm land.
Guin.: Timbo (Oct.) *Pobéguin* 1749! 1749*bis*! Baffing R. (Nov.) *Pobéguin* 1832! Mamou (Nov.) *Chev.* 34893! **S.L.:** Musaia (Dec.) *Deighton* 4464! Bumbuna (Oct.) *Thomas* 3210! Waterloo (Dec.) *Deighton* 3281! Sefadu (Dec.) *Deighton* 3565! Mange (Nov.) *Jordan* 820! **Lib.:** Ganta (Nov.) *Harley* 1719! **Ghana:** Amedzofe and Vane (Nov., Dec.) *Morton* A4048! *Rose Innes* GC 31175! 31179! Shiare (Dec.) *Hall & Jenik* CC1080! **N. Nig.:** Rafin Bauna North F.R. (Oct.) *Hepper* 1174! Vom (Oct.) *Gambles* 15! Riyom (Nov.) *Clayton* 1444! Gembu (Nov.) *Magaji & Tuley* 1855! **S. Nig.:** Ogurude (Jan.) *Holland* 286! Ikom (Oct.) *Jones* 6770! Ezillo (Oct.) *Tuley* 943! Nsukka (June) *Ariwaodu* 1431! Also in south tropical Africa, with isolated specimens from Tanzania, Ethiopia, Congo Republic and E. Cameroun.

Additional species.

1. **E. barrelieri** *Daveau* in Journ. de Bot. 8: 289 (1894). Closely resembles *E. minor*, differing in the narrower unequal glumes, longer appressed lemmas and eglandular leaf margins. Recorded from Aïr by De Miré & Gillet in J. Agric. Trop. 3: 732 (1956).
2. **E. trichophora** *Coss. & Dur.* in Bull. Soc. Bot. Fr. 2: 311 (1855). *E. atherstonei* Stapf (1900). Similar to *E. cylindriflora* but perennial. A North and South African species recorded from Aïr by De Miré & Gillet in J. Agric. Trop. 3: 733 (1956).

38. **TRIPOGON** Roem. & Schult., Syst. Veg. 2: 34 (1817); Phillips & Launert in Kew Bull. 25: 301 (1971).

Spikelets 10–12 mm. long, dark green; glumes subequal, sharply pointed, 7–8 mm. long; lemmas 5 mm. long with a stout awn-point 1–2 mm. long .. .. 1. *major*
Spikelets 2·5–4 mm. long, closely appressed to the rhachis; upper glume 2–3 mm. long, the lower half as long; lemmas 1–2 mm. long, minutely mucronate ..2. *minimus*

1. **T. major** *Hook. f.* in J. Linn. Soc. 7: 230 (1864), incl. subsp. *jaegeranus* Gledhill & subsp. *deflexus* Gledhill in Bol. Soc. Brot. 41: 166 (1967). *T. unisetus Pilger* (1932). *T. liebenbergii* C. E. Hubbard (1934). *T. snowdenii* C. E. Hubbard (1934). *T. jaegeranus* A. Camus in J. Agric. Trop. 1: 212 (1954). Densely tufted perennial about 30 cm. high, with narrow involute leaf-blades up to 15 cm. long; mountain grassland.
S.L.: Bintumane Peak 5,000–6,390 ft. (Jan.–May) *Jaeger* 389! *T. S. Jones* 108! 154! Mt. Loma (Oct.) *Jaeger* 7860! Tingi Mts. (Dec.) *Morton & Gledhill* SL 3112! **N. Nig.:** Chappal Waddi *Jackson, Magaji & Tuley* 2027! **W. Cam.:** Mann's Springs (Dec.) *Boughey* GC 12512! Onyanga (Jan.) *Steele* 72! Cam. Mt. 7,000–8,000 ft. (Dec.) *Mann* 2098! *Maitland* 870! 1220! Sudan and Ethiopia to Malawi.
2. **T. minimus** (*A. Rich.*) *Hochst. ex Steud.* Syn. Pl. Glum 1: 301 (1854); A. Chev. in Rev. Bot. Appliq. 14: 181; Berhaut, Fl. Sén. ed. 2, 388; Jacques-Félix in J. Agric. Trop. 2: 521. *Festuca minima* A. Rich., Tent. Fl. Abyss. 2: 436 (1851). Densely tufted perennial about 15 cm. high, with fine convolute leaf-blades 2·5–5 cm. long; shallow soils around rock outcrops.
Maur.: 15°40′ N, 9°30′ W *Rossetti* 61/301! **Sen.:** *Heudelot* 290! **Mali:** Dioura (Sept.) *Farrow* 59! **Iv. C.:** Sifié (Oct.) *Aké Assi* 5754! Séguéla to Vavoua (Oct.) *Adjanohoun* 389*a*! **Ghana:** Bole to Wa (Sept.) *Rose Innes* GC 30179! Damongo Scarp F.R. (Nov.) *Rose Innes* GC 30832! Kwahu Tafo (Dec.) *Adams* 4923! Gradaw, Banda Hills (Oct.) *Rose Innes* GC 31622! Kintampo to Bamboi (Oct.) *Rose Innes* GC 30624! **Niger:** Toukounous *Bartha* 92! **N. Nig.:** Sokoto (Aug.) *Dalz.* 498! Bindawa (Aug.) *Grove* 48! Kufena Hill, Zaria (Sept.) *Clayton* 1321! Anara F.R. (July) *Clayton* 1223! Pansham (Aug.) *Lawlor & Hall* 617! **S. Nig.:** Lagos (Feb.) *W. MacGregor* 232! Igboora (Oct.) *Haines* 331! Fashola Rock, Oyo (Nov.) *Gledhill* 679! Throughout tropical and South Africa. (See Useful Plants.)

39. **TRICHONEURA** Anderss. in Vet. Akad. Handl. Stockh. 1853: 148 (1855); Ekman in Ark. Bot. 11, 9: 1 (1912).

Leaf-sheaths pilose with tubercle-based hairs; blades 3–6 cm. long and up to 4 mm. broad; ligule membranous; inflorescence 8–18 cm. long, with up to 30 ascending spikes 2–5 cm. long; axis angular, scabrid-pubescent; glumes linear-lanceolate, acuminate, 1-nerved, 5–6 mm. long, slightly exceeding the 4–5 florets; lemmas 2–3 mm. long, conspicuously ciliate on the margins, sparsely hispidulous elsewhere, the apex shortly bilobed with an awn-point 0·5 mm. long from the sinus; anthers 0·4 mm. long .. .. *mollis*

T. mollis (*Kunth*) *Ekman* in Ark. Bot. 11, 9: 10 (1912); Berhaut, Fl. Sén. ed. 2, 401. *Leptochloa mollis* Kunth, Rev. Gram. 2: 443 (1831). *L. longiglumis* Hitchc. in Proc. Biol. Soc. Wash. 43: 91 (1930). *Triodia mollis* (Kunth) Dur. & Schinz (1894). *Crossotropis mollis* (Kunth) Stapf in Hook., Ic. Pl. t. 2609 (1899). *Uralepis ciliata* Steud., Syn. Pl. Glum. 1: 247 (1854). *Trichoneura arenaria* (Hochst. & Steud.) Ekman (1912). Annual about 30 cm. high; dry soils.
Maur.: Dahr of Néma *Rossetti* 61/180! 61/197*a*! **Sen.:** *Roger*! Walo (Sept.) *Leprieur* 17! Tivaouane (Dec.) *Berhaut* 1437! **Mali:** Labbézenga (Sept.) *Hagerup* 453! **Niger:** Toukounous *Bartha* 82! **N. Nig.:** Sokoto (Oct.) *Moiser* 135!

40. **DINEBRA** Jacq., Fragm. 77 (1809).

Leaf-blades flat, broadly linear, up to 15 cm. long and 1 cm. broad; ligule membranous; inflorescence stiff, up to 25 cm. long; racemes 0·5–6 cm. long, at first erect, at length

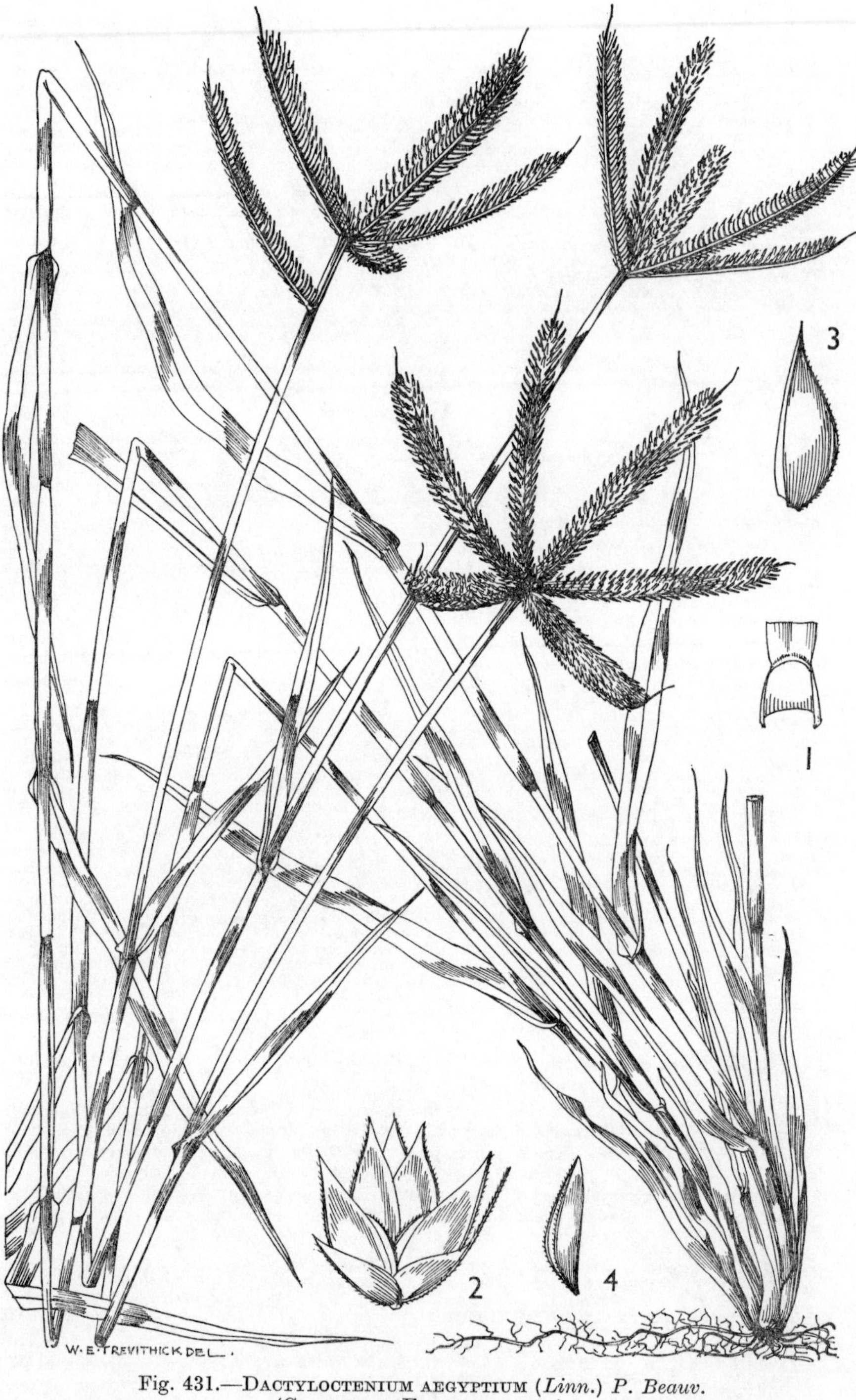

Fig. 431.—DACTYLOCTENIUM AEGYPTIUM (*Linn.*) *P. Beauv.* (GRAMINEAE-ERAGROSTIDEAE).

1, ligule. 2, spikelet. 3, lemma. 4, palea.

deflexed; glumes 6–8 mm. long, long attenuate to an awn-like tip, exceeding and concealing 2–3 florets; lemmas 2–3 mm. long, acute .. .. .. *retroflexa*

D. retroflexa (*Vahl*) *Panz.* in Denkschr. Acad. Wiss. München 270, t. 12 (1814); Berhaut, Fl. Sén. ed. 2, 401. *Cynosurus retroflexus* Vahl, Symb. Bot. 2: 20 (1791). *Dinebra arabica* Jacq., Fragm. 77 (1807); Chev. Bot. 746, and in Rev. Bot. Appliq. 14: 130. *Leptochloa arabica* (Jacq.) Kunth (1829). Coarse annual up to 60 cm. high; on dry soils.
Sen.: *Roger* 11! 32! **N. Nig.:** Marte (Nov.) *Golding* 40! Kalkala *Gwynn* 132*a*! Logomani (Nov.) *Johnston* N89! Ngala, Dikwa (Mar.) *Tuley* 122! Talasse, Bauchi Prov. (Oct.) *De Leeuw & Magaji* 1990! Throughout tropical and S. Africa, extending eastwards through Egypt, Iraq and India. (See Useful Plants.)

41. DACTYLOCTENIUM Willd., Enum. Hort. Berol. 1029 (1809).

Leaf-blades broadly linear, up to 20 cm. long and 7 mm. broad, tapered to a fine point, more or less pilose with tubercle based hairs; spikes 2–6 (rarely solitary), 2–5 cm. long, ending in a point 2–5 mm. long; spikelets 3 mm. long, 3–5-flowered, glabrous; upper glume with a short oblique awn; lemmas ovate in side view, acuminate or with a short awn-point; grain coarsely rugose .. .. *aegyptium*

D. aegyptium (*Linn.*) *P. Beauv.* Ess. Agrost. t. 15, f. 2 (1812); A. Chev. in Rev. Bot. Appliq. 14: 129; Berhaut, Fl. Sén. ed. 2, 389, 398; De Miré & Gillett in J. Agric. Trop. 3: 730. *Cynosurus aegyptius* Linn., Sp. Pl. 72 (1753). *Chloris guineensis* Schumach., Beskr. Guin. Pl. 75 (1827). *Dactyloctenium mucronatum* (Michx.) Willd. (1809). *D. prostratum* Willd. (1809)—A. Chev. in Rev. Bot. Appliq. 14: 130. *D. aegyptiacum* Willd. (1809)—Chev. Bot. 747. *Eleusine aegyptia* (Linn.) Roberty in Bull. I.F.A.N. 17: 50 (1955). Loosely tufted or stoloniferous annual, prostrate or ascending, 15–60 cm. high; a weed, often on trampled pathsides.
Maur.: Gujaf *Rossetti* 61/212! **Sen.:** *Roger* 11! Dakar (May) *Baldwin* 5733! Bargny (Aug.) *Adam* 14926*b*! Djembering (Sept.) *Broadbent* 144! Oussouye (Sept.) *Broadbent* 109! **Gam.:** *Saunders* 24! *Ruxton* 16! Yundum (Aug.) *Corbett* 7! *Frith* 127! Genieri (July) *Fox* 129! **Mali:** Timbuktu (Aug.) *Hagerup* 250! Kara (Aug.) *Farrow* 119! Kersani (Aug.) *Farrow* 120! **Port. G.:** Bolama *Rodriguez & Carvalho*! **Guin.:** Kouroussa (July) *Pobéguin* 477! **S.L.:** Rokupr (Sept.) *Jordan* 111! Freetown (Oct.) *Deighton* 2156! Aberdeen (Nov.) *Small* 793! Magbile (Dec.) *Thomas* 5961! Benduma *Deighton* 1969! **Lib.:** Rocktown (Mar.) *Baldwin* 11638! Zwedru (Aug.) *Baldwin* 7043! Cape Palmas (June) *Baldwin* 5967! **Iv. C.:** Lamto (May) *Aké Assi* 10602! Divo F.R. (Aug.) *Boughey* GC 14648! Abidjan (June) *Adjanohoun* 424*a*! Grand Bassam (Mar.) *Leeuwenberg* 3145! Tiebissou (May) *Leeuwenberg* 4262! **U. Volta:** Banfora (June) *Leeuwenberg* 4431! Ouagadougou (July, Sept.) *Mordant* 349! *Scholz* 5! **Ghana:** Navrongo (May) *Vigne* FH 4547! Tamale (Mar.) *Williams* 604! K'bong (Mar.) *Johnson* 1014! Winneba (Apr.) *Morton* A1986! Tema (Jan.) *Hall* 2160! **Togo Rep.:** Atakpame to Sokode (Oct.) *Rose Innes* GC 31380! Palime *Stage* 78! **Dah.:** Cotonou (Apr.) *Chev.* 23472! **Niger:** Toukounous *Bartha* 7! Agades (Feb.) *Bradley* 20! **N. Nig.:** Kworre (Sept.) *Palmer* 2! Biu (Sept.) *Kennedy* FHI 8014! Naraguta (July) *Lely* 416! Nupe *Barter* 1388! Ilorin (July) *Ward* 83! **S. Nig.:** Fashola *Thorold*! Ibadan (Nov.) *Meikle* 657! Ijokodo (Sept.) *Hambler* 392! Lagos I. (Jan.) *Killick* 88! Throughout the Old World tropics, introduced into America. (See Useful Plants.)

42. ACRACHNE Wight & Arn. ex Chiov. in Ann. Ist. Bot. Roma 8: 361 (1908).

Racemes 3–10 cm. long, in several whorls along a central axis up to 5 cm. long or more; spikelets compressed, mostly 5–7 mm. long, several flowered; glumes narrowly lanceolate, aristate-acuminate; lemmas glabrous, ovate in profile, truncate-emarginate at the tip with the midrib excurrent into a short awn-point; ligule hyaline and densely ciliate .. .. *racemosa*

A. racemosa (*Heyne ex Roem. & Schult.*) *Ohwi* in Bull. Tokyo Sci. Mus. 18: 1 (1947). *Eleusine racemosa* Heyne ex Roem. & Schult., Syst. Veg. 2: 583 (1817). *E. verticillata* Roxb., Fl. Ind. 1: 346 (1820); F.W.T.A., ed. 1, 2: 518; Berhaut, Fl. Sén. ed. 2, 299. *Leptochloa racemosa* (Heyne ex Roem. & Schult.) Kunth (1829). *L. verticillata* (Roxb.) Kunth (1829). *L. schimperana* Hochst. (1855). *Acrachne verticillata* (Roxb.) Chiov. (1908). Annual 30–60 cm. high; disturbed soils.
Sen.: Banks of Senegal R. *Heudelot* 309! **Mali:** Timbuktu (Aug.) *Hagerup* 282! Tropics of the Old World, and in the West Indies.

43. ELEUSINE Gaertn., Fruct. 1: 7, t. 1 (1789).

Spikes 3–6 mm. broad, slender, straight; grain concealed within the floret; lemma 2·7–3(–3·5) mm. long; spikelets 4–5·5 mm. long; rhachis 0·5–1 mm. broad 1. *indica*
Spikes 9–13 mm. broad, stout, incurved or rarely straight; grain frequently exposed 2. *coracana*

1. **E. indica** (*Linn.*) *Gaertn.* Fruct. 1: 8 (1789); Chev. Bot. 747, and in Rev. Bot. Appliq. 14: 129; Berhaut, Fl. Sén. ed. 2, 399. *Cynosurus indicus* Linn., Sp. Pl. 72 (1753). *Eleusine glabra* Schumach., Beskr. Guin. Pl. 53 (1827). Eerect annual up to 60 cm. high; common wayside weed.
Sen.: Dakar (Apr.) *Clayton* 4002! 4072! Cape Verde (Jan., Nov.) *Naegelé* 36! 1708! Oussouye to Ziguinchor (Sept.) *Broadbent* 103! **Gam.:** *Hancock* 27! *Ruxton* 21! *Saunders* 25! **Mali:** Kayes (Oct.) *Hepper* 3736! Pétal, Macina (Sept.) *Davey* 50! **Guin.:** Timbo (Oct.) *Pobéguin* 1700! Kouroussa (July) *Pobéguin* 544! Kankan (Mar.) *Chev.* 563! Kouria *Chev.* 15078! **S.L.:** King Tom *Deighton* 2004! Rokupr (July) *Jordan* 455! Kamasigi to Kanike (Oct.) *Glanville* 95! Makute (June) *Thomas* 457! Bonthe (Nov.) *Deighton* 2268! **Lib.:** Monrovia (May) *Baldwin* 5823! Paynesville (Sept.) *van Dillewijn* 91! Nanakru (Mar.) *Baldwin* 11576! Harper (June) *Baldwin* 5971! Gbanga (Sept.) *Linder* 510! **Iv. C.:** Sassandra (Feb.) *Leeuwenberg* 2898! Dabou (Nov.) *Leeuwenberg* 1945! Cocody (June) *Adjanohoun* 426*a*! Abidjan (Jan.) *Hédin*! **U. Volta:** Ouagadougou (Sept.) *Scholz* 6! **Ghana:** Tamale *Williams* 3! Kintampo (Dec.) *Vigne* FH 3198! Aburi (Sept.) *Johnson* 834! Cape Coast (July) *T. Vogel* 55! Achimota (Apr.) *Ankrah* GC 20134! **Togo Rep.:** Palime (Mar.) *Stage* 3! **Dah.:** Cotonou to Allada (Oct.) *Risopoulos* 1226! **Niger:** Gaya *White* 26! **N. Nig.:** Zaria Dist. *C. B. Taylor* 31! Samaru (July) *Clayton* 1238! Naraguta F.R., Jos (July) *Lawlor & Hall* FHI 46546! Ilorin (July) *Ward* 86! Nupe *Barter* 1386! **S. Nig.:** Ibadan (Nov.) *Clayton* 561! Lagos (Feb.) *W. MacGregor* 45! Benin (Dec.) *Brenan* 8633! Obudu Plateau, 5,000 ft. (Dec.) *Tuley* 45! Opobo (Apr.) *Jeffreys* 11! **W. Cam.:** Bambui Farm (May) *Pedder* 27! Buea (Nov., Jan.) *Migeod* 94! *Maitland* 350! Bambili (Feb.) *Bauer* 6! Nkambe (Feb.) *Charter* FHI 36988! **F. Po:** *T. Vogel*! Throughout the tropics. (See Useful Plants.)

Fig. 432.—ELEUSINE INDICA (*Linn.*) *Gaertn.* (GRAMINEAE-ERAGROSTIDEAE).

A, junction of sheath and blade of leaf. B, ligule. C, portion of spike. D, spikelet. D_1, lower glume. D_2, upper glume. D_3, lemma. E, flower. F, grain.

2. **E. coracana** (*Linn.*) *Gaertn.* Fruct. 1: 8 (1789); Chev. Bot. 747, and in Rev. Bot. Appliq. 14: 129; Berhaut, Fl. Sén. ed. 2, 399; Portères in Bull. I.F.A.N. 13: 1 (1951). *Cynosurus coracanus* Linn., Syst. Nat. ed. 10, 2: 875 (1759). A crop plant cultivated in Nigeria, particularly Jos Plateau, N. Cameroons and Lake Chad area. Also in E. Africa from Ethiopia to Natal and in India.

[Note: *E. africana* J. Kennedy-O'Byrne in Kew Bull. 12: 65 (1957), is a tetraploid species resembling diploid *E. indica*, with the lemmas (3·5–)4–5 mm. long, the spikelets 6–7·5 mm. broad, and the rhachis 1·5–2 mm. broad. Although some plants from our area encroach upon these dimensions, typical specimens of *E. africana* seem to be absent from our area.]

44. COELACHYRUM Hochst. & Nees in Linnaea 16: 221 (1842).

Leaf-blades 2–6 cm. long and up to 5 mm. broad; inflorescence 4–5 cm. long, 3–5 branched; spikelets 3·5–4 mm. long, 6–8-flowered; lemmas broadly obtuse, yellowish green with 3 prominent dark green nerves, glabrous or hispidulous; grain more or less discoid, radiately grooved on one side and deeply hollowed on the other *brevifolium*

C. brevifolium *Hochst. & Nees* in Linnaea 16: 221 (1842). *C. oligobrachiatum* A. Camus in Bull. Mus. Hist. Nat., sér. 2, 3: 546 (1931); F.W.T.A., ed. 1, 2: 518; De Miré & Gillet in J. Agric. Trop. 3: 701. *Eleusine brevifolia* (Hochst. & Nees) Steud. (1854). *Eragrostis coelachyrum* Benth. (1881). Annual, 5–45 cm. high, the culms ascending from a loose tuft; a desert grass.
Mali: Tin Zaoutène (Sept.) *Rossetti* 191/59! **N. Nig.:** Arege (Oct.) *Golding* 37! North Africa from Mauretania to Egypt and the Sudan, and in Arabia.

45. BEWSIA Goossens in S. Afr. J. Sci. 37: 183 (1941).

Spikelets 2–4-flowered; glumes lanceolate, 2·5–8 mm. long; lemmas 4–6 mm. long, pubescent on either side of the mid-nerve below, obtuse or notched at the tip; awn up to 8 mm. long, arising about ⅓ the length of the lemma from its tip; grain terete *biflora*

B. biflora (*Hack.*) *Goossens* in S. Afr. J. Sci. 37: 184 (1941). *Diplachne biflora* Hack. in Bull. Herb. Boiss. 3: 387 (1895). Tufted perennial 45–90 cm. high, the spikelets reddish brown; in woodland.
Dah.: Natitingou to Tanguieta (Aug.) *Adjanohoun* 461! **N. Nig.:** Jos Plateau (Sept.) *Tuley* 1643! Tanzania to South Africa.

46. POGONARTHRIA Stapf in Fl. Capensis 7: 316 (1898).

Inflorescence linear, up to 20 cm. long and 3 cm. broad; racemes 1–2 cm. long, curved, divergent; spikelets 4–10 mm. long, 5–12-flowered, laterally compressed; glumes lanceolate, the lower 1 mm. long, the upper 1·5–2 mm.; lemmas 2–2·5 mm. long, keeled .. .. *squarrosa*

P. squarrosa (*Licht. ex Roem. & Schult.*) *Pilger* in Notizbl. Bot. Gart. Berl. 5: 149 (1910). *Poa squarrosa* Licht. ex Roem. & Schult., Syst. Veg. 2: 553 (1817). Erect perennial 45–90 cm. high; bare stony and disturbed soils.
Ghana: Nungua (June) *Ankrah* GC 20152! **Dah.:** Agouagon (May) *Chev.* 23547! **N. Nig.:** Gombe to Bauchi (Dec.) *Magaji* MG 100! Sara Hills (Oct.) *Hepper* 1127! Isanlu Makatu (Jan.) *Clayton* 710! Yola to Biu (Nov.) *Magaji & Tuley* 1846! **S. Nig.:** Oyo (Nov.) *Clayton* 562! Throughout tropical and S. Africa.

47. LEPTOCHLOA P. Beauv., Ess. Agrost. 71 (1812).

Leaf-blades broad, oblong-lanceolate, 4–12 cm. long and 8–18 mm. broad, thin, flaccid, abruptly acute; glumes 2 mm. long, lanceolate, acuminate, exceeding the single floret; lemma 1·5 mm. long, acute .. .. 1. *uniflora*

Leaf-blades linear, 2–8 mm. broad, long-attenuate at the tip; spikelets 2–6-flowered; lemmas obtuse to emarginate:

Racemes flexuous; culms typically 10–20-noded, hard, decumbent and rooting at the lower nodes; spikelets 2·5–3 mm. long, 3–6-flowered; lower glume narrowly lanceolate, 0·75–1 mm. long; upper glume narrowly oblong, 1·2–1·7 mm. long, obtuse or emarginate, mucronate; lemmas 1·5–2 mm. long, obscurely ciliolate on the nerves 2. *caerulescens*

Racemes straight; culms about 4-noded, erect:

Upper glume linear to narrowly lanceolate, acute, 1·5–1·8 mm. long; lower glume 1·2–1·5 mm. long; spikelets 2–3 mm. long, 2–5-flowered; lemmas 1–1·5 mm. long, glabrous .. .. 3. *filiformis*

Upper glume narrowly oblong, obtuse and mucronate, 1–1·3 mm. long; lower glume 0·8–1 mm. long; spikelets 1·5–2 mm. long, 2–3-flowered; lemmas 1 mm. long, minutely mealy hairy .. .. 4. *panicea*

1. **L. uniflora** *Hochst. ex A. Rich.* Tent. Fl. Abyss. 2: 409 (1851). *Craspedorhachis uniflora* (Hochst. ex A. Rich.) Chippendall in Meredith, Grasses & Pastures of S. Africa 205 (1955). Erect or trailing annual, 30–60 cm. high; pathsides and fallow land.
Ghana: New Tafo (Sept.) *Lovi* WACRI 3909! Ampam (Dec.) *Adams* 5030! Aburi (Mar., June, Sept.) *Deighton* 613! *Ankrah* GC 20050! *Johnson* 833! **S. Nig.:** *Thomas* 1764! Ibadan (May) *Townrow* 8! Idanre (Jan.) *Brenan & E. W. Jones* 8731! Aguku Dist. *Thomas* 1000! 1045! Throughout tropical and S. Africa, and in India and Ceylon.

2. **L. caerulescens** *Steud.* Syn. Pl. Glum. 1: 209 (1854); Berhaut, Fl. Sén. ed. 2, 401. Annual up to 1 m. high, sometimes erect but usually the culms decumbent and stolon-like at the base; near water.
Sen.: *Roger* 24! **Mali:** El Oualadji (Mar.) *Chev.* 43897! Koli-Koli Stream (July) *Lean* 50! Tondigamé (Oct.) *Farrow* 24! Labbézenga (Sept.) *Hagerup* 444! **Port. G.:** Nabijã (Dec.) *Guerra* 512! **S.L.:** Rokupr (Oct.) *Jordan* 126! Makump (Sept.) *Glanville* 353! Batkanu (Apr.) *Thomas* 11! Njala (May) *Deighton* 1731!

L. Mabessi (Apr.) *Deighton* 4755! **Iv. C.:** Vavoua (May) *Adjanohoun* 268a! Soubré (Nov.) *Leeuwenberg* 3223! *de Wilde* 3327! Gansé to Dabakala (June) *Aké Assi* 10303! Réserve de Bouna, Kakpin (June) *Aké Assi* 10255! **U. Volta:** Namenda (June) *Scholz* 114! **Ghana:** Senchi (Apr.) *Irvine* 2875! Kpandai (Sept.) *Ankrah* GC 20232! Bamboi to Kintampo (Oct.) *Rose Innes* GC 30626! Keta (Quittah) (Sept.) *Akpabla* 332! Nungua (Apr.) *Rose Innes* GC 30009! **N. Nig.:** Arege (Oct.) *Golding* 33! Sherifuri (June) *Thornewill* 172! Nupe *Barter* 160! Jebba I. (Mar.) *Meikle* 1282! Lokoja I. (Mar.) *Maggs* OK48! **S. Nig.:** Omo F.R. (Aug.) *Okeke* FHI 30131! Lagos *Dawodu* 214! Inoma Anambra F.R. (Feb.) *Jones* 2763! Afikpo (Apr.) *Jones* 1416! (FHI 7407!) Amasama F.R., Brass Dist. (Dec.) *Daramola* FHI 46938! Extends eastwards to the Sudan, and southwards through the Congo to Angola and Zambia.

3. **L. filiformis** (*Lam.*) *P. Beauv.* Ess. Agrost. 71, 161, 166 (1812). *Festuca filiformis* Lam., Tab. Encycl. 1: 191 (1791). Erect annual 30–60 cm. high; weedy places.
Ghana: Ashanti (July) *Akpabla* 3856! **N. Nig.:** *Ward* 28! **S. Nig.:** Ibadan *West*! Moor Plantation, Ibadan (Mar.) *Clayton* 493! A species of warm temperate and tropical America, introduced into W. Africa and Angola.

4. **L. panicea** (*Retz.*) *Ohwi* in Bot. Mag. Tokyo 55: 311 (1941). *Poa panicea* Retz., Obs. Bot. 3: 11 (1783). *Leptochloa chinensis* of F.W.T.A., ed. 1, 2: 517; of Chev. Bot. 748, and Rev. Bot. Appliq. 14: 121, not of (Linn.) Nees (1824). Erect or ascending annual about 60 cm. high with slender racemes; old farmland and weedy places.
Ghana: Takoradi (Mar.) *Deighton* 559! Cape Coast (July) *Hall* 1515! Kpong (Dec.) *Gyadu* GC 20530! Achimota *Akpabla*! Nungua (Mar.) *Clayton* 950! A species from tropical Asia; perhaps introduced, but now widely distributed in tropical Africa.

48. DIPLACHNE P. Beauv., Ess. Agrost. 80 (1812).

Leaf-blades narrow, linear, tapering to a filiform point; racemes erect, with slightly flexuous branches up to 17 cm. long; glumes lanceolate, the lower 1–2 mm., the upper 2·5–4 mm. long; lemmas 2–3 mm. long, narrowly oblong, scarcely keeled, silky ciliolate on the lateral nerves .. .. *fusca*

D. fusca (*Linn.*) *P. Beauv. ex Stapf* in Fl. Capensis 7: 591 (1900); Chev. Bot. 748, and in Rev. Bot. Appliq. 14: 121; Berhaut, Fl. Sén. ed. 2, 407; De Miré & Gillet in J. Agric. Trop. 3: 731. *Festuca fusca* Linn., Sp. Pl. ed. 2, 109 (1762). *Poa senegalensis* Desf., Cat. Pl. Paris ed. 3: 21, 387 (1829), without description. *Uralepis alba* Steud., Syn. Pl. Glum. 1: 248 (1854). Tufted perennial with ascending culms 60–150 cm. high; spikelets olive green, becoming pallid; in swamps.
Sen.: *Perrottet* 926! *Heudelot* 499! Nouk (Jan.) *Roger* 59! Walo (Sept.) *Perrottet* 18! Seleki (Jan.) *Chev.* 2404! **Niger:** Zinder (Oct.) *Hagerup* 576! Agades (Dec.) *de Wailly* 4940! Maradi (Aug.) *Hall* 392! **N. Nig.:** Lake Chad (Dec.–June) *Elliott* 155! *Jackson* 2579! (FHI 59163!) *Golding* 85! *Gwynn* 129! *De Leeuw* 1896! Throughout Africa, extending eastwards through the Orient and India, to China and Australia.

49. CTENIUM Panz. in Denkschr. Acad. Muench. 1813: 288, t. 13 (1814), *nom. cons.*; Pilger in Notizbl. Bot. Gart. Berl. 9: 114 (1924); Clayton in Kew Bull. 16: 471 (1963).

Spikes solitary:
Lemmas of both lower florets and palea of fertile floret with a line of yellow glands along the margin; second and third lemmas with 2–8 long bristles from side nerves near apex; annual, with flat leaf-blades 1–3 mm. broad; spike 10–30 cm. long, the rhachis shortly ciliate along the margins; lateral awn of the upper glume shorter than glume, rarely vestigial .. .. 1. *elegans*
Lemmas without glands or bristles:
Rhachis of spike villous along the margins (hairs 1–2 mm. long); annual, with flat leaf-blades up to 2 mm. broad; sterile lemmas subequal, keels glabrous, the ciliate margins of the lower forming a short, very dense, crest; fertile lemma 3·5–4 mm. long; spike 2–10 cm. long, coiled in a flat spiral.. .. .. .. 2. *villosum*
Rhachis of spike shortly pubescent along the margins; perennial with narrow convolute leaf-blades; sterile lemmas unequal, the lower 1·5–2 mm. the upper 2–2·5 mm. long, margins and sometimes keels ciliate but not crested; fertile lemma 2·5–3 mm. long; spike (5–)7–20(–30) cm. long, almost straight or coiled in a lax corkscrew spiral .. .. 3. *newtonii*
Spikes 2–5, digitate (rarely solitary):
Upper glume 6·5–8 mm. long:
Back of rhachis glabrous, the edges with a ciliate fringe; first lemma irregularly 3-toothed; second lemma entire, glabrous; margins of fertile lemma long ciliate 4. *canescens*
Back of rhachis pubescent; sterile lemmas bilobed, the margins ciliate 5. *ledermannii*
Upper glume 3–4 mm. long; first lemma vestigial, or represented by an awn; second lemma 1·5–2 mm. long; fertile floret with entire lemma and 3-toothed palea 6. *sesquiflorum*.

1. **C. elegans** *Kunth* Rev. Gram. 1: 295 (1830); Pilger in Notizbl. Bot. Gart. Berl. 9: 115; A. Chev. in Rev Bot. Appliq. 14: 128 (incl. var. *longispicatum*); Berhaut, Fl. Sén. ed. 2, 390. *Chloris elegans* (Kunth) Roberty in Bull. I.F.A.N. sér. A, 17: 51 (1955), not of Kunth (1816). *Ctenium serpentinum* Steud., Syn. Pl. Glum. 1: 202 (1854). Up to about 120 cm. high, with aromatic leaves; on dry sandy soils.
Sen.: *Heudelot* 143! Kaolack (Nov.) *Chev.* 28453! **Gam.:** (July) *Brooks* 9! *Saunders* 79! **Mali:** Koréra (Nov.) *Rossetti* 59/217! **Port. G.:** Pirada, Gabu (Nov.) *Esp. Santo* 3156! **U. Volta**: Samandeni (Nov.) *Kmoch* 176! Yonde (Nov.) *Scholz* 121! **Ghana:** *Beal* 16! Tamale (Dec.) *Williams* 462! Nakpanduri to Bunkpurugu (Oct.) *Rose Innes* GC 30746! Navrongo to Naga (Nov.) *Rose Innes* GC 32525! Nakpanduri to Gambaga (Nov.) *Rose Innes* GC 32513! **Niger:** Niamey (Oct.) *Hagerup* 480! 21 miles N. of Zinder (Nov.) *Money-Kyrle* Fe 18! Wadi Dollol *White* 27! Toukounous *Bartha* 65! **N. Nig.:** Zaria (Nov.) *Dandaura* FHI 7001! Sokoto (Sept.) *Dalz.* 505! Katsina (Oct.) *Clayton* 1342! Katagum *Dalz.* 247! Maiduguri (Oct.) *Johnston* W32! Also in the Sudan. (see Useful Plants.)

2. **C. villosum** *Berhaut* in Mém. Soc. Bot. Fr. 1953–54: 10 (1954), and Fl. Sén. ed. 2, 390. Up to about 45 cm high; on shallow soils.
Sen.: Ouassadou (Sept.) *Berhaut* 647! Tambacounda (Sept.) *Berhaut* 3188! **Mali:** Sikasso to Bobo-Dioulasso (Sept.) *Adam* 15102! Ségou *Lécard* 258! **Guin.:** Youkounkoun (Oct.) *Pitot*! **Ghana:** Yendi to Gambaga (Sept.) *Ankrah* GC 20263! Yendi to Chereponi (Sept.) *Rose Innes* GC 30472! Navrongo to Lawra (Oct.) *Rose Innes* GC 30279! Bawku to Nakpanduri (Oct.) *Ankrah* GC 20494! Bawku to Bolgatanga (Oct.) *Rose Innes* GC 31443! **Togo Rep.:** Kande (Oct.) *Rose Innes* GC 31411!

3. **C. newtonii** *Hack.* in Bol. Soc. Brot. 5: 229 (1887); Pilger l.c. 115 (incl. vars. *majusculum* Pilger & *productum* Pilger); Berhaut, Fl. Sén. ed. 2, 390; Clayton in Kew Bull. 16: 472; Aké Assi, Contrib. 2: 280. *C. schweinfurthii* Pilger l.c. 116 (1924). *C. elegans* of Chev. Bot. 745, and of Rev. Bot. Appliq. 14: 128, partly (*Chev.* 2333). *C. camposum* A. Chev. in Rev. Bot. Appliq. 14: 128 (1934). Up to 1·2 m. high, with aromatic leaves; usually on poor or shallow soils. A very variable grass, the types of *C. newtonii* and *C. schweinfurthii* representing two extremes which appear to be linked by a full range of intermediates.
Sen.: Kédougou (Nov.) *Adam* 20013! Nétéboulou (Oct.) *Diallo* 548! **Mali:** Koulikoro (Oct.) *Chev.* 2333! Sikasso (Sept.) *Adam* 15433! Banguita (Sept.) *Farrow* 118! **Port G.:** Barro to Sedengal (Nov.) *Esp. Santo* 3125! **Guin.:** Timbo (Oct.) *Pobéguin* 1769! Kouroussa (Aug.) *Pobéguin* 496! Kindia *Jac.-Fél.* 175! Friguiagbé (Aug.) *Chillou* 603! Dalaba (Oct.) *Adames* 391! **S.L.:** Matamba to Manyakoi (Oct.) *Glanville* 93! Tabe (Sept.) *Deighton* 3036! Bumbuna (Oct.) *Thomas* 3930! Kono, Tingi Mts. (Dec.) *Morton & Gledhill* SL 3168! **Lib.:** Duport (Oct.) *Barker* 1447! **Iv. C.:** Tiebiessou (Oct.) *Chev.* 34286! Toumodi to Yamounokro (Oct.) *Adjanohoun* 325a! Sokrogbo to Ahouaho (Sept.) *Adjanohoun* 263a! Sifié (Oct.) *Adjanohoun* 262a! Bouna (Oct.) *Rose Innes* GC 31545! **U. Volta:** Po to Leo (Oct.) *Rose Innes* GC 31483! Leo to Ouassa (Oct.) *Rose Innes* GC 31498! Doutié (Sept.) *Scholz* 2b! Koumbili (Oct.) *Scholz* 179! **Ghana:** Accra Plains (Oct.) *Ankrah* GC 20087! *Brown* 426! Bole to Sawla (Oct.) *Rose Innes* GC 30665! Red Volta bridge, Bolgatanga to Bawku (Oct.) *Rose Innes* GC 30247! Togo Plateau (Nov.) *Morton* A3599! **Togo Rep.:** Atakpamé to Sokodé (Oct.) *Rose Innes* GC 31384! Badou to Atakpamé (Sept.) *Rose Innes* GC 31361! Sansanné Mango (Oct.) *Rose Innes* GC 31420! Palimé *Stage* 52! **Dah.:** Zumbodji (Sept.) *Newton* 7! Porto Novo (Dec.) *Froment* 1078! **N. Nig.:** Ilorin (Oct.) *Ward* 0054! Wuya ferry (Sept.) *Clayton* 359! Jebba *Barter*! Naraguta (Oct.) *Kennedy* FHI 8048! Anara F.R., Zaria (Oct.) *Keay* FHI 5481! Vogel Peak area (Nov.) *Hepper* 1365! **S. Nig.:** Lagos (Oct.) *Dalz.* 1120! Olokemeji (Nov.) *Onochie* FHI 8114! Sobo Plains (Sept.) *Butler-Cole* 3! Nanka, Awka (Nov.) *Keay* FHI 21527! Ogoja *Rosevear* 14/30a! **W. Cam.:** Bamenda (Mar.) *Brunt* 1049! **F. Po:** *Milne*! Also in Congo, Angola, Uganda and Sudan.

4. **C. canescens** *Benth.* in Fl. Nigrit. 566 (1849); Pilger l.c. 118; A. Chev. in Rev. Bot. Appliq. 14: 128. *Chloris pulchra* Schumach. (1827), illegitimate name. Perennial up to 1·2 m. high, with characteristic short curved rhizomes covered in white scales; on sandy soils.
Iv. C.: Salaye to Wolobidé (Dec.) *Adjanohoun* 438a! Bouna to Sayé (Nov.) *Aké Assi* 9305! **U. Volta:** Kapori (Oct.) *Scholz* 173! **Ghana:** Accra Plains (Sept.) *Ankrah* GC 20026! Kete Krachi to Attebubu (Oct.) *Rose Innes* GC 30606! Sunyani to Wenchi (Sept.) *Rose Innes* GC 30148! Tamale to Yendi (Sept.) *Ankrah* GC 20248! 33 miles S. of Lawra (Oct.) *Rose Innes* GC 30303! **Togo Rep.:** Sansanne Mango (Oct.) *Rose Innes* GC 31419! **Dah.:** Zumbodji (Sept.) *Newton* 8! Cocotomey (Jan.) *Froment* 1118! Cotonou (Dec.) *Froment* 1017! **N. Nig.:** Kontagora (Dec.) *Hubbard* S581! Lokoja *Barter*! Gindiri, Jos Plateau (Oct.) *Hepper* 1137! Minna (Dec.) *Keay* FHI 37301! Biliri, Bauchi Prov. (Oct.) *Magaji* 1987!

5. **C. ledermannii** *Pilger* l.c. 119 (1924). Perennial 60–90 cm. high, with dark green paired spikes.
N. Nig.: Chappal Waddi (Nov.) *Jackson, Magaji & Tuley* 2053! **S. Nig.:** Obudu Plateau (Feb.) *Tuley* 492! 1019! **W. Cam.:** Bayango, Bamenda (Sept.) *Ujor* FHI 30215! Also in E. Cameroun and C. Afr. Rep.

6. **C. sesquiflorum** *W.D. Clayton* in Kew Bull. 14: 239 (1960). About 30 cm. high, decumbent and rooting at the lower nodes.
Guin.: Ditinn to Diaguissa (Apr.) *Chev.* 12860! Ditinn (Apr.) *Pitot*!

50. TETRAPOGON Desf., Fl. Atlant. 2: 388, t. 255 (1799).

Leaf-sheaths keeled, flabellate; ligule very short, ciliolate; blades linear, glaucous, about 6 cm. long, ciliate with long hairs towards the base; upper sheath dilated-spathaceous, with or without a short blade, enclosing the inflorescence; raceme solitary, one-sided, 4–5 cm. long; spikelets densely overlapping, truncate, with about 4 fertile florets, and 4 sterile florets above; glumes membranous, acuminate, the upper 7–9 mm. long; lemmas broadly ovate, villous, awned from just below the bilobed tip, the margins with a membranous wing above .. .. *cenchriformis*

T. cenchriformis (*A. Rich.*) *W.D. Clayton* in Kew Bull. 16: 250 (1962). *Lepidopironia cenchriformis* A. Rich., Tent. Fl. Abyss. 2: 442, t. 101 (1851). *Chloris cenchriformis* (A. Rich.) Baill. (1893). *C. savatieri* Baill. (1893). *C. spathacea* Hochst. ex Steud. (1854). *Tetrapogon spathaceus* (Hochst. ex Steud.) Hack. ex Dur. & Schinz (1895)—F.W.T.A., ed. 1, 2: 520; Chev. Bot. 746, and in Rev. Bot. Appliq. 14: 131; Berhaut, Fl. Sén. ed. 2, 422; De Miré & Gillet in J. Agric. Trop. 3: 736. Annual, 10–60 cm. high; dry sandy or gravelly soils.
Maur.: El Beyed *Rossetti* 61/325! **Sen.:** Dahra (Nov.) *Raynal* 7642 *bis*. **Mali:** Tabankort (Jan.) *Chev.*! Boulel (Oct.) *Davey* 055! Dorey (Aug.) *Rossetti* 59/125! **Niger:** Tanout (Apr.) *Guile*! Also in Cape Verde Islands, and eastwards to Kenya and Arabia. (See Useful Plants.)

51. CHLORIS Sw., Prod. Veg. Ind. Occ. 25 (1788).

Lemma shortly awned, awn less than $1\frac{1}{2}$ times as long as lemma:
 Lemmas pilose with hairs 2 mm. long:
  Spikes 1–4 (usually 2), 4–7 cm. long, embraced below by the spathaceous upper leaf-sheath; upper glume 6 mm. long; lemma boat-shaped, dark brown, with an awn 4·5 mm. long from near the tip .. .. 1. *lampropaia*
  Spikes 15–30, 10–15 cm. long, digitate or in tiers, the inflorescence silky hairy; upper glume 3 mm. long; lemma linear-lanceolate, pilose on the nerves with white hairs 2. *robusta*
 Lemmas shortly ciliate:
  Spikelet conspicuously cuneate; upper lemmas 2, truncate, empty; awns 2, the longest about 3 mm. long .. .. 3. *pilosa*
  Spikelet not cuneate; upper lemmas 1 or 2, subacute in profile, accompanied by paleas; rhachilla terminating in a little clavate appendage .. .. 4. *gayana*

Lemma long awned, awn over twice as long as lemma:
Spikelets (1–)2-awned:
Lemma nearly glabrous; leaf-blades abruptly narrowed to a sub-acute point 5. *pycnothrix*
Lemma long pilose on margins above (hairs 2 mm. long); leaf-blades gradually tapering to an acute point .. .. 6. *virgata*
Spikelets 3–6-awned:
Spikelets 3-awned, awns 5–7 mm. long; fertile lemma acute, ciliate towards apex (hairs 2 mm. long), without glands; sterile florets truncate .. .. 7. *barbata*
Spikelets 4–6-awned; fertile lemma lanceolate, shortly ciliate above, with a row of glands down each side .. .. 8. *prieurii*

1. **C. lampropraria** *Stapf* in Mém. Soc. Bot. Fr. 2, 8: 220 (1912); Chev. Bot. 746, and in Rev. Bot. Appliq. 14: 126. Tufted annual 30–60 cm. high; on dry bare soils.
Mali: Baguirmi *Chev.* 9653 *bis*! Bore (Aug.) *Demange* 14/1957! **U. Volta:** Ouagadougou (Aug.) *Chev.* 24674! **N. Nig.:** Garin Basan (Aug.) *De Leeuw* 1103! Damaturu (Aug.) *De Leeuw* 1097*a*! Zaria *Taylor* 7! Samaru, Zaria (Aug.) *Sawyer* K13! Dan Tashi (Aug.) *Hall* 315! Also in the Sudan and E. Africa (See Useful Plants.)

2. **C. robusta** *Stapf* l.c. 221 (1912); A. Chev. in Rev. Bot. Appliq. 14: 127; Berhaut, Fl. Sén. ed. 2, 398. Robust perennial 1–2 m. high, with glaucous leaves and with the culms often thick and woody below; in sandy river beds.
S.L.: Mafindo Falls, Kailahun (Feb.) *Deighton* 4004! **Iv. C.:** Marabadiassa (Dec.) *Adjanohoun* 178*a*! *Aké Assi* 7178! Bouna (Oct.) *Rose Innes* GC 31553! Sanlo to Kakpin (Nov.) *Aké Assi* 9291! **Ghana:** Nakpanduri to Bawku (Oct.) *Rose Innes* GC 30750! Yendi (Dec.) *Adams & Akpabla* 4101! Volta R., near Daboya (Nov.) *Thorold* 288! Navrongo (Dec.) *Adams & Akpabla* 4361! Sakogu to Shishe (Oct.) *Ankrah* GC 20507! **Dah.:** Lake Azri to Zagnanado (Feb.) *Chev.* 23054! Ouerne (Dec.) *Froment* 1074! **N. Nig.:** Jebba (Dec.) *Hagerup* 692! Okene (Nov.) *Clayton* 579! Minna (Dec.) *Meikle* 736! Sokoto (Oct.) *Dalz.* 500! Gurum, Vogel Peak area (Nov.) *Hepper* 1255! Yola (Dec.) *Dalz.* 269! **S. Nig.:** Lagos (Oct.) *Foster* 2! Awba Hills F.R. (Jan.) *Onochie & Jones* FHI 14666! Ogun R., Oyo (Feb.) *Brenan* 8959! Fashola, Oyo (Jan.) *Keay* FHI 25642! Nzam Anambra F.R., Onitsha (Feb.) *Jones* FHI 6888! Also in the Sudan, Uganda, and Congo.

3. **C. pilosa** *Schumach.* Beskr. Guin. Pl. 55 (1827); Berhaut, Fl. Sén. ed. 2, 399. *C. breviseta* Benth. in Fl. Nigrit. 566 (1849); Chev. Bot. 746, and in Rev. Bot. Appliq. 14: 125. *C. nigra* Hack. (1906). Annual, 30–60 cm. high; common on roadsides and old farmland.
Maur.: Néma *Rossetti* 61/172! **Sen.:** *Roger* 22! Dakar (May) *Baldwin* 5714! Djembering (Sept.) *Broadbent* 161! Ziguinchor (Sept.) *Broadbent* 79! **Gam.:** *Saunders* 54! Sankuli-Kunda (Sept.) *Pirie* 12/33! **Mali:** Dioura (Nov.) *Davey* 167! Kara (Aug.) *Farrow* 98! Bouta (Feb.) *Davey* 31! **Guin.:** Kouroussa (July) *Pobéguin* 546! Baffing R. (Nov.) *Pobéguin* 1824! Nzérékoré (Oct.) *Baldwin* 9722! **S.L.:** Freetown (Sept.) *Hepper* 919*a*! Newton (Nov.) *Deighton* 1462! Mano (May) *Deighton* 674! Makump (Oct.) *Glanville* 75! Musaia (Apr.) *Deighton* 5386! **Lib.:** Timbo (Mar.) *Baldwin* 11223! Sasstown (Mar.) *Baldwin* 11603! Harper (June) *Baldwin* 5993! 10 miles inland from River Cess (Mar.) *Baldwin* 11228! Gbanga (Sept.) *Linder* 520! **Iv. C.:** Kouibly (Oct.) *Adjanohoun* 413*a*! Tehini (Aug.) *Oldeman* 353! *de Wilde* 819! Bouaké (Sept.) *Oldeman* 411! **U. Volta:** Ouagadougou (Sept.) *Scholz* 31*a*! 31*b*! 120! **Ghana:** Cape Coast *T. Vogel* 41! Nungua, Accra (June) *Ankrah* GC 20150! Navrongo (June) *Vigne* FH 286! Mampong (Oct.) *Baldwin* 13519! Bawku (Oct.) *Rose Innes* GC 30250! **Togo Rep.:** Palime *Stage* 73! **Dah.:** *Newton* 24! Parakou (Sept.) *Risopoulos* 1197! **N. Nig.:** Jebba (Dec.) *Hagerup* 739! Nupe *Barter* 496! Kworre, Sokoto (Sept.) *Palmer* 9! Fika (Oct.) *Daggash* FHI 24890! Samaru, Zaria (Sept.) *Freeman* S116! Gurum, Vogel Peak area (Nov.) *Hepper* 1291! Barra, Dikwa (Dec.) *McClintock* 89! **S. Nig.:** Lagos (May, Oct.) *Denton* 41! *Dalz.* 1124! Burutu *Parsons* 4! Ibadan (Apr.) *Brenan* 9601! Awka *Thomas*! **W. Cam.:** Victoria (July) *Maitland* 9! **F. Po:** *Mann* 118! Also in Sudan, Congo, Uganda and Tanzania. (See Useful Plants.)

4. **C. gayana** *Kunth* Rev. Gram. 1: 293, t. 58 (1830); Chev. Bot. 746, and in Rev. Bot. Appliq. 14: 125; Berhaut, Fl. Sén. ed. 2, 399. Perennial up to 1·2 m. high, apparently indigenous in the northern part of our area, but probably introduced elsewhere.
Sen.: *Döllinger* 21! *Roger* 40! M'Bidjem (June) *Berhaut* 1096! **Mali:** Timbuktu (Aug.) *Hagerup* 276! El Oualadji (Mar.) *Chev.* 43927! Kadigué to Banguita (Jan.) *Davey* 5! Banguita *Farrow* 97! Saredina (May) *Davey* 91! Koubita (Jan.) *Davey* 251! **Guin.:** Paraheolle *Heudelot* 393! **Ghana:** Aiyinase (July) *Irvine* 5061! **Niger:** Wadi Dollol *White* 25! **N. Nig.:** Baga (Jan.) *Daggash* FHI 24970! Gusko (Oct.) *Golding* 21! Kalkala *Golding* 78! Wulgo (Feb.) *Davey* FHI 27175! Tilum (Apr.) *Davey* FHI 27151! Widely spread in E. Africa and S. Africa. (See Useful Plants.)

5. **C. pycnothrix** *Trin.* Gram. Unifl. 234 (1824); A. Chev. in Rev. Bot. Appliq. 14: 127. *C. intermedia* A. Rich., Tent. Fl. Abyss. 2: 407 (1851). Stoloniferous perennial up to 30 cm. high; inflorescence tinged with pink; fallow land and lawns.
Guin.: Timbo *Pobéguin* 1707! Mamou (Nov.) *Chev.* 34887! Nzérékoré (Oct.) *Baldwin* 9725! **S.L.:** Kabala *Gledhill* 224! **Iv. C.:** Kouibly (Oct.) *Adjanohoun* 301*a*! Pakobo (June) *Adjanohoun* 384*a*! **Ghana:** Wenchi (Oct.) *Rose Innes* GC 30519! Ejura (Aug.) *Irvine* 5156! Sampa (Dec.) *Morton* A1053! **N. Nig.:** Zaria (Oct., Dec.) *Taylor* 19! *Clayton* 1332! Jos Plateau (Sept.) *Lely* P769! Naraguta (Dec.) *Kennedy* FHI 1139! Vodni *Saunders* 20! **S. Nig.:** Obudu Plateau (Feb., Apr.) *Tuley* 485! *Haines* 139! **W. Cam.:** Buea (Jan., Nov.) *Maitland* 351! *Migeod* 91! Bamenda (Jan.) *Migeod* 290! Bambui (July) *Bumpus* 19! Nkambe (Feb.) *Hepper & Charter* 1938! Throughout Africa, and also in Arabia and Brazil.

6. **C. virgata** *Sw.* Fl. Ind. Occ. 1: 203 (1797); Chev. Bot. 746, and in Rev. Bot. Appliq. 14: 127. *C. meccana* Hochst. & Steud. ex Schlechtend. (1844)—A. Chev. in Rev. Bot. Appliq. 14: 125. *C. multiradiata* Hochst. (1855); A. Chev. in Rev. Bot. Appliq. 14: 126. *C. rogeoni* A. Chev. in Rev. Bot. Appliq. 14: 127 (1934). Annual up to 60 cm. high.
Mali: *Leclercq* 42716! El Oualadji (Mar.) *Chev.* 43937! Diré (Mar.) *Chev.* 43871! *Leclercq* 42357! **Niger:** Agades (Feb.) *Bradley* 5! **N. Nig.:** *Ward* 19! Gudumbali, Bornu (Oct.) *Blair-Rains*! Arege (Oct.) *De Leeuw* 2072! Widely distributed in the tropics.

7. **C. barbata** *Sw.* Fl. Ind. Occ. 1: 200 (1797); De Miré & Gillet in J. Agric. Trop. 3: 729; Aké Assi, Contrib. 2: 280. *Andropogon barbatus* Linn. (1771), not of Linn. (1754). *C. inflata* Link (1821). Annual, 30–60 cm. high, often stoloniferous; roadsides and waste lands.
Iv. C.: Abidjan (May) *Adjanohoun* 428*a*! **Ghana:** Takoradi (Mar.) *Deighton* 560! Accra Plains (Apr., Oct.) *Rose Innes* GC 30008! *Baldwin* 13439! Kpeshi (Dec.) *Adams* 3606! Pong Tamale (July) *Williams* 832! Near Lomé (Nov.) *Clayton* 394! **S. Nig.:** Lagos (Oct.) *Haines* 160! *Ward* 9! Benin (July)*Unknown Coll.* FHI 54119! Widely distributed in the tropics.

8. **C. prieurii** *Kunth* Rev. Gram. 2: 441 (1831); Chev. Bot. 746, and in Rev. Bot. Appliq. 14: 126; Berhaut, Fl. Sén. ed. 2, 397; De Miré & Gillet in J. Agric. Trop. 3: 729. Annual about 45 cm. high.
Sen.: *Roger* 23! *Perrottet* 924! *Roberty* 16817! Almadies (Nov.) *I.F.A.N.*! Dagana (Oct.) *Hepper* 3636! **Gam.:** *Skues*! **Mali:** Sansanding (Sept.) *Chev.* 24958! Kabarah (Aug.) *Chev.* 1323! Sarédina (May) *Davey* 86! Timbuktu (Aug.) *Lean* 75! Kersani (Aug.) *Farrow* 100! **U. Volta:** Gourma (Aug.) *Rossetti* 59/110! **Ghana:** Accra (Mar.) *Deighton* 573! Kpeshi lagoon (May) *Adams* 4239! 4299! **N. Nig.:** Damaturu

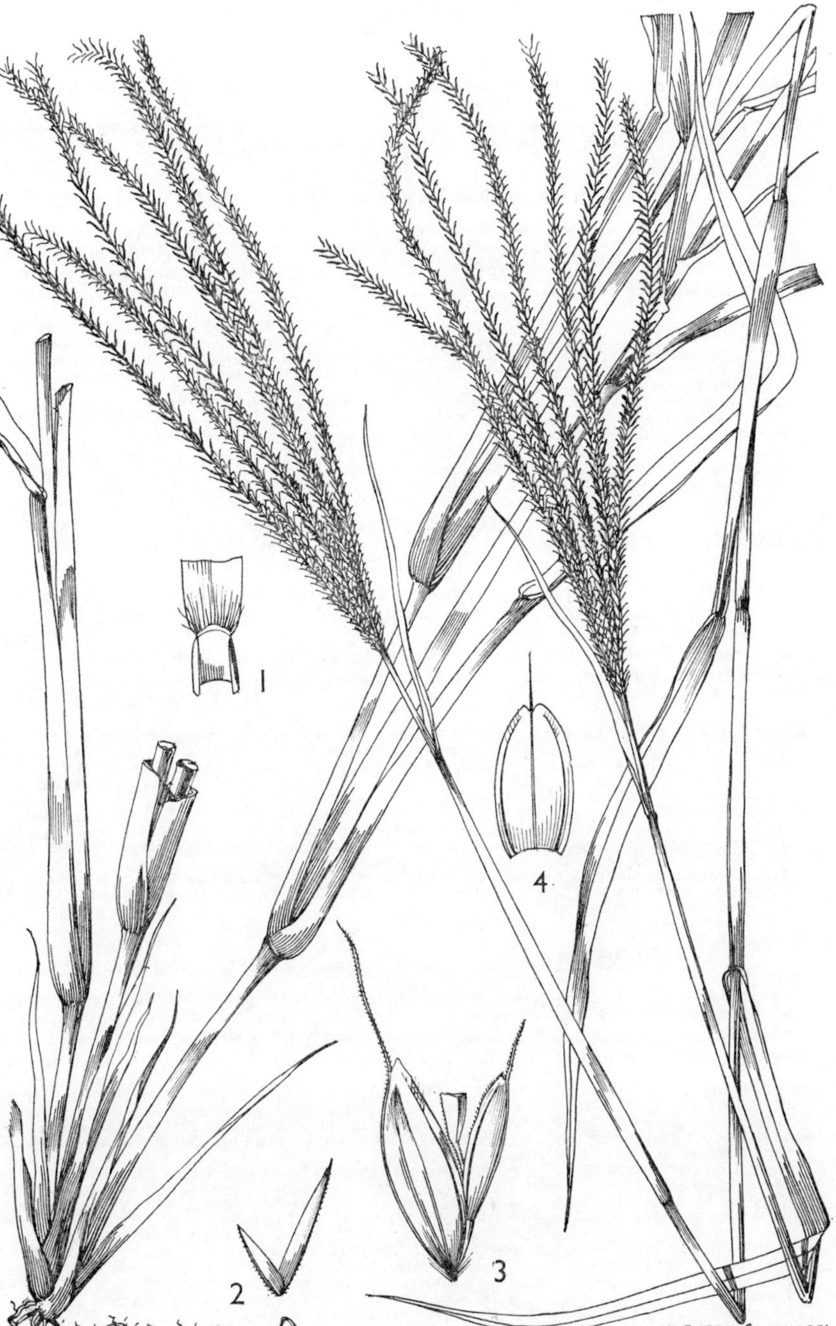

FIG. 433.—CHLORIS GAYANA *Kunth* (GRAMINEAE-CHLORIDEAE).
1, ligule. 2, lower and upper glumes. 3, florets. 4, lemma of fertile floret flattened out.

5

(Aug.) *De Leeuw* 1150! Katagum *Dalz.* 279! Kauwa (Sept.) *Gwynn* 127! Kauwa to Kingowa (Oct.) *Golding* 20! Also in Cape Verde Is., Arabia, Sudan and S.W. Africa.

Imperfectly known species

1. **C. subtriflora** *Steud.* Syn. Pl. Glum. 1: 208 (1854). Senegambia *Leprieur.* The description fits *C. pilosa*, but the number of spikes (15) is higher than usual.

2. **C. parva** *Mimeur* in Bull. Mus. Hist. Nat. sér. 2, 22: 128 (1950). Senegal *Talmy* 86. From the description this species appears to be the same as *C. prieurii*; the type is missing.

52. CHRYSOCHLOA Swallen in Proc. Biol. Soc. Wash. 54: 44 (1941).

Stoloniferous, or sometimes caespitose, annual or perennial; leaf-blades 1–15 cm. long and up to 5 mm. broad, rounded at the tip; spikes 2–10; spikelets 3–4 mm. long, laterally compressed and keeled; lower lemma long ciliate on margins and keel, shortly awned from below the tip .. .. *hindsii*

C. hindsii *C. E. Hubbard* in Kew Bull. 4: 349 (1949). *C. annua* C. E. Hubbard in Kew Bull. 12: 59 (1957). *C. caespitosa* W. D. Clayton in Kew Bull. 14: 239 (1960). *C. subaequigluma* of Berhaut, Fl. Sén. ed. 2, 398, not of (Rendle) Swallen. *Bracteola subaequigluma* of A. Chev. in Rev. Bot. Appliq. 14: 127 (1934), and of F.W.T.A., ed. 1, 2: 522, not of (Rendle) C. E. Hubbard. Culms 10–75 cm. high, the habit very variable; seasonally flooded shallow pans and clayey soils.
Sen.: M'Bouné (Aug.) *Fotius* 538! **Mali:** Boré (Sept.) *Demange* 5/1957! **U. Volta:** Mossi (Aug.) *Chev.* 24754! Ouagadougou (Sept.) *Scholz* 55! Boromo (Sept.) *Aké Assi* 6464! Fada to Koupéla (July) *Chev.* 24569! Batié to Gaoua (Oct.) *Rose Innes* GC 31537! **Ghana:** Damongo *Hinds* 16! Kintampo (Oct.) *Rose Innes* GC 30625! Yapei ferry (Oct.) *Ankrah* GC 20417! Windenaba (Aug.) *Rose Innes* GC 32004! Tamale (Sept.) *Yamoah* SLUS 804! **Togo Rep.:** Palime, Lomé *Stage* 17! **N. Nig.:** Maichi Gusau (Sept.) *De Leeuw* 1573! Also in E. Cameroun, Tanzania, Zambia and Malawi.

53. ENTEROPOGON Nees in Lindl., Introd. Nat. Syst. ed. 2, 448 (1836); Clayton in Kew Bull 21: 105 (1967).

Awn of fertile floret 1–5 mm. long, the lemma 4·5–8 mm. long; florets 2, sometimes 3; culms wiry, often bare below, the basal buds clad in short white scales; leaf-blades broadly linear .. .. 1. *rupestris*
Awn of fertile floret 10–18 mm. long, the lemma 7–10 mm. long; florets 3; culms arising from a basal tussock; leaf-blades narrowly linear, often folded
2. *macrostachyus*

1. **E. rupestris** (*J. A. Schmidt*) *A. Chev.* in Rev. Bot. Appliq. 15: 1048 (1935). *Ctenium rupestre* J. A. Schmidt, Beitr. Fl. Cap. Verd. Ins. 149 (1852). *Enteropogon somalensis* Chiov. in Ann. Istit. Bot. Roma 6: 170 (1896). *E. ruspolianus* Chiov. l.c. 7: 72 (1897). Caespitose perennial about 60 cm. high; dry places.
Maur.: Néma *Rossetti* 61/171! **Niger:** Godo (Sept.) *P. de Fabrègues* 8075! Also in the Cape Verde Is.; Kenya, Sudan, Somalia and Ethiopia; and in SW. Africa and Botswana.
2. **E. macrostachyus** (*Hochst. ex A. Rich.*) *Munro ex Benth.* in J. Linn. Soc. 19: 101 (1881). *Chloris macrostachya* Hochst. ex A. Rich., Tent. Fl. Abyss. 2: 408 (1851). *C. simplex* Schumach., Beskr. Guin. Pl. 54 (1827), illegitimate name. *Enteropogon simplex* (Schumach.) A. Chev. in Rev. Bot. Appliq. 14: 125 (1934). Perennial up to about 1 m. high, with long curved spikes.
Ghana: Accra to Ada (Dec.) *Ankrah* GC 20330! Salt Pond (Oct.) *Baldwin* 13613! Achimota (Apr.) *Irvine* 652! Accra Plains (June, Nov.) *Dalz.* 166! *Clayton* 388! Tropical Africa from the Sudan to the Transvaal and SW. Africa.

54. SCHOENEFELDIA Kunth, Rev. Gram. 1: 283, t. 53 (1830).

Spikes 1–4, digitate, slender, up to 15 cm. long; spikelets densely 2-seriate on one side of the flattened rhachis, 1-flowered; glumes subequal, subulate-acuminate, compressed and keeled, the upper 3–5 mm. long, pilose, shortly bifid with a slender curved awn about 3 cm. long arising from between the lobes .. .. *gracilis*

S. gracilis *Kunth* Rev. Gram. 1: 283, t. 53 (1830); Chev. Bot. 745, and in Rev. Bot. Appliq. 14: 122; Berhaut, Fl. Sén. ed. 2, 389, 397; De Miré & Gillet in J. Agric. Trop. 3: 735. *S. ramosa* Trin., Sp. Gram. Ic. 3: t. 359 (1836). *S. pallida* Edgew. (1852). *S. nutans* Steud., Syn. Pl. Glum. 1: 202 (1854). *S. stricta* Steud. l.c. (1854). Tufted annual 15–90 cm. high, recognised by the beautiful braiding of the awns; dry soils, particularly hardpans and depressions.
Maur.: Goberni *Rossetti* 61/307! **Sen.:** *Heudelot* 389! Borboni Desert (Aug.) *Roger* 52! M'Bambey (Nov.) *Chev.* 33858! Fatick (Jan.) *Adam*! **Mali:** Sompi (Aug.) *Chev.* 2389! Bongo to Thio (Aug.) *Chev.* 24875! Gao (Sept.) *Hagerup* 342! Tondidarou (Oct.) *Farrow* 47! Sarédina (May) *Davey* 84! **U. Volta:** Koudougou (Sept.) *Aké Assi* 6454! Ouagadougou (Oct.) *Scholz* 53*a*! **Ghana:** Accra to Ada (Dec.) *Ankrah* GC 20331! Kete Krachi (Sept.) *Rose Innes* GC 30430! Mirigu (Oct.) *Vigne* FH 4644! White Volta, Bawku to Bolgatanga (Oct.) *Rose Innes* GC 30508! Lawra *Hinds* 3848! **Togo Rep.:** Atakpamé to Sokodé (Oct.) *Rose Innes* GC 31372! **Niger:** 162 miles south of Agades (Nov.) *Money-Kyrle* Fe9! Agades (Feb.) *Bradley* 19*b*! Toukounous *Bartha* 93! **N. Nig.:** Sokoto (Oct.) *Dalz.* 506! Kano (Nov.) *Taylor* 5! Fika (Oct.) *Daggash* FHI 24885! L. Chad (Oct.) *Golding* 7! Sherifuri (Sept.) *Thornewill* 186! Extends eastwards to Sudan, Eritrea and Ethiopia; also in India. (See Useful Plants.)

55. BRACHYACHNE Stapf in Hook. Ic. Pl. 31: t. 3099 (1922); Van der Veken in Bull. Jard. Bot. Brux. 28: 83 (1958).

Leaf-blades acicular, about 2 cm. long, glabrous; spike slender, 3–6 cm. long; glumes subequal, obtuse, keeled, glabrous, 2·3 mm. long, golden-brown when dry; lemma widely emarginate, delicate, 3-nerved, nerves long pilose; palea pilose on the keels
obtusiflora

B. obtusiflora (*Benth.*) *C. E. Hubbard* in Kew Bull. 1933: 503. *Microchloa obtusiflora* Benth. in Fl. Nigrit. 565 (1849); Chev. Bot. 744. Tiny tufted annual 10–20 cm. high, growing in crevices on rock and ironstone outcrops.
Sen: Messira-Kouba (Oct.) *Boudet* 4102 (*fide* Lebrun in Bull. Soc. Bot. Fr. 116: 258 (1969)). **Mali:** *Duong*! **S.L.:** Kambia, Magbema (Nov.) *Jordan* 664! Port Loko (Oct.) *Jordan* 807! **Iv C.:** Sifié (Oct.) *Adjanohoun* 239a! *Aké Assi* 5752! **Ghana:** Burufo (Oct.) *Hinds* 5014! Damongo Scarp (Nov.) *Harris*! Jirapa to Lawra (Oct.) *Ankrah* GC 20456! Yendi to Tamale (Sept.) *Rose Innes* GC 32249! GC 32463! **N. Nig.:** Patti Lokoja *Barter* 537! *T. Vogel* 189! Also in the C. African Rep.

56. CYNODON L. Rich., Syn. Pl. 1: 85 (1805); Clayton & Harlan in Kew Bull. 24: 185 (1970); *nom. cons.*

Underground rhizomes present (in addition to surface stolons); spikes 4–6, digitate, 1·5–6(–8) cm. long; glumes $\frac{1}{2}$–$\frac{3}{4}$ as long as the lemma .. .. 1. *dactylon*
Underground rhizomes absent (but with surface stolons):
Culms very robust, woody; racemes in 2–5 whorls (rarely 1), stiff, red or purple 2. *aethiopicus*
Culms soft, not woody; racemes in 1 whorl (occasionally 2), slender, green or reddish 3. *nlemfuensis*
Racemes 4–7 cm. long; culms moderately robust to somewhat wiry, 1–1·5 mm. diam. .. .. 3a. var. *nlemfuensis*
Racemes 6–10 cm. long; culms robust, 2–3 mm. diam. .. .. 3b. var. *robustus*

1. **C. dactylon** (*Linn.*) *Pers.* l.c. (1805); Chev. Bot. 745 and in Rev. Bot. Appliq. 14: 124; Berhaut, Fl. Sén. ed. 2, 399; De Miré & Gillet in J. Agric. Trop. 3: 730. *Panicum dactylon* Linn. (1753). Stems creeping, rooting at the nodes and producing from them fascicles of barren shoots and flowering culms; 7–40 cm. high. Very variable in vegetative, and to a lesser extent spikelet, characters throughout its range, although specimens from a limited area are usually fairly constant. A common grass in lawns and on paths, and often a pest in agricultural land.
Sen.: *Roger* 113! Dakar (May) *Boivin* 390! *Baldwin* 5725! Hann (Aug.) *Berhaut* 308! Mbao (Aug.) *Broadbent* 7! **Mali:** Macina (May, Sept.) *Chev.* 24931! *Lean* 13! Timbuktu (June) *Hagerup* 108! Dogo (May) *Davey* 79! Mopti (Sept.) *Matthes* 3! **S.L.:** Rokupr (June) *Jordan* 39! **Lib.:** Ganta (Jan.) *Harley* 1741! **U. Volta:** Sindou (May) *Chev.* 872! Ouagadougou (June) *Scholz* 106a! **Ghana:** Axim (Mar.) *Chipp* 412! Ogogo *Williams* 279! Kumasi (Mar.) *La-Danso* 23! Yendi (Mar.) *Hepper & Morton* A3083! Nungua (June) *Ankrah* GC 20148! **Dah.:** Sô valley (Dec.) *Froment* 1033! **Niger:** Agades (Mar.) *Bradley* 24a! **N. Nig.:** Ilorin (Nov.) *Ward* S700! Borgu *Barter* 861! Zaria (Dec.) *Taylor* 37! Sokoto Prov. (Oct.) *Dalz.* 497! Baga Seyoram to Kalkala (Oct.) *Golding* 19! **S. Nig.:** Lagos (Mar.) *Maitland* 171a! Ibadan (Apr.) *Keay* FHI 25701! *Brenan* 9607! Port Harcourt *Maitland*! Calabar (Apr.) *Holland* 122! Distributed throughout tropical and warm temperate regions. (See Useful Plants.)
2. **C. aethiopicus** *Clayton & Harlan* in Kew Bull. 24: 187 (1970). Up to 1 m. high; a species from eastern Africa introduced as a fodder grass.
N. Nig.: Mokwa cattle ranch (Aug.) *Scholz* 108a! 108b! Ethiopia to South Africa.
3. **C. nlemfuensis** *Vanderyst* in Bull. Agric. Congo Belge 13: 342 (1922). 30–60 cm. high; an East African species introduced to West Africa both fortuitously and as a pasture grass.
3a. **C. nlemfuensis** *Vanderyst* var. **nlemfuensis.**
S. Nig.: Ilora (Oct.) *Gledhill* 647! Ethiopia to Rhodesia.
3b. **C. nlemfuensis** var. **robustus** *Clayton & Harlan* in Kew Bull. 24: 189 (1970).
Ghana: Cape Coast, Ankaful (Apr.) *Hall* 3451! Ethiopia to Rhodesia.

57. MICROCHLOA R. Br., Prod. Fl. Nov. Holl. 208 (1810).

Annual; culms slender, often branched; leaf-blades up to 3 cm. long, linear-setaceous; spikes slender, curved, up to 15 cm. long; rhachis tough, minutely ciliate; glumes lanceolate-oblong, acute, 2·2–3 mm. long, the lower keeled, the upper flattened on the back; lemma 2 mm. long, abruptly acuminate, ciliate on the nerves .. 1. *indica*
Perennial, the base clothed with fibrous remains of leaf-sheaths; otherwise similar to the preceding .. .. 2. *kunthii*

1. **M. indica** (*Linn.f.*) *P. Beauv.* Ess. Agrost. 13, t.20, f.8 (1812); Hack. in Fedde Rep. 7: 373; Merrill in Phillip. J. Sci. Bot. 7: 74; Chev. Bot. 744, and in Rev. Bot. Appliq. 14: 123; Berhaut, Fl. Sén. ed. 2, 388. *Nardus indica* Linn. f., Suppl. 105 (1781). *Rottboellia setacea* Roxb. (1798). *Microchloa setacea* (Roxb.) R. Br. (1810). 5–26 cm. high, weakly ascending or straggling; shallow and overgrazed soils.
Mali: Mopti (Sept.) *Chev.* 24922! Mahina *Lécard*! Dioura (Sept.) *Davey* 037! Labbézenga (Sept.) *Hagerup* 455! **Port. G.:** Bafata (Oct.) *Esp. Santo* 2803! **Guin.:** Dalaba *Adames* 337! **S.L.:** Falaba (Apr.) *Morton & Jarr* SL 2093! **Iv. C.:** Dimbokro to Bocanda (June) *Adjanohoun* 228a! N'zi to Bocanda (June) *Adjanohoun* 227a! Téhini (Aug.) *de Wilde* 775! **U. Volta:** Ouagadougou (Sept.) *Scholz* 60a! Bobo-Dioulasso (Sept.) *Aké Assi* 6393! Nouna to Dedougou (Oct.) *Giessler* 60! Gourma (July) *Chev.* 24554! **Ghana:** Yapei (Sept.) *Goodall* 15787! Sunyani to Wenchi (Sept.) *Rose Innes* GC 30147! Tamale (Oct.) *Baldwin* 13545! Banda hills (Oct.) *Rose Innes* GC 31625! Tindagia (Apr.) *Williams* 1035! **Togo Rep.:** Atakpamé to Sokodé (Oct.) *Rose Innes* GC 31373! Kande (Oct.) *Rose Innes* GC 31418! **N. Nig.:** Jebba *Barter*! Patti Lokoja (Sept.) *Jones* FHI 3655! Panbegwa (July) *Keay* FHI 37085! Vodni, Plateau *Saunders* 70! Sokoto (Oct.) *Dalz.* 499! **S. Nig.:** Lagos *Dawodu* 226! **W. Cam.:** Cam. Mt. (Dec.) *Mann* 2095! Throughout the tropics. (See Useful Plants.)
2. **M. kunthii** *Desv.* Opusc. 75 (1831). *M. abyssinica* Hochst. ex A. Rich., Tent. Fl. Abyss. 2: 404 (1851); F.W.T.A., ed. 1, 2: 524. Up to 30 cm. high, caespitose; generally a more robust plant than the preceding, but the size range overlaps considerably; shallow and overgrazed soils.
Iv. C.: Sifié (Oct.) *Aké Assi* 5753! **Ghana:** Accra plains *Rose Innes* GC 30020! Agomeda (Nov.) *Adams* 4387! Kpandu to Kete Krachi (Sept.) *Rose Innes* GC 30421! Andanso (Nov.) *Harris*! Kete Krachi to Kpetchu (Oct.) *Ankrah* GC 20374! **N. Nig.:** Samaru (Aug.) *Thatcher* S491! Naraguta (July) *Lely* P434! *Lawlor & Hall* 110! Vodni, Plateau *Saunders* 61! Panshanu (Aug.) *Lawlor & Hall* 208! **S. Nig.:** Sadibo, Owo (Aug.) *Symington* FHI 5631! **W. Cam.:** *Unwin* 10180! Bali (Mar.) *Brunt* 1087! Bamenda (Mar.) *Brunt* 1050! Throughout the tropics.

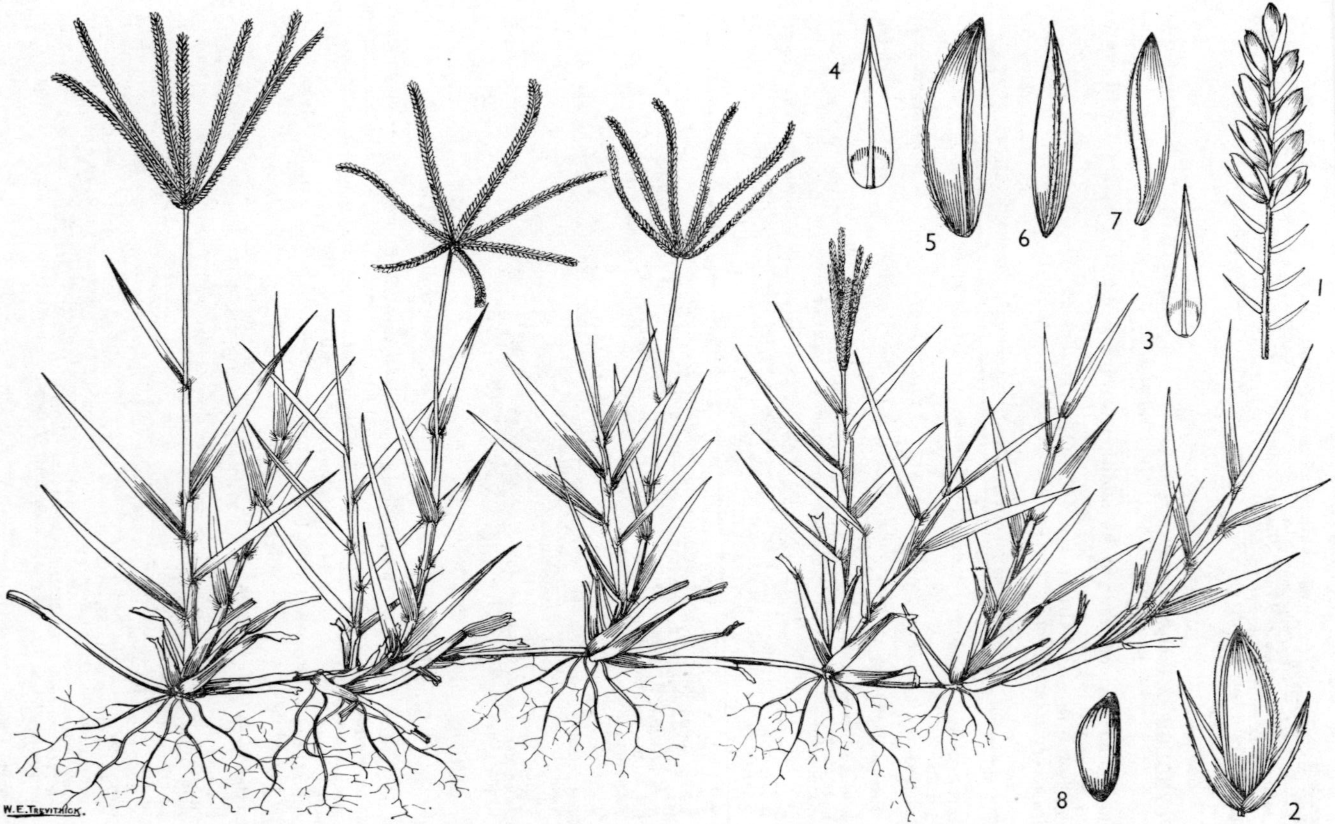

Fig. 434.—CYNODON DACTYLON (*Linn.*) *Pers.* (GRAMINEAE-CHLORIDEAE).
1, portion of spike with persistent lower glumes. 2, spikelet. 3, lower glume. 4, upper glume. 5 and 6, side and back views of lemma. 7, palea. 8, grain.

58. OROPETIUM Trin., Fund. Agrost. 98, t. 3 (1820).

Upper glume 2–3 mm. long, acute; lemma hyaline, 1·5–2 mm. long, mucronate; rhachis ± flattened, tardily disarticulating .. .. 1. *capense*
Upper glume 7·5 mm. long, long-acuminate; lemma hyaline, 3–3·5 mm. long, bearing an awn 2–2·5 mm. long; rhachis cylindrical, fragile .. .. 2. *aristatum*

1. **O. capense** *Stapf* in Fl. Capensis 7: 742 (1900). Tiny perennial up to 6 in. high, with narrow-rolled leaf-blades.
Mali: Ansongo (Sept.) *Hagerup* 3996! (This sheets bears a mixed gathering of *O. capense* and *Tripogon minimus*). Eastwards to Somali and Tanzania; also in Rhodesia and South Africa.
2. **O. aristatum** (*Stapf*) *Pilger* in Engl., Bot. Jahrb. 74: 14 (1947); Jac.-Fél. in J. Agric. Trop. 2: 524; Gillet & Quézel in J. Agric. Trop. 6: 43. *Lepturella aristata* Stapf in Mém. Soc. Bot. Fr. 2, 8: 222 (1907); Chev. Bot. 751, and in Rev. Bot. Appliq. 14: 123; Berhaut, Fl. Sén. ed. 2, 388; F.W.T.A. ed. 1, 2: 524. Tiny annual 5–10 cm. high with involute leaf-blades; shallow soil pockets on rock or ironstone outcrops.
Mali: Mahina *Lecard*! **Iv. C.:** Yérébodi to Sanlo (Nov.) *Aké Assi* 9741! **U. Volta:** Near Koupéla (July) *Chev.* 24598! Gaoua (Oct.) *Rose Innes* GC 31521! Poa (Sept.) *Aké Assi* 6438! **Ghana:** Damongo Scarp F.R. (Nov.) *Rose Innes* GC 30836! Yendi to Tamale (Oct.) *Rose Innes* GC 30686! Babile *Hinds* 3810! Lawra (Oct.) *Rose Innes* GC 31043! Jirapa to Lawra (Oct.) *Ankrah* GC 20455!

[The genus is taken to include *Kralikia* Coss. & Dur. (= *Kralikella* Coss. & Dur.) and *Chaetostichium* C. E. Hubbard]

59. SPOROBOLUS R. Br., Prod. 169 (1810).

Branches of the inflorescence verticillate, sometimes merely subverticillate but then the spikelets almost black; panicle open:
Perennials:
Panicle linear or narrowly lanceolate, usually 20–40 cm. long, of 10–20 whorls; spikelets subsessile along the primary branches:
Spikelets 2–2·2 mm. long; branches 1–2 cm. long, arcuate, incurved, successive whorls not usually overlapping though sometimes with a few of the lowest branches longer and straggling; leaf-blades flat, up to 7 mm. broad .. 1. *dinklagei*
Spikelets 2·7–3 mm. long; branches mostly 1·5–6 cm. long, spreading or ascending, usually longer than the distance between successive whorls; leaf-blades up to 7 mm. broad, but usually much narrower and often involute; spikelets dark reddish in colour .. .. 2. *sanguineus*
Panicle pyramidal, ovate, up to about 15 cm. long:
Spikelets 2–2·2 mm. long; panicle delicate, diffuse, the branches very fine and tinged with red; spikelets grey .. .. 3. *ioclados*
Spikelets 3–4 mm. long, rarely longer, clustered towards the ends of the filiform spreading branches of the inflorescence; panicle of 3–4 whorls, sometimes subverticillate:
Basal sheaths broad, papery or horny and yellowish; leaf-blades linear, involute, 15–30 cm. long and about 3 mm. broad, sometimes shorter (down to 7 cm.) but then linear-lanceolate and 3–5 mm. broad; spikelets greyish-green, 3–4 mm. long .. .. 4. *mauritianus*
Basal sheaths more or less fibrous; leaf-blades mostly 2–7 cm. long, involute, about 2 mm. broad; spikelets very dark green, 3·8–4 mm. long or more 5. *montanus*
Annuals; leaf-blades conspicuously pectinate-ciliate on the margins:
Grain spherical; lower glume half as long as the spikelet:
Grain 0·3–0·6 mm. diam.; spikelets 0·9–1·6 mm. long .. .. .. 6. *stolzii*
Grain 0·8–1 mm. diam.; spikelets 1·5–2 mm. long .. .. 7. *subglobosus*
Grain ellipsoid or obovoid, ± laterally flattened:
Lower glume lanceolate, acute to acuminate; spikelets reddish to dark red; panicle lanceolate; stamens 3 .. .. 8. *paniculatus*
Lower glume ovate, obtuse, often ± suppressed; spikelets greenish or greyish; panicle ovate:
Spikelets 1·6–2 mm. long; anthers 3, 0·7–1·1 mm. long.. .. 9. *cordofanus*
Spikelets 1–1·2 mm. long; anthers 2, 0·3–0·6 mm. long.. .. 10. *microprotus*
Branches of the inflorescence not whorled; spikelets never almost black:
Inflorescence dense or fairly dense, spike-like or with the spikelets clustered about the whole length of the primary branches; branches and pedicels not capillary:
Upper glume as long as the spikelet, or almost so:
Plants annual; panicle spiciform, slender, 3–6 mm. broad; leaf-blades flat, or convolute when dry; glumes thin, membranous, acute, the lower $\frac{2}{3}$ as long as the spikelet .. .. 11. *piliferus*
Plants perennial:
Leaf-blades convolute:
Panicle spiciform, the primary branches closely appressed to the rhachis; leaf-blades pungent; culms 1–2 mm. diam.:
Lower glume $\frac{1}{4}$–$\frac{1}{3}$ as long as the spikelet, hyaline, nerveless; panicle slender, 2–4 mm. broad, the primary branches 1–5 mm. long .. 12. *spicatus*
Lower glume $\frac{2}{3}$–$\frac{4}{5}$ as long as the spikelet, chartaceous, 1-nerved; panicle usually

6–7 mm. broad, the primary branches 5–15 mm. long; leaf-blades 2–10 cm. long .. .. 13. *virginicus*

Panicle distinctly branched, 10–40 mm. broad, the primary branches 20–50 mm. long and ± spreading; leaf-blades 10–20 cm. long, tapering to a filiform tip (rarely with the culm-leaves flat and up to 7 mm. broad in the most robust specimens); culms 2–3 mm. diam. at the base; glumes acute, the lower typically $\frac{4}{5}$ as long as the spikelet, but rather variable .. .. .. 14. *robustus*

Leaf-blades flat, 2–15 cm. long and 2–4 mm. broad, strongly glaucous; glumes acuminate, the lower $\frac{4}{5}$ to as long as the spikelet; panicle usually distinctly branched but sometimes subspiciform, 5–20 mm. broad, the branches 5–20 mm. long; culms wiry, 1 mm. diam. .. .. 15. *helvolus*

Upper glume up to $\frac{2}{3}$ as long as the spikelet (but if panicle spiciform and leaf-blades pungent see *S. spicatus*):

Spikelets markedly acuminate, 1·7–2 mm. long; lower glume a tiny hyaline scale; upper glume abruptly acuminate to a short awn-point up to 0·5 mm. long; palea $\frac{1}{2}$–$\frac{2}{3}$ as long as the lemma; panicle branches 5–25 mm. long .. 16. *molleri*

Spikelets not acuminate:

Panicle pyramidal, 3–4 cm. long, the spikelets clustered at the ends of the branches; grain ellipsoid .. .. 17. *tourneuxii*

Panicle linear, 10–40 cm. long, open or contracted, the spikelets not clustered; grain truncate:

Upper glume obtuse, $\frac{1}{4}$–$\frac{1}{3}$ as long as the spikelet; basal leaf-sheaths narrow; panicle not spiciform .. .. 18. *pyramidalis*

Upper glume acute, $\frac{1}{2}$–$\frac{2}{3}$ as long as the spikelet:

Panicle lax with spreading branches; basal leaf-sheaths breaking into fibres at maturity .. .. 19. *pellucidus*

Panicle narrow to subspiciform, the branches appressed and mostly 1–2 cm. long; basal leaf-sheaths broad and membranous .. .. .. 20. *africanus*

Inflorescence very delicate and lax, with filiform branches and often very slender pedicels, or with the spikelets clustered at the tips of the branchlets:

Perennials:

Inflorescence with long fine white hairs from the axils of the branches; spikelets 2–2·5 mm. long; glumes acuminate, both $\frac{2}{3}$ as long as the spikelet; rhachilla produced for about 1 mm. above the base of the floret .. .. 21. *subtilis*

Inflorescence not hairy in the axils; rhachilla not produced:

Spikelets clustered at the tips of the branchlets, 1·5–2·1 mm. long; upper glume $\frac{2}{3}$–$\frac{4}{5}$ as long as the spikelet, acute; basal sheaths fibrous or somewhat indurated 22. *nervosus*

Spikelets borne in a diffuse panicle:

Base fibrous, the fibres woolly tomentose within; spikelets 1·4–2·1 mm. long; upper glume acute .. .. 23. *stapfianus*

Base ± fibrous, but never woolly:

Upper glume acute; spikelets 1–1·5 mm. long .. .. 24. *festivus*

Upper glume mucronate, abruptly acuminate, or with the single nerve excurrent; spikelets 1·6–2·6 mm. long .. .. 25. *myrianthus*

Annuals:

Spikelets 2·3 mm. long; inflorescence of 10–20 spikelets, more or less racemose on the main axis, or the lowest paired; glumes subequal, about $\frac{1}{2}$ as long as the spikelet; grain ovoid-elliptic .. .. 26. *pauciflorus*

Spikelets at most 2 mm. long; inflorescence a branching panicle with numerous spikelets:

Upper glume cuspidate-acuminate, the central nerve excurrent; lower glume small (0·1–0·2 mm. in our area); spikelet 0·8–1·2(–2) mm. long; grain elliptic; leaf-blades seldom over 6 cm. long .. .. 27. *pectinellus*

Upper glume subacute to acute:

Spikelets 1·3–1·7 mm. long (exceptionally 0·8–2 mm.); grain elliptic; leaf-blades mostly 1–5 cm. long; panicle 2–12 cm. long .. .. .. 28. *infirmus*

Spikelets 0·8–1 mm. long; grain obovoid, truncate; leaf-blades mostly 10–20 cm. long; panicle 10–40 cm. long .. .. 29. *tenuissimus*

1. **S. dinklagei** *Mez* in Fedde Rep. 17: 298 (1921); A. Chev. in Rev. Bot. Appliq. 14: 112. Caespitose perennial 5–10 cm. high; sandy soils and roadsides. Closely allied to *S. sanguineus*, with which it intergrades. **S.L.:** Freetown (Sept.) *Deighton* 2003! Ronietta (Nov.) *Thomas* 5369! Njala (Oct.) *Deighton* 1420! Gbap (Oct.) *Jordan* 610! Sembehun (Oct.) *Jordan* 626! **Lib.:** Monrovia (Nov.) *Barker* 1454! Buchanan (Nov.) *Adam* 15996! Tawata (Nov.) *Baldwin* 10324! Sanokwele (Nov.) *Harley* 1617! Webo, Nyaake (June) *Baldwin* 6155! **Iv. C.:** Moossou (Nov.) *Adjanohoun* 453a!

2. **S. sanguineus** *Rendle* Cat. Welw. 2: 209 (1899). *S. oxylepis* Mez in Fedde Rep. 17: 298 (1921); F.W.T.A., ed. 1, 2: 527. *S. marginatus* of F.W.T.A., ed. 1, 2: 527. *S. schweinfurthii* Stapf (1908). *S. rhodesiensis* Stent & Rattray (1933). *S. schliebenii* Pilger (1932). Tufted perennial about 60 cm. high; pockets and crevices in rock outcrops (in southern Africa it occurs on deeper soils in *Brachystegia* woodland). **Guin.:** Dalaba (Oct.) *Adam* 12582! **Iv. C.:** Mont Mafa (Nov.) *Aké Assi* 10367! Séguélo (Oct.) *Aké Assi* 6601! Samatigela (Oct.) *Boudet* 3324! **Ghana:** Damongo Scarp (Sept.) *Rose Innes* GC 30207! Nsemre F. R. (Dec.) *Adams* 5239! Krobo Hill, Mampong Scarp (Nov.) *Rose Innes* GC 30862! Kwahu Tafo (Nov.)

Rose Innes GC 31654! Tonogo, Togo Plateau (Nov.) *Morton* A3496! **N. Nig.:** (Sept.) *Lely* P765! **S. Nig.:** Nsukka (July) *Tuley* 802! Lokoja (Oct.) *Dalz.* 294! Naraguta (Aug.) *Lely* 448! *Lawlor & Hall* 103! Panshanu (Aug.) *Lawlor & Hall* 544! Throughout tropical Africa.

[Exceedingly variable in inflorescence characters, but apparently without sufficient correlation between these characters to justify the recognition of more than one species. The variation in spikelet length is much greater (2–3·5 mm.) in southern Africa.]

3. **S. ioclados** *Nees* Fl. Afr. Austr. 161 (1841). *S. marginatus* Hochst. ex A. Rich., Tent. Fl. Abyss. 2: 397 (1851); A. Chev. in Rev. Bot. Appliq. 14: 113; Berhaut, Fl. Sén. ed. 2, 412. *Vilfa marginata* (Hochst. ex A. Rich.) Steud. (1854). *V. pallida* Nees ex Trin. (1840). *Sporobolus pallidus* (Nees ex Trin.) Boiss. (1884), not of Lindl. (1848). *S. laetevirens* Coss. (1889). Caespitose perennial 30–60 cm. high; seasonally moist, usually alkaline, soils in dry regions.

Maur.: Néma *Rossetti* 61/167! **Mali:** Toguère of Sarédina (May) *Davey* 87! Goumbomba, Koubita (Dec.) *Davey* 26! Dogo (Apr.) *Davey* 560! Banguita (Sept.) *Farrow* 9! **Niger:** Zinder to Médik (Aug.) *P. de Fabrègues* 2031! Throughout Africa, and in India.

4. **S. mauritianus** (*Steud.*) *Dur. & Schinz* Consp. Fl. Afr. 5: 822 (1894). *Vilfa mauritiana* Steud., Syn. Pl. Glum. 1: 157 (1854). *Sporobolus ledermannii* Mez in Fedde Rep. 17: 296 (1921). *S. hubbardii* A. Chev. in Bull. Mus. Hist. Nat. sér. 2, 20: 469 (1948), and in Rev. Bot. Appliq. 29: 126. Densely caespitose perennial, 30–60 cm. high.

Guin.: Diédédou to Niossomoridou (Feb.) *Chev.* 20855! Macenta (Jan., Apr.) *Jac.-Fél.* 1590! *Adam* 3359! 3644! 4249! **S.L.:** Mt. Loma (Jan., Feb., June) *Jaeger* 9161! 9391! 9774! Tingi Mts (Apr., Dec.) *Morton & Gledhill* SL 1872! SL 3057! **Ghana:** Amedzofe (July) *Thorold* 124! Gabi Mt., Amedzofe (Dec.) *Rose Innes* GC 31181! Maliato, Togo Plateau (Dec.) *Morton* A3837! Shiare (Dec.) *Jeník & Hall* CC 1040! **N. Nig.:** Kakara, Mambila Plateau (Jan.) *Hepper* 1789! **W. Cam.:** Bamenda (Mar.) *Daramola* FHI 40533! Bambui (Mar.) *Jackson* 2505! Mbakakeka Mt., Bafut-Ngemba F.R. (Feb.) *Hepper* 2125! Kishong, Bamenda (Mar.) *Brunt* 1026! Throughout tropical and S. Africa.

5. **S. montanus** *Engl.* in Abh. Preuss. Akad. Wiss. 2: 127 (1892). *Vilfa montana* Hook.f. in J. Linn. Soc. 7: 228 (1864), not of (R.Br.) P. Beauv. (1812). Densely caespitose perennial about 30 cm. high, with short rolled leaves.

W. Cam.: Cam. Mt. (Jan., Dec.) *Boughey* GC 12543! *Maitland* 1269! *Hinds* C19! *Mildbr.* 10895! *Mann* 2088!

6. **S. stolzii** *Mez* in Fedde Rep. 17: 297 (1921). *S. granularis* Mez in Fedde Rep. 17: 297 (1921); Berhaut, Fl. Sén. ed. 2, 410; F.W.T.A., ed. 1, 2: 527. Closely allied to *S. capillaris* Miq. (1851) from India, which is a more robust plant with wider leaf-blades. Annual, up to 60 cm. high.

Sen.: *Heudelot* 440! Fatick (Sept.) *Adam* 18019! Tambacounda (Oct.) *Adam* 12721! 15867! Cambérène (Sept.) *Berhaut* 2884! **Mali:** Bamako (Sept.) *Adam* 15365! Sotuba (Sept.) *Boudet* 331! Koulikoro *Chev.* 2223! Boré (Aug.) *Demange* 17/1957! **Guin.:** Pita (Sept.) *Adames* 343! **Niger:** Niamey (Oct.) *P. de Fabrègues* 322! Ethiopia to Rhodesia.

7. **S. subglobosus** *A. Chev.* in Bull. Mus. Hist. Nat., sér. 2, 20: 469 (1948). Annual, 30–90 cm. high; along roadsides.

Iv. C.: Bouaké (July) *Chev.* 22075! Koébonou to Varalé (Sept.) *Aké Assi* 6507! Bouaké to Béoumi (June) *Adjanohoun* 455a! **Ghana:** Tamale (Oct.) *Baldwin* 13549! Gambaga (Aug.) *Amensor* 581! Kanvile to Yamalkaraga (Aug.) *Rose Innes* GC 31841! Tumu (June) *Rose Innes* GC 31705! Lawra (Sept.) *Irvine* 4736! Nakpanduri (Sept.) *Rose Innes* GC 32114! **N. Nig.:** Nupe *Barter* 1395! Jebba (June) *Clayton* 1159! Ogbomosho to Ilorin (July) *Sawyer* K9! Zongon Daji (Nov.) *Clayton* 599! Zungeru to Bida (Nov.) *Freeman* S192! **S. Nig.:** Igbetti (July) *Stanfield*!

[Differs from *S. paniculatus* only in grain shape, which may not justify its acceptance as a distinct species W.D.C.]

8. **S. paniculatus** (*Trin.*) *Dur. & Schinz* Consp. Fl. Afr. 5: 823 (1895); Clayton in Kew Bull. 19: 294. *Vilfa paniculata* Trin. in Mém. Acad. Sci. Petersb. sér. 6, 5, 2: 67 (1840). *Triachyrum micranthum* Steud., Syn. Pl. Glum. 1: 176 (1854). *Sporobolus micranthus* (Steud.) Dur. & Schinz, Consp. Fl. Afr. 5: 822 (1895). *S. strictus* Franch. in Bull. Soc. Hist. Nat. Autun 8: 368 (1895); A. Chev. in Rev. Bot. Appliq. 14: 115. *S. patulus* Hack. in Oesterr. Bot. Zeitschr. 52: 58 (1902); F.W.T.A., ed. 1, 2: 527. *S. regularis* Mez in Fedde Rep. 17: 299 (1921); Clayton in Kew Bull. 19: 294. *S. granularis* of F.W.T.A., ed. 1, 2: 527, partly. *S. myxosperma* Stapf ex Hutch. & Dalz., F.W.T.A., ed. 1, 2: 527 (1936), name only; Berhaut, Fl. Sén. ed. 2, 412. *S. polycyclus* Berhaut, Fl. Sén. 253 (1954), name only. Up to 60 cm. high, with dark red spikelets; rock outcrops and hillsides.

Sen.: Ouassadou (Dec.) *Berhaut* 1408! **Guin.:** Dalaba-Diaguissa Plateau (Oct.) *Chev.* 18767! 18878! Timbo *Pobéguin* 1719! Nzérékoré (Oct.) *Baldwin* 9692! **S.L.:** Hill Station (Oct.) *Deighton* 157! Regent (Oct.) *Jordan* 949! Picket Hill (Nov.) *T. S. Jones* 223! Sefadu (Nov., Dec.) *Deighton* 3564! 4667! **Lib.:** Tawata (Nov.) *Baldwin* 10352! Nimba (Nov., Dec.) *Adam* 16488! 20070! 20140! *Adames* 734! **Ghana:** Damongo (Aug.) *Rose Innes* GC 31983! Ejura (Sept.) *Asare* 10132! Wenchi to Bamboi (Dec.) *Adams & Akpabla* 4503! **N. Nig.:** Kaduna R., Mokwa to Bida (Aug.) *Sawyer* K10! Sha (Sept.) *J. Hall* K12! Naraguta (Sept.) *Olorunfemi* FHI 55012! Vom (Oct.) *De Leeuw* 1605! Njawai (Nov.) *Jackson, Magaji & Tuley* 1910! **S. Nig.:** Quorra, Atta (Sept.) *T. Vogel* 89! Okelifi, Ondo (Nov.) *Onochie* FHI 34222! Mt. Orosun (= Carter Peak), Idanre (Oct.) *Keay* FHI 22591! Obudu Plateau (Apr., Nov.) *Tuley* 1008! *Haines* 363! **W. Cam.:** Bum, Bamenda (May) *Maitland* 7a! Ekwe, Bamenda (May) *Maitland* 6a! Ndop (Nov.) *Gillett* 10! Bamenda (Jan.) *Migeod* 348! Throughout tropical Africa; also in Madagascar and Mexico.

9. **S. cordofanus** (*Steud.*) *Coss.* in Bull. Soc. Bot. Fr. 36: 253 (1889). *Triachyrum cordofanum* Steud., Syn. Pl. Glum. 1: 176 (1854). *Sporobolus humifusus var. cordofanus* (Steud.) Massey, Sudan Grasses 43 (1926). *S. pyramidatus* of Lebrun in Bull. Soc. Bot. Fr. 116: 265 (1969), not of (Lam.) Hitchc. Annual or short-lived perennial 30–60 cm. high; sandy soils.

Sen.: Thioump (Aug.) *Audru* 2403! Bétio (Aug.) *Audru* 2074! 2076! 2175! Boslabal (Sept.) *Mosnier* 2355! **Niger:** Diffa (Sept.) *P. de Fabrègues* 2146! **N. Nig.:** Sokoto (Sept.) *De Leeuw* 2030a! Bornu (Oct.) *De Leeuw* 2061! Kaura Namoda (July) *Latilo* FHI 62594! Sudan Republic to Rhodesia.

[There is no clear separation between *S. cordofanus* and *S. ioclados*, but I hesitate to associate the weedy annual and the tufted perennial in the same species.]

10. **S. microprotus** *Stapf* in Mém Soc. Bot. Fr. 2, 8: 218 (1912); Clayton in Kew Bull. 19: 293. *S. scabriflorus* Stapf ex Massey, Sudan Grasses 42 (1926). *S. aequiglumis* Stapf ex A. Chev. in Rev. Bot. Appliq. 14: 111 (1934). *S. coromandelianus* of F.W.T.A., ed. 1, 2: 527, including var. *senegambiae* A. Chev. in Bull. Mus. Hist. Nat., sér. 2, 20: 469 (1948); Berhaut, Fl. Sén. ed. 2, 410. *S. humifusus* var. *cordofanus* of A. Chev. in Rev. Bot. Appliq. 14: 113. About 30 cm. high, panicle ovate; roadsides and waste places.

Sen.: *Roger* 13! M'bouné (Aug.) *Fotius* 541! Yoff (Oct.) *Berhaut* 458! **Mali:** Dioura (Sept.) *Davey* 71! Boré (July) *Davey* 15/1957! Douentza (Aug.) *Davey* 16/1957! Ansongo (Sept.) *Hagerup* 390! Sanga *Dieterlen*! **Iv. C.:** Béoumi (Aug.) *Aké Assi* 6843! Ferkessédougou (Sept.) *Aké Assi* 6301! Ouagadougou to Yako (Aug.) *Chev.* 24762! **U. Volta:** Bobo-Dioulasso (Sept.) *Aké Assi* 6394! Ouagadougou (Sept.) *Scholz* 57! Ouagadougou to Sikasso (Sept.) *Adam* 15441! Boromo (Sept.) *Aké Assi* 6463! **Ghana:** Bawku Town (Oct.) *Rose Innes* GC 31440! Morago R. bridge, Sakogu to Shishe (Oct.) *Ankrah* GC 20505! Pong Tamale to Nabogo (Sept.) *Oppong* 193! Lawra *Hinds* 3547! Grupe, Sawla to Damongo (Dec.) *Rose Innes* GC 31016! **Niger:** Toukounous (Feb.) *P. de Fabrègues* 370! **N. Nig.:** Fika (Oct.) *Daggash* FHI 24886! Biu Dist. (Sept.) *Kennedy* FHI 8005! Biliri, Bauchi Prov. (Oct.) *De Leeuw & Magaji* 1985! Damagum (Sept.) *De Leeuw* 1242! Extends eastwards through C. African Rep. and Sudan to Eritrea and Kenya.

11. **S. piliferus** (*Trin.*) *Kunth* Enum. Pl. 1: 211 (1833). *Vilfa pilifera* Trin., Diss. Bot. 157 (1824). *Sporobolus stachydanthus* A. Rich., Tent. Fl. Abyss. 2: 394 (1851). *S. praecox* A. Chev. in Rev. Bot. Appliq. 14:

113 (1934), and 29: 124, also in Bull. Mus. Hist. Nat. sér. 2, 20: 468 (1948). Small annual, 15–30 cm. high; shallow soils.

Guin.: Tossékré, Fouta Djalon (Oct.) *Adam* 12756! Labé (Nov.) *Chev.* 34907! Mali (Nov.) *Chev.* 34337! Mt. Loura (Nov.) *Chev.* 34916! **N. Nig.:** Panshanu (Aug.) *Lawlor & Hall* 336! 438! Throughout tropical Africa, extending eastwards through India to Malesia.

12. **S. spicatus** (*Vahl*) *Kunth* Rev. Gram. 1: 67 (1829); Chev. Bot. 743, and in Rev. Bot. Appliq. 14: 115; Berhaut, Fl. Sén. ed.2, 392; De Miré & Gillet in J. Agric. Trop. 3: 736. *Agrostis spicata* Vahl, Symb. Bot. 1: 9 (1790). Stoloniferous perennial 30–60 cm. high, the leaves painfully pungent; alkaline or saline soils.

Maur.: Nouakchott to Tanit *Rossetti* 59/20! **Sen.:** *Roger* 58! St. Louis (Oct., Nov.) *Chev.* 33899! Dakar (Jan., Dec.) *Talmy*! *Berhaut* 1161! Djembering (Sept.) *Broadbent* 175! **Mali:** Sarédina (Mar., May) *Davey* 52! 92! *LeClercq* 42639! Timbuktu (June, Aug.) *Hagerup* 124! *Lean* 72! **N. Nig.:** L. Chad (Dec.) *Golding* 18! *Elliott* 154! Baga Seyoram (Jan., Feb.) *Daggash* FHI 24968! *Davey* FHI 27166! 70 mi. north of Maiduguri (Mar.) *De Leeuw* 1914! Throughout the drier regions of Africa from Morocco to the Transvaal, extending through the Middle East into India. (See Useful Plants.)

13. **S. virginicus** (*Linn.*) *Kunth* Rev. Gram. 1: 67 (1829); A. Chev. in Rev. Bot. Appliq. 14: 116; Berhaut, Fl. Sén. ed.2, 392; Clayton in Kew Bull. 19: 295. *Agrostis virginica* Linn., Sp. Pl. 63 (1753). *Vilfa virginica* (Linn.) P. Beauv. (1812). *V. conferta* (J. A. Schmidt) Steud. (1854). *Agrostis congener* Schumach., Beskr. Guin. Pl. 46 (1827). *Sporobolus pungens* of Chev. Bot. 742, and in Rev. Bot. Appliq. 14: 114, not of Kunth. *S. confertus* J. A. Schmidt, Fl. Cape Verd. Ins.: 142 (1852)—*fide* A. Chev., Fl. Cap Vert.: 1044 (1935). *S. littoralis* (Lam.) Kunth (1829). Rhizomatous perennial up to about 60 cm. high, often with distichous leaf-blades; sandy sea-shores.

Sen.: *Adanson*! *Heudelot* 499! Hann (Oct.) *Berhaut* 2792! Djembering (Sept.) *Broadbent* 178! Oussouye (Sept.) *Broadbent* 117! **S.L.:** Lumley to Aberdeen (Aug.) *Melville & Hooker* 145! Sussex (May) *Deighton* 2680! Mesima (Apr.) *Adames* 35! Freetown (Aug.) *Hooper* 119! Juba (May) *Morton* SL 1330! **Lib.:** Monrovia (May, Oct.) *Baldwin* 5812! *Harley* 1671! Sinkor (July) *Barker* 1386! Harper (June) *Baldwin* 5966! Garaway (Mar.) *Baldwin* 11632*a*! **Iv. C.:** Sassandra (May) *Chev.* 17941! **Ghana:** Anomabu (June) *Irvine* 1464! Elmina (Mar.) *Rose Innes* GC 30887! Axim (Apr.) *Thorold* 44! Kpeshi lagoon (Sept.) *Milne-Redhead* 5139! Old Ningo (May) *Rose Innes* GC 31262! **Dah.:** Cotonou (Apr.) *Chev.* 23475! Tropical sea-shores generally.

14. **S. robustus** *Kunth* Rev. Gram. 2: 425 (1832); A. Chev. in Rev. Bot. Appliq. 14: 114; Berhaut, Fl. Sén. ed.2, 414; Clayton in Kew Bull. 19: 296. *S. assakae* Cab. (1936). *S. insulanus* Parl. (1846)—Hooker, Fl. Nigrit. 187 (1849), as "*S. insularis*". *S. littoralis* (Lam.) Kunth var. *elongatus* Dur. & Schinz, Consp. Fl. Afr. 5: 821 (1895). *S. senegalensis* Chiov., Pl. Nov. Minus Not. Aeth. 26 (1928). *Vilfa robusta* (Kunth) Trin. (1840). *V. insulanus* (Parl.) Steud. (1854). *Agrostis abyssinica* Ehrenb. & Hempr. ex Trin. (1840), in synonymy. *A. barbata* Pers. var. *senegalensis* Pers., Syn. Pl. 1: 76 (1805), illegitimate name. Stoloniferous perennial 60–90 cm. high; saline soils of the sea coast, usually on the landward side of *S. virginicus*.

Sen.: (Feb.) *Dollinger* 19! M'bour to Rufisque (Mar.) *Chev.* 44129! Mbao (Oct.) *Berhaut* 2882! Almadies (Aug.) *Broadbent* 26! **Gam.:** Brumen ferry (July) *Rhind & Deighton* 5586! Balingho ferry (July) *Fox* 154! **S.L.:** Martin (Jan.) *Deighton* 3543! Mabatu (June) *Deighton* 4023! Mortail (Feb.) *Deighton* 4116! **Lib.:** Grand Bassa (July) *T. Vogel* 31! **Ghana:** Accra (May) *Dalz.* 170! Kpeshi lagoon, Labadi (Mar.) *Adams* 3778! Keta (May) *Thorold* 62! Elmina (Mar.) *Rose Innes* GC 30888! Tema (June) *Rose Innes* GC 30358! **Togo Rep.:** Lomé *Warnecke* 316! **W. Cam.:** Ambas Bay, Victoria (Apr.) *Brenan* 9588! Victoria (July) *Maitland* 7! W. African coast from Rio de Oro to Angola. (See Useful Plants.)

15. **S. helvolus** (*Trin.*) *Dur. & Schinz* Consp. Fl. Afr. 5: 820 (1894); Berhaut, Fl. Sén. ed.2, 392, 415; De Miré & Gillet in J. Agric. Trop. 3: 736. *Vilfa helvola* Trin. in Mém. Acad. Sci. Petersb. sér.6, livr. 1–2: 52 (1840). *V. glauca* Trin. (1840), in synon. *V. glaucifolia* Steud., Syn. Pl. Glum. 1: 154 (1854). *Sporobolus glaucifolius* (Steud.) Hochst. ex Dur. & Schinz, Consp. Fl. Afr. 5: 820 (1894); Chev. Bot. 742, and in Rev. Bot. Appliq. 14: 112. *Agrostis aequalis* Ehrenb. & Hempr. ex Trin. (1840), in synon. Perennial, about 45 cm. high; seasonally flooded land in dry regions.

Maur.: *Sadio* 1521! Keur Masséne *Rossetti* 59/6! **Sen.:** *Adanson*! Richard-Tol (Feb.) *Döllinger* 20! Ross (Feb.) *Berhaut* 1382! **Mali:** *Chev.* 43897! Bandiagara (Sept.) *Rogeon* 209! Koubita (Jan.) *Davey* 257! Banguita (Sept.) *Farrow* 20! Baragungu *Hagerup* 314! **N. Nig.:** Rann (Feb.) *Davey* FHI 27171! Arege (Oct.) *Golding* 38! Ngala (Oct.) *Gwynn* 115! 60 m. north of Maiduguri (Mar.) *De Leeuw* 1907! Malamfaturi (Jan.) *De Leeuw* 2065! Extends eastwards into Tanzania, Kenya, Sudan and Somaliland, thence through Arabia to India. (See Useful Plants.)

16. **S. molleri** *Hack.* in Bol. Soc. Brot. 5: 213 (1887); A. Chev. in Rev. Bot. Appliq. 14: 113. *S. mayumbensis* Franch. in Bull. Soc. Hist. Nat. Autun 8: 367 (1895). Small annual about 30 cm. high; cultivated land and pathsides.

Guin.: Kindia (Oct.) *Adam* 12585! Dalaba (Oct.) *Adames* 386! **N. Nig.:** Tiba Plateau (Dec.) *Peter & Tuley* 67! Gembu (Sept., Nov.) *Oche* MRK 80! *Magaji & Tuley* 1851! Chappal Waddi (Nov.) *Jackson, Magaji & Tuley* 2074! **S. Nig.:** Obudu Plateau (Nov.) *Tuley* 990! **W. Cam.:** Cam. Mt. (Mar.) *Brenan* 9262! Buea (Jan., Nov., Dec.) *Maitland* 349! *Migeod* 101! *Hinds* C73! Bamenda (June) *Daramola* FHI 41560! **F. Po:** Moka (Jan., Dec.) *Guinea* 2064! 2066! *Boughey* 141! Through the Congo to Tanzania and Rhodesia.

17. **S. tourneuxii** *Coss.* in Bull. Soc. Bot. Fr. 36: 250 (1889). *S. sindicus* Stapf ex T. Cooke, Fl. Bombay 2: 1018 (1908). Perennial up to about 30 cm. high; culms tough, almost woody, at the base; leaf-blades 3–4 cm. long; saline soils.

Maur.: Nouakchott to Akjoujit (Nov.) *Adam* 19416! Also in Tunisia, Somali Republic, Arabia & Pakistan.

18. **S. pyramidalis** *P. Beauv.* Fl. Oware 2: 36 (1812); Chev. Bot. 732; Berhaut, Fl. Sén. ed.2, 415; Clayton in Kew Bull. 19: 289. *Vilfa pyramidalis* (P. Beauv.) Trin. ex Steud., Nom. Bot. ed.2, 2: 786 (1841). *Agrostis extensa* Schumach., Beskr. Guin. Pl. 45 (1827). *A. owarensis* Roem. & Schult., Syst. 2, Mant. 199 (1824). *Sporobolus wombaliensis* Vanderyst (1920), in synon. *S. indicus* of A. Chev. in Rev. Bot. Appliq. 14: 113, not of Linn. Densely tufted perennial; it belongs to a difficult group of closely related species allied to the American *S. indicus* Linn. (see also *S. pellucidus* and *S. africanus*).

Gam.: Yundum, Kombo North (Aug.) *Frith* 128! **Mali:** Ségou (Apr.) *Roberty* 2689! **Port. G.:** Banjara Farim (Aug.) *Esp. Santo* 3062! Contubo el Canhamine (Sept.) *Esp. Santo* 2759! Piche (Sept.) *Pereira* 3188! 3252! **Guin.:** Kindia *Jac.-Fél.* 128! Kouroussa (July) *Pobéguin* 549! Baffing R. (Nov.) *Pobéguin* 1821! Konkouré (Mar.) *Pitot* 206! **S.L.:** Kambia (Dec.) *Deighton* 813! Magburaka (Oct.) *Glanville* 72! Mano *Thomas* 10381! Rokupr, Magbema (July) *Jordan* 454! Bunebu Bumpe (Apr.) *Deighton* 1689! **Lib.:** Monrovia (May) *Barker* 1306! *Baldwin* 5814! Sanokwele (Sept.) *Baldwin* 9500! Suakoko (Apr.) *Blickenstaff* 44! Gletown, Tchien Dist. (July) *Baldwin* 6783! **Iv. C.:** Sassandra valley (Feb., Apr.) *Portères* 619! *Leeuwenberg* 2851! Cocody (June) *Adjanohoun* 422*a*! **U. Volta:** Mangodara (June) *Scholz* 10! Samandeni (Sept.) *Kmoch* 113! Banfora (June) *Leeuwenberg* 4432! **Ghana:** Kintampo (Dec.) *Vigne* FH 3203! Tamale (May) *Williams* 818! Kumasi (Apr.) *La Danso* 37! Aburi (Sept.) *Johnson* 841! Dayi R. bridge, Kpandu to Accra (Sept.) *Ankrah* GC 20211! **Togo Rep.:** Palime (Mar.) *Stage* 9! **Dah.:** Allada (Mar.) *Chev.* 23382! **N. Nig.:** Katagum Dist. *Dalz.* 286! Bida (Mar.) *Meikle* 1319! Naraguta (July) *Lely* P410! Yola (Sept.) *Kennedy* FHI 7296! Vogel Peak (Nov.) *Hepper* 1452*a*! Mambila Dist., Sardauna Prov. *Hill*! **S. Nig.:** Lagos (June) *Dalz.* 1309! Ibadan (Apr.) *Meikle* 1412! Aguku Dist. *Thomas* 1062! Opobo *Jeffreys* 3! Oban *Talbot* 845! **W. Cam.:** (Dec.) *Unwin* 9039! Victoria (Nov.) *Maitland* 173! Ekuse, Bamenda (May) *Maitland* 5*a*! Baba, Ndop (May) *Brunt* 428! Tropical and S. Africa, extending to Yemen, Madagascar and Mauritius. (See Useful Plants.)

Note. Robust plants typical of *S. pyramidalis* do not occur in tropical America, and the smaller local

Fig. 435.—SPOROBOLUS AFRICANUS (*Poir.*) *Roberty & Tournay* (GRAMINEAE-SPOROBOLEAE).

1, ligule. 2, portion of branch with spikelets. 3, spikelet. 4, lower glume. 5, upper glume. 6, lemma. 7, palea.

species has therefore been regarded as distinct and named *S. jacquemontii*. Identical plants are found in West Africa, suggesting the introduction of the American species. Unfortunately it intergrades with depauperate specimens of *S. pyramidalis*, so that in Africa the two species are doubtfully separable:-

Culms 2–5 mm. diam. at base, usually 90–160 cm. high; leaf-blades 3–10 mm. broad; panicle branches spreading horizontally at anthesis, the lower mostly 5–10 cm. long; grain 0·8–1 mm. long, obovoid, strongly truncate (see above) .. .. *S. pyramidalis*

Culms 1–2 mm. diam. at base, usually 50–70 cm. high; leaf-blades 1·5–2·5 mm. broad; panicle branches loosely appressed to the main axis, the lower 3–4 cm. long; grain 1–1·1 mm. long, obovoid-elliptic, slightly truncate at the top .. .. *S. jacquemontii*

[**S. jacquemontii** *Kunth* Rev. Gram. 2: 427 (1831); Clayton in Kew Bull. 19: 288. *Vilfa jacquemontii* (Kunth) Trin. (1840). *S. pyramidalis* P. Beauv. var. *jacquemontii* (Kunth) Jov. & Guéd. in Bull. Cent. Etud. Rech. Sci., Biarritz 7: 60 (1968), invalidly published.
S.L.: Freetown (May) *Baldwin* 5796! **Lib.:** Ganta (Sept.) *Harley* 275! Sarbo (July) *Baldwin* 6380! **Iv. C.:** Adiopodoumé (Nov.) *Leeuwenberg* 1885! **Ghana:** Aburi (Sept.) *Johnson* 842! Legon Hill (Oct.) *Adams* 3360! **N. Nig.:** Oturkpo, Benue Prov. (Oct.) *Kennedy* 8027! **S. Nig.:** Lagos (Mar.) *Maitland* 172! **W. Cam.:** Buea (July) *Maitland* 18!]

19. **S. pellucidus** *Hochst.* in Flora 38: 201 (1855). Perennial about 60 cm. high.
Niger: Guéré (Aug.) *P. de Fabrègues* 2024! Also from Ethiopia to Zambia.

20. **S. africanus** (*Poir.*) *Robyns & Tournay* in Bull. Jard. Bot. Brux. 25: 242 (1955); Clayton in Kew Bull. 19: 291. *Agrostis africana* Poir. in Lam., Encycl. Meth. Bot. suppl. 1: 254 (1810). *A. capensis* Willd., Sp. Pl. ed. 4, 1: 372 (1798), not of Lam. (1783). *Vilfa capensis* P. Beauv., Ess. Agrost. 16, 147, 181 (1812). *Sporobolus capensis* (P. Beauv.) Kunth, Enum. Pl. 1: 212 (1833); F.W.T.A., ed. 1, 2: 527. *S. batesii* A. Chev. in Bull. Mus. Hist. Nat. sér. 2, 20: 469 (1948). Perennial, up to about 60 cm. high; montane grasslands.
N. Nig.: Mambila Plateau: Maisamari (Aug.) *De Leeuw* 1735! Gembu (Dec.) *Daramola* FHI 62425! Chappal Waddi (Nov.) *Jackson, Magaji & Tuley* 2072! **S. Nig.:** Obudu Plateau (Feb.-May) *Tuley* 482! 621! *Haines* 144! **W. Cam.:** Cam. Mt. (Jan.) *Maitland* 927! W. of Mann's Springs (Dec.) *Morton* K856! Bambui (Dec.) *Boughey* 10749! Bamenda road (Dec.) *Boughey* 10450! Ndu, Nkambe Div. (Feb.) *Hepper* 1946! South Africa, extending northwards into the highlands of tropical Africa; also in Mauritius, Ceylon, Australia, New Zealand, Hawaii and Easter Island. [Some intermediate specimens (such as *Bumpus* 16 from Bambui) have a long acute upper glume, but a spreading, pyramidal panicle. They have been described as *S. natalensis* (Steud.) Dur. & Schinz (1894) and *S. indicus* var. *laxus* (Nees) Stapf (1900), and may be of hybrid origin.]

21. **S. subtilis** *Kunth* Rev. Gram. 2: 421 (1831); A. Chev. in Rev. Bot. Appliq. 14: 115. *Vilfa subtilis* (Kunth) Trin. in Mém. Acad. Sci. Petersb. sér. 6, 2: 88 (1840). *Sporobolus barbigerus* Franch. in Bull. Soc. Hist. Nat. Autun 8: 371 (1895). Densely tufted perennial 30–45 cm. high, with coarsely fibrous base.
S.L.: Mt. Loma (Feb.) *Jaeger* 9400! 9419! Tingi Mts. *Morton & Gledhill* SL 3060! **Lib.:** Monrovia (June, Aug.) *Baldwin* 5918! 9194! Paynesville (Feb.-Mar.) *Harley* 2117! *Voorhoeve* 862*b*! Duport (Oct.) *Barker* 1445! **S. Nig.:** Obudu Plateau (Feb., Apr.) *Tuley* 483! *Haines* 359! Extends through the Congo and Tanzania to Zambia, Rhodesia, Angola, Mozambique and S. Africa; also in Madagascar.

22. **S. nervosus** *Hochst.* in Flora 38: 202 (1855). *S. longibrachiatus* Stapf in Kew Bull. 1907: 219 (1907). Tufted perennial 15–60 cm. high.
Maur.: Nouakchott (Oct.) *Adam* 21775! Ethiopia to Tanzania, and in S.W. Africa.

23. **S. stapfianus** *Gandoger* in Bull. Soc. Bot. Fr. 66: 302 (1920); De Miré & Gillet in J. Agric. Trop. 3: 736. *S. festivus* var. *stuppeus* Stapf in Fl. Capensis 7: 582 (1900). *S. stuppeus* (Stapf) Stent in Bothalia 2: 264 (1927). Very similar to the following, but with a less delicate inflorescence.
N. Nig.: Zaranda Mt. (May) *Lely* 196! Tropical and S. Africa generally.

24. **S. festivus** *Hochst. ex A. Rich.* Tent. Fl. Abyss. 2: 398 (1850); Chev. Bot. 742, and in Rev. Bot. Appliq. 14: 112 (incl. var. *fibrosus* Stapf ex Stent in Bothalia 2: 264 (1927)); Berhaut, Fl. Sén. ed. 2, 412. *Vilfa festiva* (Hochst. ex A. Rich.) Steud., Syn. Pl. Glum. 1: 158 (1854). Densely tufted perennial up to 60 cm. high, with strongly fibrous base and very delicate, often reddish, inflorescence; crevices and hollows in rock outcrops and shallow soils.
Maur.: Néma *Rossetti* 61/190! **Gam.:** *Saunders* 51! Genieri to Massambe (July) *Fox* 172! **Mali:** Kersani (Aug.) *Farrow* 75! Dioura (Sept.) *Farrow* 74! **Guin.:** Tambailo (Aug.) *Adam* 14838! **Iv. C.:** Séguéla (Apr.) *Leeuwenberg* 3267! **U. Volta:** Bougoura (June) *Scholz* 8! Barmoussa (June) *Scholz* 8*a*! Kain (July) *Scholz* 8*b*! **Ghana:** Navrongo (June) *Vigne* FH 4545! Zuarungu (June) *Williams* 509! Berekum to Sampa (Apr.) *Morton* A3239! Gradaw, Banda Hills (Oct.) *Rose Innes* GC 31626! Kintampo to Bamboi (Oct.) *Rose Innes* GC 30623! **N. Nig.:** Anara F.R., Zaria Prov. (July) *Keay & Okeke* FHI 43505! Bakura-Tureta F.R. near Bimasa, Sokoto Prov. (July) *Keay* FHI 37790! Gaya, Kano Prov. (July) *Daggash* FHI 22387! Katagum Dist. *Dalz.* 253! Nupe *Barter* 1392! **S. Nig.:** Ipapo, Lagos (May) *Denton* 30! Ago-Are F.R., Oyo Prov. (May) *Keay* FHI 37753! Aku Rock (Apr.) *Hambler* 462! Owo (Apr.) *Haines* 149! **W. Cam.:** (Dec.) *Unwin* 9703! 9753! Throughout tropical and S. Africa, and in Madagascar. (See Useful Plants.)

25. **S. myrianthus** *Benth.* in Fl. Nigrit. 565 (1849). Loosely tufted perennial with short leaves and slender culms 30–90 cm. high, the basal sheaths somewhat indurated and more or less fibrous.
N. Nig.: Patti Lokoja (Sept.) *T. Vogel* 190! Jos Plateau *Batten-Poole* 126! Panshanu F.R., Bauchi Prov. (Aug.) *Drew*! *Lawlor & Hall* 85! 398! Sha (Sept.) *Hall* K47! Eastwards to Kenya and south to Rhodesia.

26. **S. pauciflorus** *A. Chev.* in Bull. Mus. Hist. Nat. Paris, sér. 2, 20: 470 (1948). A tiny annual up to 12 cm. high; on ironpans.
Guin.: Nimba (Aug.) *Schnell* 3437!

27. **S. pectinellus** *Mez* in Fedde Rep. 17: 295 (1921); Berhaut, Fl. Sén. ed. 2, 412. *S. vogelii* Stapf ex A. Chev. in Rev. Bot. Appliq. 14: 116 (1934); Chev. Bot. 743. Delicate annual about 30 cm. high; shallow soils.
Sen.: *Heudelot* 497! Oussadou (Sept.) *Berhaut* 615! **Mali:** Bamako (Sept.) *Adam* 14966! **Port. G.:** Piche (Sept.) *Pereira* 3249! **Guin.:** Ditinn (Sept.) *Schnell* 7405! Iribéléya to Timbo *Chev.* 18283! Kouroussa (Aug., Sept.) *Pobéguin* 506! 1095! Tossekré, Fouta Djalon (Oct.) *Adam* 12757! **S.L.:** Musaia (Sept.) *Small* 274! Makene (Oct.) *Glanville* 80! Koinadugu Distr. (Aug.) *Mead* 255! Jui peninsular (Oct.) *Morton & Jarr* SL2333! **Iv. C.:** Bouaflé to Zuenoula (Oct.) *Adjanohoun* 311*a*! **Ghana:** N. of Bole, Wa road (Sept.) *Rose Innes* GC 30175! Damongo Scarp F.R. (Nov.) *Rose Innes* GC 30838! Atonso (Oct.) *Baldwin* 13486! Gambaga F.R. (Oct.) *Ankrah* GC 20272! Kete Krachi (Sept.) *Ankrah* GC 20224! **Togo Rep.:** Atacora Mt. pass, Sokode to Lama Kara (Oct.) *Rose Innes* GC 31393! **N. Nig.:** R. Niger *McWilliam*! Anara F.R. (Sept.) *Jackson* 619968! Oturkpo, Idoma Div. (Sept.) *Kennedy* 8019! Panshanu (Aug.) *Lawlor & Hall* 416! Sha (Sept.) *Hall* K16! **S. Nig.:** Lagos *MacGregor* 140! Also in Uganda and the Congo.

28. **S. infirmus** *Mez* in Fedde Rep. 17: 294 (1921). *S. tenuis* Stapf ex A. Chev. in Bull. Mus. Hist. Nat. sér. 2, 20: 469 (1948), and in Rev. Bot. Appliq. 29: 125. *S. schnellii* A. Chev. l.c. 470 (1948), and in Rev. Bot. Appliq. 29: 126. Slender annual 10–30 cm. high, rather variable; rock outcrops and ironstone exposures.
Guin.: Mt. Loura, Fouta Djalon (Sept.) *Schnell* 7121! Macenta (Oct.) *Baldwin* 9753! Mt. Koiré, Nzérékoré (Sept.) *Baldwin* 13283! Friguiagbé (July) *Chillou* 562! **S.L.:** Wonkifu (Oct.) *Jordan* 937! Yakala (Sept.) *Thomas* 2359! **Sefadu** (Sept.) *Jordan* 521! *Adames* 270! Rolip (Sept.) *Jordan* 536! **Iv. C.:** Mid-Cavally to Mid-Sassandra (Sept.) *Portères* 1001! **Ghana:** Burufo (Oct.) *Hinds* 5009! Krobo Hill, Mampong (Oct.) *Rose Innes* GC 31155! Kwahu Tafo (Nov., Dec.) *Rose Innes* GC 31651! *Adams* 4921! Volta R., Kete

Krachi (Oct.) *Rose Innes* GC 30928! **N. Nig.:** (Sept.) *Lely* P754! Panshanu (Aug.) *Lawlor & Hall* 428! *Hall* 508! Jebba *Barter*! **S. Nig.:** Lagos *MacGregor* 47! Igboora (Oct.) *Haines* 332! Iseyin (Oct.) *Jackson* 2360! **W. Cam.:** Ndop (July) *Boughey* GC 10962*a*! Bambui (Mar.) *Brunt* 72! Also in the C. African Rep., Congo and Uganda.

29. **S. tenuissimus** (*Schrank*) *O. Ktze.* Rev. Gen. Pl. 3: 369 (1893). *Panicum tenuissimum* Schrank in Denkschr. Bot. Ges. Regensb. 2: 26 (1822). *Vilfa minutiflora* Trin., Diss. Bot. 158 (1824). *Sporobolus minutiflorus* (Trin.) Link, Hort. Berol. 1: 88 (1827); F.W.T.A., ed. 1, 2: 527; Berhaut, Fl. Sén. ed. 2, 412. *S. bibansiensis* Vanderyst (1920), in synonymy. Annual 30–90 cm. high.
Sen.: Dakar (May) *Baldwin* 5726! Sangalkam (June) *Raynal* 5958! **Guin.:** Conakry (Jan., June) *Dalz.* 8418! 8419! *Martine* 6! **Lib.:** Nimba (Sept., Dec.) *Adam* 20282! *Adames* 544! **S. Nig.:** *Boughey* GC 11400! Lagos (Oct.) *Haines* 180! **F. Po:** Santa Isabel *Swarbrick* 2910! Occurs in Gabon, Kenya and Tanzania, extending eastwards through Arabia and Mauritius to India; also in tropical America.

60. CRYPSIS Ait., Hort. Kew. 1: 48 (1789); C. E. Hubbard in Hook., Ic. Pl. 35: t. 3457 (1947); Lorch in Bull. Res. Counc. Israel 11D: 91 (1962); *nom. cons.*

Inflorescence an ovoid head of densely crowded spikelets, up to 2 cm. long and 0·8 cm. broad, embraced below by two inflated leaf-sheaths; spikelets 2·5–4 mm. long; glumes shortly ciliate on the keel .. .. *schoenoides*

C. schoenoides (*Linn.*) *Lam.* Tab. Encycl. Méth. Bot. 1: 166 (1791); Lebrun in Bull. Soc. Bot. Fr. 166: 258 (1969). *Phleum schoenoides* Linn., Sp. Pl. 60 (1753). *Heleochloa schoenoides* (Linn.) Host (1801)—F.W.T.A., ed.1, 2: 530. *Crypsis compacta* Steud., Syn. Pl. Glum. 1: 151 (1854). Annual, with prostrate stems radiating from the root.
Sen.: *Leprieur*! *Roger* 31! Extends northwards into Mediterranean Europe, and eastwards through Ethiopia and the Middle East to N.W. India; isolated records southwards to Malawi.

61. PEROTIS Ait., Hort. Kew. 1: 85 (1789).

Base of spikelets abruptly rounded, without a definite callus; spikelets 1·7–2·7 mm. long, crowded, usually spreading; awns 9–17 mm. long .. .. .. 1. *patens*
Base of spikelets drawn out into a distinct stipitiform callus:
Spikelets 1·5–2·5 mm. long (excl. callus); awns 5–15 mm. long; glumes, and their slender mid-nerve, uniformly scaberulous or hispidulous; grain terete .. 2. *indica*
Spikelets 2·5–3·5 mm. long:
Spike lax, the spikelets distant by several times their own width; mid-nerve of upper glume depressed and spinulose; glumes irregularly hispidulous; awn 5–15 mm. long; grain strongly flattened .. .. 3. *hildebrandtii*
Spike dense, the spikelets contiguous; mid-nerve of upper glume neither depressed nor markedly spinulose; glumes hispidulous in lines along their length; awn 20–35 mm. long; grain slightly flattened .. .. 4. *scabra*

1. **P. patens** *Gandoger* in Bull. Soc. Bot. Fr. 66: 301 (1920). *P. transvaalensis* Gandoger l.c. (1920). Glaucous annual or short-lived perennial up to 60 cm. high, with dense spikes; roadsides and weedy places.
Iv. C.: Pakobo Nord (June) *Adjanohoun* 378*a*! Abouabou (Jan.) *Leeuwenberg* 2391! Bouaké (Sept.) *de Wilde* 977! **U. Volta:** Sindou (June) *Leeuwenberg* 4332! **Ghana:** Accra *Brown* 385! Kete Krachi to Kpetchu (Oct.) *Ankrah* GC 20378! Tamale (Dec.) *Williams* 457! Damongo (Mar.) *Bannerman-Bruce*! Navrongo (June) *Vigne* FH 4544! **Togo Rep.:** Lomé *Warnecke* 314! Sokode to Atakpame (Oct.) *Rose Innes* GC 31382! **Dah.:** Porto Novo (Mar.) *Etève* 64! **N. Nig.:** Ilorin *Ward* 53! Nupe *Barter* 754! Okene (Mar.) *Maggs* OK52! Naraguta (Aug.) *Lely* P437! Lafia (Oct.) *Kennedy* FHI 8028! Yola (Sept.) *Dalz.* 281! **S. Nig.:** Lagos (Mar.) *Maitland* 162! *Dalz.* 1119! Ibadan (Jan.) *Meikle* 999! Shaki (Aug.) *Latilo, Taylor & Leafers* FHI 43242! Okigwi to Awgu (Mar.) *Tuley* 613! Tropical and S. Africa, and Madagascar.
2. **P. indica** (*Linn.*) *O. Ktze.* Rev. Gen. Pl. 2: 787 (1891).* *Anthoxanthum indicum* Linn. Sp. Pl. 28 (1753). *Perotis spicata* (Linn.) Th. & H. Durand (1909). *P. latifolia* Ait. (1789), superfluous name. Annual about 30 cm. high, the spike rather loose in W. African specimens; roadsides and weedy places.
S.L.: Hastings (Aug.) *Jordan* 2167! Mahera (Nov.) *Deighton* 5682! Newton (Feb., Aug.) *Jordan* 197! *Deighton* 3761! **S. Nig.:** Ilaro, Abeokuta (Apr.) *Daramola* FHI 46012! Ibadan (Feb.) *Townrow* 7! Okomu F.R. (Dec.) *Brenan* 8626! Benin (Apr., Sept.) *Maggs* 9! *Haines* 365! Also in E. Cameroun, C. African Rep., Congo and Tanzania; and in India and SE. Asia. (See Useful Plants.)
3. **P. hildebrandtii** *Mez* in Fedde Rep. 17: 145 (1921). Annual, 15–45 cm. high, geniculately ascending, with purplish or reddish " bottle brush " inflorescence; roadsides and lawns.
S.L.: Freetown (Jan., Aug., Sept.) *Hooper* 123! *Harvey* 107! *Gledhill* 144! Lumley to Aberdeen (Aug.) *Melville & Hooker* 151! **Lib.:** Sangwin, Sinoe Co. (Mar.) *Baldwin* 11322! **Ghana:** Half-Assini (Dec.) *Rose Innes* GC 30544! Legon (Apr.) *Morton* A4160! Biakpa (Dec.) *Rose Innes* GC 31184! Jasikan to Abotoase (Oct.) *Rose Innes* GC 30923! Ajena to Akwamu West (June) *Adams* 4797! **Dah.:** Cotonou (Dec.) *Froment* 1011! **S. Nig.:** Lagos (Mar., Oct.) *Maitland* 169! *Haines* 161! Ibadan (Sept.) *Latilo* FHI 54899! Owo (Apr.) *Maggs* OK61! Oyo, Fashola *Thorold* 4! Also in Kenya, Tanzania, and the Seychelles.
4. **P. scabra** *Willd. ex Trin.* Gram. Unifl. 172 (1824). Annual up to about 60 cm. high, decumbent at the base; roadsides and lawns.
Sen.: (Sept.) *Farmer* 72! *Heudelot* 650! M'Bao (Nov.) *I.F.A.N.*! **Gam.:** (July) *Brooks* 3! **S.L.:** Mambolo, Kichom (Oct.) *Glanville* 425! Pamaro (Nov.) *T. S. Jones* 44!

62. TRAGUS Haller, Hist. Stirp. Helv. 2: 203 (1768); Schweickerdt in Ann. Natal Mus. 10: 15 (1941); *nom. cons.*

Upper glume of lowermost spikelet of each cluster 7-nerved, 3·5–5 mm. long; clusters composed of 2–4 fertile spikelets and 1–2 reduced sterile spikelets; anthers 0·6–0·8 mm. long .. .. 1. *racemosus*

* This name was adopted by F.W.T.A., ed. 1, and other West African authors, but it was applied to the other 3 species of *Perotis* and not to *P. indica* proper.

Fig. 436.—PEROTIS PATENS *Gandoger* (GRAMINEAE-ZOYSIEAE).
A, spikelet.

Upper glume of lowermost spikelet of each cluster 5-nerved, 2–3 mm. long; clusters composed of 2 spikelets, the upper smaller and borne upon a short internode; anthers 0·4–0·6 mm. long .. .. 2. *berteronianus*

1. **T. racemosus** (*Linn.*) *All.* Fl. Ped. 2: 241 (1785); Chev. Bot. 721, and in Rev. Bot. Appliq. 14: 110; Berhaut Fl. Sén. ed. 2, 389, 393; De Miré & Gillet in J. Agric. Trop. 3: 737. *Cenchrus racemosus* Linn., Sp. Pl. 1049 (1753). *Tragus decipiens* (Fig. & De Not.) Boiss. (1884). *T. paucispina* Hack. (1901). *T. racemosus* var. *erectus* Mut. (1837), var. *decipiens* (Fig. & De Not.) Dur. & Schinz (1894), and var. *paucispina* (Hack.) Maire (1941). Annual, up to about 30 cm. high; roadsides and waste places on dry sandy soils.
Maur.: Atar (Sept.) *Kesby* 41! **Mali:** Sahel Canal *Leclercq* 42757! Southern Europe, N. Africa and C. African Rep., extending through Ethiopia and Somaliland into Arabia; also in S. Africa.

2. **T. berteronianus** *Schult.* Mant. 2: 205 (1824); Lebrun in Bull. Soc. Bot. Fr. 116: 266 (1969). *T. racemosus* var. *berteronianus* (Schult.) Hack. in Osterr. Bot. Zeit. 51: 195 (1901). *T. ciliatus* Leprieur ex Kunth, Rev. Gram. 2: 413 (1831), in synon. *Lappago racemosa* (Linn.) Honck. var. *erecta* Kunth, Rev. Gram. 2: 413, t. 120 (1831).
Mali: Oualata (July, Aug.) *Boëry*! Goundam (Aug.) *Chev.* 2334! Timbuktu (July, Aug.) *Hagerup* 172! *Lean* 74! **U. Volta:** Gourma (Aug.) *Rossetti* 93/1959! Throughout Africa and the warmer parts of America; also in Arabia, Afghanistan and China.

63. LEPTOTHRIUM Kunth, Rev. Gram. 1: 156 (1829). *Latipes* Kunth, Rev. Gram. 1: 261 (1830).

Leaf-blades linear, flat or rolled, 2–10 cm. long; inflorescence up to 15 cm. long or more, the rhachis tough and wavy; spikelets single or paired, often purplish, 3–5·5 mm. long, persistent on hard flattened cuneate peduncles and falling with them; one glume in each cluster drawn out into a recurved flattened tail up to 8 mm. long and ciliate with hooked hairs; other glumes thickened and tuberculate-spinulose on the back *senegalense*

L. senegalense (*Kunth*) *W. D. Clayton* in Kew Bull. ined. (1972). *Latipes senegalensis* Kunth, Rev. Gram. 1: 261 (1830); F.W.T.A., ed. 1, 2: 537; Chev. Bot. 722, and in Rev. Bot. Appliq. 14: 111; Berhaut, Fl. Sén. ed. 2, 389; De Miré & Gillet in J. Agric. Trop. 3: 734. *Lappago latipes* Steud. (1854). *Tragus senegalensis* Gay ex Kunth, in synon. (1830). Perennial, forming tough wiry bunches on dry soils.
Maur.: Tagent *Coll. Admin.* 3473! **Sen.:** (Nov.) *Roger* 19! *Heudelot* 494! **Mali:** Goundam (Aug.) *Chev.* 2331! Bandiagara to Douentza (July) *Rogeon* 449! Timbuktu (July, Aug.) *Hagerup* 206! *Lean*! Oualata (Sept., Oct.) *Boëry*! **Ghana:** Prampram (Feb.) *Darko* 559! Also in E. Africa from Egypt to Tanzania, extending through Arabia to Persia and Pakistan.

Additional genus.

Zoysia matrella (*Linn.*) *Merr.*, widely used as a lawn grass in tropical Asia, has been introduced to Ife, Nigeria for the same purpose. *Z. tenuifolia* Trin., a finer leaved species from Korea & Japan, has been used for lawns in Lomé, Togo (*Charter* FHI 40822!).

64. TRISTACHYA Nees, Agrost. Bras. 458 (1829); F.T.A. 10: 52 (1937); C. E. Hubbard in Kew Bull. 1936: 321; Conert in Engl., Bot. Jahrb. 77: 296 (1957); Phipps in Kirkia 4: 101 (1964); Clayton in Kew Bull. 21: 120 (1967).

Pedicels connate; triads in a narrow raceme, with short erect peduncles; lower glume beset with tubercle-based hairs .. .. 1. *thollonii*
Pedicels free, up to 6 cm. long; spikelets in a narrow panicle; lower glume glabrous 2. *superba*

1. **T. thollonii** *Franch.* in Bull. Soc. Hist. Nat. Autun 8: 374 (1895): F.T.A. 10: 62; Conert l.c. 306. *Apochaete thollonii* (Franch.) Phipps l.c. 105 (1964). Tufted perennial up to 1·2 m. high; seasonally damp soils.
N. Nig.: Shika, Zaria (Oct.) *Magaji* 1631! Abinsi (Dec.) *Dalz.* 882! 886! Also extends from the Congo to Zambia and Malawi.

2. **T. superba** (*De Not.*) *Schweinf. & Aschers.* Beitr. Fl. Aethiop. 302 (1867); Conert l.c. 299; Phipps l.c. 102. *Loudetia superba* De Not. in Ind. Sem. Hort. Genuens. 24 (1852) & in Ann. Sci. Nat. sér. 3, 19: 369 (1853); F.T.A. 10: 47; Berhaut, Fl. Sén. ed. 2, 416. Stout perennial 1·5–3·5 m. high; base of culm swollen and clad in silky hairy leaf-sheaths.
Mali: Bamako (Sept.) *Adam* 15353! Sikasso (Sept.) *Adam* 15049! **U. Volta:** Takalédougou, Banfora (Sept.) *Aké Assi* 6316! Samandeni, Bobo-Dioulasso (Oct.) *Kmoch* 138! **Ghana:** Yapei Ferry, White Volta R. (Aug.) *Rose Innes* GC 31992! Tamale to Yendi (Sept.) *Rose Innes* GC 30451! Kulpeni R., Yendi *Hinds* 8! Daka R., Yendi (Nov.) *Addei* SLUS 734! Damongo scarp (Sept.) *Hall* CC 841! **N. Nig.:** Michika, Bornu (Aug.) *De Leeuw* 1814! Mando F.R., Birnin Gwari Dist. (Sept.) *Keay* FHI 28032! Gombe to Bauchi (Dec.) *Oche* 96! Yankari Game Reserve (Sept.) *Oche & Tuley* 1693! Extends through tropical Africa.
[Although *Loudetia* was not properly described until 1854, the epithet *superba* was validly described in combination with the earlier illegitimate generic name.]

65. ARUNDINELLA Raddi, Agrost. Bras. 37, t. 1, fig. 3 (1823); F.T.A. 10: 1 (1937); Hubbard in Kew Bull. 1936: 319; Conert in Engl., Bot. Jahrb. 77: 329 (1967); Phipps in Kirkia 4: 97 (1964) and in Can. J. Bot. 45: 1047 (1967); Clayton in Kew Bull. 21: 121 (1967).

Perennial up to 3 m. high; culms hard; leaf-blades stiff, linear; panicle oblong, 30–50 cm. long, the spikelets arising singly or in pairs from the primary branches, rarely from secondary branches; spikelets 4·5–5·5 mm. long .. 1. *nepalensis*
Annual; culms soft; leaf-blades flaccid, narrowly lanceolate; panicle effuse, obovate; up to 25 cm. long; spikelets 1·5–2·5 mm. long .. .. 2. *pumila*

1. **A. nepalensis** *Trin.* Gram. Panic. 62 (1826). *A. ecklonii* Nees (1841), incl. var. *major* C. E. Hubbard (1934); F.T.A. 10: 2; F.W.T.A., ed. 1, 2: 546; A. Chev. in Rev. Bot. Appliq. 14: 42; Berhaut, Fl. Sén. ed. 2, 417. An extremely variable grass, the culms 0·6–3·5 m. high, and the panicles 10–50 cm. long, but with the spikelets relatively constant; the West African specimens fall into the upper part of this range.
Sen.: Niokolo-Koba (Feb., Apr., Oct.) *Berhaut* 1596! *Adam* 15714! 17437! **Mali:** Marigot Farako (Feb.) *Duong* 2102! **Guin.:** Labé to Mali (Nov.) *Chev.* 34593! Mali to Mt. Laura *Chev.* 34601 *bis*! Tropical and South Africa, extending through India to China and Australia.

2. **A. pumila** (*Hochst. ex A. Rich.*) *Steud.* Syn. Pl. Glum. 1: 114 (1854); F.T.A. 10: 3; Conert in Engl., Bot. Jahrb. 77: 340; Bor in J. Ind. Bot. Soc. 27: 62. *Acratherum pumilum* Hochst. ex A. Rich., Tent. Fl. Abyss. 2: 414 (1851). *Arundinella effusa* C. E. Hubbard in Kew Bull. 12: 63 (1957).
Guin.: Mali *Jac.-Fél.* 1988! Massif du Benna (Oct.) *Jac.-Fél.* 7163! Kindia (Jan.) *Chillou* 1806! Ditinn (Apr.) *Pitot*! **S.L.:** Tingi Mts., Kono (Dec.) *Morton & Gledhill* SL 2930! **Ghana:** Boti Falls, Huhunya (Oct.) *Hall* 1619! **N. Nig.:** Wana, Lafia (Nov.) *Hepburn* 157! **W. Cam.:** Metschum R. Falls, Bamenda (Jan.) *Keay & Russell* FHI 28524! Also in NE. Africa, extending through India to Indonesia.

66. LOUDETIOPSIS Conert in Engl., Bot. Jahrb. 77: 277 (1957); Clayton in Kew Bull. 21: 122 (1967).

Lobes of upper lemma awned, with a tuft of hairs below each lobe; stamens 3:
- Pedicels 5–25 mm. long, the spikelets in loose groups of 2 or 3; spikelets 6–10 mm. long .. .. 1. *occidentalis*
- Pedicels 0·5–1 mm. long, the spikelets in close clusters of 3:
  - Annual; spikelets 5 mm. long; lower glume 3–3·5 mm. long; panicle dense, of 40 or more triads .. .. 2. *pobeguinii*
  - Perennial; spikelets 6–12 mm. long; lower glume 4–11 mm. long; panicle loose, of 4–35 triads .. .. 3. *tristachyoides*

Lobes of upper lemma awnless, the lemma uniformly hairy or glabrous on the back; stamens usually 2:
- Annual; spikelets 14–17 mm. long with, or rarely without, pale yellow tubercle-based hairs; callus of upper floret 2-toothed; awns 7–12 cm. long .. 4. *kerstingii*
- Perennials:
  - Awns 4·5–7 cm. long:
    - Callus of upper floret obliquely 1-toothed or truncate; spikelets 13–20 mm. long; lower glume 8–13 mm. long, narrowly lanceolate, tapering to an acute or acuminate tip, the lateral nerves encrusted with a continuous line of tubercles; peduncle sharply recurved .. .. 5. *chrysothrix*
    - Callus of upper floret 2-toothed; spikelets 9–13 mm. long; lower glume 5–8 mm. long, lanceolate, narrowing to an obtuse tip, the lateral nerves usually with isolated tubercles, rarely glabrous; peduncle straight or flexuous, rarely recurved 6. *scaëttae*
  - Awns 0·3–3 cm. long:
    - Peduncles sinuous or sharply recurved, and densely hairy beneath the triads:
      - Awns 20–30 mm. long; spikelets 6–7 mm. long; leaf-blades setaceous, involute, up to 1 mm. broad; panicle branches pilose to villous; lower lemma 3-nerved; stamens 2 .. .. 7. *capillipes*
      - Awns 3–5 mm. long; triads deciduous as a whole; lower lemma 7-nerved; stamens 3:
        - Column of awn lacking; upper lemma sparsely pubescent, with an oblong truncate callus .. .. 8. *baldwinii*
        - Column of awn 1·5 mm. long, twisted; upper lemma densely pilose, with a very short rounded callus .. .. 9. *falcipes*
    - Peduncles straight or almost so, and glabrous beneath triads (though sometimes with hairy pedicels):
      - Lower glume $\frac{3}{4}$ to almost as long as the upper; glumes glabrous or with a few tubercle-based hairs; lower lemma 5–9-nerved; stamens 3 .. 10. *trigemina*
      - Lower glume $\frac{1}{2}$–$\frac{2}{3}$ as long as upper; stamens 2:
        - Basal leaf-sheaths tomentose below; glumes and pedicels glabrous, or the lower glume pubescent towards the apex; lower glume acute to acuminate; lower lemma 3–5-nerved .. .. 11. *glabrata*
        - Basal leaf-sheaths glabrous below; glumes and pedicels usually setose with tubercle-based hairs; lower lemma 3-nerved:
          - Culm glabrous and smooth below panicle; panicle contracted and rather dense, 2·5–11 cm. broad, branches erect or ascending; lower glume obtuse to sub-acute .. .. 12. *ambiens*
          - Culm scabrid and rough to the touch below panicle; panicle open and very loose, 6–25 cm. broad, branches widely spreading and often conspicuously whorled; lower glume truncate to obtuse .. .. 13. *thoroldii*

1. **L. occidentalis** (*Jac.-Fél.*) *W. D. Clayton* in Kew Bull. 21: 123 (1967). *Dilophotriche occidentalis* Jac.-Fél. in J. Agric. Trop. 7: 408 (1960). *Danthoniopsis occidentalis* Jac.-Fél. in Rev. Bot. Appliq. 30: 424 (1950), without latin descr. Perennial, 60–90 cm. high.
Guin.: Conakry (Dec.) *Adam* 12617! Kindia *Jac.-Fél.* 1852! Friguiagbé (Sept.-Dec.) *Chillou* 25! 105! Kanea (Oct.) *Adam* 12630! (The tufts of hair on the upper lemma are lacking in this specimen).

2. **L. pobeguinii** (*Jac.-Fél.*) *W. D. Clayton* in Kew Bull. 21: 123 (1967). *Dilophotriche pobeguinii* Jac.-Fél. in J. Agric. Trop. 7: 408 (1960). *Danthoniopsis pobeguinii* Jac.-Fél. in Rev. Bot. Appliq. 30: 423 (1950), without latin descr. About 60 cm. high; moist places.
Sen.: Niokolo-Koba (Jan.) *Berhaut* 4672! Kanéméré (Aug.) *Fotius* 369 *ter* (*fide* Lebrun in Bull. Soc. Bot.

Fr. 116: 263 (1969)). **Port. G.:** Madina de Boé, Gabú (Dec.) *Esp. Santo* 2848! **Guin.:** Seriba to Kiffaya (Oct.) *Adam* 12703! Touba to Kadé (June) *Pobéguin* 2035!

3. **L. tristachyoides** (*Trin.*) *Conert* in Engl., Bot. Jahrb. 77: 291 (1957). *Panicum tristachyoides* Trin. in Bull. Sci. Acad. Petersb. 1: 71 (1836). *Tristachya tristachyoides* (Trin.) C. E. Hubbard in Kew Bull. 1935: 309 (1935); F.T.A. 10: 59; F.W.T.A., ed. 1, 2: 546. *T. purpurea* C. E. Hubbard l.c. 308; F.T.A. 10: 60; F.W.T.A., ed. 1, 2: 546. *T. multinodis* C.E. Hubbard l.c. 308; F.T.A. 10: 59; F.T.W.A., ed. 1, 2: 546. *T. microstachya* Nees ex Steud. (1854). *T. tuberculata* Stapf (1897). *T. minuta* A. Chev. in Rev. Bot. Appliq. 14: 43 (1934), partly, without proper descr. *Danthoniopsis tristachyoides* (Trin.) Jac.-Fél. in Rev. Bot. Appliq. 30: 424 (1950). *D. tuberculata* (Stapf) Jac.-Fél. l.c. 423 (1950); Berhaut, Fl. Sén. ed.2, 417. *D. multinodis* (C. E. Hubbard) Jac.-Fél. l.c. 423 (1950). *D. purpurea* (C. E. Hubbard) Jac.-Fél. l.c. 424 (1950). *Arundinella tristachyoides* (Trin.) Roberty in Bull. I.F.A.N. 17: 56 (1955). *Loudetiopsis purpurea* (C. E. Hubbard) Conert in Engl., Bot. Jahrb. 77: 292 (1957). *Dilophotriche purpurea* (C. E. Hubbard) Jac.-Fél. in J. Agric. Trop. 7: 408 (1960). *D. tristachyoides* (Trin.) Jac.-Fél. l.c. (1960). *D. tuberculata* (Stapf) Jac.-Fél. l.c. (1960). Up to 1·2 m. high; damp pockets on rock outcrops.
Sen.: *Heudelot* 141! Kanéméré (Aug.) *Fotius* K369 (*fide* Lebrun in Bull. Soc. Bot. Fr. 116: 263 (1969)). **Mali:** (Aug.) *Duong*! Bamako (Sept.) *Adam* 14954! Sikasso (Sept.) *Adam* 15107! **Guin.:** Conakry (Oct.) *Adam* 12612! Kouria to Trébéléya (Sept.) *Chev.* 18278! Mamou (Sept.) *Schnell* 6783! Pita *Adames* 360! Timbo (Sept.) *Pobéguin* 1796! **S.L.:** Brookfields (Oct.) *Deighton* 2176! Sugar Loaf Mt. (Oct.) *T. S. Jones* 249! Kambia, Magbema (Jan.) *Jordan* 661! Karina (Oct.) *Glanville* 82! Bumbuna (Oct.) *Thomas* 3117! **U. Volta:** Cascade de Toussiana (Oct.) *Scholz* 68!

4. **L. kerstingii** (*Pilger*) *Conert* in Engl., Bot. Jahrb. 77: 289 (1957). *Trichopteryx kerstingii* Pilger in Engl., Bot. Jahrb. 34: 128 (1904). *Tristachya kerstingii* (Pilger) C. E. Hubbard in Kew Bull. 1934: 435; F.T.A. 10: 57; F.W.T.A., ed. 1, 2: 546; Aké Assi, Contrib. 2: 283. *T. glabrinodis* C. E. Hubbard in Kew Bull. 4: 358 (1949). *Diandrostachya kerstingii* (Pilger) Jac.-Fél. in J. Agric. Trop. 7: 408 (1960). *D. glabrinodis* (C. E. Hubbard) Phipps in Bol. Soc. Brot. 41: 202 (1967). Up to 1 m. high; panicles erect, peduncles straight or stiffly flexuous; shallow soils over ironstone.
Mali: Sindou Koroni (Sept.) *Adam* 15187! **Iv. C.:** Bassawa (Oct.) *Adjanohoun* 274a! Toupé to Saléya (Jan.) *Adjanohoun* 222a! Bouna to Batié (Oct.) *Rose Innes* GC 31543! Yérébodi to Sanlo (Nov.) *Aké Assi* 9271! **U. Volta:** Lafègué (Sept.) *Adam* 15145! Baoulé (Oct.) *Kmoch* 124! Takalédougou, Banfora (Sept.) *Aké Assi* 6334! Batié to Bouna (Oct.) *Rose Innes* GC 31538! Cascade de Toussiana (Oct.) *Scholz* 75! **Ghana:** Sampa (Dec.) *Vigne* FH 3491! Lawra *Hinds* 3853! Bamboi to Bole (Oct.) *Rose Innes* GC 30649! Gambaga to Yendi (Sept.) *Ankrah* GC 20265! Chereponi to Yendi (Oct.) *Rose Innes* GC 30972! **N. Nig.:** Nupe *Barter* 1385! Mokwa (Nov.) *De Leeuw & Oche* MRK 96! Lokoja (Sept.) *H. B. Jones* FHI 3656!

5. **L. chrysothrix** (*Nees*) *Conert* in Engl., Bot. Jahrb. 77: 285 (1957). *Tristachya chrysothrix* Nees, Agrost. Bras. 460 (1829); F.T.A. 10: 56; F.W.T.A., ed. 1, 2: 546. *T. fulva* C. E. Hubbard in Kew Bull. 1934: 434; F.T.A. 10: 56; F.W.T.A., ed. 1, 2: 546. *Loudetiopsis fulva* (C. E. Hubbard) Conert in Engl., Bot. Jahrb. 77: 287 (1957). *Diandrostachya chrysothrix* (Nees) Jac.-Fél. in J. Agric. Trop. 7: 408 (1960). *D. fulva* (C. E. Hubbard) Jac.-Fél. l.c. (1960). Up to 1·2 m. high, with panicles of nodding triads of spikelets covered in long brown hairs; dry soils.
Guin.: Timbo (Sept.) *Pobéguin* 1758! **S.L.:** Bintumane Peak, 5,000–6,000 ft. (Jan., Nov.) *T. S. Jones* 110! *Jaeger* 404! Sankan Biriwa, 6,080 ft. (Jan.) *Cole* 156! Tingi Mts. (Dec.) *Morton* SL 2961! **Togo Rep.:** Badou to Atakpame (Sept.) *Rose Innes* GC 31360! **N. Nig.:** Jos Plateau (Aug.) *Lely* P493! Jos to Bukuru (Aug.) *Keay* FHI 20078! William Camp, Jos Plateau, 4,200 ft. (Sept.) *Lawlor & Hall* 646! Sha (Sept.) *Hall* K2! Hoss (Apr.) *Tuley* 1521! Also in Brazil and Paraguay.

6. **L. scaëttae** (*A. Camus*) *W. D. Clayton* in Kew Bull. 21: 123 (1967). *Tristachya scaëttae* A. Camus in Bull. Soc. Bot. Fr. 85: 556 (1939). *Diandrostachya scaëttae* (A. Camus) Phipps in Bol. Soc. Brot. 41: 202 (1967). Up to 2 m. high; shallow soils over rock or ironstone.
Guin.: Tinka Mt., Dalaba (Nov.) *Schnell* 2255! Dalaba-Diaguissa Plateau, 1000–1300 m. (Sept.–Oct.) *Chev.* 18373! **S.L.:** Bintimani (Nov.) *Morton* SL 2667! **Iv. C.:** Ferkessédougou to Niangbo (Dec.) *Adjanohoun* 221a! Bouna to Bondoukou (Oct.) *Rose Innes* GC 31546! **U. Volta:** Sindou (Sept.) *Aké Assi* 6368! *Adam* 15171! *Scaëtta* 2303! Batié to Bouna (Oct.) *Rose Innes* GC 31541! **Ghana:** Sawla to Wa (Oct.) *Rose Innes* GC 32399! Mole Game Reserve (Oct.) *Rose Innes* GC 32413! Damongo-Tamale-Kintampo junction (Oct.) *Rose Innes* GC 31119! Zan, Tamale to Yendi (Nov.) *Rose Innes* GC 32466! Nakpanduri to Yendi (Sept.) *Rose Innes* GC 32154! **Togo Rep.:** Kande to Sansanné Mango (Oct.) *Rose Innes* GC 31408! Sansanné Mango to Dapongo (Oct.) *Rose Innes* GC 31432! **Dah.:** Tanguieta to Kobli (Nov.) *Risopoulos* 1273! **N. Nig.:** Ilorin to Jebba (Nov.) *Jackson* 2411!
[Intergrades with the preceding species, the specimens from Guinée differing from *L. chrysothrix* only in the possession of a 2-toothed callus.]

7. **L. capillipes** (*C. E. Hubbard*) *Conert* in Engl., Bot. Jahrb. 77: 284 (1957). *Loudetia capillipes* C. E. Hubbard in Kew Bull. 1934: 432; F.T.A. 10: 45; F.W.T.A., ed. 1, 2: 544. *Tristachya minuta* A. Chev. in Rev. Bot. Appliq. 14: 43 (1934), partly, without proper descr. Slender perennial up to 75 cm. high; shallow soils over ironstone.
Guin.: *Pobéguin* 1757! Madina Tossékré (Oct.) *Adam* 12547! Tombo, Labé (July) *Adames* 315! Pita (Sept.) *Adames* 359! **Iv. C.:** Séguélo (Oct.) *Aké Assi* 6591!

8. **L. baldwinii** (*C. E. Hubbard*) *Phipps* in Kirkia 5: 249 (1966). *Loudetia baldwinii* C. E. Hubbard in Kew Bull. 4: 356 (1949). Loosely tufted, about 60 cm. high; panicle 2·5–5 cm. broad.
Guin.: Macenta (Oct.) *Baldwin* 9757!

9. **L. falcipes** (*C. E. Hubbard*) *Phipps* in Kirkia 5: 249 (1966). *Loudetia falcipes* C. E. Hubbard in Kew Bull. 12: 61 (1957). Loosely tufted, about 75 cm. high; panicle up to 10 cm. broad.
Lib.: Genne-Loffa, Kolahun Dist. (Nov.) *Baldwin* 10085! Soplima, Vonjama Dist. (Nov.) *Baldwin* 10058!

10. **L. trigemina** (*C. E. Hubbard*) *Conert* in Engl., Bot. Jahrb. 77: 283 (1957). *Loudetia trigemina* C. E. Hubbard in Kew Bull. 1934: 432; F.T.A. 10: 47; F.W.T.A., ed. 1, 2: 544. Tufted perennial 60 cm. high; panicles up to 10 cm. long; rocky hills.
N. Nig.: Riyom (Nov.) *Clayton* 1409! Shene Mts. (Aug.) *J. Hall* 537! Sha (Sept.) *J. Hall* K25! Panshanu Pass (Aug.) *J. Hall* 491! Wana *Hepburn*! **W. Cam.:** Bafut-Ngemba F. R. (July) *Brunt* 775a!

11. **L. glabrata** (*K. Schum.*) *Conert* in Engl., Bot. Jahrb. 77: 287 (1957). *Trichopteryx glabrata* K. Schum. in Engl., Bot. Jahrb. 24: 336 (1897). *Loudetia glabrata* (K. Schum.) C. E. Hubbard in Kew Bull. 1934: 431; F.T.A. 10: 46; F.W.T.A., ed. 1, 2: 544. *L. villosipes* C. E. Hubbard in Kew Bull. 4: 355 (1949). *Loudetiopsis villosipes* (C. E. Hubbard) Conert in Engl., Bot. Jahrb. 77: 283 (1957). Caespitose grass about 1 m. high; crevices in rock outcrops.
Ghana: Aketebua (Jan.) *A. S. Thomas* D51! Kwahu Tafo (Nov.) *Rose Innes* GC 31655! **S. Nig.:** Okelifi, Ondo Dist. (Nov.) *Onochie* FHI 34223! Akure F. R. (Oct.) *Keay* FHI 22666! Mt. Orosun, Idanre (Oct.) *Keay & Onochie* FHI 21563! Also in E. Cameroun.

12. **L. ambiens** (*K. Schum.*) *Conert* in Engl., Bot. Jahrb. 77: 281 (1957). *Trichopteryx ambiens* K. Schum. in Engl., Bot. Jahrb. 24: 335 (1897). *T. ternata* Stapf in Morot, Journ. Bot. 19: 106 (1905). *Loudetia ambiens* (K. Schum.) C. E. Hubbard in Kew Bull. 1934: 431; F.T.A. 10: 43; F.W.T.A., ed. 1, 2: 544. *L. ternata* (Stapf) C. E. Hubbard l.c. (1934); F.T.A. 10: 43; F.W.T.A., ed. 1, 2: 544. *Loudetiopsis ternata* (Stapf) Conert in Engl., Bot. Jahrb. 77: 282 (1957). *Tristachya parviflora* Hack. (1902). Tussocky perennial up to 2·1 m. high; seasonally marshy soils.
Guin.: Kouroussa (Sept.) *Pobéguin* 910! **Iv. C.:** Kodio Koffi Dist., Tiégouakro to Kodio Koffi (Aug.) *Chev.* 22332! Serébissou (May) *Adjanohoun* 218a! Singrobo (Oct.) *Adjanohoun* 344a! 359a! Dabou

(Nov.) *Leeuwenberg* 1938! **Ghana:** Kintampo to Bamboi Ferry (Oct.) *Rose Innes* GC 31136! Krobo Hill, Mampong Scarp (Nov.) *Rose Innes* GC 30865! Attebubu to Kete Krachi (Oct.) *Rose Innes* GC 30595! Maliato, Togo Plateau (Nov.) *Morton* A3539! Tiasi, Afram Plains (Aug.) *Hall* CC 162! **Togo Rep.:** Palime *Stage* 70! Lama Kara to Kande (Oct.) *Rose Innes* GC 31400! **Dah.:** Parakou (Nov.) *Risopoulos* 1289! **N. Nig.:** Auna to Wara (Feb., Oct.) *Magaji* 112! *Ayuba* 60! Abinsi (Oct.) *Dalz.* 885! Also in E. Cameroun & Angola.

13. **L. thoroldii** (*C. E. Hubbard*) *Phipps* in Kirkia 5: 249 (1966). *Loudetia thoroldii* C. E. Hubbard in Kew Bull. 12: 62 (1957). Tussocky perennial up to 2·1 m. high; poorly drained soils.
U. Volta: Pendjori R. near Dahomey frontier (Dec.) *Scholz* 194! **Ghana:** Tumu to Walembele (Oct.) *Ankrah* GC 20475! Damongo (Nov.) *Thorold* 285! Tamale (Oct.) *Baldwin* 13565! Yendi to Gambaga (Sept.) *Ankrah* GC 20528! Bimbila to Kpandai (Nov.) *Rose Innes* GC 32475!

67. LOUDETIA Hochst. ex Steud., Syn. Pl. Glum. 1: 238 (1854); A. Br. in Flora 24: 713 (1841), *nom. superfl.*;* F.T.A. 10: 12 (1937); C. E. Hubbard in Kew Bull. 1936: 319; Conert in Engl., Bot. Jahrb. 77: 247 (1957); Phipps in Kirkia 4: 98 (1964); Clayton in Kew Bull. 21: 122 (1967).

Spikelets 22–27 mm. long, in groups of 3 on pedicels up to 2 cm. long; lower floret barren and reduced to the 5-nerved lemma; anthers 2, their tips with a minute tuft of hairs; awn 9–20 cm. long; callus unequally 2-toothed .. .. .. 1. *togoensis*
Spikelets up to 17 mm. long (rarely up to 22 mm.) paired or solitary; lower floret with a palea, the lemma 3-nerved; anther tips glabrous:
Upper lemma loosely pilose with hairs up to 2 mm. long; leaf-blades up to 1 m. long and 2 cm. broad; panicle 30–60 cm. long, contracted and dense; spikelets 6–7 mm. long; callus of upper floret truncate; awn 12–18 mm. long, including a column 2–3 mm. long .. .. 2. *phragmitoides*
Upper lemma pubescent or glabrescent:
Annuals; awn 5–12 cm. long; spikelets 14–17 mm. long or rarely longer; lower lemma $\frac{1}{2}$–$\frac{2}{3}$ as long as the spikelet:
Callus of upper floret with two equal teeth; panicle elliptic or oblong, typically 12–20 cm. long and 3–4 cm. broad (excluding awns) .. .. .. 3. *annua*
Callus of upper floret obliquely 1-toothed, pungent, sometimes with a reduced second tooth; panicle linear, contracted, typically 15–30 cm. long and 1–2 cm. broad (excluding awns) .. .. 4. *hordeiformis*
Perennials, usually densely tufted; awn up to 5·5 cm. long; lower lemma from $\frac{3}{4}$ to as long as the spikelet:
Panicle very dense and spike-like, up to 15 cm. long and 6–14 mm. broad; branches very short, up to 6 mm. long, 1–4 spiculate; lower leaf-sheaths densely pilose at the base; lower glume narrowly lanceolate, acute, 7 mm. long, dorsally flattened, setulose from a single line of black tubercles along each lateral nerve 5. *coarctata*
Panicle loose or contracted but not spike-like, up to 60 cm. long and 15 cm. broad:
Glumes acuminate, tapering to a setaceous tip, glabrous or sparsely pilose with tubercle-based hairs; callus narrowly truncate:
Culms stout, erect; leaf-blades 20–40 cm. long; panicle dense, containing numerous spikelets; spikelets 8–15 mm. long; awn 2·5–3 cm. long; stamens 3 6. *flavida*
Culms slender, straggling; leaf-blades 2–5 cm. long; panicle scanty, containing up to 15 spikelets; spikelets 8–9 mm. long; awn 1·5–2 cm. long; stamens 2 7. *jaegerana*
Glumes, especially the lower, obtuse or subacute at the tip; stamens 2:
Callus of upper floret truncate or slightly emarginate; panicle robust, 20–60 cm. long, the branches conspicuously whorled; leaf-blades 30–70 cm. long and 6–15 mm. broad; basal sheaths silky pubescent but scarcely tomentose; nodes glabrous or bearded .. .. 8. *arundinacea*
Callus of upper floret 2-toothed:
Upper lemma obscurely 2-lobed, the lobes up to 0·2 mm. long; lower glume subacute, about half as long as the spikelet; basal leaf-sheaths glabrous or at most pubescent; spikelets 8–10 mm. long, always hairy .. 9. *kagerensis*
Upper lemma acutely 2-lobed, the lobes 0·3–1 mm. long; lower glume obtuse, usually about $\frac{1}{3}$ as long as the spikelet; basal leaf-sheaths usually woolly tomentose in our area; spikelets 8–14 mm. long, mostly glabrous 10. *simplex*

1. **L. togoensis** (*Pilger*) *C. E. Hubbard* in Kew Bull. 1934: 431; F.T.A. 10: 51; Conert l.c. 276; Berhaut, Fl. Sén. ed. 2, 416. *Trichopteryx togoensis* Pilger in Engl., Bot. Jahrb. 37: 128 (1904). *T. crinita* Stapf in Morot, Journ. Bot. 19: 106 (1905); Chev. Bot. 743, and in Rev. Bot. Appliq. 14: 42. *T. figarii* Chiov. (1917). *Arundinella togoensis* (Pilger) Roberty in Bull. I.F.A.N., sér. A, 17: 56 (1955). Annual up to 1 m. high, with long curved awns; on dry or shallow soils.
Maur.: 15°40′N: 9°30′W. *Rossetti* 61/296! **Sen.:** *Heudelot* 305! **Gam.:** *Ruxton* 131! **Mali:** Koulikoro (Oct.) *Chev.* 2372 (partly)! Ségou to Sansanding (Sept.) *Chev.* 2375! Pého Dioura (Jan.) *Davey* 216! Kanikomboli, Macina (Mar.) *Chev.* 24889! In Tillit (Aug.) *Rossetti* 135/59! **Port. G.:** Piche, Canquelifá (Sept.) *Pereira* 3261! **Guin.:** Fetoré Plain, Passo, Fouta Djalon (Oct.) *Adames* 399! Kassagui to Kinde, Fouta Djalon (Nov.) *Adames* 414! Kouroussa (Aug.–Sept.) *Pobéguin* 492! 505! **Iv. C.:** Gaoua (July) *Adjanohoun* 52*a*! Ferkessédougou to Niangbo (Dec.) *Adjanohoun* 203*a*! Bouna to Bondoukou (Oct.) *Rose Innes* GC 31551! **U. Volta:** Samandeni (Oct.) *Kmoch* 174! Bobo-Dioulasso (Sept.) *Aké Assi* 6385!

*Conservation of this name is under consideration.

Tougan (Sept.) *Aké Assi* 6418! Ouassa to Leo (Oct.) *Rose Innes* GC 31504! Ouagadougou (Sept.) *Scholz* 67! **Ghana:** Bole to Bamboi (Oct.) *Rose Innes* GC 30650! Wa to Dorimon (Oct.) *Rose Innes* GC 31029! Chuchilliga, Laura to Navrongo (Oct.) *Rose Innes* GC 30277! Bolgatanga (Aug.) *Vigne* FH 4624! Yendi to Chereponi (Oct.) *Rose Innes* GC 30966! **Togo Rep.:** Sansanne Mango to Dapango (Oct.) *Rose Innes* GC 31431! Sokodé to Lama Kara (Oct.) *Rose Innes* GC 31394! **Dah.:** Boukombe (Nov.) *Risopoulos* 1255! **Niger:** Niamey (Oct.) *Hagerup* 495! Dasso *White* 7! Dogondoutch *White* 42! **N. Nig.:** Samaru, Zaria Prov. *Thatcher* S548! Kufena hill, Zaria (Sept.) *Clayton* 1310! Zaria (Aug.) *Keay* FHI 28018! Damaturu (July) *De Leeuw* 309! Maiduguri (Oct.) *Johnston* N30! Extends eastwards to the Sudan Republic. (See Useful Plants.)

2. **L. phragmitoides** (*Peter*) *C. E. Hubbard* in Kew Bull. 1934: 428; F.T.A. 10: 18; Berhaut, Fl. Sén. ed. 2, 417. *Trichopteryx phragmitoides* Peter in Fedde Rep., Beih. 40: 1, Anh.: 96 (1930). *T. flammida* of Chev. Bot. 744. *Arundinella flammida* of A. Chev. in Rev. Bot. Appliq. 14: 41 (1934). *Loudetia flammida* of Conert in Engl., Bot. Jahrb. 77: 250 (1957), partly. A pampas-like perennial up to 4·5 m. high, forming dense tussocks on marshy ground. Closely allied to the S. American species *L. flammida* (Trin.) C. E. Hubbard.

Sen.: Niokolo (Nov.) *Adam* 15919! **Mali:** Oyako Valley, Bamako (Dec.) *Duong* 2077! Sikasso to Bobo-Dioulasso (Sept.) *Adam* 15090! **Guin.:** Baffing R. (Oct.) *Pobéguin* 1801! Bambaya (Oct.) *Jaeger* 2131! **S.L.:** Mapotolon to Suribulomia (Nov.) *T. S. Jones* 41! Benekoro (Nov.) *Glanville* 321! Tingi Mts. *Morton & Gledhill* SL 2950! **Lib.:** Zigida, Vonjama Dist. (Oct.) *Baldwin* 10015! Palilah, Gbanga Dist. (Feb.) *Baldwin* 11036! Ganta, Sanokwele Dist. (Sept.-Oct.) *Baldwin* 12016! *Harley* 1018! Sanokwele (Sept.) *Baldwin* 9490! **Iv. C.:** Singrobo (Oct., Nov.) *Adjanohoun* 342*a*! Abouabou Forest *de Wilde* 3163! *Oldeman* 689! **U. Volta:** Cascade de Toussiana (Oct.) *Scholz* 69! **Ghana:** Bole to Bamboi Ferry (Oct.) *Rose Innes* GC 32405! Damongo (Nov.) *Thorold* 304! Yapei Ferry to Kintampo (Nov.) *Rose Innes* GC 30853! Bimbila to Kpandai (Nov.) *Rose Innes* GC 32480! Golokuati to Hohoe (Sept.) *Rose Innes* GC 30915! **N. Nig.:** Ilorin (Oct.) *Ward* 94! Jebba *Barter* 1028! Badeggi (Sept.) *Clayton* 365! Jos Plateau *Batten-Poole* 389! Vom (Nov.) *Gambles* 7! **S. Nig.:** Oyo to Iseyin (Oct.) *Onochie* FHI 34930! Lagos (Feb.) *W. MacGregor* 162! Enugu Extension F.R., Udi Dist. (Sept.) *Onochie* FHI 34103! Agolo *Thomas* 293! Obudu Plateau road (Feb.) *Tuley* 578! **W. Cam.:** Baba, Ndop Plain (Mar.) *Brunt* 122! Throughout tropical Africa. (See Useful Plants.)

3. **L. annua** (*Stapf*) *C. E. Hubbard* in Kew Bull. 1934: 429; F.T.A. 10: 40; Conert l.c. 271; Berhaut, Fl. Sén. ed. 2, 416; Lebrun in Bull. Soc. Bot. Fr. 116: 262 (1969). *Trichopteryx annua* Stapf in Kew Bull. 1897: 295. *T. thorbeckei* Pilger (1914). *Loudetia bidentata* Berhaut in Mém. Soc. Bot. Fr. 1952–54: 11 (1954). Annual, up to 1·2 m. high; humid soils flanking ditches.

Sen.: Kaolak (Oct.) *Berhaut* 5667! Niokolo-Koba, Tambacounda (Oct.) *Adam* 15488! Elinkine, Casamance (Oct.) *Adam* 18261! Forêt de Boudié, Casamance (Nov.) *Berhaut* 6379! **Port. G.:** Boruntuma (Dec.) *Pereira & Correia* 2175! Caium R., Gabu Dist. (Oct.) *Esp. Santo* 3110! **Guin.:** Koundara (Oct.) *Adam* 12759! **U. Volta:** Kapori (Oct.) *Scholz* 75*a*! **Ghana:** Bolgatanga to Bawku (Oct.) *Rose Innes* GC 31100! Bawku to Nakpanduri (Oct.) *Ankrah* GC 20493! **N. Nig.:** *Lely* 761! Zaria Dist. *C. B. Taylor* 30! Jos Plateau (Aug.-Oct.) *Lely* P650! P808! Vodni, Pankshin Div. *Saunders* 51! Extends to Sudan Republic. (See Useful Plants.)

4. **L. hordeiformis** (*Stapf*) *C. E. Hubbard* in Kew Bull. 1934: 431; F.T.A. 10: 41; Conert l.c. 272. *Trichopteryx hordeiformis* Stapf in Kew Bull. 1897: 297 (1897); Chev. Bot. 744, and in Rev. Bot. Appliq. 14: 42; Berhaut, Fl. Sén. ed. 2, 395. *Arundinella hordeiformis* (Stapf) Roberty in Bull. I.F.A.N., sér. A, 17: 56 (1955). Annual, up to 1·5 m. high; disturbed sandy soils.

Sen.: Goulonnibo to Tambacounda (Oct.) *Pitot*! Niokolo-Koba (Nov.) *Adam* 15919! Casamance (Sept.-Oct.) *Adam* 18165! 18274! Tièl (Oct.) *Raynal* 7673. **Gam.:** Yoroberi Kunda (Sept.) *Pirie* 47/33! **Mali:** Koulikoro (Oct.) *Chev.* 2372 *bis*! Ségou to Sansanding (Sept.) *Chev.* 2375! Keribane (Oct.) *Davey* 84! **Port. G.:** Praia Varela, Suzana (Nov.) *Esp. Santo* 3140! **Iv. C.:** Séguéla to Vavoua (Oct.) *Adjanohoun* 312*a*! Bouaké to Langouassou (July) *Chev.* 22149! Bouaké to M'Bayakro (Oct.) *Adjanohoun* 350*a*! Bouna to Bondoukou (Oct.) *Rose Innes* GC 31557! Dabakala to Katiola (Nov.) *Aké Assi* 9233! **U. Volta:** Sindou (Sept.) *Aké Assi* 6364! Ouassa to Leo (Oct.) *Rose Innes* GC 31492! Nankoum (Oct.) *Scholz* 168! Linguékoro (Nov.) *Scholz* 86! **Ghana:** Lawra to Navrongo (Oct.) *Rose Innes* GC 30275! Sawla to Damongo (Oct.) *Rose Innes* GC 30987! Damongo (Nov.) *Thorold* 284! Tamale (Nov.) *Williams* 1023! Gambaga Scarp, Nakpanduri (Oct.) *Rose Innes* GC 30738! **Togo Rep.:** Dapango to Sansanné Mango (Oct.) *Rose Innes* GC 31433! **Dah.:** Abomey to Boguila (Feb.) *Chev.* 23183! Boukombe (Nov.) *Risopoulos* 1264! **Niger:** Zinder (Nov.) *Money-Kyrle* Fe21! Niamey (Oct.) *Hagerup* 502*a*! Dasso *White* 18! 12! **N. Nig.:** Sokoto (Sept.) *Dalz.* 504! Katsina (Oct.) *Clayton* 1340! Jebba *Barter* 954! Acharane F.R., Okura Dist. *Latilo* FHI 34013! Abinsi (Oct.) *Dalz.* 883! **S. Nig.:** Enugu Extension F.R., Udi Dist. (Sept.) *Onochie* FHI 34105! Ngwo, Udi Dist. (Oct.) *Jones* 6777! Igbete (Nov.) *Gledhill* 709! Iseyin (Oct.) *Gledhill* 1022! Extends east to Sudan Republic. (See Useful Plants.)

5. **L. coarctata** (*A. Camus*) *C. E. Hubbard* in Kew Bull. 1934: 428; F.T.A. 10: 37; Conert l.c. 268. *Tristachya coarctata* A. Camus in Bull. Soc. Bot. Fr. 53: 774 (1933). *T. triticoides* A. Camus & C. E. Hubbard ex A. Chev. in Rev. Bot. Appliq. 14: 43 (1934). Densely tufted perennial, 1–1·2 m. high; poorly drained soils.

Guin.: Timbi Madina (Jan.) *Langdale-Brown* 2551! Dalaba *Chev.* 34898! (Oct.) *Adames* 324! Timbo to Ditinn (Sept.) *Chev.* 18445! **N. Nig.:** Anara F.R., Zaria (Oct.) *Keay* FHI 5447! Auna to Warra (Feb.) *Magaji* 111! Vom, Jos Plateau (Oct.) *Gambles* 57! Sugu (Dec.) *Peter & Tuley* 74! Also in the Congo, Sudan & Zambia.

6. **L. flavida** (*Stapf*) *C. E. Hubbard* in Kew Bull. 1934: 429; F.T.A. 10: 34; Conert l.c. 264. *Trichopteryx flavida* Stapf in Kew Bull. 1897: 298. *T. acuminata* Stapf l.c. 297 (1897). *T. glabra* Hack. (1901). *Loudetia acuminata* (Stapf) C. E. Hubbard in Kew Bull. 1934: 429; F.T.A. 10: 33; F.W.T.A., ed. 1, 2: 544; Conert l.c. 263. Coarse tufted perennial 1·2–1·5 m. high with golden brown panicles; shallow soils in open savanna woodland.

U. Volta: Tumu to Leo (Oct.) *Rose Innes* GC 31489! Zoaga (Oct.) *Scholz* 158! **Ghana:** Wa to Lawra (Oct.) *Ankrah* GC 20446! Tumu to Walembele (Oct.) *Rose Innes* GC 32269! Yapei Ferry (Oct.) *Rose Innes* GC 30675! Bawku to Nakpanduri (Oct.) *Ankrah* GC 20486! Gambaga F.R. (Oct.) *Ankrah* GC 20270! **Togo Rep.:** Dapango to Timbou (Oct.) *Rose Innes* GC 31437! **N. Nig.:** Biu (Nov.) *Magaji & Oche* MG 51! Jebba *Barter* 953! Lokoja (Oct.) *Dalz.* 291! Bogalo, Benue R. *Macleod* 80! Vom, 4,000 ft. (Oct.) *Gambles* 34! Tropical Africa and Transvaal.

7. **L. jaegerana** *A. Camus* in J. Agric. Trop. 1: 212 (1954). Slender straggling perennial with culms up to about 30 cm. long.

S.L.: Mt. Loma (Sept.) *Jaeger* 1705! 7669! Tingi Mts. (Jan.) *Morton & Gledhill* SL 3036!

8. **L. arundinacea** (*Hochst. ex A. Rich.*) *Steud.* Syn. Pl. Glum. 1: 238 (1854); F.T.A. 10: 20; Conert l.c. 254. *Tristachya arundinacea* Hochst. ex A. Rich., Tent. Fl. Abyss. 2: 417 (1851). *Trichopteryx arundinacea* (Hochst. ex A. Rich.) Hack. ex Engl. (1891); Chev. Bot. 743, and in Rev. Bot. Appliq. 14: 42. *Loudetia eriopoda* C. E. Hubbard (1934). Tufted perennial up to 3 m. high; dry soils, commonly associated with inselbergs and rock outcrops.

Sen.: Niokolo-Koba (Nov.) *Adam* 15919! **Guin.:** Kindia *Jac.-Fél.* 181! Dalaba to Songueta *Chev.* 20171! Timbo (Oct.) *Pobéguin* 1729! Macenta (Oct.) *Baldwin* 9751! **S.L.:** Kambia (Dec.) *Deighton* 814! Binle (Oct.) *Adames* 255! Mabonto (Oct.) *Thomas* 3601! Matamba to Manyakoi (Oct.) *Glanville* 64! Musaia (Mar.) *Deighton* 5398! **Lib.:** Karmadhun, Kolahun Dist. (Nov.) *Baldwin* 10222! **Iv. C.:** Séguélo (Oct.) *Aké Assi* 6586! Séguéla to Vavoua (Oct.) *Adjanohoun* 271*a*! Séguéla to Bouaké (Oct.) *Adjanohoun* 275*a*! Dimbokro to Toumodi (Sept.) *Adjanohoun* 211*a*! Bouna to Bondoukou (Oct.) *Rose Innes* GC

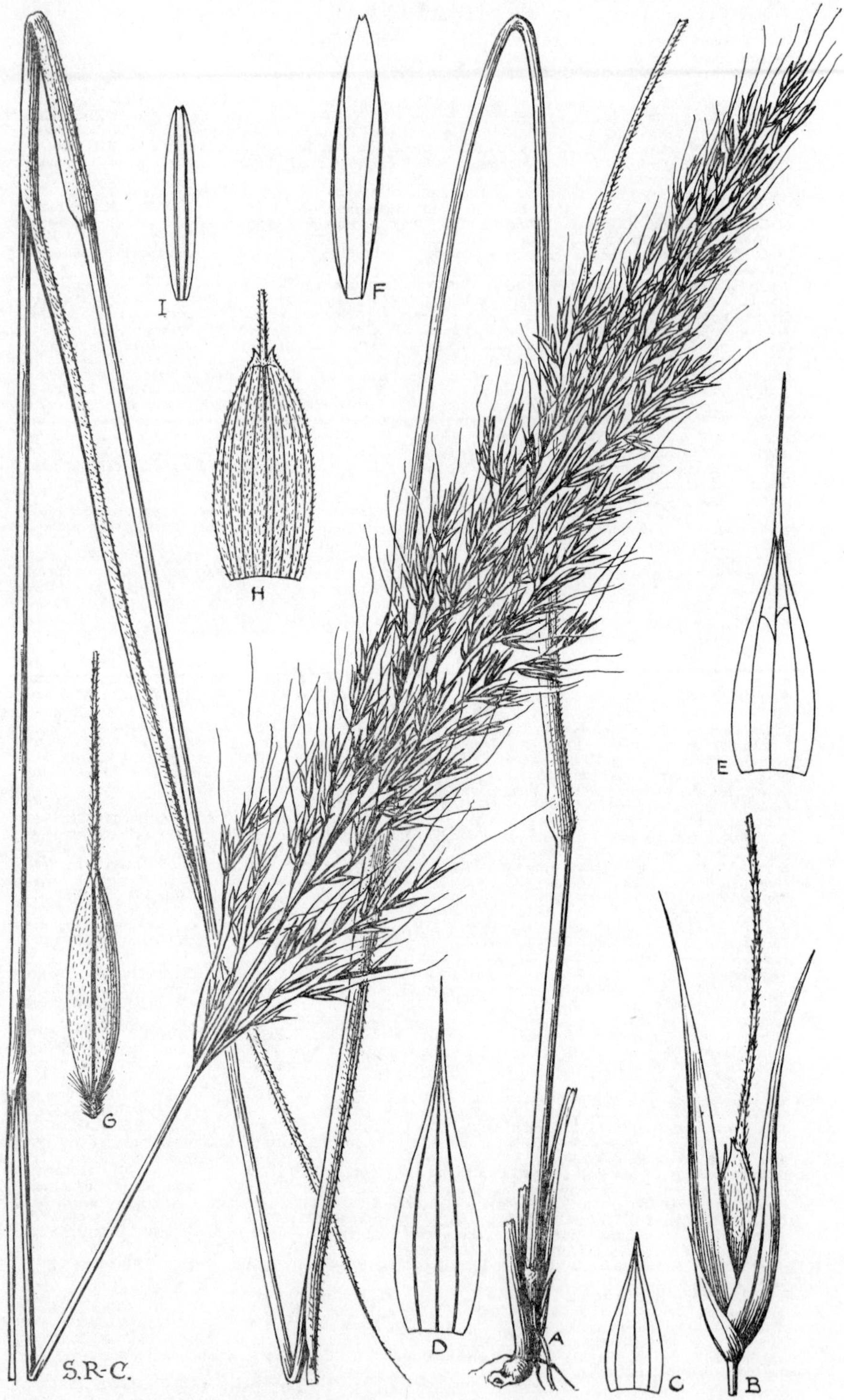

Fig. 437.—LOUDETIA FLAVIDA (*Stapf*) *C. E. Hubbard*
(GRAMINEAE-ARUNDINELLEAE).

A. plant. B, spikelet. C, lower glume. D, upper glume. E, lower lemma. F, lower palea. G, upper floret. H, upper lemma. I, upper palea.

31549! **U. Volta:** Takalédougou (Sept.) *Aké Assi* 6352! Chutes Comoé (Sept.) *Adam* 15298! Banfora (Sept.) *Adam* 15148! Cascade de Toussiana (Oct.) *Scholz* 76! **Ghana:** Nkwanta, Buem-Krachi Dist. (Apr.) *Hall* 1393! Damongo (Sept.) *Goodall* GC 15914! Kintampo to Bamboi Ferry (Oct.) *Ankrah* GC 20319! Bimbila to Yendi (Sept.) *Rose Innes* GC 32195! Maliato, Togo Plateau (Nov.) *Morton* A3527! **Togo Rep.:** Kande to Sansanné Mango (Oct.) *Rose Innes* GC 31409! **N. Nig.:** Kabba *Parsons*! Jos Plateau (Aug.) *Lely* P498! Fadan Karshi, Wamba Div. (Oct.) *Kennedy* FHI 8035! Naraguta F.R., Jos (Dec.) *Kennedy* FHI 1154! Dakemi, Adamawa Div. (Nov.) *Hepper* 1373! **S. Nig.:** Lagos *W. MacGregor* 240! Usonigbe F.R., Sobo Plains (Oct.) *Keay & Onochie* FHI 21596! Agulu, Awka Dist. (Nov.) *Keay* FHI 22282! Ifon F.R., Owo Dist. (Oct.) *Adebusuyi* FHI 43581! Oban *Talbot* 843! **W. Cam.:** Banso, 5,800 ft. (June) *Brunt* 586! Nyen, 5,000 ft. (May) *Brunt* 1127! Mbenguri, 4,000 ft. (Mar.) *Brunt* 1036! Bamenda, 6,000 ft. (Jan.) *Migeod* 347! Kambo *Boughey*! South to Angola & Mozambique. (See Useful Plants.)

[The glumes and lower lemma are usually glabrous, but a few specimens bear a tuft of tubercle-based hairs at the tip of the lower lemma. In var. *hensii* (De Wild.) C. E. Hubbard ex Pichi-Sermolli (1951) (var. *trichantha* (Peter) C. E. Hubbard ex Hutch. in F.W.T.A., ed. 1, 2: 544) an East African variety recorded from Sierra Leone in our area, the glumes are pilose with tubercle-based hairs.]

9. **L. kagerensis** (*K. Schum.*) *C. E. Hubbard ex Hutch.* in F.W.T.A., ed. 1, 2: 544 (1936); F.T.A. 10: 28; Conert l.c. 255; Aké Assi, Contrib. 2: 282. *Trichopteryx kagerensis* K. Schum. in Engl., Pflanzenw. Ost.-Afr. C: 109 (1895). *T. reflexa* Pilger in Engl., Bot. Jahrb. 33: 52 (1902). *Tristachya kagerensis* (K. Schum.) A. Chev. in Rev. Bot. Appliq. 14: 43 (1934). Up to 1 m. high with wiry culms and fine leaves; shallow soils.

Guin.: Friguiagbé (Aug.) *Chillou* 601! Dalaba (Sept.) *Schnell* 6892! Dalaba to Diaguissa (Sept.-Oct.) *Chev.* 18861! Timbo (Oct.) *Pobéguin* 1728! **S.L.:** Freetown (Dec.) *Deighton* 796! Waterloo (Aug.) *Melville & Hooker* 305! Brookfields (Oct.) *Deighton* 2168! Sankan Biriwa, 4,500 ft. (Jan.) *Cole* 129! Bintumane Peak, 5,000 ft. (Jan.) *T. S. Jones* 152! **Iv. C.:** Zuénoula to Bouaflé (Oct.) *Adjanohoun* 323a! Dougba, Oumé to Hiré (Sept.) *Adjanohoun* 207a! Singrobo (Oct.) *Adjanohoun* 340a! **Ghana:** Buipe Ferry (Oct.) *Rose Innes* GC 31123! Yinahin Range, Kumasi (Nov.) *Rose Innes* GC 30528! Kwahu Tafo (Dec.) *Adams* 4927! Pepease to Nkwantanang (Apr.) *Morton* A777! Yendi to Kete Krachi (Oct.) *Ankrah* GC 20414! **N. Nig.:** Mambila (Aug.) *De Leeuw* 1789! Mayo Daga (Nov.) *Magaji & Tuley* 1877! East Africa, Congo and Angola.

10. **L. simplex** (*Nees*) *C. E. Hubbard* in Kew Bull. 1934: 431; F.T.A. 10: 25; Conert l.c. 259; Aké Assi, Contrib. 2: 282. *Tristachya simplex* Nees, Fl. Afr. Austr. 269 (1841). *Trichopteryx nigritiana* Stapf in Kew Bull. 1897: 297. *T. camerunensis* Stapf in Kew Bull. 1897: 296. *Loudetia camerunensis* (Stapf) C. E. Hubbard in Kew Bull. 1934: 431; F.T.A. 10: 24; F.W.T.A. ed. 1, 2: 544; Conert l.c.: 258. *Arundinella simplex* (Nees) Roberty in Bull. I.F.A.N., sér. A, 17: 56 (1955). Tufted perennial 0·3–1·5 m. high; a common grass on shallow soils.

Sen.: Niokolo-Koba (Oct.) *Adam* 15615! Kanéméré (July) *Fotius* K228 (*fide* Lebrun in Bull. Soc. Bot. Fr. 116: 263 (1969)). **Mali:** Bamako (Sept.) *Adam* 14991! Bougouni to Sikasso (Sept.) *Adam* 15046! **Guin.:** Bandakure, Fouta Djalon *Adames* 283! Timbo (Oct.) *Pobéguin* 1725! Kouroussa (Aug.) *Pobéguin* 504! Fon Massif, 1,000–1,100 m. (Aug.) *Schnell* 6616! **S.L.:** Russell (Dec.) *Deighton* 3285! Sakasakala (Sept.) *Deighton* 5141! Mt. Loma (Sept.) *Jaeger* 1674! **Lib.:** Nimba, Sanokwele Dist. (Sept., Dec.) *Adam* 16459! 20351! *Adames* 481! **Iv. C.:** Upper Sassandra and Cavally (Dec.-May) *Portères* 403! 549! Tehini (Aug.) *Oldeman* 289! Bouaflé to Zuénoula (Oct.) *Adjanohoun* 302a! Fetekro to Tiebissou (July) *Chev.* 22192! **U. Volta:** Samandeni (Oct.) *Kmoch* 173! Gaoua to Batié (Oct.) *Rose Innes* GC 31524! Ouagadougou (Sept.) *Irvine* 4685! Doutié (Sept.) *Scholz* 70! **Ghana:** Bole (Sept.) *Irvine* 4754! Navrongo (Sept.) *Vigne* FH 4628! Salaga to Tamale (Oct.) *Rose Innes* GC 30313! Vane to Hohoe (Sept.) *Ankrah* GC 20220! Wiawia, Hohoe Dist. (Oct.) *St. Cl.-Thompson* FH 1535! **Togo Rep.:** Mountain top, Ahamansu, Togo frontier (Sept.) *Rose Innes* GC 31327! Badou to Atakpame (Sept.) *Rose Innes* GC 31347! Sokodé to Lama Kara (Oct.) *Rose Innes* GC 31392! **Dah.:** Bambereka to Ina (Sept.) *Risopoulos* 1203! **N. Nig.:** Jebba *Barter*! Naraguta F.R., Jos Plateau (Aug.) *Keay* FHI 20056! Samaru, Zaria (Sept.) *Clayton* 1277! Cheranchi, Katsina (Oct.) *Clayton* 1346! Abinsi (Oct.) *Dalz.* 867! Vogel Peak, 5,500 ft. (Nov.) *Hepper* 2752! Obudu Plateau (Dec.) *Tuley* 58!* **S. Nig.:** Opi, Nsukka Div. (Aug.) *Jones* 6439! Nsukka (Jan.) *Okigbo* 124! **W. Cam.:** Ndop, 3,800 ft. (May) *Brunt* 1135! Santa, 6,000 ft. (July) *Brunt* 854! 856!* Bambui (Dec.) *Pedder* 6! Bamenda, 5,000 ft. (Dec.) *Baldwin* 13856!* Cam. Mt., 7,000–8,000 ft. (Nov.-Feb.) *Mann* 2080!* *Migeod* 217!* *Maitland* 1036!* Tropical and South Africa. (See Useful Plants.)

[A form from Cameroun with large dark brown spikelets has been separated as *L. camerunensis*. These specimens (marked*) are quite distinct in our area, but similar plants occur in East Africa where they merge into the main body of the species. It seems that *L. camerunensis* should be included in *L. simplex*, of which it is merely an extreme variant.—W.D.C.]

68. DANTHONIOPSIS Stapf in Hook., Ic. Pl. 31: t. 3075 (1916); F.T.A. 10: 73 (1937); Hubbard in Kew Bull. 1936: 319 (1936); Jacques-Félix in Rev. Bot. Appliq. 10: 422 (1950); Conert in Engl., Bot. Jahrb. 77: 317 (1957); Phipps in Kirkia 4: 113 (1964); Clayton in Kew Bull. 21: 121 (1967).

Lower leaf-sheaths densely and shortly silky-villous at the base; blades broadly linear, up to 60 cm. long and 2 cm. broad; panicle narrowly ovate, about 20 cm. long, loose; spikelets 8–10 mm. long, variegated with purple or brown; glumes glabrous, 3-nerved, the lower ovate, 5–6 mm. long, the upper as long as the spikelet; lower lemma 5-nerved; stamens 3; upper lemma sparsely pilose, with a transverse line of hairs below each lateral lobe; keels of the upper palea narrowly winged throughout; awn 14–16 mm. long, flattened below .. .. *chevalieri*

D. chevalieri *A. Camus & C. E. Hubbard* in Rev. Bot. Appliq. 14: 780 (1934); F.T.A. 10: 76; A. Chev. in Rev. Bot. Appliq. 14: 44 (1934); Jacques-Félix l.c. 424; Conert l.c. 320; Berhaut, Fl. Sén. ed 2, 417. *Arundinella chevalieri* (A. Camus & C. E. Hubbard) Roberty in Bull. I.F.A.N. 17: 55 (1955). Tufted perennial 1·8–2·7 m. high.

Guin.: Labé to Mali (Nov.) *Chev.* 34603! Dalaba to Souguéta *Chev.* 20190! Timbo (Oct.) *Pobéguin* 1727! Ditinn (Oct.) *Pitot*! Sériba (Oct.) *Pitot*! **S.L.:** Serikudi (Nov.) *Glanville* 313! Musaia (Dec.) *Deighton* 4523!

69. TRICHOPTERYX Nees in Lindl., Nat. Syst. ed. 2: 449 (1836); F.T.A. 10: 4; Hubbard in Kew Bull. 1936: 319 (1936); Conert in Engl., Bot. Jahrb. 77: 238 (1957); Phipps in Kirkia 4: 119 (1964); Clayton in Kew Bull. 21: 120 (1967).

Annual; culms erect or geniculately ascending; glumes glabrous or bearing stiff tubercle-based hairs .. .. 1. *elegantula*

Perennial; culms wiry and trailing; glumes minutely pubescent .. 2. *marungensis*

1. **T. elegantula** (*Hook. f.*) *Stapf* in Hook., Ic. Pl. 24: t. 2394 (1895); F.T.A. 10: 11; Conert l.c. 246. *Arundinella elegantula* Hook. f. in J. Linn. Soc. 7: 233 (1864). *Trichopteryx glanvillei* C. E. Hubbard in Kew Bull. 1934: 426; F.T.A. 10: 11; F.W.T.A., ed. 1, 2: 545. A delicate annual, 5–17 cm. high.
S.L.: Bintumane (Nov., Jan.) *Glanville* 335! *T. S. Jones* 100! 147! *Morton* SL2660! Mt. Loma (Oct.) *Jaeger* 7861! 8312! **N. Nig.:** Njawai (Nov.) *Jackson, Magaji & Tuley* 1908! Chappal Waddi (Nov.) *Jackson, Magaji & Tuley* 2026! **S. Nig.:** Obudu Plateau (Nov.) *Tuley* 978! **W. Cam.:** Cam. Mt., 6,000–8,500 ft. (Nov., Dec.) *Migeod* 204! *Mann* 2092! Banso to Bamenda (Oct.) *Tamajong* FHI 23489! L. Bambuluwe (Sept.) *Ujor* FHI 30209! Bambui *Bumpus* Bam/11! South to Malawi and Zambia.
2. **T. marungensis** *Chiov.* in Nuovo Giorn. Bot. Ital. n.s. 26: 67 (1919); F.T.A. 10: 9 (1937). *T. gracillima* C. E. Hubbard in Kew Bull. 1934: 425; F.T.A. 10: 8; Phipps l.c. 120; Conert l.c. 243. *T. decumbens* C. E. Hubbard in Kew Bull. 4: 350 (1949); Conert l.c. 244. Trailing wiry culms 30–60 cm. long; swampy places.
N. Nig.: Tugan (Dec.) *Daramola* FHI 62430! Mayodaga, Mambila Plateau (Apr.) *Gbile & Daramola* FHI 62892! **W. Cam.:** Kumbo to Oku (Feb.) *Hepper & Charter* 2019! Nkambe (Feb.) *Brunt* 991! Mbiami (June) *Brunt* 751! Also in Uganda, and southwards to Rhodesia.

70. ISACHNE R. Br., Prodr. Fl. Nov. Holl. 196 (1810); F.T.A. 9: 1090 (1934).

Lemmas similar in shape and texture, and more or less in size; florets almost contiguous:
Florets subrotund, indurated; panicle diffuse; leaf-blades linear-lanceolate:
Lemmas minutely pubescent; glumes slightly shorter than the florets 1. *buettneri*
Lemmas glabrous; glumes as long as or slightly longer than the florets 2. *mauritianus*
Florets elliptic, subacute, thinly coriaceous, usually glabrous; panicle contracted about the primary branches; leaf-blades linear .. .. .. 3. *angolensis*
Lemmas dissimilar, the lower larger, thinner, glabrous, the upper smaller, minutely hairy; florets separated by a minute rhachilla-joint; spikelets contracted about the primary branches:
Leaf-blades linear-lanceolate, 6–12 cm. long, glabrescent below; spikelets about 2 mm. long .. .. 4. *guineensis*
Leaf-blades lanceolate, 3–6 cm. long, pubescent below; spikelets about 1·5 mm. long, rarely longer .. .. 5. *kiyalaensis*

1. **I. buettneri** *Hack.* in Verh. Bot. Ver. Prov. Brandenb. 31: 69 (1889); F.T.A. 9: 1091; Chev. Bot. 724, and in Rev. Bot. Appliq. 14: 40. Ascending or half-climbing in forest shade.
Port. G.: S. Domingo (Aug.) *Pereira* 3131! **S.L.:** Kasewe Hills F.R. (June) *Deighton* 5838! Potoru (Apr.) *Deighton* 1661! Heddle's Farm, Freetown (May) *Deighton* 1194! Kuntaia (June) *Thomas* 411! Makali (July) *Deighton* 4398! **Lib.:** Reppue's Town *Linder* 370! Ganta (Sept.) *Baldwin* 12177! Gretown (July) *Baldwin* 6760! Diebla (July) *Baldwin* 6303! Cape Palmas *T. Vogel*! **Iv. C.:** Cavally R. *Chev.* 19547! Abidjan (Sept., Oct.) *Leeuwenberg* 1809! *de Wilde* 886! Yapo Forest (Oct.) *Adjanohoun* 403*a*! **Ghana:** Shiare (Apr.) *Hall* 1407! Mpraeso Scarp (Dec.) *Adams* 5178! Aburi (June) *Ankrah* GC 20047! Kade (Jan.) *Ankrah* GC 20332! Aiyinase (July) *Irvine* 5068! **N. Nig.:** Nupe *Barter* 1362! **S. Nig.:** Benin (Sept.) *Maggs* 7! Opobo *Jeffreys* 7! Oban *Talbot* 769! Ogbokum (May) *Jones & Onochie* FHI 18882! Obudu (Apr.) *Haines* 360! **W. Cam.:** Kumba (Apr.) *Ejiofor* FHI 29346! **F. Po:** Monte Balea (Dec.) *Guinea* 354! Also in E. Cameroun, the Congos, Uganda and Zambia.
2. **I. mauritiana** *Kunth* Rev. Gram. 1: 243 (1830). *I. aethiopica* Stapf & C. E. Hubbard in Kew Bull. 1933: 300; F.T.A. 9: 1092. In forest shade.
S. Nig.: Obudu Plateau (Nov.) *Tuley* 1033! **W. Cam.:** Buea (Mar.) *Boughey* GC 6762! Cam. Mt. (Dec.) *Maitland* 879! *Hinds* C71! Esele (Feb.) *Maitland* 1032! Mimbia (Mar.) *Brenan* 9547! **F. Po:** (Dec.) *Boughey* GC 10819! Moka (Jan., Sept.) *Guinea* 1967*a*! *Wrigley* 655! Pico Serrano (Jan.) *Guinea* 2013! Also in S. Tomé, and throughout E. Africa, extending through the Congo, Rhodesia and Malawi to Mozambique, Madagascar and Mauritius.
3. **I. angolensis** *Rendle* Cat. Welw. 2: 166 (1899). *Coelachne occidentalis* Jac.-Fél. in Adansonia, n.s. 6: 533 (1966). Weakly erect perennial 30–45 cm. high; damp shady places.
N. Nig.: Tugan, Mambila plateau (Dec.) *Daramola* FHI 62431! Also in E. Cameroun, Zambia, Malawi and Angola.
4. **I. guineensis** *Stapf & C. E. Hubbard* in Kew Bull. 1933: 302; F.T.A. 9: 1094; A. Chev. in Rev. Bot. Appliq. 14: 41. Culms ascending, rooting at the lower nodes.
Guin.: Mali to Mt. Loura (Nov.) *Chev.* 37601! Timbi (Oct.) *Adam* 12520!
5. **I. kiyalaensis** *Robyns* in Bull. Jard. Bot. Brux. 9: 199 (1932); F.T.A. 9: 1096. *Panicum kiyalaense* Vanderyst in Bull. Agric. Congo Belge 10: 246 (1919), provisional name. *Isachne caillei* A. Chev. in Rev. Bot. Appliq. 14: 41 (1934). Weakly ascending, up to about 45 cm. high, culms slender; marshy places in shade.
Guin.: Dalaba (Apr.) *Chev.* 18139! Kindia (Dec.) *Jac.-Fél.* 7443! Koba (Nov.) *Jac.-Fél.* 7268! Timbi (Oct.) *Adam* 12512! **S.L.:** Giema (Apr.) *Deighton* 1640! Regent (Oct.) *Deighton* 2183! Yetaya (Sept.) *Thomas* 2446! Port Loko (Dec.) *Thomas* 6521! Kenema (Jan.) *Thomas* 7542! **Lib.:** Ganta (Oct.) *Harley* 2052! Monrovia (Oct.) *Cooper* 26! Kandessu (Dec.) *Baldwin* 10667! Mecca (Nov.) *Baldwin* 10434! Taurazon (Mar.) *Baldwin* 11482! **Iv. C.:** Abouabou (Nov.) *Adjanohoun* 452*a*! Grabo to Olodio (Feb.) *Aké Assi* 9429! **Ghana:** Awuabo to Krisin (Mar.) *Rose Innes* GC 30880! **S. Nig.:** Jamieson R. (Jan.) *Stanfield & Lowe* 1947! Sapoba (Feb.) *Onochie* FHI 31944! Udi (Sept.) *Onochie* FHI 34089! Aguku *Thomas* 631! Also in Congo and Angola.

71. HETERANTHOECIA Stapf in Hook., Ic. Pl. 30: t. 2927 (1911); F.T.A. 9: 1098 (1934).

Leaf-blades lanceolate, short, 1·5–3 cm. long, 3–5 mm. broad; ligule a line of long fine hairs; spikelets borne in short one-sided racemes 0·5–1 cm. long scattered along a main axis; raceme rhachis ciliate on the margins, tapering to an acute tip; spikelets 1·5–2 mm. long; glumes subequal, a little shorter than the spikelet; lemmas chartaceous, the lower as long as the spikelet, subacuminate, glabrous except at the base; the upper half as long, pubescent .. .. *guineensis*

H. guineensis (*Franch.*) *Robyns* in Bull. Jard. Bot. Brux. 9: 201 (1932); F.T.A. 9: 1099; Berhaut, Fl. Sén. ed. 2, 403. *Dinebra guineensis* Franch. in Bull. Soc. Hist. Nat. Autun 8: 376 (1895). *Heteranthoecia isachnoides* Stapf (1911)—A. Chev. in Rev. Bot. Appliq. 14: 41. Annual, up to 60 cm. high; culms ascending, rooting at the lower nodes; in swamps or pond margins.
Guin.: Timbi (Dec.) *Jac.-Fél.* 7421! Nimba Mts. (Apr.) *Schnell* 5038! **S.L.:** Rowala (July) *Thomas* 1091! Pateya to Njagbema (July) *Deighton* 5791! Lunsar (Mar.) *Jordan* 26! Tebor (July) *Deighton* 5808! Kissy Tungi (Sept.) *Jordan* 528! **Lib.:** *Harley* 1750! Ganta (Jan.) *Harley* 2104! Nimba (Jan.) *Adam* 20667! **N. Nig.:** Nupe *Barter* 1348! Kontagora to Ibeto (Jan.) *Meikle* 1091! Jos (Oct.) *Hepper* 1185! **S. Nig.:** Enugu *Jones* FHI 6226! Extends eastwards to Uganda and Tanzania, and southwards to the Congo and Angola.

72. SETARIA P. Beauv., Ess. Agrost. 51 (1812); F.T.A. 9: 768 (1920–30); *nom. cons.*

Inflorescence a cylindrical spike-like panicle:
- Bristles retrorsely barbellate, clinging tenaciously to clothing; upper glume as long as the spikelet, concealing the smooth upper lemma .. .. 1. *verticillata*
- Bristles antrorsely barbellate or smooth:
  - Annual; upper glume up to ½ as long as the spikelet, exposing the conspicuously rugose upper lemma; spikes 2–8 cm. long, rarely longer, with tawny bristles 6–8 mm. long; spikelets 2–2·3 mm. long .. .. 2. *pallide-fusca*
  - Perennials:
    - Basal sheaths pubescent with silky white hairs below; upper glume ⅘ to almost as long as the spikelet, concealing the obscurely rugose upper lemma; spikelets turgid; spikes 5–12 cm. long, yellowish green .. .. .. 3. *ciliolata*
    - Basal sheaths glabrous below; upper glume ¼–¾ as long as the spikelet, exposing the conspicuously rugose upper lemma; spikelets more or less dorsally compressed:
      - Culm 5–10-noded, 3–6 mm. diam. at base, compressed; basal sheaths strongly keeled and often flabellately interleaved, but falling away in older specimens; leaf-blades 4–10 mm. broad, glabrous; spikes 10–30 cm. long .. 4. *anceps*
      - Culms 2–4-noded, 1–3 mm. diam. at base, not compressed; basal sheaths more or less rounded on the back or weakly keeled, not flabellately folded; leaf-blades 2–6 mm. wide; spikes 5–20 cm. long:
        - Old leaf-sheaths breaking into a dense mass of fibres at the base; leaf-blades glabrous or sparsely hairy .. .. 5. *aurea*
        - Old leaf-sheaths not becoming fibrous at the base; leaf-blades glabrous 6. *sphacelata*

Inflorescence a branched panicle:
- Leaf-blades flat, not pleated; panicle branches glabrous or minutely scabrid, except for sparse tubercle-based hairs:
  - Upper lemma strongly transversely rugose; upper glume ½–⅔ as long as the spikelet; leaf-blades linear, firm, without false petioles .. .. .. 7. *longiseta*
  - Upper lemma smooth; upper glume ¾ as long as the spikelet; leaf-blades lanceolate-linear, flaccid, the lower often narrowed at the base into a false petiole 1–8 cm. long 8. *gracilipes*
- Leaf-blades pleated fan-wise at the base; panicle branches pubescent, with or without tubercle-based hairs:
  - Annual, up to 1 m. high, ascending or geniculate at the base; leaf-blades up to about 1·5 cm. wide; upper lemma strongly rugose .. .. 9. *barbata*
  - Perennials, with stout culms up to 3·5 m. high and broad leaf-blades up to 10 cm. wide; upper lemma obscurely rugose:
    - Spikelets 3–4 mm. long; lower lemma terminating in a short incurved beak, and exceeding the upper lemma by 0·2–0·5 mm.; panicle stiff, erect ..10. *caudula*
    - Spikelets 2·5–3 mm. long; lower lemma equalling or very shortly exceeding the upper lemma:
      - Panicle stiff, erect, dense; the branches rigid, projecting from the main axis at an angle of 45°–90° or even slightly deflexed, decreasing gradually in length to form a pagoda-like spire; culm up to 4 mm. diam. at base of panicle; palea of lower floret as long as the lemma; leaf-sheaths glabrous; leaf-blades up to 10 cm. broad .. .. 11. *megaphylla*
      - Panicle slender, more or less drooping, lax; the branches flexuous, ascending and often appressed to the main axis, of irregular lengths; culm up to 2 mm. diam. at base of panicle; palea of lower floret usually absent, less often up to ¾ as long as the lemma, rarely of the same length; leaf-sheaths usually pilose with tubercle-based hairs, less often glabrous; leaf-blades usually under 5 cm. broad 12. *chevalieri*

1. **S. verticillata** (*Linn.*) *P. Beauv.* Ess Agrost. 51, 178 (1812); F.T.A. 9: 824; Chev. Bot. 734, and in Rev. Bot. Appliq. 14: 31; Berhaut, Fl. Sén. ed. 2, 393; De Miré & Gillet in J. Agric. Trop. 3: 735. *Panicum verticillatum* Linn., Sp. Pl. ed. 2, 82 (1762). *P. adhaerens* Forsk. (1775). *P. aparine* Steud. (1854). *Setaria adhaerens* (Forsk.) Chiov. in Nuov. Giorn. Bot. Ital. n.s. 26: 77 (1919). Ascending annual up to 1 m. high; weedy places.
Maur.: Gani (Oct.) *Hepper* 3650! **Sen.:** *Roger*! **Gam.:** *Ruxton* 140! **Mali:** Sarédina (May) *Davey* 83! Dioura (Oct.) *Davey* 063! **U. Volta:** Ouahigouya (Aug.) *Chev.* 24871! Zoaga (Oct.) *Scholz* 155! Niangoloko (July) *Aké Assi* 10757! **Ghana:** Achimota (Oct.) *Irvine* 829! Christiansborg, Accra (Apr.) *Johnson* 1022! Burufu (May) *Agric. Dept.* 17! Daboya (Nov.) *Thorold* 286! Damongo to Sawla (Oct.) *Rose Innes* GC

Fig. 438.—SETARIA PALLIDE-FUSCA (*Schumach.*) *Stapf & C. E. Hubbard* (GRAMINEAE-PANICEAE).

A, ligule. B, spikelet with bristles showing sides of lower and upper glumes and lower and upper lemmas. C, D, lemma of upper floret from front and back.

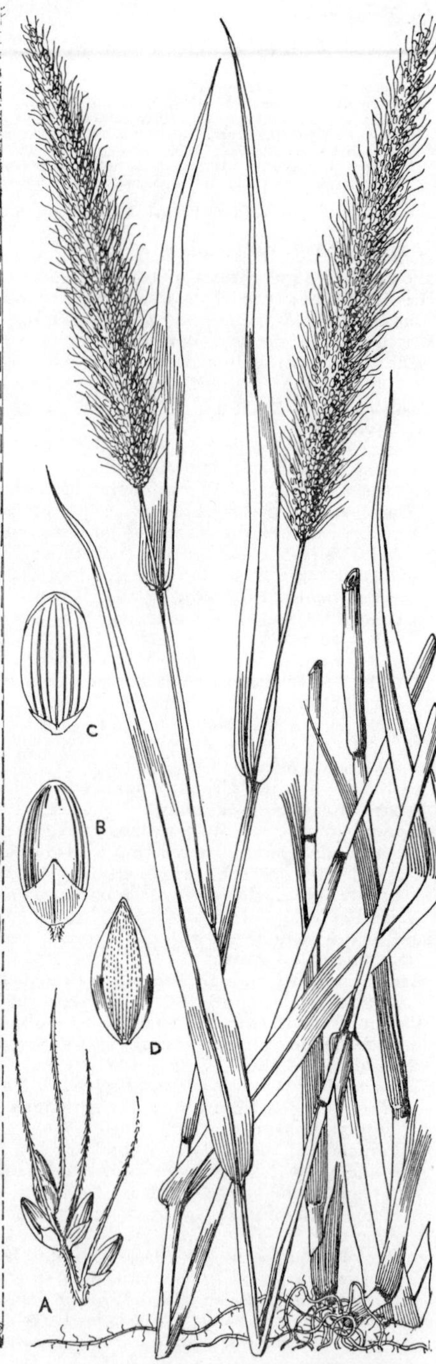

Fig. 439.—SETARIA VERTICILLATA (*Linn.*) *P. Beauv.* (GRAMINEAE-PANICEAE).

A, spikelet with bristles. B, spikelet showing lower glume and lower lemma. C, spikelet showing upper glume. D, upper floret showing palea and margins of lemma.

30998! **Niger:** Zinder (Oct.) *Hagerup* 579! **N. Nig.:** Nupe *Barter* 1356! Sokoto (Sept.) *Palmer* 25! Katagum *Dalz.* 287! Fika (Oct.) *Daggash* FHI 24899! Mongonu (Oct.) *Golding* 12! **S. Nig.:** Lagos *W. Macgregor* 129! Widespread in tropical and warm temperate regions. (See Useful Plants.)

[It has been stated that *S. verticillata* is a temperate species distinct from the tropical *S. adhaerens*. These however are only two of several overlapping geographical races which scarcely warrant specific rank].

2. **S. pallide-fusca** (*Schumach.*) *Stapf & C. E. Hubbard* in Kew Bull. 1930: 259; F.T.A. 9: 815; A. Chev. in Rev. Bot. Appliq. 14: 31; Berhaut, Fl. Sén. ed. 2, 396; De Miré & Gillet in J. Agric. Trop. 3: 735; Aké Assi, Contrib. 2: 289. *Panicum pallide-fuscum* Schumach., Beskr. Guin. Pl. 58 (1827). *Setaria rubiginosa* (Steud.) Miq. (1857). *S. imberbis* of Chev. Bot. 734. *S. glauca* var. *pallide-fusca* (Schumach.) Koyama in J. Jap. Bot. 37: 237 (1962). Ascending, up to about 60 cm. high. Closely allied to *S. glauca* (Linn.) P. Beauv. of warm temperate regions, which has larger spikelets 3 mm. long.

Sen.: Dakar (May, Aug.) *Baldwin* 5713! *Berhaut* 2819! Djembering (Sept.) *Broadbent* 163! Mt. Roland (Aug.) *Broadbent* 58! Ziguinchor (Sept.) *Broadbent* 77! **Gam.:** *Saunders* 13! *Ruxton* 160! Sankuli-Kunda (Sept.) *Pirie* 2/33! Wallikunda (Sept.) *Macluskie* 13! Yundum *Frith* 129! **Mali:** Bamako (Oct.) *Hepper* 3772! Dioura (Aug.) *Davey* 009! Banguita (Sept.) *Farrow* 49! Kouli (Aug.) *Farrow* 50! **Port. G.:** Pirada (June) *Esp. Santo* 3036! Piche (Sept.) *Pereira* 3182! **Guin.:** Timbo (Nov.) *Pobéguin* 1827! Tamba to Sambaïlo (Sept.) *Pitot*! **S.L.:** Freetown to Hastings (Aug.) *Hooper* 133! Koinadugu Distr. (Aug.) *Mead* 221! **Iv. C.:** Batié (Oct.) *Rose Innes* GC 31539! Tehini (Aug.) *Oldeman* 303! *de Wilde* 823! **U. Volta:** Banfora (June) *Leeuwenberg* 4425! Ouagadougou (Sept.) *Scholz* 59! Nassougou (June) *Scholz* 59*a*! **Ghana:** Achimota (May) *Rose Innes* GC 30355! Sapoba to Wapuli (Sept.) *Rose Innes* GC 30469! Tamale *Williams* 1083! Nangodi (July) *Williams* 535! Lawra (Aug.) *Thorold* 219! **Togo Rep.:** Badou to Atakpame (Sept.) *Rose Innes* GC 31359! **Niger:** Dogondoutch *White* 40! **N. Nig.:** Jebba *Barter* 1359! Gindiri, Plateau (Oct.) *Hepper* 1148! Sokoto (Sept.) *Palmer* 36! Kano (July) *Onwudinjoh* FHI 22358! Biu (Sept.) *Kennedy* FHI 8006! Gurum, Vogel Peak (Nov.) *Hepper* 1311! **S. Nig.:** Lagos *W. MacGregor* 143! Onitsha (Aug.) *Jones* FHI 6689! Tropical regions in Africa, Asia and Australasia. (See Useful Plants.)

3. **S. ciliolata** *Stapf & C. E. Hubbard* in F.T.A. 9: 807 (1930). About 60 cm. high, with narrow glaucous leaf-blades; heavy clay.

N. Nig.: Biu (Oct.) *De Leeuw* 1315! Kaltungo, Bauchi Prov. (Oct.) *De Leeuw & Magaji* 1992! Biu (Oct.) *Oche* MRK125! Also in Sudan, Kenya, Uganda and Rhodesia.

4. **S. anceps** *Stapf ex Massey* Sudan Grasses 33 (1926); F.T.A. 9: 793; A. Chev. in Rev. Bot. Appliq. 14: 29; Clayton in Kew Bull. 20: 263; Aké Assi, Contrib. 2: 288. *S. tenuispica* Stapf & C. E. Hubbard in F.T.A. 9: 805 (1930); A. Chev. in Rev. Bot. Appliq. 14: 30. Tufted perennial up to 2 m. high; on moist soils.

Maur.: Lac Rkiz (Sept.) *Sadio*! **Sen.:** *Richard*! Richard Tol (Feb.) *Döllinger* 17! Dakar (Oct.) *Adam* 18499! Bambey (Aug.) *Adam* 14938! **Mali:** Dari (Sept.) *Farrow* 51! Kara (Aug.) *Farrow* 53! Mopti (Sept.) *Lean* 88! San (Sept.) *Chev.* 2257 *bis*! **Port. G.:** S. Domingo (Aug.) *Pereira* 3136! **Guin.:** Fouta Djalon (Oct.) *Adam* 12567! **S.L.:** Messima (June) *Fisher* 2013! Kagberi (Sept.) *Jordan* 544! Kichom (July) *Deighton* 3758! Njala (June) *Deighton* 1759! Kittam (June) *T. S. Jones* 163! **Iv. C.:** Bassam to Yakassé (Jan.) *Hedin*! Bouaké to Katiola (Sept.) *de Wilde* 987! Dimbokro to Bocanda (May) *Adjanohoun* 252*a*! **U. Volta:** Mossi (Aug.) *Chev.* 24592! 24765! Farakoba (June) *Scholz* 95! Aragio (Nov.) *Scholz* 189! **Ghana:** Takoradi (Oct.) *Howes* 983! Keta (May) *La Danso* 54! Somanya to Akuse (May) *Rose Innes* GC 30354! Kete Krachi to Yendi (Sept.) *Rose Innes* GC 30428! Funsi to Tumu (Oct.) *Rose Innes* GC 30793! **Togo Rep.:** Lomé *Warnecke* 318! Atakpamé to Sokodé (Oct.) *Rose Innes* GC 31366! Palimé *Stage* 37! 81! **Dah.:** Cotonou (Nov.) *Clayton* 397! Sémé (Sept.) *Adjanohoun* 533! **N. Nig.:** Ilorin (Aug.) *Ward* 14! W. Yagba (May) *Ward* 1! Dogon Kurmi, Jemaa (Apr.) *E. W. Jones* 278! Zonkwa (Nov.) *Freeman* S173! Nupe *Barter* 1378! **S. Nig.:** Lagos *W. MacGregor* 170! Sobo plains, Warri Prov. (Sept.) *Cole* 29! Oyo (Aug.) *J. B. Gillett* 1541! Obubra (May) *Jones* 1581! Obudu Plateau (Dec.) *Tuley* 47! **W. Cam.:** above Bamenda, 8,000 ft. (Jan.) *Migeod* 329! Bambui (Dec.) *Pedder* 5! Jakiri *Hepper* 2070! Nkambe (Nov.) *Charter* FHI 36985! Tropical Africa.

[A variable species, the largest plants grading imperceptibly into the giant *S. splendida* Stapf (1930). These hybridize readily with the following two species; Hacker (in Austr. J. Bot. 16: 539, 551 (1968)) finds several polyploid levels in each, and little evidence of cytological barriers between them. Morphological distinctions are equally blurred, and it may be better to treat them as a single polymorphic species.—W. D. C.]

5. **S. aurea** *Hochst. ex A. Br.* in Flora 24: 276 (1841). *Pennisetun aureum* Hochst. ex A. Rich., Tent. Fl. Abyss. 2: 378 (1851). Densely caespitose, up to 1·2 m. high.

Iv. C.: Touba to Waninou (June) *Adjanohoun* 253*a*! **Ghana:** Kwahu Tafo (Apr.) *Morton* A687! Kwahu Tafo to Adawso (Oct.) *Rose Innes* GC 30568! Banda Ravine (Apr.) *Morton* A3271! **Togo Rep.:** Badou to Atakpamé (Sept.) *Rose Innes* GC 31352! **N. Nig.:** Naraguta, Plateau (Oct.) *Kennedy* FHI 8049! Darazo F.R., Bauchi (Oct.) *De Leeuw & Magaji* 1997! Sha (Sept.) *Hall* K37! **S. Nig.:** Iseyin (May) *Dawodu* 46! **W. Cam.:** Ndop Plain (Mar.) *Brunt* 261*b*! Jakiri (June) *Brunt* 690! Also in Kenya, Uganda and Tanzania, extending into the Sudan and Congo.

[*S. aurea* is apparently conspecific with the common East African species *S. trinervia* Stapf (1930), the latter being imperfectly distinguished by its 3-nerved upper glume.—W. D. C.]

6. **S. sphacelata** (*Schumach.*) *Stapf & C. E. Hubbard ex M. B. Moss* in Kew Bull. 1929: 195 (1929); F.T.A. 9: 795, partly; A. Chev. in Rev. Bot. Appliq. 14: 30; Clayton in Kew Bull. 20: 262; Berhaut, Fl. Sén. ed. 2, 396. *Panicum sphacelatum* Schumach., Beskr. Guin. Pl. 78 (1827). *Setaria laxispica* Stapf in F.T.A. 9: 802 (1930). Caespitose perennial about 1 m. high, arising from a short rhizome.

Sen.: Niokolo-Koba (June) *Adam* 14191! **Guin.:** Sambaïlo (Aug.) *Adam* 14844! **U. Volta:** Ouagadougou (July) *Scholz* 123! Po (July) *Scholz* 32*a*! **Ghana:** Winnebah plains *Dalz.* 8424! Nungua, Accra plains (Oct.) *Gyadu*! Legon Hill (Dec.) *Adams* 3593! Anum (Apr.) *Rose Innes* GC 30338! Kete Krachi to Yendi (Sept.) *Rose Innes* GC 30433! **Togo Rep.:** Palime, Lomé *Stage* 26! 77! **N. Nig.:** Patti Lokoja (May) *Richardson*! Naraguta F.R. (July) *Lawlor & Hall* FHI 45594! FHI 45595! Zaria *Taylor* 24! Vogel Peak (Nov.) *Hepper* 1426! Tropical Africa generally. (See Useful Plants.)

[*S. sphacelata* strictly refers to short tufted plants with narrow glaucous leaf-blades from the Accra plains. Similar plants are widely distributed in tropical Africa, and the boundaries of the species are ill-defined. The name has often been applied indiscriminately to various members of the complex.]

7. **S. longiseta** *P. Beauv.* Fl. Oware 2: 81, t. 110, fig. 2 (1819); F.T.A. 9: 836; A. Chev. in Rev. Bot. Appliq. 14: 31; Aké Assi, Contrib. 2: 289. *Panicum longifolium* Schumach. (1827). *P. rescissum* Trin. (1834). Perennial, usually 60–90 cm. high; shady places on pathsides and fallow land.

Guin.: Tinkisso *Pobéguin*! **S.L.:** *T. Vogel* 110! Freetown (Dec.) *Deighton* 800! Mt. Loma (Nov., Dec.) *Jaeger* 485! *Morton* SL421! SL2615! **Iv. C.:** Toumodi (Oct.) *Adjanohoun* 335*a*! **Ghana:** Legon (Oct.) *Adams* 3472! Anum to Asikuma (Sept.) *Rose Innes* GC 30385! Mampong (Oct.) *Baldwin* 13525! Kumasi (Oct.) *Baldwin* 13459! Assuantsi Agric. Stn. (Feb.) *Dalz.* 8422! **Togo Rep.:** Badou to Atakpame (Sept.) *Rose Innes* GC 31343! **Dah.:** Porto Novo (Aug.) *Adjanohoun* 27! **N. Nig.:** Nupe *Barter* 915! Naraguta, Plateau (Aug.) *Lely* P440! Gindiri, Plateau (Oct.) *Hepper* 1108! Wuse to Bukum (Oct.) *Oche & Tuley* 1668! Gembu (Dec.) *Daramola* FHI 62469! **S. Nig.:** Lagos (June) *Dalz.* 1321! *W. MacGregor* 73! Ibadan (Mar.) *Meikle* 1305! Akure (Dec.) *Brenan* 8637! Aguku *Thomas* 673! Throughout tropical Africa. (See Useful Plants.)

8. **S. gracilipes** *C. E. Hubbard* in Kew Bull. 4: 362 (1949). A sprawling annual 60–90 cm. high; leaf-blades occasionally with indistinct pleats; damp places in shade.

Iv. C.: Groumania (July) *Aké Assi* 9097! **Ghana:** Kete Krachi to Kpandu (Sept.) *Rose Innes* GC 30407!

Dah.: Pobé (Sept.) *Adjanohoun* 66! **N. Nig.:** Kwal, Jos plateau (Aug.) *Tuley* 1579! **S. Nig.:** Igboora (Oct.) *Haines* 346! Obokum, Ogoja Prov. (May) *Jones & Onochie* FHI 18878! Ekosoro, Obubra Div. (Apr.) *Jones* 1527!

9. **S. barbata** (*Lam.*) *Kunth* Rev. Gram. 1: 47 (1829); F.T.A. 9: 854; Chev. Bot. 734, and in Rev. Bot. Appliq. 14: 32; Berhaut, Fl. Sén. ed. 2, 401. *Panicum barbatum* Lam., Tab. Encycl. Meth. Bot. 1: 171 (1797). *P. lineatum* Schumach., Beskr. Guin. Pl. 61 (1827). *P. basisetum* Steud. (1854)—Chev. Bot. 724. A weed on disturbed land.

Sen.: *Roger* 17! M'Bidjem *Thierry* 8! Bignona (Sept.) *Adam* 18032! Ziguinchor (Sept.) *Broadbent* 88! **Mali:** Yatenga (Aug.) *Chev.* 24877! Ségou *Lécard* 245! **Guin.:** Kindia *Jac.-Fél.* 174! *Kranz* 291! Timbo (Oct.) *Pobéguin* 1732! Madina (Oct.) *Adam* 12553! Pita (Sept.) *Adames* 338! **S.L.:** Musaia (Oct.) *Thomas* 2666! Yakala (Sept.) *Thomas* 3284! **Iv. C.:** Toumodi *Chev.* B22257! Singrobo (Oct.) *Adjanohoun* 353*a*! Diebougou to Gaoua (Oct.) *Rose Innes* GC 31517! Bouaké (Sept.) *de Wilde* 957! **U. Volta:** Doutié (Sept.) *Scholz* 48*b*! **Ghana:** Atwabo *Fishlock* 38! Kumasi (Apr.) *La Danso* 52! Wenchi (Sept.) *Rose Innes* GC 30152! Ajena (Nov.) *Ankrah* GC 20027! Bamboi Ferry (Oct.) *Rose Innes* GC 31146! **Togo Rep.:** Loma Kara to Kande (Oct.) *Rose Innes* GC 31403! **Dah.:** Allada (Mar.) *Chev.* 23374! **N. Nig.:** Bida (Sept.) *Clayton* 361! Oturkpo (Oct.) *Kennedy* FHI 8026! Naraguta, Plateau (Aug.) *Lely* P441! Ruma, Katsina (Aug.) *Onwudinjoh* FHI 24007! Biu (Sept.) *Kennedy* FHI 8017! **S. Nig.:** Lagos *Dalz.* 1134! Abeokuta *Irving*! Ibadan (Oct.) *Onochie* FHI 7943! Onitsha (Aug.) *Jones* FHI 6684! *Barter* 1746! **W. Cam.:** Victoria (Jan., Nov.) *Maitland* 78! 149! Buea (Dec.) *Maitland* 859! Cam. Mt., 7,500 ft. (Nov.) *Migeod* 218*a*! Bamenda (Aug.) *Brunt* 1261! **F. Po:** *T. Vogel* 34! Santa Isabel (Dec.) *Guinea* 170! Also in Sudan, Congo and Angola, and in Madagascar, tropical Asia and America. (See Useful Plants.)

10. **S. caudula** *Stapf* in F.T.A. 9: 845 (1930). Tufted perennial 1·2–1·8 m. high, with broad plicate leaf-blades.

N. Nig.: Chappal Waddi (Nov.) *Jackson, Magaji & Tuley* 2059! **S. Nig.:** Obudu Plateau (Feb.) *Tuley* 486! **W. Cam.:** Cam. Mt., 2,000–3,000 ft. (Dec.) *Mann* 2102! Buea (Dec.) *Maitland* 107! Masaka Camp, Cam. Mt., 6,000 ft. (Dec.) *Maitland* 885! Bamenda (Oct.) *Agric. Off.* Bam/18! Ndop, Bamenda (Dec.) *Boughey* GC 11128! **F. Po:** (Dec.) *Boughey* GC 10823! Also in the Sudan, Congo, Uganda, Tanzania and Zambia.

11. **S. megaphylla** (*Steud.*) *Dur. & Schinz* Consp. Fl. Afr. 5: 773 (1894); F.T.A. 9: 840; A. Chev. in Rev. Bot. Appliq. 14: 32. *Panicum megaphyllum* Steud., Syn. Pl. Glum. 1: 53 (1854). *P. phyllomacrum* Steud. (1854). *Setaria sulcata* of Chev. Bot. 734. Coarse perennial up to 3 m. high; marshy places in forest.

Port. G.: Bedanda (Jan.) *Pereira* 2822! **S.L.:** Gorahun (Nov.) *Deighton* 354! Baiima (Sept.) *Deighton* 3073! Falaba (Oct.) *Small* 431! Musaia (Oct.) *Thomas* 2650! Yonibana (Oct.) *Thomas* 3988! **Lib.:** *Reynolds*! Soplima, Vonjama (Nov.) *Baldwin* 10067! Javajai, Boporo (Nov.) *Baldwin* 10270! Gbanga (Sept.) *Linder* 509! Suacoco, Gbanga (Oct.) *Daniel* 23! **Iv. C.:** Abidjan (Aug.) *de Wilde* 659! **Ghana:** Accra to Ho (Sept.) *Rose Innes* GC 30397! Akwapim (Jan.) *Irvine* 2601! Mampong (Sept.) *Darko* 716! Suhum (Sept.) *Ankrah* GC 20071! Togo Plateau (Sept.) *Thompson* 1448! **Togo Rep.:** Badou (Sept.) *Rose Innes* GC 31341! **N. Nig.:** Abuja (Dec.) *Meikle* 739! Jemaa (Apr.) *E. W. Jones* 38! Wamba to Assob (Oct.) *Oche & Tuley* 1686! **S. Nig.:** Lagos *W. MacGregor* 233! Awba Hills F.R. (Oct.) *Jones* FHI 7098! Old Oyo F.R. (Sept.) *Latilo* FHI 23550! Oban Rubber Estate (July) *Tuley* 668! Ikom (Sept.) *Olorunfemi & Adetunji* FHI 45482! **W. Cam.:** Victoria *Maitland*! Buea (Oct.) *Migeod* 1! Kumba (Oct.) *Ejiofor* FHI 8452! **F. Po:** (Dec.) *Mann* 110! Extending to E. and C. Africa.

12. **S. chevalieri** *Stapf* in F.T.A. 9: 842 (1930); A. Chev. in Rev. Bot. Appliq. 14: 32. *S. megaphylla* (Steud.) Dur. & Schinz var. *chevalieri* (Stapf) Berhaut, Fl. Sén. ed. 2, 401 (1954). Coarse perennial up to 3 m. high; forest margins.

Sen.: *Heudelot* 701! **Gam.:** *Ingram*! **Guin.:** Timbo (Oct.) *Pobéguin* 1733! Baffing R. (Oct.) *Pobéguin* 1733*bis*! 1736! **S.L.:** Njala (Oct., Dec.) *Deighton* 1341! 1505! Kenema *Lane-Poole* 451! Joru (Dec.) *Pyne* 47! Kamabai to Kabala (Dec.) *Deighton* 5689! **Lib.:** Bumbuna (Nov.) *Linder* 1330! Tappita (Aug.) *Baldwin* 9054! Sanokwele (Nov.) *Harley* 1613! Nimba (Dec.) *Adam* 16501! Ganta (Oct.) *Harley* 1034! **Iv. C.:** Nimba (Aug.) *Boughey* GC 18167! Anguédédou (Sept.) *de Wilde* 3103! Tonkoui (Aug.) *Boughey* GC 18337! **Ghana:** Aburi (Nov.) *Howes* 1013! Wenchi to Bamboi (Oct.) *Rose Innes* GC 31149! Mampong (Sept.) *Rose Innes* GC 30138! Otrokpe (Oct.) *Vigne* FH 4016! Kumasi (Aug.) *La Danso* 129! **Togo Rep.:** Badou (Sept.) *Rose Innes* GC 31342! **N. Nig.:** Dagoul, R. Niger *T. Vogel*! Jemaa (Oct., Nov.) *Clayton* 1416! *Gambles* 62! Baissa (Feb.) *De Leeuw* AS10! Njawai to Mayo Daga (Nov.) *Jackson, Magaji & Tuley* 2107! **S. Nig.:** Olokemeji (Nov.) *Onochie* FHI 8137! Okomu F.R., Benin (Dec.) *Brenan* 8451! Akure (Nov.) *Clayton* 559! Awka *Thomas* 45! Opobo *Jeffreys* 13! **W. Cam.:** Victoria *Maitland* 82! Buea (Jan., Feb.) *Maitland* 106! 334! 336! *Dalz.* 8427! **F. Po:** Clarence Peak (Oct.) *T. Vogel* 44! Throughout tropical Africa. (See Useful Plants.)

[A variable grass in which forms with slender erect panicles and glabrous leaf-sheaths approach very closely to *S. megaphylla*, with which it may hybridize. It is closely related to *S. paniculifera* (Steud.) Fourn. of tropical America, which is distinguished by its narrower more acute spikelets.]

Additional species.

S. viridis (*Linn.*) *P. Beauv.* Ess. Agrost. 51, 171, 181 (1812). *Panicum viride* Linn., Syst. Nat. ed. 10, 2: 870 (1759). Annual, inflorescence spiciform; upper glume as long as the spikelet. Warm temperate regions. Recorded at Aïr, Niger, by De Miré & Gillet in J. Agric. Trop. 3: 736 (1956).

73. SACCIOLEPIS Nash in Britton, Man. Fl. North. U.S. 89 (1901); F.T.A. 9: 747 (1920).

False spikes much attenuated downwards, bearing there imperfect tightly appressed spikelets (these sometimes absent in the smallest spikes); branches fused with the axis of the false spike; spikelets 2·5–3·5 mm. long; lower glume with a single broad median nerve and hyaline margins; lower lemma with a U-shaped fringe of hairs on the back .. .. 1. *ciliocincta*

False spikes cylindrical, without imperfect florets; branches not or imperfectly fused with the axis; lower glume 3- or more-nerved; lower lemma glabrous or uniformly hairy:

Spikelets 0·7–1 mm. long, glabrous .. .. 2. *micrococca*

Spikelets 1·8–4 mm. long:

Spikelets dorsally slightly compressed, obtuse to subacute, quite glabrous, light green, 2·5–3 mm. long; false spikes stout, 10–30 cm. long, rarely shorter .. 3. *africana*

Spikelets more or less laterally compressed, acute to acuminate:

Spikelets 1·8–2·5 mm. long; leaf-sheaths without auricles:

Lower lemma conspicuously gibbous, the nerves thickened on the bulge; spikelets glabrous, strongly laterally compressed, 2·2–2·5 mm. long; spikes up to 40 cm. long .. .. 4. *cymbiandra*

Lower lemma not markedly gibbous, unthickened, the nerves distinct; spikelets nearly always hairy, slightly laterally compressed, 1·8–2·2(–2·5) mm. long; spikes up to 12 cm. long, rarely longer .. .. 5. *chevalieri*

Spikelets 3–4 mm. long (rarely less); leaf-sheaths with auricles 0·5–5 mm. long at the mouth:

Annual, ascending from a decumbent rambling base emitting long aerial roots from its nodes; culms up to about 1·5 mm. diam. at the base; leaf-blades seldom over 15 cm. long; spikelets 3–3·5 mm. long, usually hairy.. .. 6. *auriculata*

Perennial, erect or ascending from a short underground rhizome; culms about 3 mm. diam. at the base; leaf-blades often 20–30 cm. long; spikelets 3·5–4 mm. long, usually glabrous .. .. 7. *rigens*

1. **S. ciliocincta** (*Pilger*) *Stapf* in F.T.A. 9: 751 (1920); Berhaut, Fl. Sén. ed. 2, 390. *Panicum ciliocinctum* Pilger in Engl., Bot. Jahrb. 33: 48 (1904). Annual up to 12 ins. high; seasonal marsh on ironstone pavement.
Sen.: Tambacounda *Berhaut* 900. **Iv. C.:** Bassawa (Oct.) *Adjanohoun* 276*a*! Boundiali to Mankono (Nov.) *Aké Assi* 8286! **U. Volta:** Bobo-Dioulasso (Sept.) *Adam* 15121! **Ghana:** Burufo (Sept.-Nov.) *Hinds* 5012! *Harris*! *Rose Innes* GC 32345! *Hall* CC 754! Sampa (Dec.) *Morton* A 1066! Also in the Congo and Sudan.

2. **S. micrococca** *Mez* in Fedde Rep. 15: 122 (1918); F.T.A. 9: 753; A. Chev. in Rev. Bot. Appliq. 14: 28; Berhaut, Fl. Sén., ed. 2, 390. Tufted annual 30–60 cm. high; marshy ground, rice fields.
Sen.: *Heudelot* 551! Gassane (Oct.) *Raynal* 7688. **Mali:** Dogo (Jan.) *Davey* 203! **Guin.:** Kouroussa *Pobéguin* 536! **S.L.:** Fintonia (Jan.) *Morton & Gledhill* SL 588! **Iv. C.:** Bassawa (Oct.) *Adjanohoun* 295*a*! Boundiali (Nov.) *Aké Assi* 8278! **Ghana:** Tamale (Nov.) *Williams* 1025! *Thorold* 292! Tamale to Bolgatanga (Oct.) *Rose Innes* GC 31112! Banda (Dec.) *Morton* GC 25096! Damongo (Dec.) *Morton* GC 25055! **N. Nig.:** Minna (Jan., Dec.) *Keay* FHI 37318! *Meikle* 725! Shendam (Nov.) *Clayton* 1460! Borgu *Barter* 759! Maiduguri (Oct.) *Johnston* N27! **S. Nig.:** Oyo *Gledhill* 740! Igboora (Oct.) *Haines* 353! Abakaliki (Nov.) *Tuley* 8! Also in Sudan, Uganda, Zambia and Malawi.

3. **S. africana** *C. E. Hubbard & Snowden* in Kew Bull. 1936: 294. *S. interrupta* of F.T.A. 9: 757; of F.W.T.A., ed. 1, 2: 555; of A. Chev. in Rev. Bot. Appliq. 14: 28; and of Berhaut, Fl. Sén. ed. 2, 390. *Panicum interruptum* of Chev. Bot. 727, not of Willd. (1798). Perennial 1.2–1.8 m. high, with thick spongy culms, rooting at the lower nodes; growing in water.
Gam.: *Ruxton* 122! Wallikunda (Oct.) *Macluskie* 30! **Mali:** Gao (Jan.) *Chev.* 43084! 43120! Tenenkou (Sept.) *Lean* 103! Banguita to Dogo (Jan.) *Davey* 10! Sarédina (May) *Davey* 123! **Guin.:** Timbo *Pobéguin* 1740! Kindia (Dec.) *Jac.-Fél.* 126! 7444! Seliori (Oct.) *Adam* 12678! Nzérékoré (Oct.) *Baldwin* 9720! **S.L.:** Bumba (Sept.) *Thomas* 1991! Mamaha (Nov.) *Thomas* 4508! Rokupr (Sept.) *Jordan* 94! Musaia (Sept.) *Small* 259! Sherbro Island (July) *Jones* 164! **Lib.:** Gbanga (Sept.) *Linder* 521! Ganta (Sept.) *Baldwin* 12021! *Harley* 1015! **Iv. C.:** Man (Aug.) *Boughey* GC 18416*b*! Fettekro to Bassawa (Oct.) *Adjanohoun* 297*a*! Bassawa (Oct.) *Adjanohoun* 282*a*! **U. Volta:** Yondé (Nov.) *Scholz* 190! **Ghana:** Anum to Asikuma (Sept.) *Rose Innes* GC 30388! Kpandai (Sept.) *Ankrah* GC 20235! Damongo to Sawla (Nov.) *Rose Innes* GC 30828! Nalerigu (Dec.) *Adams & Akpabla*! Lungiri (Nov.) *Thorold* 269! **N. Nig.:** Bida (Aug.) *Sawyer* K34! Naraguta, Plateau (June) *Lely* 327! Fodama (Nov.) *Moiser* 224! Sokoto *Dalz.* 478! Gajibo, Bornu (Nov.) *Johnston* N98! **S. Nig.:** Lagos *Rowland*! Akinjare, Owo (Apr.) *Jones* FHI 3122! Aligbeta (Aug.) *Onochie* FHI 33291! Opobo *Jeffreys*! Awka (Aug.) *Jones* FHI 6693! Extends eastwards to Ethiopia and Tanzania, and southwards to Malawi and Angola. (See Useful Plants.)
[Closely related to *S. interrupta* (Willd.) Stapf, an Asiatic species with more acute spikelets and shorter upper lemma.]

4. **S. cymbiandra** *Stapf* in F.T.A. 9: 758 (1920); A. Chev. in Rev. Bot. Appliq. 14: 29; Berhaut, Fl. Sén. ed. 2, 390. Erect perennial up to 1 m. high, with spongy culm; growing in water and marshy places.
Port. G.: Nova Lamego to Pixe (Jan.) *Raimundo & Guerra* 897! Pitche (Dec.) *Pereira & Correia* 2378! **Guin.:** Baffing R. *Pobéguin* 1746! **S.L.:** Kambia (Dec.) *Deighton* 804! Mabould (Oct.) *Thomas* 3555! Rokupr (Oct.) *Jordan* 129! **Iv. C.:** Ferkessédougou (Nov.) *Leeuwenberg* 2023! **U. Volta:** Ouagadougou (Oct.) *Adam* 12741! **Ghana:** *Beal* 45! Nassia (Dec.) *Vigne* FH 4671! Menji to Banda (Dec.) *Adams* 3106! Sawla (Nov.) *Harris*! Lawra (Oct.) *Ankrah* GC 20463! **N. Nig.:** Edozhigi, Bida (Jan.) *Clayton* 642! Badeggi (Dec.) *Haines* 109! **S. Nig.:** Abakaliki (Nov.) *Tuley* 9!

5. **S. chevalieri** *Stapf* in F.T.A. 9: 754 (1920); A. Chev. in Rev. Bot. Appliq. 14: 28; Berhaut, Fl. Sén. ed. 2, 390; Aké Assi, Contrib. 2: 288. *Panicum chevalieri* of Chev. Bot. 725, name only. Erect perennial up to 60 cm. high growing from a short rhizome, the culms soft below; marshy places.
Sen.: Kanémére (Oct.) *Fotius* K867. **Mali:** Dogo (Apr.) *Davey* 553! Niger "delta" *Duong*! **Guin.:** Labé (Oct.) *Adam* 12566! Koulountou (Mar.) *Adam* 17892! **Iv. C.:** Toumodi to Yamoussokro (Oct.) *Adjanohoun* 330*a*! N. Korhogo (Oct.) *Boudet* 3121! **U. Volta:** Ouagadougou (Oct.) *Adam* 12743! Toussiana (June) *Scholz* 101! **Ghana:** Tiasi (Aug.) *Hall* CC 156! Attebubu to Kete Krachi (Oct.) *Rose Innes* GC 30597! Wulugu (Oct.) *Rose Innes* GC 30777! Burufo, Lawra (Apr.) *Adams* 4059! Salaga to Bimbila (Oct.) *Rose Innes* GC 30960! **Dah.:** Mt. Atacora (June) *Chev.* 24186! **N. Nig.:** Vodni *Saunders* 12! Vom (Nov.) *Gambles* 5! Anara F.R. (Oct.) *Keay* FHI 5445! Vogel Peak (Nov.) *Hepper* 1335! Wamba (Dec.) *Haines* 107! **W. Cam.:** Jakiri, Bamenda (Feb.) *Hepper* 2072! Also in Kenya, Uganda, Zambia and Malawi.
[The boundaries of this species are probably too widely drawn, but little improvement can be made until its indistinctly differentiated allies in east and central Africa have been revised.]

6. **S. auriculata** *Stapf* in F.T.A. 9: 762 (1920); A. Chev. in Rev. Bot. Appliq. 14: 29. Straggling grass of wet places, with light green spikelets.
Guin.: Baffing R. (Sept.) *Pobéguin* 1745! Bilima (Dec.) *Jac.-Fél.* 7448! Friguiagbé (Dec.) *Chillou* 1108! **S.L.:** Njala (Nov.) *Deighton* 1430! Kamaron (Nov.) *Morton* SL 3348! Moria to Fintonia (Jan.) *Morton & Gledhill* SL 589! Sefadu (Nov.) *Deighton* 4661! Makump (Dec.) *Glanville* 98! **Ghana:** Vane to Dzolokpuita (Nov.) *Morton* GC 9358! **N. Nig.:** Borgu *Barter* 752! Minna (Dec.) *Meikle* 726! Kogigiri, Jos (Oct.) *Hepper* 1049! Kufena Hill, Zaria (Sept.) *Clayton* 1307! Maska (Nov.) *Thatcher* S512! **S. Nig.:** Lagos *W. MacGregor* 157! Olokemeji to Iseyin (Nov.) *Onochie* FHI 8161! Ezillo (Oct.) *Tuley* 938! Igbete Rock (Nov.) *Gledhill* 756! Also in Sudan, Uganda, Congo, Zambia and Angola.

7. **S. rigens** *A. Chev.* in Rev. Bot. Appliq. 14: 29 (1934). *Panicum rigens* Mez in Engl., Bot. Jahrb. 34: 141 (1903), not of Swartz (1788); F.T.A. 9: 767. *S. auriculata* of F.W.T.A., ed. 1, 2: 555, partly. Up to 1·2 m. high, with light brown spikelets; on moist soils.
Ghana: Kete Krachi to Kpandu (Sept.) *Rose Innes* GC 30411! Kpandai (Sept.) *Ankrah* GC 20236! Yendi to Zabzugu (Oct.) *Rose Innes* GC 30695! Kpetchu to Kete Krachi (Oct.) *Ankrah* GC 20384! Tamale to Salaga (Aug.) *Rose Innes* GC 31875! **Togo Rep.:** Sokode to Basari *Kersting* 595!

74. CYRTOCOCCUM Stapf in F.T.A. 9: 745 (1920).

Leaf-blades narrowly lanceolate to broadly linear, up to 12 cm. long and 1·5 cm. broad, very acute, thin, prominently ciliate near the base; panicles delicate, broadly oblong to obcuneate in outline, somewhat flat-topped, the base enclosed in the uppermost leaf-sheath; pedicels long and slender, loosely beset with long white hairs; spikelets about 2 mm. long, dark brown; glumes thinly pilose .. .. *chaetophoron*

C. chaetophoron (*Roem. & Schult.*) *Dandy* in J. Bot. 69: 55 (1931); Berhaut, Fl. Sén. ed. 2, 413. *Panicum chaetophoron* Roem. & Schult. in Linn., Syst. Veg., ed. 9, 2: 884 (1817). *Cyrtococcum setigerum* Stapf in F.T.A. 9: 746 (1920); F.W.T.A., ed. 1, 2: 555; A. Chev. in Rev. Bot. Appliq. 14: 27. *Panicum setigerum* P. Beauv. (1806), not of Retz. (1786). *Panicum jardinii* Steud. (1854). *Isachne jardinii* (Steud.) Dur. & Schinz (1894). Perennial up to 1 m. high; culms slender, ascending; forest shade.
Sen.: Oussouye (Nov.) *Berhaut* 6603! **S.L.:** Freetown (Nov.) *Gledhill* 235! Njala (Nov.) *Deighton* 1495! **Lib.:** Kakatown *Whyte*! Zuie (Nov.) *Baldwin* 12033! Peahtah (Dec.) *Baldwin* 10599! Ganta (Oct., Nov.) *Harley* 1713! 2065! **Iv. C.:** Brafouédi (Oct.) *Adjanohoun* 435a! Abidjan (Dec.) *Schnell* 3903! **Ghana:** Aburi hills (Nov.) *Johnson* 286! Aiyinasi (Dec.) *Rose Innes* GC 30327! Bunsu to Anyinam (Nov.) *Adams* 4434! Asuansi (Dec.) *Rose Innes* GC 30532! Yinahin Range, Kumasi (Nov.) *Rose Innes* GC 30524! **N. Nig.:** Kabba road (Oct.) *Parsons* 28! Abuja (June) *E. W. Jones* 164! Kagoro to Jemaa (Oct.) *Magaji & Tuley* 1776! Kubacha (Oct.) *Magaji & Tuley* 1744! **S. Nig.:** Lagos *W. MacGregor* 312! Otta (Oct.) *Haines* 113! Mamu F.R. (Dec.) *Onochie* FHI 7492! Ife (Jan.) *Sandford* 216! Gambari (Nov.) *Jackson* 3231168! Also in E. Cameroun, Congo, Angola and the Sudan.

75. PANICUM Linn., Sp. Plant. 55 (1753), and Gen. Pl. ed. 5, 29 (1754); F.T.A. 9: 638 (1920).

Inflorescence consisting of several more or less secund compound racemes, the secondary branches very short (3–5 mm. long) and closely appressed to the spreading primary branches; leaf-blades broadly linear, 2–8 mm. broad; spikelets glabrous, elliptic, 1–1·5 mm. long; lower glume about $\frac{1}{3}$ as long as the spikelet; lower lemma 3-nerved; upper lemma smooth .. .. 1. *laxum*
Inflorescence an open panicle, very rarely with the spikelets loosely contracted about the primary branches:
 Spikelets slightly oblique to gibbous in profile, diffusely scattered in delicate open panicles; leaf-blades thin, ovate to lanceolate; upper glume and lower lemma 5-nerved:
  Lower glume as long as the spikelet, lanceolate; spikelets 1·2–1·5 mm. long, glabrous or sometimes pubescent; leaf-blades ovate to lanceolate, amplexicaul at the base:
   Panicle branchlets viscid from brown patches of glandular tissue; upper lemma minutely papillose; rhachilla internode not distinctly swollen .. 2. *heterostachyum*
   Panicle branchlets neither viscid nor glandular; upper lemma smooth; second rhachilla internode swollen, and covered by a small bulge at the base of the upper glume .. .. 3. *brevifolium*
  Lower glume half as long as the spikelet, broadly ovate; spikelets 1–1·2 mm. long, obscurely puberulous; leaf-blades narrowly lanceolate .. .. 4. *trichoides*
 Spikelets symmetrical in profile:
  Upper glume and lower lemma pectinate at the tip, with 3 long and two smaller teeth; lower glume trifid (in West African specimens almost to the base, with the lateral lobes often suppressed); spikelets 3–3·5 mm. long; inflorescence axis often bearing clavate hairs .. .. 5. *ecklonii*
  Upper glume and lower lemma entire:
   Axis and branches of the inflorescence bearing clavellate-tipped hairs:
    Lower glume 1-nerved, lanceolate; spikelets 2–3 mm. long; lower lemma 9-nerved; leaf-blades 2–8 cm. long and 2–8 mm. broad .. .. .. 6. *hymeniochilum*
    Lower glume 5-nerved, ovate, $\frac{1}{2}$–$\frac{2}{3}$ as long as the spikelet; spikelets 3 mm. long, oblong, acute; lower lemma 11-nerved; leaf-blades up to 12 cm. long and 20 mm. broad .. .. 7. *sadinii*
   Axis and branches of the inflorescence glabrous, or if hairy the hairs not clavellate-tipped:
    Upper lemma transversely rugose; spikelets oblong, 3–3·5 mm. long; lower glume obtuse, about $\frac{1}{3}$ as long as the spikelet; panicle up to 50 cm. long, the lower branches usually whorled .. .. 8. *maximum*
    Upper lemma not transversely rugose:
     Upper lemma verrucose; spikelets broadly elliptic to subglobose in outline, obtuse to subacute at the apex, tending to break up at maturity; lower glume usually $\frac{2}{3}$–$\frac{4}{5}$ as long as the spikelet (to p. 427):
      Perennials:
       Leaf-blades lanceolate, 5–13 cm. long, 5–17 mm. broad, glabrous, cordate at base; panicle loose, obovate; spikelets 1·5–2 mm. long, glabrous 9. *dinklagei*
       Leaf-blades linear or involute:
        Spikelets 1–1·5 mm. long, greyish, pubescent; blades involute, sometimes flattened but rarely much over 2 mm. broad:
         Culms usually 7- or more-noded, leafy; innovations extravaginal, the buds

clad in short scales; leaf-blades involute or flattened, 5–15 cm. long; lower lemma truncate to emarginate at the tip with prominent hyaline wings; lower glume $\frac{3}{4}$–$\frac{4}{5}$ as long as the spikelet .. .. .. 10. *pubiglume*

Culms 2–4-noded, the leaves mostly basal; innovations intravaginal, without scales; leaf-blades acicular, 2–12 cm. long; lower lemma obtuse at the tip without wings; lower glume $\frac{1}{2}$–$\frac{2}{3}$ as long as the spikelet 11. *filifolium*

Spikelets 1·5–2·4 mm. long; leaf-blades linear, flat, 3 mm. or more broad:

Spikelets 1·5–1·8 mm. long, glabrous or pubescent, dull, yellowish grey; upper glume emarginate at the tip when flattened, the margins expanded into delicate narrow hyaline wings; panicle branches spreading; culms 1–3 mm. diam. at base; leaf-blades 3–10 mm. broad, 5–20 cm. long 12. *strictissimum*

Spikelets 1·7–2·5 mm. long, with a metallic bronze lustre; upper glume obtuse at the tip, without hyaline wings; panicle branches ascending:

Culms stout, 2–4 mm. diam. at base; leaf-blades 5–10 mm. broad, 10–30 cm. long; spikelets usually glabrous .. .. .. 13. *praealtum*

Culms slender, up to 1 mm. diam. at base; leaf-blades 3–5 mm. broad, 5–20 cm. long; spikelets glabrous or pubescent .. 14. *baumannii*

Annuals:

Upper lemma with deep longitudinal striations, verrucose only towards the tip; spikelets 2·5 mm. long, glabrous; leaf-blades 3–13 cm. long, 3 mm. broad 15. *calocarpum*

Upper lemma without incised striations:

Leaf-blades linear, 2–7 cm. long, 1–3 mm. broad; spikelets (1–)1·4–1·7(–2) mm. long, grey to yellow, pubescent to villous, rarely glabrous; upper lemma warts hemispherical .. .. 16. *lindleyanum*

Leaf-blades lanceolate to linear-lanceolate:

Blades 1–2(–3) cm. long, 2–3(–7) mm. broad; upper lemma warts longitudinally elongated; spikelets 1–1·5 mm. long, glabrous or pubescent 17. *glaucocladum*

Blades 4–6 cm. long, 4–6 mm. broad; upper lemma warts hemispherical, prominent; spikelets 1·8–2·2 mm. long, glabrous .. 18. *gracilicaule*

Upper lemma smooth; spikelets acuminate or distinctly acute (except *P. parvifolium*), usually not plump (except *P. turgidum*):

Spikelets 5–6 mm. long, glabrous; upper glume and lower lemma acuminate, 7-nerved .. .. 19. *nigerense*

Spikelets up to 4·5 mm. long, usually much less:

Spikelets gaping open early and permanently, the tips of upper glume and lower lemma usually acuminate and turned slightly outwards:

Perennials; lower glume mostly $\frac{1}{2}$–$\frac{2}{3}$ as long as the spikelet:

Upper glume and lower lemma 7–9-nerved:

Leaf-blades glaucous, glabrous, rarely over 25 cm. long, 4–6 mm. broad; spikelets 3·6–4·3 mm. long; panicle sparse, with 5–10 primary branches occurring singly .. .. 20. *anabaptistum*

Leaf-blades dark green, glabrous or pubescent, 40–70 cm. long, 6–10 mm. broad or more; spikelets 2·8–3·8 mm. long; panicle dense, with 15–25 or more primary branches usually occurring in whorls of 3–5 21. *phragmitoides*

Upper glume and lower lemma 5-nerved, or the former 7-nerved:

Panicle with long spreading hairs from the branchlets, or at least from the pedicels; culms slender, 1·5 mm. diam. and compactly caespitose at the base; leaf-blades glabrous or hirsute, up to 3 mm. broad; spikelets 2–3 mm. long, the glumes acuminate and drawn out into awnlets 0·5–1 mm. long; roots pinkish .. .. 22. *pilgeri*

Panicle branches glabrous:

Plants compactly caespitose, the slender culm up to 2 mm. diam. and coated with persistent leaf-sheaths which form a dense erect mass at the base of the plant; panicle 10–15 cm. long:

Leaf-blades up to 3 mm. broad, flat or often involute; basal sheaths silvery-tomentose below; spikelets 2–2·8 mm. long .. 23. *dregeanum*

Leaf-blades 0·5 mm. diam., setaceous, glabrous; basal sheaths glabrous or sometimes pubescent; spikelets 1·5–1·7 mm. long .. 24. *congoense*

Plants neither compactly caespitose nor with a dense coat of persistent leaf-bases; culms 3–8 mm. diam. at base, rarely less; leaf-blades flat, 4–10 mm. broad; panicle 20–30 cm. long; spikelets 2–2·7 mm. long 25. *fluviicola*

Annuals; lower glume usually $\frac{3}{4}$ as long as the spikelet or more; lower lemma 5-nerved; leaf-blades up to 20 cm. long, but usually less than 10 cm.:

Branchlets, or at least the pedicel tips, bearing stiff white hairs up to 2 mm. long; spikelets pilose with tubercle-based hairs, rarely glabrous, 2–2·7 mm. long; whole plant softly pilose .. .. 26. *griffonii*

Branchlets, pedicels and spikelets glabrous:
Glumes gaping widely, separated by an angle of 90°; spikelets 1·2–1·8 mm. long; lower glume long-acuminate, with an awn-point 0·3–0·5 mm. long 27. *afzelii*
Glumes not widely gaping, separated by an angle of about 45°:
Spikelets 1·7–2·1 mm. long; lower glume at most acuminate, with a mucro up to 0·2 mm. long; panicle untidily ovate .. .. ..28. *walense*
Spikelets 2·5–2·7 mm. long; lower glume drawn out into an awnlet 0·3–1 mm. long; panicle compactly lanceolate .. .. 29. *paucinode*
Spikelets not conspicuously and permanently gaping or only when mature, the tips of the upper glume and lower lemma straight or slightly incurved:
Upper glume pilose above, and bearing 2–4 bristles as long as the spikelet; leaf-blades lanceolate, 3–8 cm. long, 8–23 mm. broad, with transverse veins; spikelets 2–2·5 mm. long; lower palea suppressed .. 30. *acrotrichum*
Upper glume glabrous or pubescent, but without long bristles:
Lower lemma 3-nerved; spikelets 2 mm. long, pubescent, lanceolate; lower glume $\frac{1}{3}$–$\frac{1}{2}$ as long as the spikelet, separated from the upper by an internode 0·1 mm. long; leaf-blades narrowly lanceolate, 3–8 cm. long 31. *mueense*
Lower lemma 5–11-nerved:
Lower glume $\frac{1}{2}$ as long as the spikelet or more, rarely less (*P. monticola*), lanceolate to ovate:
Suffrutescent, with glaucous woody culms and hard pungent leaf-blades; spikelets (3–)3·5–4 mm. long, plump, acute at the tip; lower glume $\frac{2}{3}$ to almost as long as the spikelet; lower lemma 9–11-nerved 32. *turgidum*
Herbaceous:
Plants perennial, rambling or creeping, the leaf-blades narrowly lanceolate to lanceolate:
Leaf-blades less than 2 cm. long, lanceolate, amplexicaul at the base; spikelets 1·2–1·8 mm. long, elliptic-oblong, subacute, glabrous; lower glume $\frac{1}{2}$–$\frac{2}{3}$ as long as the spikelet, 3-nerved .. .. 33. *parvifolium*
Leaf-blades mostly over 4 cm. long, not amplexicaul:
Lower glume 3-nerved, $\frac{2}{3}$–$\frac{3}{4}$ as long as the spikelet, lanceolate; leaf-blades mostly 5–12 cm. long, gradually tapering from base to tip, reflexed on the stem; spikelets 2–2·3 mm. long, glabrous or rarely scabrid-hairy, the nerves prominent and often forming raised ribs; lower palea present; inflorescence with or without long glistening white hairs on the axis .. .. 34. *hochstetteri*
Lower glume 1-nerved; lower palea absent:
Lower glume ovate, $\frac{1}{3}$–$\frac{1}{2}$ as long as the spikelet; spikelets 2·5–3 mm. long, glabrous; leaf-blades lanceolate (if linear see *P. coloratum*) 35. *monticola*
Lower glume lanceolate, $\frac{1}{2}$–$\frac{2}{3}$ as long as the spikelet; spikelets 2–2·5 mm. long, glabrous or sparsely hispid .. .. 36. *calvum*
Plants annual; lower palea present:
Leaf-blades 1–2 cm. long, or sometimes a little longer, narrowly lanceolate, not amplexicaul; culms prostrate; spikelets 1·7–2 mm. long, glabrous or pubescent; lower glume 3-nerved, $\frac{3}{4}$ the length of the spikelet; upper lemma smooth .. .. 37. *pusillum*
Leaf-blades 5–45 cm. long, linear; culms erect, robust; spikelets 2·5–3 mm. long, glabrous; lower glume 5-nerved:
Spikelets borne in contiguous pairs, distant on long branchlets; lower glume sharply acute at the tip, $\frac{1}{2}$–$\frac{2}{3}$ as long as the spikelet; palea about half as long as the lower lemma; leaf-blades softly hairy above and below .. ..38. *pansum*
Spikelets neither paired nor contiguous, evenly spaced on the branches; lower glume obtuse to subacute at the tip, $\frac{1}{2}$ as long as the spikelet or a little less; palea as long as the lower lemma; leaf-blades glabrous and glaucous, or stiffly pilose below near the base.. .. 39. *laetum*
Lower glume $\frac{1}{3}$ as long as the spikelet or less, very broadly ovate, clasping the base of the spikelet:
Leaf-blades linear, firm, not narrowed at the base:
Plants creeping, with long wiry underground rhizomes; barren shoots distichously leafy, the blades more or less involute, rigid and glaucous; lower lemma 9-nerved, with a palea .. .. 40. *repens*
Plants not creeping:
Spikelets cuspidate acuminate at the tip, 2·5–3·5 mm. long, narrowly ovate; lower lemma 7-nerved, accompanied by a palea; panicle broadly ovate, loose, the branches spreading .. 41. *porphyrrhizos*
Spikelets acute at the tip:
Lower floret without a palea (in most W. African plants), its lemma

9-nerved or very rarely 7-nerved; spikelets ovate-elliptic, 2·5–3 mm. long; panicle terminal, narrowly oblong, dense, the primary branches ascending with the spikelets loosely contracted about them
42. *subalbidum*

Lower floret with a palea, its lemma 7–9-nerved; spikelets ovate, plump, 2–3 mm. long, prominently nerved; panicle broadly ovate, more or less spreading .. .. 43. *coloratum*

Leaf-blades narrowly lanceolate, thin and delicate with fine transverse veins, up to 15 cm. long, narrowed to the midrib at the base:

Upper glume 3-nerved; lower glume very broadly ovate, subacute
44. *comorense*

Upper glume 5-nerved; lower glume ovate-elliptic, usually obtusely rounded .. .. 35. *monticola*

1. **P. laxum** *Sw.* Prod. Veg. Ind. Occ. 23 (1788); Hitchc. & Chase in Contrib. U. S. Nat. Herb. 15: 115; Aké Assi, Contrib. 2: 286. Annual, 30–60 cm. high, the inflorescence superficially resembling that of a *Brachiaria*; roadsides and clearings, particularly on damp soils.
Guin.: Séguéa (Oct.) *Adam* 12641! **S.L.:** Freetown (May) *Baldwin* 5797! Njala (June) *Deighton* 732! Musaia (Mar.) *Deighton* 5439! Gbangbama to Senge (Nov.) *Deighton* 2343! Mt. Bintimani, 5,600 ft. (July) *Bakshi* 303! **Lib.:** Bushrod I. (Apr.) *Barker* 1292! Suacoco (Apr.) *Blickenstaff* 43! Blazie (Nov.) *Adam* 16039! Monrovia (Oct.) *van Harten* 178! Nimba Mt. (July) *Leeuwenberg & Voorhoeve* 4607! 4666! **Iv. C.:** Séguéla (July) *Adjanohoun* 240*a*! Ayamé to Bianoua (Nov.) *Aké Assi* 9763! **Ghana:** Elmina (July) *Hall* 1533! Awuabo to Krisin (Mar.) *Rose Innes* GC 30889! Aiyinasi (Dec.) *Rose Innes* GC 30545! Kade (Jan.) *Ankrah* GC 20333! Princestown (Mar.) *Morton* A316! **N. Nig.:** Zaria (Aug.) *Sawyer* K41! **S. Nig.:** Lagos (Oct.) *Haines* 187! A native of tropical America, introduced to W. Africa; first collected at Njala in 1927, it has now become naturalized.

2. **P. heterostachyum** *Hack.* in Oest. Bot. Zeitschr. 51: 430 (1901). Erect or ascending annual up to 60 cm high; in shade.
Niger: Maradi to Takieta (Nov.) *P. de Fabrègues* 1217! Widespread in Africa.

3. **P. brevifolium** *Linn.* Sp. Pl. 59 (1753); F.T.A. 9: 731; Chev. Bot. 724, and in Rev. Bot. Appliq. 14: 26; Berhaut, Fl. Sén. ed. 2, 416. *P. plantagineum* Schumach., Beskr. Guin. Pl. 64 (1827), not of Link (1827). *P. subobliquum* Stapf in F.T.A. 9: 723 (1920); F.W.T.A. ed. 1, 2: 554; A. Chev. in Rev. Bot. Appliq. 14: 26. *P. ovalifolium* Poir. (1816). *P. guineense* Desv. ex Poir. (1816). *P. amplexicaule* Poir. ex P. Beauv. (1819). *P. trichopiptum* Steud. (1854). A rambling annual 30–60 cm. high, with long aerial roots from the nodes; forest shade. Very rarely the outer spikelets of the panicle are long-hairy and apparently surrounded by a delicate halo.
Sen.: Sédhiou (Oct.) *Adam* 18445! *Berhaut* 6410! **Port. G.:** Bricama (Dec.) *Pereira* 3582! **Guin:** Timbo *Pobéguin* 1722! Fouta Djalon (Oct.) *Chev.* 18755! *Adam* 12751! **S.L.:** Rokupr (Sept.) *Jordan* 312! Bumbuna (Oct.) *Thomas* 3397! Bafodeya (Apr.) *Sc. Elliot* 1892! Mabala (Oct.) *Glanville* 61*a*! Bintumane, 3,200 ft. (Jan.) *T. S. Jones* 123! **Lib.:** Kakatown *Johnston*! Monrovia (Nov.) *Dinklage* 3269! Buchanan (Nov.) *Adam* 15999! Wanau (Nov.) *Harley* 2057! Ganta (Sept.) *Baldwin* 12017! **Iv. C.:** Abidjan (Nov.) Dec.) *Leeuwenberg* 1878! *Boughey* GC 13508! Adiopodoumé (Oct.) *Leeuwenberg* 1711! Dabou (June) *Leeuwenberg* 4469! Guéyo (Mar.) *Leeuwenberg* 3803! **Ghana:** Aiyinasi (Dec.) *Rose Innes* GC 30326! Kintampo to Bamboi (Oct.) *Rose Innes* GC 31141! Kwamang to Mampong (Nov.) *Rose Innes* GC 30872! Kumasi (Apr.) *La Danso* 34! Vane (Dec.) *Adams* 4723! **N. Nig.:** Gusau (Oct.) *De Leeuw & Magaji* 1955! **S. Nig.:** Lagos (Oct.) *Dalz.* 1121! Ibadan (Oct.) *Keay* FHI 13761! Okomu F.R., Benin (Dec.) *Brenan* 8420! Onitsha *Barter* 1781! Port Harcourt (Oct.) *Stubbings* 63! **W. Cam.:** Buea (Jan., July) *Maitland* 20! 326! Bamenda (May) *Maitland* 8*a*! 9*a*! Mensham R. (Feb.) *Brunt* 951! **F. Po:** *T. Vogel* 85! *Mann* 116! Bahia de Venus (Dec.) *Guinea* 250! Throughout the Old World tropics.

4. **P. trichoides** *Sw.* Prodr. Veg. Ind. Occ. 24 (1788); F.T.A. 9: 730. Ascending annual about 30 cm. high.
Ghana: Mansu *Cummins* 11! Kumasi (Aug.) *Irvine* 4521! **S. Nig.:** *Thomas* 1760! Benin (Sept.) *Maggs* 8! Epe (Aug.) *Gillett* 15265! Degema (Sept.) *Holland* 135! Umudike (July) *Tuley* 791! Throughout tropical America, and also in the Malayan islands; probably introduced to Africa.

5. **P. ecklonii** *Nees* Fl. Afr. Austr. 43 (1841). A densely caespitose perennial about 30 cm. high.
S.L.: Mt. Loma (Feb.) *Jaeger* 9371! Bintumane (June) *Jaeger* 9773! *Morton & Gledhill* SL 1091*a*! Tingi Mts (Apr., Dec.) *Morton & Gledhill* SL 1859! SL 3102! **Lib.:** Nimba (Feb.) *Adam* 20884! **Iv. C.:** Nimba Mts. (Aug.) *Boughey* GC 18067! **N. Nig.:** Chappal Waddi (Apr.) *Gbile & Daramola FHI* 63281! **S. Nig.:** Obudu (Apr.) *Haines* 143! **W. Cam.:** Bamenda (Jan., Oct.) *Daramola* FHI 40599! *Agric. Off.* 11! Bafut-Ngemba F.R. (Feb., Mar.) *Hepper* 2116! *Coombe* 215! Bambui (Mar.) *Brunt* 77! Tanzania and southwards to South Africa.

6. **P. hymeniochilum** *Nees* Fl. Afr. Austr.: 46 (1841). *P. kisantuense* Vanderyst ex Robyns in Mém. Inst. Col. Belge 1, 6: 25 (1932). *P. djalonense* A. Chev. in Rev. Bot. Appliq. 14: 27 (1934); F.W.T.A. ed. 1, 2: 554. *P. snowdenii* C. E. Hubbard in Kew Bull. 1928: 132. Prostrate perennial, rarely with the clavellate hairs absent; moist places.
Guin.: Dalaba (Apr.) *Caille* in *Hb. Chev.* 18139 *bis*! **Iv. C.:** Pakobo Nord (Oct.) *Adjanohoun* 369*a*! **N. Nig.:** Nguroje (Nov.) *Magaji & Tuley* 1868! 1872*a*! Mambila (Aug.) *De Leeuw* 1747! 1792! **W. Cam.:** ?Cam. Mt. *Boughey* GC 12694! Kumbo (Feb.) *C.N.A.D.* 1789! Throughout tropical Africa.

7. **P. sadinii** (*Vanderyst*) *Renvoize* in Kew Bull. 22: 485 (1968). *Brachiaria sadinii* Vanderyst in Bull. Agric. Congo Belge 16: 665 (1925). *Panicum lineatum* Trin., Sp. Gram. 2: t.233 (1829), not of Schumach. (1827); F.T.A. 9: 653; F.W.T.A. ed. 1, 2: 551; Aké Assi, Contrib. 2: 286. *P. scandens* Mez in Engl., Bot. Jahrb. 57: 190 (1921), not of (Schrad. ex Roem. & Schult.) Trin. (1826). Ascending perennial 60–90 cm. high, the base decumbent and rooting at the nodes; very rarely with the clavellate hairs reduced to stumps or absent; shady places.
Guin.: Pela (Sept.) *Baldwin* 13273! Tossekré (Oct.) *Adam* 12669! Macenta (Apr.) *Pitot*! **S.L.:** Yonibana (Oct.) *Thomas* 4049! Kanya (Oct.) *Thomas* 3015! Mabaka (Oct.) *Glanville* 71! Moselelo (Nov.) *Deighton* 2376! Picket Hill (Nov.) *T. S. Jones* 215! **Lib.:** Gola Forest (Dec.) *Baldwin* 10717! Nimba (Dec.) *Adam* 16497! Banga (Oct.) *Linder* 1191! Ganta (Sept.) *Baldwin* 9287! Grand Bassa (Mar.) *Baldwin* 11233! Also in the Congo, Zambia and Angola.

8. **P. maximum** *Jacq.* Ic. Pl. Rar. 1: 2, t.13 (1781); F.T.A. 9: 655; Chev. Bot. 727, and in Rev. Bot. Appliq. 14: 20; Berhaut, Fl. Sén. ed. 2, 412. *P. praelongum* Steud., Syn. Pl. Glum. 1: 73 (1854). *P. sparsum* Schumach., Beskr. Guin. Pl. 64 (1827). A robust perennial up to 3 m. high, densely tufted from a stout rhizome; moist places and roadside ditches.
Sen.: Sangalkam (Oct.) *Berhaut* 2897! **S.L.:** Freetown (Oct.) *Deighton* 2147! *T. Vogel*! Wellington (Jan.) *Morton & Gledhill* SL 1634! **Lib.:** Ganta (Feb.) *Harley* 1801! Harper (Mar., June) *Baldwin* 5983! 5999! 11640! **Iv. C.:** Abidjan to Dabou (Oct.) *de Wilde* 3109! Adiopodoumé (Nov.) *Leeuwenberg* 1887! Cocody (June) *Adjanohoun* 409*a*! Abidjan (Aug.) *de Wilde* 653! **Ghana:** Kumasi (Mar.) *La Danso* 18! Adawso (Oct.) *Rose Innes* GC 30583! Accra Plains (Oct.) *Baldwin* 13430! Ahamansu to Jasikan (Oct.) *Rose Innes* GC 30920! Kpeve (Mar.) *Lloyd Williams* 71! **Dah.:** Cotonou (Apr.) *Chev.* 4469! **N. Nig.:** Ilorin (Oct.) *Ward* 56! Lokoja (Oct.) *Dalz.* 285! Abinsi (June) *Dalz.* 904! Okene (Oct.) *Magaji*

Fig. 440.—Panicum maximum *Jacq.* (Gramineae-Paniceae).

1, ligule. 2, spikelet. 3, lower glume. 4, upper glume. 5, upper floret showing lemma. 6, flower.

MG 255! Makurdi (Aug.) *De Leeuw* 1712! **S. Nig.:** Lagos (July) *Dalz.* 1311! Ibadan (Nov.) *Meikle* 649! Tapa (Apr.) *Hambler* 466! Ala (Nov.) *Thomas* 1972! Opobo *Jeffreys* 1! **W. Cam.:** Victoria (Feb., July) *Maitland* 15! 152! *Brunt* 1040! *Kalu* FHI 13984! Buea (Jan.) *Maitland* 332! Throughout the tropics.
[Although very variable elsewhere in Africa, the W. African material is fairly uniform and is closely similar to specimens from the type locality in the West Indies.] (See Useful Plants.)

9. **P. dinklagei** *Mez* in Engl., Bot. Jahrb. 57: 190 (1921). Robust perennial, with firm dark green leaves and thin cane-like culms scrambling over other vegetation.
Guin.: Kindia *Jac.-Fél.* 209! Macenta (Oct.) *Baldwin* 9789! Nzérékoré to Macenta (Oct.) *Baldwin* 9743! Dyeke (Oct.) *Baldwin* 9656! Ségućа (Oct.) *Adam* 12638! **S.L.:** Mabala (Oct.) *Glanville* 61! Regent (Oct.) *Pyne* 137! Picket Hill (Nov.) *T. S. Jones* 226! Sembehun (Oct.) *Jordan* 625! Kenema (Jan.) *Thomas* 7642! **Lib.:** Vonjama (Oct.) *Baldwin* 9949! Monrovia (May) *Baldwin* 5831! Ganta (Oct.) *Harley* 1712! Moala (Nov.) *Linder* 1363! Nimba (Dec.) *Adam* 16462! **Ghana:** Maliato (Nov.) *Morton* A3556! Shiare (Dec.) *Hall* CC 1129!

10. **P. pubiglume** *Stapf* in F.T.A. 9: 670 (1920); Chev. Bot. 729, and in Rev. Bot. Appliq. 14: 20; Berhaut, Fl. Sén. ed. 2, 413. Loosely tufted perennial about 60 cm. high, with weak culms; on wet or swampy soils.
Sen.: Kanéméré (July) *Fotius* K306! Kabrousse to Oussouye *Adam* 18363! Marécage des Bayottes (Aug.) *Berhaut* 6332! **Mali:** Sikasso (May) *Chev.* 805! **Guin.:** Dalaba–Diaguissa Plateau (Oct.) *Chev.* 18766! Friguiagbé (Sept.) *Chillou* 1689! Chutes Komoré (Sept.) *Adam* 15251! Guésservolé (June) *Adam* 14655! **S.L.:** Gbap (June) *Adames* 50! **Iv. C.:** Singrobo (Oct.) *Adjanohoun* 306*a*! **Ghana:** Kintampo to Bamboi (Oct.) *Rose Innes* GC 31139! Damongo scarp (Sept.) *Rose Innes* GC 30202! Ho to Senchi (May) *Rose Innes* GC 30376! Nakpanduri to Gambaga (Oct.) *Ankrah* GC 20501! Wa (Oct.) *Rose Innes* GC 32376! **Togo Rep.:** Palimé *Stage* 58! **N. Nig.:** Nupe *Barter* 1013! Baro (Oct.) *Parsons* 13! Lokoja *Richardson*! Abinsi (June) *Dalz.* 899! Kontagora (Nov.) *De Leeuw & Magaji* 2002! Mundar, Jos (Sept.) *Lawlor & Hall* 634! **S. Nig.:** Igboora (Oct.) *Haines* 328! Also in the Congo.

11. **P. filifolium** *W. D. Clayton* in Kew Bull. 21: 111 (1967). Densely tufted perennial about 30 cm. high, with slender erect culms; in sandy savanna.
Lib.: Monrovia (Aug.) *Baldwin* 9184! 9185! 13020! Robertsfield (Mar.) *Harley* 2119! Buchanan (Nov.) *Adam* 16055!

12. **P. strictissimum** *Afzel. ex Sw.* Adnot. Bot. 4 (1829); F.T.A. 9: 666; A. Chev. in Rev. Bot. Appliq. 14: 20. *Isachne sapinii* Vanderyst (1925). *Panicum sapinii* (Vanderyst) Robyns (1932). Culms erect or ascending, 60–90 cm. high; in swamps.
Guin.: Baffing (Oct.) *Adam* 12665! **S.L.:** Kambia (Sept.) *Jordan* 308! Rokon (July) *Jordan* 485! Madina (Aug.) *Jordan* 915! Tabe to Jama (Sept.) *Deighton* 3026! Mabum (Aug.) *Thomas* 1540! **S. Nig.:** Mamu R. (Aug.) *Jones* 6691! Also in the Congo.
[Closely related to the two following species, but with a preference for moist habitats.]

13. **P. praealtum** *Afzel. ex Sw.* Adnot. Bot. 5 (1829); F.T.A. 9: 667; A Chev. in Rev. Bot. Appliq. 14: 20; Berhaut, Fl. Sén. ed. 2, 414; Schnell in Ic. Pl. Afr. IFAN. 5: 117 (1960). *P. caillei* A. Chev., Bot. 725 (1920), name only. Robust perennial 60–120 cm. high; sandy soils in savanna.
Sen.: Bakor (Oct.) *Adam* 18494! Kédougou (Nov.) *Fotius* K696! **Guin.:** Kouria *Caille* 15010! Kouria to Trebeleya (Sept.) *Chev.* 18229! Friguiagbé (Oct.) *Chillou* 1743! **S.L.:** Mando (Sept.) *Adames* 134! Bubuya (Sept.) *Jordan* 318! Binkolo (Aug.) *Thomas* 1878! Musaia (Aug.) *Deighton* 4399! Falaba (Oct.) *Small* 448! **Ghana:** summit E. of Ahamansu village (Sept.) *Rose Innes* GC 31326! **N. Nig.:** Gembu (Dec.) *Daramola* FHI 62320! **S. Nig.:** Udi Plateau (Sept.) *Tuley* 870! Abakaliki (Aug.) *Nwauzo* 767! Enugu (July, Sept.) *Opara* 805! *Onochie* FHI 34084! Also in E. Cameroun.
[A more robust plant than the following species with which it intergrades, and apparently limited to W. Africa.]

14. **P. baumannii** *K. Schum.* in Engl., Bot. Jahrb. 24: 331 (1897). *P. fulgens* Stapf in F.T.A. 9: 668 (1920); F.W.T.A., ed. 1, 2: 553; Aké Assi, Contrib. 2: 285. *P. globulosum* Mez in Engl., Bot. Jahrb. 57: 188 (1921). Erect or weakly ascending perennial 60–90 cm. high; often on shallow stony soils.
Guin.: Manea (June) *Chillou* 2619! **S.L.:** Tingi Mts. (Dec.) *Morton & Gledhill* SL 3039! **Lib.:** Nimba Mts. (Nov.) *Adam* 24715! **Iv. C.:** Singrobo (Oct.) *Adjanohoun* 339*a*! Dabou (Mar., Nov.) *Leeuwenberg* 1951! 3126! Séguélo (Oct.) *Aké Assi* 6603! Koébonou to Varalé (Sept.) *Aké Assi* 6506! **Ghana:** Sawla to Wa (Oct.) *Rose Innes* GC 31028! Kpandu to Accra (Sept.) *Ankrah* GC 20214! Kpandu to Kete Krachi (Sept.) *Rose Innes* GC 31314! Ahamansu to Jasikan (Oct.) *Rose Innes* GC 30919! Golokuati (Sept.) *Rose Innes* GC 20354! **N. Nig.:** Anara F.R., Kanjimi (Nov.) *Latilo* FHI 37968! Afaka F.R. (Nov.) *Clayton* 1390! Wuya Ferry (Sept.) *Clayton* 363! Samaru (Oct.) *Freeman* S178! Zaria Dist. *Taylor* 3! **S. Nig.:** Ezillo (Oct.) *Tuley* 964! **W. Cam.:** Bambui (Sept.) *Nditapah* 52! Also in the Congo and Angola, extending eastward to the Sudan, Tanzania and Zambia.

15. **P. calocarpum** *Berhaut* in Bull. Soc. Bot. Fr. Mém. 1953–54: 11 (1954), and Fl. Sén. ed. 2, 414. About 75 cm. high; swamps.
Sen.: Badi (Dec.) *Berhaut* 767!

16. **P. lindleyanum** *Nees ex Steud.* Syn. Pl. Glum. 1: 91 (1854); F.T.A. 9: 673; A. Chev. in Rev. Bot. Appliq. 14: 21; Berhaut, Fl. Sén. ed. 2, 413. *P. hystrix* Steud. l.c. 95 (1854); F.W.T.A., ed. 1, 2: 553; Chev. Bot. 726, and in Rev. Bot. Appliq. 14: 21. *P. drosocarpum* Stapf (1905). *P. viciniflorum* Stapf in F.T.A. 9: 675 (1920); F.W.T.A., ed. 1, 2: 553; A. Chev. in Rev. Bot. Appliq. 14: 21. *P. filicaule* Stapf in F.T.A. 9: 676 (1920); F.W.T.A., ed. 1, 2: 553; A. Chev. in Rev. Bot. Appliq. 14: 21. *P. trochainii* (A. Camus) A. Camus in Bull. Mus. Hist. Nat., sér. 2, 6: 99 (1934); F.W.T.A., ed. 1, 2: 553. *Isachne trochainii* A. Camus in Bull. Mus. Hist. Nat., sér. 2, 5: 250 (1933); F.T.A. 9: 1098; A. Chev. in Rev. Bot. Appliq. 14: 40. *Panicum gemmeum* C. E. Hubbard in Kew Bull. 4: 361 (1949). A delicate tufted or weakly ascending annual from 10 cm to about 60 cm. high; shallow pools and wet flushes on rock outcrops and ironstone pans.
Sen.: *Heudelot* 583! Janu de Fayil (Oct.) *Berhaut* 5666! Bignona (Feb.) *Trochain* 1438! Kabrousse (Oct.) *Adam* 18290! Oussouye (Mar.) *Berhaut* 7252! **Mali:** *Lécard* 250! Koulikoro (Oct.) *Chev.* 2178! **Port. G.:** Orango Grande (Jan.) *Raimundo & Guerra* 928! **Guin.:** Kouroussa (Dec.) *Pobéguin* 541! Mamou to Trebeleya (Sept.) *Chev.* 8622 *bis*! Nzérékoré (Oct.) *Baldwin* 9689! Macenta (Oct.) *Baldwin* 9795! Filicoundji (Oct.) *Adam* 12590! **S.L.:** Rokupr (Nov.) *Jordan* 690! Sugar Loaf Mt. (Oct.) *T. S. Jones* 257! Tombo (Jan.) *Deighton* 995! Brookfields (Oct.) *Deighton* 2174! Mapatolon (Nov.) *T. S. Jones* 40! **Lib.:** Monrovia (Nov.) *Barker* 1464! Buchanan (Nov.) *Adam* 16056! Tawata (Nov.) *Baldwin* 10349! Vonjama (Oct.) *Baldwin* 9861! Genne-Loffa (Nov.) *Baldwin* 10084! **Iv. C.:** Sifié (Oct.) *Aké Assi* 5716! **Ghana:** Nakpanduri to Yendi (Oct.) *Rose Innes* GC 30735! Berekum to Sampa (Dec.) *Adams* 5273! Kwahu Tafo (Dec.) *Harris*! Hohoe to Dafo (Sept.) *Rose Innes* GC 31319! Yinahin range (Nov.) *Rose Innes* GC 30530! **Togo Rep.:** Mango (Nov.) *Bille* JG 339! **N. Nig.:** Anara F.R. (Oct.) *Keay* FHI 5500! Jos (Oct.) *Hepper* 1059! Okene (Oct.) *Magaji* MG 278! **S. Nig.:** Idanre (Jan.) *Brenan & Keay* 8658! Ife (Feb.) *Hall* 64! Oyo (April) *Gledhill* 986! Also in Central African Rep., Congo and Zambia.
[Although usually a more slender plant than the perennial species, *P. lindleyanum* is difficult to separate from them unless the base is well represented. It displays a wide range of variation, but there are no obvious discontinuities nor any clear distinguishing features. It seems best, for the present, to treat it as a single species—W. D. C.]

17. **P. glaucocladum** *C. E. Hubbard* in Kew Bull. 1933: 501. *P. pusillum* var. *glabriglumatum* Schnell in Rev. Gen. Bot. 57: 290 (1950). Slender decumbent annual; damp places in shade.
Guin.: Nimba (Sept.) *Schnell* 3684! **S.L.:** Mt. Loma (Oct., Nov.) *Jaeger* 7862! *Morton* SL 2715! Bintumane (Jan.) *T. S. Jones* 84! Tingi hills (Dec.) *Morton & Gledhill* SL 3042! Toma, Bonthe I. (Nov.) *Deighton* 2282! **Lib.:** Fisherman's lake (Dec.) *Baldwin* 10898!

18. **P. gracilicaule** *Rendle* in Cat. Welw. 2: 179 (1899); F.T.A. 9: 672; Berhaut, Fl. Sén. ed. 2, 415. *P sublaetum* Stapf in F.T.A. 9: 671 (1920); F.W.T.A., ed. 1, 2: 553; A. Chev. in Rev. Bot. Appliq. 14: 21. *P. membranaceum* Robyns (1932). *P. tambacoundense* Berhaut in Mém. Soc. Bot. Fr. 1953–54: 11 (1954), and Fl. Sén. ed. 2, 414. Up to 60 cm. high, with flaccid light green leaves; roadsides and weedy places.
Sen.: Niokolo-Koba (Oct.) *Adam* 15675! Barkédédji (Oct.) *Berhaut* 3867! Tambacounda (Oct.) *Adam* 12722! Sefa (Oct.) *Adam* 18416! Bougnadou (Oct.) *Adam* 18478! **Gam.:** Yundum (Dec.) *Austin* 21! **Port. G.:** Mansoa (Nov.) *Pereira & Correia* 1983! Susana (Dec.) *Raimundo & Guerra* 262! 307! Bissau I. (Nov.) *Raimundo & Guerra* 45! Nova Lamego (Dec.) *Pereira & Correia* 2427! **Guin.:** Baffing R. (Nov.) *Pobéguin* 1723! 1822! Kouroussa (Oct.) *Pobéguin* 523! Sériba to Kiffaya (Oct.) *Adam* 12698! **S.L.:** Musaia (Dec.) *Deighton* 4478! **N. Nig.:** Riyom (Nov.) *Clayton* 1410! Shendam (Nov.) *Clayton* 1452! Plateau Prov. (Aug.) *Lely* P496! Mundar, Jos (Sept.) *Lawlor & Hall* 636! Abinsi *Dalz.* 892! Extends through the Congo into Malawi and Zambia.

19. **P. nigerense** *Hitchc.* in Proc. Biol. Soc. Wash. 43: 90 (1930). Annual up to 1·5 m. high, with a large panicle of reddish-brown spikelets; on dry sandy soils.
Mali: Macina (Sept.) *Chev.* 24911! **U. Volta:** Sindou (Sept.) *Adam* 15174! Toussiana to Banfora *Adam* 15219! **Niger:** Niamey (Oct.) *Hagerup* 481! Zinder (Apr.) *Guile*! **N. Nig.:** Katsina (Oct.) *Clayton* 1338! *Ward* 63! Damaturu (Sept.) *De Leeuw* 1162!

20. **P. anabaptistum** *Steud.* Syn. Pl. Glum. 1: 75 (1854); F.T.A. 9: 678; A. Chev. in Rev. Bot. Appliq. 14: 22; Berhaut, Fl. Sén. ed 2., 413. *P. oxyanthum* Steud. (1854). *P. cinereo-viride* Mez in Engl., Bot. Jahrb. 57: 190 (1921). *P. glaucifolium* Hitchc. in Proc. Biol. Soc. Wash. 43: 90 (1930). *Garnotia africana* Janowski in Fedde Rep. 17: 86 (1921). Up to about 1·5 m. high; on flood plains.
Maur.: Lake Rkiz (Sept.) *Sadio* 10! **Sen.:** Richard-Tol (Feb.) *Döllinger* 18! Walo (Dec.) *Roger* 51! St. Louis (Oct.) *Chev.* 33893! *Heudelot* 286! **Mali:** Zingueni to Sompi (Aug.) *Chev.* 2289! Toguère of Dogo (May) *Davey* 80! Djenné (Sept) *Chev.* 2274! Diré *Rogeon* 257! Kabara (Sept.) *Hagerup* 304! **Iv. C.:** Koro to Kaboro-Kendé (Sept.) *Chev.* 24887! **Ghana:** Gushiago to Pigu (Oct.) *Rose Innes* GC 30709! Kpasinkpe (Oct.) *Rose Innes* GC 30778! Bawku to Bolgatanga (Aug.) *Rose Innes* GC 32092! Lawra (Sept.) *Hall* CC 766! **Niger:** Zinder (Nov.) *Hagerup* 616! Dassa *White* 17! Toukounous *Bartha* 37! **N. Nig.:** Nupe *Barter* 838! Sokoto (Sept.) *Palmer* 16! Hadejia (Dec.) *Clayton* 474! Bornu Prov. (Oct.) *Johnston* N 48! Katagum Dist. *Dalz.* 274! **W. Cam.:** Bama (Dec.) *McClintock* 104! Also in E. Cameroun and the Sudan. (See Useful Plants.)

21. **P. phragmitoides** *Stapf* in F.T.A. 9: 677 (1920); A. Chev. in Rev. Bot. Appliq. 14: 21. *P. klingii* Mez in Engl., Bot. Jahrb. 57: 188 (1921). *P. afrum* Mez l.c. Robust reed-like grass up to 2 m. high; in savanna, on dry soils.
Guin.: Pita (Sept.) *Adames* 358! **Iv. C.:** Bouna to Bondoukou (Oct.) *Rose Innes* GC 31550! Bouaflé to Zuénoula (Oct.) *Adjanohoun* 320*a*! Bocanda (June) *Adjanohoun* 249*a*! Bouaké (Sept.) *de Wilde* 942! **U. Volta:** Sikasso to Bobo-Dioulasso (Sept.) *Adam* 15124! Samandéni (Oct.) *Kmoch* 133! 156! Doutié (Sept.) *Scholz* 15*a*! Gaoua to Batié (Oct.) *Rose Innes* GC 31529! Ouassa to Leo (Oct.) *Rose Innes* GC 31490! **Ghana:** Sawla Junction (Oct.) *Rose Innes* GC 30306! Sawla to Tamale (Oct.) *Rose Innes* GC 30673! Gambaga (Oct.) *Rose Innes* GC 30266! Kete Krachi (Oct.) *Rose Innes* GC 30610! Kpevi to Anum (Sept.) *Rose Innes* GC 30899! **Togo Rep.:** Badou to Atakpamé (Sept.) *Rose Innes* GC 31348! Sokode to Lama Kara (Oct.) *Rose Innes* GC 31388! Bismarckburg *Kling* 243! 254! **N. Nig.:** Shika (Sept.) *Clayton* 1271! Zaria Dist. (Sept.) *Freeman* S21! *Taylor* 45! Naraguta (Oct.) *Kennedy* FHI 8045! Gindiri (Oct.) *Hepper* 1141! **S. Nig.:** Obudu Plateau (Nov.) *Tuley* 1021! 1022! **W. Cam.:** Bamenda (Dec.) *Agric. Off.*! *Boughey* 11148! Ndop (Mar., Nov.) *Brunt* 22! *Gillett* 13! Maibinka (Feb.) *Brunt* 985! Extends through Cameroun and the Congos to Angola and Malawi.

22. **P. pilgeri** *Mez* in Engl., Bot. Jahrb. 34: 146 (1904); F.T.A. 9: 686. *P. lasiopodum* Stapf (1905). Densely tufted perennial up to about 1 m. high; wet and marshy places.
Guin.: Kouroussa (Aug.) *Pobéguin* 500! Gali to Hinde (July) *Adames* 311! Kindia (Oct.) *Kent* 44! **S.L.:** Waterloo (June) *Deighton* 2742! Bumban (Oct., Nov.) *Deighton* 3276! *Morton* SL 2902! Tingi Mts. (Dec.) *Morton & Gledhill* SL 2952! SL 3173! **Iv. C.:** Mankono (July) *Chev.* 21981! Singrobo (Oct.) *Adjanohoun* 307*a*! Brobo (Sept.) *de Wilde* 941! Bouaké (Sept.) *Oldeman* 410! Toumodi to Yamoussokro (Oct.) *Adjanohoun* 328*a*! **U. Volta:** Sikasso to Bobo-Dioulasso (Sept.) *Adam* 15083! **Ghana:** Kwahu Tafo (Aug., Oct.) *Rose Innes* GC 30569! *Hall* CC 123! Amentin (Oct.) *Rose Innes* GC 30588! 30589! Kete Krachi to Yendi (Sept.) *Rose Innes* GC 30438! Velempele, Wa (Aug.) *Addei* SLUS 7! **N. Nig.:** Nupe *Barter* 1377! Vom (Oct., Nov.) *Gambles* 4! 53! Yola to Makridi (Aug.) *De Leeuw* 1720!

23. **P. dregeanum** *Nees* Fl. Afr. Austr. 42 (1841); F.T.A. 9: 684; A. Chev. in Rev. Bot. Appliq. 14: 22; Berhaut, Fl. Sén. ed. 2, 414. *P. amethystinum* Franch. (1895). *P. chilianthum* A. Chev., Sudania 62, 132 (1911), name only. *P. poecilanthum* Stapf in F.T.A. 9: 684 (1920); F.W.T.A., ed. 1, 2: 553. Tufted, erect, 60–90 cm. high; seasonally moist soils.
Sen.: Boukitingo (Oct.) *Adam* 18323! Oussouye (Aug.) *Berhaut* 6255! **Guin.:** Friguiagbé (Sept.) *Chillou* 704! Sambaïlo (Sept.) *Pitot*! **S.L.:** Tingi Mts. (Dec.) *Morton & Gledhill* SL 3092! Loma Mt. (Dec.) *Morton* SL 395! **Lib.:** Paynesville (Sept.) *Van Dillewijn* 79! **N. Nig.:** Anara F.R., Zaria (May) *Keay* FHI 22870! Naraguta (July) *Lely* 423! Abinsi (July) *Dalz.* 897! **W. Cam.:** (Dec.) *Unwin* 9185! Throughout tropical and S. Africa.

24. **P. congoense** *Franch.* in Bull. Soc. Hist. Nat. Autun 8: 342 (1895); F.T.A. 9: 682; Aké Assi, Contrib. 2: 285. *P. mitophyllum* Pilger (1903). Very densely tufted grass 30–60 cm. high, with tiny spikelets; in marshes.
Sen.: Salikénie (Oct.) *Adam* 18523! **S.L.:** Waterloo (Aug.) *Melville & Hooker* 301*b*! Rokupr (Mar.) *Jordan* 421! Bonthe Isl. (Nov.) *Deighton* 2299! Tabe (Sept.) *Deighton* 3037! Kitchom (Jan.) *Deighton* 937! **Lib.:** Paynesville (Sept.) *van Dillewijn* 79! Monrovia (Aug.) *Baldwin* 9196! Duport (Nov.) *Linder* 1461! Nimba (Feb., Oct.) *Adam* 20878! *Adames* 703! **Ghana:** Essiama (Oct., Dec.) *Hall* 2658! *Rose Innes* GC 30551! Awiebo to Krisin (Dec.) *Rose Innes* GC 30540! Also in the Congos and Angola.

25. **P. fluviicola** *Steud.* Syn. Pl. Glum. 1: 89 (1854); F.T.A. 9: 689. *P. aphanoneurum* Stapf in F.T.A. 9: 687 (1920); F.W.T.A., ed. 1, 2: 553; Chev. Bot. 724, and in Rev. Bot. Appliq. 14: 22; Berhaut, Fl. Sén. ed. 2, 414 (incl. var. *silvicola* Berhaut, Fl. Sén. ed. 1, 254 (1954), without latin descr.); Aké Assi, Contrib. 2: 284. *P. rowlandii* Stapf in F.T.A. 9: 688 (1920); F.W.T.A., ed. 1, 2: 553. *P. purpurascens* Mez in Engl., Bot. Jahrb. 57: 188 (1921), not of Raddi (1823). *P. sparmannii* Mez l.c. 188. *P. radicosum* Mez l.c. 189. A reed-like grass 1–2 m. high; basal sheaths glabrous or sometimes pubescent; streamsides, flood plains and moist soils.
Maur.: Lake Rkiz (Sept.) *Sadio* 121! **Sen.:** Niokolo-Koba (Oct.) *Adam* 15642! 15646! 15651! **Gam.:** *Don*! *Hancock* 21! **Mali:** San (Jan.) *Demange* 1018! Macina (Sept.) *Chev.* 24950! Sarédina (Mar.) *Davey* 69! Mopti (Sept.) *Matthes* 6! Dari (Sept.) *Farrow* 40! **Port. G.:** Bafata (Dec.) *Pereira & Correia* 2521! Bissau Is. (Nov.) *Raimundo & Guerra* 117! Bubaque (Jan.) *Raimundo & Guerra* 994! Farim (Oct.) *Pereira* 3387! **Guin.:** Baffing R. *Pobéguin* 1720! 1721! 1730! Kindia *Jac.-Fél.* 256! Sita (Oct.) *Adam* 12685! **S.L.:** Batkanu (Dec.) *Deighton* 5693! Musaia (Dec.) *Deighton* 4489! Kamiototo (Nov.) *Glanville* 311! Masuri (Oct.) *Glanville* 84! Makeni (Dec.) *Morton & Gledhill* SL 310! **Iv. C.:** Nambonkaha (Nov.) *Leeuwenberg* 2046! Toumodi (Oct.) *Adjanohoun* 348*a*! Zuenoula (Oct.) *Adjanohoun* 316*a*! **U. Volta:** Leo to Po (Oct.) *Rose Innes* GC 31470! Fada-N'Gourma to Pama (Aug.) *Scholz* 142! **Ghana:** Morago R. (Oct.) *Rose Innes* GC 30753! Wulugu (Oct.) *Rose Innes* GC 30776! Kpandai (Oct.) *Rose Innes* GC 30953! Tonogo (Nov.) *Morton* A3505! Anum (Sept.) *Rose Innes* GC 30390! **Togo Rep.:** Sokode (Oct.) *Kersting* 254! *Schroeder* 146! Dapango (Oct.) *Rose Innes* GC 31425! Palime *Stage* 59! **Niger:** Dallol (Jan.) *Virgo* 76! **N. Nig.:** Katagum Dist. *Dalz.* 258! Ekan (Sept.) *Ward* 17! Naraguta (Dec.) *Kennedy*

12! Minna Div. (Dec.) *Meikle* 737! Abinsi *Dalz.* 905! **S. Nig.:** Lagos (Oct.) *Dalz.* 1328! Ago-Are F.R. (Nov.) *Keay* FHI 37697! Jesse (Sept.) *Butler-Cole* 2! Aboh *Barter* 343! Enugu (Sept.) *Onochie* FHI 34104! Throughout tropical Africa.

26. **P. griffonii** *Franch.* in Bull. Soc. Hist. Nat. Autun 8: 342 (1895); F.T.A. 9: 691; A. Chev. in Rev. Bot. Appliq. 14: 22; Berhaut, Fl. Sén. ed. 2, 413; Aké Assi, Contrib. 2: 285. Straggling or weakly ascending, 15–30 cm. high; shallow soil on rock outcrops and inselbergs.
Sen.: Bougnadou *Adam* 18483! Sédhiou (Nov.) *Berhaut* 6404! **Guin.:** Macenta (Oct.) *Baldwin* 9760! **S.L.:** N. Prov. (Oct.) *Glanville* 94! Rokon (Oct.) *Jordan* 575! Njala (July) *Deighton* 1167! Makeni (Jan.) *Deighton* 4149! Mabould (Oct.) *Thomas* 3662! **Lib.:** Vonjama (Oct.) *Baldwin* 9876! Palilah (Aug.) *Baldwin* 9149! Nimba (Dec.) *Adam* 16458! Zorzor (Oct.) *Baldwin* 10027! Gbau (Sept.) *Baldwin* 9449! **Iv. C.:** Vaoua (Oct.) *Adjanohoun* 314*a*! Séguéla (Oct.) *Adjanohoun* 242*a*! Odienné (Oct.) *Aké Assi* 8236! **Ghana:** Gradaw (Oct.) *Rose Innes* GC 31624! Krobo Hill (Oct.) *Rose Innes* GC 31152! Tonogo (Nov.) *Morton* A3493! Atonso (Oct.) *Baldwin* 13535! Kumasi (Dec.) *Adam* 4451! **N. Nig.:** Vodni *Saunders* 1! Riyom (Oct., Nov.) *Clayton* 1404! *De Leeuw* 1626! Sha (Sept.) *Hall* K28! **S. Nig.:** Lagos *Dawodu* 186! Ebute Metta (Nov.) *Millen* 13! Igboora (Oct.) *Haines* 330! Ado Rock (Oct.) *Savory & Keay* FHI 25447! Mt. Orosun, Idanre Hills (Oct.) *Keay* FHI 22597! **W. Cam.:** Bamenda (Dec.) *Baldwin* 13837*a*! Extends southwards through E. Cameroun and Congo to Angola, and eastwards to Uganda and Tanzania.

27. **P. afzelii** *Sw.* Adnot Bot. 5 (1829); F.T.A. 9: 695; A. Chev. in Rev. Bot. Appliq. 14: 23; Berhaut, Fl. Sén. ed. 2, 414. *P. muscarium* Trin. (1829). Delicate grass about 30 cm. high; very closely related to the two following species; shallow or disturbed soils in moist places.
Sen.: Ziguinchor (Dec.) *Berhaut* 6707! Kanéméré (Sept.) *Fotius* K489! **Port. G.:** Bissau Is. (Nov., Dec.) *Pereira & Correia* 2096! 2129! *Raimundo & Guerra* 207! **Guin.:** Conakry (Oct.) *Adam* 12603! 12614! Filicoundji (Oct.) *Adam* 12589! Dalaba (Nov.) *Chev.* 34594! Timbo *Pobéguin* 1717! Manéa (Dec.) *Chillou* 2301! **S.L.:** Regent (Dec.) *Sc. Elliot* 4115! Kambia (Jan.) *Deighton* 914! Roboli (Nov.) *Jordan* 170! Magbile (Dec.) *Thomas* 6467! Foni flats, Brookfields (Oct.) *Deighton* 2143! **Iv. C.:** Bassawa (Oct.) *Adjanohoun* 270*a*! Ouango-Fitini (Dec.) *Aké Assi* 7527! Dabakala to Comoë (Nov.) *Aké Assi* 9247! **U. Volta:** Samandeni (Oct.) *Kmoch* 175! Cascade de Toussiana (Oct.) *Scholz* 77! **Ghana:** Burufu (Oct.) *Rose Innes* GC 31042! 32348! *Hinds* 5015! Mole Reserve (Nov.) *Hall & Enti* GC 35831!

28. **P. walense** *Mez* in Engl., Bot. Jahrb. 34: 146 (1904), as "*watense*"; Clayton in Kew Bull. 20: 264; Lebrun in Bull. Soc. Bot. Fr. 116: 263. *P. humile* Nees ex Steud., Syn. Pl. Glum. 1: 84 (1854), not of Thunb. ex Trin. (1825); F.T.A. 9: 693; F.W.T.A., ed. 1, 2: 554; Chev. Bot. 726, and in Rev. Bot. Appliq. 14: 23; Berhaut, Fl. Sén. ed. 2, 415. *P. austro-asiaticum* Ohwi in Act. Phytotax. & Geobot. 11: 45 (1942). *P. vescum* Stewart (1945). Annual 30–60 cm. high, with a delicate bushy inflorescence; often forming a dense cover over shallow or disturbed sites on moist soils.
Sen.: Matam (Dec.) *Chev.* 2235! Tamboukané (Dec.) *Chev.* 2234! Kala (Jan.) *Davey* 201! Walo (Mar.) *Leprieur* 52! Sefa (Oct.) *Adam* 18429! **Gam.:** *Saunders* 80! **Mali:** Macina (Nov.) *Duong* 1186! Dioura (Sept.) *Farrow* 29! Sorme (Sept.) *Farrow* 43! Zangoula (Oct.) *Farrow* 64! **Guin.:** Madina (Oct.) *Adam* 12511! Kinkon to Pita (Oct.) *Pitot*! Labé (Nov.) *Adames* 413! **S.L.:** Fonima (Oct.) *Fisher* 71! **U. Volta:** Koumbili (Oct.) *Scholz* 181! **Niger:** (Sept.) *P. de Fabrègues* 2114! **N. Nig.:** Hadejia (Dec.) *Clayton* 473! Kano (Dec.) *Hagerup* 666! Naraguta F.R. (Oct.) *Hepper* 1015! *Moiser* 201! Birnin Kebbi (Nov.) *Vaillant* 2774! **S. Nig.:** Olokemeji (Nov.) *Jackson* 581169! Also in Sudan, India, China and Malesia.

29. **P. paucinode** *Stapf* in F.T.A. 9: 692 (1920). 60–90 cm. high, rather more robust than the preceding; shallow or disturbed sites on moist soils.
Guin.: Timbo *Pobéguin* 1718! **Ghana:** Burufu (Sept.) *Hall* CC 751! Yendi to Gambaga (Oct.) *Rose Innes* GC 30480! Tamale to Bolgatanga (Nov.) *Hall & Enti* 35967! **N. Nig.:** Kufena Hill, Zaria Prov. (Sept.) *Clayton* 1308! Shendam (Nov.) *Clayton* 1451! Yola (Aug., Nov.) *Dalz.* 270! *Hepper* 1202! *Lely* 786! Also in E. Cameroun.

30. **P. acrotrichum** *Hook. f.* in J. Linn. Soc. 7: 226 (1864); F.T.A. 9: 721. A straggling grass, rooting from the lower nodes; in forest shade.
W. Cam.: Cam. Mt. (Dec., Feb., Mar.) *Mann* 2100! *Maitland* 851! 1040! *Brenan* 9533! Oku, Bamenda Div. (Feb.) *Hepper* 2028! **F. Po:** El Pico (Dec.) *Boughey* 119! 134!

31. **P. mueense** *Vanderyst* in Bull. Agric. Congo Belge 10: 248 (1919). *P. microthyrsum* Stapf in F.T.A. 9: 729 (1920); F.W.T.A., ed. 1, 2: 544. *Ichnanthus mueensis* (Vanderyst) Vanderyst l.c. 16: 681 (1925). Perennial, with culms decumbent and rooting at the nodes; gallery forest.
S. Nig.: Okomu F.R., Benin (Dec., Jan., Feb.) *Richards* 5058! *Brenan* 8397! 8896! **W. Cam.:** Kindong R., Mbalange (Jan.) *Binuyo & Daramola* FHI 35496! Also in E. Cameroun and Congo.

32. **P. turgidum** *Forsk.* Fl. Aegypt.-Arab. 18 (1775); F.T.A. 9: 706; Chev. Bot. 730, and in Rev. Bot. Appliq. 14: 24; Berhaut, Fl. Sén. ed. 2, 415; De Miré & Gillet in J. Agric. Trop. 3: 734. Perennial up to 1·2 m. high, with stout roots, and sturdy stems branching untidily in all directions; dry sandy soils.
Sen.: *Heudelot* 393! Saint Louis (Oct.) *Berhaut* 857! **Mali:** *Leclercq* 42488! Ségou (Mar.) *Chev.* 44000! Timbuktu (June, Aug.) *Hagerup* 97! *Chev.* 1221! *Lean* 65! **Niger:** Agadez (Feb., Apr.) *Bradley* 6! *Guile*! Through N. Africa and the Sudan to Arabia, India and Iraq. (See Useful Plants.)

33. **P. parvifolium** *Lam.* Tab. Encycl. 1: 173 (1791); F.T.A. 9: 726; A. Chev. in Rev. Bot. Appliq. 14: 26; Berhaut, Fl. Sén. ed. 2, 414. *P. raripilum* Kunth (1831). Creeping perennial with long wiry stolons, and rigid glaucous leaves; swamps.
Sen.: (June) *Leprieur*! **Mali:** Sikasso (Feb.) *Laferrère* 19! **S.L.:** Mambolo (Apr.) *Glanville* 225! Rokupr (Mar.) *Jordan* 19! Songo (Nov.) *T. S. Jones* 291! Njala (Sept.) *Deighton* 2099! Mayoso (Aug.) *Thomas* 1503! **Lib.:** Monrovia (June) *Baldwin* 5854! Buchanan (Nov.) *Adam* 16006! Suacoco (Aug.) *Barker* 1393! Duport (Nov.) *Linder* 1458! Ganta (Oct.) *Harley* 2053! **Ghana:** Atuabo (Aug.) *Thorold* 220! Eikwe (Dec.) *Rose Innes* GC 30546! Esiama (Mar.) *Rose Innes* GC 30877! Gulumpe (Nov.) *Rose Innes* GC 30855! Nakpanduri to Gambaga (Oct.) *Ankrah* GC 20500! **Dah.:** Sémé (Feb.) *Raynal* 13539! **S. Nig.:** Lagos (Sept.) *Dalz.* 1319! Emo R. (July) *Jeffreys* 17! Sapoba (Sept.) *Onochie* FHI 34276! Eket Dist. (May) *Onochie* FHI 32927! Mamu R. flood plain (Aug.) *Jones* FHI 6695! Tropical Africa generally, Madagascar, and tropical America.

34. **P. hochstetteri** *Steud.* Syn. Pl. Glum. 1: 90 (1854); F.T.A. 9: 724. *P. mokaense* Mez in Engl., Bot. Jahrb. 57: 189 (1921). *Isachne refracta* Hook. f. (1864). Loosely tufted perennial up to 1 m. high with weak stems; in forest, and in grassland just above the forest limit.
S.L.: Loma Mt. (Nov., Dec.) *Morton* SL 2716! *Jaeger* 8232! 8317! Bintumane (Jan.) *T. S. Jones* 122! **Ghana:** Shiare (Dec.) *Jenik & Hall* CC 1035! **N. Nig.:** Gembu (Dec.) *Daramola* FHI 62476! Chappal Waddi (Nov.) *Jackson, Magaji & Tuley* 2025! 2031! **S. Nig.:** Obudu Plateau (Nov.) *Tuley* 1028! **W. Cam.:** Bamenda (Oct.) *Agric. Off.* Bam. 16! Ashong (Mar.) *Brunt* 1091! Cam. Mt. (Dec., Jan.) *Mann* 2082! *Boughey* 12591! *Maitland* 849! **F. Po:** Moka (Nov.) *Mildbr.* 7086! Musola (Jan.) *Guinea* 1011! 1013! Also in the Congo, extending eastwards through Uganda to Kenya and Ethiopia.

35. **P. monticola** *Hook. f.* in J. Linn. Soc. 7: 226 (1864); F.T.A. 9: 722. *P. mannii* Mez in Engl., Bot. Jahrb. 34: 143 (1904). *P. macrophyllum* Guinea ex W. D. Clayton in Kew Bull. 21: 111 (1967).
S. Nig.: Obudu Plateau (Nov.) *Tuley* 1031! **W. Cam.:** Cam. Mt. *Mann* 1353! *Hinds* C67! Mann's Spring (Jan.) *Steele* 93! Buea (Jan.) *Maitland* 331! 1266! **F. Po:** Moka (Dec.) *Boughey* 63! Musola to Balacha (Jan.) *Guinea* 1470! 1472! Mountain tops throughout tropical Africa.

36. **P. calvum** *Stapf* in F.T.A. 9: 723 (1920); Aké Assi, Contrib. 2: 284. Closely allied to the preceding.
Iv. C.: Mt. Tonkoui (Nov.) *Aké Assi*! **W. Cam.:** Cam. Mt. (Dec.) *Hinds* C63! Buea (Nov., Dec., Jan.) *Migeod* 60! *Maitland* 104! 327! 850! Also in the Congo, extending eastwards to Kenya and Tanzania.

37. **P. pusillum** *Hook. f.* in J. Linn. Soc. Bot. 7: 227 (1864); F.T.A. 9: 275. A weak prostrate annual growing in loose mats; wet flushes on rock and ironstone outcrops.

S.L.: Mt. Loma (Oct.) *Jaeger* 7799! **N. Nig.:** Vom (Sept.) *Magaji & Tuley* 1803! Njawai (Nov.) *Jackson, Magaji & Tuley* 1909! Chappal Waddi (Nov.) *Jackson, Magaji & Tuley* 2023! **W. Cam.:** Bafut-Ngemba F.R. (Feb.) *Hepper* 2124! Nyango Camp (Dec.) *Maitland* 1218! Cam. Mt. (Sept., Dec.) *Boughey* A307! *Mann* 2090! Also in the Congo, extending eastwards through Uganda to Kenya and Ethiopia.

[The inclusion of the *Jaeger* plant, a tiny annual with 1-nerved lower glume, is open to question.—W.D.C.]

38. **P. pansum** *Rendle* Cat. Welw. 2: 117 (1899); F.T.A. 9: 700; Lebrun in Bull. Soc. Bot. Fr. 116: 263; *P. kerstingii* Mez in Engl., Bot. Jahrb. 34: 145 (1904); F.T.A. 9: 699; F.W.T.A., ed. 1, 2: 554; Chev. Bot. 727, and in Rev. Bot. Appliq. 14: 23; Berhaut, Fl. Sén. ed. 2, 412, 415; Aké Assi, Contrib. 2: 286. 60–90 cm. high; a weed of waste places.

Sen.: Kanéméré (Sept.) *Fotius* K409! Vélingara (Nov.) *Boudet* 4605! Bougnadou (Oct.) *Adam* 18470! **Guin.:** Baffing (Oct.) *Pobéguin* 1823! **Iv. C.:** Sampa to Bondoukou (Oct.) *Rose Innes* GC 31563! Bouaké (May) *de Wilde* 53! Bocanda (June) *Adjanohoun* 250*a*! Lomo (Oct.) *Aké Assi* 6544! **U. Volta:** Bama (Oct.) *Kmoch* 143! Folonto (Oct.) *Scholz* 90! Sindou (Sept.) *Aké Assi* 6365! **Ghana:** Mankuma (Oct.) *Rose Innes* GC 30663! Tamale (Oct.) *Baldwin* 13567! Kete Krachi (Oct.) *Rose Innes* GC 30924! Kpandu (Sept.) *Ankrah* GC 20360! Adome Bridge to Asikuma (Nov.) *Adams* 5207! **Togo Rep.:** Lama-Kara Kande (Oct.) *Rose Innes* GC 31401! Dapango (Oct.) *Bille* TG 69! **N. Nig.:** Daddara (Nov.) *Latilo* FHI 43763! Fika, Bornu Prov. (Oct.) *Daggash* FHI 24896! Nupe *Barter* 1374! Yola (Sept.) *Kennedy* FHI 7281! Gurum, Vogel Peak area (Nov.) *Hepper* 1230! **S. Nig.:** Lagos *W. MacGregor* 155! Ibadan (Oct.) *Onochie* FHI 7942! Igboora (Oct.) *Haines* 349! Aguku Dist. *Thomas* 623! 660! Extends southwards through the Congo to Angola, and eastwards into the Sudan.

39. **P. laetum** *Kunth* Rev. Gram. 2: 399 (1831); F.T.A. 9: 700; A. Chev. in Rev. Bot. Appliq. 14: 23; Berhaut, Fl. Sén. ed. 2, 416; De Miré & Gillet in J. Agric. Trop. 3: 734. *P. psilopodium* var. *afrum* A. Chev., Sudania 34, 160 (1911); Chev. Bot. 728; name only. *P. albidulum* Steud., Syn. Pl. Glum. 1: 69 (1854). 30–60 cm. high; moist soils.

Maur.: Kiffa *Jumelle* 6! Néma *Rossetti* 61/265! Gani (Oct.) *Hepper* 3652! **Sen.:** Dakar (Oct.) *Berhaut* 2867! Almadies (Aug.) *Broadbent* 25! Badion (Sept.) *Boudet* 3887! **Gam.:** *Ruxton* 62! **Mali:** Oualata *Jumelle* 7! Dioura (Oct.) *Davey* 68! *Leclerq* 42648! Labézenga (Sept.) *Hagerup* 4416! Takadji (Aug.) *Chev.* 2281! **U. Volta:** Ouagadougou (Sept.) *Scholz* 58! 124! Mossi (July) *Chev.* 24599! **Ghana:** Bawku (Aug.) *Hall* CC 552! **Niger:** Toukounous *Bartha* 57! **N. Nig.:** Bornu (Sept.) *Goss* 4! Gajibo (Aug., Dec.) *Grove* 52! *Johnston* N2! *Davey* FHI 27121! Katagum Dist. *Dalz.* 262! Also in the Sudan, Ethiopia and Eritrea. (See Useful Plants.)

40. **P. repens** *Linn.* Sp. Pl. ed. 2, 87 (1762); F.T.A. 9: 708; Chev. Bot. 729, and in Rev. Bot. Appliq. 14: 25; Berhaut, Fl. Sén. ed. 2, 415. *P. convolutum* P. Beauv. ex Spreng. (1852). Rhizomatous perennial about 60 cm. high; coastal and riverside sands.

Sen.: *Heudelot* 979! *Adanson* 126! Djembering (Sept.) *Broadbent* 179! **Mali:** Goundam (Aug.) *Chev.* 2247! Diré *Leon* 82! Faguibine (Sept.) *Chev.* 2248! *Leclerq* 42961! Ansongo (Sept.) *Hagerup* 373! **Guin.:** Dabola (Apr.) *Pitot*! **S.L.:** Rokupr (Oct.) *Jordan* 563! Mahera (Mar.) *Glanville* 233! Bonthe (Jan.) *Dalz.* 941! Limba (Apr.) *Deighton* 1690! Mungeru (Jan.) *Thomas* 7258! **Lib.:** Robertsport (Dec.) *Baldwin* 10889! Monrovia (May, Nov., Dec.) *Baldwin* 5834! 10491! *Linder* 1417! Sinkor (July) *Barker* 1378! **Iv. C.:** Ferkessédougou (Nov.) *Leeuwenberg* 2015! Mbrago (Jan.) *Hédin*! Cocody (June) *Adjanohoun* 429*a*! **Ghana:** Takoradi (June) *Morton* A3431! Elmina (July) *Hall* 1531! Cape Coast (June) *Hall* 1466! Sakumo Lagoon (Sept.) *Ankrah* GC 20550! Keta (June) *Thorold* 92! **Niger:** Gaya *White* 24! **N. Nig.:** Baga Seyoram (Oct.) *Golding* 26! Chad (Mar.) *Gwynn* 105! Naraguta (July) *Lely* 401! Ilorin (Oct.) *Ward* 57! Nupe *Barter* 1355! **S. Nig.:** Lagos *W. MacGregor* 215! Badagry (Aug.) *Onochie* FHI 33465! Ibadan (July) *Onochie* FHI 33239! Onitsha (July) *Jones* FHI 6261! Big Town, Eket Dist. (May) *Onochie* FHI 32930! **W. Cam.:** Victoria (Nov.) *Maitland* 84! 85! Throughout the tropics and subtropics. (See Useful Plants.)

41. **P. porphyrrhizos** *Steud.* Syn. Pl. Glum. 1: 72 (1854); F.T.A. 9: 712. Caespitose perennial up to 1 m. high; by ditches and on moist soils.

Sen.: Dakar (May) *Baldwin* 5719! 5736! **Ghana:** Cape Coast (June) *Hall* 3056! Weija (June) *Hall* 3101! Accra Plains (Nov.) *Ankrah* GC 20032! Kpong (Dec.) *Gyadu* GC 20535! Achimota (June) *Irvine* 5027! **Niger:** Biffa (Sept.) *P. de Fabrègues* 2156! Also in Sudan, Ethiopia, Uganda, Tanzania and Zambia.

[*P. porphyrrhizos* intergrades with both *P. subalbidum* and *P. coloratum*; its status will remain unclear until material from southern Africa has been properly revised.]

42. **P. subalbidum** *Kunth* Rev. Gram. 2: 397, t. 112 (1831); Chev. Bot. 730; Berhaut, Fl. Sén. ed. 2, 415. *P. glabrescens* Steud., Syn. Pl. Glum. 1: 71 (1854); A. Chev. in Rev. Bot. Appliq. 14: 25. *P. longijubatum* Stapf in F.T.A. 9: 718 (1920); F.W.T.A., ed. 1, 2: 554; Aké Assi, Contrib. 2: 287. *P. kermesinum* Mez in Engl., Bot. Jahrb. 57: 189 (1921). *P. paludosum* of Berhaut, Fl. Sén. ed. 2, 415, not of Roxb. 0·6–2 m. high with spongy culms and broad lush leaves; swamps.

Maur.: Nima (Oct.) *Hepper* 3710! Rosso (Oct.) *Hepper* 3618! **Sen.:** Richard-Tol (Sept.) *Martine* 132! Fété Faouron (Aug.) *Fotius* 500! Rong (Sept.) *Audru* 2532! Bambilor (May) *Raynal* 5871! Gadiaga, Kayar (July) *Raynal* 7178! **Gam.:** *Ruxton* 27! Sankuli Kunda (Sept.) *Pirie* 6/33! 13/33! 15/33! Wallikunda (Sept.) *Macluskie* 20! **Mali:** Kerdial (Sept.) *Davey* 198! Bamako (Aug.) *Waterlot* 1258! Macina (Sept.) *Chev.* 24891! Mopti (Sept.) *Lean* 91! Ansongo (Sept.) *Hagerup* 376! **Guin.:** Kouroussa (July) *Pobéguin* 488! Friguiagbé (Aug., Sept.) *Chillou* 1650! 1688! **S.L.:** Kambia (July) *Jordan* 53! Rokupr (Feb.) *Deighton* 4184! Bumba (Sept.) *Thomas* 1938! Makene (Aug.) *Deighton* 5032! Katakera (Apr.) *Glanville* 231! **Iv. C.:** Sikensi to N'douci (June) *Adjanohoun* 381*a*! **U. Volta:** Bobo-Dioulasso (Sept.) *Aké Assi* 6392! Ouagadougou (July, Sept.) *Scholz* 64! 64*a*! Lediga (July) *Scholz* 122! **Ghana:** Bawku (Oct.) *Rose Innes* GC 30248! Yapei (Sept.) *Goodall* 15762! Aburi (May) *La Danso* 63! Elmina (July) *Hall* 1532! Christiansborg (Apr.) *Johnson* 1028! **Niger:** Toukounous *Bartha* 48! Maradi (Aug.) *Hall* 393! Abalak (Sept.) *P. de Fabrègues* 45! **N. Nig.:** Sokoto (Aug.) *Dalz.* 483! Kaita (Aug.) *Onwudinjo* FHI 24002! Ilorin (Aug.) *Ward* 15! Naraguta (July) *Lely* 426! Samaru (July) *Taylor* 28! **S. Nig.:** Lagos (Sept.) *Haines* 343! Throughout tropical Africa.

[Most West African plants have no lower palea, a short ligule, smooth upper leaf surface and erect culms. In eastern Africa this form is mixed with another form having a palea in the lower floret, a long ligule, scabrid upper leaf surface and decumbent culms; there are also intermediates in which the character correlation breaks down. The taxonomic significance of these forms is at present not clear. The species is closely allied to *P. dichotomiflorum* Michx. of America; the upper glume and lower lemma of the latter are nearly always 7-nerved, and it commonly bears both axillary and terminal inflorescences.]

43. **P. coloratum** *Linn.* Mant. 1: 30 (1767); F.T.A. 9: 713; incl. var. *minus* Stapf ex Chiov., Res. Sci. Miss. Stef.-Paoli 1, Bot. Coll. 183, 226 (1916); F.T.A. 9: 715. Tufted perennial up to 1·2 m. high.

N. Nig.: Arege, Bornu (Oct.) *De Leeuw* 2058! Egypt to South Africa; introduced as a fodder grass to West Africa and other tropical countries.

[A polymorphic species from southern Africa. The specimen cited is a depauperate form known as var. *minus*.]

44. **P. comorense** *Mez* in Engl., Bot. Jahrb. 57: 185 (1921). Weakly ascending annual, 30–60 cm. high, decumbent at the base; forest shade.

Ghana: Kpandu to Kete Krachi (Sept.) *Rose Innes* GC 30406! **N. Nig.:** Jemaa (Oct.) *Gambles* 64! **S. Nig.:** Olokemeji F.R. (Nov.) *Jones, Keay & Onochie* FHI 14572! *Jackson* 1181168! Throughout tropical Africa.

Imperfectly known species.

P. deustum *Thunb.* Prodr. 1: 19 (1794); Berhaut, Fl. Sén. ed. 2, 404, 415.
Sen.: *Leprieur*!
Apparently a mislabelled duplicate from Quartin Dillon's gathering of this species in Ethiopia.

P. ozogonum *Steud.* Syn. Pl. Glum. 1: 68 (1854).
Sen.: *Leprieur.*
Probably the same as *Brachiaria ramosa.*

76. STENOTAPHRUM Trin., Fund. Agrost. 175 (1820); F.T.A. 9: 578 (1920).

False spike winged, the herbaceous wings forming chambers on alternate sides of a wavy midrib and produced upwards into a broad acute tooth; racemes of 3–5 spikelets, sunk in chambers on the anterior face of the spike .. .. 1. *dimidiatum*

False spike subcylindrical, corky, entire; racemes bearing a solitary spikelet (rarely more) with a short barren rhachis appressed to its outer side, deeply embedded in the anterior face of the spike .. .. 2. *secundatum*

1. **S. dimidiatum** (*Linn.*) *Brongn.* in Duperr., Bot. Voy. Coquille 127 (1831); F.T.A. 9: 580. *Panicum dimidiatum* Linn., Sp. Pl. 57 (1753). *Stenotaphrum glabrum* Trin. (1820). Stoloniferous perennial of coastal regions, with obtuse leaf-tips and compressed basal sheaths.
U. Volta: Ouagadougou (June) *Scholz* 113! **Ghana:** Accra (July) *Rose Innes* GC 30082! Ohawu (July) *Irvine* 4936! Aiyinase (July) *Irvine* 5103! Cape Coast (Sept.) *Hall* 2378! Introduced into W. Africa; also in E. Africa, Madagascar and India.

2. **S. secundatum** (*Walt.*) *Kuntze* Rev. Gen. Pl. 2: 794 (1891); F.T.A. 9: 579 (excl. syn. *S. glabrum*); A. Chev. in Rev. Bot. Appliq. 13: 891. *Ischaemum secundatum* Walt., Fl. Carol. 249 (1788). *Rottboellia tripsacoides* Lam., Tab. Encycl. 1: 205 (1791). *Stenotaphrum glabrum* of Chev. Bot. 737, not of Trin. *S. americanum* Schrank (1820)—Berhaut, Fl. Sén. ed. 2, 389. A strongly stoloniferous perennial of seashores, with obtuse leaf tips and compressed basal sheaths.
Sen.: Dakar (Dec.) *Clayton* 4073! **S.L.:** *Smeathman*! Freetown (Dec.) *Deighton* 465! Yele, Turtle Is. (Nov.) *Deighton* 2311! **Lib.:** Monrovia (Jan., Oct., Nov.) *Linder* 1426! *Harley* 1670! 1082*a*! Reppue's Town (July) *Harley* 1082! Garaway (Mar.) *Baldwin* 11632! **Iv. C.:** Port Bouet (Mar., Dec.) *Raynal* 13582! *Aké Assi* 9858! Grand Bassam (Mar.) *Leeuwenberg* 3136! **Ghana:** Half Assinie (Jan., July) *Chipp* 274! *Thorold* 294! Axim (Apr.) *Thorold* 42! Atwabo *Fishlock* 10! Essiama *Boughey* GC 10264! **S. Nig.:** Nun R. (Sept.) *Mann* 529! Opobo (July) *Jeffreys* 18! **W. Cam.:** Victoria (July) *Maitland* 25! The west coast of Africa from Sierra Leone to S. Africa, the east coast of America, the Pacific generally and Australia. (See Useful Plants.)

77. ACROCERAS Stapf in F.T.A. 9: 621 (1920).

Leaf-blades lanceolate, up to 20 mm. broad, rounded and contracted at the base; spikelets oblong, 5–5·5 mm. long; tips of keels of upper palea reflexed 1. *zizanioides*

Leaf-blades linear or narrowly linear-lanceolate, about 5 mm. broad, rounded- subcordate at the base and amplexicaul; spikelets lanceolate, 6 mm. long; tips of keels of upper palea reflexed .. .. 2. *amplectens*

1. **A. zizanioides** (*Kunth*) *Dandy* in J. Bot. 69: 54 (1931); Berhaut, Fl. Sén. ed. 2, 404. *Panicum zizanioides* Kunth in Humb. & Bonpl., Nov. Gen. & Sp. Pl. 1: 100 (1816); Chev. Bot. 731 (excl. var.). *Acroceras oryzoides* Stapf in F.T.A. 9: 622 (1920); A. Chev. in Rev. Bot. Appliq. 14: 18. *Panicum oryzoides* Sw. (1788), not of Ard. (1763). *Echinochloa zizanioides* (Kunth) Roberty (1955). Rambling perennial 60–120 cm. high, rooting at lower nodes; moist and shady soils, or growing in shallow water.
Sen.: Sédhiou (Nov.) *Berhaut* 6410 *bis*! **Guin.:** Kindia *Jac.-Fél.* 169! Timbo (Mar.) *Pobéguin* 1712! Nzérékoré (Sept., Oct.) *Baldwin* 9728! 13282! **S.L.:** Falaba (Mar.) *Sc. Elliot* 5097! Kanya (Oct.) *Thomas* 2943! Matotoka (July) *Thomas* 1355! Mahera (Nov.) *Deighton* 5668! Gebwema (Aug.) *Deighton* 5569! **Lib.:** Monrovia (Aug.) *Baldwin* 12544! Gbanga (Sept.) *Linder* 519! Gletown (July) *Baldwin* 6769! Ganta (Aug.) *Harley* 1823! Harbel (Sept.) *Baldwin* 9221! **Iv. C.:** Boka (Aug.) *Schnell* 6515! Adiopodoumé (Oct.) *Leeuwenberg* 1802! Ferkessédougou (Nov.) *Leeuwenberg* 2017! Abidjan to Dabou (Dec.) *Boughey* GC 14459! Agboville forest (Sept.) *Adjanohoun* 396*a*! **Ghana:** Pokoase (Sept.) *Ankrah* GC 20075! Gambaga to Bawku (Oct.) *Rose Innes* GC 30754! Damongo to Sawla (Nov.) *Rose Innes* GC 30827! Asuansi (Dec.) *Rose Innes* GC 30534! Accra to Ho (Sept.) *Rose Innes* 30398! **N. Nig.:** Jebba *Barter*! Akpata (Dec.) *Clayton* 617! Anara (Sept.) *Jackson* 219968! Jemaa (Oct.) *Gambles* 60! Maraba (Oct.) *Oche & Tuley* 1680! **S. Nig.:** Lagos *Dawodu* 94! Abeokuta *Irving*! Sapoba (Nov.) *Keay & Meikle* 551! Umuduru (Aug.) *Jones* FHI 6445! Orem (Jan.) *Onochie* FHI 36049*x*! **W. Cam.:** Victoria (Jan.) *Maitland* 1291! Victoria to Buea (Aug.) *Maitland* 65! Cam. Mt. (Dec.) *Maitland* 880! Likomba (Oct.) *Mildbr.* 10578! **F. Po:** *Mann* 1176! Extends southward through E. Cameroun and the Congos to Angola; also in India and tropical America. (See Useful Plants.)

2. **A. amplectens** *Stapf* in F.T.A. 9: 625 (1920); A. Chev. in Rev. Bot. Appliq. 14: 18; Berhaut, Fl. Sén. ed. 2, 404. *Panicum zizanioides* var. *angustatum* A. Chev., Bot. 731 (1920), name only. *Acroceras basicladum* Stapf in F.T.A. 9: 626 (1920). *Neohusnotia amplectens* (Stapf) C. Hsu in Journ. Fac. Sci. Univ. Tokyo, Bot. 9: 94 (1965). Scrambling annual 60–90 cm. high, rooting at the lower nodes; marshy places or shallow water.
Sen.: Casamance (Jan.) *Chev.* 2188! Bignona (Sept.) *Adam* 18054! Djembering (Sept.) *Broadbent* 142! **Gam.:** Wallikunda (Aug.) *Macluskie* 8! Sankuli Kunda (Sept.) *Pirie* 52/33! **Mali:** Koulikoro *Chev.* 2187! Ségou *Lecard* 246! Mopti (Oct.) *Davey* 197! Kara (Aug.) *Davey* 189! Dogo (Mar.) *Davey* 017! **Guin.:** Timbo (Oct.) *Pobéguin* 1713! Baffing R. *Pobéguin* 1713 *bis*! **S.L.:** Makete (Oct.) *Glanville* 86! Musaia (Aug.) *Haswell* 104! Kambia (Dec.) *Deighton* 803! Subu (Jan.) *T. S. Jones* 383! Mange Bure (Oct.) *Deighton* 5604! **Iv. C.:** Bouaké (Sept.) *Oldeman* 409! Bouaké to Katiola (Sept.) *de Wilde* 985! **U. Volta:** Tindangou to Porga (Aug.) *Scholz* 139! Banfora (Sept.) *Aké Assi* 6313! Koupéla *Chev.* 24600! Ouagadougou (Sept.) *Scholz* 139*a*! **Ghana:** Yendi (Nov.) *Thorold* 275! Kpandai (Sept.) *Ankrah* GC 20233! Damongo to Tamale (Oct.) *Rose Innes* GC 30219! Bolgatanga to Navrongo (Oct.) *Rose Innes* GC 30244! Walewale to Gambaga (Oct.) *Rose Innes* GC 31109! **Togo Rep.:** Sansanne Mango (Oct.) *Rose Innes* GC 31417! **N. Nig.:** Jebba (Dec.) *Hagerup* 677! Badeggi (Sept.) *Clayton* 302! Naraguta F.R. (July) *Lawlor & Hall* FHI 46608! Maska (Oct.) *Freeman* S144! Zaria (Aug.) *Sawyer* K33! **S. Nig.:** Lagos *W. MacGregor* 219! Eruwa (Oct.) *Haines* 326! Okomu F.R., Benin (Dec.) *Brenan & E. W. Jones* 8592! Uguoka, Awka (Aug.) *Jones* FHI 6697! Ogoja to Abakaliki (Aug.) *Tuley* 761! Also in the Sudan, C. African Rep. and Congo. (See Useful Plants.)

78. MICROCALAMUS Franch. in Journ. de Bot. 3: 282 (1889), not of Gamble (1890); F.T.A. 9: 490 (1919).

Leaf-blades mostly 5–8 cm. long and 2–3·5 cm. broad, glabrous, much constricted at the base; sheaths and nodes pilose, with a transverse ciliate rim on the outside at junction of sheath and blade; spikelets 6–6·5 mm. long, slightly compressed laterally; lower glume adaxial, about a quarter the length of the spikelet; upper lemma laterally flattened at the tip, its margins inrolled over the prominent keels of the palea; callus of the upper floret rounded .. .. *barbinodis*

M. barbinodis *Franch.* in Journ. de Bot. 3: 282 (1889); F.T.A. 9: 491. Ascending perennial about 30 cm. high with wiry culms; forest shade.
W. Cam.: Banga, S. Bakundu F.R. (Mar.) *Brenan* 9442! Also in Middle Congo.

79. COMMELINIDIUM Stapf in F.T.A. 9: 627 (1920).

Leaf-blades mostly 5–8 cm. long and 1·5–3·5 cm. broad, glabrous, minutely pubescent or sparsely pilose, much constricted at the base, acuminate at the tip; spikelets 5–5·5 mm. long, dorsally flattened; lower glume abaxial, about half the length of the spikelet, separated from the upper by an internode 1 mm. long; upper lemma laterally flattened at the tip, and with a rounded callus .. .. *gabunense*

C. gabunense (*Hack.*) *Stapf* in F.T.A. 9: 629 (1920). *Panicum gabunense* Hack. in Verhandl. Bot. Ver. Brandenb. 31: 70 (1889). *Commelinidium nervosum* Stapf in F.T.A. 9: 628 (1920); F.W.T.A., ed. 1, 2: 558; A. Chev. in Rev. Bot. Appliq. 14: 19. *C. mayumbense* (Franch.) Stapf l.c.; *Panicum mayumbense* Franch. in Bull. Soc. Hist. Nat. Autun 8: 343 (1895). *Echinochloa nervosa* (Stapf) Roberty (1955). A scrambling perennial emitting long aerial roots from the nodes; forest shade.
Lib.: Soplima (Nov.) *Baldwin* 10069! Wanau (Jan., Dec.) *Harley* 1727! 2105! Bahtown (Aug.) *Baldwin* 8033! Zeahtown (July) *Baldwin* 6932! **Iv. C.:** Guidéko *Chev.* 16396! Assikasso (Dec.) *Chev.* 22598! Bouaké (Oct.) *Adjanohoun* 404a! Issia (Aug.) *Boughey* GC 14678! **Ghana:** Cape Coast (July) *Hall* 1512! Jachi, Kumasi (July) *Thorold* CB 113! Yaya F.R. (Dec.) *Adams* 5231! Kumasi (Oct.) *Baldwin* 13472! Volta River F.R. (Nov.) *Morton* GC 6069! **N. Nig.:** Acharane F.R., Kabba Prov. (Nov.) *Latilo* FHI 34016! **S. Nig.:** Olokemeji (Nov.) *Jackson* 181169! **W. Cam.:** Mamfe (Mar.) *Coombe* 197! Also in E. Cameroun, the Congos, Uganda and Angola.

80. ICHNANTHUS P. Beauv., Ess. Agrost. 56 (1812); F.T.A. 9: 743 (1920).

Leaf-blades broadly lanceolate, oblique and rounded on one side at the base, 5–6 cm. long and 1·5–2 cm. broad; inflorescence lax, the upper primary branches racemose, the lower usually with secondary branches; spikelets 4·5–5 mm. long, strongly nerved, on pedicels up to 1 cm. long; lower glume abaxial; callus of the upper floret drawn into a very short stipe with small lateral hyaline wings .. .. .. *vicinus*

I. vicinus (*Bail.*) *Merr.* Enum. Phillip. Fl. Pl. 1: 70 (1923). *Panicum vicinum* Bail., Syn. Queensl. Fl. Suppl. 3: 82 (1890). *Ichnanthus pallens* of F.W.T.A., ed. 1, 2: 554, of F.T.A. 9: 744, and of A. Chev. in Rev. Bot. Appliq. 14: 27, not of Munro. Perennial, up to 60 cm. high, rooting at the lower nodes.
S.L.: Tingi Mts. (Dec.) *Morton & Gledhill* SL 3013! Mamaha (Nov.) *Thomas* 4635! 4571! **Lib.:** Nimba (Dec.) *Adames* 807! Also in E. Cameroun, throughout tropical Asia, and in Australia.
[Closely allied to *I. pallens* (Sw.) Munro from tropical America, which has shorter spikelets, and a more compact panicle].

81. CHLOACHNE Stapf in Hook., Ic. Pl. t. 3072 (1916); F.T.A. 9: 489 (1919).

Leaf-blades lanceolate, 6–12 cm. long and 5–15 mm. broad, acuminate, asymetrical at the base; racemes distant; spikelets 7 mm. long; lower glume abaxial, about two thirds as long as the spikelet .. .. *oplismenoides*

C. oplismenoides (*Hack.*) *Stapf ex Robyns* in Bull. Jard. Brux. 9: 173 (1932). *Panicum oplismenoides* Hack. in Bol. Soc. Brot. 6: 141 (1888). *Chloachne secunda* Stapf in Hook., Ic. Pl. t. 3072 (1916); F.T.A. 9: 489. *Poecilostachys flaccidula* Stapf ex Rendle in J. Linn. Soc. 40: 231 (1911). *Oplismenus anomalus* Mez ex Peter in Fedde, Rep., Beih. 40, 1: 220, 222 (1931). A trailing grass with long aerial roots from the lower nodes; in montane forest shade.
S. Nig.: Sonkwala, 5,200 ft, Obudu Div., (Dec.) *Savory & Keay* FHI 25257! **W. Cam.:** Cam. Mt., 4,000 ft. (Feb.) *Mann* 1354! Liwanga, 5,500 ft. (Mar.) *Brenan* 9551! Esele, 5,200 ft. (Feb.) *Maitland* 1031! Buea (Jan.) *Maitland* 340! Bafut-Ngemba F.R., 7,000 ft. (Feb.) *Hepper* 2197! **F. Po:** El Pico, 4,800 ft. (Dec.) *Boughey* 167! In montane forests of Sudan, Ethiopia, Kenya, Tanzania, Rhodesia and Malawi.

82. PSEUDECHINOLAENA Stapf in F.T.A. 9: 494 (1919).

Leaf-blades lanceolate, 1–6 cm. long; inflorescence slender, of several lax racemes; spikelets about 4 mm. long, obliquely ovoid, gaping; lower glume almost as long as the spikelet, adaxial; upper glume shorter, gibbous, usually armed with conspicuous hooked bristles but these sometimes short and appressed; lower lemma chartaceous, with membranous margins .. .. *polystachya*

P. polystachya (*Kunth*) *Stapf* in F.T.A. 9: 495 (1919); A. Chev. in Rev. Bot. Appliq. 13: 884. *Echinolaena polystachya* Kunth in Humb. & Bonpl., Nov. Gen. & Sp. 1: 119 (1816). *Echinochloa polystachya* (Kunth) Roberty (1955), not of Hitchc. (1920). *Panicum uncinnatum* Raddi (1823)—Chev. Bot. 730. Culms slender, ascending, the prostrate portion rooting from the nodes and often forming a dense carpet; a forest grass in damp shady places.

Guin.: Dalaba-Diaguissa Plateau *Chev.* 18570! Dyecke (Sept., Oct.) *Baldwin* 9667! 13323! **S.L.:** Potoru (Apr.) *Deighton* 1662! Kenema (Nov.) *Deighton* 416! Musaia (Dec.) *Deighton* 5699! Gebwema (Oct.) *Deighton* 5193! *Pyne* 15! **Lib.:** Nimba (Dec.) *Adam* 16500! Wohmen (Oct.) *Baldwin* 12029! Gbanga (Sept.) *Linder* 705! Dubo (Sept.) *Baldwin* 9484! Wanau (Nov.) *Harley* 2056! **Iv. C.:** Nimba (Aug.) *Boughey* GC 18173! Mt. Tonkoui (Oct.) *Adjanohoun* 398*a*! **Ghana:** Bunsu (July) *Thorold* 110! Aburi (June) *Ankrah* GC 20051! Amentia (Aug.) *Vigne* FH 3998! Kwahu (Dec.) *Adams* 5115! Kumasi (Oct.) *Baldwin* 13471! **S. Nig.:** Oyo (Apr.) *Haines* 200! Okomu F.R. (June, Dec.) *Onochie* FHI 34615! *Brenan* 8506! Aboh *Barter* 1354! Oban *Talbot* 851! **W. Cam.:** Esele (Feb.) *Maitland* 1030! Buea (Jan., Feb., Dec.) *Keay* FHI 28577! *Hinds* C1! *Maitland* 147! Ndop (Dec.) *Boughey* GC 1112! **F. Po:** Monte Balea (Dec.) *Guinea* 347! Moka (Dec.) *Boughey* 151! Occurs throughout the tropics.

83. ERIOCHLOA Kunth in Humb. & Bonpl., Nov. Gen. & Sp. 1: 94 (1815); F.T.A. 9: 497 (1919).

Lower glume suppressed, more or less adnate to the swollen lowest rhachilla internode, and terminating in an obscure rim above it; racemes simple, the spikelets pedicelled in pairs; spikelets mostly 3–3·5 mm. long, sparsely appressed-pilose, lanceolate, acuminate; upper glume produced into an awn-point 0·2–1 mm. long or more; lower floret reduced to a barren lemma; upper lemma obtuse, and bearing a mucro 0·1–0·6 mm. long.. .. .. 1. *nubica*

Lower glume developed, about 0·5 mm. long, obscurely and obtusely 3-lobed, adnate below to the swollen internode; racemes usually compound, the rhachis triquetrous, and the spikelets borne upon short side branches; side-branches commonly appressed to the rhachis, but sometimes spreading and paniculate; spikelets 3–3·8 mm. long, glabrous or pilose, ovate-lanceolate, acute; upper glume acute; lower floret with a well developed palea; upper lemma obtuse, with a minute mucro at the apex 2. *meyerana*

1. **E. nubica** (*Steud.*) *Hack. & Stapf ex Thell.* in Vierteljahrschr. Nat. Ges. Zürich 64: 697 (1919); Chev. Bot. 723; Lebrun in Bull. Soc. Bot. Fr. 116: 262 (1969). *Helopus nubicus* Steud., Syn. Pl. Glum. 1: 100 (1854). *Eriochloa acrotricha* (Steud.) Hack. ex Thell. op. cit. 52: 435 (1907), not of Hack. (1905); F.W.T.A., ed. 1, 2: 564; F.T.A. 9: 499; A. Chev. in Rev. Bot. Appliq. 13: 884; Berhaut, Fl. Sén. ed. 2, 404. *Helopus acrotrichus* Steud. (1854). *Digitaria acrotricha* (Steud.) Roberty (1955). Tufted annual 30–60 cm. high.
Maur.: Gani (Oct.) *Hepper* 3653! **Sen.:** Dakar (Jan.) *Chev.* 2289! Richard-Tol *Roger* 17! **Mali:** Boré (Aug.) *Demange* 18/1957! **Ghana:** Bawku to Gambaga (Oct.) *Ankrah* GC 20283! Cape Coast (Mar.) *Hall* 1316! Nungua, Accra (Feb., Oct.) *Rose Innes* GC 30333! *Ankrah* GC 20091! Keta (June) *Thorold* 88! **N. Nig.:** Kano (Nov.) *Taylor* 4! Bumsa (Aug.) *De Leeuw* 1129! Ngala *Gwynn* 113! Maiduguri to Dikwa (Oct.) *De Leeuw* 1830! South of Arege (Oct.) *De Leeuw* 2073! Widely spread in tropical and S. Africa, extending into Arabia. (See Useful Plants.)

2. **E. meyerana** (*Nees*) *Pilger* in Engl. & Prantl, Pflanzenfam., Aufl. 2, 14, e: 56 (1940). *Panicum meyeranum* Nees, Fl. Afr. Austr. 32 (1841); F.T.A. 9: 650; F.W.T.A. ed. 1, 2: 551. *Eriochloa borumensis* Hack. (1901), not of Stapf in F.T.A. 9: 500 (1919). Perennial, 90–120 cm. high. The species is almost exactly intermediate between *Brachiaria mutica* and *Eriochloa borumensis* of Stapf (not of Hack.), some specimens approaching very closely to these extremes.
Ghana: *Irvine* 1427! Cape Coast (May) *Hall* 2975! Christiansborg (Apr.) *Johnson* 1036! Labadi, Accra (Mar.) *Morton* GC 6654! Nungua, Accra (Nov.) *Ankrah* GC 20033! Throughout tropical and S. Africa, extending into Arabia; also in Thailand.

84. OPLISMENUS P. Beauv., Fl. Oware 2: 14 (1808); F.T.A. 9: 630 (1920).

Perennial; awns viscid, smooth, stiff, the tips truncate; leaf-blades narrowly ovate to lanceolate-linear, 1·5–12 cm. long and 7–25 mm. wide, rounded at the base, the tip acuminate; racemes 0·5–3 cm. long; spikelets 2–4 mm. long, with awns up to 15 mm. long .. .. 1. *hirtellus*

Annual; awns antrorsely barbellate, slender, flexuous, tapering; leaf-blades narrowly ovate to lanceolate, 1·5–4 cm. long and 3–11 mm. broad, rounded at the base, the tip acuminate; racemes 0·8–2 cm. long; spikelets 2–2·5 mm. long, with awns up to 10 mm. long .. .. 2. *burmannii*

1. **O. hirtellus** (*Linn.*) *P. Beauv.* Ess. Agrost. 54, 170 (1812); F.T.A. 9: 631; Chev. Bot. 733, and in Rev. Bot. Appliq. 14: 19; Berhaut, Fl. Sén. ed. 2, 401. *Panicum hirtellum* Linn., Syst. Nat., ed. 10, 2: 870 (1759). *P. incanum* Schumach. (1827). *Oplismenus africanus* P. Beauv. (1808). Straggling, often with long aerial roots from the nodes; forest shade. Rather variable, particularly in the size, shape and hairiness of its leaf-blades, and also in panicle characters.
Sen.: Tanma (July) *Berhaut* 2779! Mbidjem (Oct.) *Raynal* 6469*b*. **Mali:** Sikasso (Dec.) *Vuillet* 506! **Guin.:** Baffing R. (Nov.) *Pobéguin* 1737! **S.L.:** Mamaha *Thomas* 4639! Njala (Nov.) *Deighton* 1496! Magkuna (Nov.) *Marmo* 78! Gola (Nov.) *Deighton* 374! Magboloko (Nov.) *Glanville* 99! **Lib.:** Kakatown *Whyte*! So (Oct.) *Linder* 1130! Bangee (Nov.) *Baldwin* 10368! Vonjama (Oct.) *Baldwin* 9867! Ganta (Dec.) *Harley* 1054! **Iv. C.:** Bouroukrou (Jan.) *Chev.* 16572! Upper Sassandra (Dec.) *Portères*! Tiébissou (Nov.) *Leeuwenberg* 2091! Kokondékro (Oct.) *Adjanohoun* 352*a*! Singrobo (Nov.) *Adjanohoun* 377*a*! **Ghana:** Cape Coast (July) *Hall* 1530! Aburi (Oct.) *Ankrah* GC 20113! Akwapim hills (Nov.) *Johnson* 781! S. Fomansu F.R. (Dec.) *Adams* 4453! Yinahin hills (Nov.) *Rose Innes* GC 30526! **Togo Rep.:** Misahöhe (Nov.) *Mildbr.* 7284! Sokodé-Basari *Schroeder* 159! **N. Nig.:** Acharane F.R., Kabba Prov. (Nov.) *Latilo* FHI 34017! Lokoja to Kabba (Oct.) *Parsons* 26! Naraguta, Plateau Prov. (Oct.) *Hepper* 1071! Jemaa (Oct.) *Gambles* 59! Nguroje, Mambila (Sept.) *Oche* 75! **S. Nig.:** Lagos (Dec.) *Dalz.* 1324! Olokemeji F.R. (Nov.) *Jones, Keay & Onochie* FHI 14576! Usonigbe F.R. (Oct.) *Keay & Onochie* FHI 21597! Bende F.R. (Nov.) *Jones* FHI 6819! Oban *Talbot* 849! **W. Cam.:** Victoria (Dec.) *Maitland* 83! Buea (Nov.) *Migeod* 24! Cam. Mt., 5,600 ft. (Feb.) *Maitland* 1037! Kumba (=Johann-Albrechtshöhe) *Staudt* 450! Bambui (July) *Brunt* 1188! **F. Po:** *T. Vogel* 140! *Mann* 112! Moka (Sept., Dec.) *Melville* 478! *Boughey* 56! Bahia de Venus (Dec.) *Guinea* 254! Widely distributed in tropical and S. Africa, Mascarenes, America and Polynesia. (See Useful Plants.)

2. **O. burmannii** (*Retz.*) *P. Beauv.* Ess. Agrost. 54, 168, 169 (1812); F.T.A. 9: 636; Chev. Bot. 733, and in Rev. Bot. Appliq. 14: 19; Berhaut, Fl. Sén. ed. 2, 395, 401. *Panicum burmannii* Retz., Obs. Bot. 3: 10 (1783). Creeping, often with long aerial roots from the nodes; forest shade.

Fig. 441.—ECHINOCHLOA COLONA *Link* (GRAMINEAE-PANICEAE).

1, spikelet, front view. 2, same, side view. 3, lemma of upper floret. 4 & 5, grains, front and side view.

Sen.: Casamance *Heudelot* 590! Kafountine (Dec.) *Adam* 18186! Hann (Mar.) *Chev.* 44093! **Guin.:** Kouria *Chev.* 14924! Dyeke (Oct.) *Baldwin* 9654! Badabon (Oct.) *Schnell* 7552! **S.L.:** Rokupr (Oct.) *Jordan* 652! Kambia (Dec.) *Deighton* 806! Bumbuna (Oct.) *Thomas* 3343! Mamaba (Nov.) *Thomas* 4414! Mabala (Oct.) *Glanville* 60! **Lib.:** Wohmen, Vonjama (Oct.) *Baldwin* 12030! **Iv. C.:** Abidjan (Dec.) *Boughey* GC 13518! Singrobo (Nov.) *Adjanohoun* 375a! **Ghana:** Kwamang to Mampong (Nov.) *Rose Innes* GC 30871! Kumasi (Nov.) *Adams* 4423! Ho (Dec.) *Phillips* GC 30552! Anum (Dec.) *Rose Innes* GC 31173! Assuantsi (July) *Irvine* 5080! **N. Nig.:** *Lely* P752! Panshanu (Aug.) *Lawlor & Hall* 223! **S. Nig.:** Lagos (Oct.) *Baldwin* 13638! Ibadan (Oct., Nov.) *Onochie* FHI 6990! *Baldwin* 13676! Ifon (Oct.) *Maggs* 22! Ndoni *Barter* 1767! **W. Cam.:** Victoria *Maitland* 69! Buea (Oct., Dec.) *Migeod* 97! *Maitland* 130! **F. Po:** *T. Vogel* 212! 246! *Guinea* 576! Throughout the tropics. (See Useful Plants.)

85. ECHINOCHLOA P. Beauv., Ess. Agrost. 53 (1812); F.T.A. 9: 604 (1920).

Ligule represented by a fringe of hairs, at least in the lower leaves:
- Spikelets obtuse, 3 mm. long (if 4 mm. long see *Brachiaria obtusiflora*), scaberulous towards the tips only; lower glume obtuse, about a third as long as the spikelet; upper glume as long as the spikelet; racemes erect, about 2 cm. long, nearly their own length apart .. .. 1. *obtusiflora*
- Spikelets acute, acuminate or awned, scabrid or ciliate on the nerves:
  - Spikelets mostly 2·5–3·5 mm. long, awnless; racemes 20 or more, up to 8 cm. long, in a dense panicle up to 30 cm. long .. .. 2. *pyramidalis*
  - Spikelets mostly 4–6 mm. long, the upper lemma usually with an awn 4–20 mm. long; racemes 5–15, up to 6 cm. long; inflorescence variable, the spikelets densely imbricate and the racemes closely overlapping, or the spikelets contiguous and secund and the racemes distant .. .. 3. *stagnina*

Ligule absent:
- Racemes simple, 1–3 cm. long, suberect, about half their own length apart, alternate or rarely in whorls of 2 or 3; spikelets 2–3 mm. long, acute or cuspidate, sometimes with the lower lemma drawn out into a short awn-point up to 1 mm. long; leaf-blades 3–7 mm. broad .. .. 4. *colona*
- Racemes compound, up to 10 cm. long, the spikelets densely clustered on short side branches; spikelets 2·5–3 mm. long, the lower lemma usually with a short awn up to 2 mm. long; leaf-blades 10–15 mm. broad .. .. .. 5. *crus-pavonis*

1. **E. obtusiflora** *Stapf* in F.T.A. 9: 606 (1920); Berhaut, Fl. Sén. ed. 2, 406. Erect annual up to 1 m. high; in shallow pools and wet places.
Niger: Wadi Dollol *White* 23! **N. Nig.:** Gajibo (Nov.) *Johnston* N61! Fika (Oct.) *Daggash* FHI 24881! Kufena Hill, Zaria (Sept.) *Clayton* 1309! Ungun Shefu, Zaria Prov. (Nov.) *Freeman* S185! Yola (Nov.) *Hepper* 1200! Also in E. Cameroun and the Sudan.

2. **E. pyramidalis** (*Lam.*) *Hitchc. & Chase* in Contrib. U.S. Nat. Herb. 18: 345 (1917); F.T.A. 9: 615; A. Chev. in Rev. Bot. Appliq. 14: 18; Berhaut, Fl. Sén. ed. 2, 405. *Panicum pyramidale* Lam., Tab. Encycl. 1: 171 (1791); Chev. Bot. 729. *Echinochloa senegalensis* Mez in Notizbl. Bot. Gart. Berl. 7: 56 (1917); Berhaut, Fl. Sén. ed. 2, 405. A robust rhizomatous perennial up to 3·5 m. high, somewhat variable in size of spikelets and density of panicle; on river banks, marshes or in open water, sometimes forming extensive meadows on flood plains.
Sen.: *Adanson* 126! *Roger* 22! *Heudelot* 443! Ziarsy (Oct.) *Berhaut* 7590! L. Tanma (July) *Adam* 14894! **Gam.:** *Ruxton* 52! Sankuli Kunda (Sept.) *Pirie* 3/33! Wallikunda (Sept.) *Macluskie* 9! 10! **Mali:** Sébi to L. Debo (Aug.) *Chev.* 2173! Sompi to L. Debo (Aug.) *Chev.* 2179! Bura (Sept.) *Hagerup* 410! Mopti (Sept.) *Matthes* 14! *Lean* 89! **Port. G.:** Ingore to Barro (Aug.) *Esp. Santo* 3082! **S.L.:** Gbap (Jan.) *Adames* 9! Gbenti (Mar.) *Adames* 160! Nongoba Bullom (Oct.) *Jordan* 602! Kibai Island (Mar.) *T. S. Jones* 399! Mano (May) *Bakshi* 157! **Iv. C.:** Cocody (June) *Adjanohoun* 427a! **U. Volta:** Mossi (July) *Chev.* 24605! Ouagadougou (Sept.) *Irvine* 4677! *Scholz* 61! Namenda (June) *Scholz* 116! **Ghana:** Cape Coast Castle (Apr.) *T. Vogel* 31! Sakumo Lagoon, Accra (July) *Irvine* 2788! Agbosome (May) *La Danso* 72! Senchi (July) *Thorold* 127! Gambaga to Bawku (Oct.) *Rose Innes* GC 30756! **Niger:** Niamey (Jan.) *Vaillant* 782! **N. Nig.:** Jebba (June) *Onochie* FHI 40246! Badeggi (Sept.) *Clayton* 339! Naraguta, Plateau Prov. (July) *Lely* P413! Sokoto (Sept.) *Dalz.* 482! Kuka (Feb.) *T. Vogel* 37! **S. Nig.:** Ibu (Aug.) *T. Vogel* 49! Lagos (May, June) *Dalz.* 1327! *Haines* 173! 179! Onitsha (May) *Onochie* FHI 7517! **W. Cam.:** Victoria (July) *Maitland* 11! Throughout tropical Africa. (See Useful Plants.)

3. **E. stagnina** (*Retz.*) *P. Beauv.* Ess. Agrost. 53, 161, 171 (1812); F.T.A. 9: 617; A. Chev. in Rev. Bot. Appliq. 14: 17; Berhaut, Fl. Sén. ed. 2, 403. *Panicum stagninum* Retz., Obs. Bot. 5: 17 (1789); Chev. Bot. 730 (incl. var. *burgu* (A. Chev.) A. Chev.). *P. scabrum* Lam. (1797)—Chev. Bot. 729. *P. scabrum* subsp. *stagninum* (Retz.) A. Chev. in Comptes Rend. Assoc. Franç. Avanc. Sci. 1900: 651 (1901); subsp. *oryzetorum* A. Chev. l.c. 651; subsp. *lelievrei* A. Chev. l.c. 651; subsp. *burgu* (A. Chev.) A. Chev. l.c. 652. *P. burgu* A. Chev. in Rev. Cult. Colon 7: 513 (1900). *P. oryzetorum* (A. Chev.) A. Chev., Bot. 728 (1920), not of Bal. (1890). *Echinochloa oryzetorum* (A. Chev.) A. Chev. in Rev. Bot. Appliq. 14: 18 (1934). *E. lelievrei* (A. Chev.) Berhaut in Mém. Soc. Bot. Fr. 1953–54: 9 (1954), & Fl. Sén. ed. 2, 403. Robust perennial up to 2 m. high (occasionally weaker and behaving as an annual), with stout, often floating rhizomes; in swamps and in water, often forming large floating mats.
Maur.: Rosso (Oct.) *Hepper* 3621! **Sen.:** *Roger* 22! 28! *Adanson* 173! *Heudelot* 300! Gae (Oct.) *Hepper* 3656! **Mali:** Ségu *Lécard* 244! Sébé to Debo (Aug.) *Chev.* 2170! Mopti (Sept.) *Lean* 90! Banguita to Dogo (Jan.) *Davey* 14! Mopti (Sept.) *Matthes* 12! **S.L.:** Mange, Bure (Nov., Dec.) *Jordan* 722! 821! **U. Volta:** Mossi (Aug.) *Chev.* 24757! Diébougou (Sept.) *Aké Assi* 6479! **Ghana:** Lawra to Navrongo (Oct.) *Rose Innes* GC 30293! Bawku (Oct.) *Ankrah* GC 20287! Tumu to Leo (Oct.) *Rose Innes* GC 31484! Navrongo (Sept.) *Irvine* 4678! Savelugu (Dec.) *Morton* GC 9889! **Niger:** Niamey (Oct.) *Hagerup* 527! S. Dallol (Jan.) *Virgo* 72! **N. Nig.:** Nupe *Barter* 843! Sokoto (Nov.) *Dalz.* 479! Kano (Dec.) *Hagerup* 673! Hadejia (Dec.) *Clayton* 467! Gajibe (Nov.) *Davey* FHI 27107! **S. Nig.:** Itu, Calabar (Jan.) *Jones* FHI 6189! Tropical regions in Asia and Africa.

4. **E. colona** (*Linn.*) *Link* in Hort. Berol. 2: 209 (1833); F.T.A. 9: 607; A. Chev. in Rev. Bot. Appliq. 13: 892; Berhaut, Fl. Sén. ed. 2, 404 (incl. specimens ascribed to *E. crus-galli*); De Miré & Gillet in J. Agric. Trop. 3: 731; Aké Assi, Contrib. 2: 291. *Panicum colonum* Linn., Syst. Nat. ed. 10, 2: 870 (1759); Chev. Bot. 725 (incl. var. *equitans* (Hochst. ex A. Rich.) Hack. ex Dur. & Schinz (1894)). *P hookeri* Parl. (1846). *P. daltoni* Parl. ex Webb (1849). *P. equitans* Hochst. ex A. Rich. (1851). *P. brachiaraeforme* Steud. (1854). *Oplismenus daltoni* (Parl. ex Webb) Schmidt (1852). *Echinochloa verticillata* Berhaut in Mém. Soc.Bot. Fr. 1953–54: 9 (1954), & Fl. Sén. ed. 2, 403. *E. equitans* (Hochst. ex A. Rich.) C. E. Hubbard ex Troupin (1956). Annual grass 30–60 cm. high, erect or sometimes rooting at the lower nodes; damp places and near water.

Maur.: Podor to Boghé (Oct.) *Hepper* 3676! Oujaf *Rossetti* 61/210! **Sen.:** *Roger* 62! *Heudelot* 534*! Dakar (May) *Baldwin* 5720! Tambacounda (Sept.) *Berhaut* 3179! Youmbeul (Dec.) *Berhaut* 5612! Kaolack (Sept.) *Adam* 18025*! Baila (Sept.) *Adam* 18116*! **Gam.:** *Saunders* 29*! 53! *Ruxton* 19! 49! Wallikunda (Aug., Sept.) *Macluskie* 12! 26! **Mali:** *Leclercq* 42755! Koulikoro (Oct.) *Chev.* 2177! Gao (Sept.) *Hagerup* 356! Diré (Aug.) *Lean* 64! Tondigame (Oct.) *Farrow* 86! **Guin.:** Kouroussa (July) *Pobéguin* 481! Mamou (Aug.) *Adames* 331! **S.L.:** Bali (Apr.) *Glanville* 232! Kambia (Dec.) *Deighton* 848! Kukuna (Oct.) *Jordan* 940! Rokupr (July) *Jordan* 486! Katema (Dec.) *Jordan* 741! **Iv. C.:** Sérébissou (May) *Adjanohoun* 223*a*! Ouango-Fitini (June) *Adjanohoun* 390*a*! **U. Volta:** Po (Oct.) *Rose Innes* GC 31466! Gaoua (Oct.) *Rose Innes* GC 31532! Ouagadougou (Sept.) *Scholz* 63! 63*a*! Tenkodogo (June) *Scholz* 115! Nassougou (June) *Scholz* 115*a*! **Ghana:** Accra (May) *Adams* 2416! Nungua (Sept.) *Ankrah* GC 20007! Zuarungu (Sept.) *Thorold* CB 201! Navrongo (July) *Vigne* FH 4715! Bawku (Oct.) *Rose Innes* GC 30258! **Niger:** Toukounous *Bartha* 15! **N. Nig.:** Jebba *Barter* 1396! Jimeta, Yola (Dec.) *Hepper* 1613! Sokoto (Sept.) *Palmer* 26! Kaita, Katsina (Aug.) *Onwudinjoh* FHI 24001! Dikwa (Oct.) *Johnston* N50! **S. Nig.:** Lagos (Jan., July) *Haines* 125! 175! *Gledhill* 801! Throughout the tropics and subtropics. (See Useful Plants.)

[*E. colona* is rather more variable in Africa than elsewhere in its range. Forms with cuspidate spikelets 2·5–3 mm. long and with 2–3 branches from the lowest node of the panicle have been separated as *E. equitans* and *E. verticillata*, but they appear to grade imperceptibly into typical *E. colona*. Our species differs from *E. crus-galli* (Linn.) P. Beauv. of warm temperate regions in its simple shorter racemes, narrower leaf-blades, and smaller awnless spikelets; I have seen no specimens entirely typical of *E. crus-galli* from our area, but those marked * approach very closely to it. The declension of the epithet depends upon whether it is regarded as a noun or as an adjective. Lexicographers differ, but the adjectival use was acceptable to those of Linnaeus' own time—W.D.C.]

5. **E. crus-pavonis** (*Kunth*) *Schult.* Mant. 2: 269 (1824); F.T.A. 9: 612; A. Chev. in Rev. Bot. Appliq. 14: 17; Aké Assi, Contrib. 2: 292. *Oplismenus crus-pavonis* Kunth in Humb. & Bonpl., Nov. Gen. & Sp. 1: 108 (1815). Erect annual 1–2 m. high, near water.

Guin.: Baffing R. (Oct.) *Pobéguin* 1735! Dyeke (Oct.) *Baldwin* 9665! Macenta (Oct.) *Baldwin* 9842! **S.L.:** Port Loko (Dec.) *Thomas* 5848! Bali (Apr.) *Glanville* 228! Musaia (Dec.) *Deighton* 4517! Gbap (Oct.) *Jordan* 617! Bandajuma (Nov.) *Deighton* 3782! **Lib.:** Vonjama (Oct.) *Baldwin* 9868! Kailahun (Nov.) *Baldwin* 10124! Javajai (Nov.) *Baldwin* 10267! Sanokwele (Sept.) *Baldwin* 9489! 9546! **Iv. C.:** Touba to Odienné (Oct.) *Aké Assi* 8222! 8224! **Ghana:** Kumasi (Oct., Nov.) *Howes* 1000! *Bannerman-Bruce* 4440! Sampa to Wenchi (Oct.) *Rose Innes* GC 31575! **N. Nig.:** Naraguta (Aug.) *Lely* P447! Shendam (Oct.) *De Leeuw* 1612! Tiba Plateau (Dec.) *Peter & Tuley* 61! Mambila Plateau (Jan., Sept.) *Hepper* 1658! *Oche* 74! **S. Nig.:** Lagos *Dawodu* 225! Ibadan (Mar., Apr.) *Meikle* 1241! 1425! Oke Eleyele (July) *Keay & Jones* FHI 19435! Oban *Talbot* 850! **W. Cam.:** Bamenda (Feb.) *Daramola* FHI 40474! Kumbo, Bamenda (Feb.) *Hepper* 1983! Wum, Bamenda (Dec.) *Boughey* GC 11588! Santa (July) *Bumpus* 30! Tropical Africa and America.

86. PASPALIDIUM Stapf in F.T.A. 9: 582 (1920).

Inflorescence up to 30 cm. long, the racemes 1–3 cm. long, regularly spaced along a common axis and more or less appressed to it; spikelets imbricate, 2·5 mm. long; lower glume a rounded scale 0·5 mm. long; upper glume ⅗ the length of the spikelet, obtuse; lower lemma acute .. .. *geminatum*

P. geminatum (*Forsk.*) *Stapf* in F.T.A. 9: 583 (1920); A. Chev. in Rev. Bot. Appliq. 13: 891; Berhaut, Fl. Sén. ed. 2, 405. *Panicum geminatum* Forsk., Fl. Aegypt.-Arab. 18 (1775). *Echinochloa geminata* (Forsk.) Roberty (1955). *Panicum fluitans* Retz. (1783)—Chev. Bot. 726. A stoloniferous perennial with stout, rather spongy, culms; damp places.

Sen.: Cambérène (Oct.) *Berhaut* 982! Mbao (Aug.) *Broadbent* 9! **Mali:** Bourem (Mar.) *Chev.* 43780! Rharous (Mar.) *Chev.* 43829! Timbuktu (July) *Chev.* 1231! Niafounké (Mar.) *Davey* 79! Toumoura (Jan.) *Davey* 24! **Iv. C.:** Abidjan (Nov.) *Adjanohoun* 395*a*! Adiopodoumé (May, Oct.) *Adjanohoun* 399*a*! *Leeuwenberg* 4186! **Ghana:** Korle Lagoon, Accra (Oct.) *Ankrah* GC 20325! Cape Coast Castle (July) *T. Vogel* 54! *Irvine* 5108! Nungua (Apr.) *Morton* A4170! **Niger:** Niamey (Oct.) *Hagerup* 550! *Aké Assi* 10391! Dosso to Niamey (Feb.) *Chev.* 43751! Guidimouni (July) *P. de Fabrègues* 485! **N. Nig.:** Sokoto (June, Nov.) *Dalz.* 484! *Moiser* 143! Kworre (Sept.) *Palmer* 21! L. Chad (Dec., Mar.) *Elliott* 160! *De Leeuw* 1900! **S. Nig.:** Ilesha (Apr.) *Haines* 138! Throughout tropical Africa, extending eastwards to Egypt, Madagascar and India, and westwards to tropical America. (See Useful Plants.)

87. UROCHLOA P. Beauv., Ess. Agrost. 52 (1812); F.T.A. 9: 586 (1920).

Leaf-blades linear-lanceolate, coarsely pilose with tubercle-based hairs; spikelets ovate, acuminate, 5 mm. long, borne singly in two rows on the raceme; lower glume irregularly truncate, narrower than the spikelet and about ¼ as long, with a tuft of tubercle-based hairs in the centre; upper glume as long as the spikelet, pubescent; lower lemma ciliate with tufts of tubercle-based hairs along the margins; upper lemma broadly ovate, transversely rugose, the mucro up to 1·5 mm. long *trichopus*

U. trichopus (*Hochst.*) *Stapf* in F.T.A. 9: 589 (1920); A. Chev. in Rev. Bot. Appliq. 13: 891. *Panicum trichopus* Hochst. in Flora 27: 254 (1844). A coarsely leaved annual about 60 cm. high.

Sen.: Dahra (Oct.) *Raynal* 7642 *bis*. Ndjarno (Sept.) *Raynal* 7565. **Mali:** Gao (Sept.) *Hagerup* 349! *Wailly* 4750 *bis*! Gourma to Haoussa (July) *Rogeon* 462! **N. Nig.:** Garin Basan (Aug.) *De Leeuw* 1095! Bama (Oct.) *De Leeuw* 1836! Extends eastwards to Eritrea and south to Mozambique. (See Useful Plants.)

88. BRACHIARIA Griseb. in Ledeb., Fl. Ross. 4: 469 (1853); F.T.A. 9: 505.

Lower glume ½ or more than ½ as long as the spikelet, obtuse, 9–13-nerved:
 Spikelets glabrous, conspicuously reticulate-nerved, 4–4·5 mm. long, sessile, borne in 1–2 rows upon a narrow rhachis with sparsely ciliate margins; leaf-blades linear-lanceolate, the margins cartilaginous and minutely undulate-serrulate 1. *stigmatisata*
 Spikelets hairy; perennials:
  Rhachis of racemes subtriquetrous, wavy, glabrous or with a few scattered hairs along the margins; spikelets 4·5 mm. long, in 1–2 rows; upper glume and lower

FIG. 442.—BRACHIARIA BRIZANTHA (*Hochst. ex A. Rich.*) *Stapf* (GRAMINEAE-PANICEAE).

1, ligule. 2, axis of raceme. 3, spikelet. 4, lower glume. 5, upper glume. 6, lemma of upper floret. 7, palea.

lemma pubescent; leaf-blades linear lanceolate, the margins cartilaginous and minutely spinulose .. .. **2. *humidicola***

Rhachis of racemes herbaceous, more or less ribbon-like, with the setulosely ciliate margins curving over the base of the spikelets; spikelets mostly 3–3·5 mm. long:

Leaf-blades convolute, glaucous, up to 4 mm. broad; racemes up to 3 cm. long, sparsely ciliate with pale hairs at most 2 mm. long .. .. .. 3. *falcifera*

Leaf-blades flat, 4–10 mm. broad; racemes up to 5 cm. long, copiously ciliate with fulvous hairs up to 4 mm. long .. .. 4. *jubata*

Lower glume at most half as long as the spikelet (rather longer in *B. xantholeuca*):

Rhachis of racemes flat, herbaceous, more or less ribbon-like; spikelets glabrous:

Spikelets 4–5 mm. long, solitary, subsessile, in 1–2 rows; margins of raceme-rhachis scabrid (if ciliate see 8. *B. brizantha*); lower glume separated from the upper by an internode 0·5 mm. long .. .. 5. *plantaginea*

Spikelets 2·5–3·5 mm. long; margins of raceme-rhachis scabrid (but pedicels may be setose); internode between glumes obscure or absent:

Spikelets sessile in 1 row; racemes 2–4, 2–6 cm. long; leaf-blades linear, up to 6 mm. broad.. .. .. 6. *distachyoides*

Spikelets paired, one of each pair on an elongated pedicel, usually irregularly 4-rowed, often borne upon short side-branches (very rarely subsessile in 2 rows); racemes 8–20, 2–8 cm. long; leaf-blades lanceolate-linear, up to 10 mm. broad 7. *mutica*

Rhachis of racemes triquetrous or filiform:

Plants perennial:

Spikelets 4·5–5·5 mm. long, in 1 row, narrowed at the base into a short stipe 0·5 mm. long, glabrous or rarely sparsely hairy; margins of raceme-rhachis ciliate 8. *brizantha*

Spikelets 2–3 mm. long, in 2 rows, silky villous, usually with a transverse apical fringe, rarely almost glabrous; basal leaf-sheaths woolly tomentose; lower lemma muticous or with an awn-point up to 2 mm. long .. .. 9. *brachylopha*

Plants annual, the basal sheaths glabrous or pubescent:

Lowest internode of rhachilla forming a bead-like swelling covered by the base of the lower glume; spikelets 3–4 mm. long, glabrous, in 2 (rarely 4) rows 10. *callopus*

Lowest internode of the rhachilla not swollen, though sometimes elongated:

Racemes densely and irregularly 4-rowed, the spikelets borne on short side branches; spikelets 3·5–4·5 mm. long, plump, glabrous, narrowed at the base into a little stipe 0·5 mm. long.. .. .. 11. *obtusiflora*

Racemes 1–2-rowed, or spikelets much shorter:

Spikelets 4–4·5 mm. long, obtuse to subacute, glabrous; racemes loosely 2-rowed, the spikelets paired with the primary pedicelled; leaf-blades glabrous to pilose below, the margins cartilaginous and serrulate .. 12. *serrifolia*

Spikelets 2–4 mm. long:

Leaf-blades velvety pubescent, not cartilaginous on the margin; spikelets 3–4 mm. long, nearly always solitary and sessile in two dense rows, hispidulous (rarely glabrous), rostrate-acuminate or cuspidate at the tip, narrowed at the base into a short stipe up to 0·5 mm. long; racemes 1–3 cm. long .. .. 13. *xantholeuca*

Leaf-blades glabrous, pilose or (in *B. distichophylla*) pubescent; spikelets 2–3 mm. long, paired with one of the spikelets more or less pedicelled, or borne on short side branches (but often solitary in *B. distichophylla*), glabrous or hispidulous, the stipe obscure or absent:

Spikelets 2–2·5 mm. long; leaf-blades 2–8 mm. wide and 1–8 cm. long, the margins cartilaginous and spinulose; racemes slender, secund:

Spikelets solitary above, paired below and sometimes on very short secondary branches at the base of the raceme, not gaping; racemes 1–4 cm. long 14. *distichophylla*

Spikelets borne on short side-branches, forming little clusters or glomerules of 2–6, frequently gaping; racemes up to 8 cm. long .. 15. *kotschyana*

Spikelets 2·5–3 mm. long; leaf-blades up to 20 mm. broad and 25 cm. long, not cartilaginous on the margins; racemes scarcely secund, 2–8 cm. long:

Pairs of spikelets very loosely scattered, distant by up to 10 mm. or more; primary pedicels up to 10 mm. or more long .. .. 16. *deflexa*

Pairs of spikelets approximate, evenly distant by less than their own length; primary pedicels not more than 2 mm. long:

Margins of leaf-blades ciliate with tubercle-based hairs towards the rounded base; spikelets mostly subsessile .. .. 17. *lata*

Margins of leaf-blades not ciliate below; one spikelet of each pair usually pedicelled .. .. **18. *ramosa***

1. **B. stigmatisata** (*Mez*) *Stapf* in F.T.A. 9: 520 (1919); A. Chev. in Rev. Bot. Appliq. 13: 885; Berhaut, Fl. Sén. ed. 2, 404; Aké Assi, Contrib. 2: 293. *Panicum stigmatisatum* Mez in Engl., Bot. Jahrb. 34: 140 (1904). Annual up to 60 cm. high; disturbed soils.

Sen.: Djembering (Sept.) *Broadbent* 129! Ziguinchor (Sept.) *Broadbent* 78! Sari Hamadi (Sept.) *Diallo* 207! **Gam.:** *Ruxton* 135! **Mali:** Bamako (Sept.) *Adam* 15324! Kaboila (Sept.) *Adam* 15072! **Guin.:** Timbo *Pobéguin* 1711! **S.L.:** Port Loko (Oct.) *Jordan* 362! Gbinle (Oct.) *Jordan* 789! Makene (Sept.) *Deighton* 3981! Kagberi (Sept.) *Jordan* 541! **Iv. C.:** N'zi to Bocanda (June) *Adjanohoun* 183*a*! Ouagadougou (June) *Scholz* 111! Tindangouto Porga (Aug.) *Scholz* 138! **U. Volta:** Ouahigouya (Sept.) *Aké Assi* 6433! Mossi (Aug.) *Chev.* 24755! **Ghana:** Abotoase to Worawora (Oct.) *Ankrah* GC 20371! Kete Krachi to Kpandu (Sept.) *Rose Innes* GC 30417! Pong Tamale (Aug.) *Addei* 38! Yendi (Sept.) *Thorold* 202! Lawra (Aug.) *Thorold* 217! **N. Nig.:** Ilorin (Aug.) *Ward* L102! Batati (Aug.) *Sawyer* K16! Jebba *Barter*! Naraguta F.R. (July) *Lawlor & Hall* FHI 45552! Zaria (Sept.) *Freeman* S126! **S. Nig.:** Umudike (June) *Tuley* 785! Also in the Sudan. (See Useful Plants.)

2. **B. humidicola** (*Rendle*) *Schweick.* in Kew Bull. 1936: 297 (1936). *Panicum humidicola* Rendle in Cat. Welw. 2: 169 (1899). *Brachiaria dictyoneura* of F.T.A. 9: 512 (partly), and of F.W.T.A., ed. 1, 2: 561. A stoloniferous perennial up to about 60 cm. high.
N. Nig.: Naraguta, Plateau Prov. (July) *Lely* P405! *Lawlor & Hall* FHI 46591! Vodni, Plateau Prov. *Saunders* 35! Hoss, Plateau Prov. (Aug.) *Keay* FHI 12694! Also in Tanzania, Rhodesia, Malawi, Angola and Natal. [Closely allied to the caespitose *B. dictyoneura* (Fig. & De Not.) Stapf].

3. **B. falcifera** (*Trin.*) *Stapf* in F.T.A. 9: 517 (1919); A. Chev. in Rev. Bot. Appliq. 13: 884. *Panicum falciferum* Trin., Gram. Pan. 127 (1826); Chev. Bot. 726. *P. collare* Schumach., Beskr. Guin. Pl. 60 (1827). *Brachiaria brevis* of F.W.T.A., ed. 1, 2: 563, partly. Tufted perennial about 60 cm. high.
Ghana: *Irvine* 2660! Christiansborg (Mar.) *Johnson* 1021! Legon Hill (Apr.) *Adams* 4274! Nungua, Accra Plains (Apr.) *Ankrah* GC 20136! 20137!

4. **B. jubata** (*Fig. & De Not.*) *Stapf* in F.T.A. 9: 563 (1919). *Panicum jubatum* Fig. & De Not. in Mem. Acc. Torin. ser. 2, 14: 331 (1854). *Brachiaria fulva* Stapf in F.T.A. 9: 518 (1919); F.W.T.A., ed. 1, 2: 563; A. Chev. in Rev. Bot. Appliq. 13: 884; Berhaut, Fl. Sén. ed. 2, 405. *B. brevis* Stapf in F.T.A. 9: 519 (1919); F.W.T.A., ed.1, 2: 563, partly; A. Chev. in Rev. Bot. Appliq. 13: 884. Tufted grass 60–120 cm. high; often on shallow soils with seasonal extremes of drought and waterlogging.
Sen.: Djembering (Sept.) *Broadbent* 127! **Gam.:** *Skues*! *Ruxton* 18! **Mali:** Klela (July) *Farrow* 90! Banguita (Sept.) *Farrow* 89! **Port. G.:** Bambadinca (July) *Pereira* 3052! **Guin.:** Kouroussa (Apr.) *Pobéguin* 699! **Iv. C.:** Dabou (June, Oct., Nov.) *Adjanohoun* 410*a*! *Roberty* 15573! *Leeuwenberg* 1959! **U. Volta:** Barmorissa (June) *Scholz* 117! Banfora (June) *Leeuwenberg* 4350! **Ghana:** Kpandu to Accra (Sept.) *Ankrah* GC 20197! Yendi to Kete Krachi (Oct.) *Rose Innes* GC 30955! Tamale (Oct.) *Baldwin* 13551! Damongo (Oct.) *Rose Innes* GC 30209! Navrongo (June) *Vigne* FH 4540! **Togo Rep.:** Lomé *Stage* 24! **Dah.:** Agouagon to Savalou (May) *Chev.* 23677! **N. Nig.:** Badeggi, Bida (Sept.) *Clayton* 307! Anara F.R. (May) *Keay* FHI 22973! Kankiya, Katsina (Aug.) *Onwudinjoh* FHI 24008! Naraguta, Plateau Prov. (June) *Lely* 289! Jemaa (Apr.) *E. W. Jones* 275! **S. Nig.:** Ogboro road (May) *Dawodu* 34! Awba Hills F.R. (Apr.) *Onochie* FHI 32079! Igbete-Kishi (Nov.) *Gledhill* 733! Throughout tropical Africa. (See Useful Plants.)

5. **B. plantaginea** (*Link*) *Hitchc.* in Contrib. U.S. Nat. Herb. 12: 212 (1909). *Panicum plantagineum* Link, Hort. Berol. 1: 206 (1827). Annual, rooting at the lower nodes; leaf-blades thin, broadly linear.
U. Volta: Wagadugu (Aug.) *Irvine* 5175! **Ghana:** Kumasi, (Mar., Apr., May) *Adams* 3844! *Darko* 5564! *La Danso* 51! *Akpabla* 3846! Navrongo (Aug.) *Irvine* 5176! **W. Cam.:** Buea (July, Dec.) *Maitland* 26! 860! Also in the Congo and in tropical America.

6. **B. distachyoides** *Stapf* in F.T.A. 9: 530 (1919). Slender annual, rooting at the lower nodes; growing in or near water.
S.L.: Port Loko (Oct.) *Jordan* 809! Njala (July) *Deighton* 1172! Tabe to Jama (Sept.) *Deighton* 3027! Kambia (Aug.) *Jordan* 299! **Iv. C.:** Zuénoula to Vavoua (Oct.) *Adjanohoun* 368*a*! Mankono (July) *Aké Assi* 6151! Boka to Lenguera (June) *Boudet* 2832! **Ghana:** Takoradi (Aug.) *Hall* 3319! Accra (Aug.) *Hall* 3265! **S. Nig.:** Lagos (July) *Dawodu* 163! *Haines* 117! 171! 319! Also in the Congo.

7. **B. mutica** (*Forsk.*) *Stapf* in F.T.A. 9: 526 (1919); A. Chev. in Rev. Bot. Appliq. 13: 885; Berhaut, Fl. Sén. ed. 2, 406. *Panicum muticum* Forsk., Fl. Aegypt.-Arab. 20 (1775); Chev. Bot. 728. *P. purpurascens* Raddi (1823). *P. numidianum* Lam. (1791). *Brachiaria purpurascens* (Raddi) Henr. in Blumea 3: 434 (1940). *B. numidianum* (Lam.) Henr. l.c. (1940). *Urochloa mutica* (Forsk.) Nguyen in Nov. Sist. Vys. Rast. 1966: 13 (1966). Perennial, often rooting at the lower nodes, 0·6–1·8 m. high; wet places.
Sen.: Tivaouane (Dec.) *Chev.* 2294! Tiarsy (Oct.) *Berhaut* 975! **Gam.:** *Boteler*! **Mali:** Kayes (June) *Chev.* 26019! Dogo (Apr.) *C.I.P.A.S.* 605! Ansongo (Sept.) *Hagerup* 415! Mopti (Sept.) *Matthes* 9! 16! **S.L.:** Mange (Nov.) *Jordan* 822! **Lib.:** Grand Bassa (July) *T. Vogel* 32! Buchanan (Mar.) *Baldwin* 11169! **U. Volta:** Samandeni (Oct.) *Kmoch* 157! **Ghana:** Nandaw (May) *Agric. Dept.*! Kumasi (May) *Darko* 5566! Tumu to Tamale (Oct.) *Rose Innes* GC 32290! Tumu to Leo (Oct.) *Ankrah* GC 20467! *Rose Innes* GC 31485! **Niger:** Niamey (Sept.) *P. de Fabrègues* 2086! **N. Nig.:** Shendam (Nov.) *Clayton* 1450! Fodama (Nov.) *Moiser* 220! Sokoto (Aug.) *Dalz.* 481! Baga (Jan.) *Daggash* FHI 24967! Wulgo (Mar.) *Johnston* N24! **S. Nig.:** Ikorodu (Oct.) *Haines* 334! **W. Cam.:** Victoria (July, Oct.) *Maitland* 10! 69! 73! Buea (Dec.) *Maitland* 105! Throughout the tropics and subtropics. (See Useful Plants.)
[Henrard (l.c.) supports Hitchcock (Man. Grasses W. Indies 247) in maintaining the American *B. purpurascens* as a separate species, but the distinguishing characters are not reliable in W. Africa. A form with the spikelets subsessile in two rows is occasionally found in W. Africa, but it is otherwise similar to the typical form.]

8. **B. brizantha** (*Hochst. ex A. Rich.*) *Stapf* in F.T.A. 9: 531 (1919); A. Chev. in Rev. Bot. Appliq. 13: 885. *Panicum brizanthum* Hochst. ex A. Rich., Tent. Fl. Abyss. 2: 363 (1851). Robust perennial up to about 2 m. high.
Guin.: Gali to Hinde (July) *Adames* 310! **S.L.:** Yenkisa (Nov.) *Glanville* 310! Sekurede (Nov.) *Jaeger* 8163! **Iv. C.:** Briankouma to Touba (Oct.) *Aké Assi* 8194! **Ghana:** Ejura to Kintampo (Nov.) *Rose Innes* GC 30858! Maliato, Togo Plateau (Nov.) *Morton* A3535! Ejura (Oct.) *Asare* 9637! **N. Nig.:** Omu Aran, Ilorin Prov. (Oct.) *Ward* L101! Sanga River F.R., Jemaa (Nov.) *Keay* FHI 37243! Kogigiri, Jos (Oct.) *Hepper* 1047! Naraguta, Plateau Prov. (July) *Lely* P422! Maska, Katsina Prov. (Aug.) *Thatcher* S456! **S. Nig.:** Olokemeji (Nov.) *Onochie* FHI 8123! Ila (Oct.) *Thomas* 1930! Ubiaja F.R., Ishan (Apr.) *Daramola* FHI 31274! Usonigbe F.R., Benin (Oct.) *Keay & Onochie* FHI 19684! Awka (June) *Jones* FHI 1776! **W. Cam.:** Bamenda (Oct.) *Agric. Off.* Bam 14! Donga Plain *Brunt* 1010! Iwo & Mencham (Feb.) *Brunt* 939! Throughout tropical Africa and in S. Africa. (See Useful Plants.)

9. **B. brachylopha** *Stapf* in F.T.A. 9: 539 (1919); A. Chev. in Rev. Bot. Appliq. 13: 886. *Panicum brachylophum* (Stapf) A. Chev., Bot. 724 (1920). *P. serratum* (Thunb.) Spreng. var. *brachylophum* (Stapf) A. Chev., Bot. 730 (1920). Tufted perennial about 60 cm. high. Very closely allied to *Brachiaria serrata* (Thunb.) Stapf of eastern and southern Africa.
Iv. C.: Dabou (June) *Adjanohoun* 412*a*! *Chev.* 17148! Toumodi (Aug.) *Boughey* GC 18554! Séguéla (Apr.) *Leeuwenberg* 3260! Asakra (Aug.) *Boughey* GC 18633! **U. Volta:** Doutié (Sept.) *Scholz* 22*a*! **Ghana:** Kete Krachi to Yendi (Sept.) *Ankrah* GC 20242! Biakpa (Sept.) *Rose Innes* GC 30905! Lawra to Navrongo (Oct.) *Rose Innes* GC 30286! Bole (May) *Vigne* FH 3829! Damongo (Mar.) *Adams* 3949! **Togo Rep.:** Sokodé to Lama Kara (Oct.) *Rose Innes* GC 31397! **N. Nig.:** Lokoja (May) *Richardson*! Nupe *Barter* 1397! Kusheriki (July, Aug.) *Clayton* 1101! *Sawyer* K18! Birnin Gwari (June) *Keay* FHI 25871!

10. **B. callopus** (*Pilg.*) *Stapf* in F.T.A. 9: 533 (1919). *Panicum callopus* Pilg. in Engl. Bot. Jahrb. 33: 46 (1902). *Brachiaria stipitata* C. E. Hubbard in Kew Bull. 1933: 498; Berhaut, Fl. Sén. ed. 2, 405, 406. Annual with spongy culms up to 1 m. high; shallow pools.
Sen.: *Bouchet*! Baïla, Casamance (Sept.) *Adam* 18136! **Mali:** Ségou (Nov.) *Roberty* 2959! **Guin.:** Mamou (Oct., Nov.) *Chev.* 34598! *Adam* 12646! **S.L.:** Kukuna (July) *Morton & Gledhill* SL 46! **Iv. C.:** Boundiali

to Mankono (Nov.) *Aké Assi* 8295! Boundiali to Tingréla (Nov.) *Aké Assi* 8284! **U. Volta:** Baoulé (Oct.) *Kmoch* 122! **N. Nig.:** Damagum (Oct.) *De Leeuw* 1233! Also in the Sudan.

11. **B. obtusiflora** (*Hochst. ex A. Rich.*) *Stapf* in F.T.A. 9: 534 (1919). *Panicum obtusiflorum* Hochst. ex A. Rich., Tent. Fl. Abyss. 2: 367 (1851). Annual, 2–4 ft. high; on clay soil with *Acacia seyal*.
N. Nig.: Damaturu (Nov.) *De Leeuw* 1171! Also in Sudan and Ethiopia.

12. **B. serrifolia** (*Hochst.*) *Stapf* in F.T.A. 9: 548 (1919). *Panicum serrifolium* Hochst. in Flora 38: 196 (1855). Annual 30–90 cm. high; in shade.
Niger: Toukounous (Mar.) *P. de Fabrègues* 1794! Extends eastwards to Ethiopia and southwards to Rhodesia.

13. **B. xantholeuca** (*Hack. ex Schinz*) *Stapf* in F.T.A. 9: 541 (1919); A. Chev. in Rev. Bot. Appliq. 13: 886; Berhaut, Fl. Sén. ed. 2, 405; De Miré & Gillet in J. Agric. Trop. 3: 728. *Panicum xantholeucum* Hack. ex Schinz in Verhandl. Bot. Ver. Brandenb. 30: 141 (1889). *P. anisotrichum* Mez in Notizbl. Bot. Gart. Berl. 7: 70 (1917). *P. orthostachys* Mez in Notizbl. Bot. Gart. Berl. 7: 66 (1917). *Brachiaria orthostachys* (Mez) W. D. Clayton in Kew Bull. 20: 265 (1966); Lebrun in Bull. Soc. Bot. Fr. 116: 257 (1969). *B. hagerupii* Hitchc. in J. Wash. Acad. Sci. 19: 303 (1929); F.W.T.A., ed. 1, 2: 564; A. Chev. in Rev. Bot. Appliq. 13: 888; Berhaut, Fl. Sén. ed. 2, 406. Annual 30–45 cm. high, growing on old farm land.
Maur.: Néma *Rossetti* 61/193! **Sen.:** *Adanson*! Walo *Leprieur* 21! Ross-Béthio (Aug.) *Audru* 2259! Dakar (Aug.) *Adam* 14934! **Gam.:** Genieri (July) *Fox* 127! **Mali:** Timbuktu (Aug.) *Lean* 76! *Hagerup* 271! Oualata to Heissa *Jumelle* 5! Gourma (Aug.) *Rossetti* 59/114! Bango to Koro (Aug.) *Chev.* 24880! Gao (Jan.) *Chev.* 43039! **Ghana:** Sinibaga (Aug.) *Rose Innes* GC 32071! Navrongo (June) *Vigne* FH 4538! *Rose Innes* GC 31673! Tili to Windenaba (Aug.) *Rose Innes* GC 32019! **Niger:** Toukounous *Bartha* 3! Gouré (Aug.) *P. de Fabrègues* 771! **N. Nig.:** Nupe *Barter* 1366! Zaria (July) *Clayton* 1254! Sokoto (July) *Dalz.* 476! 476*a*! Bindawa (Aug.) *Grove* 16! Throughout tropical Africa.
[In a few specimens the racemes are rather lax with the spikelets paired, the primary on a pedicel; otherwise the spikelets are similar to the typical form.]

14. **B. distichophylla** (*Trin.*) *Stapf* in F.T.A. 9: 557 (1919); A. Chev. in Rev. Bot. Appliq. 13: 888; Berhaut, Fl. Sén. ed. 2, 405. *Panicum distichophyllum* Trin., Gram. Pan. 147 (1826); Chev. Bot. 725. *P. serrulatum* Schumach., Beskr. Guin. Pl. 62 (1827). *P. viviparum* Schumach. (1827). *P. despreauxii* Steud. (1854). *P. distichophylloides* Mez (1905). Slender annual up to about 45 cm. high; sometimes displays a vegetative proliferation of the spikelets; disturbed soils.
Maur.: Goberni *Rossetti* 61/332! **Sen.:** Hann (Oct.) *Berhaut* 2765! Cayor *Heudelot* 398! Ziguinchor (Sept.) *Broadbent* 67! **Gam.:** *Saunders* 12! 34! 50! Sankuli-Kunda (Sept.) *Pirie* 11/33! Yundum *Ashrif* 29! **Mali:** Sindou (Oct.) *Chev.* 875! Macina (Sept.) *Davey* 40! Dioura (Sept.) *Farrow* 87! In Tillit (Aug.) *Rossetti* 59/132! Ansongo (Sept.) *Hagerup* 413! **Guin.:** Timbo (Oct.) *Pobéguin* 1734! Kouroussa (July) *Pobéguin* 476! 486! Kindia *Jac.-Fél.* 137! *Kranz* 281! **S.L.:** Freetown (Aug.) *Deighton* 2032! Mabum (Aug.) *Thomas* 1577! Kitchom (Jan.) *Deighton* 930! Kumrabai-Mamila (Nov.) *Deighton* 4936! Kambia (July) *Jordan* 466! **Lib.:** Harper (June) *Baldwin* 6000! **Iv. C.:** Pakobo (June) *Adjanohoun* 385*a*! Tehini (Aug.) *Oldeman* 356! **U. Volta:** Sindou (June) *Leeuwenberg* 4317! Banfora (June) *Leeuwenberg* 4347! **Ghana:** Legon (Jan.) *Adams* 3653! Kete Krachi (Oct.) *Rose Innes* GC 30929! Half Assinie (Dec.) *Rose Innes* GC 30541! Avume (Aug.) *Thorold* 120! Navrongo (June) *Vigne* FH 4537! **Togo Rep.:** Lomé *Warnecke* 169! Atakpamé to Sokodé (Oct.) *Rose Innes* GC 31379! **Dah.:** Sémé (Sept.) *Adjanohoun* 532! Cotonou (Dec.) *Froment* 1049! **Niger:** Niamey (Oct.) *Hagerup* 484! Toukounous *Bartha* 4! Wadi Dollol *White* 28! **N. Nig.:** Nupe *Barter* 367! Bida (Jan.) *Meikle* 1103! Lafia (Oct.) *Kennedy* FHI 8029! Katsina (Oct.) *Clayton* 1355! Sokoto (July) *Dalz.* 475! Gurum, Vogel Peak (Nov.) *Hepper* 2783! **S. Nig.:** Lagos *Dalz.* 1135! Ibadan (Apr.) *Brenan* 9600! Aku Rock (Apr.) *Hambler* 465! Opobo *Jeffreys*! Awka (June) *Jones* FHI 1784! Probably throughout tropical Africa, but the taxonomy of the species in east and central Africa is confused as it tends to intergrade with the following. (See Useful Plants.)

15. **B. kotschyana** (*Hochst. ex Steud.*) *Stapf* in F.T.A. 9: 559 (1919); A. Chev. in Rev. Bot. Appliq. 13: 889. *Panicum kotschyanum* Hochst. ex Steud., Syn. Pl. Glum. 1: 68 (1854); Chev. Bot. 727. Slender annual about 60 cm. high.
Sen.: Kayar (Sept.) *Raynal* 7347! **N. Nig.:** Yola (July, Sept.) *Dalz.* 275! *Kennedy* FHI 7286! Ibaji Ojoko F.R., Kabba Prov. (June) *Latilo* FHI 47753! Gembu, Mambila (Sept.) *Oche* 87! **S. Nig.:** Aguku *Thomas* 1390! **W. Cam.:** Wum, Bamenda (Dec.) *Boughey* GC 11568! Extends eastwards to Eritrea, and south to Angola and the Congo. (See Useful Plants.)

16. **B. deflexa** (*Schumach.*) *C. E. Hubbard ex Robyns* in Bull. Jard. Bot. Brux. 9: 181 (1932); Berhaut, Fl. Sén. ed. 2, 413; incl. *var. sativa* Portères in Agron. Trop. 6: 38 (1951), without latin descr. *Panicum deflexum* Schumach., Beskr. Guin. Pl. 63 (1827). *Brachiaria regularis* (Nees) Stapf in F.T.A. 9: 544 (1919); A. Chev. in Rev. Bot. Appliq. 13: 886; De Miré & Gillet in J. Agric. Trop. 3: 728. *Panicum regulare* Nees, Fl. Afr. Austr. 41 (1841). *P. nudiglume* Hochst. (1844)—Chev. Bot. 728. *Pseudobrachiaria deflexa* (Schumach.) Launert in Mitt. Bot. München 8: 158 (1970). Annual up to 45 cm. high; the inflorescence, with its distant spikelets and long pedicels, simulates a panicle, and the plant is commonly mistaken for a species of *Panicum*; a weed of cultivated land and disturbed soils.
Gam.: Dipa Kunda *Fox* 157! **Mali:** Sompi (Aug.) *Chev.* 2282! Boré (Aug.) *Demange* 19/1957! **Port G.:** Contubo (Aug.) *Pereira* 3113! **Iv. C.:** Adiopodoumé (Nov.) *Leeuwenberg* 1937! Bouaké (May) *de Wilde* 58! **U. Volta:** Mossi (Aug.) *Chev.* 24752! **Ghana:** Achimota (May) *Irvine* 1633! Aburi (Aug.) *Johnston* 1004! Otrokpe (Oct.) *Vigne* FH 4015! Cape St. Paul (Apr.) *Rose Innes* GC 31228! Yamoransa (Jan.) *Thorold* 28! **Togo Rep.:** Palime *Stage* 79! **Dah.:** Niaouli (Oct.) *Risopoulos* 1238! **N. Nig.:** Nupe *Barter* 1367*b*! Kalawala, Bornu (Aug.) *De Leeuw* 1088! Badeggi to Lapai (June) *E. W. Jones* 186! **S. Nig.:** Lagos *Dawodu* 70! Ibadan (Mar.) *Meikle* 1260! Olokemeji F.R. (Nov.) *Olafusi* FHI 14264! Fashola *Thorold* 3! Benin (Dec.) *Brenan* 8630! Throughout tropical Africa, extending into Arabia, Madagascar and S. Africa. (See Useful Plants.)

17. **B. lata** (*Schumach.*) *C. E. Hubbard* in Hook., Ic. Pl. t. 3363 (1938); Berhaut, Fl. Sén. ed. 2, 406; Aké Assi, Contrib. 2: 293. *Panicum latum* Schumach., Beskr. Guin. Pl. 61 (1827). *Urochloa lata* (Schumach.) C. E. Hubbard in Kew Bull. 1934: 112 (1934); F.W.T.A., ed. 1, 2: 560; De Miré & Gillet in J. Agric. Trop. 3: 740. *U. insculpta* (Steud.) Stapf in F.T.A. 9: 599 (1920); A. Chev. in Rev. Bot. Appliq. 13: 892. *Panicum insculptum* Steud. (1854)—Chev. Bot. 727. *P. exasperatum* Nees ex Steud. (1854). A coarse-leaved annual, usually 30–60 cm. high, but sometimes much higher; a weed of cultivated land and disturbed soil.
Maur.: Nema *Rossetti* 61/175! **Sen.:** Cambérène (Oct.) *Berhaut* 1107! Bargny (Aug.) *Adam* 14924! 14927! 14928! Mt. Roland (Aug.) *Broadbent* 50! **Gam.:** *Ruxton* 49! 58! **Mali:** Djenné (Sept.) *Chev.* 2196! Sarédina (Mar.) *Davey* 57! Boré (July) *Demange* 6/1957! Dioura (Oct.) *Davey* 067! **Guin.:** Kouroussa (July) *Pobéguin* 478! Conakry (Aug.) *Chillou* 2738! **Iv. C.:** N'Douci (Apr.) *Leeuwenberg* 3324! Minankro (Nov.) *Boudet* 1457! **U. Volta:** Bougoura (June) *Scholz* 28! Ouagadougou (July) *Scholz* 126! **Ghana:** Christiansborg (Mar.) *Johnson* 1013! Takoradi (Apr.) *Brenan* 9613! Kumasi (June) *La Danso* 69! Damongo (Sept.) *Rose Innes* GC 30199! Navrongo (June) *Vigne* FH 4539! **Dah.:** Parakou (Sept.) *Risopoulos* 1198! Ajuda *Newton* 2-23! Cotonou (Apr.) *Chev.* 23476! *Froment* 1197! Niaouli (Oct.) *Risopoulos* 1239! **N. Nig.:** Mokwa (June) *Clayton* 1193! Sokoto (July) *Dalz.* 477! Maska (July) *Thatcher* S553! Biu (Sept.) *Kennedy* FHI 8012! Kalkala, Chad (July) *Gwynne* 131! **S. Nig.:** Lagos *W. MacGregor* 109! Ibadan (Jan., Mar., Sept.) *Deighton* 541! *Keay* FHI 25415! *Gledhill* 795! Extending east to Ethiopia and Arabia.
B. lata var. *pubescens* C. E. Hubbard in Kew Bull. 4: 125 (1949), distinguished by its pubescent spikelets, is occasionally found in our area.

18. **B. ramosa** (*Linn.*) *Stapf* in F.T.A. 9:542 (1919); A. Chev. in Rev. Bot. Appliq. 13: 886; Berhaut, Fl. Sén.

ed. 2, 405. *Panicum ramosum* Linn., Mant. 1: 29 (1767); Chev. Bot. 729. *Echinochloa ramosa* (Linn.) Roberty in Bull. I.F.A.N. 17: 64 (1955). *Panicum arvense* Kunth (1831). *P. brachylachrum* Steud. (1854). *P. cognatissimum* Steud. (1854). *Urochloa ramosa* (Linn.) Nguyen in Nov. Sist. Vys. Rast. 1966: 13 (1966). Annual 30–60 cm. high.
Maur.: Gani (Oct.) *Hepper* 3655! **Sen.:** *Roger* 15! Richard-Tol (Aug.) *Roger* 34! Bargny (Aug.) *Adam* 14925! Dakar (Apr.) *Clayton* 4003! **Gam.:** Yundum (Sept.) *Wallace* 7! **Mali:** Bamako (Sept.) *Adam* 15367! El Oualadji (Mar.) *Chev.* 43915! Diré (May) *Rogeon & Leclercq* 268! Bore (Aug.) *Demange* 20/1957! Timbuktu (Aug.) *Hagerup* 259! **Port. G.:** Pirada (June) *Esp. Santo* 3037! **N. Nig.:** Kalkala, Chad *Golding* 78! Also in eastern Tropical Africa, extending to India and Thailand.

89. ENTOLASIA Stapf in F.T.A. 9: 739 (1920).

Leaf-blades lanceolate, up to 18 mm. broad, with faint transverse veins beneath; panicle up to 15 cm. long, the racemes densely distributed along a central axis, bearing the spikelets upon short (6–10 mm. long) appressed secondary racemes; spikelets lanceolate, 2·5 mm. long; lower glume a short hyaline scale; upper glume and lower lemma as long as the spikelet; upper lemma villous .. .. *olivacea*

E. olivacea Stapf in F.T.A. 9: 740 (1920). *Panicum olivaceum* Stapf ex A. Chev., Sudania 60, 90 (1911), name only.
Guin.: Ziama (Feb.) *Adam* 3549 (*fide* Lebrun in Bull. Soc. Bot. Fr. 116: 370 (1969)). Extends eastwards through Chad and Cameroun to Congo.

90. PASPALUM Linn., Syst. Nat. ed. 10, 2: 855 (1759); F.T.A. 9: 568 (1919).

Spikelets with a ciliate fringe from the margins of the upper glume:
Plants stoloniferous; spikelets orbicular, greenish yellow, 1·5 mm. long, closely appressed to the rhachis of paired slender racemes; basal leaf-sheaths flattened 1. *conjugatum*
Plants caespitose; spikelets 2–4 mm. long, green to brownish:
Racemes 3–5; spikelets ovate, acute, 3–4 mm. long .. .. 2. *dilatatum*
Racemes 10–18, from a long main axis; spikelets elliptic, 2–3 mm. long 3. *urvillei*
Spikelets glabrous or minutely pubescent, without a ciliate fringe:
Spikelets oblong-ovate, acute, rather flattened, 3–3·5 mm. long, glabrous; racemes usually paired, up to 5 cm. long; leaf-blades linear, acute, up to about 10 cm. long, the sheaths imbricate .. .. 4. *vaginatum*
Spikelets orbicular to broadly elliptic, conspicuously plano-convex:
Spikelets minutely pubescent, dark brown, 1·5 mm. long; racemes 20–30, fastigiate 5. *paniculatum*
Spikelets quite glabrous:
Leaf-blades linear-lanceolate, more or less rounded at the base, up to 20 mm. broad and 10–20 cm. long; spikelets 2·5–3 mm. long; racemes 3–4(–10), up to 7 cm. long .. .. 6. *auriculatum*
Leaf-blades linear, not rounded at the base, up to 12 mm. broad; spikelets 1·8–2·5 mm. long, rarely more:
Racemes 1–5, rarely more, the longest up to about 8 cm. long; stems slender, 1–3 mm. diam. at the base, usually tufted, rarely with thin stolons or roots from the lower nodes; basal sheaths glabrous or pilose .. .. 7. *orbiculare*
Racemes 3–5 the longest over 8 cm. long, or more numerous; stems stout, 3–7 mm. diam. at the base, rooting at the lower nodes; basal sheaths nearly always glabrous .. .. 8. *polystachyum*

1. **P. conjugatum** *Berg.* in Act. Helv. Phys. Math. 7: 129, t. 8 (1772); F.T.A. 9: 569; Chev. Bot. 722, and in Rev. Bot. Appliq. 13: 889. *P. africanum* Poir. (1816). Creeping perennial up to about 60 cm. high, sometimes more; open places in forests.
S.L.: Rokupr (June) *Jordan* 452! Njala (Sept.) *Deighton* 2086! Kamakuia (May) *Thomas* 296! Bumbuna (Oct.) *Thomas* 3342! Makump (Oct.) *Glanville* 57! **Lib.:** Jabrocca (Dec.) *Baldwin* 10870! Timbo (Mar.) *Baldwin* 11225! Gbanga (Sept.) *Linder* 511! Ganta (May) *Harley* 305! Gbawia (July) *Baldwin* 6711! **Iv. C.:** Adiopodoumé (Sept.) *Adjanohoun* 431*a*! Abidjan (Oct., Dec.) *Boughey* GC 14458! *Leeuwenberg* 1705! Gagnoa (Aug.) *Boughey* GC 14651! Bianouan (Apr.) *Leeuwenberg* 3967! **Ghana:** Dawa (May) *Howes* 907! Mampong (Sept.) *Rose Innes* GC 30134! Assin-Yan-Kumasi *Cummins* 166! Aburi (June) *Ankrah* GC 20044! Amedzofe (Oct.) *Morton* 6044! **Dah.:** Porto Novo (Aug.) *Adjanohoun* 25! **N. Nig.:** Baissa (Mar.) *Ayuba* AS14! Kagoro to Jemaa (Oct.) *Magaji & Tuley* 1774! **S. Nig.:** Lagos (May) *Dalz.* 1314! Ibadan (Dec.) *Meikle* 913! Benin (Dec.) *Brenan* 8533! Opobo (Apr.) *Jeffreys* 31! Calabar (Apr.) *Holland* 116! **W. Cam.:** Victoria (June) *Rosevear*! Buea (Feb.) *Maitland* 151! Ndop, Bamenda (Dec.) *Boughey* GC 11145! Mamfe (Mar.) *Coombe* 227! Bambui (July) *Brunt* 1189! **F. Po:** *T. Vogel* 14! *Mann* 119! Moka (Sept., Dec.) *Wrigley* 596! *Boughey* 113! S. Isabel (Dec.) *Guinea* 179! Throughout the tropics. (See Useful Plants.)
2. **P. dilatatum** *Poir.* in Lam., Encycl. Méth. Bot. 5: 35 (1804). Perennial 0·6–1·5 m. high.
Ghana: Kumasi (June) *La Danso* 73! Kade (Jan.) *Agyakwah* GC 31202! A S. American grass, introduced throughout the tropics.
3. **P. urvillei** *Steud.* Syn. Pl. Glum. 1: 24 (1854). Tufted perennial 1–2 m. high.
Lib.: Robertsport (Dec.) *Baldwin* 10887! A S. American grass, now found in most tropical countries.
4. **P. vaginatum.** *Sw.* Prod. Veg. Ind. Occ. 21 (1788); F.W.T.A., ed. 1, 2: 561; F.T.A. 9: 570; A. Chev. in Rev. Bot. Appliq. 13: 889; Berhaut, Fl. Sén. ed. 2, 399. *P. distichum* Linn., Syst. Nat. ed. 10, 2: 855 (1759), rejected name. *P. squamatum* Steud. (1854). A stoloniferous creeping grass 30–60 cm. high; sea-shores, near the beach or on tidal flats.
Sen.: *Roger*! *Heudelot* 451! Malika (Aug.) *Broadbent* 47! Oussouye (Sept.) *Broadbent* 118! St. Louis (Oct.) *Hepper* 3609! **Gam.:** Balingho (July) *Fox* 155! **S.L.:** Freetown (May) *Deighton* 1177! Mambolo (Jan.) *Deighton* 977! Rokupr (July) *Jordan* 459! Petifu Creek (Aug.) *Pelly* 139! Tumbu (Apr.) *Glanville*

229! **Iv. C.:** Grand Lahou (May) *Adjanohoun* 432*a*! Sassandra (Dec.) *Leeuwenberg* 2237! **Ghana:** Sakumo Lagoon (Sept., Nov.) *Morton* A4117! *Ankrah* GC 20540! Anyanuwi (Feb.) *La Danso* 16! Anyako (July) *Irvine* 4934! Keta Lagoon (Aug.) *Rose Innes* GC 30083! **Dah.:** Abomey (Dec.) *Froment* 1000! Cotonou to Allada (Oct.) *Risopoulos* 1217! **S. Nig.:** Lagos (Jan.) *Baldwin* 14010! *Barber*! Nun R. *T. Vogel* 60! **W. Cam.:** Victoria (Mar., July) *Brunt* 1042! *Maitland* 8! Bota (Aug.) *Chuml* CDC 249! **F. Po:** *Mann* 441! Tropics and subtropics generally. (See Useful Plants.)

[Owing to a mistake in typification the name *P. distichum* has been consistently misapplied to another species, and should now be rejected as a *nomen confusum*.]

5. **P. paniculatum** *Linn.* Syst. Nat. ed. 10, 2: 855 (1759); F.T.A. 9: 577; A. Chev. in Rev. Bot. Appliq. 13: 890. *P. multispica* Steud. (1854). Caespitose perennial up to 1·2 m. high.
Lib.: Sarbo (July) *Baldwin* 6389! Jabroke (July) *Baldwin* 6494! **W. Cam.:** Tiko (Feb.) *Maitland* 976! Buea (Jan., Mar., July) *Maitland* 4! 352! *Brenan* 9332! Victoria to Buea (Mar.) *Brunt* 1046! **F. Po:** El Pico (Dec.) *Boughey* 185! Estrada farm (Jan.) *Akpabla* GC 11494! Santa Isabel *Swarbrick* 2922! Also in E. Cameroun, Gabon, S. Tomé, Principé, Uganda, Mauritius, New Guinea, Australia, Polynesia and Tropical America.

6. **P. auriculatum** *J. S. Presl* Rel. Haenk. 217 (1830); F.T.A. 9: 572; A. Chev. in Rev. Bot. Appliq. 13: 890; Aké Assi, Contrib. 2: 292. Culms erect or geniculate and rooting at the lower nodes, up to 1·5 m. high, 4–5 mm. diam. at the base; marshy places.
Sen.: Mahon (Oct.) *Boudet* 3995 (*fide* Lebrun in Bull. Soc. Bot. Fr. 116: 264 (1969)). **S.L.:** Bumban (Sept.) *Thomas* 1967! Kaballa (Sept.) *Thomas* 2229! Jigaya (Sept.) *Thomas* 2837! Kamakwie (Feb.) *Deighton* 4045! **Lib.:** Kailahun (Nov.) *Baldwin* 10133! Ganta (Jan., Sept.) *Harley* 2103! 2136! *Baldwin* 12022! Gbanga (Sept.) *Linder* 706! **W. Cam.:** Bamenda (Oct.) *Agric. Off.* Bam. 19! Throughout the moister parts of tropical Africa and in Java, Tonkin and the Phillipines.

7. **P. orbiculare** *Forst.* Fl. Insul. Austr. Prodr. 7 (1786). *P. commersonii* Lam., Tab. Encycl. Méth. Bot. 1: 175 (1791). *P. scrobiculatum* of Berhaut, Fl. Sén. ed. 2, 398, not of Linn. *P. scrobiculatum* Linn. var. *commersonii* (Lam.) Stapf in F.T.A. 9: 573 (1919); F.W.T.A., ed. 1, 2: 561; A. Chev. in Rev. Bot. Appliq. 13: 890; var. *deightonii* C. E. Hubbard in Kew Bull. 1928: 40; F.W.T.A., ed. 1, 2: 561. *P. barbatum* Schumach., Beskr. Guin. Pl. 53 (1827), not of (Trin.) Schult. (1827). *P. ledermannii* Mez in Fedde Rep. 15: 65 (1917). Perennial, more or less tufted or rarely stoloniferous, up to about 60 cm. high; usually in moist or shady places.
Maur.: Nima (Oct.) *Hepper* 3711! **Sen.:** *Adanson* 158! Matam (Dec.) *Chev.* 2249! Mbao (Oct.) *Berhaut* 2767! Djembering (Sept.) *Broadbent* 143! Ziguinchor (Sept.) *Broadbent* 66! **Gam.:** *Saunders* 55! Yundum (Dec.) *Austin* 15! *Ashrif* 20! 41! Genieri *Fox* 144! **Mali:** Mopti (Sept.) *Matthes* 8! Labézenga (Sept.) *Hagerup* 443! Diafarabé (Sept.) *Lean* 96! **Guin.:** Timbo (Oct.) *Pobéguin* 1704! Friguiagbé (Sept.) *Chillou* 3890! **S.L.:** Freetown (Aug.) *Deighton* 778! Rokupr (July) *Jordan* 277! Njala (May) *Deighton* 649! Musaia (Dec.) *Deighton* 4427! Serabu (Apr.) *Deighton* 1680! **Lib.:** Jabrocca (Dec.) *Baldwin* 10872! Monrovia (May) *Baldwin* 5822! Gbanga (Sept.) *Linder* 514! Ganta (May) *Harley* 955! Bahtown (Aug.) *Baldwin* 8039! **Iv. C.:** Toumodi (Dec.) *Boughey* GC 14428! Asakra (Aug.) *Boughey* GC 18623! Bengasso (May) *Adjanohoun* 266*a*! Cocody (May) *Adjanohoun* 420*a*! Singrobo (Oct.) *Adjanohoun* 354*a*! **U. Volta:** Samandéni (Oct.) *Kmoch* 169! Ouagadougou (June) *Scholz* 112! Bobo-Dioulasso (June) *Scholz* 29*a*! **Ghana:** Cape Coast *Rose Innes* GC 30015! Labadi (Mar.) *Deighton* 577! Mampong (Sept.) *Rose Innes* GC 30136! Navrongo (June) *Vigne* FH 4546! Bawku (Oct.) *Rose Innes* GC 30249! **Togo Rep.:** Badou to Atakpame (Sept.) *Rose Innes* GC 31354! **Dah.:** Cotonou to Allada (Oct.) *Risopoulos* 1219! **N. Nig.:** Jebba (Feb.) *Meikle* 1180! Abinsi (Sept.) *Dalz.* 872! Naraguta *Lely* P406! Zaria *Taylor* 33! Sokoto (Sept.) *Palmer* 29! **S. Nig.:** Lagos (Mar.) *Hepper* 2263! Ibadan (Feb.) *Brenan* 8975! Benin (Dec.) *Brenan* 8632! Opobo *Jeffreys* 25! Onitsha (Oct.) *Jones* FHI 6759! **W. Cam.:** Victoria (July) *Maitland* 5! Buea (Nov.) *Migeod* 109! Kumba (Nov.) *Thorold* TN96! Nkambe (Nov.) *Charter* FHI 36987! Bamenda (Oct.) *Agric. Off.* Bam. 15! **F. Po:** *Mann* 119 (partly)! El Bokoko de San Carlos (Jan.) *Guinea* 898! Old World tropics generally. (See Useful Plants.)

[A very variable species in which the customary division between *P. orbiculare* and *P. commersonii* (see Jansen in Reinwardtia 2: 319 (1953)) breaks down in W. Africa. It probably represents a swarm of apomictic segregates with as many subdivisions as one cares to find characters, and I have deemed it best to treat the species in a wide sense. The closely allied *P. scrobiculatum* Linn. is an annual crop plant from India.]

8. **P. polystachyum** *R. Br.* Prod. Fl. Nov. Holl. 188 (1810). *P. scrobiculatum* Linn. var. *polystachyum* (R. Br.) Stapf in F.T.A. 9: 576 (1919); F.W.T.A., ed. 1, 2: 561; Chev. Bot. 723, and in Rev. Bot. Appliq. 13: 890. A very variable grass, distinguished (perhaps arbitrarily) from *P. orbiculare* by its robust habit with large inflorescences and stout stems rooting at the nodes, but there are many intermediate forms; wet places.
Gam.: *Ruxton* 88! Wallikunda (Sept.) *Macluskie* 16! 29! **Mali:** Nafadie (Oct.) *Chev.* 2250! Banguita (Sept.) *Farrow* 44! Dari (Sept.) *Farrow* 45! **Guin.:** Kouroussa (Aug.) *Pobéguin* 487! Kindia *Jac.-Fél.* 177! **S.L.:** Freetown (Aug.) *Melville & Hooker* 212! Njala (Aug.) *Deighton* 774! 2914! Kambia (July) *Jordan* 461! Makump (July) *Thomas* 966! Rokupr (Aug.) *Hooper* 91! **Lib.:** Paynesville (Sept.) *van Dillewijin* 90! Monrovia (May, June) *Barker* 1319! *Baldwin* 5944! Ganta (Oct.) *Harley* 1040! **Iv. C.:** Ferkessédougou (Nov.) *Leeuwenberg* 2028! Abidjan (Oct.) *Leeuwenberg* 1803! Lamto *Adjanohoun* 411*a*! Assanou (Aug.) *Aké Assi* 6848! **U. Volta:** Baoulé (Oct.) *Kmoch* 121! Latourgo to Varalé (Sept.) *Aké Assi* 6515! Konpiénga (June) *Scholz* 119! **Ghana:** Kpandu to Accra (Sept.) *Ankrah* GC 20200! Nsawam (Jan.) *Irvine* 1939! Senchi (July) *Thorold* 126! Tamale to Yendi (Oct.) *Rose Innes* GC 30680! Lawra to Navrongo (Oct.) *Rose Innes* GC 30287! **Togo Rep.:** Palimé, Lomé *Stage* 28! **N. Nig.:** Ajidogari, Ilorin (July) *Ward* 11! Badeggi, Bida (Sept.) *Clayton* 314! Abinsi (Jan.) *Dalz.* 871! Naraguta, Plateau Prov. (July) *Lely* 332! Bindawa (Aug.) *Grove* 35! **S. Nig.:** Olokemeji (Sept.) *Jones* FHI 20264! Onitsha (Aug.) *Jones* FHI 6696! Buruto (Sept.) *Parsons* 1! Nun R. (Sept.) *Mann* 535! Port Harcourt (Sept.) *Holland* 132! **W. Cam.:** Sabonari, Mbaw Plain (May) *Brunt* 397! Throughout Africa, and in the Mascarenes, Australia and New Guinea. (See Useful Plants.)

91. AXONOPUS P. Beauv., Ess. Agrost. 12, 154 (1812); Black. in Adv. Front. Pl. Sci. 5 (1963); F.T.A. 9: 565 (1919); Gledhill in Bol. Soc. Brot. 40: 125 (1966), and in J. West Afr. Sci. Ass. 11: 20 (1966).

Upper lemma almost as long as the spikelet, obscurely hairy at the tip; leaf-blades 1·5–7 mm. broad; spikelets 1·6–2·1 mm. long; nodes glabrous .. .. 1. *affinis*
Upper lemma distinctly shorter than the spikelet; leaf-blades 8–16 mm. broad:
Spikelets 2–2·4 mm. long; nodes pubescent; upper lemma about $\frac{4}{5}$ as long as the spikelet, bearing a short tuft of hairs at the apex .. .. 2. *compressus*
Spikelets 2·8–4 mm. long; nodes glabrous; upper lemma about $\frac{2}{3}$ as long as the spikelet, glabrous .. .. 3. *flexuosus*

1. **A. affinis** *Chase* in J. Wash. Acad. Sci. 28: 180 (1938). A creeping lawn grass.
S.L.: Fourah Bay *Gledhill* 62! **Lib.:** Nimba (Jan.) *Adam* 20576! A species from tropical and subtropical America, introduced as a lawn grass.

Fig. 443.—PASPALUM ORBICULARE *Forst.* (GRAMINEAE-PANICEAE).

1, ligule. 2, portion of raceme, front view showing spikelets and stalks. 2a, 2b, flattened axis. 3, portion of raceme, back view. 4, upper glume. 5, lower lemma. 6, upper lemma. 7, palea. 8, grain.

2. **A. compressus** (*Sw.*) *P. Beauv.* Ess. Agrost. 12, 154, 167 (1812); F.T.A. 9: 566; A. Chev. in Rev. Bot. Appliq. 13: 889; Jac.-Fél. in J. Agric. Trop. 4: 139; including subsp. *brevipedunculatus* Gledhill in Phytomorph. 12: 413 (1962). *Milium compressum* Sw., Prodr. Veg. Ind. Occ. 24 (1788). *Paspalum platicaulon* Poir. (1804)—Chev. Bot. 722. *Axonopus kisantuensis* Vanderyst in Bull. Agric. Congo Belge 16: 667 (1925). *A. brevipedunculatus* Gledhill in Bol. Soc. Brot. 40: 127 (1966). *Echinochloa compressa* (Sw.) Roberty in Bull. I.F.A.N. 17: 60 (1955). A strongly stoloniferous perennial up to 60 cm. high; on pathsides and lawns in the forest zone.
Guin.: Kindia (Dec.) *Jac.-Fél.* 7441! **S.L.:** Freetown (Aug.) *Deighton* 2030! Makump (Aug.) *Deighton* 1364! Mabum (Aug.) *Thomas* 1599! Rokupr (July) *Jordan* 456! Aureol (Jan.) *Gledhill* 45! **Lib.:** Vahun, Kolahun Dist. (Nov.) *Baldwin* 10231! Bomi Hills (Feb.) *Baldwin* 14079! Cavally (June) *Baldwin* 6013! *Harley* 1740! Nimba (Dec.) *Adam* 20067! **Ghana:** Cape Coast (July) *Hall* 1506! Kumasi (Mar., June) *Adams* 3843! *La Danso* 74! Aburi (May, June) *La Danso* 67! *Ankrah* GC 20045! **S. Nig.:** Lagos (Mar.) *Maitland* 164! Agege (Oct.) *Haines* 185! Ibadan (May, Nov.) *Townrow* 21! *Clayton* 560! **W. Cam.:** Victoria (Mar., July) *Brunt* 1038! *Maitland* 17! Cam. Mt. *Maitland*! Kumba (Nov.) *Thorold* TN97! A native of tropical America, now introduced to most tropical countries. (See Useful Plants.)

3. **A. flexuosus** (*Peter*) *C. E. Hubbard ex Troupin*, Fl. Garamba 18 (1956); Jac.-Fél. in J. Agric. Trop. 4: 139; Aké Assi, Contrib. 2: 294. *Digitaria flexuosa* Peter in Fedde, Rep. Beih. 40, 1: Anh. 60 (1930). *Axonopus compressus* of F.T.A. 9: 566, and of F.W.T.A., ed. 1, 2: 564 (partly). *A. compressus* subsp. *congoensis* Henr. in Blumea 5: 529 (1945); var. *congoensis* (Henr.) G. A. Black in Adv. Front. Pl. Sci. 5: 31 (1963). *A. arenosus* Gledhill in Bol. Soc. Brot. 40: 130 (1966). A robust stoloniferous perennial up to 1 m. high; swampy places.
Guin.: Forécariah (Dec.) *Jac.-Fél.* 7403! Kindia *Jac.-Fél.* 127! Dalaba (Nov.) *Chev.* 34899! Timbo (Oct.) *Pobéguin* 1703! Madina Tossékré (Oct.) *Adam* 12542! **S.L.:** Njala (July, Sept.) *Deighton* 734! 750! 2075! Kichom (July) *Deighton* 3765! **Lib.:** Kailahun (Nov.) *Baldwin* 10126! Roberts Field (July) *Baldwin* 6644! Harbel (June) *Baldwin* 5924! 5925! Gbau (Sept.) *Baldwin* 9452! **Iv. C.:** Cavally basin *Chev.* 19762! **Ghana:** Eikwe village (Dec.) *Rose Innes* GC 30547! Esiama (Feb., Mar.) *Rose Innes* GC 30875! *Irvine* 2340! Atuabo (Apr.) *Thorold* CB 41! Ankaful (Mar.) *Hall* 1296! **Dah.:** Ouidah (Aug.) *Adjanohoun* 219! Porto Novo *Adjanohoun* 255! Sémé (Feb.) *Raynal* 13540! **N. Nig.:** Sha (Sept.) *Hall* K34! **S. Nig.:** Ikoyi Plains, Lagos (Oct., July) *Dalz.* 1313! *Jones* FHI 19419! Brass R. *Barter* 1842! Mamu R. flood plain (Aug.) *Jones* FHI 6698! Opobo *Jeffreys* 29! **F. Po:** Moka (Sept.) *Wrigley & Melville* 595! Throughout tropical Africa.

92. ACRITOCHAETE Pilger in Engl., Bot. Jahrb. 32: 53 (1903); F.T.A. 9: 481 (1919).

Leaf-blades narrowly lanceolate, sparsely pilose, narrowed at the base, drawn out into a filiform tip; inflorescence of 1–4 slender racemes appressed to the main axis; spikelets secund, lanceolate, 6 mm. long, on pedicels 2 mm. long; lower glume an ovate scale 1 mm. long; upper glume and lower lemma hispidulous, with a few scattered bristles .. .. *volkensii*

A. volkensii *Pilger* in Engl., Bot. Jahrb. 32: 53 (1903). Perennial, with weak trailing culms up to 1 m. long.
N. Nig.: Chappal Waddi, Sardauna Prov. (Nov.) *Magaji, Jackson & Tuley* 1968! **W. Cam.:** Cam. Mt., 6,000 ft. (Dec.) *Hinds* C65! Mokanda, 6,000 ft. (Feb.) *Maitland* 1339! Mann's Spring, 7,000–7,500 ft. (Dec.) *Boughey* GC 10716! GC 12623! Also on the mountains of east and north-east Africa.

93. ALLOTEROPSIS J. S. Presl. ex C. B. Presl, Rel. Haenk. 343 (1830); emend. Hitchc. in Contrib. U.S. Nat. Herb. 12: 210 (1909); F.T.A. 9: 482 (1919); Butzin in Willdenowia 5: 123 (1968).

Leaf-blades linear, narrow at the base:
Spikelets 5–7 mm. long; upper lemma with an awn 3 mm. long; basal sheaths silky hairy; racemes 2–5, digitate; margins of upper glume ciliate, and sometimes extended into a broad fimbriate wing; lower lemma sometimes transversely barred with purple bands .. .. 1. *semialata*
Spikelets 3–4 mm. long; upper lemma mucronate; basal sheaths at most pubescent; racemes 2(−3); margins of upper glume shortly ciliate .. 2. *angusta*
Leaf-blades lanceolate, rounded-cordate at the base; basal sheaths pubescent:
Spikelets lanceolate, 3·5–4 mm. long; upper glume with a narrow fringe of short appressed hairs along the margins, and tipped with an awn-point up to 1 mm. long; upper lemma with an awn 3–5 mm. long; margins of leaf-blades cartilaginous and crisped-wavy, usually glabrous; racemes 10–20, in 2 or more whorls 3. *paniculata*
Spikelets ovate, 4 mm. long; upper glume with a dense conspicuous band of sub-appressed hairs along the margins, and tipped with an awn point up to 1 mm. long; upper lemma with an awn 2–5 mm. long; margins of leaf-blades cartilaginous and pectinate-ciliate; racemes 5–10, in 1–2 whorls .. .. 4. *cimicina*

1. **A. semialata** (*R. Br.*) *Hitchc.* in Contrib. U.S. Nat. Herb. 12: 210 (1909); F.T.A. 9: 483; A. Chev. in Rev. Bot. Appliq. 13: 883. *Panicum semialatum* R. Br., Prodr. 192 (1810). *Axonopus semialatus* (R. Br.) Hook. f. (1896)—Chev. Bot. 723. *Coridochloa semialata* (R. Br.) Nees (1833). Tufted perennial 60–90 cm. high with erect culms; in savanna.
Sen.: Siminti (June) *Adam* 14185. Kanéméré (June) *Fotius* K124 (*fide* Lebrun in Bull. Soc. Bot. Fr. 116: 257 (1969). **Guin.:** Timbo *Pobéguin* 1833! Mamou to Kindia *Chev.* 13580! **S.L.:** Port Loko (May) *Deighton* 4783! *Jordan* 869! **Ghana:** Tamale to Yendi (May) *Akpabla* 508! Han road, Wa (July) *Adams* 916! Adamsu (Mar.) *Morton* A3243a! **Dah.:** Atacora Mts. *Chev.* 24183! **N. Nig.:** Zaranda Mt. (May) *Lely* 195! Jos Plateau (Apr.) *Lely* P242! Kaduna (May) *Keay* FHI 25797! Anara F.R., Zaria (May) *Keay* FHI 22948! **W. Cam.:** Bangola, Ndop (Mar.) *Brunt* 279! Throughout the tropics of Africa and Asia.

2. **A. angusta** *Stapf* in F.T.A. 9: 485 (1919). Perennial 60–90 cm. high, with slender ascending culms; swampy places.
N. Nig.: Gembu to Donga Valley, Sardauna Prov. (Aug.) *De Leeuw* 1808! Also in Uganda, Kenya, Zambia and Angola.

3. **A. paniculata** (*Benth.*) *Stapf* in F.T.A. 9: 486 (1919); A. Chev. in Rev. Bot. Appliq. 13: 883; Berhaut, Fl. Sén. ed. 2, 417. *Urochloa paniculata* Benth., Fl. Nigrit. 558 (1849). *Axonopus paniculatus* (Benth.) A. Chev., Bot. 723 (1920). *Echinochloa paniculata* (Benth.) Roberty in Bull. I.F.A.N. 17: 67 (1955). *Panicum benthamii* Steud. (1854). *P. aubertii* Mez (1905). *Mezochloa aubertii* (Mez) Butzin in Willdenowia 4: 211 (1960). Annual up to about 1·2 m. high; moist soils and marshy places.
Mali: San (June, Nov.) *Chev.* 1092! 2291! **Guin.:** Kindia *Jac.-Fél.* 260! **S.L.:** Mano *Thomas* 10646! Rombe (Apr.) *Glanville* 227! Kambia (Dec.) *Deighton* 845! Rotebana (Mar.) *Adames* 154! Rokupr (May) *Jordan* 38! **Lib.:** Gbeidin (Apr., May) *Harley* 1807! 2034! Baila (May) *Harley* 1415! **Iv. C.:** Soubré (Apr.) *Leeuwenberg* 3224! **Ghana:** Kpong (June) *Ankrah* GC 20055! Amedica, R. Volta (May) *Irvine* 3006! **N. Nig.:** Lokoja *Ansell*! **S. Nig.:** Ajagbodudu, Benin (Aug.) *Wright* 59! Onitsha *Barter* 1756! Afikpo (June) *Stone* 10! Abakaliki (Apr.) *Haines* 140! Also in C. African Rep., Congo, Malawi, Zanzibar and Mauritius. (See Useful Plants.)

4. **A. cimicina** (*Linn.*) *Stapf* in F.T.A. 9: 487 (1919). *Milium cimicinum* Linn., Mant. Alt. 184 (1771). *Panicum cimicinum* (Linn.) Retz. (1783). *Coridochloa cimicina* (Linn.) Nees (1833). Annual 60–90 cm. high; moist soils.
N. Nig.: Maska, Katsina Prov. (Nov.) *Thatcher* S551! Damaturu (July) *De Leeuw* 1031! Throughout tropical Asia and Africa.

94. DIGITARIA Heist. ex Fabric., Enum. Pl. Hort. Helmst. 207 (1759); F.T.A. 9: 422 (1919); Henrard, Monograph of the genus Digitaria (1950).

Pedicel tips bearing white setulae as long as the spikelets, the latter quite glabrous; lower glume suppressed, the upper reduced to a small scale and exposing much of the chestnut-brown upper lemma; lower lemma 3-nerved:
Perennial, with densely silky base; spikelets 1·7–2·1 mm. long, the upper glume distinct and up to 0·5 mm. long .. .. 1a. *diagonalis* var. *hirsuta*
Annual; spikelets 0·7–1 mm. long, the upper glume represented by a minute scale 2. *minutiflora*
Pedicel tips glabrous or hispidulous:
Racemes in a succession of 6 or more whorls on a slender common axis about 20 cm. long, the whorls each of 10–12 racemes; raceme rhachis triquetrous, 3–5 cm. long; spikelets 1·8 mm. long; upper glume 3-nerved, as long as the spikelet; leaf-blades lanceolate from a broad subcordate base .. .. 3. *perrottetii*
Racemes digitate or subdigitate, sometimes with an elongated common axis but then with the racemes longer than the axis and few in number:
Upper glume shorter than the lower lemma, 3-nerved, the back of the upper lemma partly exposed:
Base of culms densely coated with fibrous remains of leaf-sheaths; rhachis triquetrous, without wings; spikelets clothed in fine pale brown hairs; fruit dark chestnut brown:
Spikelets 2–2·5 mm. long .. .. 4. *homblei*
Spikelets 3–3·5 mm. long .. .. 5a. *phaeotricha* var. *patens*
Base of culms not fibrous:
Upper lemma with an awn-point 1·3 mm. long; spikelets 1·7 mm. long, obscurely puberulous with verrucose hairs; racemes digitate, 2–4 cm. long, the rhachis flat and herbaceous with a low rounded midrib .. .. 6. *aristulata*
Upper lemma acute or acuminate, but not aristulate:
Spikelets ternate in the middle part of the raceme (sometimes the longest pedicel is partly adnate to the midrib and the spikelets appear to be alternately paired and solitary, or the lowest spikelet is aborted):
Rhachis of racemes triquetrous, without lateral wings:
Hairs of spikelets usually short, appressed-clavate with acute tips, not mixed with stout bristles; spikelets 1·7–2·2 mm. long; fruit black 7. *delicatula*
Hairs of spikelets mixed with stiff glistening bristles; spikelets 1·8–2 mm. long; fruit black .. .. 8. *lecardii*
Rhachis of racemes flat, laterally winged, the wings about the same width as the rounded or triquetrous midrib:
Perennial; tips of pedicels glabrous; spikelets 2–2·2 mm. long, obscurely hairy with appressed truncate hairs; upper glume obtuse, about ½ as long as spikelet; fruit light chestnut brown .. .. .. 9. *seminuda*
Annuals; tips of pedicels with a corona of short hairs up to 0·3 mm. long:
Spikelets with dense lines of short appressed hairs between the nerves, the hairs clavate with obtuse tips; spikelets 1·8–2 mm. long; upper glume acute, ¾ as long as spikelet; fruit black .. .. 10. *ternata*
Spikelets quite glabrous, 2 mm. long, dark brown to black, the 3 central nerves of the lower lemma close together; midrib of rhachis triquetrous 11. *iburua*
Spikelets paired in the middle part of the raceme; midrib of rhachis triquetrous and usually winged:
Lower lemma apparently 3-nerved in face view (actually 5-nerved, with the lateral nerves contiguous); spikelets 3 mm. long, narrowly lanceolate; midrib of rhachis triquetrous, 3-winged; fruit yellowish-brown .. 12. *timorensis*
Lower lemma 5-nerved on the face, the nerves spaced:
Spikelets acuminate at the tip, 3–4 mm. long, with or without stiff brown

bristle-hairs along the margins of the lower lemma; racemes 15–25 cm. long, stiff, straight; fruit bluish-grey .. .. .. 13. *acuminatissima*

Spikelets acute at the tip, up to 3 mm. long:

Spikelets quite glabrous, 1·8–2 mm. long, about 3 times as long as broad; racemes triquetrous, scarcely winged, 3–6 cm. long on a short common axis up to 3 cm. long, spreading almost horizontally, the spikelets borne rather laxly on one side; lower glume a green ovate scale 0·3 mm. long; upper glume $\frac{2}{3}$–$\frac{3}{4}$ as long as the spikelet; fruit bluish grey .. 14. *abyssinica*

Spikelets hairy, though often obscurely so, very rarely glabrous; racemes narrowly winged, the spikelets appressed; lower glume varying from an obscure rim to a distinct scale:

Spikelets 3 times as long as broad, mostly 2–2·3 mm. long; racemes 5–11 cm. long, digitate, stiffly ascending; spikelets imbricate, overlapping by at least half their own length; upper glume usually half as long as the spikelet or more; pedicelled spikelet often with a rufous ciliate fringe at maturity 15. *ciliaris*

Spikelets 5 times as long as broad, mostly 2–2·5 mm. long; racemes 8–14 cm. long, digitate or more often on a common axis up to 6 cm. long, more or less spreading; spikelets less dense, overlapping by less than half their length at least in parts of the raceme; upper glume usually less than half as long as the spikelet; pedicelled spikelet without a conspicuous ciliate fringe .. .. 16. *horizontalis*

Upper glume as long as the lower lemma, 5- or rarely 3-nerved, concealing the back of the upper lemma:

Rhachis of racemes flat, ribbon-like, with a low rounded midrib:

Spikelets pubescent or hirsute with verrucose hairs:

Spikelets 2 mm. long, clothed in fleecy white hairs drawn out into a tuft about 1 mm. beyond the end of the spikelet; lower lemma with a prominent glabrous middle nerve; racemes paired, 8–15 cm. long .. .. 17. *argyrotricha*

Spikelets 1·2–1·5 mm. long, with short appressed hairs not extending beyond the spikelet; racemes typically paired, 2–5 cm. long .. .. 18. *longiflora*

Spikelets quite glabrous:

Plant creeping, up to about 20 cm. high; spikelets 1·5 mm. long; racemes typically paired, 2–6 cm. long; fruit light brown, the tip distinctly protruding beyond the lower lemma .. .. 19. *fuscescens*

Plant geniculately ascending, 40–80 cm. high; spikelets 1·7–2 mm. long; racemes 3–5, 4–15 cm. long; fruit bluish grey, not exceeding the lower lemma 20. *exilis*

Rhachis of racemes triquetrous, sometimes narrowly winged but then the spikelets attached to a sharp-edged midrib:

Spikelets densely clothed in long silvery or purplish hairs produced into a long tuft 3–4 mm. beyond the end of the spikelet; spikelets 3–4 mm. long, ternate; lower glume a truncate, nerveless, hyaline scale 0·75 mm. long; upper glume 3-nerved; upper lemma acuminate, rostrate .. .. 21. *gayana*

Spikelets glabrous or hairy, the hairs not extending beyond the tip of the spikelet; spikelets paired:

Upper glume 0·5 mm. longer than the lower lemma, caudate-acuminate at the tip, 7-nerved; spikelets 2·5–3 mm. long, narrowly lanceolate, obscurely puberulous; lower glume a membranous collar separated from the rest of the spikelet by a distinct internode 0·1–0·2 mm. long .. .. 22. *debilis*

Upper glume and lower lemma subequal, acute:

Lower glume a broad truncate scale, $\frac{1}{4}$–$\frac{1}{3}$ the length of the spikelet, with a snow white fringe of hairs protruding from underneath; spikelets 3 mm. long, pubescent with verrucose hairs .. .. 23. *patagiata*

Lower glume an obscure glabrous rim at the base of the spikelet:

Spikelets 3·7–4 mm. long, minutely pubescent; upper glume 3-nerved 24. *gentilis*

Spikelets 1·7–2·2 mm. long:

Spikelets appressed-pubescent, the hairs very fine with curled tips, 1·6–2 mm. long; upper glume 5-nerved; fruit greyish purple .. 25. *leptorhachis*

Spikelets quite glabrous, 2–2·2 mm. long; upper glume 3-5-nerved; fruit grey .. .. 26. *barbinodis*

1. **D. diagonalis** (*Nees*) *Stapf* in Fl. Capensis 7: 381 (1898). *Panicum diagonale* Nees, Fl. Afr. Austr. 23 (1841). *P. uniglume* Hochst. ex A. Rich., Tent. Fl. Abyss. 2: 370 (1851). *Digitaria uniglumis* (Hochst. ex A. Rich.) Stapf in F.T.A. 9: 474 (1919).

1a. **D. diagonalis** var. **hirsuta** (*De Wild. & Th. Dur.*) *Troupin* Fl. Garamba 1: 29 (1956). *Panicum diagonale* var. *hirsutum* De Wild. & Th. Dur., Pl. Thonn. Congol. 4 (1900). *Digitaria uniglumis* var. *hirsuta* (De Wild. & Th. Dur.) Robyns (1955). *D. uniglumis* var. *major* Stapf in F.T.A. 9: 476 (1919); F.W.T.A., ed. 1, 2: 569; Henrard, Monogr. Digitaria 767; A. Chev. in Rev. Bot. Appliq. 13: 883; Aké Assi, Contrib. 2: 296. Robust erect perennial, 1–3 m. high, with handsome spreading inflorescence.
Sén.: Kanéméré (July) *Fotius* K307! **S.L.:** Bintumane Peak, 5,000 ft. (Jan.) *T. S. Jones* 150! Mt. Loma (Nov.) *Jaeger* 503! *Morton* SL 2684! **Iv. C.:** Bouaflé to Zuénoula (Oct.) *Adjanohoun* 331*a*! Odiénné

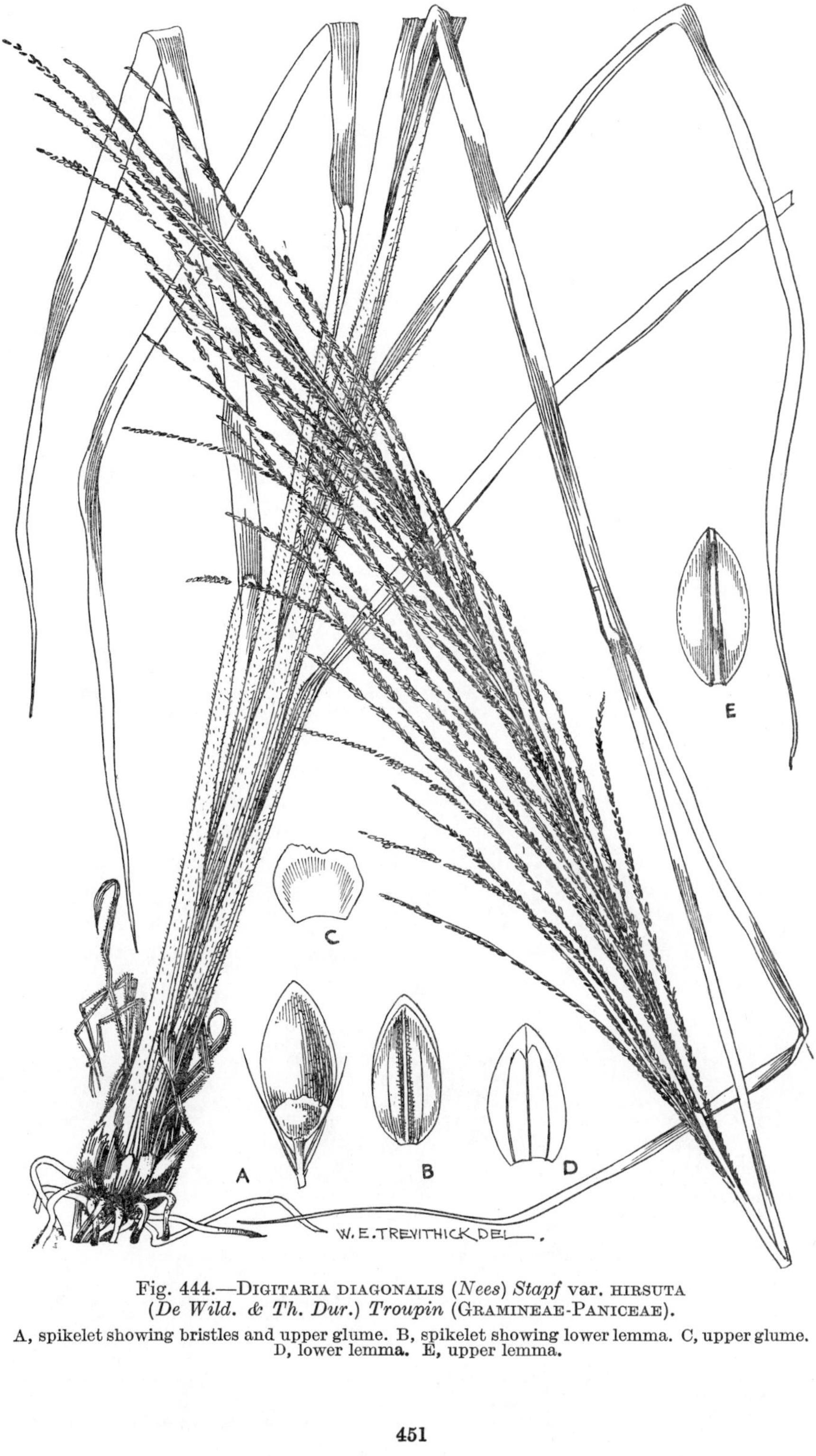

Fig. 444.—Digitaria diagonalis (*Nees*) *Stapf* var. hirsuta (*De Wild. & Th. Dur.*) *Troupin* (Gramineae-Paniceae).

A, spikelet showing bristles and upper glume. B, spikelet showing lower lemma. C, upper glume. D, lower lemma. E, upper lemma.

to Boundiali (July) *Adjanohoun* 287*a*! **Ghana:** Bosomoa F.R. (June) *Vigne* FH 3054! Sawla to Wa (Nov.) *Rose Innes* GC 30825! Yapei ferry to Kintampo (Nov.) *Rose Innes* GC 30851! Ejura to Kintampo (Oct.) *Rose Innes* GC 30516! Banda (Apr.) *Morton* A3286! **Togo Rep.:** Sokode to Lama Kara (Oct.) *Rose Innes* GC 31389! **N. Nig.:** Gwada, Niger Prov. (Sept.) *Lamb* FHI 3164! Gwari, Niger Prov. (Aug.) *Onochie* FHI 35701! 35713! Mando F.R., Zaria (June) *Keay* FHI 37078! Vogel Peak (Nov.) *Hepper* 1500! Maisamari, Mambila (Sept.) *Oche* MRK57! **S. Nig.:** Lagos *W. MacGregor* 208! Ago-Are F.R., Shaki (Oct.) *Sofoluwe* FHI 38180! Akure to Owo (Oct.) *Maggs* 20! Mamu River F.R., Awka (Sept.) *Onochie* FHI 34075! Ikem, Nsukka (Oct.) *Jones* FHI 6774! **W. Cam.:** Bafut-Ngemba F.R. (Feb.) *Hepper* 2206! Bambui (Dec.) *Pedder* 3! *Boughey* GC 10751! Kishong (June) *Brunt* 713! Throughout tropical Africa.

2. **D. minutiflora** (*Hochst. ex A. Rich.*) *Stapf* in F.T.A. 9: 476 (1919); Henr. l.c. 421; A. Chev. in Rev. Bot. Appliq. 13: 883. *Panicum minutiflorum* Hochst. ex A. Rich., Tent. Fl. Abyss. 2: 362 (1851). Erect grass about 60 cm. high.
Guin.: Timbo (Oct.) *Pobéguin* 1701! Madina Tossékré (Oct.) *Adam* 12560! **S.L.:** Benekoro (Sept.) *Glanville* 328! Bintumane (Nov.) *Jaeger* 8223! Loma Mt. (Nov.) *Morton* SL 2685! Extends eastwards to Ethiopia and Tanzania, southwards through the Congo to Zambia.

3. **D. perrottetii** (*Kunth*) *Stapf* in F.T.A. 9: 435 (1919); Henr. l.c. 547; A. Chev. in Rev. Bot. Appliq. 13: 879; Berhaut, Fl. Sén. ed. 2, 403. *Panicum perrottetii* Kunth, Rev. Gram. 2: 395 (1829). *P. pseudagrostis* Steud., Syn. Pl. Glum. 1: 71 (1854). *Milium minutiflorum* Trin. (1834). Erect annual with coarse leaves, 1–1·2 m. high.
Sen.: *Leprieur*! *Roger*! *Heudelot* 448! Also in Angola, Rhodesia, Malawi, Mozambique, Tanzania and Madagascar.

4. **D. homblei** *Robyns* in Inst. Roy. Col. Belge, Mém. 1: 27 (1931). Caespitose perennial 60 cm. high.
N. Nig.: 30mi. W. of Kaduna (June) *Clayton* 1098! Anara F.R., Zaria (May) *Keay* FHI 22865! Zaria *Taylor* 32! Iyatawa (July) *Keay* FHI 25933! Also in Katanga and Zambia.

5. **D. phaeotricha** (*Chiov.*) *Robyns* in Inst. Roy. Col. Belge, Mém. 1: 28 (1931). *D. parlatorei* (Steud.) Chiov. var. *phaeotricha* Chiov. in Ann. Bot. Roma 13: 41 (1914).

5a. **D. phaeotricha** var. **patens** *W. D. Clayton* in Kew Bull. 23: 294 (1969). Caespitose perennial 60 cm. high.
S.L.: Mt. Loma (Mar., Apr.) *Jaeger* 9677! 9791!

6. **D. aristulata** (*Steud.*) *Stapf* in F.T.A. 9: 471 (1919); Henr. l.c. 50; Chev. Bot. 731, and in Rev. Bot. Appliq. 13: 882; Berhaut, Fl. Sén. ed. 2, 400. *Panicum aristulatum* Steud. Syn. Pl. Glum. 1: 42 (1854). Creeping *Cynodon*-like annual up to about 15 cm. high.
Sen.: *Leprieur*! **Mali:** San (June) *Chev.* 1094! Djenné (June) *Chev.* 1121! San to Djenné (June) *Lean* 40!

7. **D. delicatula** *Stapf* in F.T.A. 9: 454 (1919); Henr. l.c. 173; A. Chev. in Rev. Bot. Appliq. 13: 881. Erect annual about 60 cm. high.
Sen.: Kanéméré (Aug.) *Fotius* K570 (*fide* Lebrun in Bull. Soc. Bot. Fr. 116: 258 (1969)). **Mali:** Banguita (Sept.) *Farrow* 109! **Iv. C.:** Bocanda (June) *Adjanohoun* 257*a*! Niangbo (July) *Aké Assi* 10758! Sipilou (Oct.) *Aké Assi* 9680! **U. Volta:** Adioukrou (Feb.) *Chev.* 17149! **Ghana:** Akwamu to Ajena (June) *Adams* 4795! Wenchi (Sept.) *Rose Innes* GC 30154! Kpandu to Accra (Sept.) *Ankrah* GC 20206! Yendi (Sept.) *Thorold* 203! Pong-Tamale (Oct.) *Adesi*! **Dah.:** Parakou (Sept.) *Risopoulos* 1193! **N. Nig.:** Panshanu (Aug.) *Lawlor & Hall* 197! 206! Damagum (Sept.) *De Leeuw* 1235! Jebba (Aug.) *Hall* 301! Guga F.R., Zaria (Aug.) *De Leeuw* 1951! **S. Nig.:** Lagos *W. MacGregor* 210! Aladidan (July) *Sawyer* K6! Igboora (Oct.) *Haines* 335!

8. **D. lecardii** (*Pilger*) *Stapf* in F.T.A. 9: 450 (1919); Henr. l.c. 382; A. Chev. in Rev. Bot. Appliq. 13: 880; Berhaut, Fl. Sén. ed. 2, 400. *Panicum sanguinale* var. *lecardii* Pilger in Engl., Bot. Jahrb. 30: 118 (1901). Erect annual about 60 cm. high; scarcely distinct from the preceding.
Mali: Ségou *Lecard* 243! **Iv. C.:** Téhini (Aug.) *de Wilde* 776! **U. Volta:** Fada, N'Gourma *Chev.* 24556! Kassandé (July) *Scholz* 39! Barmoussa (June) *Scholz* 121! **Ghana:** Damongo (Sept.) *Goodall* 15917! Navrongo (Aug.) *Williams* 552! Gambaga (Oct.) *Vigne* FH 4637! Tamale (Aug.) *Irvine* 4536! Damongo to Tamale (Aug.) *Rose Innes* GC 31926! **Dah.:** Tanguieta to Boukombe (Nov.) *Risopoulos* 1272! **N. Nig.:** Nupe *Barter* 1358! Vodni, Plateau *Saunders* 59! Kuduru (Aug.) *Sawyer* K3! Samaru (Aug.) *Thatcher* S478! Yola (Sept.) *Dalz.* 274! **S. Nig.:** Igbetti (July) *Stanfield*!

9. **D. seminuda** *Stapf* in F.T.A. 9: 446 (1919); Henr. l.c. 675; A. Chev. in Rev. Bot. Appliq. 13: 880; Aké Assi, Contrib. 2: 296. Tufted perennial 60–90 cm. high.
Guin.: Kouria (July) *Caille in Hb. Chev.* 15020! **S.L.:** Mabum (Aug.) *Thomas* 1516! Tingi Mts. (Dec.) *Morton & Gledhill* SL 3127! Njala (June, July) *Deighton* 733! 3232! Madina (Aug.) *Jordan* 914! **Ghana:** Mankessim (May) *Hall* 2968! Kpandai (Sept.) *Ankrah* GC 20240! Cape Coast (June) *Hall* 3053! Elmina (June) *Hall* 1485! Kete Krachi to Yendi (Oct.) *Rose Innes* GC 30951! **N. Nig.:** Shika, Zaria (Sept.) *De Leeuw & Magaji* 1951*a*! Gembu, Mambila (Aug.) *De Leeuw* 1786! **S. Nig.:** Lagos *Dalz.* 1425! *Haines* 314! Opobo *Jeffreys* 24! Ogoni (July) *Tuley* 781!

10. **D. ternata** (*A. Rich.*) *Stapf* in Fl. Capensis 7: 376 (1898); F.T.A. 9: 452; Henr. l.c. 737; A. Chev. in Rev. Bot. Appliq. 13: 881. *Cynodon ternatus* A. Rich., Tent. Fl. Abyss. 2: 405 (1851). Tufted annual up to 60 cm. high.
Mali: Man to Bamako (Sept.) *Adam* 15005! **Guin.:** Timbo *Pobéguin*! **S.L.:** Mabonto (Aug.) *Deighton* 1301! **Ghana:** Lawra (Sept.) *Hall* CC 636! **N. Nig.:** Naraguta F.R. (July) *Lawlor & Hall* 112! Riyom, Plateau (Aug.) *Sawyer* K5! Hoss, Plateau (Aug.) *Keay* FHI 12693! Samaru, Zaria (Aug., Sept.) *Thatcher* S484! Biu (Oct.) *De Leeuw* 1317! **S. Nig.:** Obudu Plateau (Nov.) *Tuley* 1025! **W. Cam.:** Bamenda (Oct.) *Agric. Off.* Bam 13! Bambui (Aug.) *Brunt* 1241! Throughout tropical and S. Africa, extending eastwards through India to Thailand and China.

11. **D. iburua** *Stapf* in Kew Bull. 1915: 382 (1915); F.T.A. 9: 455; Henr. l.c. 338; Portères in J. Agric. Trop. 2: 359; A. Chev. in Rev. Bot. Appliq. 13: 881. Annual crop plant up to about 1·4 m. high.
Iv. C.: Azaguié (Oct.) *Aké Assi* 10331! **Togo Rep.:** Kandé, Kpessede, Atalоté (*fide* Portères). **Dah.:** Birni to Natitingou (*fide* Portères). **N. Nig.:** Zaria (Sept.) *Lamb* 54! Jos (Oct., Nov.) *Hepper* 2867! *Keay* FHI 21429! Riyom (Nov.) *Philcox* 1011! 1012! (See Useful Plants.)

12. **D. timorensis** (*Kunth*) *Bal.* in Morot, Journ. de Bot. 4: 138 (1890); Henr. l.c. 745. *Panicum timorense* Kunth, Enum. Pl. 1: 83 (1833). Slender creeping annual about 23 cm. high, probably introduced.
W. Cam.: Buea (July) *Maitland* 2! Also in Mauritius, and extending eastwards from India to Malesia and Polynesia.

13. **D. acuminatissima** *Stapf* in F.T.A. 9: 441 (1919); Henr. l.c. 4 (incl. var. *conformis* Henr. and var. *typica* subvar. *grandiflora* Henr.); A. Chev. in Rev. Bot. Appliq. 13: 880; Berhaut, Fl. Sén. ed. 2, 400. *D. sanguinalis* var. *subacuminata* A. Chev., Sudania 30, 33 (1911); Chev. Bot. 732; name only. Annual 60–90 cm. high with long stiff racemes; on riverside flats subject to seasonal flooding.
Maur.: L. Rkiz (Oct.) *Adam* 21818! **Mali:** Goundam to Faginbirg (Aug.) *Chev.* 2284! San (Sept.) *Chev.* 2218! Mopti (Sept.) *Chev.* 24957! Gao (Aug.) *de Wailly* 5132! **Ghana:** Pong-Tamale (June) *Addei* 757! Ningo Mt. (July) *Hall* 3148! Amedica (Aug.) *Hall* 3301! Lawra (Sept.) *Hall* CC 772! **N. Nig.:** Lokoja (Mar.) *Maggs* OK49! Abinsi (May) *Dalz.* 907! Nguru (July) *Golding* 42! Sokoto (Sept.) *De Leeuw* 1556! Bornu (Oct.) *De Leeuw* 2045! Extends southward to the Congo and Angola.

14. **D. abyssinica** (*Hochst. ex A. Rich.*) *Stapf* in Kew Bull. 1907: 213; F.T.A. 9: 460; Henr. l.c. 1. *Panicum abyssinicum* Hochst. ex A. Rich., Tent. Fl. Abyss. 2: 360 (1851). A slender creeping perennial with culms up to about 30 cm. high.
N. Nig.: Gembu, Mambila Plateau (Jan., Sept., Nov.) *Hepper* 1841! *Oche* MRK 81! *Magaji & Tuley* 1849! Mayo Daga to Njawai (Nov.) *Jackson, Magaji & Tuley* 1892! **S. Nig.:** Obudu Plateau (Nov.

Tuley 1032! **W. Cam.:** Bambui (May) *Pedder* 24! Ndop (Apr.) *Brunt* 374! Nkambe (Feb.) *Hepper & Charter* 1936! Bamenda to Santa *Akpabla* GC 10562! Buea (Mar.) *Boughey* GC 7018! Also in east Africa from Ethiopia to Tanzania.

15. **D. ciliaris** (*Retz.*) *Koel.* Descr. Gram. 27 (1802). *Panicum ciliare* Retz., Obs. Bot. 4: 16 (1786). *P. adscendens* Kunth in Humb. & Bonpl., Nov. Gen. et Sp. Pl. 1: 97 (1816). *Digitaria adscendens* (Kunth) Henr. in Blumea 1: 92 (1934); Henr., Monogr. Digitaria 9 & 998 (incl. subsp. *chrysoblephora* (Fig. & De Not.) Henr., subsp. *marginata* (Link) Henr. and var. *rhachiseta* Henr.); De Miré & Gillet in J. Agric. Trop. 3: 731. *D. marginata* Link (1821); F.T.A. 9: 439 (incl. var. *linkii* Stapf, and var. *fimbriata* (Link) Stapf; A. Chev. in Rev. Bot. Appliq. 13: 879; Aké Assi, Contrib. 2: 295. *D. sanguinalis* of Chev. Bot. 732, not of Scop. [this closely allied species of temperate regions has shorter spikelets and upper glumes, see Ebinger in Brittonia 14: 248 (1962)]. Creeping or ascending annual up to about 60 cm. high; common on roadsides, old farmland, and waste ground.
Maur.: Néma *Rossetti* 61/173! Rosso (Oct.) *Hepper* 3620! **Sen.:** Ziguinchor (Jan.) *Chev.* 2188 *bis*! Dakar (May, Oct.) *Baldwin* 5740! *Adam* 14929! Diégoune (Sept.) *Adam* 18064! Sefa (Oct.) *Adam* 18427! **Gam.:** *Saunders* 26! *Ruxton* 59! Yundum (Aug., Dec.) *Frith* 17! 126! Mandingo *Pirie* 1*a*! **Mali:** Macina (Sept.) *Chev.* 24884! Bamba (Mar.) *Chev.* 43813! Timbuktu (July, Aug.) *Hagerup* 207! *Lean* 66! Dioura (Aug.) *Davey* 62! **Port. G.:** S. Domingos (Aug.) *Pereira* 3130! **Guin.:** Timbo *Pobéguin* 1703! 1705! Kouroussa (Aug.) *Pobéguin* 498! **S.L.:** Freetown (Dec., Mar.) *Deighton* 798! *Gledhill* 46! Mane (May) *Deighton* 2879! Newton (Aug.) *Deighton* 2202! Kissy (Nov.) *Deighton* 4388! **Lib.:** Monrovia (Dec.) *Adam* 16521! Paynesville (Sept.) *van Dillewijn* 89! Nimba (May) *Adam* 25517! **U. Volta:** Koudougou (Sept.) *Aké Assi* 6455! **Ghana:** Accra (June) *Dalz.* 169! Damongo (Mar.) *Adams* 3970! Mamprusi *Hepper & Morton* A3152! Bolgatanga (Oct.) *Rose Innes* GC 30241! Bawku (Oct.) *Rose Innes* GC 30253! **Niger:** Niamey (Oct.) *Hagerup* 515! Toukounous *Bartha* 28! **N. Nig.:** Ilorin (Mar.) *Meikle* 1316! Nupe *Barter* 1365! Naraguta *Lely* P402! Sokoto (Sept.) *Palmer* 4! Fika (Oct.) *Daggash* FHI 24891! Throughout the tropics.

[This and the following species are closely related and are both variable. The most reliable distinguishing features are the shape and disposition of the spikelets and the attitude of the racemes, features which, unfortunately, are difficult to describe in words.]

16. **D. horizontalis** *Willd.* Enum. Hort. Berol. 92 (1809); F.T.A. 9: 436; Henr. l.c. 329; Chev. Bot. 732, and in Rev. Bot. Appliq. 13: 879. *D. horizontalis* var. *porrantha* (Steud.) Henr. ex Hubbard & Vaughan, The grasses of Mauritius and Rodriguez 88 (1940); Henr., Monogr. Digitaria 333. *Panicum porranthum* Steud. (1854)—Chev. Bot. 728. *P. diamesum* Steud. (1854). *Digitaria diamesum* (Steud.) Henr. l.c. 177 (1950). *D. reflexa* Schumach. (1827)—Henr. l.c. 621. *D. nuda* Schumach. (1827)—Henr. l.c. 500. *D. velutina* of F.W.T.A., ed. 1, 2: 567; and of Berhaut, Fl. Sén. ed. 2, 400. (This closely allied species is confined to East Africa.) Creeping or ascending annual 30–60 cm. high; waste places.
Sen.: *Perrottet* 909! Kafountine (Sept.) *Adam* 18174! **Gam.:** Sankuli-Kunda (Sept.) *Pirie* 8! Yundum *Ashrif* 33! **Port. G.:** Piche (Sept.) *Pereira* 3174! **Guin.:** Baffing R. *Pobéguin* 1705*bis*! Kouroussa (July) *Pobéguin* 549! Kindia *Jac.-Fél.* 261! **S.L.:** Mabum (Aug.) *Thomas* 1545! Royanka (Sept.) *Melville & Hooker* 638! Rokupr (June) *Jordan* 274! Makump (Aug.) *Deighton* 1366! Musaia (Apr.) *Deighton* 5457! **Lib.:** Bushrod I. (Apr.) *Barker* 1291! Du R. (Aug.) *Linder* 223! Monrovia (May) *Baldwin* 5819! Ganta (July) *Harley* 1809! Gretown (July) *Baldwin* 6795! **Iv. C.:** Mbrabo (Jan.) *Hédin*! Tonkoui (Aug.) *Boughey* GC 18324! Lamé, Abidjan (Oct.) *Leeuwenberg* 1738! Cocody (June) *Adjanohoun* 416*a*! Gagnoa (Aug.) *Boughey* GC 14647! **U. Volta:** Banfora (July) *Scholz* 34! Ouagadougou (July) *Scholz* 34*b*! **Ghana:** Achimota, Accra (Apr.) *Ankrah* GC 20117! Damongo (Sept.) *Rose Innes* GC 30197! Mampong (Oct.) *Baldwin* 13532! Half Assinie (Dec.) *Rose Innes* GC 30542! Aburi (Apr.) *Johnson* 1010! **Togo Rep.:** Lomé *Warnecke* 340! Palime *Stage* 57! **Dah.:** Sémé (Sept.) *Adjanohoun* 531! Cotonou (July) *Chev.* 4477! **N. Nig.:** Ilorin (Oct.) *Ward* 45! Lokoja *Richardson*! Abinsi *Dalz.* 880! Acharane F.R., Kabba Prov. (June) *J. K. Jackson* 5163! **S. Nig.:** Lagos *W. MacGregor* 6! Ibadan (Nov.) *Meikle* 658! Okomu F.R. (Dec.) *Brenan* 8623! Umuahia (July) *Jones* FHI 6214! Opobo (Oct.) *Jeffreys* 8! **W. Cam.:** Buea (July, Oct.) *Migeod* 90! *Maitland* 3! 24! 203! Bamenda to Buli (Mar.) *Brunt* 1084! **F. Po:** Clarence Peak (Oct.) *T. Vogel*! Throughout tropical Africa and America. (See Useful Plants.)

17. **D. argyrotricha** (*Anderss.*) *Chiov.* in Result. Sci. Miss. Stef.-Paoli 1: 183 (1916); F.T.A. 9: 468; Henr. l.c. 49. *Panicum argyrotrichum* Anderss. in Peters, Nat. Reise Mossamb. 2: 548 (1864). Tufted or trailing annual.
Ghana: Legon (Jan., Apr.) *Adams* 3647! *Morton* A4159! Keta (July, Dec.) *Thorold* 273! *Hall* 3263! Cape St. Paul (June) *Rose Innes* GC 31224! Also in east Africa from the Sudan to Mozambique.

18. **D. longiflora** (*Retz.*) *Pers.* Syn. Pl. 1: 85 (1805); F.T.A. 9: 469; Henr. l.c. 409; A. Chev. in Rev. Bot. Appliq. 13: 882; Berhaut, Fl. Sén. ed. 2, 399, 400; Aké Assi, Contrib. 2: 294. *Paspalum longiflorum* Retz., Obs. Bot. 4: 15 (1786). Creeping stoloniferous annual or short-lived perennial up to about 30 cm. high; roadsides, fallow land and waste places.
Sen.: Niokolo-Koba (Oct.) *Adam* 15568 *bis*! Diégoune (Sept.) *Adam* 18066! Oussouyé (Sept.) *Broadbent* 105! Ziguinchor (Sept.) *Broadbent* 64! Djembering (Sept.) *Broadbent* 152! **Gam.:** *Pirie*! *Ruxton* 137! Genieri (July) *Fox* 173! **Mali:** Ségou *Lécard* 252 (partly)! Kara (Aug.) *Farrow* 112! Banguita (Sept.) *Farrow* 114! Kouli (Aug.) *Farrow* 116! **Guin.:** Timbo (Oct.) *Pobéguin* 1702 *bis*! 1706! 1828! Madina Tossekré (Oct.) *Adam* 12549! Dalaba (June) *Adames* 274! **S.L.:** Fwendu *Deighton* 1589! Njala (May, Sept.) *Deighton* 130! *Small* 20! Kabala (July) *Glanville* 238! Kambia (July) *Jordan* 465! **Lib.:** Monrovia (Nov.) *Dinklage* 3270! Timbo (Mar.) *Baldwin* 11226! Ganta (Jan.) *Harley* 1743! Sinkor (June) *Barker* 1369! Buchanan (Nov.) *Adam* 16028! **Iv. C.:** Touba to Odienné (Oct.) *Aké Assi* 8307! **U. Volta:** Ouahigouya (July) *Scholz* 125! **Ghana:** Kpeshi (Apr.) *Adams* 4258! Kpandu to Accra (Sept.) *Ankrah* GC 20199! Wenchi (Sept.) *Rose Innes* GC 30150! Jasikan (July) *Thorold* CB119! Mampong (Oct.) *Baldwin* 13520! **Dah.:** Porto Novo (Dec.) *Froment* 1093! **N. Nig.:** Ilorin (Oct.) *Ward* 85! Naraguta F.R. (July) *Lawlor & Hall* FHI 46599! Panshanu (Aug.) *Lawlor & Hall* 182! Zaria (Aug., Dec.) *Clayton* 1287! *Taylor*! **S. Nig.:** Lagos *MacGregor* 39! *Dalz.* 1427! Benin (Sept.) *Maggs* 6! Opobo *Jeffreys* 5! Sapoba (Aug.) *Keay* FHI 46321! Throughout the tropics.

19. **D. fuscescens** (*Presl*) *Henr.* in Med. Rijks Herb. 61: 8 (1930). *Paspalum fuscescens* Presl, Rel. Haenk. 213 (1830). Stoloniferous annual resembling the previous species.
Lib.: Sinkor (July) *Barker* 1387! **Iv. C.:** Cocody (June) *Adjanohoun* 418*a*! **Ghana:** Accra (Aug.) *Hall* 3304! 3460! Nyakpala (Nov.) *Thorold* 289*a*! **S. Nig.:** Lagos (June, Oct.) *Hall* 3474! *Haines* 176! **W. Cam.:** Mukonji *Maitland* 1175! Throughout tropical Asia, and in Tanzania.

20. **D. exilis** (*Kippist*) *Stapf* in Kew Bull. 1915: 385; F.T.A. 9: 470; Henr., Monogr. Digitaria 237; Portères in J. Agric. Trop. 2: 365 (with key to varieties on p. 498); Chev. Bot. 731, and in Rev. Bot. Appliq. 13: 882; Berhaut, Fl. Sén. ed. 2, 400. *Paspalum exile* Kippist in Clarke in Proc. Linn. Soc. 1: 157 (1842). *Panicum exile* (Kippist) A. Chev., Bot. 726 (1920). Annual crop plant 30–75 cm. high.
Gam.: *Dir. of Agric.*! *Dawe* 50! Mandingo *Pirie* 1! **Mali:** San (Sept.) *Chev.* 2216! 2217! **Port. G.:** Piche (Sept.) *Pereira* 3175! **Guin.:** Labé to Mali (Nov.) *Chev.* 34905! Timbo (Oct.) *Pobéguin* 1702! Kouroussa (July, Aug.) *Pobéguin* 490! 1088! **S.L.:** Mayoso (Aug.) *Thomas* 1485! Makump (Sept.) *Glanville* 398! Roruks (July) *Deighton* 1324! Musaia (July) *Deighton* 4823! Rokupr (Oct.) *Jordan* 564! **U. Volta:** Banfora (Sept.) *Aké Assi* 6358! **Ghana:** Lawra *Agric. Dept.*! **Togo Rep.:** Difalu *Kersting* 193! Badou to Atakpamé (Sept.) *Rose Innes* GC 31358! **N. Nig.:** Nassarawa (Sept.) *Elliot* 105! Plateau Prov. (Sept.) *Lely* P755! Jal, Plateau Prov. (Nov.) *Philcox* 1016! Kwakwi, Plateau Prov. (Nov.) *Philcox* 1006! Zaria (Sept.) *Lamb* 53! (See Useful Plants.)

21. **D. gayana** (*Kunth*) *Stapf ex A. Chev.* Sudania 1: 163 (1911); F.T.A. 9: 449; Henr. l.c. 278; Chev. Bot. 732, and in Rev. Bot. Appliq. 13: 880; Berhaut, Fl. Sén. ed. 2, 398. *Panicum gayanum* Kunth, Rev.

Gram. 1: 239 (1829). *Panicum dimidiostachyum* Steud. (1854). Tufted annual 30–90 cm. high, with a distinctive silky inflorescence; fallow land and disturbed soils.
Maur.: Goberni *Rossetti* 61/313! **Sen.:** *Heudelot* 446! Kouma (Aug.) *Roger* 56! Sangalkam (Sept.) *Berhaut* 2817! **Gam.:** Sankuli Kunda (Sept.) *Pirie* 7/33! **Mali:** Dioura (Aug.) *Davey* 64! *Farrow* 110*a*! Dorey (Aug.) *Rossetti* 59/128! **Port. G.:** Praia de Varela (Aug.) *Esp. Santo* 3072! Piche (Sept.) *Pereira* 3212! **Guin.:** Kindia *Jac.-Fél.* 258! **U. Volta:** Banfora (June) *Leeuwenberg* 4429! Tenkodogo (June) *Scholz* 96! Bobo-Dioulasso (Oct.) *Scholz* 9*a*! **Ghana:** Ejura (June) *Thorold* 80! Bawku to Gambaga (Oct.) *Ankrah* GC 20275! Bawku (Oct.) *Rose Innes* GC 30254! Tamale (Dec.) *Williams* 458! Navrongo (June) *Vigne* FH 4541! **Togo Rep.:** Sokodé to Atakpamé (Oct.) *Rose Innes* GC 31383! Sansanné Mango to Dapango (Oct.) *Rose Innes* GC 31426! **Dah.:** Porto Novo (Dec.) *Froment* 1091! **Niger:** Niamey (Oct.) *Hagerup* 482! Toukounous *Bartha* 61! **N. Nig.:** Nupe *Barter* 718! Stirling *T. Vogel* 163! Zaria (Aug.) *Keay* FHI 28027! Bindawa, Katsina (Aug.) *Grove* 46! Sokoto (Aug.) *Dalz.* 501! **S. Nig.:** Lagos (Mar.) *Dalz.* 1128! *Maitland* 157! Igboora (June) *Haines* 371! Oban (July) *Tuley* 812! Throughout tropical Africa. (See Useful Plants.)

22. **D. debilis** (*Desf.*) *Willd.* Enum. Hort. Berol. 91 (1809); F.T.A. 9: 464; Henr. l.c. 165 (incl. var. *reimarioides* (Anderss.) Henr. l.c. 168); Chev. Bot. 731, and in Rev. Bot. Appliq. 13: 882; Berhaut, Fl. Sén. ed. 2. 400. *Panicum debile* Desf., Fl. Atlant. 1: 59 (1798). Annual, geniculately ascending, up to 60 cm. high.
Sen.: Pont-Diak (July) *Berhaut* 2784! Mbidjem (July) *Raynal* 6028! **Mali:** San (June) *Chev.* 1095! **S.L.:** Makump (May) *Deighton* 1717! **U. Volta:** Tengréla (July) *Aké Assi* 10756! **Ghana:** Akuse (May, Aug.) *Irvine* 3009! *Hall* 3299! Pong Tamale (June) *Addei* 756! Kete Krachi to Yendi (Oct.) *Rose Innes* GC 30952! **N. Nig.:** Abinsi (May) *Dalz.* 879! Riyom (Aug.) *Sawyer* K37! Shaganu (July) *Cook* 381! Katagum *Dalz.* 254! Sokoto (July) *Dalz.* 496! **S. Nig.:** Lagos *Foster* 7*a*! Obudu Plateau (Apr., May) *Tuley* 626! *Haines* 368! **W. Cam.:** Bamenda (Jan., May) *Migeod* 406! *Maitland* 8! Bambui (May) *Pedder* 25! Ndop (May) *Brunt* 431! Batibo (Mar.) *Brunt* 1063! Throughout tropical Africa, extending to the Mediterranean countries and S. Africa. (See Useful Plants.)

23. **D. patagiata** *Henr.* l.c. 525 (1950); Berhaut, Fl. Sén. ed. 2, 400. Creeping; inflorescence a solitary raceme; moist soils.
Sen.: Casamance (Oct.) *Adam* 18259! **Guin.:** Kindia *Jac.-Fél.* 258! 2131! Friguiagbé (July) *Chillou* 1634!

24. **D. gentilis** *Henr.* l.c. 284 (1950); Berhaut, Fl. Sén. ed. 2, 400.
Sen.: Hann (Sept.) *Trochain* 570.

25. **D. leptorhachis** (*Pilger*) *Stapf* in F.T.A. 9: 462 (1919); Henr. l.c. 393; A. Chev. in Rev. Bot. Appliq. 13: 882. *Panicum leptorhachis* Pilger in Engl., Bot. Jahrb. 30: 119 (Mar. 1901). *Digitaria nigritiana* (Hack.) Stapf in F.T.A. 9: 463 (1919); Henr. l.c. 494. *Panicum nigritianum* Hack. in Oest. Bot. Zeitschr. 51: 293 (Aug. 1901). *Digitaria chevalieri* Stapf in F.T.A. 9: 458 (1919), as to the perennial form; Henr. l.c. 119; A. Chev. in Rev. Bot. Appliq. 13: 881; Berhaut, Fl. Sén. ed. 2, 400. *D. richardsonii* Mez in Engl., Bot. Jahrb. 57: 193 (1921); Henr. l.c. 625. Decumbent or ascending annual or short-lived perennial up to about 60 cm. high; racemes often bare at the base, but with traces of arrested spikelets, and sometimes clothed in spreading white hairs.
Sen.: Baila (Sept.) *Adam* 18131! Djembering (Sept.) *Broadbent* 146! 171! **Mali:** Ségou *Lécard* 252! Bamako (Nov.) *Adam* 15351! **Iv. C.:** Bingerville (Dec.) *Chev.* 16003! **U. Volta:** Sindoukoroni (Sept.) *Adam* 15197! **Ghana:** Cape Coast (June) *Hall* 3122! Wenchi to Kumasi (Oct.) *Rose Innes* GC 30523! Mampong (Oct.) *Baldwin* 13516! Kwahu Tafo (Nov.) *Rose Innes* GC 31652! Adidome (Nov.) *Ankrah* GC 20510! **Togo Rep.:** Sokode to Atakpame (Oct.) *Rose Innes* GC 31370! Lomé *Stage* 27! **Dah.:** Cotonou to Allada (Oct.) *Risopoulos* 1214! **N. Nig.:** Lokoja (June) *Richardson*! Naraguta, Jos (July) *Lely* P407! Vodni *Saunders* 23! 49! Abukur, Katsina (Oct.) *Clayton* 1356! **S. Nig.:** Lagos (Mar., June) *Dalz.* 1123! 1426! *Maitland* 165! Brass R. (Jan.) *Barter* 27! Awka (May) *Jones* FHI 6508! Also in the Congo.

26. **D. barbinodis** *Henr.* l.c. 67 (1950).
Mali: *Chev.* 43090! **N. Nig.:** Kwakwi, Plateau (Nov.) *Philcox* 1005 [mixed with *D. exilis*]! Rahoss, Riyom (Nov.) *Philcox* 1013 [mixed with *D. iburua*]!

95. RHYNCHELYTRUM Nees in Lindley, Nat. Syst. ed. 2, 446 (1836); F.T.A. 9: 869 (1930).

Lower glume up to 3 mm. long, separated from the rest of the spikelet by a conspicuous internode nearly 1 mm. long; spikelets 5–12 mm. long; upper glume and lower lemma shortly and obtusely bilobed, awned or awnless, the former conspicuously gibbous, tapering above into a narrow glabrous beak .. .. 1. *villosum*

Lower glume about 1 mm. long, separated from the upper by a short internode up to 0·3 mm. long:

Palea of lower floret ciliate on the keels; upper glume and lower lemma bilobed or emarginate, shortly awned, gibbous and clothed in silky hairs below, narrowing above into a glabrous beak a quarter to a half the length of the spikelet; spikelets 2·5–6 mm. long .. .. 2. *repens*

Palea of lower floret glabrous on the keels; upper glume and lower lemma bilobed and awned, more or less straight on the back and not narrowed into a beak, hairy on keel and margins or margins only, otherwise glabrous; spikelets 4–6 mm. long 3. *bellespicatum*

1. **R. villosum** (*Parl.*) *Chiov.* in Ann. Ist. Bot. Roma 8: 310 (1908). *Monachyron villosum* Parl. in Hook., Fl. Nigrit. 191 (1849). Annual or short-lived perennial, 45–100 cm. high. The size of the spikelets is remarkably variable.
Ghana: Zabzugu (Apr.) *Morton* GC 9065! Kumasi (May) *Darko* 5568! **Niger:** Aïr (*fide* De Miré & Gillet in J. Agric. Trop. 3: 735). Also in the Cape Verde Is., most of tropical Africa outside our area, Madagascar, Arabia and India.

2. **R. repens** (*Willd.*) *C. E. Hubbard* in Kew Bull. 1934: 110. *Saccharum repens* Willd., Sp. Pl. 1: 322 (1798). *Tricholaena repens* (Willd.) Hitchc. in U.S. Dept. Agric. Misc. Pub. No. 243: 331 (1936). *T. sphacelata* Benth. (1849). *T. rosea* Nees (1835)—Chev. Bot. 733. *Rhynchelytrum roseum* (Nees) Stapf & C. E. Hubbard ex Bews (1929)—F.T.A. 9: 880; A. Chev. in Rev. Bot. Appliq. 14: 33. *Melinis argentea* Mez (1921). *M. brachyrhynchus* Mez (1921). Annual 60–90 cm. high; panicles fluffy, silvery white to pink. A weed of old farmland.
S.L.: Freetown (Dec.) *Deighton* 3333! Tower Hill (Feb., Apr.) *Gledhill* 110! 468! **Iv. C.:** Abidjan (Apr.) *Leeuwenberg* 3919! Dihabo to Bouaké (July) *Chev.* 22057! Cocody (June) *Adjanohoun* 421*a*! Téhini (Aug.) *de Wilde* 764! Bouaké (May) *de Wilde* 54! **U. Volta:** Banfora (June) *Leeuwenberg* 4419! Kassandé (July) *Scholz* 37*a*! **Ghana:** Cape St. Paul (Feb.) *Thorold* 5! Legon, Accra (Oct.) *Adams* 3347! Ejura (Aug.) *Andoh* 5058! Tamale (Dec.) *Williams* 451! Navrongo (June) *Vigne* FH 4543! **Togo Rep.:** *Mahoux*! Lomé *Warnecke* 167! Palime *Stage* 72! Atakpame to Sokode (Oct.) *Rose Innes* GC 31377! **Dah.:** Cotonou

Debeaux 157! Cotonou to Allada (Oct.) *Risopoulos* 1212! **N. Nig.:** Nupe *Barter* 756! Lokoja *Richardson* 5! Gindiri, Plateau (Oct.) *Hepper* 1145! Yola (June) *Dalz.* 278! Biu (Sept.) *Kennedy* FHI 8013! **S. Nig.:** Lagos (Mar.) *Dalz.* 1125! *Maitland* 170! Ibadan (Jan.) *Meikle* 943! Nalende, Ibadan (Nov.) *Onochie* FHI 6986! Abeokuta (Apr.) *Baldwin* 11982! **W. Cam.:** Buea (Jan.) *Maitland* 324! *Kalu* FHI 13981! Babessi (Jan.) *Migeod* 285! Bambui (Dec.) *Pedder* 9! Bamenda (Jan.) *Migeod* 313! Throughout tropical Africa, and introduced to most other tropical countries. (See Useful Plants.)

3. **R. bellespicatum** (*Rendle*) *Stapf & C. E. Hubbard* in F.T.A. 9: 900 (1930). *Tricholaena bellespicata* Rendle in Cat. Welw. 2: 196 (1899). About 30 cm. high.
N. Nig.: Jos (July, Oct.) *Blair Rains*! *De Leeuw* 1624*a*! Naraguta F.R. (July) *Lawlor & Hall* FHI 45574! Also in Angola and Malawi.

96. TRICHOLAENA Schrad. in Roem. & Schult., Syst. Veg. 2, Mant. 163 (1824); F.T.A. 9: 908 (1930).

Spikelets 1·7–2·2 mm. long; lower glume a minute hyaline rim; upper lemma glabrescent to pubescent on the margins; panicle loose, up to 15 cm. long *monachne*

T. monachne (*Trin.*) *Stapf & C. E. Hubbard* in F.T.A. 9: 909 (1930); A. Chev. in Rev. Bot. Appliq. 14: 33. *Panicum monachne* Trin. in Spreng., Neue Entd. 2: 86 (1821). *Xylochlaena monachne* (Trin.) Stapf (1922). *Tricholaena delicatula* Stapf & C. E. Hubbard in F.T.A. 9: 911 (1930); F.W.T.A., ed. 1, 2: 569. *T. bicolor* (Schumach.) C. E. Hubbard in F.T.A. 10: 88 (1937). *Aira bicolor* Schumach., Beskr. Guin. Pl. 65 (1827). Decumbent perennial 30–60 cm. high, rooting at the lower nodes; seashore sands, often drenched by salt spray.
Ghana: Saltpond (Oct.) *Baldwin* 13614! Labadi (Mar.) *Deighton* 574! Teshi (Nov.) *Irvine* 803! Tema (June) *Rose Innes* GC 30360! Nungua (June) *Rose Innes* GC 30079! Tropical and S. Africa and Mascarene Islands, mostly on the coast but also from some inland stations.

97. MELINIS P. Beauv., Ess. Agrost. 54 (1812); F.T.A. 9: 916 (1930).

Spikelets deeply grooved between the prominent nerves, glabrous; leaf-blades linear, softly pilose with sticky hairs; panicle narrow, compact; upper lemma usually long awned, but sometimes with the awn short or absent .. 1. *minutiflora*
Pedicels scaberulous, without long hairs; lower glume an ovate scale 0·2–0·3 mm. long .. .. 1a. *minutiflora* var. *minutiflora*
Pedicels bearing a few white hairs towards the apex; lower glume a scarious rim up to 0·1 mm. long .. .. 1b. *minutiflora* var. *setigera*
Spikelets not grooved, finely nerved:
Pedicels without long hairs, scaberulous; leaf-blades linear, sparsely pilose with sticky hairs; panicle rather dense with short pedicels; spikelets usually glabrous; lower glume reduced to an obscure rim; upper glume 7-nerved, distinctly bilobed with denticulate lobes; lower lemma with long conspicuous awn, bifid for ¼ of its length, the lobes acute .. .. 2. *macrochaeta*
Pedicels bearing long white hairs towards the apex; leaf-blades linear-lanceolate, not sticky, glabrous to softly pubescent; spikelets puberulous to sparsely hairy; upper glume truncate or obscurely lobed; lower lemma shortly bilobed, usually awned:
Lower glume reduced to an obscure rim; panicle delicate, open, with capillary branchlets; pedicels smooth between the bristles; upper glume 5-nerved 3. *tenuissima*
Lower glume a tiny ovate scale 0·25 mm. long; panicle moderately dense, the branches ascending; pedicels scaberulous; upper glume 7-nerved .. .. 4. *effusa*

1. **M. minutiflora** *P. Beauv.* Ess. Agrost. 54, t.11, f. 4 (1812); F.T.A. 9: 931; Chev. Bot. 722, and in Rev. Bot. Appliq. 14: 34; including var. *inermis* Rendle, Cat. Welw. 2: 200 (1899), and var. *mutica* Hack. in Sitz. Acad. Wiss. Wien. 89: 126 (1884); Berhaut, Fl. Sén. ed. 2, 419. Perennial, erect or ascending from a prostrate base, the culms up to 2 m. long; leaves covered in sticky hairs and smelling strongly of molasses or linseed oil.

1a. **M. minutiflora** *P. Beauv.* var. **minutiflora**
Guin.: Dalaba (Nov.) *Chev.* 31900! Kindia *Jac.-Fél.* 237! **S.L.:** Pujehun (Dec.) *Deighton* 256! Bintumane (Nov., Jan.) *Glanville* 332! 478! *T. S. Jones* 111! Mt. Loma (Nov.) *Jaeger* 525! **Lib.:** Sanoquelle (Nov.) *Harley* 1615! Nimba (Dec.) *Adam* 20142! *Adames* 822! **Iv. C.:** Dabou (Feb.) *Chev.* 17255! **Ghana:** Kumasi (Nov.) *Vigne* FH 4084! Krobo Hill (Nov.) *Irvine* 1707! Mt. Demi (Dec.) *Howes* 1059! Togo Plateau (Nov.) *St. Clair-Thompson* 1624! *Morton* A3481! **N. Nig.:** Ekan, Ilorin (Oct.) *Ward* 64! Maisamari, Mambila Plateau (Jan.) *Hepper* 1646! **S. Nig.:** Lagos (Dec.) *Dalz.* 1326! Newi, Onitsha (Oct.) *Jones* FHI 6222! Ikwette, Ogoja (Dec.) *Savory & Keay* FHI 25233! Obudu (Apr.) *Haines* 362! **W. Cam.:** Ndop (Dec.) *Boughey* GC 11130! Ndu (Jan.) *Keay & Russell* FHI 28434! Bamenda (Dec., Jan.) *Baldwin* 13840! *Migeod* 409! Tropical Africa and America; introduced into most other tropical regions. (See Useful Plants.)

1b. **M. minutiflora** var. **setigera** *W. D. Clayton* in Kew Bull. 21: 113 (1967).
W. Cam.: Buea (Nov.-Feb.) *Maitland* 102! 102*a*! 148! *Migeod* 223! *Kalu* FHI 13982! Also in E. Cameroun.

2. **M. macrochaeta** *Stapf & C. E. Hubbard* in Kew Bull. 1926: 443; F.T.A. 9: 927; A. Chev. in Rev. Bot. Appliq. 14: 34. Annual, erect or ascending, up to 1·2 m. high, often with stilt roots from the lower nodes; leaves less sticky and odorous than *M. minutiflora*.
Iv. C.: Bouaké (May) *de Wilde* 129! **N. Nig.:** *Lely* P841! Kabba (Oct.) *Parsons* 23! Riyom, Plateau (Nov.) *Clayton* 1408! Jos (Nov.) *Keay* FHI 21428! Naraguta, Plateau (Oct.) *Kennedy* FHI 8044! Vom (Dec.) *Haines* 153! (this specimen has setose pedicels; ? hybrid with *M. tenuissima*). **S. Nig.:** Ago-Are, Shaki (Nov.) *Olorunfemi* FHI 40305! Yala Hill, Oyo Prov. (May) *Hambler* 464! Olokemeji (Nov.) *Clayton* 569! (this specimen has an open panicle and scale-like lower glume; ? hybrid with *M. minutiflora*.) Throughout tropical Africa.

3. **M. tenuissima** *Stapf* in Hook., Ic. Pl. 27: t. 2660 (1900); F.T.A. 9; 926. Perennial, erect or geniculately ascending, up to 1·5 m. long; leaves usually pubescent in West African specimens, but not markedly sticky.
Iv. C.: Béoumi (Dec.) *Adjanohoun* 454*a*! **Ghana:** Kumasi (Dec.) *Adams* 4448! Manhia (Nov.) *Irvine* 2555!

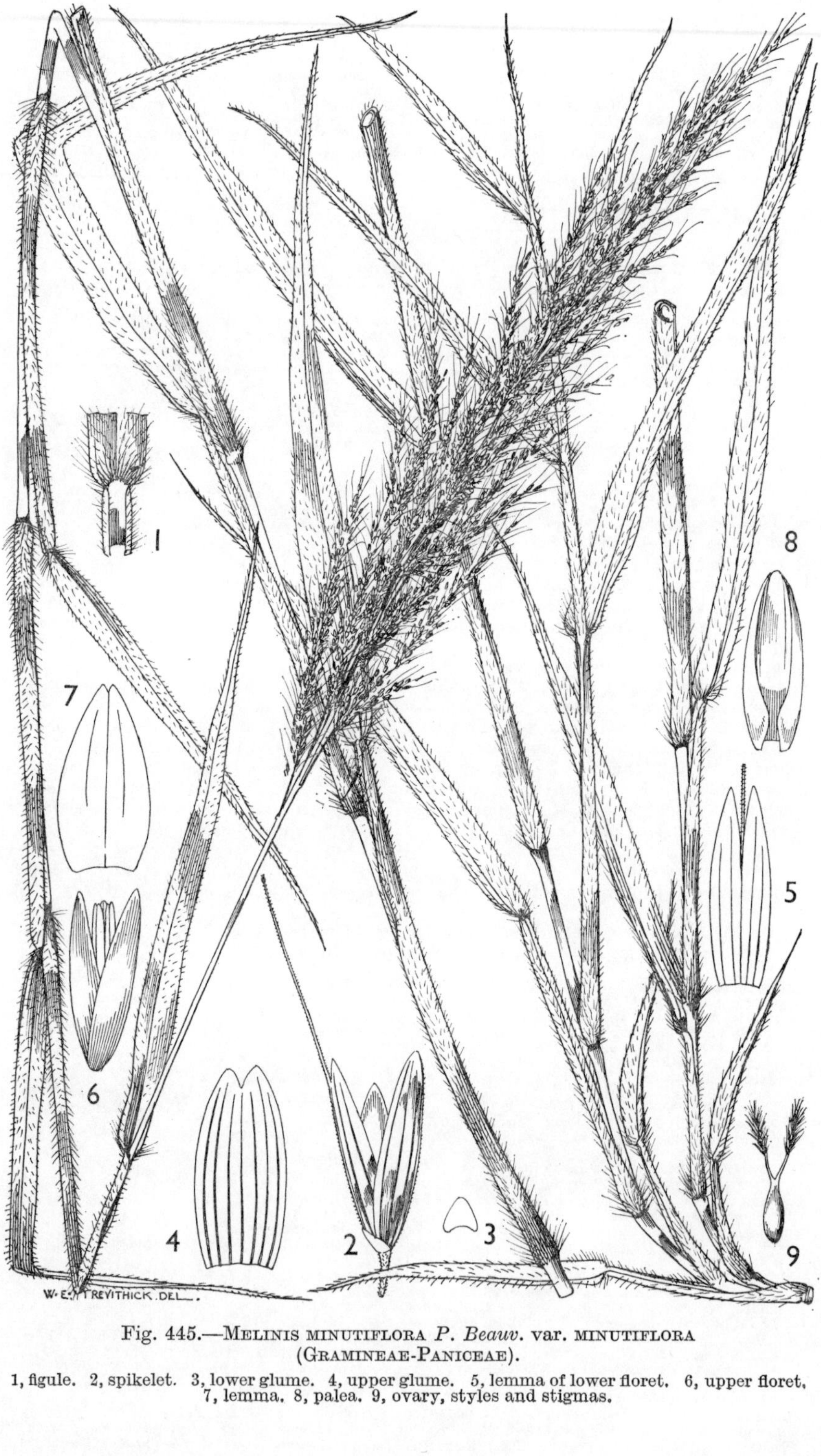

Fig. 445.—MELINIS MINUTIFLORA *P. Beauv.* var. MINUTIFLORA (GRAMINEAE-PANICEAE).

1, ligule. 2, spikelet. 3, lower glume. 4, upper glume. 5, lemma of lower floret. 6, upper floret. 7, lemma. 8, palea. 9, ovary, styles and stigmas.

Shiare (Nov., Dec.) *Morton* A4097! *Jenik & Hall* CC 1110! Ejura (Dec.) *Morton* GC 9600! **N. Nig.:** Tiba Plateau (Dec.) *Peter & Tuley* 69! **S. Nig.:** Ibadan (Mar., Nov.) *Deighton* 546! *Stanfield* FHI 55610! Ibadan to Abeokuta (Mar.) *Schlechter* 12338! **W. Cam.:** Bamenda (Dec.) *Baldwin* 13839! Throughout tropical Africa.

4. **M. effusa** (*Rendle*) *Stapf* in Kew Bull. 1922: 310, and l.c. 1926: 444; F.T.A. 9: 925. *M. minutiflora* var. *effusa* Rendle, Cat. Welw. 2: 200 (1899). Perennial with ascending culms 60–90 cm. high.
Ghana: Tonogo, Togo Plateau (Nov.) *Morton* A3507! **N. Nig.:** Naraguta (Dec.) *Kennedy* 21! 27! Gembu (Dec.) *Daramola* FHI 62472! **W. Cam.:** Ndop (Dec.) *Boughey* GC 11129! Extends south to the Congo, Malawi and Angola; east to Uganda and Kenya.

98. ANTHEPHORA Schreb., Beschr. Gräser 2: 105 (1810); F.T.A. 9: 933 (1930).

Clusters of spikelets broadly conical, the lower glumes united for one third of their length to form a disc-like rimmed base 3–4 mm. diam; spikelets glabrous; spikes 3–7 cm. long .. .. 1. *cristata*
Clusters of spikelets oblong, the lower glumes free almost to the base:
Lower glume scaberulous, narrowly ovate to lanceolate, acuminate, the nerves not anastomosing; lower lemma glabrous or rarely ciliate at the tip; spikelets 4–6 mm. long; spikes dense, cylindric, 8–17 cm. long .. .. .. 2. *ampullacea*
Lower glume pubescent, becoming shortly villous at the base; lower lemma ciliate towards the tip:
Lower glume lanceolate, setaceously acuminate, 4–8 mm. long, the nerves not anastomosing; spikes very dense, 10–20 cm. long, the clusters closely imbricate 3. *hochstetteri*
Lower glume elliptic-oblong, shortly acute, 4–6 mm. long, the nerves anastomosing on the inner side; spikes dense, cylindric, 13–21 cm. long .. 4. *nigritana*

1. **A. cristata** (*Doell*) *Hack. ex De Wild. & Dur.* Reliq. Dewevr. 255 (1901); F.T.A. 9: 935; Chev. Bot. 721, and in Rev. Bot. Appliq. 14: 34. *A. elegans* var. *cristata* Doell in Mart., Fl. Bras. 2, 2: 314 (1877). Annual, erect or ascending, up to 1m. high.
Guin.: Conakry (Aug.) *Jac.-Fél.* 7001! **Iv. C.:** Cocody (May) *Adjanohoun* 419*a*! **Ghana:** Cape Coast (May) *Morton* A3666! Anomabu to Cape Coast (June) *Irvine* 1466! Saltpond (Jan.) *Thorold* C.B. 29! Elmina (May) *Morton* A906! **Dah.:** Gebé (May) *Newton* 14*a*! Cotonou (Mar.) *Chev.* 23486*bis*! **S. Nig.:** Lagos (Oct.) *Dawodu* 17! *Dalz.* 1129! Badagry (Feb.) *Haines* 133! Abeokuta (Feb.) *Burtt* 31! Also in the Congos, Angola and Brazil.
2. **A. ampullacea** *Stapf & C. E. Hubbard* in F.T.A. 9: 939 (1930). *A. acuminata* (Rendle) Robyns ex Stapf & C. E. Hubbard l.c. 940 (1930). *A. elegans* var. *acuminata* Rendle, Cat. Welw. 2: 193 (1899). Perennial up to 1 m. high. Specimens from Guinée have the lower lemma ciliate at the tip.
Guin.: Kindia (Oct.) *Jac.-Fél.* 483! Benna Plateau (Oct.) *Jac.-Fél.* 7175! **N. Nig.:** Aboh *Barter* 463! Ankpa to Oturkpo (June) *Daramola* FHI 38027! Idah (Mar.) *Clayton* 860! **S. Nig.:** Enugu (Sept.) *Onochie* FHI 34116! Udi (May, Nov., Dec.) *Jones & Rosevear* FHI 18852! *Onyeagocha* FHI 7774! *Keay* FHI 22277! Awgu to Oji (Mar.) *Tuley* 613! Also in Congo and Angola.
3. **A. hochstetteri** *Nees* in Flora 27: 249 (1844); F.T.A. 9: 944. Tufted, glaucous perennial up to 1 m. high.
Mali: Kidal (Sept.) *Rossetti* 59/152! Extends eastwards to Kenya, Uganda, Sudan, Ethiopia and Somalia.
4. **A. nigritana** *Stapf & C. E. Hubbard* in F.T.A. 9: 937 (1930). Perennial, up to 75 cm. high.
Niger: Toukounous *Bartha* 46! **N. Nig.:** Nupe *Barter* 1380! Sokoto (July) *Dalz.* 507! Kano (July, Nov.) *Onwudinjoh* FHI 22384*a*! *Taylor* 6! Katagum *Dalz.* 260! Also in the Sudan.

99. PARATHERIA Griseb., Cat. Pl. Cuba. 236 (1866); F.T.A. 9: 1084 (1934).

Nodes bearded; leaf-blades glabrous or pubescent; branches of the inflorescence and bristles scabrid; peduncles 2–5 mm. long; bristles (15—)25–40 mm. long; spikelets (6—)10–12 mm. long, glabrous except for a little tuft of hair surrounding the base; glumes ovate, obtuse, 1–2 mm. long .. .. 1. *prostrata*
Nodes glabrous; leaf-blades glabrous; branches of the inflorescence and bristles quite smooth; peduncles 10–15 mm. long; bristles 15–25 mm. long; spikelets 10–12 mm. long, glabrous; glumes ovate, acute, 2–3 mm. long .. 2. *glaberrima*

1. **P. prostrata** *Griseb.* Cat. Pl. Cuba 236 (1866); F.T.A. 9: 1085; Berhaut, Fl. Sén. ed. 2, 396. Prostrate perennial, growing in water; cleistogamous spikelets, resembling normal spikelets except for the absence of the bristle, occur at the base of the raceme and also in the uppermost leaf-sheaths.
Sen.: Oussouye (Apr.) *Berhaut* 5835! **Guin.:** Soumbalako Maounde (June) *Adames* 280! **S.L.:** Njala (May, July) *Deighton* 704! 737! Mabum (Aug.) *Thomas* 1515! Luti (July) *Jordan* 283! Rosino (July) *Jordan* 43! **U. Volta:** Banfora (July) *Adjanohoun* 192*a*! **S. Nig.:** Yaba (June) *Dalz.* 1423! Also in the Congo, Zambia, SW. Africa, Madagascar and tropical America.
2. **P. glaberrima** *C. E. Hubbard* in Kew Bull. 4: 365 (1949). Prostrate perennial, growing in water; cleistogamous spikelets as in preceding species.
S.L.: Luti (July) *Deighton* 4337!

100. BECKEROPSIS Fig. & De Not. in Mem. Acad. Sci. Torino ser. 2, 14: 365 (1854); F.T.A. 9: 948 (1934); Clayton in Hook., Ic. Pl. 37; t.3643 (1967).

Perennial; leaf-blades firm, up to 60 cm. long and 20 mm. broad, usually with a slender petiole-like base; apex of leaf-sheath usually bearded with a line of hairs on the abaxial side; raceme rhachis and peduncles pubescent, the spikelets overlapping by ¾ their own length or more; spikelets acute, 2·5–3 mm. long; glumes subequal, up to 0·5 long; lower lemma hispidulous, the upper scaberulous .. .. 1. *uniseta*
Annual; leaf-blades thin, flaccid, up to 10(—25) cm. long and 6(—14) mm. broad, the lower narrowed into slender filiform false petioles; apex of leaf-sheath without hairs on the abaxial side; raceme rhachis and peduncles scaberulous, spikelets overlapping

Fig. 446.—BECKEROPSIS LAXIOR *W. D. Clayton* (GRAMINEAE-PANICEAE).

1, habit, × 1. 2, raceme, × 2. 3, spikelet, × 10. 4, lower glume, × 10. 5, upper glume, × 10. 6, lower sterile lemma (inner aspect), × 10. 7, lower lemma (outer aspect), × 10. 8, upper lemma (inner aspect), × 10. 9, upper lemma (outer aspect), × 10. 10, flower, × 10. 11, lodicule, × 10. 12, palea, × 10. 13, grain inside palea, × 10. From *Keay* FHI 22678.

by ½ their own length; spikelets acuminate, mostly 3·5–4 mm. long; glumes subequal, the upper 0·75–1 mm. long; both lemmas spinulose-hispidulous above 2. *laxior*

1. **B. uniseta** (*Nees*) *K. Schum.* in Engl., Pflanzenw. Ost-Afr. B: 52 (1895); F.T.A. 9: 949; A. Chev. in Rev. Bot. Appliq. 14: 34; Berhaut, Fl. Sén. ed. 2, 421. *Gymnothrix uniseta* Nees, Fl. Afr. Austr. 66 (1841). A tall perennial up to 3·5 m. high; in woodland shade.
Sen.: Oussouye (Oct.) *Adam* 18420! **Port. G.:** São Domingo (Jan.) *Pereira* 1491! **Guin.:** Baffing R. (Nov.) *Pobéguin* 1815! **S.L.:** Musaia (Dec.) *Deighton* 4432! Sefadu to Jagbwema (Nov.) *Deighton* 4668! Kamiototo (Nov.) *Glanville* 312! Bintumane Peak, 5,300 ft. (Jan.) *T. S. Jones* 67! *Jaeger* 484! **Lib.:** Yekepa (Dec.) *Adames* 825! **Iv. C.:** Singrobo (Oct.) *Adjanohoun* 356a. Marahoué to Séguéla (Oct.) *Aké Assi* 6654! **U. Volta:** Koumbili (Oct.) *Scholz* 178! Leo to Po (Oct.) *Rose Innes* GC 31479! Samandeni (Oct.) *Kmoch* 163! **Ghana:** Kwahu Tafo (Oct.) *Rose Innes* GC 30573! Togo Plateau F.R. (Nov.) *St. Clair-Thompson* FH 1643! Sekodumasi (Oct.) *Vigne* FH 3403! Damongo to Sawla (Oct.) *Rose Innes* GC 30988! Banda hills *Rose Innes* GC 31611! **Dah.:** Konanda to Ketou (Nov.) *Risopoulos* 1278! **N. Nig.:** Ilorin (Oct.) *Ward* 93! Kaiama (Oct.) *Ward* 28! Okene (Oct.) *Ward* L151! Plateau Prov. (Sept.) *Lely* P760! Anara F.R., Zaria (Oct.) *Keay* FHI 5439! Vogel Peak area (Nov.) *Hepper* 1451! **S. Nig.:** Olokemeji F.R. (Nov.) *Latilo* FHI 36689! Akure to Owo (Oct.) *Maggs* 16! Ila (Oct.) *Thomas* 1934! Ogoja Prov. *Rosevear* 8/30a! Enugu (Nov.) *Jones* FHI 6234! **W. Cam.:** Bambuko F.R. (Nov.) *Argent* 974! Ndop (Dec.) *Boughey* GC 11111! Bambui (July) *Bumpus* Bam 25! Wum (Nov.) *Brunt* 874! Kuk-we (Nov.) *Brunt* 879! Throughout tropical Africa. (See Useful Plants.)

2. **B. laxior** *W. D. Clayton* in Bull. Brit. Mus. Bot. 3: 118 (1963). *B. nubica* of F.T.A. 9: 952 (partly), and of F.W.T.A. ed. 1, 2: 571, not of (Hochst.) Fig. & De Not. Thin-leaved annual up to 80 cm. high; exposed rock domes and inselbergs.
Ghana: Mpraeso Scarp (Nov.) *Morton* A2764! **N. Nig.:** (Sept.) *Lely* P751! **S. Nig.:** Igboora (Oct.) *Haines* 342! Ado Rock (Oct.) *Savory & Keay* FHI 25448! Idanre (Oct.) *Keay* FHI 22678! Okelifi, Ondo (Nov.) *Onochie* FHI 34226! Igbete (Nov.) *Gledhill* 753! **W. Cam.:** Victoria to Bota (Nov.) *Maitland* 82! Buea (Nov.) *Migeod* 107! *Maitland*! Also in E. Cameroun, S. Tomé and Annobon.

101. PENNISETUM L. Rich. in Pers., Syn. Pl. 1: 72 (1805); F.T.A. 9: 954 (1934).

Inflorescence reduced to a cluster of 2–4 spikelets, subsessile and enclosed in the uppermost leaf-sheath; spikelets 10–20 mm. long; leaf-blades up to about 12 cm. long, subobtuse at the tip .. .. 1. *clandestinum*
Inflorescence a false spike of numerous spikelets, conspicuously exserted:
Rhachis angular with sharp-edged decurrent wings below the scars of the fallen involucres, glabrous or at most puberulous (but see *P. pedicellatum*); peduncles deciduous; upper glume as long as the spikelet:
Upper glume and lower lemma acuminate, 6–7 mm. long; upper lemma acuminate, chartaceous, glabrous, not readily separating from the rest of the spikelet; spikelets solitary, sessile; involucral bristles glabrous, in a single whorl, up to 2 cm. long; basal sheaths laterally compressed .. .. 2. *ramosum*
Upper glume and lower lemma more or less 3-lobed, 2·5–6 mm. long; upper lemma obtuse, thinly coriaceous, ciliolate at the apex, readily disarticulating from the rest of the spikelet:
Spikelets in clusters of 1–5 within the involucre, at least one of the spikelets upon a pedicel 1–3 mm. long; bristles densely woolly plumose, forming a fluffy ovate involucre 0·5–1 cm. long; spikelets 4–6 mm. long; rhachis puberulous, very rarely pubescent .. .. 3. *pedicellatum*
Spikelets solitary and sessile within the involucre:
Bristles, or at least the inner ones, plumose in the lower half:
Spikelets 3–5 mm. long; false spike (6–)8–10(–15) mm. broad, excluding the bristles; longest bristle 15–25 mm. long, the other inner bristles usually more than twice as long as the spikelet, rarely less.. .. .. 4. *polystachion*
Spikelets 2·5–3 mm. long; false spike 4–7 mm. broad, excluding the bristles; longest bristle 6–12 mm. long, the other bristles less than twice as long as the spikelet .. .. 5. *subangustum*
Bristles glabrous, or the longer obscurely ciliate:
Longest bristle 11–16 mm. long, the others irregularly 2-whorled and up to twice as long as the spikelet or more; spikelets 3–4 mm. long; false spike 7 mm. broad, excluding the bristles .. .. 6. *atrichum*
Longest bristle 5–6 (–8) mm. long, the others in 1 whorl of 6–8 and subequal to the spikelet or up to 1½ times as long, rarely longer; spikelets 2·5–3 mm. long; false spike 4–6 mm. broad, excluding the bristles .. .. 7. *hordeoides*
Rhachis cylindric or with rounded ribs, glabrous or hairy:
Plants perennial:
Involucres borne upon a peduncle 1–3 mm. long, and falling with it at maturity; bristles plumose, up to about 3 cm. long; spikelets in clusters of 1–3, 4·5–6·5 mm. long; anther-tips smooth; leaf-blades rigid, about 3 mm. broad, usually convolute or folded .. .. 8. *setaceum*
Involucres subsessile, or disarticulating from the peduncle at maturity with the latter remaining upon the rhachis; bristles glabrous (sometimes obscurely ciliate in *P. purpureum*):
Rhachis densely pubescent to loosely pilose; culm pubescent just below spike; bristles unequal:
Spikes 14–30 mm. broad (excl. bristles); leaf-blades up to 40 mm. broad; anther tips with a tiny tuft of hairs; rhachis densely pubescent; persistent peduncles

obscure, involucres subsessile; spikelets 4–7 mm. long; lower floret variable 9. *purpureum*

Spikes 6–9 mm. broad (excl. bristles); leaf-blades up to 10 mm. broad, but usually narrower; anther-tips smooth; rhachis shortly and loosely pilose, at least in the lower part and beneath the spike; persistent peduncles conspicuous, up to 1 mm. long, appressed to the rhachis; spikelets 5 mm. long; lower floret reduced to an empty lemma as long as the spikelet .. .. .. 10. *monostigma*

Rhachis glabrous or hispidulous; culm glabrous below spike; anther tips glabrous:

Spikelets 6–7 mm. long; spikes terminal and axillary; lower floret as long as the spikelet, usually barren; leaf-blades up to 20 mm. broad 11. *trachyphyllum*

Spikelets 3–5 mm. long; spikes terminal only:

Terminal bristle of the involucre not much longer than the rest, up to 8 mm. long; lower floret barren, almost as long as the spikelet .. .. 12. *glaucocladum*

Terminal bristle of the involucre twice as long as the rest, up to 18 mm. long:

Lower floret male, as long as the spikelet .. .. .. 13. *giganteum*

Lower floret barren, and reduced to a lemma ½ as long as the spikelet 14. *glabrum*

Plants annual; rhachis pubescent or pilose; anther tips with a tiny tuft of hairs;

Clusters readily deciduous; grain not exserted from the spikelet; wild grasses or degenerate escapes from cultivation:

Upper lemma glabrous on the margins, narrowly lanceolate, acuminate, thinly coriaceous, readily deciduous; rhachis slender, more or less flexuous; leaf-blades and sheaths mostly hirsute; involucres 5–10 mm. long, sometimes with 1 bristle up to 8 mm. longer than the rest:

Involucral bristles plumose and fluffy .. .. 15. *violaceum*

Involucral bristles merely ciliate .. .. 16. *fallax*

Upper lemma ciliate on the margins, narrowly ovate, acute, chartaceous, tardily deciduous; rhachis stiff, straight; leaf-blades and sheath mostly glabrous:

Involucral bristles plumose .. .. 17. *stenostachyum*

Involucral bristles glabrous .. .. 18. *dalzielii*

Clusters persistent; grain usually exserted from the gaping spikelet; cultivated plants .. .. 19. *americanum*

1. **P. clandestinum** *Hochst. ex Chiov.* in Ann. Ist. Bot. Roma 8: 41 (1903); F.T.A. 9: 1009. A creeping perennial with stout stolons forming a dense mat; the silvery filaments of the stamens are conspicuously exserted from the leaf-sheath when in flower.
S. Nig.: Obudu Plateau (Feb., Apr.) *Tuley* 478! *Haines* 366! **W. Cam.:** Bafut-Ngemba F.R. (Jan.) *Daramola* FHI 40629! Bamenda to Bali (Mar.) *Brunt* 1082! Bambui (June) *Brunt* 1181! Introduced in 1946, and now naturalized; native in the E. African highlands, and widely introduced to upland regions throughout the tropics.

2. **P. ramosum** (*Hochst.*) *Schweinf.* Beitr. Fl. Aethiop. 301 (1867); F.T.A. 9: 976. *Gymnothrix ramosa* Hochst. in Flora 27: 252 (1844). Annual, up to 1·2 m. high; on seasonally swampy heavy soils.
N. Nig.: Ngala (Oct.) *Gwynne* 114! Gumari (Nov.) *Davey* FHI 27111! Logomani (Nov.) *Johnston* W74! Biu (Sept.) *Kennedy* FHI 8001! Kingowa (Oct.) *Golding* 27! Extends east to Eritrea, south to Uganda and Tanzania. (See Useful Plants.)

3. **P. pedicellatum** *Trin.* in Mém. Acad. Sci. Petersb. sér. 6, 3: 184 (1834); F.T.A. 9: 1065; Chev. Bot. 736, and in Rev. Bot. Appliq. 14: 38; Berhaut, Fl. Sén. ed. 2, 395; De Miré & Gillet in J. Agric. Trop. 3: 734; incl. var. *pubirhachis* Berhaut in Mém. Soc. Bot. Fr. 1953–54: 10 (1954), and Fl. Sén. ed. 2, 394 (the variety is distinguished by its pubescent rhachis). *P. implicatum* Steud. (1854). Annual up to about 1 m. high, with fluffy white or purple inflorescences; often dominant upon fallow land in the drier savanna.
Maur.: Néma *Rossetti* 61/249! Nima (Oct.) *Hepper* 3715! **Sen.:** *Farmar* 17! Walo (Aug.) *Roger* 50! Matam (Dec.) *Chev.* 2262*bis*! M'Bambey (Nov.) *Chev.* 33819! Ngazobile (Nov.) *Chev.* 33925! **Gam.:** *Saunders* 117! *Brooks* 12! Yoroberi Kunda (Sept.) *Pirie* 48/33! Yundum *Corbett* 1! 11! **Mali:** Diré *Rogeon* 282! Bamako (June) *Chev.* 25987! San (Sept.) *Chev.* 2264! Ansongo (Sept.) *Hagerup* 376*a*! Dioura (Jan.) *Davey* 215! **Guin.:** Kouroussa (Nov.) *Pobéguin* 1089! Timbi Madina (Jan.) *Langdale-Brown* 2547! **Iv. C.:** Haut Sassandra & Haut Cavally (Nov.) *Portères* 386! Singrobo (Sept.) *Adjanohoun* 393*a*! Béoumi (Aug.) *Aké Assi* 6844! **U. Volta:** Ouahigouya (Sept.) *Aké Assi* 6434! Ouagadougou (Sept.) *Scholz* 41*a*! **Ghana:** Nungua, Accra (Sept.) *Ankrah* GC 20079! Yendi to Chereponi (Oct.) *Rose Innes* GC 30964! Tamale (Dec.) *Williams* 448! Damongo to Tamale (Oct.) *Rose Innes* GC 30212! Navrongo (Sept.) *Vigne* FH 4632! **Togo Rep.:** Anécho *Mahoux* 16! Sansanné Mango to Dapango (Oct.) *Rose Innes* GC 31434! **Dah.:** Boukombe (Nov.) *Risopoulos* 1260! **Niger:** Niamey (Jan.) *Chev.* 43210! Dasso *White* 8! 15! Tessaoura (Feb.) *Chev.* 43687! Toukounous *Bartha* 55! **N. Nig.:** Ilorin (Aug.) *Ward* 13! Nupe *Barter* 1371! Samaru, Zaria (Dec.) *Thatcher* S559! Sokoto (Oct.) *Moiser* 153! Katagum *Dalz.* 282! Gwoza, Dikwa (Dec.) *McClintock* 81! **S. Nig.:** Lagos (Sept.) *W. MacGregor* 1! 51! *Tuley* 869! Ibadan (May) *Latilo* FHI 44091! Also in E. Cameroun, and extending eastwards to Ethiopia, the Sudan and India. (See Useful Plants.)
[This species may hybridize with *P. polystachion;* some specimens have the narrow plumose involucre of the latter but containing a single pedicelled spikelet.]

4. **P. polystachion** (*Linn.*) *Schult.* Syst. Veg. Mant. 2: 146 (1824); F.T.A. 9: 1057; A. Chev. in Rev. Bot. Appliq. 14: 39; Berhaut, Fl. Sén. ed. 2, 395. *Panicum polystachion* Linn., Syst. Nat., ed. 10, 2: 870 (1759). *Pennisetum setosum* (Sw.) L. Rich. in Pers., Syn. Pl. 1: 72 (1805); Chev. Bot. 736, and in Rev. Bot. Appliq. 14: 38. *Cenchrus setosus* Sw., Prodr. Veg. Ind. Occ. 26 (1788). *Panicum cauda-ratti* Schumach. (1827). *Pennisetum cauda-ratti* (Schumach.) Franch. (1895). *P. tenuispiculata* Steud. (1854)—A. Chev. in Rev. Bot. Appliq. 14: 39. *P. elegans* Nees ex Steud. (1854). Annual or perennial, 0·3–1·8 m. high, varying greatly in the stiffness and colouring of the inflorescence; common on fallow land and disturbed soils.
Sen.: Charoi-Kow, Cape Verde *Brunner* 119! Ngazobib (Nov.) *Chev.* 33927! **Gam.:** *Boteler*! Yundum *Corbett* 12! **Mali:** Diafarabé (Jan.) *Davey* 403! Bamako (Sept.) *Adam* 15372! **Port. G.:** Pitche (Dec.) *Pereira & Correia* 2397! Bissau (Dec.) *Pereira & Correia* 2152! **Guin.:** Labé to Mali (Nov.) *Chev.* 34906! Timbo (Oct.) *Pobéguin* 1741! 1762! Mont Loura (Dec.) *Schnell* 2376! Timbi Madina (Jan.) *Langdale-Brown* 2556! **S.L.:** Mandu (Apr.) *Jordan* 224! Newton (Nov.) *Deighton* 1467! Regent (Dec.) *Sc. Elliot* 4047! Samaia (May) *Thomas* 230! Freetown (Mar.) *Dawe* 800! **Lib.:** Harper (June) *Baldwin* 5982!

Monrovia (Aug.) *Baldwin* 9201! Vonjama (Oct.) *Baldwin* 9943! Soplima (Nov.) *Baldwin* 10063! Duport (Sept.) *van Dillewijn* 98! **Iv. C.:** Haut Sassandra & Haut Cavally (May) *Chev.* 711! Dimbokro (Nov.) *Roberty* 12530! Cocody (June) *Adjanohoun* 406*a*! Cosrou (May) *Leeuwenberg* 4251! Yérébodi to Sanlo (Nov.) *Aké Assi* 9270! **U. Volta:** Farako-Ba (Nov.) *Kmoch* 181! Tondoura (Oct.) *Scholz* 80! Banan-kélédaga (Oct.) *Kmoch* 146! Po (Oct.) *Rose Innes* GC 31078! Po to Leo (Oct.) *Rose Innes* GC 31482! **Ghana:** Elmina (Jan.) *Rose Innes* GC 31198! Achimota (Apr.) *Irvine* 2861! Yendi (Oct.) *Rose Innes* GC 30976! Mampong (Oct.) *Rose Innes* GC 31151! Kintampo (Dec.) *Vigne* FH 3199! **Togo Rep.:** Sokode to Bassari (Jan.) *Schroeder* 139! **Dah.:** *Burton*! **Niger:** Niamey (Oct.) *Hagerup* 557! **N. Nig.:** Nupe *Barter* 979! Zaria (Dec.) *Thatcher* S560! Jos (Nov.) *Keay* FHI 21025! Kano (Dec.) *Hagerup* 657! Katagum *Dalz.* 283! Vogel Peak (Nov.) *Hepper* 1227! 1353! **S. Nig.:** Lagos (Oct.) *Dalz.* 1331! Ibadan (Nov.) *Keay* FHI 28137! Iseyin (May) *Denton* 27! Benin (Oct.) *Maggs* 12! Opobo *Jeffreys* 10! **W. Cam.:** Mbaw plain (May) *Brunt* 487! Bamenda (Aug.) *Brunt* 1268! Throughout the tropics. (See Useful Plants.)

It is not at present clear whether the annual and perennial forms merit recognition as distinct [species, if so, then the name *P. polystachion* should be retained for the annuals and *P. setosum* be applied to the perennials.]

5. **P. subangustum** (*Schumach.*) *Stapf & C. E. Hubbard* in Kew Bull. 1933: 271; F.T.A. 9: 1062; Berhaut, Fl. Sén. ed. 2, 395. *Panicum subangustum* Schumach., Beskr. Guin. Pl. 59 (1827). *Pennisetum gracile* Benth. (1849). Annual, up to about 1·2 m. high, sometimes higher; on fallow land and disturbed soils. **Sen.:** *Farmar* 153! Bargny (Nov.) *Berhaut* 2955! Niokolo-Koba (Oct.) *Adam* 15519! **Gam.:** *Boteler*! Yoroberi Kunda (Sept.) *Pirie* 48A/33! Yundum *Corbett* 3! '6! **Port. G.:** Cacine (Jan.) *Pereira & Guerra* 2911! Susana (Dec.) *Raimundo & Guerra* 392! Bissau Is. (Nov.) *Raimundo & Guerra* 62! 102! Safim (Nov.) *Raimundo & Guerra* 7! **Guin.:** Conakry *Rattray*! Baffing R. (Oct.) *Pobéguin* 1743! Timbo (Oct.) *Pobéguin* 1760! Farana (Apr.) *Pitot*! **S.L.:** Mabonto *Thomas* 3505! Newton (Nov.) *Deighton* 1466! Makump (Oct.) *Glanville* 54! Rokon (Oct.) *Jordan* 568! Madagbo (Feb.) *Bakshi* 10! **Lib.:** Monrovia (June, Sept., Nov.) *Baldwin* 5947! *Wrigley & Melville* 702! *Barker* 1468! Mecca (Nov.) *Baldwin* 10423! **Iv. C.:** Abidjan (Dec.) *Boughey* GC 14457! **U. Volta:** Ouagadougou (Sept.) *Scholz* 41*b*! **Ghana:** Otrokpe (Oct.) *Vigne* FH 4014! Tarkwa (Dec.) *Morton* A3613! New Tafo *Lovi* 3960! Bawku to Bolgatanga (Oct.) *Rose Innes* GC 31464! Tumu to Leo (Oct.) *Rose Innes* GC 31488! **Togo Rep.:** Palime *Stage* 64! **N. Nig.:** Jemaa *Clayton* 1421! **S. Nig.:** Lagos (Oct.) *Dalz.* 1333! Ibadan (Nov.) *Russell* FHI 14954! Burutu (Sept.) *Parsons* 5! Orlu (July) *Jones* FHI 6659! Oban *Talbot* 847! **W. Cam.:** Victoria *Maitland* 153! Wum (Nov.) *Brunt* 877! Also occurs in the Congo. (See Useful Plants.)

6. **P. atrichum** *Stapf & C. E. Hubbard* in Kew Bull. 1933: 282; F.T.A. 9: 1061; Berhaut, Fl. Sén. ed. 2, 394. *P. parviflorum* var. *majus* Chev., Bot. 735, name only. Perennial, up to 2 m. high. **Sen.:** *Heudelot* 305! **Gam.:** Sankuli-kunda (Sept.) *Pirie* 35/33! **Mali:** Bamako (Sept.) *Adam* 15016! **Port. G.:** Catio (Jan.) *Raimundo & Guerra* 791! Boruntuma (Dec.) *Pereira & Correia* 2319! **Guin.:** Timbo (Oct.) *Pobéguin* 1761 (partly)! Wellia (Feb.) *Sc. Elliot* 1892! Friguiagbé (Jan.) *Chillou* 3437! Pita (Oct.) *Adam* 12556! **S.L.:** Magbema (Oct.) *Jordan* 802! Makump (July) *Thomas* 918! **Iv. C.:** Abidjan (Aug.) *Monod*! **N. Nig.:** Fakun, Borgu *Barter* 842! Naraguta, Plateau Prov. (Aug.) *Lely* P439! Extends through the Congo to Kenya, Malawi and Zambia.

7. **P. hordeoides** (*Lam.*) *Steud.* Syn. Pl. Glum. 1: 103 (1854); F.T.A. 9: 1063; Berhaut, Fl. Sén. ed. 2, 394. *Panicum hordeoides* Lam., Tab. Encycl. Meth. Bot. 1: 170 (1791). *Pennisetum parviflorum* Trin., Gram. Pan. 65 (1826); Chev. Bot. 735, and in Rev. Bot. Appliq. 14: 39. *P. siguirense* Mimeur in Rev. Bot. Appliq. 30: 314 (1950). Annual, up to 1·2 m. high; disturbed soils. **Sen.:** Niokolo-Koba *Sadio* 5! Ziguinchor (Sept.) *Broadbent* 89! **Gam.:** *Ruxton* 42! Sankuli-kunda (Sept.) *Pirie* 5/33! **Mali:** Ségou (Sept.) *Chev.* 2259! **Guin.:** Timbo (Oct.) *Pobéguin* 1140 (in part)! Nzérékoré (Oct.) *Baldwin* 9688! Dyéké (Oct.) *Baldwin* 9662! Siguiri (Nov.) *Jac.-Fél.* 1414! Dalaba (Oct.) *Adames* 387! **S.L.:** *Barter*! Pujahun *Thomas* 8041! Kenema (Jan.) *Thomas* 7807! Bumbuna (Oct.) *Thomas* 3806! Musaia (Dec.) *Deighton* 4551! **Lib.:** Grand Bassa (July) *T. Vogel* 93! Gbanga (Oct.) *Linder* 1197! Ganta (Dec.) *Harley* 2101! Tappita (Aug.) *Baldwin* 9084! Vonjama (Oct.) *Baldwin* 9942! **Iv. C.:** Fonvonou-Attienkro (Jan.) *Aké Assi* 5928! **Ghana:** Kpeve (Dec.) *Rose Innes* GC 31194! Biakpa (Dec.) *Rose Innes* GC 31190! Ho to Anyirawasi (Dec.) *Blair* GC 31204! Kumasi (Apr., Dec.) *La Danso* 31! *Adams* 4452! **N. Nig.:** *Lely* 782! Kaiama (Oct.) *Ward* 21! Jemaa (Nov.) *Clayton* 1431! Plateau Prov. (Oct.) *Lely* P850! Vom (Oct.) *Gambles* 13! **S. Nig.:** Lagos *Millen* 81! Ikom (Apr.) *Jones* FHI 1555! Calabar *Thollon*! **W. Cam.:** *Unwin* 8645! Sopo (Dec.) *Maitland* 877! Ndop (Dec.) *Boughey* GC 11105! Mbaw plain (May) *Brunt* 471! Mbem (Feb.) *Brunt* 976! **F. Po:** *Mann* 111! Bahia de Venus (Dec.) *Guinea* 228*bis*! Extends south to the Congo; also in India. (See Useful Plants.)

8. **P. setaceum** (*Forsk.*) *Chiov.* in Bull. Soc. Bot. Ital. 1923: 113 (1923); F.T.A. 9: 1013. *Phalaris setacea* Forsk., Fl. Aegypt.-Arab. 17 (1775). *Pennisetum asperifolium* (Desf.) Kunth (1829)—Berhaut, Fl. Sén. ed. 2, 394. *Cenchrus asperifolius* Desf. (1800). Densely tufted perennial with hard glaucous leaf-blades. **Sen.:** *Heudelot* 484 (partly)! Extends from Tunisia to Israel, thence southwards to Tanzania; also grown as an ornamental.

9. **P. purpureum** *Schumach.* Beskr. Guin. Pl. 44 (1827); F.T.A. 9: 1016; Chev. Bot. 736, and in Rev. Bot. Appliq. 14: 36; Berhaut, Fl. Sén. ed. 2, 394. *P. benthamii* Steud. (1854). *P. pallescens* Leeke (1907). A robust perennial up to 8 m. high and 2.5 cm. diam. at the base; commonly occurs near the banks of streams.
Guin.: Baffing R. (Nov.) *Pobéguin* 1825! **S.L.:** Talla Hills *Sc. Elliot* 4936! Kabala (Dec.) *Deighton* 4410! Tumbu (Oct.) *Deighton* 3272! Juring (Dec.) *Deighton* 296! Kamabai (Jan.) *Glanville* 131! **Lib.:** Kakatown *Whyte*! Gondolahun (Nov.) *Baldwin* 10118! Kailahun (Nov.) *Baldwin* 10136! Kitoma (Jan.) *Baldwin* 11033! Nimba (Oct.) *Adames* 718! **Iv. C.:** Abidjan (Dec.) *Leeuwenberg* 2152! Dabou to Sikensi (Jan.) *Aké Assi* 6860! Adiopodoumé (Sept.) *de Wilde* 1000! **Ghana:** Krepi (Jan.) *Johnson* 541! Chirra (Dec.) *Chipp* 781! Ho (Feb.) *Howes* 1113! Abetifi to Pepease (Jan.) *Irvine* 1667! Brufu Edru *Cummins* 12*b*! **Togo Rep.:** *Kersting* A119! **N. Nig.:** Ilorin (Nov.) *Ward* L107! Nupe *Barter* 1390! Abinsi (Dec.) *Dalz.* 888! Gindiri, Plateau Prov. (Oct.) *Hepper* 1112! Mambila Plateau (Feb.) *Hepper* 1852! **S. Nig.:** Lagos (Oct.) *Dalz.* 1335! Ute, Owo (Aug.) *Symington* FHI 5644! Brass *Barter* 1836! Aboh *Barter* 1353! Awka *Thomas* 46! **W. Cam.:** Buea (Nov.) *Migeod* 3! Bamenda (Jan.) *Migeod* 311! **F. Po:** *T. Vogel*! *Mann* 115! Moka (Aug.) *Wrigley* 517! Throughout tropical Africa, and widely introduced elsewhere in the tropics. (See Useful Plants.)

10. **P. monostigma** *Pilger* in Engl., Bot. Jahrb. 30: 120 (1901); F.T.A. 9: 991. *P. jacquesii* Mimeur in Rev Bot. Appliq. 30: 314 (1950). *P. kamerunense* Mez (1917). Densely tufted perennial up to 1 m. high. **S.L.:** Bintumane Peak (Jan., Oct.) *Glanville* 474! *T. S. Jones* 113! Mt. Loma (Oct.) *Jaeger* 303! 7889! *Nichols* 7! **N. Nig.:** Chappal Waddi, Mambila Plateau (Nov.) *Jackson, Magaji & Tuley* 2098! **W. Cam.:** Mann's Spring, Cam. Mt., 7,500 ft. (Mar., Dec.) *Brenan & Onochie* 9535! *Boughey* GC 10600! Cam. Mt., 9,000 ft. (Jan., Nov., Dec.) *Maitland* 1229! *Migeod* 169! *Mildbraed* 10842! **F. Po:** *Boughey* GC 10795! *Mann* 1463! Lago Biao, 5,000 ft. (Sept.) *Wrigley* 672! Pico de Santa Isabel (Mar.) *Guinea* 2781! Also in E. Cameroun.

11. **P. trachyphyllum** *Pilger* in Engl., Bot. Jahrb. 30: 122 (1901); F.T.A. 9: 967. Perennial up to 2 m. high. **W. Cam.:** Buea (Jan.) *Maitland* 341! Bamenda (Oct.) *Agric. officer* Bam/20! Also in eastern Africa, from the mountains of Ethiopia south to Kilimanjaro.

12. **P. glaucocladum** *Stapf & C. E. Hubbard* in Kew Bull. 1933: 276; F.T.A. 9: 984. Stout perennial, 1·2–2·4 m. high; streamsides.
N. Nig.: Vom (Apr.) *Morton* K323! Also in Tanzania, Zambia, Rhodesia, Malawi and Botswana.

13. **P. giganteum** *A. Rich.* Tent. Fl. Abyss. 2: 382 (1851); F.T.A. 9: 979. *P. togoense* Mez in Engl., Bot. Jahrb. 57: 191 (1921); F.T.A. 9: 988. Stout perennial up to 5 m. high; streambanks.

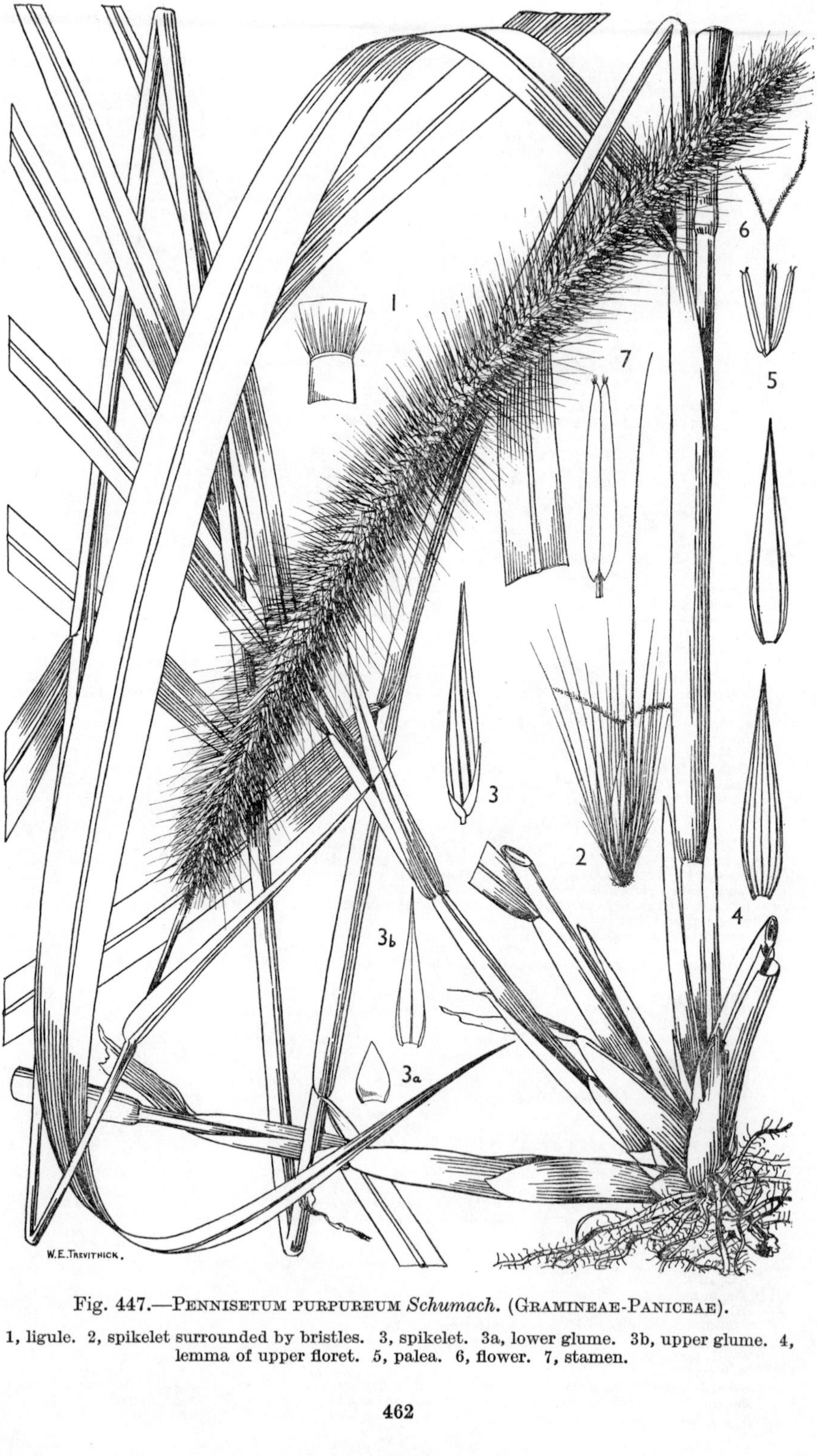

Fig. 447.—PENNISETUM PURPUREUM *Schumach.* (GRAMINEAE-PANICEAE).

1, ligule. 2, spikelet surrounded by bristles. 3, spikelet. 3a, lower glume. 3b, upper glume. 4, lemma of upper floret. 5, palea. 6, flower. 7, stamen.

Guin.: Dabola *Jac.-Fél.* 550 (*fide* Mimeur in Rev. Bot. Appliq. 30: 313). **Togo Rep.:** Sokode to Basari *Kersting* A698! **N. Nig.:** Yankari (Sept.) *Magaji & Tuley* 1798! Kwagiri, Gashaka (Nov.) *Latilo & Daramola* FHI 28762! Nguroge (Nov.) *Magaji & Tuley* 1867! Chappal Waddi, Sardauna Prov. (Nov.) *Jackson, Magaji & Tuley* 2096! Extends eastwards to the Sudan and Kenya.

[*P. giganteum* belongs to a cluster of poorly differentiated species, mainly from eastern Africa, whose taxonomy requires revision. The name is here used in a wide sense].

14. **P. glabrum** *Steud.* Syn. Pl. Glum. 1: 104 (1854); F.T.A. 9: 995. Perennial with slender culms 30–90 cm. high; damp places.
N. Nig.: Chappal Waddi, Sardauna Prov. (Nov.) *Jackson, Magaji & Tuley* 2099! Extends eastwards to the Sudan and southwards to Zambia.

15. **P. violaceum** (*Lam.*) *L. Rich.* in Pers., Syn. Pl. 1: 72 (1805); F.T.A. 9: 1027; Berhaut, Fl. Sén. ed. 2, 394; De Miré & Gillet in J. Agric. Trop. 3: 734; incl. vars. *chudeaui* (Maire) Maire and *monodii* (Maire) Maire in Fl. Afr. Nord 1: 337 (1952). *Panicum violaceum* Lam., Illustr. Gen. 1: 169 (1791). *Pennisetum mollissimum* Hochst. in Flora 27: 253 (1844); F.T.A. 9: 1022; F.W.T.A., ed. 1, 2: 574; Chev. Bot. 735, and in Rev. Bot. Appliq. 14: 35. *P. ramosissimum* Steud., Syn. Pl. Glum. 1: 105 (1854). *P. cognatum* Steud. l.c. 107. *P. subeglume* Trabut in Bull. Soc. Hist. Nat. Nord Afr. 10: 187 (1919). *P. molle* Hitchc. in Proc. Biol. Soc. Wash. 43: 91 (1930). *P. chudeaui* Maire in Bull. Mus. Hist. Nat. Paris sér. 2, 3: 252 (1931); incl. subsp. *monodii* Maire l.c.; A. Chev. in Rev. Bot. Appliq. 14: 35. *P. rogeri* Stapf & C. E. Hubbard in Kew Bull. 1933: 285; F.T.A. 9: 1023; F.W.T.A., ed. 1, 2: 574; A. Chev. in Rev. Bot. Appliq. 14: 37. *P. barteri* Stapf & C. E. Hubbard in Kew Bull. 1933: 287 (1933); F.T.A. 9: 1026; F.W.T.A., ed. 1, 2: 574. *P. darfuricum* Stapf & C. E. Hubbard in Kew Bull. 1933: 283 (1933); F.T.A. 9: 1020; A. Chev. in Rev. Bot. Appliq. 14: 37. Annual, 30–120 cm. high; a very variable species, divisible into a profusion of indistinct subordinate taxa; a weed of roadsides and fallow land.
Maur.: Oujaf *Rossetti* 61/244! **Sen.:** *Adanson*! Dakar (May.) *Rattray*! *Baldwin* 5739! Kaolack (Nov.) *Adam* 19906! Louga (Oct.) *Chev.* 34027! **Mali:** Gourma (July) *Rogeon* 448! Gao (Sept.) *Hagerup* 340! Tin Zaoutene (Sept.) *Rossetti* 59/190! In Tillit (Aug.) *Rossetti* 59/123! Kidal (Sept.) *Rossetti* 59/163! **Niger:** Tahoma to In Gall (Feb.) *Chev.* 43321! **N. Nig.:** Wurge, Chad (Oct.) *Gwynn* 122! Logomani, Bornu (Nov.) *Johnston* N84! Nupe *Barter* 1376! Also in Central African and Sudan Republics. (See Useful Plants.)

16. **P. fallax** (*Fig. & De Not.*) *Stapf & C. E. Hubbard* in Kew Bull. 1933: 270; F.T.A. 9: 1019. *Penicillaria fallax* Fig. & De Not. in Mem. Accad. Sci. Torino ser. 2, 14: 371 (1854). *Pennisetum ochrops* Stapf & C. E. Hubbard in Kew Bull. 1933: 284; F.T.A. 9: 1021; F.W.T.A., ed. 1, 2: 574; A. Chev. in Rev. Bot. Appliq. 14: 38. Annual weed 60–120 cm. high; perhaps not distinct from the preceding.
Maur.: Néma *Rossetti* 61/202! **Sen.:** Galam *Heudelot* 299! M'Bidjem (Aug.) *Thierry* 66! **N. Nig.:** Gushko, Chad (Oct.) *Golding* 22! Also in the Sudan and Eritrea. (See Useful Plants.)

17. **P. stenostachyum** (*Klotzsch ex A. Br.*) *Stapf & C. E. Hubbard* in Kew Bull. 1933: 270; F.T.A. 9: 1031. *Penicillaria stenostachya* Klotzsch ex A. Br. in Ind. Sem. Hort. Berol. 1855 App. 25 (1855). *Pennisetum sclerocladum* Stapf & C. E. Hubbard in Kew Bull. 1933: 288; F.T.A. 9: 1033; F.W.T.A., ed. 1, 2: 574. *P. sampsonii* Stapf & C. E. Hubbard l.c. 289; A. Chev. in Rev. Bot. Appliq. 14: 37. *P. niloticum* Stapf & C. E. Hubbard l.c. 293. Coarse annuals 1–1·5 m. high, usually in proximity to the cultivated species. This taxon is probably no more than an assemblage of worthless segregates from the cultivated plants.
Sen.: Walo *Perrottet* 76! Thiès to St. Louis (Nov.) *Chev.* 2266! St. Louis *Roger*! Kaolack (Nov.) *Adam* 19905! **Gam.:** *Boteler*! Georgetown (Jan.) *Sampson* 56! **N. Nig.:** Ilorin (Dec.) *Sampson* 44! Also in Egypt and Sudan. (See Useful Plants.)

18. **P. dalzielii** *Stapf & C. E. Hubbard* in Kew Bull. 1933: 290; F.T.A. 9: 1030. Very doubtfully distinct from the preceding.
N. Nig.: Sokoto *Dalz.* 511! 512! 513! (See Useful Plants.)

19. **P. americanum** (*Linn.*) *K. Schum.* in Engl., Pflanzenw. Ost-Afr. B: 51 (1895). *Panicum americanum* Linn., Sp. Pl. 56 (1753). *Pennisetum typhoides* (Burm.) Stapf & C. E. Hubbard (1933). *P. pycnostachyum* (Steud.) Stapf & C. E. Hubbard (1933); Berhaut, Fl. Sén. ed, 2, 393. *P. nigritarum* (Schlechtd.) Dur. & Schinz (1894). *P. anchylochaete* Stapf & C. E. Hubbard (1933). *P. cinereum* Stapf & C. E. Hubbard (1933). *P. gibbosum* Stapf & C. E. Hubbard (1933). *P. maiwa* Stapf & C. E. Hubbard (1933). *P. gambiense* Stapf & C. E. Hubbard (1933); Berhaut, Fl. Sén. ed. 2, 393. *P. leonis* Stapf & C. E. Hubbard (1933).

[Many species of "Bulrush Millet" were accepted in F.T.A. and F.W.T.A. ed. 1. It is becoming apparent, however, that these do not form satisfactory units, either morphologically, genetically or agriculturally. The solution lies in uniting them into a single species, within which cultivars may be recognised according to the provisions of the International Code of Nomenclature for Cultivated Plants.—W.D.C.].

Additional species.

Pennisetum divisum (*Forsk. ex Gmel.*) *Henr.* in Blumea 2: 162 (1938). *Panicum divisum* Forsk. ex Gmel., Syst. Nat. 2: 156 (1791). *Pennisetum dichotomum* (Forsk.) Del. (1813). Rhachis glabrous, inconspicuously angular; spikelets 7–8 mm. long, the lower glume half as long. A glabrous bushy perennial with stout woody stems, from N. Africa, Arabia and India.
Recorded at Aïr, Niger, by De Miré & Gillet in J. Agric. Trop. 3: 734 (1956).

P. villosum *Fresen.* in Mus. Senck. 2: 134 (1837). *P. longistylum* Hochst. ex A. Rich. (1851); Berhaut, Fl. Sén. ed. 2, 393. Perennial, with handsome plumose involucres up to 7 cm. long. An Ethiopian species sometimes grown as an ornamental.

Imperfectly known species.

P. senegalense *Steud. ex Jard.* Herbor. Côte Occid. Afr. 10 (1851), name only. Probably one of the wild annual species growing near Dakar.

102. CENCHRUS Linn., Sp. Pl. 1049 (1753), and Gen. Pl. ed. 5, 470 (1754); De Lisle in Iowa State J. Sci. 37: 259 (1963); F.T.A. 9: 1070 (1934).

Bristles of the involucre retrorsely barbellate and adhering to clothing, rigid, prickly, 4–8 mm. long; outer bristles present:
- Bristles connate at the base into a shallow elliptic disc up to about 3 mm. across, but free above the rim of the disc .. .. 1. *biflorus*
- Bristles connate for about half their length to form a deep cup .. 2. *echinatus*

Bristles of the involucre antrorsely barbellate, not prickly:
- Inner bristles flattened, rigid, 2·5–5 mm. long, ciliolate on the inside, with the margins fused for $\frac{1}{3}$ of their length above the rim of the basal disc; outer bristles suppressed or reduced to minute points .. .. 3. *setigerus*

Inner bristles slender, 9 mm. long or more, free above the basal disc; outer bristles numerous:
Involucral bristles connate for a short distance from the base to form a shallow elliptic disc up to 3 mm. across; inner bristles stiff, subequal, 15–20 mm. long, scantily ciliate in the lower half .. .. 4. *prieuri*
Involucral bristles connate only at the very base to form a disc up to 1 mm. diam.; inner bristles flexible, unequal, one longer than the others, 9–12 mm. long, densely ciliate in the lower half .. .. 5. *ciliaris*

1. **C. biflorus** *Roxb.* Fl. Ind. 1: 238 (1820); Berhaut, Fl. Sén. ed. 2, 393. *C. barbatus* Schumach., Beskr. Guin. Pl. 43 (1827); F.T.A. 9: 1079. *C. catharticus* Del. (1839)—Chev. Bot. 734, and in Rev. Bot. Appliq. 14: 39; De Miré & Gillet in J. Agric. Trop. 3: 728. *C. leptacanthus* A. Camus in Bull. Soc. Bot. Fr. 80: 774 (1934). Annual, geniculately ascending, about 60 cm. high; the burrs vary somewhat in size and spinescence, and in a few cases the acicular outer bristles are suppressed simulating *C. setigerus*; old farmland and waste places.
Maur.: Bafréchié (Sept.) *Sadio* 39! 234! Goberni *Rossetti* 61/317! **Sen.:** (July) *Roger*! Sar I. *Brunner* 29! Dakar (May) *Baldwin* 5728! Kouma (Aug.) *Roger* 63! Djembering (Sept.) *Broadbent* 125! **Gam.:** *Saunders* 31! 112! *Ruxton* 159! Yundum *Frith* 132! Genieri (July) *Fox* 125! **Mali:** Timbuktu (July) *Chev.* 1229! *Hagerup* 146! Divé (Aug.) *Lean* 63! Sarédina (Mar.) *Davey* 51! Dioura (Jan.) *Davey* 242! **Port. G.:** Buruntuma (July) *Esp. Santo* 2714! Piche to Orebode (June) *Esp. Santo* 3025! **U. Volta:** Nadjogo to Nadiokan (June) *Scholz* 4*b*! **Ghana:** Accra *Brown* 427! Ada (Jan.) *De Wit & Morton* A2991! Damongo (Mar.) *Adams* 3973! Tamale (July) *Vigne* FH 4549! Yeji (Apr.) *Williams* 226! **Dah.:** Cotonou (Apr.) *Debeaux*! Kraké (Oct.) *Risopoulos* 1230! **Niger:** Zinder (Nov.) *Money-Kyrle* Fe19! Dasso *White* 11! D'in-Abbangarit (Nov.) *Money-Kyrle* Sah39! Toukounous *Bartha* 85! 87! **N. Nig.:** Jebba (Dec.) *Meikle* 684! Ankpa (June) *Daramola* FHI 38028! Sokoto (Oct.) *Moiser* 138! Kano (July) *Onwudinjoh* FHI 22383! Mongonu (Oct.) *Golding* 13! **S. Nig.:** Lagos (Sept., Nov., Jan.) *Dalz.* 1130! *Maitland* 159! *Stubbings* 73! *Bels* 45! *Birket-Smith* 83! Extending south to Zambia and Rhodesia and east to Eritrea and India. (See Useful Plants.)
2. **C. echinatus** *Linn.* Sp. Pl. 1050 (1753); F.T.A. 9: 1082. Annual 30–90 cm. high.
Ghana: Cape Coast (May) *Hall* 3039! Accra (Oct.) *Morton*! Achimota (Apr.) *Ankrah* GC 20129! Accra to Aburi (June) *Oduro* 11! **S. Nig.:** Lagos (Oct.) *Haines* 183! A native of tropical America, occurring as an adventive in most tropical countries.
3. **C. setigerus** *Vahl* Enum. Pl. 2: 395 (1806); F.T.A. 9: 1077; Berhaut, Fl. Sén. ed. 2, 393; De Miré & Gillet in J. Agric. Trop. 3: 729. Perennial.
Ghana: Legon (Mar.) *Morton* A4156! Aiyinase (July) *Irvine* 5060! Also in Sudan, Ethiopia, Kenya, Arabia and India, and introduced elsewhere as a fodder plant.
4. **C. prieuri** (*Kunth*) *Maire* in Bull. Mus. Hist. Nat. Paris, ser. 2, 3: 523 (1931); F.T.A. 9: 1071; Berhaut, Fl. Sén. ed. 2, 395; De Miré & Gillet in J. Agric. Trop. 3: 729. *Pennisetum prieuri* Kunth, Rev. Gram. 2: 411 (1831). *P. breviflorum* Steud. (1854). *Cenchrus macrostachys* Hochst. ex Steud. (1854)—A. Chev. in Rev. Bot. Appliq. 14: 40. Annual, culms ascending, up to about 75 cm. high.
Maur.: Bafréchié (Sept.) *Sadio* 235! **Sen.:** *Adanson*! *Roger*! *Leprieur*! **Mali:** Timbuktu (Aug.) *Hagerup* 277! *Lean* 67! Dioura (Sept.) *Davey* 69! Gao (Aug.) *de Wailly* 4725! Kidal (Sept.) *Rossetti* 59/168! **Niger:** Aïr *Trabut*! Zinder (Nov.) *Hagerup* 611*a*! **N. Nig.:** Kafinsoli (Sept.) *Butler* S107! Extending eastwards to the Sudan and into India. (See Useful Plants.)
5. **C. ciliaris** *Linn.* Mant. Alt. 302 (1771); F.T.A. 9: 1072; Berhaut, Fl. Sén. ed. 2, 394; De Miré & Gillet in J. Agric. Trop. 3: 729. *Pennisetum ciliare* (Linn.) Link (1827); A. Chev. in Rev. Bot. Appliq. 14: 36. *P. cenchroides* Rich. (1805). Perennial, culms ascending, up to about 1 m. high; indigenous in the drier areas, but probably introduced in the south where it occurs on rubbish tips, roadsides, etc.
Maur.: Bafréchié (Sept.) *Sadio* 250! Oualata *Rossetti* 61/140! **Sen.:** *Leprieur*! *Richard*! Walo (Aug.) *Roger* 61! Richard-Tol (Apr.) *Chev.* 25779! St. Louis (Feb.) *Berhaut* 689! **Mali:** Sompi (Aug.) *Chev.* 2166! Bas el Ma (Aug.) *Chev.* 2165! Sarédina (Feb.) *Davey* 19! Bamba (Sept.) *Hagerup* 322! Tabankort to Kidal *Leclercq* 42687! **Ghana:** Cape Coast (May) *Morton* A3673! Accra (June, Oct.) *Thorold* 95! *Baldwin* 13441! Ada (Feb.) *Adams* 5193! Kpong (Jan.) *Gyadu* GC 31212! **Niger:** Aïr (Feb.) *Chev.* 43388! Toukounous *Bartha* 86! **N. Nig.:** Baga, Chad (Jan.) *Daggash* FHI 24963! Baga Seyoram, Kukuwa (Feb.) *Davey* FHI 27162! Throughout tropical Africa, extending into the Middle East and India; introduced to many other countries as a fodder grass. (See Useful Plants.)

103. IMPERATA Cyr., Pl. Rar. Neap. 2: 26 (1792); F.T.A. 9: 87 (1917).

Culms up to 1·25 m. high, from extensively creeping rhizomes; leaf-blades usually erect, sword-like; inflorescence cylindrical, spiciform, silky, with long white hairs from glumes, pedicels and branches; spikelets paired, both pedicelled, the pedicels unequal; glumes membranous; stamens 2 .. .. *cylindrica*
Leaf-blades narrowly linear, 3–6 mm. broad, involute or convolute, rigid, glaucous; panicles very dense and tightly cylindrical .. a. *cylindrica* var. *cylindrica*
Leaf-blades linear to lanceolate-linear, 4–18 mm. broad, flat, firm; panicles rather less dense .. .. b. *cylindrica* var. *africana*

I. cylindrica (*Linn.*) *P. Beauv.* Ess. Agrost. 165, 166 (1812); F.T.A. 9: 87; A. Chev. in Rev. Bot. Appliq. 13: 855; Berhaut, Fl. Sén. ed. 2, 396. *Lagurus cylindricus* Linn., Syst. Nat. ed. 10, 878 (1759). *Saccharum cylindricum* (Linn.) Lam. (1785).
a. **I. cylindrica** (*Linn.*) *P. Beauv.* var. **cylindrica.** *I. arundinacea* Cyr., Pl. Rar. Neap. 2: 26 (1792). *I. arundinacea* var. *europaea* Anderss. in Öfvers. K. Vet. Akad. Frh. Stock. 1855: 159 (1856). *I. cylindrica* var. *europaea* (Anderss.) Asch. & Graeb., Syn. Mittleleurop. Fl. 2: 37 (1898). Specimens from West Africa and the Sahara tend to have smaller spikelets (4–4·5 mm. long) than those from the Mediterranean area, and have been separated as *I. cylindrica* var. *parviflora* Batt. & Trab. in Bull. Soc. Bot. Fr. 53: xxxii (1907), and as *I. robustior* A. Chev. in Rev. Bot. Appliq. 29: 114 (1949).
N. Nig.: Kworre, Sokoto (Sept.) *Palmer* 27! Baga Seyoram, Kukuwa Dist. (Feb.) *Davey* FHI 27174! Kalkala, L. Chad (May) *Golding* 80! Extending to Hoggar, shores of the Mediterranean, and eastwards through Iraq, Iran and Afghanistan to southern U.S.S.R. (See Useful Plants.)
b. **I. cylindrica** var. **africana** (*Anderss.*) *C. E. Hubbard* in Imp. Agric. Bur., Joint Pub. 7: 10 (1944). *I. arundinacea* var. *africana* Anderss. l.c. (1856). *Saccharum thunbergii* Retz., Obs. Bot. 5: 17 (1789). *Imperata thunbergii* (Retz.) Nees, Fl. Afr. Austr. 89 (1841). *I. arundinacea* var. *thunbergii* (Retz.) Hack. in Bol. Soc. Brot. 6: 142 (1888); Chev. Bot. 712. *I. cylindrica* var. *thunbergii* (Retz.) Dur. & Schinz., Consp. Fl. Afr. 5: 693 (1895); Chev. Bot. 713. Usually distinguished fairly readily from var. *cylindrica* by the broader, flat leaf-blades, but in the northern part of our area the two varieties tend to merge and cannot easily be separated.

Fig. 448.—IMPERATA CYLINDRICA (*Linn.*) *P. Beauv.* var AFRICANA (*Anderss.*) *C. E. Hubbard* (GRAMINEAE-ANDROPOGONEAE).

A, ligule. B, portion of flowering branch. C, spikelet enveloped by long hairs. C_1, spikelet. D, internode.

Sen.: Dakar (May) *Baldwin* 5723! Joual to Pordutal (Feb.) *Dollinger*! **Gam.:** *Hayes* 563! **Mali:** Dio (Jan.) *Chev.* 166! Koulikoro Dist. (Nov.) *Chev.* 329! Toguèré de Sabouniaka (Feb.) *Davey* 17! **Guin.:** Kouroussa (Jan.) *Pobéguin* 633! Bambaya to Kamaro (Jan.) *Jaeger* 3892! **S.L.:** Freetown (May) *Baldwin* 5793! Njala (June) *Deighton* 2771! Mabum (Aug.) *Thomas* 1649! Bumban (Sept.) *Thomas* 1961! Kora R., Scarcies (Feb.) *Sc. Elliot* 4590! **Lib.:** Duport (Nov.) *Linder* 1445! Palilah, Gbanga Dist. (Dec.) *Baldwin* 11039! **Iv. C.:** (Jan.) *Hédin*! Boundoukou to Abengourou (Mar.) *de Wilde* 3512! Béréby (Nov.) *Oldeman* 499! Bigulaye (Jan.) *Oldeman* 881! **U. Volta:** Bobo-Dioulasso (Aug.) *Scholz* 94! **Ghana:** Kpong (Jan.) *Gyadu* GC 31205! Aburi Hills (Nov.) *Johnson* 291! Kumasi (Mar.) *La Danso* 25! Kpedsu (Dec.) *Howes* 1064! Ho (Mar.) *Williams* 60! **Togo Rep.:** Palime (Mar.) *Stage* 4! **Dah.:** *Poisson*! **N. Nig.:** Nupe *Barter* 1393! Fadau Karshi, Wamba Div. (Oct.) *Kennedy* FHI 8043! Vodni, Pankshin Div. *Saunders* 43! Kano (Dec.) *Hagerup* 662! Yola (Sept.) *Dalz.* 307! **S. Nig.:** Lagos (Sept.) *Stubbings* 1! Ibadan (Feb.) *Richards* 5079! Opobo (Oct.) *Jeffreys* 33! Port Harcourt (Dec.) *Maitland*! Okuni-Inkum (Jan.) *Holland* 279! **W. Cam.:** Buea (Jan.) *Maitland* 325! Fonfuka, Bamenda Dist. (June) *Maitland* 1771! Bamenda (Jan.) *Migeod* 346! Jakiri to Mbawver (June) *Brunt* 679! Bambili (Feb.) *Bauer* 24! Throughout tropical and S. Africa and in Madagascar.

[A third variety, var. *major* (Nees) C. E. Hubbard (1940), extends from eastern tropical Africa, through tropical Asia, to Australia. It is much more aggressively rhizomatous than our varieties, and constitutes a most serious pest of agricultural land.]

104. SACCHARUM Linn., Sp. Pl. 54 (1753), and Gen. Pl. ed. 5: 28 (1754); F.T.A. 9: 94 (1917).

Culm hairy below the panicle, the latter narrowly oblong, up to 45 cm. long; spikelets 4–6 mm. long; callus hairs very fine and white silky, about twice as long as the spikelets; leaf-blades 5–10 mm. broad; ligule crescentic
1. *spontaneum* var. *aegyptiacum*

Culm glabrous below the panicle, the latter broadly pyramidal, up to 1 m. long or more; spikelets up to 4 mm. long; callus hairs shorter, scantier and off-white; leaf-blades up to 4 cm. broad .. .. 2. *officinarum*

1. **S. spontaneum** *Linn.* Mant. Alt. 183 (1771); Panje in Ind. J. Agric. Sci. 3: 1013 (1933); Artschwager, U.S. Dept. Agric. Tech. Bull. 811 (1942).

1a. **S. spontaneum var. aegyptiacum** (*Willd.*) *Hack.* in DC., Monogr. Phan. 6: 115 (1889); F.T.A. 9: 95. *S. aegyptiacum* Willd., Enum. Hort. Berol. 1: 82 (1809). *S. biflorum* Forsk. (1775)—A. Chev. in Rev. Bot. Appliq. 13: 856. *S. punctatum* Schumach. (1827). *S. palisotii* Tausch. (1836). Perennial with culms up to 3·5 m. high; sandy river banks.
U. Volta: Tenkodogo (Nov.) *Scholz* 188! **Ghana:** Gambaga to Bawku (Oct.) *Rose Innes* GC 30274! Bawku to Nakpanduri (Oct., Nov.) *Ankrah* GC 20499! *Rose Innes* GC 32507! **N. Nig.:** Ilorin (Sept.) *Ward* L130! Wuya Ferry, Kaduna R. (Dec.) *Clayton* 408! Dogon Kurmi, Jemaa (Sept.) *Killick* 73! Naraguta, Plateau Prov. (Aug.) *Lely* P 443! Katagum Dist. *Dalz.* 277! Gurum, Vogel Peak (Nov.) *Hepper* 1326! **S. Nig.:** Nun R. (Sept.) *T. Vogel* 33! *Mann* 536! Asaba (Sept.) *Tuley* 865! Throughout tropical Africa, extending to Egypt and the N. African coast, and through Arabia to the Middle East and Afghanistan. (See Useful Plants.)
[The species is found throughout the Old World tropics. It is polymorphic, and its many forms are commonly grouped into 2 varieties. Var. *spontaneum* differs in having a deltoid ligule, and the leaf-blade narrowed to the midrib at its insertion on the sheath.]

2. **S. officinarum** *Linn.* Sp. Pl. 54 (1753); F.T.A. 9: 96; A. Chev. in Rev. Bot. Appliq. 13: 856; Berhaut, Fl. Sén. ed. 2, 419; Artschwager & Brandes in U.S. Dept. Agric., Handbook 122 (1958). Culms up to 6 m. high. Cultivated sugar cane. Grown, particularly as a chewing cane, on moist soils throughout our area. (See Useful Plants.)

105. ERIOCHRYSIS P. Beauv., Ess. Agrost. 8, t. 4, f. 11 (1812); F.T.A. 9: 91 (1917).

Callus hairs ⅓ as long as the sessile spikelet, rarely up to ⅔ as long, ferruginous; sessile spikelet bisexual, awnless; pedicelled similar but female and a little smaller; lower glume coriaceous, shining .. .. 1. *brachypogon*

Callus hairs exceeding the sessile spikelet, yellowish tawny; otherwise similar to above .. .. 2. *pallida*

1. **E. brachypogon** (*Stapf*) *Stapf* F.T.A. 9: 93 (1917); A. Chev. in Rev. Bot. Appliq. 13: 856; Berhaut, Fl. Sén. ed. 2, 401. *Saccharum brachypogon* Stapf in Mém. Soc. Bot. Fr. 2, 8: 97 (1908); Chev. Bot. 713. Perennial, about 1·2 m. high; swampy soils.
Sen.: Niokolo-Koba *Adam* 17504. **Ghana:** Sampa (Dec.) *Morton* A2629! **N. Nig.:** Nupe *Barter* 1351! Gembu, Mambila (Aug.) *De Leeuw* 1790! Through C. African Rep., Congo, Uganda and Tanzania to Rhodesia.

2. **E. pallida** *Munro* in Harv., Gen. S. Afr. Pl. ed. 2: 440 (1868); F.T.A. 9: 93; A. Chev. in Rev. Bot. Appliq. 13: 856. *Saccharum pallidum* (Munro) Benth. (1881). *S. munroanum* Hack. (1889). Perennial, about 1·2 m. high; swampy soils.
Guin.: Timbi (Dec.) *Jac.-Fél.* 7420! **N. Nig.:** Mundar, Jos (Sept.) *Lawlor & Hall* 638! Maisamari (Aug.) *De Leeuw* 1752! Occurs in C. African Rep., Ethiopia, Tanzania, Zambia, Rhodesia, Malawi and Angola.

106. SORGHUM Moench, Meth. 207 (1794); F.T.A. 9: 104 (1917); Snowden in J. Linn. Soc. 55: 191 (1955); *nom. cons.*

Nodes bearded; sessile spikelets black, densely clothed in tawny or purplish hairs
1. *purpureo-sericeum*

Nodes pubescent:
  Perennials with well-developed creeping rhizomes; spikelets 4·5–5 mm. long, obtuse, deciduous when mature; lower glume distinctly 3-toothed; leaf-blades 0·5–2 cm. broad .. .. 2. *halepense*
  Annuals or tufted perennials without rhizomes:
    Mature sessile spikelets persistent; grains large, often exceeding the glumes in size, usually much exposed; cultivated .. .. 3. *bicolor*

Mature sessile spikelets deciduous; grains enclosed by the glumes; wild:
Sessile spikelets lanceolate; awns 5–16 mm. long; leaf-blades 0·5–2 cm. broad:
Panicle ovate-oblong to lanceolate, 20–40 cm. long and up to 15 cm. broad, the branches numerous and soon spreading; sessile spikelets 6–8 mm. long; awn rather stout; grain 2–2·5 mm. broad .. .. 4. *lanceolatum*
Panicle long and narrow, 15–60 cm. long and 1–5 cm. broad, the branches scanty and suberect; sessile spikelets 6·5–7 mm. long; awn slender; grain 1·5–2 mm. broad .. .. 5. *virgatum*
Sessile spikelets narrowly ovate to ovate or elliptic:
Sessile spikelets 8–9 mm. long, 2·5–3·5 mm. broad; acuminate, awnless or more often with an awn 8–16 mm. long; panicle large, loose, 15–45 cm. long, 5–25 cm. broad; leaf-blades 1–7 cm. broad .. .. 6. *vogelianum*
Sessile spikelets 6–8 mm. long:
Upper lemma of sessile spikelet with an awn 15–30 mm. long; sessile spikelets 2·5–3·5 mm. broad, usually densely and persistently tomentose with mostly whitish silky hairs; panicle very narrow, 10–40 cm. long and 3–10 cm. broad, with suberect branches; leaf-blades 0·5–3 cm. broad .. 7. *aethiopicum*
Upper lemma of sessile spikelet awnless or with an awn 5–10 mm. long; sessile spikelets 2–2·5 mm. broad; panicle large, loose and much-branched, 20–60 cm. long and 10–25 cm. broad; leaf-blades mostly 3–6 cm. broad 8. *arundinaceum*

1. **S. purpureo-sericeum** (*Hochst. ex A. Rich.*) *Aschers. & Schweinf.* Beitr. Fl. Aethiop. 302, 310 (1867); F.T.A. 9: 139. *Andropogon purpureo-sericeus* Hochst. ex A. Rich., Tent. Fl. Abyss. 2: 469 (1851). Annual up to 1 m. high; clay soils.
N. Nig.: Biu (Oct.) *Oche* MRK126! Tallase, Bauchi Prov. *Magaji* Mg192! Eastwards to Eritrea and southwards to Tanzania.

2. **S. halepense** (*Linn.*) *Pers.* Syn. 1: 101 (1805); Snowden in J. Linn. Soc. 55: 191. *Holcus halepensis* Linn., Sp. Pl. 1047 (1753). *Andropogon halepensis* (Linn.) Brot. (1804). 0·3–1·5 m. high; old farmland, roadsides and waste places.
Sen.: Dakar (Apr.) *Clayton* 4005! **Dah.:** Cotonou (Mar.) *Aké Assi* 11137! Mediterranean region, but now found in most of the warmer parts of the world.

3. **S. bicolor** (*Linn.*) *Moench.* Méth. 207 (1794). *Holcus bicolor* Linn., Mant. Alt. 301 (1771). Cultivated throughout tropical Africa, with notable centres of diversification in W. Africa and the Sudan. Grown to a lesser extent elsewhere in the Old World tropics and subtropics, and introduced to the New World.
[Snowden (The cultivated races of Sorghum, 1936) divides the cultivated Sorghums into 31 species and numerous lesser taxa. However, there is mounting evidence that these species are almost completely interfertile, and should be regarded as the cultivars of a single species. I have adopted that position here, and those requiring a more elaborate classification are referred to Snowden's book. Curtis (in Exp. Agric. 3: 275–286 (1967)) has classified the cultivated Sorghums of Nigeria into agronomic races.
There is a small group of species (or introgression products?) which share the properties of both wild and cultivated plants; they have more or less tough racemes and tardily deciduous sessile spikelets, but occur spontaneously. The following have been introduced to, or recorded in, West Africa: *S. hewisonii* (Piper) Longley, *S. niloticum* (Stapf ex Piper) Snowden, and *S. sudanense* (Piper) Stapf. For further details see Snowden in J. Linn. Soc. Bot. 55: 191 (1955).—W.D.C.]

4. **S. lanceolatum** *Stapf* in F.T.A. 9: 112 (1917); Snowden in J. Linn. Soc. 55: 216; A. Chev. in Rev. Bot. Appliq. 13: 857. Including *S. virgatum* of F.W.T.A. ed. 1, 2: 580, not of (Hack.) Stapf. Annual or short-lived perennial up to 2·4 m. high.
Maur.: L. Rkiz (Oct.) *Adam* 21824! **Sen.:** *Heudelot* 544! 654! **Mali:** El Oualadji (Mar.) *Chev.* 43947! **Ghana:** Daboya (Nov.) *Thorold* 264! Singa (Aug.) *Rose Innes* GC 31859! **Niger:** Kolo to Niamey (Dec.) *P. de Fabrègues* 2769! **N. Nig.:** Kwarre (Nov.) *Sampson* 27! Katagum Dist. *Dalz.* 293! **S. Nig.:** Onitsha (Mar.) *Jones* FHI 6277!

5. **S. virgatum** (*Hack.*) *Stapf* in F.T.A. 9: 111 (1917). *Andropogon sorghum* var. *virgatus* Hack. in DC., Monogr. Phan. 6: 504 (1889). Annual up to 1·5 m. high.
Maur.: Matmata, Tagant (Apr.) *Monod* 157! **Sen.:** Niokolo-Koba (Feb.) *Adam* 17508! **Niger:** Illeba (Sept.) *P. de Fabrègues* 2528! Extends eastwards into Egypt and the Sudan.

6. **S. vogelianum** (*Piper*) *Stapf* in F.T.A. 9: 116 (1917); Snowden l.c. 219; A. Chev. in Rev. Bot. Appliq. 13: 857. *Andropogon sorghum* subsp. *vogelianus* Piper in Proc. Biol. Soc. Wash. 28: 34 (1915). Loosely tufted annual, or sometimes perennial, 1–5 m. high.
Mali: Gao (Aug.) *de Wailly* 5142! **Ghana:** Lawra (Sept.) *Hall* CC 770! Kpandai (July) *Rose Innes* GC 32576! **N. Nig.:** (Dec.) *Lely* 778! Makurdi (Nov.) *Sampson* 37! Ajaokuta (Aug.) *Ward* L135! Tungan Bandekwoi, Jebba (June) *Onochie* FHI 40248! **S. Nig.:** Nun R. *Vogel* 11! Idu (Sept.) *Holland* 153! Onitsha (Nov.) *Baldwin* 13733! Yenagoa (Sept.) *Tuley* 887! **W. Cam.:** Cam. R. (Jan.) *Mann* 2109!

7. **S. aethiopicum** (*Hack.*) *Rupr. ex Stapf* in F.T.A. 9: 119 (1917); Snowden l.c. 221; A. Chev. in Rev. Bot. Appliq. 13: 857. Annual, up to 3 m. high.
Sen.: N'Tiago (Sept.) *Audru* 2505! **Mali:** (Jan.) *Chev.* 43027! Kidal (Sept.) *Rossetti* 59/164! **S.L.:** Njala (Aug.) *Deighton* 772! **Niger:** Agades (Apr.) *P. de Fabrègues* 1482! Zinder (Oct.) *P. de Fabrègues* 2267! Toukounous (May) *Bartha* 11! *P. de Fabrègues* 1558! Tanout (Apr.) *P. de Fabrègues* 1426! **N. Nig.:** Gajibe (Nov.) *Johnston* N72! Tilum, Rann Dist. (Apr.) *Davey* FHI 27152! Wulgo (Feb., Mar.) *Davey* FHI 27157! *Johnston* N26! Kalkala *Golding* 77! Extends eastward to the Sudan and Somalia.

8. **S. arundinaceum** (*Desv.*) *Stapf* in F.T.A. 9: 114 (1917); Snowden l.c. 227; A. Chev. in Rev. Bot. Appliq. 13: 857. *Rhaphis arundinacea* Desv., Opusc. 69 (1831). *Andropogon arundinaceus* Willd., Sp. Pl. 4: 906 (1805), not of Berg. (1767). Annual or perennial 2–3·5 m. high.
Sen.: Niokolo-Koba (Oct.) *Adam* 15711! M'bout (Nov.) *Roberty* 10150! Kanémèré (July) *Fotius* K311! **Gam.:** Yundum, W. Div. (Dec.) *Austin* 18! **Guin.:** Nzérékoré (Oct.) *Baldwin* 9726! **S.L.:** Kenema (Jan.) *Thomas* 7714! Segbwema *Deighton* 3473! Musaia (Oct., Nov.) *Glanville* 317! *Thomas* 2637! **Lib.:** Kle, Boporo Dist. (Dec.) *Baldwin* 10578! Gbanga (Sept.) *Linder* 508! Sanokwele (Sept.) *Baldwin* 9495! Cess R. at Tappita trail (Aug.) *Baldwin* 9020! Blazié (Nov.) *Adam* 16078! **Iv. C.:** Abidjan (June) *Adjanohoun* 408a! Toumodi to Nzaakro (Aug.) *Chev.* 22424! Kouibly, Man (Oct.) *Adjanohoun* 300a! Ferkessédougou (Nov.) *Leeuwenberg* 2016! Beyo (Feb.) *Leeuwenberg* 2740! **U. Volta:** Gonaba (Dec.) *Scholz* 193! Konpiénga (Aug.) *Scholz* 146! **Ghana:** Winneba Plains (Feb.) *Dalz.* 8423! Accra to Dodowah (June) *Rose Innes* GC 30081! Kumasi (Apr.) *La Danso* 35! Odumase (Aug.) *Howes* 1203! Yapei (Aug.) *Williams* 844! **Dah.:** Logdo (Feb.) *Raynal* 13513! **N. Nig.:** Nupe *Barter* 1379! Katsina Ala (Aug.) *De Leeuw* 1713! **S. Nig.:** Lagos (May) *Dalz.* 1322! Oke Ado, Ibadan Dist. (Oct.) *Onochie* FHI 7940! Eruwa (Nov.) *Onochie* FHI 8143! Benin (Oct.) *Maggs* 21! Nun R. (Aug.) *T. Vogel* 50! **W. Cam.:** Victoria *Maitland* 977! Batibo to Mambe (June) *Brunt* 1172! **F. Po:** (Dec.) *Mann* 114! Extends south into the Congo.

[Preliminary investigations suggest that the wild species hybridize rather freely, both among each other and with the cultivated forms. It seems clear that the status of the present species will have to be reconsidered, but until more is known of their breeding behaviour it would be premature to attempt it.]

107. SORGHASTRUM Nash in Britton, Man. Fl. North U.S. 71 (1901).

Awns up to 8 mm. long, rarely up to 25 mm. (specimens from Guinée); racemes bearing 1–3 spikelets, therefore at least some of the spikelets truly sessile and accompanied by only 1 pedicel .. .. 1. *trichopus*
Awns 30–40 mm. long; racemes each bearing only 1 spikelet, therefore every spikelet borne upon a pedicel-like branch and accompanied by 2 pedicels 2. *bipennatum*

1. **S. trichopus** (*Stapf*) *Pilger* in E. & P., Pflanzenfam., Aufl. 2, 14e: 142 (1940). *Andropogon trichopus* Stap in Kew Bull. 1897: 287 (1897); Chev. Bot. 721. *Sorghum trichopus* (Stapf) Stapf in F.T.A. 9: 141 (1917); F.W.T.A., ed. 1, 2: 582; A. Chev. in Rev. Bot. Appliq. 13: 858; Berhaut, Fl. Sén. ed. 2, 417. Perennial, up to 1·5 m. high; spikelets light brown to straw coloured; river flood plains.
Sen.: Ouassadou (Dec.) *Berhaut* 1688! Niokolo-Koba *Adam* 16880! **Gam.:** *Ruxton* 30! **Mali:** Diafarabé (Sept.) *Lean* 93! Dari (Sept.) *Farrow* 54*a*! Macina (Nov.) *Duong* 1300! **Guin.:** Baffing R. (Nov.) *Pobéguin* 1731! Kouroussa (Sept.) *Pobéguin* 507! **N. Nig.:** Nupe *Barter* 1375! Bida to Badeggi (Aug.) *Onochie* FHI 35423! Badeggi (Sept.) *Clayton* 303! Damaturu (Aug.) *De Leeuw* 1133!
2. **S. bipennatum** (*Hack.*) *Pilger* in Notiz. Bot. Gart. Berl. 14: 96 (1938); Aké Assi, Contrib. 2: 297. *Andropogon bipennatus* Hack. in Flora 68: 142 (1885). *Sorghum bipennatum* (Hack.) O. Ktze., Rev. Gen. Pl. 2: 791 (1891); F.T.A. 9: 144 (1917); F.W.T.A., ed. 1, 2: 582; A. Chev. in Rev. Bot. Appliq. 13: 858; Berhaut, Fl. Sén. ed. 2, 416. Annual, up to 1·2 m. high; spikelets dark brown, with black bigeniculate awns; disturbed sites on moist soils.
Port. G.: Contubo to Djécoi, Bafata Reg. (Nov.) *Esp. Santo* 2830! Quaina, Gabú Reg. (Nov.) *Esp. Santo* 3159! Boruntuma (Dec.) *Pereira & Correia* 2232! **Guin.:** Kouroussa (Nov.) *Pobéguin* 530! **S.L.:** Dankawali, N. Prov. (Nov.) *Glanville* 324! Musaia (Dec.) *Deighton* 4434! Kurubonla (Nov.) *Morton* SL 3347! Sokurella (Nov.) *Morton* SL 2827! **Iv. C.:** Féttékro to Bassawa (Oct.) *Adjanohoun* 291*a*! Tangaoussou to Féttékro *Chev.* 22159! Séguéla to Kouroukourounga (Oct.) *Adjanohoun* 292*a*! Ouorossantiakara, Ferkessédougou (Nov.) *Leeuwenberg* 2011! Nambonkaha, Ferkessédougou (Nov.) *Leeuwenberg* 2047! **U. Volta:** Samandeni (Oct.) *Kmoch* 139! Tondoura (Oct.) *Scholz* 82! **Ghana:** Asikuma to Tsito (Nov.) *Adam* 5205! Bole (Oct.) *Rose Innes* GC 31019! Pong Tamale (Nov.) *Williams* 853! Hamale to Lawra (Oct. *Rose Innes* GC 31055! Navrongo (Oct.) *Vigne* FH 4641! **Togo Rep.:** Hnie, Sokode to Atakpame (Oct. *Rose Innes* GC 31369! **Dah.:** Boukombe (Nov.) *Risopoulos* 1257! **Niger:** Wadi Dollol *White* 22! **N. Nig.:** Lokoja (Nov.) *Dalz.* 286! Ilorin (Nov.) *Keay* FHI 28102! Jemaa (Oct.) *Gambles* 66! Fika (Oct.) *Musa Daggash* FHI 24882! Maska, Katsina Prov. (Jan.) *Thatcher* 509! **S. Nig.:** Olokemeji, Egba Dist. (Nov.) *Onochie* FHI 8145! Olokemeji F.R. (Nov.) *Latilo* FHI 36682! *Jackson* 5161168! **W. Cam.:** Bambuko F.R. (Nov.) *Argent* 976! Kuk escarpment (Nov.) *Brunt* 880! Extends southwards through Cameroun and the Congos to Angola, thence east to Zambia, Malawi, Mozambique and Madagascar.

108. CHRYSOPOGON Trin., Fund. Agrost. 187 (1820); F.T.A. 9: 159 (1917); *nom. cons.*

Stems erect, woody at the base, fastigiately branched; awns 2 cm. long or more; panicle open, the triads borne on long flexuous peduncles
1a. *aucheri* var. *quinqueplumis*
Stems creeping, with very close nodes bearing short leafy shoots; awns up to 0·5 cm. long; panicle dense, the peduncles appressed to the axis .. .. 2. *aciculatus*

1. **C. aucheri** (*Boiss.*) *Stapf* in Kew Bull. 1907: 21. *Andropogon aucheri* Boiss., Diagn. sér. 1, 5: 77 (1844).
1a. **C. aucheri var. quinqueplumis** (*A. Rich.*) *Stapf* l.c.; F.T.A. 9: 160; A. Chev. in Rev. Bot. Appliq. 13: 860. *C. quinqueplumis* A. Rich., Tent. Fl. Abyss. 2: 450 (1851). A wiry perennial of semi-arid regions.
Niger: Aïr (*fide* De Miré & Gillet in J. Agric. Trop. 3: 729). Extends eastwards to Uganda, Sudan, Ethiopia and Somalia, thence south to Tanzania.
2. **C. aciculatus** (*Retz.*) *Trin.* Fund. Agrost. 188 (1820); A. Chev. in Rev. Bot. Appliq. 13: 860. *Andropogon aciculatus* Retz., Obs. Bot. 5: 22 (1789); Chev. Bot. 715. A sward-forming perennial, sometimes used for lawns. The very long (3–4 mm.) needle-like callus is a menace, readily penetrating the mouths of cattle and the paws of dogs.
S.L.: New England (Oct.) *Morton & Jarr* SL 1441! **Iv. C.:** Adiopodoumé (Apr.) *Leeuwenberg* 3826! Port Bouet (Mar.) *Raynal* 13589! **U. Volta:** Ouagadougou (Sept.) *Scholz* 151! **Ghana:** Legon Garden (May) *Cudjoe*! Tafo (Feb.) *Rose Innes* GC 30332! **S. Nig.:** Ibadan (May, Oct.) *Townrow* 20! *Gledhill* 628! Owo (Apr.) *Haines* 198! Bori to Ogoni (July) *Nwanzo* 729! **W. Cam.:** Victoria (July, Nov.) *Maitland* 16! 84! 152! Kumba (Mar., Nov.) *Brenan* 9459! *Thorold* TN100! A species from tropical Asia, introduced to W. Africa and now well established in some localities.

Additional genus.

Polytrias amaura (*Büse*) *O. Ktze.* Rev. Gen. Pl. 788 (1891). *Andropogon amaurus* Büse in Miq., Pl. Junghn. 360 (1854). Prostrate and mat-forming; the inflorescence a solitary raceme of numerous spikelets; spikelets all alike, in threes, 2 sessile and 1 pedicelled. Commonly used as a lawn grass in the Asiatic tropics, and successfully established in the lawns of the Botanic Gardens, Victoria, W. Cam.: *Maitland* 98! 155! *Keay* FHI 28661!

109. VETIVERIA Lem.-Lisanc. in Bull. Soc. Philom. 1822: 43 (1822); F.T.A. 9: 156 (1917).

Sessile spikelets quite awnless, the callus glabrous; glumes spinulosely muricate
1. *zizanioides*
Sessile spikelets awned, the callus bearded:
Awn a short curved bristle (2)–5–(12) mm. long, enclosed or more or less exserted; callus beard up to 1 mm. long, pallid-brown .. .. .. 2. *nigritana*
Awn geniculate, up to 20 mm. long, differentiated into column and limb, the column exserted; callus beard up to 1·5 mm. long, fulvous .. .. 3. *fulvibarbis*

Fig. 449.—SORGHUM ARUNDINACEUM (*Desv.*) *Stapf* (GRAMINEAE-ANDROPOGONEAE).

1, pedicelled and sessile spikelets. 2, lower glume, and 3, upper glume from sessile spikelet. 4, grain.

1. **V. zizanioides** (*Linn.*) *Nash* in Small, Fl. Sth.-east. U.S. 67 (1903); F.T.A. 9: 157; A. Chev. in Rev. Bot. Appliq. 13: 858. *Phalaris zizanioides* Linn., Mant. Pl. 2: 183 (1771). *Chrysopogon zizanioides* (Linn.) Roberty (1960). Cultivated for its aromatic roots, and sometimes maintained as an ornamental.
S.L.: Freetown (Dec.) *Deighton* 3322! Rokupr (July) *Jordan* 475! **Lib.:** Monrovia (Sept.) *Baldwin* 9210! **Ghana:** Accra (June) *Thorold* CB91! Achimota (May, June) *Irvine* 1638! *Akpabla* 565! An Asiatic species, now introduced into most tropical countries. (See Useful Plants.)
2. **V. nigritana** (*Benth.*) *Stapf* in F.T.A. 9: 157 (1917); A. Chev. in Rev. Bot. Appliq. 13: 858; Berhaut, Fl. Sén. ed. 2, 403. *Andropogon nigritanus* Benth. in Hook., Fl. Nigrit. 573 (1849). *A. squarrosus* var. *nigritanus* (Benth.) Hack. (1889); Chev. Bot. 720. *A. zizanioides* var. *nigritanus* (Benth.) A. Chev., Bot. 721 (1920). *Lenormandia insignis* Steud. (1850). *Mandelorna insignis* Steud. (1854). Robust perennial 1·2–2·4 m. high, growing in clumps; stream sides and swampy flood plains.
Maur.: Diouk (Nov.) *Rossetti* 59/231! **Sen.:** Senegal R. *Adanson* 159! *Heudelot* 294! *Roger* 24! Bafing plains *Lecard* 123! **Gam.:** *Ruxton* 145! **Mali:** *Davey* 18! *Rogeon* 239! Soulei, Macina (May) *Lean* 9! Mopti (Sept.) *Lean* 87! Kabara (Aug.) *Chev.* 1353! **Port. G.:** Canhamina to Cambajú (Nov.) *Esp. Santo* 2969! **Guin.:** Bafing R. (Mar.) *Pobéguin* 1739! Farana (Mar.) *Sc. Elliot* 5349! **S.L.:** Batkanu (Jan.) *Glanville* 127! Katuna, Little Scarcies (Jan.) *Deighton* 968! Pujahun *Thomas* 8204! Mesima (Apr.) *Deighton* 3694! Gombahun (Sept.) *Jordan* 553! **Iv. C.:** Sérébissou, Dimbokro (May) *Adjanohoun* 231*a*! Pakobo Nord, Lamto (June) *Adjanohoun* 379*a*! Lamto (Dec.) *Aké Assi* 9814! Soubré (Nov.) *de Wilde* 3308! **U. Volta:** Baoulé (Oct.) *Kmoch* 119! Touroukoro (Nov.) *Scholz* 17*a*! **Ghana:** Accra *T. Vogel*! Kpong (June) *Ankrah* GC 20056! Kete Krachi (Dec.) *Adams* 4588! Damongo to Tamale (Oct.) *Rose Innes* GC 30218! Navrongo (Sept.) *Vigne* FH 4626! **Togo Rep.:** Koukomba plain *Kersting* A661! **Dah.:** Cotonou *Chev.* 4455! Porto Novo (Sept.) *Adjanohoun* 263! **Niger:** Sabon Birni (Feb.) *Virgo* 6! S. Dallol (Jan.) *Virgo* 74! **N. Nig.:** Badeggi (Sept.) *Clayton* 299! Kworre, Sokoto (Sept.) *Palmer* 23! Zaria (Oct.) *Keay* FHI 21421! Jimeta, Yola (Dec.) *Hepper* 1612! Kumshe road, Golumba Dist. (Dec.) *Davey* FHI 27128! **S. Nig.:** Oyan R. (June) *Holland* 9! Olokemeji F.R., Oyan R. (Nov.) *Jones, Keay & Onochie* FHI 14557! Nun R. (Aug.) *T. Vogel* 18! Niger R., Onitsha (May, Aug.) *Jones* 6686! *Onochie* FHI 7522! Tropical and S. Africa generally. (See Useful Plants.)
3. **V. fulvibarbis** (*Trin.*) *Stapf* in F.T.A. 9: 158 (1917); A. Chev. in Rev. Bot. Appliq. 13: 860; Berhaut, Fl. Sén. ed. 2, 403; Aké Assi, Contrib. 2: 297. *Andropogon fulvibarbis* Trin. in Mém. Acad. Sci. St. Petersb. sér. 6, 2: 287 (1832); Chev. Bot. 717. *A. verticillatus* Schumach. (1827), not of Roxb. (1820). Robust perennial up to 2 m. high; flood plains.
Mali: San (Sept.) *Chev.* 2340! 2342! Méma (Oct.) *Vintrebert* 9/1957! **Ghana:** Accra *T. Vogel*! Agomeda to Somanya (Nov.) *Adams* 4411! Asukawkaw R., Kete-Krachi to Kpandu (Sept.) *Rose Innes* GC 30413! Bamboi Ferry (Oct.) *Rose Innes* GC 30629! Pong Tamale (July) *Williams* 840! **Togo Rep.:** Barkoissi (Nov.) *Bille* 405!

110. ARTHRAXON P. Beauv., Ess. Agrost. 111, t. 11, f. 6 (1812); F.T.A. 9: 162 (1917).

Pedicelled spikelets, at least in the upper part of the raceme, developed; racemes villous, the silky hairs on internodes and pedicels 1–2 mm. long; sessile spikelets 2·5–3 mm. long; leaf-blades 1–2 cm. long and 3–7 mm. broad .. 1. *lancifolius*
Pedicelled spikelets not developed, the pedicels represented by a little subulate point 0·25 mm. long; racemes glabrous or inconspicuously hairy, hairs on the internodes up to 0·5 mm. long; spikelets 3–4 mm. long; leaf-blades up to 7 cm. long and 20 mm. broad .. .. 2. *quartinianus*

1. **A. lancifolius** (*Trin.*) *Hochst.* in Flora 39: 188 (1856); F.T.A. 9: 165; A. Chev. in Rev. Bot. Appliq. 13: 860; Berhaut, Fl. Sén. ed. 2, 397. *Andropogon lancifolius* Trin. in Mém. Acad. Sci. Petersb. sér. 6, 2: 271 (1832). *Arthraxon microphyllus* (Trin.) Hochst. (1856)—Chev. Bot. 715. A delicate trailing annual with slender culms; among rocks and boulders.
Sen.: Niokolo-Koba *Adam* 15706. **Mali:** Koulikoro (Oct.) *Chev.* 2237! 2238! **Guin.:** Dalaba-Diaguissa Plateau, Fouta Djalon *Chev.* 18877! Mt. Lansa, Fouta Djalon (Sept.) *Schnell* 7249! **S.L.:** Bintumane (Nov.) *Jaeger* 448! *Morton* SL 2652! **Ghana:** Nakpanduri (Nov.) *Hall & Enti* GC 35709! **N. Nig.:** Jos Plateau (Sept.) *Lely* P753! Wana (Nov.) *Hepburn* 159! Zaria (Sept.) *Clayton* 1305! Kereng, Jos Plateau (Sept.) *Oche & Tuley* 1627! Also in Ethiopia, Tanzania, Zambia, Rhodesia, Malawi, extending eastwards through India to China and Indonesia.
2. **A. quartinianus** (*A. Rich.*) *Nash* in N. Amer. Fl. 17: 99 (1912); F.T.A. 9: 166; A. Chev. in Rev. Bot. Appliq. 13: 861. *Alectoridia quartiniana* A. Rich., Tent. Fl. Abyss. 2: 448 (1851). Decumbent annual with tough wiry stems rooting at the lower nodes; shady clearings.
Guin.: Dalaba (Nov.) *Chev.* 34596! Timbo *Pobéguin* 1709! Labe (Oct.) *Adam* 12668! **S.L.:** Freetown (Dec.) *Gledhill* SL237! Musaia *Glanville* 316! Mt. Loma (Nov.) *Jaeger* 8140! **N. Nig.:** Vom (Oct.) *Gambles* 42! 75! *Oche* MRK132! Manchok (Oct.) *De Leeuw, Magaji & Tuley* 1719! Gembu (Dec.) *Daramola* FHI 62464! **S. Nig.:** Obudu Plateau (Nov.) *Tuley* 1029! **W. Cam.:** Buea (Dec.) *Maitland* 103! Bambui *Bumpus* 13! Oku, Bamenda Div. (Feb.) *Hepper* 2045! Tropical Africa, West Indies, India and Indonesia.

111. BOTHRIOCHLOA O. Ktze., Rev. Gen. Pl. 762 (1891).

Racemes usually simple or sometimes the lower divided, borne upon an elongated central axis 2·5–10 cm. long, rarely subdigitate; leaf-blades linear, long tapering to a setaceous point .. .. *bladhii*

B. bladhii (*Retz.*) *S. T. Blake* in Proc. Roy. Soc. Queensland 80: 62 (1969). *Andropogon bladhii* Retz., Obs. Bot. 2: 27 (1781). *A. intermedius* R. Br., Prodr. 202 (1810). *Amphilophis intermedia* (R. Br.) Stapf in Agric. News W. Ind. 15: 179 (1916), incl. var. *acidula* Stapf in F.T.A. 9: 174. *A. glabra* of A. Chev. in Rev. Bot. Appliq. 13: 861, not of (Roxb.) Stapf. *Bothriochloa intermedia* (R. Br.) A. Camus in Ann. Soc. Linn. Lyon 76: 164 (1931); incl. var. *acidula* (Stapf) C. E. Hubbard in Kew Bull. 1934: 109; F.W.T.A., ed. 1, 2: 583; Berhaut, Fl. Sén. ed. 2, 397. *B. glabra* of F.W.T.A., ed. 1, 2: 583, and of Berhaut, Fl. Sén. ed. 2, 397, not of (Roxb.) A. Camus. Tufted perennial up to 1 m. high.
Sen.: Dakar (Jan.) *Hagerup* 774! L. Tanma (Sept.) *Adam* 18644! Fann (Nov.) *Adam* 17771! Almadies (Aug.) *Broadbent* 16! Mt. Roland (Aug.) *Broadbent* 63! **U. Volta:** Gaoua (July) *Aké Assi* 6148! Gaoua to Ouangofétini (July) *Adjanohoun* 456*a*! Pontieba (Aug.) *Winkoun*! **Ghana:** Apam (June) *Hall* 1775! Christiansborg (Mar.) *Johnson* 1017! Nungua, Accra (Apr., Nov.) *Rose Innes* GC 30014! *Clayton* 391! Labadi to Teshi (Apr.) *Irvine* 2860! **N. Nig.:** Katagum *Dalz.* 257! Damaturu (Sept.) *De Leeuw* 1220*b*! Also in W. Indies, and extending through India to China and Australia.
[*Bothriochloa* is currently the object of intensive cytological studies which suggest that all the W. African specimens should be included in *B. bladhii*. This conclusion may need to be reviewed when investigations reach a more advanced state.]

112. DICHANTHIUM Willemet in Neue Ann. Bot. 18: 11 (1796); F.T.A. 9: 177 (1917).

Inflorescence subdigitate with 3–9 racemes; sessile spikelets imbricate, oblong, 3–4 mm. long, sparsely ciliate, with a geniculate awn 16–20 mm. long; pedicelled spikelet about as long as the sessile, awnless; anthers all about 1 mm. long .. *annulatum*

D. annulatum (*Forsk.*) *Stapf* in F.T.A. 9: 178 (1917); A. Chev. in Rev. Bot. Appliq. 13: 861; De Miré & Gillet in J. Agric. Trop. 3: 731; Berhaut, Fl. Sén. ed. 2, 398. *Andropogon annulatus* Forsk., Fl. Aegypt.-Arab. 173 (1775). Tufted perennial up to 1 m. high.
Sen.: Ross (Feb.) *Berhaut* 1438! Bargny (May) *Adam* 17672! **Mali:** Kidal (Sept.) *Rossetti* 59/153! Fairly generally distributed in the tropics and subtropics of the Old World. (See Useful Plants.)

113. EUCLASTA Franch. in Bull. Soc. Hist. Nat. Autun 8: 335 (1895); F.T.A. 9: 180 (1917).

Panicle umbelliform, pilose, of up to 15 nodding racemes 5–8 cm. long; axis of the panicle short, long-pilose at the nodes; lower glume of sessile spikelet 5 mm. long, strongly nerved; awns 3–4 cm. long; anthers of sessile spikelet normally 0·5 mm. long, those of pedicelled spikelet about 2 mm. long .. .. *condylotricha*

E. condylotricha (*Hochst. ex Steud.*) *Stapf* in F.T.A. 9: 181 (1917); A. Chev. in Rev. Bot. Appliq. 13: 861; Berhaut, Fl. Sén. ed. 2, 398. *Andropogon condylotrichus* Hochst. ex Steud., Syn. Pl. Glum. 1: 377 (1854); Chev. Bot. 717. *Euclasta glumacea* Franch. (1895). *E. graminea* Th. & Hél. Dur. (1909). *Dichanthium condylotrichum* (Hochst. ex Steud.) Roberty, Boissiera 9: 163 (1960). Annual; culms slender, weak, geniculately ascending, up to 1·5 m. high; nodes strongly bearded; in partial shade.
Sen.: Niokolo-Koba (Oct.) *Adam* 15590! Toubacouta (Oct.) *Monod* 8486! **Mali:** Koulikoro (Oct.) *Chev.* 2346! **Guin.:** Timbo (Oct.) *Pobéguin* 1775! **S.L.:** Musaia (Dec.) *Deighton* 4435! Kamba (Nov.) *Glanville* 315! **Iv. C.:** Féttékro to Bassawa (Oct.) *Adjanohoun* 281*a*! Ferkessédougou (Nov.) *Leeuwenberg* 2012! **U. Volta:** Ouagadougou (Oct.) *Scholz* 163! **Ghana:** Kayoro (Oct.) *Vigne* FH 4649! Pong Tamale (Nov.) *Williams* 855! Navrongo to Lawra (Oct.) *Rose Innes* GC 30280! Ho (Dec.) *Howes* 1041! Bame Pass (Dec.) *Adams* 4481! **Togo Rep.:** Sokode to Lama Kara (Oct.) *Rose Innes* GC 31386! **Dah.:** Boukombe (Nov.) *Risopoulos* 1263! **N. Nig.:** Jos Plateau (Oct.) *Lely* P806! Lokoja *Dalz.*! Kaiama (Oct.) *Ward* 22! Abinsi (Oct.) *Dalz.* 887! Kaduna (Oct.) *Thatcher* S540! Gurum, Vogel Peak (Nov.) *Hepper* 1306! **S. Nig.:** Olokemeji F.R. (Nov.) *Clayton* 563! *Onochie* FHI 8107! Ibadan (Nov.) *Meikle* 650! Zalende, Ibadan (Nov.) *Onochie* FHI 6988! Ago-Are (Nov.) *Olorunfemi* FHI 40302! Tropical Africa and America.

114. EREMOPOGON Stapf in F.T.A. 9: 182 (1917).

Racemes solitary, 3–4 cm. long, on filiform peduncles exserted from a spathe-like sheath; sessile spikelets lanceolate, 4 mm. long, pale and shining with reddish tips; pedicelled spikelets lanceolate-oblong, reddish, unpitted .. .. .. *foveolatus*

E. foveolatus (*Del.*) *Stapf* in F.T.A. 9: 183 (1917); A. Chev. in Rev. Bot. Appliq. 13: 861. *Andropogon foveolatus* Del., Fl. Egypte 16 (1812); Chev. Bot. 717. Densely tufted perennial up to 60 cm. high with wiry culms and glaucous leaf-blades.
Maur.: Oualata to Oujaf (Aug.) *Rossetti* 218/61! **Mali:** Bouressa to Tessalit (Sept.) *Rossetti* 196/59! Extends from Mauretania to Egypt, Sudan and Kenya, thence through Iran and Arabia to India. (See Useful Plants.)

115. THEMEDA Forsk., Fl. Aegypt.-Arab. 178 (1775); F.T.A. 9: 415 (1919).

Racemes solitary, each supported by a boat-shaped spatheole up to 5 cm. long, clustered into loose heads; the two lowest pairs of spikelets in each raceme sterile, similar, sessile, glabrous or with a few tubercle-based bristles, and forming an involucre; fertile spikelet 5–6 mm. long, chestnut brown, with a pungent rufously bearded callus; awn 4–6 cm. long .. .. *triandra*

T. triandra *Forsk.* Fl. Aegypt.-Arab. cxxiii & 178 (1775); F.T.A. 9: 416; A. Chev. in Rev. Bot. Appliq. 13: 878; Berhaut, Fl. Sén. ed. 2, 420. *T. forskalii* Hack. (1889)—Chev. Bot. 721. Tufted perennial 1·2–1·8 m. high with rusty brown spatheoles.
Sen.: Ibel *Adam* 18849. Kédougou (Nov.) *Adam* 19930! **Mali:** San (Nov.) *Chev.* 2362! **Guin.:** Ditinn (Dec.) *Jac.-Fél.* 7427! Timbo to Ditinn *Chev.* 18503! **U. Volta:** Ouassa to Leo (Oct.) *Rose Innes* GC 31505! Bobo-Dioulasso (Sept.) *Adam* 15135! **Ghana:** Anyaboni (Jan.) *A. S. Thomas* D113! Yapei Ferry (Oct.) *Rose Innes* GC 30222! Lawra to Wa (Oct.) *Rose Innes* GC 30300! Lawra to Tamale (Oct.) *Rose Innes* GC 31052! Daboya (July) *Rose Innes* GC 31821! **N. Nig.:** Belli to Jalingo, Mambila Plateau (Sept.) *Oche* MRK89! Kereng, Jos Plateau (Sept.) *Oche & Tuley* 1626! Old World tropics generally. A most important grazing grass in eastern and southern Africa, where it is often a dominant constituent of grasslands. (See Useful Plants.)
[Rather variable, but the varieties described do not seem to show much correlation with cytological investigations.]

116. TRACHYPOGON Nees, Agrost. Bras. 341 (1829); F.T.A. 9: 400 (1919).

Annual; sessile spikelet neuter, awnless or with an awn up to 3 cm. long, the pedicelled hermaphrodite with an awn about 6 cm. long; pedicels 1·5 mm. long, with very oblique disarticulation; callus 2 mm. long, pungent, fulvously bearded
1. *chevalieri*

Perennial; sessile spikelet male and awnless, the pedicelled hermaphrodite with an awn 4–7 cm. long; pedicels 3 mm. long, obliquely disarticulating; callus 2 mm. long, pungent, white-bearded .. .. 2. *spicatus*

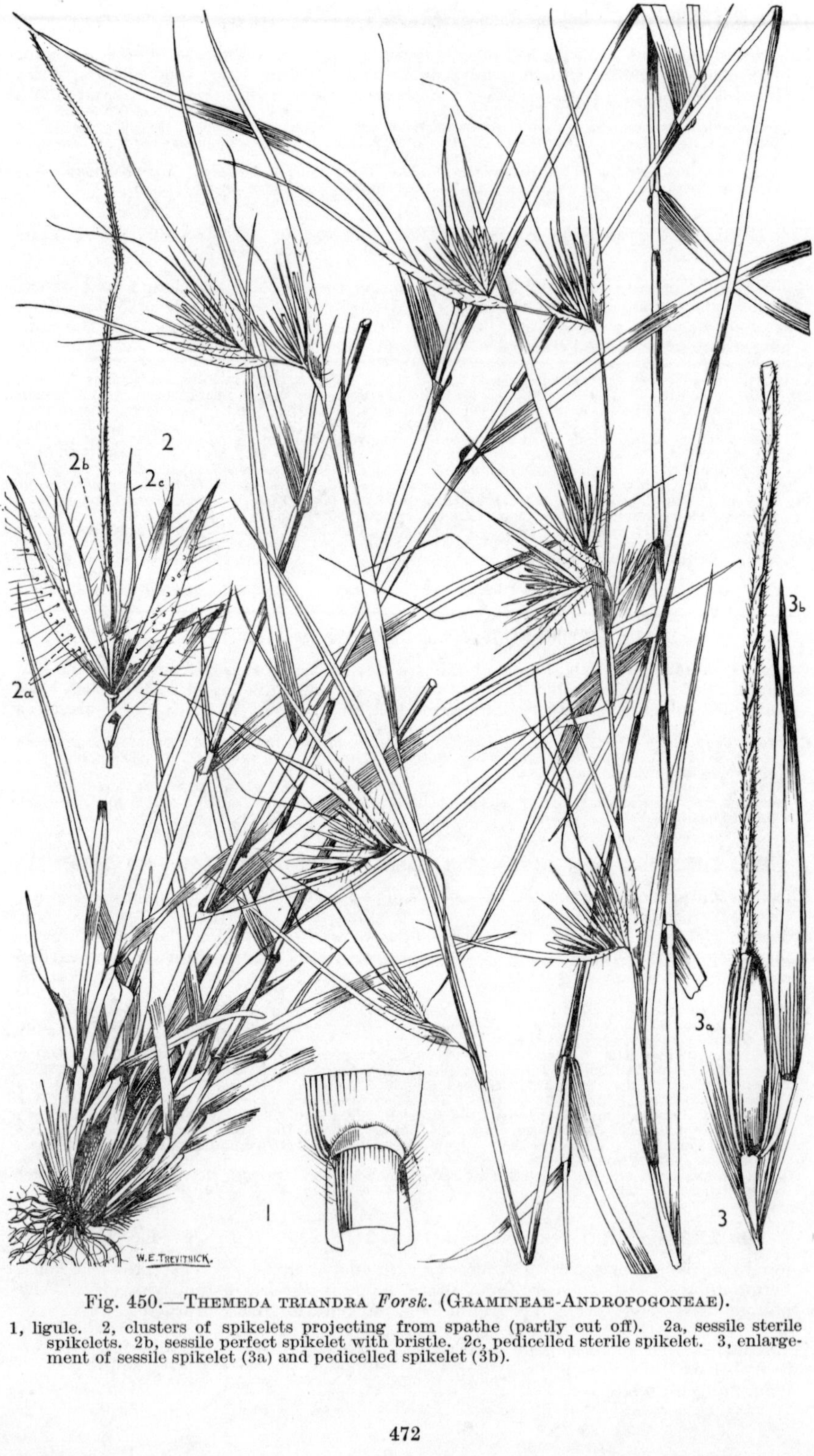

Fig. 450.—THEMEDA TRIANDRA *Forsk.* (GRAMINEAE-ANDROPOGONEAE).

1, ligule. 2, clusters of spikelets projecting from spathe (partly cut off). 2a, sessile sterile spikelets. 2b, sessile perfect spikelet with bristle. 2c, pedicelled sterile spikelet. 3, enlargement of sessile spikelet (3a) and pedicelled spikelet (3b).

1. **T. chevalieri** (*Stapf*) *Jac.-Fél.* in J. Agric. Trop. 1: 60 (1954). *Homopogon chevalieri* Stapf in Mém. Soc. Bot. Fr. 1, 8: 103 (1908); F.T.A. 9: 409; A. Chev. in Rev. Bot. Appliq. 13: 878. *Trachypogon ledermannii* Pilger in Engl., Bot. Jahrb. 48: 342 (1912); F.T.A. 9: 402; F.W.T.A., ed. 1, 2: 594. Up to 45 cm. high with flat leaf-blades and solitary racemes; ironstone outcrops.
N. Nig.: Naraguta (Aug.) *Lely* P449! Sho, Plateau Prov. (Nov.) *Keay* FHI 21020! Manchok, Mambila Plateau (Oct.) *Magaji & Tuley* 1769! Also in E. Cameroun, Congo, Tanzania and Zambia.

2. **T. spicatus** (*Linn. f.*) *O. Ktze.* Rev. Gen. Pl. 794 (1891). *Stipa spicata* Linn. f., Suppl. 111 (1781). *Trachypogon plumosus* (Willd.) Nees (1829)—F.T.A. 9: 403. *Andropogon plumosus* Willd. (1806). Wiry caespitose perennial up to 1·2 m. high; leaf-blades very narrow, usually convolute; racemes usually solitary, sometimes 2 (very rarely more in Africa).
Iv. C.: Odienné (Oct.) *Aké Assi* 8241! **U. Volta:** Sikasso to Bobo-Dioulasso (Sept.) *Adam* 15127! **Ghana:** Yendi to Nakpanduri (Oct.) *Rose Innes* GC 30737! Gambaga to Yendi (Oct.) *Rose Innes* GC 30491! Walewale to Gambaga (Oct.) *Rose Innes* GC 30265! Gambaga (Oct.) *Vigne* FH 4635! Gambaga to Bawku (Oct.) *Ankrah* GC 20276! **Togo Rep.:** Lama Kara to Kande (Oct.) *Rose Innes* GC 31404! **Dah.:** Taneka to Djougou (Nov.) *Risopoulos* 1280! **N. Nig.:** Jos Plateau (Sept.) *Lely* P767! Vodni, Pankshin Div. *Saunders* 64! Afaka F.R., Kaduna (Nov.) *Clayton* 1388! Samaru, Zaria Prov. (Sept.) *Freeman* S115! Anara F.R., Zaria Prov. (Oct.) *Keay* FHI 20108! Tropical and subtropical Africa and America.
[Very variable in number of racemes, length of ligule and hairiness of spikelets, and some authorities divide it into several species.]

117. HETEROPOGON Pers., Syn. Pl. 2: 533 (1807); F.T.A. 9: 410 (1919).

Lower glume of pedicelled spikelets acuminate, 15–25 mm. long, glabrous, with a line of glandular pits down the middle, keels narrowly winged; racemes with 1–3 awnless pairs at the base; awns 8–15 cm. long .. .. 1. *melanocarpus*
Lower glume of pedicelled spikelets acute, 8–10 mm. long, with tubercle-based hairs near the margin, without a line of pits, keels broadly and unequally winged; racemes with 3–10 awnless pairs at the base; awns 5–10 cm. long.. .. 2. *contortus*

1. **H. melanocarpus** (*Ell.*) *Benth.* in J. Linn. Soc. 19: 71 (1881); F.T.A. 9: 413; A. Chev. in Rev. Bot. Appliq. 13: 878; Berhaut, Fl. Sén. ed. 2, 421. *Andropogon melanocarpus* Ell., Sketch Bot. S. Carol. 1: 146 (1816). Annual, up to 2 m. or more; callus pungent, 5 mm. long.
Sen.: Messira (Nov.) *Berhaut* 680! 2726! **Gam.:** *Saunders* 107! **Dah.:** Boguila to Echitopa (Feb.) *Chev.* 23209! **N. Nig.:** Darazo F.R., Bauchi Prov. (Oct.) *De Leeuw & Magaji* 1994! A tropical American species, occurring sporadically in Africa and India.

2. **H. contortus** (*Linn.*) *P. Beauv. ex Roem. & Schult.* Syst. Veg. 2: 836 (1817); F.T.A. 9: 411; A. Chev. in Rev. Bot. Appliq. 13: 878; Berhaut, Fl. Sén. ed. 2, 421. *Andropogon contortus* Linn., Sp. Pl. 1045 (1753); Chev. Bot. 717. Perennial, up to about 1 m. high; awns twisted into a spire when young, and forming a malignant tangled knot at maturity; callus painfully pungent.
S.L.: Mt. Aureol (Aug., Sept.) *Melville & Hooker* 324! *Hooper* 147! Fourah Bay (Dec.) *Cole* 17! **Iv. C.:** Tehini (Aug.) *de Wilde* 811! Bouaké (May) *de Wilde* 63! **U. Volta:** Fada to Koupéla (July) *Chev.* 24574! Pama to Diabiga (Aug.) *Scholz* 148! **Ghana:** Christiansborg (Mar.) *Johnson* 1020! Nungua, Accra (Sept.) *Ankrah* GC 20042! Tamale (Oct.) *Baldwin* 13552! Navrongo (Aug.) *Vigne* FH 4621! Zuarungu *Lynn* 1077! **Togo Rep.:** Sokode to Atakpame (Oct.) *Rose Innes* GC 31371! **Dah.:** Porto Novo (Sept.) *Adjanohoun* 257! Agouagon (May) *Chev.* 23487! Cotonou (July, Nov.) *Debeaux* 333! *Chev.* 4464! *Clayton* 399! **N. Nig.:** Ilorin (Oct.) *Ward* 78! Lokoja (Nov.) *Keay* FHI 28094! Sokoto (Sept.) *Dalz.* 488! Yola (Apr.) *Dalz.* 271! Biu (Sept.) *Kennedy* FHI 8003! Throughout the tropics. (See Useful Plants.)

118. THELEPOGON Roth ex Roem. & Schult., Syst. Veg. 2: 46 (1817); F.T.A. 9: 34 (1917).

Leaf-blades lanceolate, broad and subcordate at the base, margins rigidly pectinate; racemes 2–17 in a bunch, up to 17 cm. long, stiff; barren pedicel linear, longer than the sessile spikelet, the latter 6–7 mm. long; lower glume prominently transversely rugose-tuberculate; lower floret male with a palea; upper lemma with a glabrous geniculate awn up to 2·5 cm. long .. .. *elegans*

T. elegans *Roth ex Roem. & Schult.* Syst. Veg. 2: 788 (1817); F.T.A. 9: 34; A. Chev. in Rev. Bot. Appliq. 13: 849; Berhaut, Fl. Sén. ed. 2, 397. Annual, up to 1 m. high, often with prop roots from the lower nodes; weed in disturbed soils.
Sen.: Népin *Adam* 18851. **Guin.:** Labé (Oct.) *Jac.-Fél.* 2016! **Ghana:** Yendi to Zabzugu (Oct.) *Rose Innes* GC 30704! Kete Krachi to Yendi (Nov.) *Rose Innes* GC 30440! Tamale (Nov.) *Thorold* 279! Gambaga (Nov.) *Vigne* FH 4679! Bawku (Oct.) *Rose Innes* GC 30252! **Dah.:** Boukombe (Nov.) *Risopoulos* 1256! **Niger:** Niamey (Oct.) *Hagerup* 538! Dogondoutch *White* 36! **N. Nig.:** Jebba to Mokwa (Oct.) *Hepper & Keay* 942! Naraguta, Plateau Prov. (Oct.) *Kennedy* FHI 8046! Sokoto (Oct.) *Moiser* 132*a*! Katsina (Oct.) *Clayton* 1354! Katagum *Dalz.* 284! Gurum, Vogel Peak area (Nov.) *Hepper* 1229! Throughout tropical Africa; also in India and Malesia. (See Useful Plants.)

119. ISCHAEMUM Linn., Sp. Pl. 1049 (1753), and Gen. Pl. ed. 5, 469 (1754); F.T.A. 9: 28 (1917).

Sessile spikelets concave on the back, 7–8 mm. long, chartaceous; internodes and pedicels clavate to cuneate, swollen; racemes 2–3; awn 7–10 mm. long 1. *afrum*
Sessile spikelets flat or convex on the back, coriaceous below and becoming herbaceous towards the tip:
Lower glume of sessile spikelet rugose or muricate; keels sharply involute for their whole length; racemes 2, firmly appressed together and interlocked when young:
Lower glume deeply rugose with sharp edged ridges right across its back; sessile spikelets 4–6 mm. long; pedicelled spikelets resembling the sessile and on a pedicel about ⅓ as long as the latter, or much reduced smooth and on a pedicel as long as the sessile spikelet .. .. 2. *rugosum*
Lower glume with rounded knobs on the keels, not, or only obscurely, joined across

Fig. 451.—HETEROPOGON CONTORTUS (*Linn.*) *P. Beauv. ex Roem. & Schult.* (GRAMINEAE-ANDROPOGONEAE).

1, ligule. 2, sessile fertile spikelet showing glumes, bristle and callus. 3, 4, lower glume, front and back view. 5, upper glume. 6, grain. 7, pedicelled sterile spikelet showing lower glume. 8, upper glume.

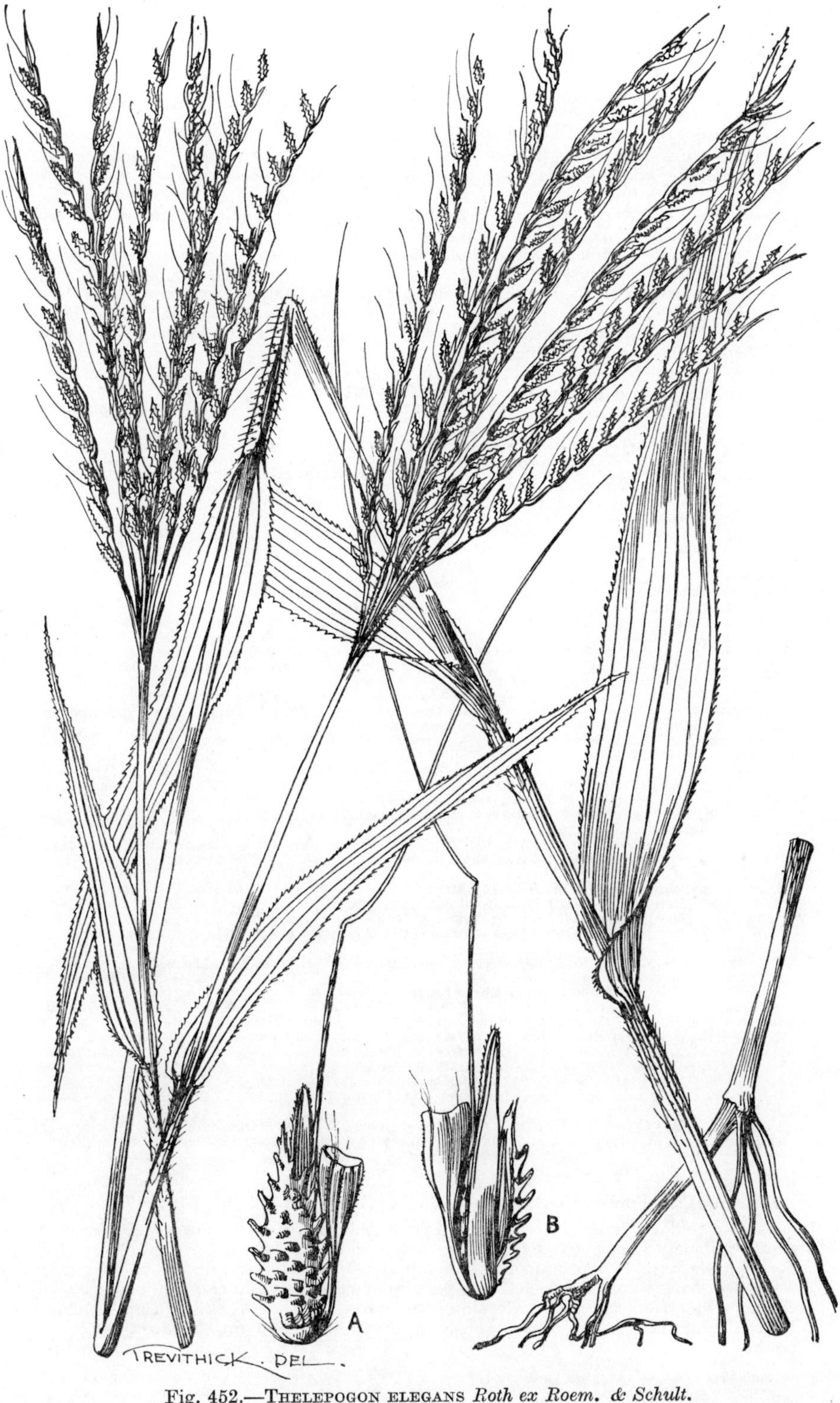

Fig. 452.—THELEPOGON ELEGANS *Roth ex Roem. & Schult.* (GRAMINEAE-ANDROPOGONEAE).

A and B, spikelets.

the back; sessile spikelets 5·5–7 mm. long; pedicelled spikelets 6 mm. long, smooth, asymmetrically winged, borne on a short pedicel 1–2 mm. long 3. *barbatum*

Lower glume of sessile spikelet smooth; sharply keeled in the upper half, more broadly involute below:

Awn 35–40 mm. long; racemes 3–4 (rarely 1–5), 12–22 cm. long, on a short common axis 1–3 cm. long, very fragile; sessile spikelets 7–10 mm. long, the keels of the lower glume winged towards the tip; spikelets, internodes and pedicels densely villous with purplish hairs .. .. 4. *amethystinum*

Awns 7–14 mm. long; racemes 2–3, digitate, up to 8 cm. long; internodes and pedicels shortly ciliate; sessile spikelets 5–6 mm. long:

Sessile spikelets with the keels of both glumes winged, the apex of the lower glume obtuse; racemes paired .. .. 5. *indicum*

Sessile spikelets with the glumes wingless, the apex of the lower glume acute:

Lower glume of sessile spikelet glabrous, ovate; culms 1 mm. diam. near base; leaf-blades narrowly lanceolate, spreading, mostly 2–10 cm. long and 4–10 mm. broad; ligule 0·5 mm. long, pilose .. .. 6. *timorense*

Lower glume of sessile spikelet pilose, lanceolate; culms 2–3 mm. diam. near base; leaf-blades broadly linear, erect, mostly 8–14 cm. long and 4–8 mm. broad; ligule 1–2 mm. long, glabrous .. .. 7. *tallanum*

1. **I. afrum** (*J. F. Gmel.*) *Dandy* in F. W. Andrews, Fl. Pl. Sudan 3: 476 (1956). *Andropogon afer* J. F. Gmel., Syst. Nat. ed. 13, 2: 166 (1791). *A. brachyatherus* Hochst. in Flora 27: 241 (1844). *A. intumescens* Pilger in Engl., Bot. Jahrb. 45: 208 (1910). *Ischaemum brachyatherum* (Hochst.) Fenzl ex Hack. in DC., Monogr. Phan. 6: 239 (1889); F.T.A. 9: 30; F.W.T.A. ed. 1, 2: 596; A. Chev. in Rev. Bot. Appliq. 13: 849. Perennial 60–90 cm. high; black clay soils. Hairiness of spikelets varies considerably, from almost glabrous to densely villous.
N. Nig.: Kworre, Sokoto (Sept.) *Palmer* 17! Shigal, Rann Dist. (Feb.) *Davey* FHI 27169! Logomani, N. Bornu (Nov.) *Johnston* N83! Ngala (Mar.) *Tuley* 1123! Yola Div. (Sept.) *Kennedy* FHI 7300! Tropical and S. Africa.
2. **I. rugosum** *Salisb.* Ic. Stirp. Rar. 1, t. 1 (1791); A. Chev. in Rev. Bot. Appliq. 13: 849; Berhaut, Fl. Sén. ed. 2, 389, 423. Straggling annual up to about 1 m. high; coastal and inland swamps.
Sen.: Kolda, Casamance (Oct.) *Adam* 18454! **Port G.:** Nova Lamego (Nov.) *Esp. Santo* 3158! **Guin.:** Kindia *Jac.-Fél.* 178! Mamou (Nov.) *Chev.* 34871! **S.L.:** Kambia (Dec.) *Deighton* 805! Rokupr (Dec.) *Jordan* 184! Freetown (Oct.) *Deighton* 2193! Mabaifu, Rokel R. (Mar.) *Glanville* 226! Sherbro I. (Nov.) *Deighton* 2302! **Lib.:** Monrovia (Nov.) *Barker* 1472! **U. Volta:** Ouagadougou (Dec.) *Aké Assi* 10445! **Ghana:** Bawku (Oct.) *Rose Innes* GC 30256! *Ankrah* GC 20292! Zowse, Bawku (Nov.) *Enti & Hall* GC 36002! An Asiatic species, now found in most tropical countries.
3. **I. barbatum** *Retz.* Obs. Bot. 6: 35 (1791). *I. goebelii* Hack. (1901). *I. imbricatum* (Munro ex Hack.) Stapf ex Ridley (1925). *Mesochium imbricatum* Munro ex Hack. (1889). Perennial about 1 m. high; swamps.
S. Nig.: Sapele (Aug.) *Onochie* FHI 33425! Tropical Asia.
4. **I. amethystinum** *J.-P. Lebrun* in J. Agric. Trop. 13: 44 (1966). *I. hirsutum* Peter in Fedde Rep., Beih. 40, 1: 356 Anh. 115 (1936), not of Spreng. (1825). Robust perennial up to 3 m. high.
U. Volta: Bouly (Nov.) *Scholz* 192! **Ghana:** Ejura to Atebubu (Oct.) *Ankrah* GC 20314! Yeji Ferry (Oct.) *Rose Innes* GC 30314! Yapei to Kintampo (Nov.) *Rose Innes* GC 30857! Pong Tamale (Oct.) *Addei* SLUS 758! Gambaga to Bawku (Oct.) *Ankrah* GC 20277! **Togo Rep.:** Koumongou (Oct.) *Bille* TG 435. **N. Nig.:** Naraguta F.R. (July) *Lawlor & Hall* FHI 46648! Zonkwa (Oct.) *De Leeuw* 1602! Also in E. Cameroun and Tanzania.
5. **I. indicum** (*Houtt.*) *Merrill* in J. Arn. Arb. 19: 320 (1938). *Phleum indicum* Houtt., Nat. Hist. II, 13: 198 (1782). *Ischaemum ciliare* Retz. (1791)—Aké Assi, Contrib. 2: 303. *I. aristatum* of authors, not of Linn. (1753).
S.L.: Mahela, Samu (Nov.) *T. S. Jones* 46! Mano (Feb.) *Deighton* 3114! Kowama, Peri (Nov.) *Deighton* 5251! **Lib.:** Robertsport (Dec.) *Baldwin* 10886! Monrovia (May, Nov.) *Baldwin* 5817! *Barker* 1473! *Adam* 16522 *bis*! *Dinklage* 3235! **Iv. C.:** Téké Forest (Mar.) *Aké Assi* 6966! **U. Volta:** Bobo-Dioulasso (June) *Scholz* 100! **S. Nig.:** Opobo (Oct.) *Maitland* 64! An Asiatic species, now distributed throughout the tropics.
6. **I. timorense** *Kunth.* Rev. Gram. 1: 369 (1830). Creeping annual; damp wayside places.
W. Cam.: Victoria (Dec.-Feb.) *Maitland* 91! 154! 154*a*! *Keay* FHI 28572! Buea (Dec.) *Maitland* 857! An Asiatic species, now found in many tropical countries.
7. **I. tallanum** *Rendle* in J. Bot. 31: 359 (1893); F.T.A. 9: 31; Chev. Bot. 715, and in Rev. Bot. Appliq. 13: 849. *Argopogon vuilletii* Mimeur in Rev. Bot. Appliq. 31: 213 (1951); Jacques-Félix in Rev. Bot. Appliq. 32: 547 (1952). A streamside grass about 1·2 m. high, decumbent and rooting at the lower nodes.
Mali: Bamako (Dec.) *Duong* 20! Sikasso to Bobo-Dioulasso (Dec.) *Vuillet* 25956! Sikasso, ravin de Sarakoro (Nov.) *Demange* 2749. Sikasso, Chutes du Farako (Nov.) *Demange* 3133. **Guin.:** Kouria *Chev.* 12860 *bis*! Benna (Oct.) *Jac.-Fél.* 7177! Dalaba *Adames* 417! Friguiagbé (Jan.) *Chillou* 1142! **S.L.:** Farangbaia (Nov.) *Lewis* 30! Ninia, Talla (Feb.) *Sc. Elliot* 4927! Loma Mt. (Mar.) *Morton & Gledhill* SL 1026!
[The species has obvious affinities to *I. fasciculatum* Brong., *I. purpurascens* Stapf and *I. timorense* var. *villosum* C. E. C. Fisher, and a final decision as to its status must await a critical revision of this difficult group.]

120. SEHIMA Forsk., Fl. Aegypt.-Arab. 178 (1775); F.T.A. 9: 35 (1917).

Racemes up to 8 cm. long; internodes and pedicels linear, densely ciliate on both sides; sessile spikelets linear, 1–1·4 cm. long; lower glume deeply hollowed between the dorsal keels with 3–5 prominent intercarinal nerves, terminating in an unequally 2-toothed, flat, membranous beak; upper glume with a straight awn 2 cm. long; awn of upper lemma distinctly ciliate along the spiral; pedicelled spikelet up to about 1·5 cm. long, lanceolate, conspicuously 7-nerved, long acuminate into 2 setaceous teeth .. .. *ischaemoides*

S. ischaemoides *Forsk.* Fl. Aegypt.-Arab. 178 (1775); F.T.A. 9: 37; A. Chev. in Rev. Bot. Appliq. 13: 849. Annual up to 60 cm. high.
Mali: *Leclercq* 42658! **Niger:** Toukounous *Bartha* 94! **N. Nig.:** Kohukuwa (Oct.) *De Leeuw* 1861! Yola (Nov.) *Lowe* 1599*a*! Tropical Africa, Arabia and India.

121. MICROSTEGIUM Nees in Lindley, Nat. Syst. ed. 2, 447 (1836).

Leaf-blades linear-lanceolate, up to about 7 cm. long, softly pilose above and beneath; racemes 2–5, subdigitate; spikelets 4·5–6 mm. long, glabrous; lower glume with a deep median channel along the back, the nerves anastomosing near the tip; lower lemma nerveless, up to 4 mm. long, or completely absent; upper lemma 0·2 mm. long, accompanied by a tiny scale-like palea; awn very short, seldom protruding beyond the tip of the spikelet .. .. *vimineum*

M. vimineum (*Trin.*) *A. Camus* in Ann. Soc. Linn. Lyon, n.s. 68: 201 (1921). *Andropogon vimineus* Trin. in Mém. Acad. Sci. Petersb., sér. 6, 2: 268 (1832). *Microstegium aristulatum* Robyns & Tournay in Bull. Jard. Bot. Brux. 25: 240 (1955). Creeping annual. Asiatic specimens have glabrous or sparsely pubescent foliage, but are otherwise similar.

W. Cam.: Litoka, Cam. Mt., 4,400 ft. (Apr.) *Maitland* 1086! Also in the Congo, and extending from India to China and Japan.

122. SCHIZACHYRIUM Nees, Agrost. Bras. 331 (1829); F.T.A. 9: 184 (1917); Jacques-Félix in Rev. Bot. Appliq. 33: 423.

Fertile lemma bifid to beyond the middle; spikelets usually laterally compressed between internode and pedicel; lower glume of sessile spikelet chartaceous to subcoriaceous, obscurely nerved, lanceolate to linear:
- Leaf-blades obtuse to abruptly acute at the apex; plants rooting at the lower nodes and creeping; pedicelled spikelets smaller than the sessile:
  - Column of awn scarcely exceeding the sessile spikelet; fertile lemma bifid almost to the base; racemes 1–2·5 cm. long:
    - Culms 3–4 mm. diam., compressed; anthers 2–3 mm. long; leaf-blades 7–17 cm. long and 4–8 mm. broad; sheaths mostly longer than the internodes, strongly compressed and keeled; spikelets 3–4·5 mm. long .. .. 1. *platyphyllum*
    - Culms up to 2 mm. diam.; anthers 0·5–1 mm. long; leaf-blades 2–6 cm. long and 2–7 mm. broad; sheaths mostly shorter than the internodes; spikelets 3–3·5 mm. long:
      - Internodes of the rhachis narrowly clavate to linear with a cupuliform tip, narrower than the spikelet; lower glume of sessile spikelet keeled in the upper half, sharply inflexed below .. .. 2. ***brevifolium***
        - Lower glume of sessile spikelet, internodes and pedicels glabrous 2a. ***brevifolium*** var. ***brevifolium***
        - Lower glume of sessile spikelet, internodes and pedicels hispid to shortly villous with white hairs .. .. 2b. *brevifolium* var. *flaccidum*
      - Internodes of the rhachis broadly clavate, swollen, wider than the spikelet; lower glume of sessile spikelet keeled to the base .. .. .. 3. *maclaudii*
  - Column of awn twice as long as the sessile spikelet or longer; lower glume of the sessile spikelet pubescent (sometimes sparsely so in *S. djalonicum*):
    - Internodes of the rhachis linear, ciliate on both sides, narrower than the sessile spikelet; lower glume of the latter narrowly lanceolate, not pinched between internode and pedicel, 3·5 mm. long; fertile lemma bifid for ⅔ of its length; racemes 2 cm. long .. .. 4. *djalonicum*
    - Internodes of the rhachis swollen, subcylindrical, densely pubescent, broader than the sessile spikelet; lower glume of the latter linear, pinched between internode and pedicel, 4 mm. long; fertile lemma bifid for ¾ of its length; racemes 4 cm. long .. .. 5. *radicosum*
- Leaf-blades gradually tapered to a fine point, sometimes abruptly acute but then the plant erect:
  - Lower glume of sessile spikelet lanceolate, not pinched between internode and pedicel, up to 6 mm. long, glabrous; pedicelled spikelets as long as or longer than the sessile:
    - Internodes and pedicels clavate to cylindrical, glabrous except for a tuft of hairs on one side at the top of the pedicel; lower glume of sessile spikelet wingless; upper glume awnless; upper lemma with an awn 10–15 mm. long; spikelets 4·5–6 mm. long .. .. 6. *penicillatum*
    - Internodes and pedicels linear or narrowly widening upwards; spikelets seldom over 4 mm. long:
      - Upper glume of sessile spikelets shortly awned; keels of lower glume wingless, rounded in the lower half; upper lemma with an awn 11–12 mm. long; pedicelled spikelet awnless .. .. 7. *lomaense*
      - Upper glume of sessile spikelets without an awn; keels of lower glume sharp, narrowly winged; upper lemma with an awn 2–2·5 cm. long; pedicelled spikelet with a short bristle 1–3 mm. long .. .. 8. *delicatum*
  - Lower glume of sessile spikelet linear, squeezed between internode and pedicel, 5–8 mm. long; pedicelled spikelets smaller than the sessile:
    - Rhachis internodes cylindrical, about 2 mm. broad at the top; racemes stout, cylindrical, 5 cm. long, at length just exserted from the spatheoles; leaf-blades

up to 5 cm. long and 2 mm. broad, acuminate at the apex; fertile lemma bifid for rather more than ½ its length; spikelets 5–7 mm. long:

Spikelets, internodes and pedicels glabrous except for the pubescent callus concealed within the joint .. .. 9. *urceolatum*

Spikelets, internodes and pedicels densely pubescent, the callus bearded and forming a conspicuous ring of hairs above each node .. 10. *nodulosum*

Rhachis internodes clavate, about 1 mm. broad at the top; racemes slender; leaf-blades up to 30 cm. long and 2–10 mm. broad:

Plants perennial; lower glume of sessile spikelet keeled throughout, glabrous or sometimes pubescent to villous; pedicelled spikelet awnless or with an awn seldom over 3 mm. long; racemes 5–12 cm. long .. .. 11. *sanguineum*

Plants annual:

Internodes, pedicels and lower glume of the sessile spikelet villous with hairs 2 mm. long; lower glume of sessile spikelet sharply inflexed on the sides, keeled only at the very tip; racemes 2–8 cm. long, usually sessile, rarely on peduncles up to 3 cm. long .. .. 12. *exile*

Internodes and pedicels shortly ciliate on one or both sides, the lower glume of the sessile spikelet glabrous though usually with ciliate margins, or rarely pubescent; lower glume of sessile spikelet keeled throughout; racemes 4–6 cm. long on peduncles 1–3 cm. long; pedicelled spikelet with an awn 3–10 mm. long .. .. 13. *ruderale*

Fertile lemma shortly bifid, the lobes not extending beyond the middle (except sometimes in *S. schweinfurthii*) or entire; spikelets usually dorsally compressed; lower glume of the sessile spikelet thinly chartaceous, finely but distinctly 3–5-nerved:

Racemes plumose with hairs 5–6 mm. long from internodes and pedicels, gathered into a terminal bunch; internodes 6–8 mm. long, the pedicels 1–2 mm. longer; awn about 6 mm. long, without a column; fertile lemma entire; leaf-blades abruptly acute 14. *pulchellum*

Racemes almost glabrous to sparsely pilose, loosely paniculate; internodes 3–5 mm. long, the pedicels subequal or shorter; awn with a distinct column:

Racemes compact, cylindrical, almost glabrous (apart from the sessile spikelet callus), long exserted from the spatheole; sessile spikelets twice as long as the internodes or more; internodes and pedicels obscurely ciliate on one side; sessile spikelets glabrous; fertile lemma bilobed to near the middle; leaf-blades narrowly linear, up to 30 cm. long and 3 mm. broad, gradually attenuate at the tip 15. *schweinfurthii*

Racemes flexuous, distinctly hairy; sessile spikelets up to 1½ times as long as the internodes; internodes and pedicels ciliate on the lower half of one side, and the upper half of the other; sessile spikelets mostly pilose:

Annual; fertile lemma entire; leaf-blades gradually acuminate 16. *scintillans*

Perennials; fertile lemma bilobed:

Leaf-blades convolute, setaceous; fertile lemma bilobed for up to ⅓ its length 17. *gresicolum*

Leaf-blades flat, shortly acute at the tip; fertile lemma bilobed for up to ¼ its length; column of awn 5–7 mm. long .. .. 18. *rupestre*

1. **S. platyphyllum** (*Franch.*) *Stapf* in F.T.A. 9: 188 (1917); A. Chev. in Rev. Bot. Appliq. 13: 862; Berhaut, Fl. Sén. ed. 2, 422. *Andropogon brevifolius* var. *platyphyllus* Franch. in Bull. Soc. Hist. Nat. Autun 8: 324 (1895). *A. platyphyllus* (Franch.) Pilger in E. & P., Pflanzenfam. ed. 2, 14e: 166 (1940). Perennial, or sometimes annual, 90–120 cm. high; grows as an understorey grass among taller species; marshy soils.
Sen.: Badé (Dec.) *Berhaut* 2734! Niokolo-Koba (Oct.) *Adam* 15720! **Guin.:** Baffing R. (Oct.) *Pobéguin* 1800! Ouassou (Nov.) *Jac.-Fél.* 7257! **S.L.:** Mano Salija (Nov.) *Deighton* 343! Yeraia (Apr.) *Glanville* 205! Mt. Loma (Nov.) *Jaeger* 490! 8202! Bintumane Peak (Jan.) *T. S. Jones* 71! **Iv. C.:** Namboukaha (Nov.) *Leeuwenberg* 2050! Beréby (Nov.) *Oldeman* 475! **U. Volta:** Samandeni (Oct.) *Kmoch* 172! Leo to Po (Oct.) *Rose Innes* GC 31082! **Ghana:** Sampa to Wenchi (Oct.) *Rose Innes* GC 31573! Damongo (Nov.) *Thorold* CB283! Tamale to Bolgatanga (Oct.) *Rose Innes* GC 30770! Lawra to Wa (Oct.) *Ankrah* GC 20444! Tumu to Leo (Oct.) *Rose Innes* GC 20469! **Togo Rep.:** Badou to Atakpame (Sept.) *Rose Innes* GC 31346! **N. Nig.:** Naraguta F.R. (Dec.) *Kennedy* 9! **S. Nig.:** Oyo (Nov.) *Gledhill* 745! **W. Cam.:** Bambuko F.R. (Nov.) *Argent* 973! Also in Sudan, Uganda, the Congos and Malawi. (See Useful Plants.)

2. **S. brevifolium** (*Sw.*) *Nees ex Büse* in Miq., Plant. Junghn. 359 (1854); F.T.A. 9: 187; A. Chev. in Rev. Bot. Appliq. 13: 862; Berhaut, Fl. Sén. ed. 2, 422. *Andropogon brevifolius* Sw., Prodr. Veg. Ind. Occ. 26 (1788); Chev. Bot. 716. *A. minimus* Rendle & C. B. Cl. ex Elliot in J. Linn. Soc. 30: 99 (1894). *Schizachyrium minutum* Gledhill in Bol. Soc. Brot. 41: 60 (1967). A delicate straggling annual 12–18 ins. high, usually found as an understorey grass beneath other species in damp places. A most variable species.

2a. **S. brevifolium** (*Sw.*) *Nees ex Büse* var. **brevifolium.**
Sen.: Bargny (Nov.) *Berhaut* 2736! **Gam.:** *Saunders* 81! **Mali:** Bamako (Jan.) *Chev.* 205! Nyamina to Koulikoro (Oct.) *Chev.* 2350! Koulikoro (Sept.) *Chev.* 25019! Somo (Jan.) *Demange* 1014! Diafarabé to Nouhoun (Feb.) *Demange* 10/1957! **Guin.:** Kindia (Oct.) *Jac.-Fél.* 7195! Timbo (Oct.) *Pobéguin* 1792! 1793! Labé *Chev.* 34875 *bis*! Kouroussa (Oct.) *Pobéguin* 522! **S.L.:** Gorango (Dec.) *Deighton* 259! Kambia (Jan., Nov.) *Deighton* 924! *Morton & Gledhill* SL 73! Mt. Loma (Dec.) *Jaeger* 8394! 8512! **Lib.:** Nimba Mts. (Nov.) *Adam* 24662! **Iv. C.:** Dabou (Nov.) *Leeuwenberg* 1964! Bassawa (Oct.) *Adjanohoun* 430a! Ferkessédougou (Nov.) *Leeuwenberg* 1990! Sipilou (Oct.) *Aké Assi* 9679! **U. Volta:** Samandeni (Oct.) *Kmoch* 161! Leo to Po (Oct.) *Rose Innes* GC 31087! **Ghana:** Kwamang to Mampong (Nov.) *Rose Innes* GC 30873! Tamale (Nov.) *Williams* 1018! Gashiago to Pigu (Oct.) *Rose Innes* GC 30712! Lawra to Wa (Oct.) *Rose Innes* GC 30301! Walembele to Tumu (Oct.) *Rose Innes* GC 30790! **N. Nig.:** Abukur, Katsina (Oct.) *Clayton* 1350! Nupe *Barter*! Minna (Dec.) *Meikle* 726a! Shika (Oct.)

Freeman S146! Zaria Dist. (Dec.) *Taylor* 41! Gurum, Vogel Peak (Nov.) *Hepper* 1327! **S. Nig.**: Igboora (Oct.) *Haines* 339! Egba (Nov.) *Latilo* FHI 36700! Ibadan (Nov.) *Latilo* FHI 54398! Ezillo (Oct.) *Tuley* 965! Obudu Plateau (Nov.) *Tuley* 1016! Throughout the tropics.

2b. **S. brevifolium** var. **flaccidum** (*A. Rich.*) *Stapf* in F.T.A. 9: 188 (1917). *Andropogon flaccidus* A. Rich., Tent. Fl. Abyss. 2: 452 (1851). *A. brevifolius* var. *flaccidus* (A. Rich.) Hack. (1889). *Schizachyrium fasciculatum* Jac.-Fél. in Rev. Bot. Appliq. 33: 434 (1953).
Guin.: Baffing R. (Nov.) *Pobéguin* 1811! Siguiri (Oct.) *Jac.-Fél.* 1337! Friguiagbé (Nov.) *Chillou* 880! **S.L.**: Rochain (Dec.) *Koroma* 6! (mixed with *S. maclaudii*). **N. Nig.**: Gembu *Magaji & Tuley* 1861! (mixed with var. *brevifolium*) **S. Nig.**: Adani (Nov.) *Tuley* 22! **W. Cam.**: Lumbe R., Victoria (Jan.) *Maitland* 1292, partly! Occasionally found throughout tropical Africa and in India.

3. **S. maclaudii** (*Jac.-Fél.*) *S. T. Blake* in Proc. Roy. Soc. Queensland 80: 76 (1969). *S. brevifolium* var. *maclaudii* Jac.-Fél. in Rev. Bot. Appliq. 33: 432 (1953).
Guin.: Kindia *Jac.-Fél.* 208! Friguiagbé (Apr.) *Chillou* 1257! **S.L.**: Kambia (Dec.) *Deighton* 831! Picket Hill (Nov.) *T. S. Jones* 224! Roruks (Nov.) *Thomas* 5699! Rokupr (Nov.) *Jordan* 678! Musaia (Dec.) *Deighton* 4472! **Lib.**: Buchanan (Dec.) *Adam* 16430! Bahtown, Tchien Dist. (Aug.) *Baldwin* 8038! Kle, Boporo Dist. (Dec.) *Baldwin* 10570! Mecca, Boporo Dist. (Nov.) *Baldwin* 10401! Ganta (Nov.) *Harley* 1718! **Ghana**: Amedzofe (Nov.) *Morton* GC 9398! Agona to Tarkwa (Aug.) *Hall* 3349! Agona to Axim (Jan.) *Hall & Enti* GC 39116! **S. Nig.**: *Thomas* 1072! **W. Cam.**: Lumbe R., Victoria (Jan.) *Maitland* 1292, partly! Also in tropical south America.

4. **S. djalonicum** *Jac.-Fél.* in Rev. Bot. Appliq. 33: 426 (1953), and in J. Agric. Trop. 5: 551. Creeping annual about 30 cm. high.
Guin.: Benna (Oct.) *Jac.-Fél.* 7170! Pita (Oct.) *Jac.-Fél.* 1956! **S.L.**: Kayihi *Glanville* 427! **Mt. Loma** (Sept., Oct.) *Jaeger* 7714! 7861 *bis*!

5. **S. radicosum** *Jac.-Fél.* in Rev. Bot. Appliq. 33: 441 (1953). Procumbent annual up to 60 cm. high.
Guin.: Tristão Is. (Nov.) *Jac.-Fél.* 7338! Kindia *Jac.-Fél.* 255!

6. **S. penicillatum** *Jac.-Fél.* in Rev. Bot. Appliq. 33: 437 (1953). Tufted perennial about 60 cm. high.
Guin.: Kindia (Sept.) *Jac.-Fél.* 448! Madina Tossekré (Oct.) *Adam* 12530! 12559! **S.L.**: Tingi Mts. (Dec.) *Morton & Gledhill* SL 3101!

7. **S. lomaense** *A. Camus* in J. Agric. Trop. 1: 210 (1954). Annual about 23 cm. high; swampy soil.
S.L.: Bintumane (Oct., Nov.) *Jaeger* 401! 7885! *Morton* SL 2666! Tingi Mts. (Dec.) *Morton & Gledhill* SL 3136!

8. **S. delicatum** *Stapf* in F.T.A. 9: 190 (1917); A. Chev. in Rev. Bot. Appliq. 13: 862. *S. alatum* Jac.-Fél. in Rev. Bot. Appliq. 33: 429 (1953). Annual about 30 cm. high; dry, shallow soils.
Guin.: Timbo (Oct.) *Pobéguin* 1791! Macenta (Oct.) *Jac.-Fél.* 1298! **Iv. C.**: Sanlo to Kakpin (Nov.) *Aké Assi* 9292! **Ghana**: Sawla to Wa (Oct.) *Rose Innes* GC 32400! **N. Nig.**: Zaria (Sept.) *Clayton* 1311! Gidan Mada, Zonkwa (Oct.) *Tuley* 1715!

9. **S. urceolatum** (*Hack.*) *Stapf* in F.T.A. 9: 190 (1917). *Andropogon urceolatus* Hack. in Flora 68: 115 (1885). *A. nodulosus* var. *glabrescens* Pilger in Engl., Bot. Jahrb. 34: 127 (1904); A. Chev. in Rev. Bot. Appliq. 13: 862. Erect annual up to about 30 cm. high, with cinnamon-coloured racemes; shallow stony soils.
Sen.: Hasirik (Oct.) *Adam* 15717! Loumbol (Oct.) *Fotius* 137 *bis*. **U. Volta**: Leo to Po (Oct.) *Rose Innes* GC 31471! Zabré (Oct.) *Scholz* 153*a*! **Ghana**: Tamale (Oct.) *Addei*! Nandom (Oct.) *Rose Innes* GC 32297! Jirapa to Lawra (Oct.) *Ankrah* GC 20458! Bolgatanga to Tamale (Oct.) *Rose Innes* 30237*a*! Salaga (Oct.) *Rose Innes* GC 32453! **Togo Rep.**: Sokode *Kersting* 260! Kande to Sansanné Mango (Oct.) *Rose Innes* GC 31407! **Niger**: Niamey (Oct.) *Hagerup* 503! **N. Nig.**: Kufena hill, Zaria (Sept.) *Clayton* 1312! Damagum (Oct.) *De Leeuw* 1264! Jos (Oct.) *Oche & Tuley* 1708! Also in Ethiopia.

10. **S. nodulosum** (*Hack.*) *Stapf* in F.T.A. 9: 194 (1919); A. Chev. in Rev. Bot. Appliq. 13: 862; Berhaut, Fl. Sén. ed. 2, 423. *Andropogon nodulosus* Hack. in Flora 68: 116 (1885); Chev. Bot. 719. Annual, about 30 cm. high; rock and ironstone outcrops or shallow gravelly soils.
Sen.: *Heudelot* 303! Niokolo-Koba (Oct.) *Adam* 15501! Casamance (Oct.) *Adam* 18612! **Mali**: Koulikoro (Oct.) *Chev.* 2401! **Iv. C.**: Odienné (Nov.) *Aké Assi* 8270! **U. Volta**: Tondoura (Oct.) *Scholz* 81! Bangaregou (Oct.) *Scholz* 153*b*! Leo to Ouassa (Oct.) *Rose Innes* GC 31503! **Ghana**: Seripe to Dokrupe (Oct.) *Ankrah* GC 20439! Damongo Scarp F.R. (Nov.) *Rose Innes* GC 30835! Yendi to Chereponi (Nov.) *Rose Innes* GC 30470! Lawra (Oct.) *Hinds* 3827! Nakpanduri to Bawku (Oct.) *Rose Innes* GC 30498! **N. Nig.**: Gombe to Mubi (Oct.) *De Leeuw* 1841! 105 miles N of Gusau (Oct.) *De Leeuw & Magaji* 1967! **S. Nig.**: Abo *Barter* 371!

11. **S. sanguineum** (*Retz.*) *Alston* Suppl. Fl. Ceylon 334 (1931); Aké Assi, Contrib. 2: 298; Berhaut, Fl. Sén. ed. 2, 423. *Rottboellia sanguinea* Retz., Obs. Bot. 3: 25 (1783). *Schizachyrium semiberbe* Nees, Agrost. Bras. 336 (1829); F.T.A. 9: 195; F.W.T.A., ed. 1, 2: 584; A. Chev. in Rev. Bot. Appliq. 13: 863; Berhaut, Fl. Sén. ed. 2, 423. *S. domingense* (Spreng. ex Schult.) Nash in N. Amer. Fl. 17: 103 (1912). *S. hirtiflorum* Nees, Agrost. Bras. 334 (1829). *S. griseum* Stapf in F.T.A. 9: 194 (1918); F.W.T.A., ed. 1, 2: 584; A. Chev. in Rev. Bot. Appliq. 13: 862. *Streptachne domingensis* Spreng. ex Schult., Mant. 2: 188 (1824). *Andropogon domingensis* (Spreng.) F. T. Hubbard (1913), not of (Roem. & Schult.) Steud. (1821). *A. semiberbis* (Nees) Kunth (1833). *A. leptostachyus* Benth. (1849). Tall tufted perennial with reddish culms up to 3 m. high; polymorphic, and occupying a wide range of habitats from shallow stony soils to wet streamsides.
Sen.: Casamance (Oct.) *Adam* 18613! Sédhiou (Nov.) *Berhaut* 6362! **Mali**: Sikasso to Bougouni (Sept.) *Adam* 15462! **Guin.**: Ségou (Sept.) *Chev.* 2354! Younkounkoun (Oct.) *Pitot*! Baffing R. (Nov.) *Pobéguin* 1774! **Iv. C.**: Toumodi (Oct.) *Adjanohoun* 386*a*! Séguéla road (Oct.) *Adjanohoun* 305*a*! Bouaké (Nov.) *Leeuwenberg* 2078! Issia (July) *Adams* 878! 879! **U. Volta**: Linguékoro (Nov.) *Scholz* 45*b*! Tondoura (Oct.) *Scholz* 45*a*! Samandeni (Oct.) *Kmoch* 151! Gaoua to Batie (Oct.) *Rose Innes* GC 31526! Leo to Ouassa (Oct.) *Rose Innes* GC 31497! **Ghana**: Accra *Brown* 428! Keta (Dec.) *Thorold* 277! Tamale (Oct.) *Baldwin* 13561! Mirigu (Oct.) *Vigne* FH 4643! Gambaga (Oct.) *Rose Innes* GC 30739! **Togo Rep.**: *Baumann* 352! Palime *Stage* 84! Atakpame to Sokode (Oct.) *Rose Innes* GC 31364! Badou to Atakpame (Sept.) *Rose Innes* GC 31345! Sansanné Mango to Dapango (Oct.) *Rose Innes* GC 31423! **Dah.**: Porto Novo (Sept.) *Adjanohoun* 258! Parakou (Nov.) *Risopoulos* 1285! **N. Nig.**: Ilorin (Oct.) *Ward* 59! Kaiama (Oct.) *Ward* 26! Abinsi (July) *Dalz.* 869! Nupe *Barter* 1360! Badeggi (Sept.) *Clayton* 315! **S. Nig.**: Lagos (Oct.) *Haines* 195! Enugu (Sept.) *Jones* FHI 6748! Mamu River F.R. *E. W. Jones* 4! Awka (Sept.) *Onochie* FHI 34071! Ogoja *Rosevear* 5/30*a*! Throughout the tropics.
[Plants with glabrous spikelets occur throughout the range; in West Africa and tropical south America the spikelets may also be pubescent or villous. The latter have been separated as *S. domingense*, but the species are otherwise identical, with many intermediates, and it seems best to unite them.—W.D.C.]

12. **S. exile** (*Hochst.*) *Pilger* in Engl., Bot. Jahrb. 54: 284 (1917); F.T.A. 9: 191; A. Chev. in Rev. Bot. Appliq. 13: 862; Berhaut, Fl. Sén. ed. 2, 423; De Miré & Gillet in J. Agric. Trop. 3: 735; Aké Assi, Contrib. 2: 298. *Andropogon exilis* Hochst. in Flora 27: 241 (1844). Annual, 60–120 cm. high, the spatheoles at length becoming rusty red; typically on disturbed gravelly soils.
Sen.: Mbao (Oct.) *Berhaut* 2729! **Mali**: Dioura (Oct.) *Davey* 091! Bore (Oct.) *Davey* 8/1957! Bouta (Feb.)*Davey* 34! Ansongo (Sept.) *Hagerup* 392! **Port G.**: Nova Lamego (Oct.) *Pereira* 3353! **Iv. C.**: Bocanda to Ndokouassikro (Sept.) *Adjanohoun* 196*a*! **U. Volta**: Ouassa to Diebougou (Oct.) *Rose Innes* GC 31511! Nouna to Dedougou (Oct.) *Giessler* 65! Ouagadougou (Sept.) *Scholz* 153! Po to Leo (Oct.) *Rose Innes* GC 31084! **Ghana**: Kpeye to Kpandu (Sept.) *Ankrah* GC 20219! Sampa (Oct.) *Rose Innes* GC 31565! White Volta Bridge (Mar.) *Hepper & Morton* A3155! Bawku to Dapango (Oct.) *Rose Innes* GC 31439! Tumu (Oct.) *Rose Innes* GC 32281! **Togo Rep.**: Atakpame to Sokode (Oct.) *Rose Innes* GC 31374! **Dah.**: Sémé (Feb.) *Raynal* 13536! Boukombe (Nov.) *Risopoulos* 1268! **Niger**: Dogondoutch

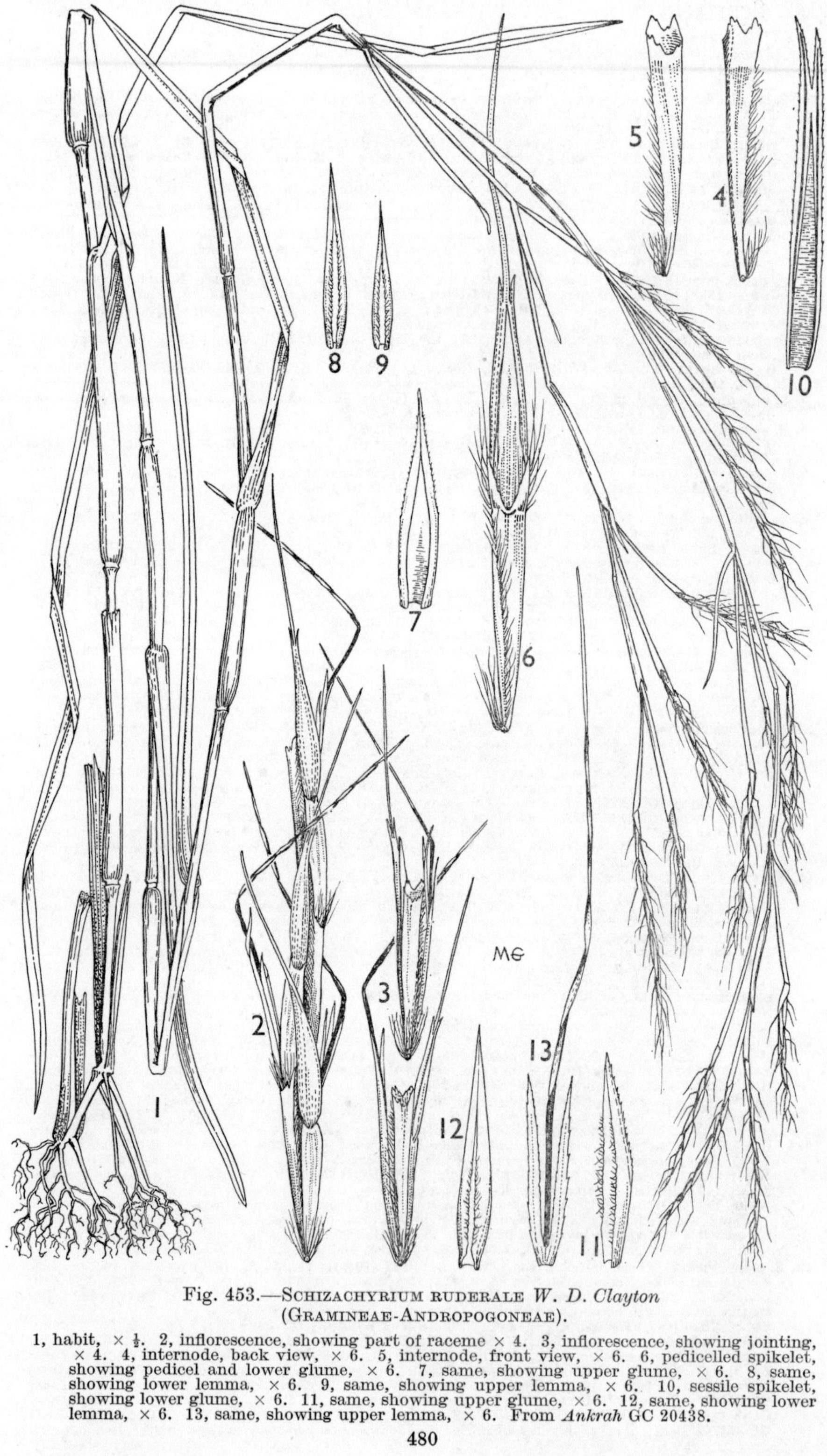

Fig. 453.—Schizachyrium ruderale *W. D. Clayton*
(Gramineae-Andropogoneae).

1, habit, × ½. 2, inflorescence, showing part of raceme × 4. 3, inflorescence, showing jointing, × 4. 4, internode, back view, × 6. 5, internode, front view, × 6. 6, pedicelled spikelet, showing pedicel and lower glume, × 6. 7, same, showing upper glume, × 6. 8, same, showing lower lemma, × 6. 9, same, showing upper lemma, × 6. 10, sessile spikelet, showing lower glume, × 6. 11, same, showing upper glume, × 6. 12, same, showing lower lemma, × 6. 13, same, showing upper lemma, × 6. From *Ankrah* GC 20438.

White 43! **N. Nig.:** Katsina (Oct.) *Clayton* 1343! Lokoja (Oct.) *Dalz.* 279! Kogigiri, Jos (Oct.) *Hepper* 1054! Kano (Dec.) *Adelodun* FHI 5406! Sokoto (Nov.) *Moiser* 142! Mongonu *Golding* 5! **S. Nig.:** Igboora (Oct.) *Haines* 338! Abakaliki (Dec.) *Baldwin* 13812! Ago-Are F.R. (Nov.) *Keay* 37687*a*! Also in the C. African Rep., Sudan, Ethiopia, Malawi and Mozambique. (See Useful Plants.)

13. **S. ruderale** *W. D. Clayton* in Kew Bull. 19: 451 (1965). Loosely tufted annual up to 2 m. high; roadside spoil heaps and new fallow land.
Sen.: Kolda (Oct.) *Boudet* 3986. Nétéboulou (Oct.) *Diallo* 556. Diamel (Nov.) *Fotius* 279. **Mali:** Sikasso (Oct.) *Demange* 2743. **Gam.:** *Saunders* 84! **Guin.:** Koundara (Oct.) *Adam* 12711! **U. Volta:** Leo to Ouassa (Oct.) *Rose Innes* GC 31495! Bangaregou (Oct.) *Scholz* 162! **Iv. C.:** Katiola to Dabakala (Nov.) *Aké Assi* 9234! Fettekro to Bassawa (Oct.) *Adjanohoun* 290*a*! **Ghana:** Tamale (Nov.) *Williams* 1021! Yendi (Oct.) *Rose Innes* GC 30978! Gambaga Scarp (Oct.) *Rose Innes* GC 30763! Navrongo to Lawra (Oct.) *Rose Innes* GC 30281! Seripe to Dokrupe (Oct.) *Ankrah* GC 20438! **Dah.:** Tanguieta to Boukombe (Nov.) *Risopoulos* 1270!

14. **S. pulchellum** (*Don ex Benth.*) *Stapf* in F.T.A. 9: 203 (1919); A. Chev. in Rev. Bot. Appliq. 13: 863; Berhaut, Fl. Sén. ed. 2, 422. *Andropogon pulchellus* Don ex Benth., Fl. Nigrit. 571 (1849); Chev. Bot. 719. *A. guineensis* P. Beauv. & Hochreut. (1898), not of Schumach. (1827). Perennial with prostrate inflorescences, the spathes becoming reddish brown, and rampant stolons; pioneer of sea-shore sands.
Sen.: Malika (July) *Adam* 17811! Cayor (Dec.) *Chev.* 2366! **S.L.:** Lumley Beach (Dec.) *Deighton* 789! Hamilton (Nov.) *Morton & Gledhill* SL 111! Mama Beach, near Kent (Dec.) *Deighton* 3287! Mahera (Nov.) *Deighton* 5670! Mano Salija (Nov.) *Deighton* 347! **Lib.:** Robertsport (Dec.) *Baldwin* 10916! 10934! Monrovia (Dec.) *Baldwin* 10490! **Iv. C.:** Grand Bassam (Mar.) *Leeuwenberg* 3137! Azuretti, Grand Bassam (Dec.) *Adjanohoun* 394*a*! Port Bouet (Dec.) *Aké Assi* 9857! **Ghana:** Aflao (May) *Thorold* CB53! Ada (Jan., June, Dec.) *T. M. Harris*! *Adams* 3622! *Morton* A1781! *Akpabla* 1879! **Dah.:** Kraké (Oct.) *Risopoulos* 1229! **S. Nig.:** Lagos (Sept., Dec.) *Dalz.* 1325! *Stubbings* 11! Tarqua Bay (Sept.-Jan.) *Birkett-Smith* 83*a*! *Ward* L141! *Haines* 152! Also in the Congo.

15. **S. schweinfurthii** (*Hack.*) *Stapf* in F.T.A. 9: 199 (1919). *Andropogon schweinfurthii* Hack. in Flora 68: 118 (1885). Erect caespitose perennial up to about 1·2 m. high with black roots; poorly drained shallow soils over hardpan.
U. Volta: Koumbili (Oct.) *Scholz* 180! Samandeni (Oct.) *Kmoch* 158! Po to Leo (Oct.) *Rose Innes* GC 31089! **Ghana:** Nungua (Nov.) *Ankrah* 43! Berekum to Sampa (Oct.) *Rose Innes* GC 31640! Kete Krachi to Yendi (Sept.) *Rose Innes* GC 30439! Gambaga to Yendi (Oct.) *Rose Innes* GC 30485! Navrongo to Tumu (Oct.) *Rose Innes* GC 30784! **Togo Rep.:** Sokode to Atakpame (Oct.) *Rose Innes* GC 31367! **N. Nig.:** Lokoja (Oct.) *Magaji* MG 263! *Dalz.* 289! Mokwa (Nov.) *Ayuba* AS104! Okene (Oct.) *Magaji* MG248! Ningi (Nov.) *Magaji & Tuley* 1823! Also in the Sudan.

16. **S. scintillans** *Stapf* in F.T.A. 9: 202 (1919); A. Chev. in Rev. Bot. Appliq. 13: 863; Berhaut, Fl. Sén. ed. 2, 422. Slender annual up to 60 cm. high; ironstone pans.
Sen.: Niokolo-Koba (Oct.) *Adam* 15631! **Guin.:** Conakry (Oct.) *Adam* 12606! Kindia (Oct.) *Jac.-Fél.* 7197! Fouta Djalon (Sept.-Oct.) *Adam* 12746! *Schnell* 7405 *ter*! *Chev.* 18664! **S.L.:** Kent (Nov.) *Deighton* 5632! No. 2 River (Nov.) *T. S. Jones* 380! Brookfields, Foni Flats (Oct.) *Deighton* 2175! Kambia (Nov.) *Jordan* 665! Lakka (Dec.) *Morton* SL 1599*a*!

17. **S. gresicolum** *Jac.-Fél.* in Rev. Bot. Appliq. 33: 446 (1953). Caespitose perennial about 30 cm. high; rock outcrops.
Mali: Kandiara R. (Oct.) *Pitot*! **Guin.:** Kindia *Jac.-Fél.* 183! Benna (Oct.) *Jac.-Fél.* 7171! **Ghana:** Amedzofe (Dec.) *Rose Innes* GC 31180! **N. Nig.:** Zonkwa (Oct.) *De Leeuw, Magaji & Tuley* 1716!

18. **S. rupestre** (*K. Schum.*) *Stapf* in F.T.A. 9: 204 (1919); Berhaut, Fl. Sén. ed. 2, 423. *Andropogon rupestris* K. Schum. in Engl., Bot. Jahrb. 24: 327 (1897). *A. compressus* Stapf in Journ. de Bot. 22: 204 (1909), incl. var. *pleiocladus* Stapf l.c.; Chev. Bot. 716. *Schizachyrium compressum* (Stapf) Stapf in F.T.A. 9: 201 (1919); F.W.T.A. ed. 1, 2: 585; A. Chev. in Rev. Bot. Appliq. 13: 863. *S. thollonii* var. *compressum* (Stapf) Jac.-Fél. in Rev. Bot. Appliq. 33: 448 (1953). *S. pratorum* C. E. Hubbard in Kew Bull. 4: 373 (1949). Caespitose perennial about 1 m. high, with black roots; stony slopes.
Sen.: Moro du Rip (Nov.) *Berhaut* 2739! Ziguinchor (Dec.) *Berhaut* 6706! Casamance R. (Jan.) *Chev.* 2348! **Mali:** Morigueya (Feb.) *Chev.* 435! **Port. G.:** Prabis, Bissau (Nov.) *Pereira & Correia* 2094! *Raimundo & Guerra* 205! Xime (Jan.) *Pereira & Correia* 2597! **S.L.:** Mateboi (Nov.) *Jordan* 834! Kasanko (Oct.) *Jordan* 647! Batkanu (Feb.) *Deighton* 4187! **Lib.:** Nimba (Oct.) *Adames* 700! *Warner* 26! 67! **Ghana:** Gradaw (Oct.) *Rose Innes* GC 31613! Banda to Bui (Oct.) *Rose Innes* GC 31599! Bamboi to Bole (Oct.) *Rose Innes* GC 30645! Dutukpene (Dec.) *Adams* 4529! Togo Plateau (Nov.) *Morton* A3484! **Togo Rep.:** Misahöhe *Baumann* 361! **N. Nig.:** Mokwa (Nov.) *De Leeuw & Magaji* 2005! **S. Nig.:** Nsukka (Jan.) *Okigbo* 117! Enugu (Sept.) *Tuley* 884!
[Closely allied to *S. thollonii* (Franch.) Stapf, which has the column of the awn only 1–2·5 mm. long.]

Imperfectly known species.

Andropogon simplex *Vahl & Schumach.* in Schumach., Beskr. Guin. Pl. 49 (1827); F.T.A. 9: 206 (1919). Apparently a species of *Schizachyrium*, but the type is lost, and the description cannot confidently be equated to any of the present species.

123. CYMBOPOGON Spreng., Pl. Pugill. 2: 14 (1815); F.T.A. 9: 265 (1919).

Lower glume of sessile spikelet lanceolate, almost flat to shallowly concave on the back with a V-shaped median groove in the lower half, laterally winged at the apex; leaf-blades with a broad (up to 2·5 cm.) rounded base, gradually tapering to a filiform tip; panicle 30 cm. long or more .. .. 1. *giganteus*

Fertile lemma awned, the awn geniculate with a twisted column 3–6 mm. long and a limb 4–7 mm. long .. .. 1a. *giganteus* var. *giganteus*

Fertile lemma awnless, or provided with a straight untwisted awn up to 6 mm. long 1b. *giganteus* var. *inermis*

Lower glume of sessile spikelet narrowly lanceolate, almost flat to deeply concave with the bottom of the depression rounded, wingless at the apex; leaf-blades linear to filiform, narrowing at the base:

Fertile lemma awnless; lower glume deeply concave between the keels; leaf-blades up to 60 cm. long or more and 15 mm. broad .. .. 2. *citratus*

Fertile lemma awned; leaf-blades up to 4 mm. broad:

Lower glume of the sessile spikelet deeply concave between the keels, these thickened and rounded in the lower half and separated by a gibbous swelling at the base; racemes scantily hairy, the hairs up to 2 mm. long; leaf-blades cauline, 2–4 mm. broad .. .. 3. *commutatus*

Lower glume of the sessile spikelet shallowly concave, the keels sharp throughout; racemes conspicuously villous with white hairs up to 4 mm. long; leaf-blades up to 3 mm. broad, mostly arising from the caespitose base .. 4. *schoenanthus*
Leaf-blades convolute, filiform, about 1 mm. broad, up to 15 cm. long; panicle scanty, the raceme-pairs borne singly or in clusters of 2–3, rarely more; old basal sheaths never curly .. .. 4a. *schoenanthus* subsp. *schoenanthus*
Leaf-blades flat and up to 3 mm. broad, but often convolute or folded in dried specimens, and up to 35 cm. long; panicle compound, the raceme-pairs aggregated into dense clusters; old basal sheaths often curly like wood-shavings
4b. *schoenanthus* subsp. *proximus*

1. **C. giganteus** *Chiov.* Gram. da Essenze 12 (1909); F.T.A. 9: 288; A. Chev. in Rev. Bot. Appliq. 13: 871; Berhaut, Fl. Sén. ed. 2, 420. *Andropogon giganteus* Hochst. in Flora 27: 242 (1844), not of Ten. (1811). Tall perennial up to 2·5 m. high with glaucous leaves and dense narrow panicles.

1a. **C. giganteus** *Chiov.* var. **giganteus**
Sen.: *Perrottet* 893! *Adanson* 47*b*! (June) *Roger*! Tambacounda (Oct., Nov.) *Chev.* 33911! M'Bidjem *Thierry* 93! **Gam.:** *Saunders* 85! *Dawe* 11! (Jan.) *Dalz.* 8415! Sukuta to Old Yundum (Feb.) *Wallace* K3! Sankuli-Kunda (Sept.) *Pirie* 51/33! **Mali:** Tiekole *Duong* 1500! Bamako (Jan.) *Duong*! Nioro to Aioun (Nov.) *Rossetti* 218/59! 219/59! Bura, Ansongo (Sept.) *Hagerup* 402! **Port. G.:** Nhacre (Nov.) *Raimundo & Guerra* 18! Mansoa (Nov.) *Pereira & Correia* 1973! **S.L.:** *T. Vogel*! **Iv. C.:** Assanou (Aug.) *Aké Assi* 6847! Séguéla (June) *Adjanohoun* 256*a*! Bouaké to M'Bayakro (Oct.) *Adjanohoun* 351*a*! Namboukaha to Ouorossantiakaha (Nov.) *Leeuwenberg* 1984! **U. Volta:** Batié to Gaoua (Oct.) *Rose Innes* GC 31534! Touroukoro (Nov.) *Scholz* 44*a*! Samandeni (Oct.) *Kmoch* 154! Kalamon (Sept.) *Aké Assi* 6489! Sindou (Sept.) *Aké Assi* 6363! **Ghana:** Accra Plains (Aug.) *Johnson* 756! Kintampo to Ejura (Oct.) *Rose Innes* GC 30619! Sawla to Bole (Oct.) *Rose Innes* GC 30662! Tamale to Bolgatanga (Oct.) *Rose Innes* GC 30230! Kugri Hill (Oct.) *Ankrah* GC 20297! **Togo Rep.:** Lomé *Warnecke* 402! Atakpame to Sokode (Oct.) *Rose Innes* GC 31365! Sansanné Mango to Dapango (Oct.) *Rose Innes* GC 31424! **Dah.:** Tanguieta to Kobli (Nov.) *Risopoulos* 1274! Kpinnou *Froment* 1149! **Niger:** Dogondoutchi *White* 54! **N. Nig.:** Kaiama (Oct.) *Ward* 37! Sokoto Prov. (July) *Dalz.* 485! Jemaa (Nov.) *Keay* FHI 37225! Gindiri, Jos Plateau (Oct.) *Hepper* 1087! Maiduguri (Oct.) *Johnston* N28! Garbabi, Sardauna Prov. (Jan.) *Latilo & Daramola* FHI 34420! **S. Nig.:** Lagos (Feb.) *W. D. MacGregor* 164! Igboora (Oct.) *Haines* 352! Ibadan (Dec.) *Fakunle & Russell* FHI 14967! Ago Are F.R., Shaki (Oct., Nov.) *Sofoluwe* FHI 38170! *Keay* FHI 37678! Throughout tropical Africa. (See Useful Plants.)

1b. **C. giganteus** var. **inermis** *W. D. Clayton* in Kew Bull. 19: 454 (1965).
Maur.: Kiffa *Cusfont*! N.W. of Nioro (Aug.) *Rossetti* 61/282! **Mali:** Macina (Nov., Dec.) *Duong* 1190! Bou Zériba (Oct.) *Monod* 4020!

2. **C. citratus** (*DC.*) *Stapf* in Kew Bull. 1906: 357; F.T.A. 9: 282; A. Chev. in Rev. Bot. Appliq. 13: 872; Berhaut, Fl. Sén. ed. 2, 420. *Andropogon citratus* DC., Cat. Hort. Monsp. 78 (1813); Chev. Bot. 716. Tall perennial 2 m. high or more, rarely flowering, the lower leaf-sheaths with a characteristic waxy bloom. Only known in the cultivated state; commonly planted along roads and in gardens.
S.L.: Njala (Apr.) *Deighton* 2495! **Lib.:** R. Cess, Grand Bassa Co. (Mar.) *Baldwin* 11229! **Iv. C.:** Tabou *Chev.* 20098! **Ghana:** Aburi (Oct.) *Howes* 997! Cultivated throughout the tropics. (See Useful Plants.)

3. **C. commutatus** (*Steud.*) *Stapf* in Kew Bull. 1907: 211; F.T.A. 9: 275. *Andropogon commutatus* Steud., Syn. Pl. Glum. 1: 387 (1854). Erect perennial 1·2–1·5 m. high.
Maur.: (Dec.) *Jumelle*! **Sen.:** Bakel (Dec.) *Berhaut* 1311! Also in Sudan and Eritrea.

4. **C. schoenanthus** (*Linn.*) *Spreng.* Pl. Pugill. 2: 15 (1815); F.T.A. 9: 268; A. Chev. in Rev. Bot. Appliq. 13: 871; Berhaut, Fl. Sén. ed. 2, 420; De Miré & Gillet in J. Agric. Trop. 3: 730; Stapf in Kew Bull. 1906: 303. *Andropogon schoenanthus* Linn., Sp. Pl. 1046 (1753). *A. laniger* Desf., Fl. Atlant. 2: 379 (1799); Chev. Bot. 718. *Cymbopogon schoenanthus* subsp. *laniger* (Desf.) Maire & Weiler, Fl. Afr. Nord. 1: 287 (1952).

4a. **C. schoenanthus** (*Linn.*) *Spreng.* subsp. **schoenanthus.** Densely caespitose glaucous perennial about 30 cm. high; in dry soils.
Mali: Ansongo (Sept.) *Hagerup* 330! **Ghana:** Saboba, Dagomba Dist. (Mar.) *Hepper & Morton* A3115! Kaleo Hill, Wa (Apr.) *Adams* 4008! N. African coast, eastwards to Arabia, the Orient and India; and southwards to the C. Sahara, Somalia and Eritrea. (See Useful Plants.)

4b. **C. schoenanthus** subsp. **proximus** (*Hochst. ex A. Rich.*) *Maire & Weiler.* Fl. Afr. Nord. 1: 287 (1952). *Andropogon proximus* Hochst. ex A. Rich., Tent. Fl. Abyss. 2: 464 (1851). *Cymbopogon proximus* (Hochst. ex A. Rich.) Stapf in F.T.A. 9: 271 (1919); F.W.T.A., ed. 1, 2: 588; A. Chev. in Rev. Bot. Appliq. 13; 871; Aké Assi, Contrib. 2: 301. *Andropogon schoenanthus* var. *proximus* (Hochst. ex A. Rich.) A. Chev., Bot. 719 (1920). Caespitose perennial 60–90 cm. high, on dry stony soils.
Maur.: Bafréchié, Trarza Circle (Sept.) *Sadio* 251! El Hagner Plateau (Aug.) *Monod* 2102! **Sen.:** *Heudelot* 304! **Mali:** Kare, near Dioura (Nov.) *Davey* 143! Douentza to Hombori (June) *Rogeon* 407! L. Horo, near Gidigada (Aug.) *Lean* 83! Ansongo to Tillaberi (Jan.) *Chev.* 43146! Gao to Ansongo (Jan.) *Chev.* 43042! **Guin.:** Falama (Mar.) *Chev.* 614! **Iv. C.:** Téhini (Aug.) *Oldeman* 328! **U. Volta:** Pontiéba (Aug.) *Winkoun*! Banfora (June) *Leeuwenberg* 4299! Ouagadougou (Oct.) *Scholz* 1*a*! Kampala (Mar.) *Scholz* 1! **Ghana:** Tong Hills (July) *Williams* 520! Tamale (May) *Saunders*! Chereponi (Oct.) *Rose Innes* GC 30975! Zuarungu (June) *Vigne* FH 4548! Bawku (Aug.) *Irvine* 4605! **Togo Rep.:** Kande (Oct.) *Rose Innes* GC 31405! **Dah.:** Atacora (June) *Chev.* 24075! **Niger:** Zinder (Nov.) *Money-Kirle* Fe 15! Tarraouaji, Aïr Mts., (Sept.) *Chopard & Villiers*! Toukounous *Bartha* 5! **N. Nig.:** Sokoto (July) *Dalz.* 486! Kworre (Sept.) *Palmer* H7535! Mongonu (Oct.) *Golding* 4! Kukawa (Mar.) *Jackson* 2597! Gujba (Aug.) *De Leeuw* 1130! Eastwards to Ethiopia, Sudan and Kenya, extending northwards into the Central Sahara. (See Useful Plants.)
[Closely allied to *C. jwarancusa* (Jones) Schult. a common species of flood plains in India.]

Various species of *Cymbopogon* are cultivated, both as ornamental garden plants and as a source of essential oils. The following have been recorded from West Africa: *C. densiflorus* (Steud.) Stapf—Gashaka Dist., Sardauna Prov., N. Nigeria; *C. nardus* (Linn.) Rendle—Hill Station, Freetown, Sierra Leone; *C. winteranus* Jowitt—Aburi, Ghana.

124. ANDROPOGON Linn., Sp. Pl. 1045 (1753), and Gen. Pl., ed. 5: 468 (1754); F.T.A. 9: 208 (1919); Clayton in Hook., Ic. Pl. 37: t. 3644 (1967).

Lower glume of sessile spikelet pitted on either side of a median groove; internodes of rhachis and pedicels linear:
Annual; inflorescence of 1–3 distant raceme-pairs partially embraced by narrow spatheoles; pits 2, circular; pedicelled spikelet with an awn 6–8 mm. long 1. *pusillus*
Perennial; inflorescence terminal, long exserted, of 2–3 digitate racemes; pits 3–8, irregularly oblong; pedicelled spikelets awnless .. .. .. 2. *lacunosus*

Lower glume not pitted:
Internodes of the rhachis and pedicels linear to filiform:
Racemes terminal on the culms; lower glume of sessile spikelet flat or slightly hollowed on the back with 5–11 intercarinal nerves, one of them median (except *A. mannii*); upper glume awned:
Base of plant silky hairy; sessile spikelet 9–11 mm. long; lower glume glabrous to puberulous, winged on the keels, with 7–11 intercarinal nerves; racemes paired .. .. 3. *distachyos*
Base of plant glabrous:
Spikelets pilose, 6–7 mm. long; lower glume with 5–7 intercarinal nerves; internodes and pedicels ciliate on both margins, glabrous on the face; leaf-blades up to 4 mm. broad, herbaceous; racemes paired .. .. .. 4. *amethystinus*
Spikelets glabrous:
Racemes usually 3–8, rarely 2 or 1; sessile spikelet 7–9 mm. long, the lower glume wingless and without intercarinal nerves; leaf-blades up to 4 mm. broad, stiff but not harshly scabrid, the midrib slender .. .. 5. *mannii*
Racemes 2; lower glume of sessile spikelet with winged keels and 5–7 intercarinal nerves:
Leaf-blades up to 2 mm. broad and 30 cm. long or more, harsh and file-like to the touch, the midrib almost ⅓ the width; sessile spikelets 6–8 mm. long; caespitose .. .. 6. *lima*
Leaf-blades up to 3 mm. broad and about 10 cm. long, herbaceous, with a slender midrib, scabrid only on the margins; sessile spikelets 7–8 mm. long; stoloniferous .. .. 7. *pratensis*
Racemes gathered into spathate false panicles; lower glume of sessile spikelet concave on the back, with no intercarinal nerves; upper glume acute:
Racemes solitary; callus of sessile spikelet 0·8 mm. long; spikelets 3–4 mm. long:
Lemma of fertile floret entire and awnless .. .. .. 8. *festuciformis*
Lemma of fertile floret bilobed and awned from the sinus .. 9. *incomptus*
Racemes paired or digitate:
Racemes 5–6; sessile spikelets about 5 mm. long, the internodes a little longer; internodes and pedicels finely filiform, arcuate, long ciliate .. 10. *tenuiberbis*
Racemes 2, rarely 3:
Leaf-blades involute, filiform; sheaths ciliate at the mouth; sessile spikelets 4·5–6·5 mm. long; awn 16–22 mm. long; rhachis internodes 4–5 mm. long 11. *curvifolius*
Leaf-blades flat, 2–4 mm. broad, abruptly acute at the tip:
Awns 5–9 mm. long; fertile lemma bifid for ⅓ its length; rhachis internodes 3 mm. long; racemes dense, the spikelets appressed; sessile spikelets 4·5–5 mm. long.. .. .. 12. *incanellus*
Awns 12–25 mm. long; fertile lemma bifid for up to ½ its length; rhachis internodes 3–4·5 mm. long; racemes loose, the spikelets divergent; sessile spikelets 4·5–6 mm. long .. .. 13. *africanus*
Internodes of the rhachis and pedicels distinctly swollen upwards, clavate or cuneate, rarely almost linear and then the back of the lower glume of the sessile spikelet flat with a conspicuous median groove (*A. tectorum*);
Lower glume of sessile spikelets linear, laterally compressed, wedged between internode and pedicel, the back deeply depressed with the keels dorsal and almost meeting over the depression; pedicels not, or obscurely, 2-lobed:
Racemes solitary, in a fastigiate panicle; upper glume of sessile spikelet with an awn about 1 cm. long; awn of fertile floret up to 5 cm. long; lower glume of pedicelled spikelet obliquely obovate-oblong, nearly 1 cm. long, many-nerved, ending in a fine bristle .. .. 14. *fastigiatus*
Racemes paired:
Upper glume of sessile spikelet and both glumes of pedicelled spikelet with fine bristle-like awns up to about 10 mm. long; inflorescence a delicate much-branched spathate panicle with conspicuous dark brown awns:
Annual; awn of fertile floret 35–40 mm. long; leaf-sheaths and blades glabrous 15. *pseudapricus*
Perennial; awn of fertile floret 20–35 mm. long; leaf-sheaths and blades glabrous or pubescent .. .. 16. *ascinodis*
Upper glume of sessile spikelet and glumes of pedicelled spikelet awnless (with awn-points up to 2 mm. long in *A. perligulatus*); raceme-pairs few on each culm, sometimes in scanty spathaceous panicles; perennials:
Racemes markedly dorsiventral, the pedicelled spikelets lying flat upon the back, imbricate, and concealing the internodes and pedicels; no auricles; internodes and pedicles ciliate on both sides, not inflated:
Back of lower glume of sessile spikelet convexly rounded and 5–6-nerved on either side of the median depression, becoming keeled only in the upper half; internodes and pedicels stout, cylindric; palea absent .. .. 17. *ivorensis*

Back of lower glume of sessile spikelet sharply 2-keeled throughout, the keels dorsal and contiguous; internodes and pedicels clavate; upper floret with a palea:
Leaf-blades linear, up to 45 cm. long and 14 mm. broad, mostly cauline; racemes 6–12 cm. long; sessile spikelets 5–7 mm. long .. .. 18. *schirensis*
Leaf-blades lanceolate-linear, 5–15 cm. long and up to 5 mm. broad, clustered towards the base; racemes 3–7 cm. long; sessile spikelets 5 mm. long 19. *dummeri*
Racemes not dorsiventral, the pedicelled spikelets more or less lateral and not covering the internodes and pedicels:
Sheath-auricles and ligules 5–12 mm. long, usually glabrous at the throat; pedicels clavate, about as wide as the gap between internode and pedicel; upper glume usually aristulate .. .. 20. *perligulatus*
Sheath-auricles and ligules up to 2 mm. long, usually setose at the throat; pedicels inflated, about twice as wide as the gap between internode and pedicel; upper glume acute .. .. 21. *canaliculatus*
Lower glume of sessile spikelet lanceolate, not wedged between internode and pedicel, flattened on the back with a shallow median groove, the keels lateral; racemes paired, in large spathate panicles:
Rhachis internodes and pedicels almost linear, slightly widening at the truncate tips, 3 mm. long, ciliate on both sides; leaf-blades thin, lanceolate, light green; panicle large, decompound; racemes 3–4 cm. long; sessile spikelet 4–5 mm. long with an awn 15–20 mm. long .. .. 22. *tectorum*
Rhachis internodes and pedicels clavate, conspicuously widening above, the pedicel more or less bilobed at the tip; leaf-blades firm; racemes 4–6 cm. long:
Leaf-blades very large, 20–40 mm. broad and up to 75 cm. long; pedicels broadly clavate, 4 mm. long, scantily ciliate on both sides and often puberulous on the face; lower glume of sessile spikelet 5–6 mm. long; awn of fertile floret 9–16 mm. long; pedicelled spikelet 4–5 mm. long, its awn up to about 1 mm. long; external ligule absent .. .. 23. *macrophyllus*
Leaf-blades usually less than 16 mm. broad, rarely up to 25 mm.:
Mouth of leaf-sheath produced into acute linear-lanceolate auricles 5–30 mm. long; lower glume of sessile spikelet with a line of hairs in the median groove; leaf-blades up to 10 mm. broad, but usually less:
Perennial, the basal sheaths compressed and keeled; internodes and pedicels glabrous or sparsely hairy on the face; sessile spikelet 5·5–7 mm. long, the awn 10–17 mm. long; upper glume glabrous to puberulous 24. *auriculatus*
Annual; internodes and pedicels pubescent to tomentose on the face; sessile spikelet 5·5–10 mm. long, the awn 18–25 mm. long; upper glume pubescent to tomentose .. .. 25. *chevalieri*
Mouth of leaf-sheath forming rounded shoulders, or rarely produced into ovate auricles up to 3 mm. long:
Sessile spikelet 9–11 mm. long, the keels of both glumes bearing a cartilaginous wing; pedicelled spikelet 8–10 mm. long, pilose; pedicels ciliate on both sides .. .. 26. *pteropholis*
Sessile spikelet up to 8 mm. long, the glumes inconspicuously winged:
Pedicels swollen, almost ovoid, about 4 mm. long and 1·8 mm. broad, ciliate both sides; sessile spikelet 5–6 mm. long, with an awn 15–22 mm. long; pedicelled spikelet glabrous or puberulous; annual .. 27. *pinguipes*
Pedicels clavate, up to 6 times as long as wide, their sides straight or only slightly convex; perennials:
Pedicels 3 mm. long, scantily ciliate on one or both sides, and often puberulous on the face; pedicelled spikelet 4–4·5 mm. long, awnless or with a mucro up to 1 mm. long; sessile spikelet 4–5 mm. long, usually with a line of hairs along the median groove; awn 9–12 mm. long; external ligule absent 28. *gabonensis*
Pedicels 4–5 mm. long, conspicuously ciliate, glabrous on the face; pedicelled spikelet 5–7 mm. long; sessile spikelet 5–8 mm. long, glabrous; external ligule often present as a membranous rim at the junction of sheath and blade on the abaxial side .. .. 29. *gayanus*
Internodes and pedicels ciliate on one side:
Pedicelled spikelet glabrous; sessile spikelet 6 mm. long; awn of sessile spikelet 10–20 mm. long, and of pedicelled spikelet 1–2 mm. 29a. *gayanus* var. *gayanus*
Pedicelled spikelet villous or pubescent; sessile spikelet 6–8 mm. long; awn of sessile spikelet 15–30 mm. long, and of pedicelled spikelet 5–10 mm. 29b. *gayanus* var. *tridentatus*
Internodes and pedicels ciliate on both sides; sessile spikelet 5–6 mm. long; awn of sessile spikelet 15–30 mm. long and of pedicelled 3–7 mm.:
Pedicelled spikelet glabrous .. .. 29c. *gayanus* var. *squamulatus*
Pedicelled spikelet villous .. .. 29d. *gayanus* var. *bisquamulatus*

1. **A. pusillus** *Hook. f.* in J. Linn. Soc. 7: 233 (1864); Clayton in Kew Bull. 19: 455. *Hyparrhenia pusilla* (Hook. f.) Stapf in F.T.A. 9: 379 (1919); F.W.T.A., ed. 1, 2: 591. Annual up to 15 cm. high.
N. Nig.: Chappal Waddi, Mambila Plateau (Nov.) *Jackson, Magaji & Tuley* 2054! **W. Cam.:** Cam. Mt., 7,000 ft. (Dec.) *Mann* 2097!
2. **A. lacunosus** *J. G. Anderson* in Bothalia 8: 113 (1962). Perennial up to 60 cm. high; wet places.
W. Cam.: Mbiami, 6,000 ft. (June) *Brunt* 746! Also in Transvaal.
3. **A. distachyos** *Linn.* Sp. Pl. 1046 (1753); F.T.A. 9: 218. Perennial, 30–75 cm. high.
W. Cam.: *Mann* 1345! Mann's Spring, Cam. Mt., 7,500 ft. (Feb., Dec.) *Boughey* GC 10621! *Mann* 2078! *Steele* 101! Mauve, 8,000 ft. (Feb.) *Maitland* 1035! Europe, N. Africa, Levant and Arabia, highlands of tropical Africa and in S. Africa.
4. **A. amethystinus** *Steud.* Syn. Pl. Glum. 1: 371 (1854); F.T.A. 9: 216, partly. *A. abyssinicus* of F.W.T.A., ed. 1, 2: 586, not of R. Br. Perennial, tufted or with short underground rhizomes, about 30 cm. high.
W. Cam.: Cam. Mt., 7,800–9,000 ft. (Nov.-Jan.) *Migeod* 170! *Hinds* C43! *Maitland*! *Keay* FHI 28587! Bafut-Ngemba F.R., 6,990 ft. (Feb.) *Hepper* 2123! **F. Po:** S. Isabel Peak (Mar.) *Guinea* 2778! Also in Ethiopia, Kenya and Uganda.
5. **A. mannii** *Hook. f.* in J. Linn. Soc. 7: 232 (1864); F.T.A. 9: 226. Densely tufted perennial 30–45 cm. high.
S.L.: Mt. Loma (Feb.-Apr.) *Jaeger* 9781! 9418! 9487! **Iv. C.:** Mt. Momi (Oct.) *Aké Assi* 9165! **W. Cam.:** Cam. Mt., 9,500–12,000 ft. (Dec., Jan.) *Mildbr.* 10892! *Maitland* 1271! *Hinds* C35! C39! *Keay* FHI 28626! **F. Po:** Clarence Peak, 9,000 ft. (Dec.) *Mann* 654!
6. **A. lima** (*Hack.*) *Stapf* in F.T.A. 9: 217 (1919). *A. amethystinus* vars. *lima* and *breviaristatus* Hack. in DC., Monogr. Phan. 6: 464 (1889). *A. amethystinus* of F.T.A. 9: 216, partly. Densely tufted perennial up to 1 m. high.
W. Cam.: Cam. Mt., 8,500 ft. (Dec.) *Mildbr.* 10856! Mauve, 8,000 ft. (Feb.) *Maitland* 1041! Ukele Camp, 8,000 ft. (Feb.) *Maitland* 1340! Johann-Albrechtshutte, 9,350 ft. (Jan., Dec.) *Breteler et al.* MC10! *Mann* 2084! Also in the Sudan, Ethiopia, the African highlands, and Malawi.
7. **A. pratensis** *Hochst. ex Hack.* in DC., Monogr. Phan. 6: 463 (1889); F.T.A. 9: 219. Stoloniferous perennial up to about 45 cm. high.
N. Nig.: Vom (Oct.) *Gambles* 26*a*! Also in Ethiopia, Kenya, Tanzania and Zambia.
8. **A. festuciformis** *Rendle* in Cat. Welw. 2: 145 (1899); Clayton in Kew Bull. 17: 469 (1964). *A. schlechteri* Hack. in Bull. Herb. Boiss., ser. 2, 6: 703 (1906). *Hypogynium schlechteri* (Hack.) Pilger (1940). *H. spathiflorum* of F.T.A. 9: 168, not of Nees. Caespitose perennial 1 m. high; flood plains and damp grasslands.
Mali: Sikasso, Chutes du Farako (Oct.) *Demange* 3039. **Guin.:** Fouta Djalon (Dec.) *Jac.-Fél.* 7430! Also in the Congo, Zambia, Angola and S. Africa.
9. **A. incomptus** *W. D. Clayton* in Kew Bull. 17: 467 (1964). Caespitose perennial up to 2·5 m. high.
Guin.: Madina Tossekré (Oct.) *Adam* 12528!
10. **A. tenuiberbis** *Hack.* in DC., Monogr. Phan. 6: 435 (1889); F.T.A. 9: 232. *A. calvescens* Stapf in F.T.A. 9: 232 (1919); F.W.T.A., ed. 1, 2: 586; A. Chev. in Rev. Bot. Appliq. 13: 863; Berhaut, Fl. Sén. ed. 2, 420. *Leptopogon tenuiberbis* (Hack.) Roberty in Boissiera 9: 203 (1960). Coarse tufted perennial up to 5 m. high, with sharply keeled and interleaved sheaths; waterlogged soils.
Sen.: Kaême (Oct.) *Kerharo & Adam* 1509. **Guin.:** Baffing R. (Oct.) *Pobéguin* 1802! **S.L.:** Gberia Timbako (Oct.) *Small* 465! Dankawali (Nov.) *Glanville* 323! **Lib.:** Siaple, Sanokwele Dist. (Sept.) *Baldwin* 9457! **Iv. C.:** Bouna to Bondoukou (Oct.) *Rose Innes* GC 31555! Boundiali to Mankono (Nov.) *Aké Assi* 8292! **Ghana:** Sawla (Oct.) *Rose Innes* GC 30667! Bole (Oct.) *Rose Innes* GC 32406! **N. Nig.:** Ilorin (Oct.) *Ward* 92! **S. Nig.:** Oyo (Oct., Nov.) *Jackson* 2335! 121168! Okigwi (Oct.) *Ariwaodu* 1408! Also in the Congo and the Sudan.
11. **A. curvifolius** *W. D. Clayton* in Kew Bull. 17: 465 (1964). Densely tufted perennial about 1 m. high; shallow pockets and clefts in rock outcrops.
Guin.: Macenta (Oct.) *Baldwin* 9825! **Lib.:** Monrovia (Aug.) *Baldwin* 9191! Genne-Loffa (Nov.) *Baldwin* 10081! **Iv. C.:** Dougba (Sept.) *Adjanohoun* 187*a*! 188*a*! Minouré (Oct.) *Adjanohoun* 364*a*! Séguéla (Oct.) *Aké Assi* 6585! Séguéla to Bouaké (Oct.) *Adjanohoun* 280*a*! **Ghana:** Krobo hill (Nov.) *Rose Innes* GC 30861! GC 31153! Wa to Ducie (Nov.) *Rose Innes* GC 30812! Kwahu Tafo (Dec.) *Adams* 4931!
12. **A. incanellus** *W. D. Clayton* in Kew Bull. 17: 466 (1964). Tufted perennial up to 1·5 m. high; seasonally flooded alluvial soils.
U. Volta: Fada N'Gourma (Aug.) *Scholz* 144! **Ghana:** Yendi (Nov.) *Addei* SLUS 741! Kete Krachi to Kpandu (Sept.) *Rose Innes* GC 30416! Dorimon (Oct.) *Rose Innes* GC 31033! Yapei ferry (Oct.) *Rose Innes* GC 31118! Lawra *Hinds* 3750!
13. **A. africanus** *Franch.* in Bull. Soc. Hist. Nat. Autun 8: 325 (1895); F.T.A. 9: 239; A. Chev. in Rev. Bot. Appliq. 13: 863; Berhaut, Fl. Sén. ed. 2, 426. *A. linearis* Stapf in F.T.A. 9: 239 (1919); F.W.T.A., ed. 1, 2: 588; A. Chev. in Rev. Bot. Appliq. 13: 864; Aké Assi, Contrib. 2: 300. *A. prolixus* Stapf (1908). *Anatherum africanum* (Franch.) Roberty in Boissiera 9: 207 (1960). Caespitose perennial 0·6–2 m. high; leaf-sheaths usually glabrous, rarely ciliate at the mouth, but specimens from Bintumane have the upper part of the sheath and lower part of the blade densely pilose; flood plains and swampy soils, less often among rock outcrops.
Maur.: L. Rkiz (Sept.) *Sadio* 5! 25! **Sen.:** Marais des Bayottes, Casamance (Aug.) *Berhaut* 6333! **Mali:** Bamako to Sotuba (Sept.) *Adam* 15327! Ségou (Nov.) *Roberty* 3344! Nouhoun (Feb.) *Davey* 420! Diafarabé (Jan.) *Demange* 1 (1957)! Klela (Oct.) *Farrow* 66! **Guin.:** Conakry (Oct.) *Adam* 12609! Labé (Oct.) *Adam* 12521! Fouta Djalon *Chev.* 18666! **S.L.:** Bumban (Oct.) *Deighton* 3278! Mt. Loma (Oct.) *Jaeger* 7963! Bintumane Peak, 5,000 ft. (Jan., Nov.) *T. S. Jones* 151! *Nichols* 4! *Glanville* 334! **Iv. C.:** Toumodi to Yamoussokro (Oct.) *Adjanohoun* 326*a*! N. Toumodi (Oct.) *Adjanohoun* 345*a*! Zuenoula (Oct.) *Adjanohoun* 310*a*! Féttékro to Bassawa (Oct.) *Adjanohoun* 284*a*! Bouna to Bondoukou (Oct.) *Rose Innes* GC 31556! **U. Volta:** Bobo-Dioulasso (Sept.) *Adam* 15091! Gaoua (Oct.) *Rose Innes* GC 31502! **Ghana:** Chache, Black Volta R. (Oct.) *Rose Innes* GC 30660! Wa to Sawla (Oct.) *Rose Innes* GC 31022! Wa to Lawra (Sept.) *Hall* 608! Tamale to Salaga (Aug.) *Rose Innes* GC 31874! **N. Nig.:** Badeggi (Sept.) *Clayton* 327! Vodni, Panshin Div. *Saunders* 31! 44! Jos Plateau (Sept.) *Lely* P758! **S. Nig.:** Ado Rock, Oyo (Oct.) *Savory & Keay* FHI 25299! Iseyin (Oct.) *Jackson* 2362! *Stanfield*! Through the Congos to Zambia and Angola; also in Kenya.
14. **A. fastigiatus** *Sw.* Prodr. Veg. Ind. Occ. 26 (1788). *Diectomis fastigiata* (Sw.) Kunth in Humb. & Bonpl., Nov. Gen. & Sp. 1: 193 (1816); F.T.A. 9: 207; F.W.T.A., ed. 1, 2: 585; A. Chev. in Rev. Bot. Appliq. 13: 863; Berhaut, Fl. Sén. ed. 2, 422. *Cymbachne fastigiata* (Sw.) Roberty in Boissiera 9: 255 (1960). Annual, up to about 1 m. high, with conspicuous purplish papery pedicelled spikelets; shallow or sandy soils, and dry fallow land.
Maur.: 15°40'N, 9°30'W. *Rossetti* 61/298! **Sen.:** Richard-Toll (Oct.) *Berhaut* 2741! **Mali:** Bore (Nov.) *Demange* 2 (1957)! Dioura (Sept.) *Farrow* 101! **Guin.:** Kindia *Jac.-Fél.* 214! Mamou (Nov.) *Chev.* 34600*bis*! Dalaba (Oct.) *Adames* 393! Puta (Jan.) *Langdale-Brown* 2564! Baffing R. (Nov.) *Pobéguin* 1814! **S.L.:** Russell (Dec.) *Deighton* 3284! Musaia (Dec.) *Deighton* 4441! Kasona (Nov.) *Glanville* 331! Lengekoro (Dec.) *Deighton* 5716! **Iv. C.:** Bouaké (Nov.) *Leeuwenberg* 2088! Féttékro to Bassawa (Oct.) *Adjanohoun* 283*a*! Bango to Kiri (Sept.) *Chev.* 24876! **U. Volta:** Baoulé (Oct.) *Kmoch* 125! Bobo-Dioulasso (Oct.) *Kmoch* 141! *Koechlin* 6946! 6968! Fada N'Gourma (Aug.) *Scholz* 145! **Ghana:** Sunyani to Wenchi (Dec.) *Adams* 5241! Seripe to Dokrupe (Oct.) *Ankrah* GC 20437! Tamale (Nov.) *Williams* 1009! Chache, Black Volta R. (Oct.) *Rose Innes* GC 30658! Kugri Hill (Oct.) *Ankrah* GC 20294!

Togo Rep.: Atacora Mt. Pass (Oct.) *Rose Innes* GC 31398! Sansanné Mango to Dapango (Oct.) *Rose Innes* GC 31429! **Dah.:** (Aug.) *Poisson*! Cotonou to Allada (Oct.) *Risopoulos* 1235! Boukombe (Nov.) *Risopoulos* 1262! **Niger:** Niamey (Oct.) *Hagerup* 500! Tahoua (Sept.) *Koechlin* 6629! Toukounous (Sept.) *Koechlin* 6563! 6909! **N. Nig.:** Okene (Nov.) *Keay* FHI 28084! Katsina (Oct.) *Clayton* 1337! Kaiama (Oct.) *Ward* 32! Abinsi (Dec.) *Dalz.* 894! Samaru (Oct.) *Thatcher* S503! **S. Nig.:** Ibadan (Nov.) *Keay* FHI 28126! Olokemeji (Nov.) *Onochie* FHI 8151! Onitsha (Dec.) *Killick* 92! Udi (Dec.) *Onyeagocha* FHI 7776! Abakaliki *Baldwin* 13809! Throughout the tropics.

15. **A. pseudapricus** *Stapf* in F.T.A. 9: 242 (1919); A. Chev. in Rev. Bot. Appliq. 13: 864; Berhaut, Fl. Sén. ed. 2, 425. *A. apricus* var. *africanus* Hack. in DC., Monogr. Phan. 6: 457 (1889); Chev. Bot. 716. Erect annual, 60–120 cm. high; shallow or sandy soils, common on fallow land.
Sen.: Vélingara to Goloumbo (Oct.) *Adam* 18611! Kaolack (Mar.) *Chev.* 28499! Tambacounda (Oct., Nov.) *Berhaut* 1761! *Adam* 12719! Bargny (Oct.) *Berhaut* 2720! **Gam.:** *Saunders* 77! *Hayes* 584! *Brooks* 11! McCarthy Is., Gambia R. (Jan.) *Dalz.* 8414! Fulladu *Frith* 149! **Mali:** Kankan *Bardau* 24! Bamako *Adam* 11241! Nyamina to Koulikoro (Oct.) *Chev.* 2347! Ségou (Oct.) *Rogeon* 243! Méma (Oct.) *Vintrebert* 11/1957! **Port. G.:** Boruntuma (Dec.) *Pereira & Correia* 2177! Farim (Dec.) *Pereira* 1039! **Guin.:** Kindia *Jac.-Fél.* 213! Mamou (Nov.) *Chev.* 34600! Bafing (Oct.) *Adam* 12664! Koundara (Oct.) *Adam* 12712! Kouroussa (Sept.) *Pobéguin* 516! **S.L.:** Freetown (Dec.) *Deighton* 797! Hill Station to Regent (Oct.) *Deighton* 2180! Mateboi (Nov.) *Jordan* 844! Rokupr (Dec.) *Jordan* 712! Musaia (Dec.) *Deighton* 4442! **U. Volta:** Samandeni (Oct.) *Kmoch* 162! Black Volta (Oct.) *Kmoch* 127! Bobo-Dioulasso (Oct.) *Kmoch* 142! Leo to Ouassa (Oct.) *Rose Innes* GC 31508! Leo to Po (Oct.) *Rose Innes* GC 31080! **Iv. C.:** Boundiali (Oct.) *Boudet* 3303! **Ghana:** Yendi to Kete Krachi (Oct.) *Ankrah* GC 20415! Tamale (Nov.) *Williams* 1016! Mirigu (Oct.) *Vigne* FH 4647! Bawku to Bolgatanga (Oct.) *Rose Innes* GC 31450! Bolgatanga to Navrongo (Oct.) *Rose Innes* GC 31098! **Togo Rep.:** Sansanné Mango to Dapango (Oct.) *Rose Innes* GC 31436! **Dah.:** Boukombe (Nov.) *Risopoulos* 1252! **N. Nig.:** Okene (Oct.) *Ward* L148! Kaiama (Oct.) *Ward* 24! Anara F.R., Zaria Prov. (Oct.) *Keay* FHI 5440! Sokoto (Oct.) *Dalz.* 490! Maiduguri, Bornu Prov. (Oct.) *Johnston* N38! **S. Nig.:** Ago Are F.R. (Nov.) *Keay* FHI 37696! Also in the Bagirmi region, Chad, and has been found in Mexico and Brazil. (See Useful Plants.)

16. **A. ascinodis** *C. B. Cl.* in J. Linn. Soc. 25: 87 (1889). *A. apricus* var. *indicus* Hack. in DC., Monogr. Phan. 6: 457 (1889). Coarse tufted perennial up to 1·5 m. high, inflorescence less dense than the preceding species; shallow stony soils and old fallow land.
Sen.: 4 km. N. of Guinea frontier (Oct.) *Boudet* 4193. Kanéméré (Sept.) *Fotius* K464. (*fide* Lebrun in Bull. Soc. Bot. Fr. 116: 257). **Guin.:** Mamai (Nov.) *Chev.* 34867*bis*! **S.L.:** Kasanko (Oct.) *Jordan* 646! Robumbe (Oct.) *Jordan* 801! Mange to Kambia (Oct.) *Adames* 259! **Iv. C.:** Béréby (Nov.) *Oldeman* 503! Toumodi (Aug.) *Boughey* GC 18542! Boka (Oct.) *Boudet* 3052! Vavoua to Séguéla (Oct.) *Adjanohoun* 405*a*! **U. Volta:** Gaoua to Batié (Oct.) *Rose Innes* GC 31523! Samandeni (Oct.) *Kmoch* 132! Bobo-Dioulasso (Oct.) *Kmoch* 115! Leo to Ouassa (Oct.) *Rose Innes* GC 31501! Tondoura (Oct.) *Scholz* 27! **Ghana:** Nungua (Nov.) *Ankrah* GC 20328! Dabala junction, Sogakope to Denu (Nov.) *Ankrah* GC 20515! Bole (Oct.) *Rose Innes* GC 30661! Kintampo to Tamale (Oct.) *Rose Innes* GC 31121! Zawse Dam (Sept.) *Ankrah* GC 20289! **Togo Rep.:** Lama Kara to Sokode (Oct.) *Rose Innes* GC 31644! **Dah.:** Parakou (Nov.) *Risopoulos* 1286! **N. Nig.:** Ilorin (Oct.) *Ward* 38! Kaiama (Oct.) *Ward* 25! Middle Belt, Niger Prov. (Dec.) *Hubbard* S576! Zaria (Oct.) *Keay* FHI 21422! Kangimi, Anara F.R. (Oct.) *Keay* FHI 5483! **S. Nig.:** Ago Are F.R., Oyo (Nov.) *Keay* FHI 37684! 37688! Igbete (Nov.) *Gledhill* 714! Also in Congo and Rhodesia, and in India, Burma and Thailand.
[Closely allied to *A. angustatus* (Presl) Steud. of tropical America, which differs in having less hairy racemes and 1-awned pedicelled spikelets.]

17. **A. ivorensis** *Adjanohoun & Clayton* in Adansonia 3: 401 (1963). Perennial up to nearly 2 m. high.
Mali: Sikasso, Chutes du Farako (Sept.) *Demange* 2736. **Iv. C.:** Kongasso (Oct.) *Adjanohoun* 388*a*! Samofigila (Oct.) *Boudet* 3319! **Togo Rep.:** Atakpamé to Sokode (Oct.) *Rose Innes* GC 31385!

18. **A. schirensis** *Hochst. ex A. Rich.* Tent. Fl. Abyss. 2: 456 (1851); F.T.A. 9: 246; Chev. Bot. 719, and in Rev. Bot. Appliq. 13: 866; Berhaut, Fl. Sén. ed. 2, 426. Erect tufted perennial up to 2 m. high; particularly on shallow soils over rock or ironstone.
Sen.: Kanéméré (July) *Fotius* K321. (*fide* Lebrun in Bull. Soc. Bot. Fr. 116: 257.) **Guin.:** Timbo (Oct.) *Pobéguin* 1780! Soumbalako to Boulivel, Fouta Djalon *Chev.* 18645! Kouroussa (Aug., Sept.) *Pobéguin* 503! 509! **S.L.:** Yonibana (Oct.) *Jordan* 591! Mamalia (Nov.) *Thomas* 4342! Bubuya (Sept.) *Jordan* 349! Bintumane (Nov.) *Glanville* 333! Musaia (Dec.) *Deighton* 4519! **Iv. C.:** Singrobo (Oct.) *Adjanohoun* 317*a*! 338*a*! Ferkessédougou (Sept.) *Aké Assi* 6302! Nimba Mts. (Aug.) *Boughey* GC 18070! Tehini (Aug.) *de Wilde* 802! **U. Volta:** Batié to Kalamon (Sept.) *Aké Assi* 6488! Gaoua to Batié (Oct.) *Rose Innes* GC 31522! Diebougou to Ouassa (Oct.) *Rose Innes* GC 31513! Ouassa to Nandom (Oct.) *Rose Innes* GC 31507! Tondoura (Oct.) *Scholz* 12*c*! **Ghana:** Accra to Kpong (Nov.) *Rose Innes* GC 31649! Amedzofe (Sept.) *Ankrah* GC 20221! Wenchi to Bole (Sept.) *Rose Innes* GC 30155! Attebubu to Ejura (Oct.) *Rose Innes* GC 30591! Gambaga (Oct.) *Vigne* FH 4634! **Togo Rep.:** Badou to Atakpamé (Sept.) *Rose Innes* GC 31350! Sokode to Atakpamé (Oct.) *Rose Innes* FHI 31381! **Dah.:** Ina to Bembereke (Sept.) *Risopoulos* 1201! **N. Nig.:** Bonu, Gwari (Sept.) *Onochie* FHI 18282! Wuya Ferry, Kaduna R. (Sept.) *Clayton* 368! Abinsi (Sept.) *Dalz.* 906! Gindiri, Jos Plateau (Oct.) *Hepper* 1138! Shika, Zaria (Oct.) *Clayton* 1330! **S. Nig.:** Lagos (Feb.) *W. MacGregor* 153! Iseyin (Oct.) *Jackson* 2366! Ago-Are F.R. (Oct.) *Sofoluwe* FHI 38183! Enugu (Sept.) *Jones* FHI 6749! Ogoja Prov. *Rosevear* 3/30*a*! **W. Cam.:** Bambui (Dec.) *Pedder* 2! Bafut-Ngemba F.R. (Feb.) *Hepper* 2207! Wum area (Nov.) *Brunt* 881! Throughout tropical and S. Africa. (See Useful Plants.)

19. **A. dummeri** *Stapf* in F.T.A. 9: 248 (1919). Densely tufted perennial up to 1 m. high, coated at the base with persistent fibrous sheaths.
S.L.: Bintumane (Oct.) *Jaeger* 7894! Loma Mts. (Nov.) *Morton* SL 2671! **W. Cam.:** Cam. Mt., 7,000–9,000 ft. (Jan.-Dec.) *Mildbr.* 10857! *Hinds* C53! *Maitland* 943! Bafut-Ngemba F.R., 7,050 ft. (July) *Brunt* 776! Bambui (Dec.) *Boughey* GC 10384! Also in Ethiopia, Kenya, Uganda and Tanzania.

20. **A. perligulatus** *Stapf* in Kew Bull. 1908: 410; F.T.A. 9: 250; A. Chev. in Rev. Bot. Appliq. 13: 866. Erect perennial 1·2–1·8 m. high; anthers red; swampy soils.
Sen.: Dinedéfélo (Nov.) *Adam* 20002! **Mali:** *Duong* 1260! Bamako (Sept.) *Duong*! *Adam* 15313! **Guin.:** Tossékré (Oct.) *Adam* 12538! 12738! Boumalol, Fouta Djalon (Oct.) *Adam* 12563! Timbo (Oct.) *Pobéguin* 1776! Fouta Djalon (Dec.) *Jac.-Fél.* 7439! **S.L.:** Waterloo (Sept.) *Melville & Hooker* 473! Materboi (Oct., Nov.) *Glanville* 78! *Jordan* 833! Sasa (Sept.) *Jordan* 336! Batkanu (Nov.) *Jordan* 847! **Iv. C.:** Singrobo (Oct.) *Adjanohoun* 318*a*! Bocanda (June) *Adjanohoun* 190*a*! Bouna to Batié (Oct.) *Rose Innes* GC 31542! **U. Volta:** Bobo-Dioulasso (Sept.) *Adam* 15089! Prata (Nov.) *Scholz* 185! **Ghana:** Togo Plateau F.R. (Sept.) *St. Clair-Thompson* 1507! Mampong Scarp (Nov.) *Rose Innes* GC 30866! Bolgatanga to Tamale (Oct.) *Rose Innes* GC 31115! Tumu to Walembele (Oct.) *Ankrah* GC 20473! Bolgatanga to Bawku (Oct.) *Rose Innes* GC 30504! **Togo Rep.:** Misahöhe *Baumann* 318! Dapango to Sansanné Mango (Oct.) *Rose Innes* GC 31435! **Dah.:** Sémé (Feb.) *Raynal* 13541! **N. Nig.:** Maska, Katsina Prov. (Nov.) *Thatcher* S508! Minna (Jan.) *Keay* FHI 37316! Wuya Ferry (Sept.) *Clayton* 369! R. Bahago (Oct.) *Keay* FHI 5444! Abinsi *Dalz.* 906*a*!

21. **A. canaliculatus** *Schumach.* Beskr. Guin. Pl. 52 (1827); F.T.A. 9: 251; Chev. Bot. 716, and in Rev. Bot. Appliq. 13: 866; Berhaut, Fl. Sén., ed. 2, 425; Aké Assi, Contrib. 2: 299; incl. var. *fastigians* Stapf in F.T.A. 9: 252 (1919). *A. eucnemis* Trin. (1832). *A. macleodiae* Stapf in F.T.A. 9: 256 (1919); F.W.T.A., ed. 1, 2: 588 (this is a rare aberrant form with solitary racemes). Tufted perennial up to nearly 2 m. high; anthers yellow; swampy soils.
Mali: Diafarabé (Sept.) *Lean* 101! Dari (Sept.) *Farrow* 72! **Guin.:** Timbo (Oct.) *Pobéguin* 1776!

Fig. 454.—ANDROPOGON SCHIRENSIS *Hochst. ex A. Rich.* (GRAMINEAE-ANDROPOGONEAE).

A, ligule. B, C, pairs of spikelets. B_1, C_1, internodes of racemes. B_2, C_4, sessile spikelet. B_3, C_3, pedicel of pedicelled spikelet. B_4, C_2, pedicelled spikelet.

Kouroussa (Jan.) *Pobéguin* 513! San (Sept.) *Chev.* 2349! **Iv. C.:** Bondoukou to Bouna (Oct.) *Rose Innes* GC 31560! Singrobo (Oct.) *Adjanohoun* 341*a*! Bouaké (Sept.) *Oldeman* 408! **U. Volta:** Tondoura (Oct.) *Scholz* 78! **Ghana:** Achimota (May) *Irvine* 1632! Kpong (Dec.) *Gyadu* GC 20524! Agomeda to Somanya (Nov.) *Adams* 4416! Ekumdipe Drift, Daka R. (Oct.) *Ankrah* GC 20403! Pong Tamale (July) *Williams* 838! **Togo Rep.:** Sokode to Atakpamé (Oct.) *Rose Innes* GC 31368! **N. Nig.:** Ajaokuta (Aug.) *Ward* L139! Badeggi (Sept.) *Clayton* 306! Extends southward to Malawi.

22. **A. tectorum** *Schum. & Thonn.* Beskr. Guin. Pl. 49 (1827); F.T.A. 9: 257; Chev. Bot. 720, and in Rev. Bot. Appliq. 13: 868; Berhaut, Fl. Sén. ed. 2, 425; incl. var. *falsopetiolatus* Reznik (1933), and var. *acutatus* Reznik (1933). *A. spectabilis* K. Schum. (1897). *A. tenuiculmis* Reznik in Bull. Mus. Hist. Nat. Paris, sér. 2, 5: 495 (1933); A. Chev. in Rev. Bot. Appliq. 13: 870. Up to about 3 m. high with cane-like culms, often supported by prop roots, and well-developed false petioles; in shade of trees.
Sen.: Bakor (Oct.) *Adam* 18493! 18496! **Mali:** Sikasso (Sept.) *Adam* 15465! **Port. G.:** Teixeira Pinto (Dec.) *Raimundo & Guerra* 550! Boruntuma (Dec.) *Pereira & Correia* 2329! Farim (Jan.) *Pereira* 1280! **Guin.:** Timbo (Oct.) *Pobéguin* 1784! 1809! Mali (Nov.) *Chev.* 34879! Kindia *Jac.-Fél.* 221! Mt. Koiré (Sept.) *Baldwin* 13281! Friguiagbé (Dec.) *Chillou* 1057! **S.L.:** Freetown (Feb., Mar.) *Johnston* 58! *Dawe* 687! New England (Dec.) *Small* 809! Kambia (Dec.) *Deighton* 856! Sieamamaia (Mar.) *Deighton* 4602! **Lib.:** Sanokwele (Jan.) *Baldwin* 14059! Bilimu Mt. (Dec.) *Harley* 1543! 1622! Nimba (Dec.) *Adam* 20214! **Iv. C.:** Bouaké (Aug.) *Pitot*! Daloa to Abidjan (Dec.) *Boughey* 599! Féttékro to Bassawa (Oct.) *Adjanohoun* 293*a*! Haut Sassandra (Dec.) *Portères* 438! **U. Volta:** Samandeni (Oct.) *Kmoch* 135! 167! Touroukoro (Nov.) *Scholz* 23*a*! **Ghana:** Amoma (Jan.) *Vigne* FH 3551! Gambaga to Bawku (Oct.) *Rose Innes* GC 30270! Seripe to Dokrupe (Oct.) *Ankrah* GC 20434! Tamale to Kintampo (Oct.) *Rose Innes* GC 31132! Tumu to Leo (Oct.) *Ankrah* GC 20468! **Togo Rep.:** *Baumann* 467! *Büttner* 301! Misahöhe (Nov.) *Mildbr.* 7208! **Dah.:** Cotonou (Nov.) *Clayton* 398! Parakou (Nov.) *Risopoulos* 1293! **N. Nig.:** Lokoja *Dalz.* 296! Jagindi (Nov.) *Keay & Onochie* FHI 21718! Kontagora (Dec.) *Hubbard* S579! Makwando, Niger Prov. (Nov.) *Ayuba* 70! **S. Nig.:** Lagos (Nov.) *Dalz.* 1310! Ibadan (Nov.) *Meikle* 654! Idanre (Jan.) *Brenan* 8645! Enugu (Nov.) *Jones* FHI 6809! Port Harcourt (Dec.) *Maitland* 714! **W. Cam.:** Bambuko F.R. (Nov.) *Argent* 977! Munkep (Jan.) *Brunt* 932! Also in C. African Republic. (See Useful Plants.)

23. **A. macrophyllus** *Stapf* in F.T.A. 9: 264 (1919); A. Chev. in Rev. Bot. Appliq. 13: 868; incl. var. *pilosus* Reznik (1933). *A. gabonensis* of F.W.T.A., ed. 1, 2: 587, not of Stapf. A robust grass up to 3·5 m. high, with broad coarse leaf-blades.
Guin.: Filicounji (Oct.) *Adam* 12592! Nimba (Feb.) *Schnell* 4260! **S.L.:** Musaia (Dec.) *Deighton* 4526! 4527! Newton (Nov.) *Deighton* 1469! Kambia (Nov.) *Jordan* 686! 687! **Lib.:** Nimba (Apr., Oct., Dec.) *Harley* 2212! *Adam* 16469! *Adames* 699! **Iv. C.:** Bondoukou to Bouna (Oct.) *Rose Innes* GC 31558! Féttékro (Oct.) *Adjanohoun* 277*a*! Sipilou Mt. (Mar.) *Adjanohoun* 691*a*! Bouaflé (Sept.) *Aké Assi* 6534! Tiebissou to Boka (Oct.) *Boudet* 3038! **Ghana:** Sekodumasi (Oct.) *Vigne* FH 3390! Kintampo to Ejura (Oct.) *Rose Innes* GC 30512! Tamale to Kintampo (Oct.) *Rose Innes* GC 31133! Sampa to Wenchi (Oct.) *Rose Innes* GC 31569! **Togo Rep.:** Badou to Atakpamé (Sept.) *Rose Innes* GC 31344! **N. Nig.:** Manchok (Oct.) *Magaji & Tuley* 1767! **S. Nig.:** Lagos *W. MacGregor* 248! Enugu (Sept.) *Jones* 6750! **F. Po:** Ureka (Feb.) *Guinea* 2453!

24. **A. auriculatus** *Stapf* in F.T.A. 9: 258 (1919); A. Chev. in Rev. Bot. Appliq. 13: 868; Berhaut, Fl. Sén. ed. 2, 425; Aké Assi, Contrib. 2: 299. *A. guineensis* Steud., Syn. Pl. Glum. 1: 371 (1854), not of Schumach. (1827). Caespitose perennial up to 1·5 m. high; sandy beaches.
Sen.: *Farmar* 45! 157! Niokolo-Koba (June) *Adam* 14203! Dionouar (Dec.) *Adam* 18916! **Gam.:** *Boteler*! **Port. G.:** Bedanda (Jan.) *Pereira & Correia* 2780! Buba (Jan.) *Pereira & Correia* 2944! **Guin.:** Koba (Nov.) *Jac.-Fél.* 7272! **S.L.:** Yoni (Mar.) *Deighton* 2480! Kitchom (Jan.) *Deighton* 1009! Hamilton (Dec.) *Gledhill* 120! Mama beach (Dec.) *Deighton* 3321! Rokupr (Dec.) *Jordan* 711! **Iv. C.:** Abouabou (Mar.) *Raynal* 13647! **S. Nig.:** Nun R. (Sept.) *Mann* 533! Opobo (Oct.) *Jeffreys* 9! Calabar (Mar.) *Brenan* 9212! Brass (Dec.) *Daramola* FHI 43287! Obudu Plateau (Feb.) *Tuley* 491! Also in E. Cameroun.

25. **A. chevalieri** *Reznik* in Bull. Mus. Hist. Nat. Paris, sér. 2, 5: 497 (1933); A. Chev. in Rev. Bot. Appliq. 13: 870. *A. felicis* Reznik l.c. 499; Berhaut, Fl. Sén. ed. 2, 425. *A. pseudauriculatus* Mimeur in Bull. Mus. Hist. Nat. 22: 128 (1950). Annual 2–2·5 m. high, apparently intergrading with the preceding species; spikelet length very variable, even in the same inflorescence; dry stony soils.
Sen.: Niokolo-Koba (Oct.) *Adam* 15600! 15892! **Mali:** Bamako (June, Sept.) *Adam* 15036! 15038! 15310! Koulouba (Sept.) *Duong*! *Vuillet* 461! **Port. G.:** Bissau (Nov.) *Raimundo & Guerra* 17! Mansoa (Nov.) *Pereira & Correia* 1971! **Guin.:** Mali (Nov.) *Chev.* 37878! Kindia *Jac.-Fél.* 244! Sériba (Oct.) *Pitot*!

26. **A. pteropholis** *W. D. Clayton* in Hook., Ic. Pl. 37: t. 3644 (1967). Perennial up to nearly 2 m. high; roadsides and stony soils.
Ghana: Gambaga to Bawku (Oct.) *Rose Innes* GC 30268! Walewale to Gambaga (Oct.) *Rose Innes* GC 31107! Nakpanduri (Oct.) *Yamoah* SLUS 514!

27. **A. pinguipes** *Stapf* in Kew Bull. 1908: 411; F.T.A. 9: 254; A. Chev. in Rev. Bot. Appliq. 13: 866; Berhaut, Fl. Sén. ed. 2, 426. Annual, 0·3–1·5 m. high.
Sen.: M'Bidjem (Aug.) *Thierry* 92! Dakar (Oct., Nov.) *Adam* 15474! 17802! Niokolo-Koba (Oct.) *Adam* 15975! Iles du Saloum (Dec.) *Adam* 18927! (See Useful Plants.)

28. **A. gabonensis** *Stapf* in Journ. de Bot. sér. 2, 2: 207 (1909); F.T.A. 9: 260. Robust perennial up to about 3·5 m. high.
W. Cam.: Bamenda (Jan., Dec.) *Migeod* 321! *Agric. Off.* 49! Ndop (Dec.) *Boughey* GC 10464! Nyen (May) *Brunt* 1128! 1130! Extends through the Congos, and into Angola.

29. **A. gayanus** *Kunth* Enum. Pl. 1: 491 (1833); F.T.A. 9: 261; Chev. Bot. 717, and in Rev. Bot. Appliq. 13: 866.

29a. **A. gayanus** *Kunth* var. **gayanus**—(as var. *genuinus*) Hack. in DC., Monogr. Phan. 6: 447 (1889); Berhaut, Fl. Sén. ed. 2, 425. *A. infrasulcatus* Reznik in Bull. Mus. Hist. Nat. Paris, ser. 2, 5: 496 (1933); A. Chev. in Rev. Bot. Appliq. 13: 870. *A. reconditus* Steud., Syn. Pl. Glum. 1: 386 (1854). A tall perennial with a somewhat glaucous appearance and a conspicuously fastigiate habit; seasonal swamps and flood plains.
Maur.: L. Rkiz (Oct.) *Sadio* 120! Rosso to Nouakchott (Oct.) *Adam* 21757! **Sen.:** *Roger* 29! Fatik to Kaolak (Oct.) *Chev.* 33938! Ouassoudou (Nov.) *Berhaut* 2719! Niokolo-Koba (Oct.) *Adam* 15669! Gouloumbou, Casamance (Oct.) *Adam* 18622! **Mali:** *Rogeon* 220! San (Sept.) *Chev.* 2361! Timbuktu *Hagerup* 109! Macina (Nov.) *Duong* 2025! Saraféré (July) *Lean* 60! **U. Volta:** Ouassa to Diebougou (Oct.) *Rose Innes* GC 31510! Pendjari R. (Aug.) *Scholz* 141! Samandeni (Oct.) *Kmoch* 171! Tindangou to Konpienga (Aug.) *Scholz* 143! **Ghana:** Kpong (Sept.) *Rose Innes* GC 30382! Kpandu to Kete Krachi (Sept.) *Rose Innes* GC 30408! Yendi to Gambaga (Sept.) *Ankrah* GC 20260! Tamale to Sawla (Oct.) *Rose Innes* GC 30674! Kpevi to Amedzofe (Sept.) *Rose Innes* GC 30900! **Togo Rep.:** Sansanné Mango to Kande (Oct.) *Rose Innes* GC 31415! Palime *Stage* 85! **Dah.:** Cotonou (July) *Chev.* 4471! Porto Novo (Sept.) *Adjanohoun* 397! **N. Nig.:** Ajaokuta (Aug.) *Ward* L134! Ibu *T. Vogel* 32! Nupe *Barter* 1383! Wuya, Bida (Aug.) *Sawyer* K30! Badeggi (Sept.) *Clayton* 298! **S. Nig.:** Nun R. *T. Vogel* 7! Aguku *Thomas* 963! Orle F.R. (Aug.) *Onochie* FHI 33281! Yenagoa (Sept.) *Tuley* 888! Also in the Sudan. (See Useful Plants.)

29b. **A. gayanus** *Kunth* var. **tridentatus** *Hack.* in DC., Monogr. Phan. 6: 449 (1889). *A. tridentatus* Hochst. in Flora 27: 246 (1844), not of Roxb. (1820). *A. tomentellus* Steud., Syn. Pl. Glum. 1: 371 (1854). Tufted perennial 1·2–1·8 m. high, with occasional solitary racemes. This variety is remarkable in that it contains a high proportion of diploids (2n = 20), whereas the other varieties are tetraploid.

Maur.: Néma *Rossetti* 61/169! **Sen.:** Dakar (May) *Baldwin* 5732! Ile de la Madeleine, Dakar (Nov.) *Adam* 17774! M'Bambey (Mar.) *Pitot*! Almadies (Nov.) *Hb. I.F.A.N.*! **Gam.:** Karawan (Jan.) *Dalz.* 8413! **Mali:** Dorey (Aug.) *Rossetti* 59/126! Kidal (Sept.) *Rossetti* 59/151! Dioura (Jan., Feb.) *Davey* 214! 253! Mederdra (Sept.) *Popov* 125! **Ghana:** Gambaga (Apr.) *Morton* 9006! **Togo Rep.:** Sansanné Mango to Dapango (Oct.) *Rose Innes* GC 31427! **Niger:** Dogondoutchi *White* 41! **N. Nig.:** Bimasa Bakusa Tureta N.A. F.R. (Nov.) *Latilo* FHI 43790! Zurmi, Sokoto Prov. (Apr.) *Keay* FHI 16166! Kano (Dec.) *Meikle* 806! Kingowa (Oct.) *Golding* 28! Kauwa (Sept.) *Gwynn* 124! Also in the Sudan.

29c. **A. gayanus** *Kunth* var. **squamulatus** (*Hochst.*) *Stapf* in F.T.A. 9: 263 (1919); Chev. Bot. 718; Berhaut, Fl. Sén. ed. 2, 425; De Miré & Gillet in J. Agric. Trop. 3: 725; Aké Assi, Contrib. 2: 300. *A. squamulatus* Hochst. in Flora 27: 244 (1844). Tufted perennial up to 3 m. high; a common savanna grass.
Maur.: L. Rkiz (Sept.) *Sadio* 119! **Sen.:** Niokolo-Koba (Oct.) *Adam* 15761! 15923! 15927! Gouloumbou (Oct.) *Adam* 12717! Richard-Toll (Oct.) *Martine* 18! **Gam.:** *Saunders* 73! **Mali:** Gao (Sept.) *Hagerup* 355! Macina (Nov.) *Duong* 1285! Ségou (Sept.) *Roberty* 3929! Kouli (Nov.) *Farrow* 110! Dari (Sept.) *Farrow* 77! **Guin.:** Bafing (Oct.) *Adam* 12660! Konkouré (Mar.) *Pitot* 230! **S.L.:** *Sc. Elliot* 4261! 5932! Dodo (Apr.) *Deighton* 3920! **Lib.:** Nimba (Dec.) *Adam* 20208! **U. Volta:** Zoaga (Oct.) *Scholz* 157! Zabré to Ziou (Oct.) *Scholz* 154! Po (Oct.) *Rose Innes* GC 31467! **Ghana:** Nungua, Accra plains (Sept.) *Ankrah* GC 20057! Kpedsu (Jan.) *Howes* 1119! Ejura to Salaga (Oct.) *Ankrah* GC 20305! Bolgatanga to Tamale (Oct.) *Rose Innes* GC 30234! Bawku (Oct.) *Ankrah* GC 20295! **Togo Rep.:** Palime *Stage* 82! Lama-Kara (Apr.) *Aké Assi* 10523! **Dah.:** *Poisson*! Cotonou to Allada (Oct.) *Risopoulos* 1206! **N. Nig.:** Okene (Oct.) *Ward* L149! Lokoja (Oct.) *Dalz.* 288! Zaria *Taylor* 16! Gindiri, Plateau Prov. (Oct.) *Hepper* 1142! Hadejia (Dec.) *Clayton* 466! Gangumi, Gashaka (Nov.) *Latilo & Daramola* FHI 28774! Gurum, Vogel Peak (Nov.) *Hepper* 1324! Donga plain (Feb.) *Brunt* 1009! **S. Nig.:** Olokemeji (Nov.) *Onochie* FHI 8113! Odinjo (Jan.) *Swarbrick* 2968! Port Harcourt to Ogoni (July) *Tuley* 778! Obudu Plateau (Nov.) *Tuley* 974! Ogoja *Rosevear* 11/30*a*! **W. Cam.:** Bali to Bamenda (Mar.) *Brunt* 1066! Extending east to the Sudan, and south to S. Africa.

29d. **A. gayanus** *Kunth* var. **bisquamulatus** (*Hochst.*) *Hack.* in DC., Monogr. Phan. 6: 448 (1889); Berhaut, Fl. Sén. ed. 2, 425. *A. bisquamulatus* Hochst. in Flora 27: 245 (1844); incl. var. *argyrophoeus* Stapf in Mém. Soc. Bot. Fr. 2, 8: 102 (1907); Chev. Bot. 717. Tufted perennial up to 3 m. high; savanna.
Maur.: L. Rkiz (Sept.) *Sadio* 15! 22! **Sen.:** Mankono-Ba, Casamance (Oct.) *Adam* 18467! **Gam.:** *Dawe* 10! Yundum *Corbett* 10! **Mali:** Ras el Mar (Oct.) *Rossetti* 59/211! Koulouba (Jan.) *Duong*! Balasoko (Nov.) *Duong*! **Port. G.:** Mansoa (Nov.) *Pereira & Correia* 1972! **Guin.:** Abudia *Krause*! Kouroussa (Dec.) *Pobéguin* 540! Kissidougou *Martine* 228! **Iv. C.:** Abidjan (June) *Adjanohoun* 423*a*! Bouaké (Dec.) *Pitot*! **U. Volta:** Bobo-Dioulasso (Oct.) *Kmoch* 148! Samandeni (Oct.) *Kmoch* 153! 180! Ouagadougou (Oct.) *Scholz* 24*a*! **Ghana:** Accra plains (Oct.) *Baldwin* 13429! Chindiri (Nov.) *Thorold* 309! Ejura (Nov.) *Vigne* FH 3468! Ejura to Yeji (Nov.) *Clayton* 386! Pong Tamale (Nov.) *Williams* 851! **Dah.:** Parakou (Nov.) *Risopoulos* 1287! **N. Nig.:** Ilorin (Dec.) *Keay* FHI 37295! Minna (Dec.) *Keay* FHI 37300! Toro, Plateau Prov. (Nov.) *Semple* 166! Zaria (Oct.) *Thatcher* S498! Maiduguri (Oct.) *Johnston* N33! **S. Nig.:** Abeokuta (Oct.) *Onochie* FHI 40288! Ago-Are F.R. (Nov.) *Keay* FHI 37685! Also in C. African Rep. and the Sudan.

Imperfectly known species.

A. gambiensis *A. Chev.* in Rev. Bot. Appliq. 13: 870 (1933). An inadequately described species from Senegal, of which the type is missing.

A. guineensis *Schumach.* Beskr. Guin. Pl. 51 (1827). *Cymbachne guineensis* (Schumach.) Roberty in Boissiera 9: 244 (1960). The description suggests *A. gayanus*, but it is not conclusive and I have been unable to trace any authentic specimens.

A. leucostachyus *Kunth* in Humb. & Bonpl., Nov. Gen. & Sp. 1: 187 (1816); F.W.T.A., ed. 1, 2: 586; A. Chev. in Rev. Bot. Appliq. 13: 866. A tropical American species recorded from Senegal; probably introduced.

A. sp. A. Apparently an annual form of *A. gayanus* var. *squamulatus*.
N. Nig.: mile 69 Yola to Biu (Nov.) *Magaji & Tuley* 1842!

125. DIHETEROPOGON Stapf in Hook., Ic. Pl. t. 3093 (1922); Clayton in Kew Bull. 20: 73 (1966).

Leaf-blades not cordate at the base; pedicelled spikelets 16–25 mm. long, laterally winged towards the tip; sessile spikelets 7–8 mm. long, excluding the pungent callus which is 4–5 mm. long; awn 9–15 cm. long, pubescent; homogamous pairs 3–9 at the base of each raceme .. .. 1. *grandiflorus*

Leaf-blades cordate and semi-amplexicaul at the base; pedicelled spikelets 9–13 mm. long, wingless; sessile spikelets 6–7 mm. long, with an acute to pungent callus 1–2 mm. long; awn 4–8 cm. long; homogamous pairs 1 at the base of the lower raceme:

Annual; awn hirtellous to hoary hirsute .. .. 2. *hagerupii*

Perennial; awn shortly pubescent to hirtellous .. 3a. *amplectens* var. *catangensis*

1. **D. grandiflorus** (*Hack.*) *Stapf* in Hook., Ic. Pl. t. 3093 (1922). *Andropogon grandiflorus* Hack. in Flora 68: 127 (1885). *Heteropogon grandiflorus* (Hack.) Roberty in Boissiera 9: 137 (1960). Perennial 1–2 m. high; poor sandy soils in drainage hollows or near streams.
N. Nig.: Nupe *Barter* 1373! Chappal Waddi, Mambila (Nov.) *Jackson, Magaji & Tuley* 2079! **S. Nig.:** Udi, Onitsha Prov. (Sept., Oct.) *Jones* 6782! *Onochie* FHI 34114! Enugu (Sept.) *Tuley* 883! Extends into the Congo, Tanzania, Zambia, Rhodesia, Malawi and Angola.

2. **D. hagerupii** *Hitchc.* in Proc. Biol. Soc. Wash. 43: 89 (1930); A. Chev. in Rev. Bot. Appliq. 13: 870; Lebrun in Bull. Soc. Bot. Fr. 116: 259. *Heteropogon hagerupii* (Hitchc.) Roberty in Boissiera 9: 137 (1960). Annual, 1–1·5 m. high; dry sandy or gravelly soils.
Maur.: 15°40′N, 9°30′W. *Rossetti* 61/299! **Sen.:** *fide* Lebrun *l.c.* **Mali:** Bara, nr. Ansongo (Sept.) *Hagerup* 401! Shirat (Nov.) *Chev.* 33830! Macina *Duong*! Dioura, Sahel (Sept.) *Davey* 20! Safoulabé, Goudan *Duong*! **Guin.:** Gouloumbo to Tambacounda (Oct.) *Pitot*! Sériba to Koundara (Oct.) *Adam* 19709! **U. Volta:** Bondoukuy (Sept.) *Aké Assi* 6405! Black Volta bridge (Sept.) *Aké Assi* 6467! Sindou (Sept.) *Aké Assi* 6373! Gaoua to Diebougou (Oct.) *Rose Innes* GC 31518! Ouassa to Leo (Oct.) *Rose Innes* GC 31506! **Ghana:** Wa to Dorimon (Oct.) *Rose Innes* GC 31032! Lawra to Wa (Oct.) *Rose Innes* GC 32356! **Niger:** Dasso *White* 19! Dogondoutchi *White* 37! Zinder (Nov.) *Money-Kyrle* Fe20! Toukounous *Bartha* 41! **N. Nig.:** Sokoto Prov. *Dalz* 489! Bimasa, Sokoto Prov. (Nov.) *Latilo* FHI 43776! Katsina (Oct.) *Clayton* 1339! *Imam* FHI 27805! Katagum Dist. *Dalz.* 267! (See Useful Plants.)

3. **D. amplectens** (*Nees*) *W. D. Clayton* in Kew Bull. 20: 75 (1966). *Andropogon amplectens* Nees, Fl. Afr. Austr. 104 (1841); F.T.A. 9: 243. *Cymbachne amplectens* (Nees) Roberty in Boissiera 9: 242 (1960).

3a. **D. amplectens** var. **catangensis** (*Chiov.*) *W. D. Clayton* l.c. (1966). *Andropogon amplectens* var. *catangensis* Chiov. in Ann. Bot. Roma 13: 38 (1914). *A. diversifolius* Rendle, Cat. Welw. 2: 148 (1899). *A. amplectens* ar. *diversifolius* (Rendle) Stapf in F.T.A. 9: 244 (1919); F.W.T.A., ed. 1, 2: 588; A. Chev. in Rev. Bot.

Appliq. 13: 864; Berhaut, Fl. Sén. ed. 2, 423. *A. subamplectens* Berhaut, Fl. Sén. ed. 2, 424, without latin descr. Perennial with short scaly rhizomes, 1·2–2·1 m. high; usually glabrous, but occasionally densely pubescent all over; poor sandy soils, or shallow rocky soils of hill slopes.
Sen.: Niokolo-Koba (Oct.) *Adam* 15930! Ziguinchor (Dec.) *Berhaut* 6685! **Guin.:** Kanca (Oct.) *Adam* 12623! Kindia *Jac.-Fél.* 187! Pita (Sept.) *Adames* 361! Timbo (Oct.) *Pobéguin* 1779! **Iv. C.:** Odienné (Oct.) *Aké Assi* 8255! **U. Volta:** Baoulé (Oct.) *Kmoch* 126! **Ghana:** Bawku to Bolgatanga (Oct.) *Rose Innes* GC 31447! Gambaga to Bawku (Oct.) *Rose Innes* GC 30272! Gambaga F.R. (Oct.) *Ankrah* GC 20269! Salaga to Tamale (Oct.) *Rose Innes* GC 30312! Banda to Bui (Oct.) *Rose Innes* GC 31603! **N. Nig.:** Gindiri, Jos Plateau (Oct.) *Hepper* 1140! Kukakowa (Oct.) *De Leeuw* 1858! Gombe, Bauchi Prov. (Oct.) *Magaji* 180! Extends from the Congo, Kenya and Tanzania southwards to the Cape. (See Useful Plants.)

126. HYPARRHENIA Fourn., Mex. Pl. 2: 51 (1886); F.T.A. 9: 291 (1919); Clayton in Kew Bull., Add. Ser. No. 2 (1969).

Callus of sessile spikelet broadly rounded, semicircular; spikelets glabrous; raceme-bases unequal, the upper more or less flattened, glabrous, 1·5–3 mm. long 1. *glabriuscula*
Callus of sessile spikelet acute to pungent, rarely obtuse but then the other characters not as above:
Upper raceme-base terete or filiform, usually much longer than the lower, glabrous or sometimes softly hirtellous, without an appendage; spatheoles narrow:
Spikelets wholly or partly rufously hairy:
Basal leaf-sheaths densely pubescent to tomentose below:
Sheath hairs white; callus of sessile spikelets 0·8–1·2 mm. long, acute 2. *nyassae*
Sheath hairs dark purplish red; callus of sessile spikelets 0·5–0·8 mm. long, obtuse to narrowly truncate at the tip .. .. 3. *smithiana*
Culms 40–100 cm. high; spikelet indumentum very dense, dark rufous to chocolate coloured .. .. 3a. *smithiana* var. *smithiana*
Culms 150–250 cm. high; spikelet indumentum rufous or fulvous 3b. *smithiana* var. *major*
Basal leaf-sheaths glabrous:
Fertile spikelets, or some of them, awnless, 12–19 per raceme-pair 4. *exarmata*
Fertile spikelets awned:
Sessile spikelets 3–5 mm. long; upper raceme-base 2–3·5 mm. long, with or without 1 homogamous pair; callus obtuse, 0·2–0·8 mm. long; racemes with (7–)9–14 awns per pair; pedicelled spikelets awnless, rarely mucronate 5. *rufa*
Sessile spikelets 5·5–7 mm. long; upper raceme-base 3·5–7 mm. long, with 1–2 homogamous pairs; callus acute to pungent, 1–2 mm. long; racemes 4–7-awned per pair; pedicelled spikelet often with an awn-point up to 2 mm. long .. .. 6. *poecilotricha*
Spikelets white-hairy or glabrous:
Upper raceme with or without 1 homogamous pair at the base:
Plants perennial:
Spikelets glabrous to hispidulous; racemes 2–6-awned per pair; callus linear, 1–2 mm. long; upper raceme-base 1·5–2·5 mm. long, hirtellous; pedicelled spikelets with an awn up to 5 mm. long .. .. .. 7. *finitima*
Spikelets pubescent to villous:
Racemes not deflexed .. .. 8. *hirta*
Racemes, or some of them, deflexed at maturity .. .. .. 9. *quarrei*
Plants annual:
Racemes 6–9-awned per pair; column of awn bearing hairs up to 1·5 mm. long; upper raceme-base 2–3 mm. long:
Awn 2·5–4 cm. long, pubescent with hairs 0·1–0·5 mm. long; callus 0·8–1·2 mm. long, acute .. .. 10. *violascens*
Awn 5–8 cm. long, hirtellous with hairs 1–1·5 mm. long; callus 1·5–2·5 mm. long, pungent .. .. 11. *bagirmica*
Racemes 2-awned per pair; awn hirsute with hairs 3–5 mm. long; upper raceme-base 4–7 mm. long .. .. 12. *barteri*
Upper raceme with 2 homogamous pairs:
Peduncle with spreading yellow hairs towards the tip; racemes deflexed; awns 3–5 per raceme-pair; pedicelled spikelet with an awn 2–10 mm. long 13. *familiaris*
Peduncle with or without white hairs towards the tip; racemes not deflexed; pedicelled spikelet with an awnlet up to 5 mm. long; awn hirtellous with hairs 0·7–1·2 mm. long; upper raceme-base typically 4·5–8 mm. long; callus 1·8–3 mm. long:
Annual; awn 5–7·5 cm. long .. .. 14. *figariana*
Perennial; awn 3–5·5 cm. long .. .. 15. *filipendula*
Spikelets glabrous; racemes 2-awned per pair.. .. 15a. var. *filipendula*
Spikelets white villous; racemes 2–4-awned per pair .. 15b. var. *pilosa*
Upper raceme-base flattened, scarcely longer than the lower:
*Raceme-bases bearded; 1 pair of homogamous spikelets at base of lower raceme only:

Raceme-bases unappendaged, though sometimes with a scarious rim:
Pedicelled spikelets glabrous:
Spatheoles 0·8–1·8 cm. long, enclosing the racemes; awns 3–5 per raceme-pair, up to 1·6 cm. long; callus almost square, broadly obtuse .. 16. *cymbaria*
Spatheoles 3·5–5 cm. long, the racemes exserted; awns 6–11 per raceme-pair, 2·8–3·4 cm. long; callus oblong to cuneate, subobtuse .. 17. *cyanescens*
Pedicelled spikelets villous; awns 4–7 per raceme-pair:
Awn 0·7–1·3 cm. long; callus rounded; spatheole 1·2–2·3 cm. long; peduncle 3–13 mm. long; pedicelled spikelet muticous or with an awn-point up to 1 mm. long; culms robust and supported by stilt roots .. .. 18. *umbrosa*
Awn over 1·5 cm. long; callus cuneate; spatheole 2–4 cm. long; peduncle 10–25 mm. long:
Culms robust, supported by stilt roots; awn 2·2–4 cm. long; pedicelled spikelets usually with a bristle 2–6 mm. long .. .. 19. *rudis*
Culms slender, without stilt roots, arising in clumps from a short underground rhizome; awn 1·5–2·5 cm. long; pedicelled spikelets with a short awn-point 1–3 mm. long .. .. 20. *collina*
Raceme-bases produced at the insertion of the lowest fertile spikelet into a scarious appendage 0·5–4 mm. long; racemes 2–4-awned per pair:
Annuals; panicle loose and leafy; awn 4–6·5 cm. long; pedicelled spikelet with a bristle 2–10 mm. long:
Awns 2 per raceme-pair; appendage narrowly oblong, 3·5–4 mm. long; sessile spikelet 7·5–10 mm. long .. .. 21. *niariensis*
Awns 3 per raceme-pair; appendage oblong, 0·5–2 mm. long; sessile spikelet 5–7 mm. long .. .. 22. *welwitschii*
Perennials:
Pedicel tip with a short obtusely triangular lobe; sessile spikelet 4–6 mm. long, the awn 1–2·5 cm. long; pedicelled spikelet muticous or with a mucro up to 1 mm. long; panicle narrow, dense .. .. 23. *bracteata*
Pedicel tip drawn out into a subulate tooth 0·2–1·5 mm. long; sessile spikelet 6–10 mm. long, the awn 2·2–5·5 cm. long; pedicelled spikelet with a bristle 1–5 mm. long; panicle loose, often scanty .. .. .. 24. *newtonii*
*Raceme-bases glabrous; 2 pairs of homogamous spikelets at the base of both racemes:
Sessile spikelets awnless .. .. 25. *mutica*
Sessile spikelets awned:
Perennials:
Column of awn pubescent with hairs 0·2–0·4(–0·5) mm. long, the awn 2–5·5 cm. long; peduncles 3–15 mm. long; spatheoles 2–4·5 cm. long 26. *diplandra*
Column of awn hirtellous with hairs 0·5–1·7 mm. long, the awn 4·5–7·5 cm. long; peduncles 10–35 mm. long; spatheoles 3–7 cm. long .. 27. *subplumosa*
Annual; awn 7–11 cm. long .. .. 28. *involucrata*
Pedicelled spikelet with an awn 8–20 mm. long; awns 4 per raceme-pair; base of spatheole and supporting ray glabrous, or sometimes bearded at the node 28a. var. *involucrata*
Pedicelled spikelet with an awn 1–5 mm. long; awns 2(–4) per raceme-pair; base of spatheole and supporting ray pilose .. .. 28b. var. *breviseta*

1. **H. glabriuscula** (*Hochst. ex A. Rich.*) *Anderss. ex Stapf* in F.T.A. 9: 372 (1919); Clayton in Kew Bull., Add. Ser. 2: 47 (1969). *Andropogon glabriusculus* Hochst. ex A. Rich., Tent. Fl. Abyss. 2: 468 (1851). *Hyparrhenia amoena* Jac.-Fél. in J. Agric. Trop. 1: 46 (1954); Berhaut, Fl. Sén. ed. 2, 426. Tufted perennial 1·2–1·8 m. high, closely resembling *Cymbopogon*; seasonally swampy soils and river flood plains.
Sen.: Niokolo-Koba (Sept., Oct.) *Adam* 15715! 15811! 17147! Kidira (Nov.) *Berhaut* 1896! **U. Volta:** *Prost*! Go (Oct.) *Scholz* 169! 170! Nanguédi (Nov.) *Scholz* 187! **Ghana:** Han to Tumu (Oct.) *Rose Innes* GC 32427! Naga to Navrongo (Nov.) *Rose Innes* GC 32524! Bawku to Bolgatanga (Oct.) *Rose Innes* GC 31099! Bamboi to Bole (Nov.) *Brand* SLUS 750! Pong Tamale (Oct.) *Rose Innes* GC 32429! **N. Nig.:** Biu (Oct.) *De Leeuw* 1314! Buratai (Oct.) *De Leeuw* 1322! Shendam (Nov.) *Clayton* 1465! Ningi (Nov.) *Magaji & Tuley* 1825! Pambegua to Kaduna (Oct.) *Oche & Tuley* 1701! Also in Ethiopia, Mozambique and Malawi.
2. **H. nyassae** (*Rendle*) *Stapf* in F.T.A. 9: 313 (1919); Clayton l.c. 53. *Andropogon nyassae* Rendle in J. Bot. 31: 358 (1893). *A. chrysargyreus* (Stapf) Stapf ex A. Chev., Sudania 1: 77 (1911). *Cymbopogon chrysargyreus* Stapf in Journ. de Bot. sér. 2, 2: 213 (1909). *Hyparrhenia chrysargyrea* (Stapf) Stapf in F.T.A. 9: 312 (1919). *H. vulpina* Stapf l.c. 310 (1919). *H. vanderystii* (De Wild.) Vanderyst (1920). *H. schmidiana* A. Camus (1955). Tufted perennial 0·6–1·5 m. high; in savanna woodland.
Ghana: Accra plains (Nov.) *Ankrah* GC 20109! Throughout tropical Africa; also in Thailand and Vietnam.
3. **H. smithiana** (*Hook.f.*) *Stapf* in F.T.A. 9: 314 (1919). *Andropogon smithianus* Hook.f. in J. Linn. Soc. Bot. 7: 232 (1864).
3a. **H. smithiana** (*Hook.f.*) *Stapf* var. **smithiana.** Tufted perennial 30–90 cm. high; mountain grasslands.
S.L.: Mt. Loma (July) *Jaeger* 6975! Bintumane (Jan., July, Nov.) *T. S. Jones* 116! *Bakshi* 289! *Jaeger* 322! *Morton* SL 2669! **W. Cam.:** Bambui (Dec.) *Boughey* GC 10752! L. Oku (Jan.) *Keay & Lightbody* FHI 28499! Bafut-Ngemba F.R. *Hepper* 2848! Bamenda (Jan.) *Migeod* 343! Cam. Mt. *Mann* 1342!
3b. **H. smithiana var. major** *W. D. Clayton* in Kew Bull., Add. Ser. 2: 57 (1969). *H. chrysargyrea* of F.W.T.A. ed. 1, 2: 591, of A. Chev. in Rev. Bot. Appliq. 13: 872, and of Aké Assi, Contrib. 2: 301, not of Stapf. Tufted perennial 1·5–2·4 m. high; lowland savanna.
Sen.: Kédougou (Sept.) *Fotius* K610! **Guin.:** Timbi Madina (Jan.) *Langdale-Brown* 2555! Dalaba (Oct.) *Adam* 12672! **S.L.:** Musaia (Dec.) *Deighton* 4522! Tingi Mts. (Apr., Dec.) *Morton & Gledhill* SL 1945! SL 3033! Loma Mts. (Nov.) *Morton* SL 2558! **Iv. C.:** Séguéla (Oct.) *Adjanohoun* 288*a*! 319*a*!

Zuenoula to Vavoua (Oct.) *Adjanohoun* 382a! Asakra (Aug.) *Boughey* GC 18628! Dabou (Jan.) *Adjanohoun & Aké Assi*! **U. Volta:** Samandeni (Oct.) *Kmoch* 164! Diebougou to Ouassa (Oct.) *Rose Innes* GC 31514! Leo to Po (Oct.) *Rose Innes* GC 31075! **Ghana:** Wa (Oct.) *Rose Innes* GC 32385! Bamboi to Bole (Oct.) *Rose Innes* GC 30646! Salaga to Tamale (Oct.) *Rose Innes* GC 32440! Gambaga (Oct.) *Ankrah* GC 20271! Kpong (May) *Rose Innes* GC 30364! **Togo Rep.:** Badou to Atakpame (Sept.) *Rose Innes* GC 31349! **N. Nig.:** Zaria (Oct., Nov.) *Freeman* S170! *Thatcher* S529! S538! Afaka F.R., Zaria (Oct.) *Olorunfemi* FHI 55087! Lokoja (Oct.) *Dalz*! Ayere (Dec.) *Clayton* 616! **S. Nig.:** Olokemeji (Nov.) *Clayton* 567! *Jackson* 7161168! Ago-Are F.R. (Nov.) *Keay* FHI 37683! Ibadan (May) *Townrow* 22! Ekpeshi (Nov.) *Keay* FHI 28083!

4. **H. exarmata** (*Stapf*) *Stapf* in F.T.A. 9: 308 (1919); A. Chev. in Rev. Bot. Appliq. 13: 872; Clayton l.c. 60. *Cymbopogon exarmatus* Stapf in Journ. de Bot. sér. 2, 2: 210 (1909). *Andropogon rufus* var. *exarmatus* (Stapf) Stapf ex A. Chev., Sudania 1: 180 (1911). Perennial, or perhaps sometimes annual, 1·5 m. high; in savanna, often bordering swampy soils. Care should be taken not to confuse this species with specimens of *H. rufa* in which the awn has been destroyed by the attack of a smut fungus.
U. Volta: Bobo-Dioulasso (Oct.) *Kmoch* 120! Also in C. African Rep., the Congo and Sudan.

5. **H. rufa** (*Nees*) *Stapf* in F.T.A. 9: 304 (1919); A. Chev. in Rev. Bot. Appliq. 13: 872; Berhaut, Fl. Sén. ed. 2, 426; Clayton l.c. 60. *Trachypogon rufus* Nees, Agrost. Bras. 345 (1829). Perennial or sometimes annual, 0·3–2·4 m. high; typically a tall savanna grass, often preferring moister sites, but dwarfed forms may be found by roadsides and other frequented places.
Maur.: L. Rkiz (Sept.) *Sadio* 116! **Sen.:** Cape Verde (Dec.) *Adam* 1841! Dienoundiella (Dec.) *Berhaut* 2713! **Gam.:** Wallikunda (Sept.) *Macluskie* 23! **Mali:** Bamako (Dec.) *Adam* 11309! 11409! Koulikoro (Sept.) *Rogeon* 228! Macina (Dec.) *Duong* 1509! Kadigue to Banguita *Davey* 8! **Port. G.:** Canquelifa (Dec.) *Pereira* 3617! **Guin.:** Seriba (Oct.) *Adam* 12710! La Kolenté (Nov.) *Chillou* 1030! Dalaba (Oct.) *Adam* 12672! Baffing (Nov.) *Pobéguin* 1816! Nzérékoré (June) *Adam* 5306! **S.L.:** Musaia (Feb., Mar.) *Deighton* 4205! 5448! Benekoro (Nov.) *Glanville* 318! Dumbaia (Jan.) *Morton & Gledhill* SL 573! Sefadu (Dec.) *Deighton* 3573! Senehun (Apr.) *Morton & Gledhill* SL 1963! **Iv. C.:** *Adam* 6594! Nambonkaha (Nov.) *Leeuwenberg* 2048! Ferkessédougou (Nov.) *Leeuwenberg* 2018! Zuenoula (Oct.) *Adjanohoun* 309a! Beriby (Nov.) *Oldeman* 474! **U. Volta:** Leo to Po (Oct.) *Rose Innes* GC 31469! Koumbili to Koumbo (July) *Scholz* 50a! Koumbili to Natiédiougou (Oct.) *Scholz* 183! Nanguédi (Nov.) *Scholz* 186! **Ghana:** Lawra to Wa (Oct.) *Rose Innes* GC 32353! Navrongo to Wiaga (Oct.) *Ankrah* GC 20480! Pong Tamale (Dec.) *Morton* GC 9834! Elmina (Nov.) *Rose Innes* GC 31172! Achimota (Dec.) *Ankrah* GC 20115! **Togo Rep.:** Lama-Kara (Apr.) *Aké Assi* 10532! Lomé *Stage* 19! **Dah.:** Gouka (Mar.) *Froment* 1186! Kpinnou (Jan.) *Froment* 1133! Boukombe (Nov.) *Risopoulos* 1258! Cotonou to Allada (Oct.) *Risopoulos* 1237! **N. Nig.:** Sokoto (Oct.) *Dalz.* 487! Gajibo (Dec.) *Davey* FHI 27123! Zaria (Nov.) *Thatcher* S523! Minna (Dec.) *Keay* FHI 37303! Gurum, Vogel Peak (Nov.) *Hepper* 1310! **S. Nig.:** Ogurude (Jan.) *Holland* 276! Olokemeji (Nov.) *Onochie* FHI 8139! Ibadan (Nov.) *Keay* FHI 28138! Ibadan to Ilorin (Nov.) *Jackson* 2421! Obudu (Dec.) *Tuley* 60! **W. Cam.:** Nkambe (Feb.) *Hepper* 1879! Wum (Nov.) *Brunt* 870! Bambui to Bamenda (June) *Brunt* 1182! Nyen (May) *Brunt* 1129! Victoria *Maitland* 150a! Throughout tropical Africa, and also in tropical America. Introduced into most other tropical countries. (See Useful Plants.)

6. **H. poecilotricha** (*Hack.*) *Stapf* in F.T.A. 9: 309 (1919); Clayton l.c. 69. *Andropogon poecilotrichus* Hack. in Bol. Soc. Brot. 3: 138 (1885). Perennial, 0·6–1·5 m. high; in savanna. Perhaps the result of hybridization between *H. rufa* and *H. filipendula*.
Guin.: Timbo (Dec.) *Adam* 2673! Nzérékoré to Beyla (May) *Adam* 5050! Labe (Nov.) *Adames* 410! **Ghana:** Gambaga (Oct.) *Rose Innes* GC 30764! Kintampo (Dec.) *Vigne* FH 3201! **N. Nig.:** Njawai (Nov.) *Jackson, Magaji & Tuley* 1923! **W. Cam.:** Bambui (June) *Brunt* 1185! Throughout tropical Africa.

7. **H. finitima** (*Hochst.*) *Anderss. ex Stapf* in F.T.A. 9: 299 (1919); Clayton l.c. 72. *Andropogon finitimus* Hochst. in sched., Schimp. Iter. Abyss. 1797 (1844). *Cymbopogon finitimus* (Hochst.) T. Thoms. in Speke, Journ. Discov. Source Nile 652 (1863). Tufted perennial, 1–2 m. high; inflorescence dense, often fastigiate; a ruderal species of roadsides, waste land and farm fallows.
S.L.: Mateboi (Nov.) *Jordan* 843! Rather widely distributed, but mainly in south tropical Africa and Uganda.

8. **H. hirta** (*Linn.*) *Stapf* in F.T.A. 9: 315 (1919); A. Chev. in Rev. Bot. Appliq. 13: 874; De Miré & Gillet in J. Agric. Trop. 3: 733; Clayton l.c. 75. *Andropogon hirtus* Linn., Sp. Pl. 1046 (1753). Perennial 30–60 cm. high with a dense basal tussock of narrowly linear or filiform leaves; dry soils in open places.
Niger: Aïr (*fide* de Miré & Gillet). Shores of the Mediterranean, extending through the Middle East to Pakistan, and southwards through Egypt and Arabia to northern Tanzania; also in South Africa. Probably introduced in Australia and Central America.

9. **H. quarrei** *Robyns* Fl. Agrost. Congo Belge 1: 171 (1929); Clayton l.c. 82. Tufted, with culms 1–2 m. high and leaf-blades up to 5 mm. broad; roadsides, clearings and disturbed sites. Perhaps the result of hybridization between *H. hirta* and *H. nyassae*.
N. Nig.: Maska (Nov.) *Thatcher* S510! Abuja (Dec.) *Thatcher* S585! Bokkos (Nov.) *Magaji & Tuley* 1816! Mayo Daga (Nov.) *Magaji & Tuley* 1879! Throughout tropical Africa.

10. **H. violascens** (*Stapf*) *W. D. Clayton* in Kew Bull., Add. Ser. 2: 88 (1969). *H. soluta* var. *violascens* Stapf in F.T.A. 9: 319 (1919); F.W.T.A., ed. 1, 2: 591. Annual about 1 m. high; roadsides and old cultivation.
N. Nig.: Bimasa Bakuro-Tureta F.R. (Nov.) *Latilo* FHI 43780! Katagum *Dalz.* 263! Fika (Oct.) *Daggash* FHI 24880! Shendam (Nov.) *Clayton* 1469! Abinsi (Nov.) *Dalz.* 891! Gurum, Vogel Peak (Nov.) *Hepper* 1282! Also in E. Cameroun and Chad. (See Useful Plants.)

11. **H. bagirmica** (*Stapf*) *Stapf* in F.T.A. 9: 319 (1919); A. Chev. in Rev. Bot. Appliq. 13: 874; Berhaut, Fl. Sén. ed. 2, 424; Clayton l.c. 89. *Cymbopogon bagirmicus* Stapf in Journ. de Bot. sér. 2, 2: 214 (1909). *C. solutus* Stapf l.c. 211 (1909). *Andropogon bagirmicus* (Stapf) A. Chev., Sudania 1: 166 (1911). *A. brachypodus* Stapf ex A. Chev., Sudania 1: 167 (1911), without descr. *Hyparrhenia soluta* (Stapf) Stapf in F.T.A. 9: 318 (1919). Annual up to nearly 2 m. high; roadsides and old cultivation.
Sen.: Niokolo-Koba (Oct.) *Adam* 15886! Tambacounda (Oct.) *Adam* 12720! Samba to Torodo (Oct.) *Adam* 18456! **N. Nig.:** Moshi, Katsina (Oct.) *Clayton* 1379! Samaru, Zaria (Oct., Nov.) *Thatcher* S497! S499! Darazo F.R., Bauchi Prov. (Oct.) *De Leeuw & Magaji* 1993! Damagum (Oct.) *De Leeuw* 1254! Also in Chad and the C. African Republic.

12. **H. barteri** (*Hack.*) *Stapf* in F.T.A. 9: 321 (1919); A. Chev. in Rev. Bot. Appliq. 13: 874; Clayton l.c. 90. *Andropogon barteri* Hack. in Flora 68: 124 (1885); Chev. Bot. 716. Annual up to nearly 2 m. high; inflorescence narrow, dense, of up to 8 fastigiate tiers; old farmland and roadsides.
Togo Rep.: Sokode to Lama Kara (Oct.) *Rose Innes* GC 31391! **N. Nig.:** Fadan Karshi, Wamba (Oct.) *Kennedy* FHI 8037! Niger Prov. (Nov.) *Freeman* S196! Gwada (Sept.) *Lamb* FHI 3159! Lokoja (Oct.) *Dalz.* 295! *Barter*! Okene (Oct.) *Ward* L145! **S. Nig.:** Awka (Oct.) *Thomas*! *Jones* FHI 6791! Ezillo (Oct.) *Tuley* 933! Also in C. African and Congo Republics, Malawi and Zambia.

13. **H. familiaris** (*Steud.*) *Stapf* in F.T.A. 9: 325 (1919); Clayton l.c. 92. *Andropogon familiaris* Steud., Syn. Pl. Glum. 1: 385 (1854). *Hyparrhenia effusa* (Bal.) A. Camus (1931). Tufted perennial 0·6–1·5 m. high; savanna grassland.
Guin.: Mamou (Oct.) *Adam* 12755! Nzérékoré to Beyla (May) *Adam* 5050a! **Lib.:** Robertsfield (June) *Adam* 21346! **Ghana:** Amedzofe (Dec.) *Rose Innes* GC 31177! **N. Nig.:** Gembu (Nov.) *Magaji & Tuley* 1856! Njawai (Nov.) *Jackson, Magaji & Tuley* 1924! **S. Nig.:** Ogoja Prov. *Roseveare* 10/30a! Obudu Plateau (Nov.) *Tuley* 775! 1030! **W. Cam.:** Bamenda (Dec.) *Baldwin* 13835! Nyen (May) *Brunt* 1126! Bali (Apr.) *Brunt* 1057! Also in the Congo and adjacent territories, and in Vietnam.

Fig. 455.—HYPARRHENIA RUFA *Stapf* (GRAMINEAE-ANDROPOGONEAE).

A, ligule. B and C, pair of spikelets. B_1, internode of raceme. B_2, C_1, pedicel of pedicelled spikelet. B_3, C_2, sessile spikelet from back and front respectively. B_4, C_3, pedicelled spikelet.

14. **H. figariana** (*Chiov.*) *W. D. Clayton* in Kew Bull., Add. Ser. 2: 94 (1969). *Cymbopogon figarianus* Chiov. in Bull. Soc. Bot. Ital. 1917: 59 (1917). *Andropogon figarianus* Chiov. l.c., in synon. *A. filipendulus* var. *calvescens* Hack. in DC., Monogr. Phan. 6: 635 (1889). *Hyparrhenia barteri* var. *calvescens* (Hack.) Stapf in F.T.A. 9: 322 (1919). Erect annual 0·6–2 m. high; inflorescence fastigiate; a coarser plant than *H. filipendula* but otherwise very similar; waste land and roadsides.
N. Nig.: Panshanu, Jos Plateau (Aug.) *Lawlor & Hall* 183! 339! Also in Congo, Sudan, Uganda and Tanzania.

15. **H. filipendula** (*Hochst.*) *Stapf* in F.T.A. 9: 323 (1919); Clayton l.c. 95. *Andropogon filipendulus* Hochst. in Flora 29: 115 (1846). Caespitose with culms 0·6–2 m. high; in savanna.

15a. **H. filipendula** (*Hochst.*) *Stapf* var. **filipendula.**
Guin.: Madina Tossekré (Oct.) *Adam* 12543! 12545! Fetoré (Oct.) *Adames* 398! Mali (Dec.) *Schnell* 2358! **N. Nig.:** Jos Plateau (Oct.) *Lely* P807! Jos (Nov.) *Clayton* 1448! Gindiri (Oct.) *Hepper* 1135! Panshanu Pass (Aug.) *Lawlor & Hall* 456! Biu (Oct.) *De Leeuw* 1308! **W. Cam.:** Jakiri (June) *Brunt* 766!

15b. **H. filipendula** (*Hochst.*) *Stapf* var. **pilosa** (*Hochst.*) *Stapf* in F.T.A. 9: 324 (1919); Clayton l.c. 97. *Andropogon filipendulus* var. *pilosus* Hochst. in Flora 29: 115 (1846).
N. Nig.: Naraguta (July) *Lely* P433! Okene (Nov.) *Clayton* 581! Throughout tropical Africa, and in E. Asia from Ceylon to Australia.

16. **H. cymbaria** (*Linn.*) *Stapf* in F.T.A. 9: 332 (1919); Clayton l.c. 110. *Andropogon cymbarius* Linn., Mant. Alt. 303 (1771). Robust perennial; culms 2–3·5 m. high, initially slender and rambling, subsequently erect and sustained by stilt roots; in savanna.
N. Nig.: Mambila Plateau: Chappal Waddi (Nov.) *Jackson, Magaji & Tuley* 1991! Mayo Daga (Nov.) *Magaji & Tuley* 1873! Maisamari (Jan., Sept.) *Hepper* 1645! *Oche* MRK 51! Ngelyaki F.R., Gembu (Nov.) *Daramola* FHI 62352! **W. Cam.:** Wum (Nov.) *Brunt* 876! Bafut-Ngemba F.R. (Mar.) *Onochie* FHI 34845! Bamenda (Jan.) *Migeod* 315! Bambui (Aug.) *Brunt* 1269! Throughout tropical Africa, and in Madagascar.

17. **H. cyanescens** (*Stapf*) *Stapf* in F.T.A. 9: 351 (1919); A. Chev. in Rev. Bot. Appliq. 13: 875; Clayton l.c. 120. *Cymbopogon cyanescens* Stapf in Journ. de Bot. sér. 2, 2: 209 (1909). *Andropogon cyanescens* (Stapf) A. Chev., Sudania 1: 35 (1911); Chev. Bot. 717. Robust perennial up to nearly 3 m. high; moist alluvial soils.
Sen.: Movisassoum (Nov.) *Boudet* 4481 *bis*! Velingara (Nov.) *Boudet* 4558! 4593! **Gam.:** *Ruxton* 32! **Mali:** Myomina to Koulikoro (Oct.) *Chev.* 2359! **Guin.:** *Scaëtta* 3152! Kouroussa (Sept.) *Pobéguin* 508! Kankan (Oct.) *Duong*! **S.L.:** Musaia (Dec.) *Deighton* 4440! **Iv. C.:** Boundiali (Oct.) *Boudet* 3312! **Ghana:** Sawla to Wa (Oct.) *Rose Innes* GC 32394! Bole (Oct.) *Rose Innes* GC 32407! Damongo (Mar.) *Morton* GC 8670! Yendi to Zabzugu (Nov.) *Rose Innes* GC 32471! Tamale to Salaga (Oct.) *Rose Innes* GC 32441! **N. Nig.:** Maska (Nov.) *Thatcher* S506! Zaria (Oct.) *Freeman* S154! Anara F.R. (Oct.) *Keay* FHI 5490! Biu (Oct.) *De Leeuw* 1328*a*! Bida (Dec.) *Hubbard* S586! Ilorin (Oct.) *Ward* 39! **S. Nig.:** Olokemeji (Nov.) *Clayton* 568! Also in Congo and Angola.

18. **H. umbrosa** (*Hochst.*) *Anderss. ex W. D. Clayton* in Kew Bull., Add. Ser 2: 127 (1969). *Andropogon umbrosus* Hochst. in sched., Schimp. Iter Abyss. 1116 (1842). *Cymbopogon umbrosus* (Hochst.) Pilger (1929). Culms stout, 1·2–1·8 m. high; highland grassland, by roadsides and in old cultivation.
N. Nig.: Mandaga, Mambila Plateau (Jan.) *Latilo & Daramola* FHI 28995! **W. Cam.:** Nkambe (Feb.) *Charter* FHI 36983! Santa (Jan.) *Brunt* 908! Bambui (Dec.) *Pedder* 4! Bamenda (June) *Brunt* 1176! Widely scattered in Africa, but not common.

19. **H. rudis** *Stapf* in F.T.A. 9: 344 (1919); Clayton l.c. 128. Coarse perennial 2–3 m. high; savanna grasslands, favouring damp or alluvial soils.
Ghana: Sawla to Wa (Oct.) *Rose Innes* GC 32395! **N. Nig.:** Zaria (Jan.) *Bumpus* B6! Vom, Jos Plateau (Oct.) *Gambles* 41! *Oche* 131! Tiba Plateau, Adamawa (Dec.) *Peter & Tuley* 52! Gembu (Nov.) *Magaji & Tuley* 1854! Throughout tropical Africa.

20. **H. collina** (*Pilger*) *Stapf* in F.T.A. 9: 337 (1919); Clayton l.c. 130. *Cymbopogon collinus* Pilger in Engl., Bot. Jahrb. 54: 287 (1917). *Andropogon collinus* Pilger (1910), not of Lojac. (1909). Loosely clumped perennial 30–120 cm. high; damp soils and upland grassland.
N. Nig.: Chappal Waddi, Mambila Plateau (Nov.) *Jackson, Magaji & Tuley* 2073! **W. Cam.:** Nkambe (June) *Brunt* 977! Throughout tropical Africa.

21. **H. niariensis** (*Franch.*) *W. D. Clayton* in Kew Bull., Add. Ser. 2: 140 (1969). *Andropogon niariensis* Franch. in Bull. Soc. Hist. Nat. Autun 8: 330 (1895). Culms 1–2 m. high; in savanna.
W. Cam.: Wum (Nov.) *Brunt* 882! Also in Central African and Congo Republics.

22. **H. welwitschii** (*Rendle*) *Stapf* in F.T.A. 9: 356 (1919); A. Chev. in Rev. Bot. Appliq. 13: 876; Clayton l.c. 142. *Cymbopogon welwitschii* Rendle, Cat. Welw. 2: 157 (1899). *Hyparrhenia gracilescens* Stapf in F.T.A. 9: 357 (1919); F.W.T.A., ed. 1, 2: 591; A. Chev. in Rev. Bot. Appliq. 13: 876. Coarsely tufted annual 0·3–3 m. high; light to moderate shade in savanna woodland, favouring deep soils with adequate moisture.
Guin.: Fouta Djalon *Chev.* 20196! Labé (Nov.) *Adames* 409! Kindia *Jac.-Fél.* 231! Bafing (Oct.) *Adam* 12658! Timbo (Oct.) *Pobéguin* 1797! **S.L.:** Falaba (Nov.) *Miszewski* 17! *Morton* SL 2857! Mamudia (Nov.) *Glanville* 330! Musaia (Feb., Dec.) *Deighton* 4279! 4433! **Iv. C.:** Touba (Oct.) *Aké Assi* 8210! Madinani (Oct.) *Boudet* 3314! Kouiby (Dec.) *Portères* 424! Séguéla (Oct.) *Adjanohoun* 286*a*! **Ghana:** Biakpa (Nov.) *Morton*! **Dah.:** Parakou (Nov.) *Risopoulos* 1296! **N. Nig.:** Anara F.R., Zaria (Oct.) *Keay* FHI 5438! Zungeru (Nov.) *Freeman* S189! Jemaa (Nov.) *Clayton* 1434! Lokoja (Oct.) *Dalz.* 292! Chappal Waddi, Mambila Plateau (Nov.) *Jackson, Magaji & Tuley* 2094! **S. Nig.:** Ogoja Prov. *Rosevear* 20/30*a*! Extends southwards around the Congo basin to Angola.

23. **H. bracteata** (*Humb. & Bonpl. ex Willd.*) *Stapf* in F.T.A. 9: 360 (1919); Clayton l.c. 144. *Andropogon bracteatus* Humb. & Bonpl. ex Willd., Sp. Pl. 4: 914 (1806). Densely tufted, 0·6–1·5 m. high; seasonally wet soils.
Iv. C.: *Bégué* 136! **U. Volta:** Bobo-Dioulasso (June) *Adam* 15093! **N. Nig.:** Vom (Oct.) *De Leeuw* 1620! Vodni *Saunders* 9! Gembu, Mambila (Aug.) *De Leeuw* 1793! Maisamari, Mambila Plateau (Aug., Sept.) *De Leeuw* 1751! *Oche* MRK 58! **S. Nig.:** Obudu Plateau (Nov.) *Tuley* 1023! **W. Cam.:** Nyen (May) *Brunt* 1132! Wum (Nov.) *Brunt* 868! Bambui (July) *Bumpus* Bam/26! Bamenda (Dec.) *Baldwin* 13833! Bamenda to Bali (Apr.) *Brunt* 1667! Extends southwards through western Africa to Rhodesia; also in S. America.

24. **H. newtonii** (*Hack.*) *Stapf* in F.T.A. 9: 363 (1919); Clayton l.c. 148. *Andropogon newtonii* Hack. in Bol. Soc. Brot. 3: 137 (1885). *Hyparrhenia lecomtei* (Franch.) Stapf in F.T.A. 9: 362 (1919). Densely tufted, 60–120 cm. high; stony hillsides.
Guin.: Labé (Nov.) *Adames* 411! **S.L.:** Loma Mts. (Nov.) *Morton* SL 2744! **N. Nig.:** Vogel Peak, Sardauna Prov. (Nov.) *Hepper* 1424! Chappal Waddi, Mambila Plateau (Nov.) *Jackson, Magaji & Tuley* 2095! **W. Cam.:** Bambui (Dec.) *Pedder* 7! Bamenda (June) *Brunt* 1178! 1179! Banso to Bamenda (Oct.) *Tamajong* FHI 23486! Extends southwards through western Africa to Rhodesia and Transvaal; also in Madagascar, Thailand, Indo-China and Indonesia.

25. **H. mutica** *W. D. Clayton* in Kew Bull., Add. Ser. 2: 161 (1969). Coarse grass 1·5–2·5 m. high; swampy places.
S.L.: Bintumane (Nov., Dec., Jan.) *Morton*! *T. S. Jones* 78! *Jaeger* 492! Loma Mts. (Nov.) *Morton* SL 2720! **Lib.:** Nimba Reserve (Nov.) *Adames* 746! **Ghana:** Essiama (Mar., Dec.) *Rose Innes* GC 30549! GC 30874! **N. Nig.:** Gangumi, Gashaka Dist. (Dec.) *Latilo & Daramola* FHI 28794! Widespread in tropical Africa, but not common.

Fig. 456.—Hyparrhenia mutica *W. D. Clayton* (Gramineae-Andropogoneae).

1, habit, basal part, × ⅔. 2, inflorescence, × ⅔. 3, raceme-pair, × 4. 4, raceme-bases, × 12. 5, single raceme spread open, × 4. 6, lower glume of sessile spikelet, including callus, × 6. 7, upper lemma of sessile spikelet, × 6. From *P. Adames* 746.

26. **H. diplandra** (*Hack.*) *Stapf* in F.T.A. 9: 368 (1919); A. Chev. in Rev. Bot. Appliq. 13: 876; Berhaut, Fl. Sén. ed. 2, 424; Clayton l.c. 166. *Andropogon diplandrus* Hack. in Flora 68: 123 (1885). *A. osikensis* Franch. (1895). *A. pachyneuros* Franch. (1895). *A. obscurus* K. Schum. (1897). *Cymbopogon diplandrus* (Hack.) De Wild. (1919)—Chev. Bot. 717. *Hyparrhenia pachystachya* Stapf in F.T.A. 9: 370 (1919). Coarse grass 2–3 m. high.
Guin.: Dalaba (Oct.) *Adam* 12673! Baffing R. (Oct.) *Pobéguin* 1803! Konkouré (Mar.) *Pitot* 279! Macenta (Oct.) *Baldwin* 9750! Nimba (Oct.) *Schnell* 3865! **S.L.:** Kambia (Dec.) *Deighton* 877! Juring (Dec.) *Deighton* 294! Mt. Loma (Nov.) *Jaeger* 499! Loma Mansa (Dec.) *Morton* SL 447! Senehun (Apr.) *Morton & Gledhill* SL 1964! **Lib.:** Gbanga (Dec.) *Baldwin* 10538! Nimba (Dec., Jan., Apr.) *Harley* 2211! *Adam* 20207! 20637! **Iv. C.:** Bingerville (Dec.) *Chev.* 20091! *Hédin*! **N. Nig.:** Mandaga, Mambila Plateau (Jan.) *Latilo & Daramola* FHI 28993! Gembu, Mambila (Sept., Nov.) *Oche* MRK 84! *Magaji & Tuley* 1858! Donga R., Mambila (Sept.) *Oche* MRK 86! **S. Nig.:** Oyo (Oct.) *Stanfield*! Enugu (May) *Jones* FHI 18862! Ogoja Prov. *Rosevear* 16/30*a*! Obudu (Apr.) *Haines* 370! **W. Cam.:** Nkambe (Feb.) *Charter* FHI 36981! Wum (Nov.) *Brunt* 875! Bafut-Ngemba F.R. (Mar.) *Hepper* 2242! Bamenda (Dec.) *Baldwin* 13838! Ndu (Jan.) *Keay* FHI 28432! Throughout tropical Africa; also in Thailand, Vietnam and Indonesia.

27. **H. subplumosa** *Stapf* in F.T.A. 9: 368 (1919); A. Chev. in Rev. Bot. Appliq. 13: 876; Berhaut, Fl. Sén. ed. 2, 424; Clayton l.c. 164. Robust perennial 2–3 m. high; savanna woodland, often on poor dry soils.
Sen.: Kanéméré (Nov.) *Fotius* K723! Vélingara (Oct.) *Adam* 18619! Niokolo-Koba (Nov.) *Adam* 17070! Tambacounda (Nov.) *Adam* 20031! Tambacounda to Niokolo-Koba (Oct.) *Adam* 15638! **Mali:** Sikasso (Sept.) *Adam* 15460! **Guin.:** Pita (Oct.) *Adam* 12555! Dalaba *Chev.* 20185! Timbo (Oct.) *Pobéguin* 1799! Baffing R. (Nov.) *Pobéguin* 1812! Siguiri (Jan.) *Jac.-Fél.* 1331! **S.L.:** Kambia (Nov.) *Jordan* 689! Mabala (Oct.) *Glanville* 62! Samaia (May) *Thomas* 227! Musaia (Dec.) *Deighton* 4444! Mamodia (Nov.) *Glanville* 329! **Lib.:** Nimba Reserve (Oct., Nov.) *Adames* 698! 748! **Iv. C.:** Ferkessédougou (Nov.) *Leeuwenberg* 1999! Séguéla to Vavoua (Oct.) *Adjanohoun* 361*a*! Zuénoula to Vavoua (Oct.) *Adjanohoun* 313*a*! Toumodi (Apr.) *Leeuwenberg* 3321! Dabou (Nov.) *Leeuwenberg* 1967! **U. Volta:** Samandéni (Oct.) *Kmoch* 128! Bobo-Dioulasso (Oct.) *Kmoch* 147! Ouassa (Oct.) *Rose Innes* GC 31509! Leo (Oct.) *Rose Innes* GC 31074! Tondoura (Oct.) *Scholz* 83! **Ghana:** Wa to Lawra *Rose Innes* GC 32354! Gambaga (Oct.) *Ankrah* GC 20285! Yendi to Zabzugu (Nov.) *Rose Innes* GC 32470! Elmina (Nov.) *Rose Innes* GC 31171! Achimota (June) *Irvine* 3056! Kpong to Somanya (Nov.) *Rose Innes* GC 31647! **Togo Rep.:** Misahöhe *Baumann* 328! **Dah.:** Parakou *Risopoulos* 1294! Cotonou to Allada (Oct.) *Risopoulos* 1234! **Niger:** Wadi Dollol *White* 21! **N. Nig.:** Zaria (Nov.) *Thatcher* S536! Anara F.R., Zaria (Oct.) *Keay* FHI 20109! Riyom (Nov.) *Clayton* 1442! Abinsi *Dalz.* 901! Kaiama (Oct.) *Ward* 18! Gangumi, Gashaka Dist. (Nov.) *Latilo & Daramola* FHI 28776! **S. Nig.:** Olokemeji (Nov.) *Clayton* 566! Ibadan (Nov.) *Ogua* FHI 7945! Iseyin (Oct.) *Jackson* 2364! Akpaka F.R. (Nov.) *Onochie* FHI 40429! Obudu Plateau (Feb.) *Tuley* 489! Also found occasionally southward to Angola and Rhodesia. (See Useful Plants.)

28. **H. involucrata** *Stapf* in F.T.A. 9: 377 (1919); Clayton l.c. 158. *Anthistiria barteri* Munro ex Oliv. (1875), without descr. *Androscoepia barteri* Anderss. ex Oliv. (1875), without descr. *Hyparrhenia notolasia* Stapf in F.T.A. 9: 377 (1919); F.W.T.A., ed. 1, 2: 591. Robust annual 2 m. high; in savanna, particularly on dry or shallow soils.

28a. **H. involucrata** *Stapf* var. **involucrata**
Ghana: Bolgatanga to Tamale (Oct.) *Ankrah* GC 20300! Yendi to Zabzugu (Nov.) *Rose Innes* GC 32472! **N. Nig.:** Funtua to Yashi (Oct.) *Keay* FHI 21147! Samaru, Zaria (Oct.) *Freeman* S168! Plateau Prov. (Oct.) *Lely* P849! Nupe *Barter* 957! Lokoja (Oct.) *Dalz.* 299! **S. Nig.:** Igbetti (Oct.) *Stanfield* FHI 55499! Ago-Are F.R. (Nov.) *Keay* FHI 37680! FHI 37693! Awba Hills F.R. (Nov.) *Jones* FHI 7322! Ngwo (Oct.) *Jones* FHI 6783! Fashola Rock (Nov.) *Gledhill* 677! Also in E. Cameroun and C. African Republic.

28b. **H. involucrata** var. **breviseta** *W. D. Clayton* in Kew Bull., Add. Ser. 2: 159 (1969).
Iv. C.: Séguélo (Oct.) *Aké Assi* 6567! **U. Volta:** Zabré (Oct.) *Scholz* 161! Leo to Po (Oct.) *Rose Innes* GC 31475! **Ghana:** Wa (Oct.) *Rose Innes* GC 32373! Navrongo (Oct.) *Vigne* FH 4640! Nakpanduri (Sept.) *Rose Innes* GC 32119! Yapei ferry (Oct.) *Rose Innes* GC 30676! Kete Krachi to Yendi (Sept.) *Rose Innes* GC 30448! **Dah.:** Boukombe (Nov.) *Risopoulos* 1259! Also in Chad and C. African Republic.

127. HYPERTHELIA W. D. Clayton in Kew Bull. 20: 438 (1966).

Racemes not deflexed; raceme-bases produced into bract-like linear to narrowly lanceolate appendages 4–11 mm. long; homogamous pairs 1 at the base of the lower raceme; sessile spikelets 10–14 mm. long, including a pungent callus 3–6 mm. long; awn 5–10 cm. long, hirtellous with yellow hairs .. .. *dissoluta*

H. dissoluta (*Nees ex Steud.*) *W. D. Clayton* in Kew Bull. 20: 441 (1966). *Anthistiria dissoluta* Nees ex Steud., Syn. Pl. Glum. 1: 400 (1854). *Hyparrhenia dissoluta* (Nees ex Steud.) C. E. Hubbard in F.W.T.A., ed. 1, 2: 591 (1936); Berhaut, Fl. Sén. ed. 2, 421. *H. ruprechtii* Fourn., Mex. Pl. Gram. 67 (1886); F.T.A. 9: 326; A. Chev. in Rev. Bot. Appliq. 13: 874. Perennial up to nearly 3 m. high, with a fastigiate yellowish inflorescence; a savanna grass typical of waysides and disturbed places.
Maur.: Gaberni *Rossetti* 61/292! **Sen.:** Dakar (Oct.) *Adam* 2141! Hann (Oct.) *Berhaut* 1078! Tambacounda (Oct.) *Adam* 18625! Niokolo-Koba *Adam* 14270! Kayar (Oct.) *Trochain* 654! **Gam.:** *Hayes* 514! **Mali:** Niger Delta *Duong*! Kadigué to Banguita (Jan.) *Davey* 6! Macina (Sept.) *Chev.* 24943! Bamako (Sept.) *Adam* 14989! *Rogeon* 221! **Guin.:** Kissidougou to Kankan (Oct.) *Jaeger* 2156! **Port. G.:** Piche (Sept., Nov.) *Pereira* 3186! 3211! 3491! Cabuca (Oct.) *Pereira* 3303! **Iv. C.:** Oussou (Mar.) *Scaëtta*! Dabou (Nov.) *Leeuwenberg* 1940! *Chev.* 17150! Dihabo to Bouaké (July) *Chev.* 22054! **U. Volta:** Bobo-Dioulasso (Oct.) *Kmoch* 140! Sindou (Sept.) *Aké Assi* 6374! Sikasso to Bobo-Dioulasso (Sept., Oct.) *Adam* 15095! *Koechlin* 6979! Po to Leo (Oct.) *Rose Innes* GC 31481! **Ghana:** Navrongo (Sept.) *Vigne* FH 4629! Bolgatanga to Tamale (Oct.) *Rose Innes* GC 30232! Gambaga to Yendi (Sept.) *Rose Innes* GC 30460! Pong Tamale (July) *Williams* 841! Legon Hill (Oct.) *Adams* 3438! **Togo Rep.:** Sansanné Mango to Dapango (Oct.) *Rose Innes* GC 31422! Djagba, Atakpame to Sokode (Oct.) *Rose Innes* GC 31363! **Dah.:** Ouidah to Adjounja (Apr.) *Chev.* 23466! Agouagon (May) *Chev.* 23499! Djougou to Pabigou (June) *Chev.* 23928! **Niger:** Filingé (Sept.) *Koechlin* 6882! Tahoua (Sept.) *Koechlin* 6771! Say (Sept.) *Koechlin* 6394! Toukounous *Bartha* 27! **N. Nig.:** Kworre, Sokoto (Sept.) *Palmer* 1! Katagum Dist. *Dalz.* 265! Anara F.R., Zaria (May) *Keay* FHI 22968! Ilorin (Oct.) *Ward* 61! Jebba *Barter* 281! **S. Nig.:** Lagos *W. MacGregor* 320! Ikoyi Plains (Mar., June) *Dalz.* 1316! *Maitland* 160*a*! Aworo (June) *Dawodu* 52! Igboora *Haines* 337! Throughout tropical Africa; also in Mexico, Paraguay and Colombia. (See Useful Plants.)

128. PARAHYPARRHENIA A. Camus in Bull. Mus. Hist. Paris, sér. 2, 22: 404 (1950); Clayton in Kew Bull. 20: 434 (1966).

Annual; awn of fertile lemma 3·5–10 cm. long .. .. 1. *annua*
Perennial; awn of fertile lemma 2·7–3·5 cm. long .. .. .. 2. *perennis*

Fig. 457.—PARAHYPARRHENIA PERENNIS *W. D. Clayton* (GRAMINEAE-ANDROPOGONEAE).

1, base of plant with leaves, × ⅔. 2, false panicle, × ⅔. 3, raceme-pair, × 4. Sessile spikelet: 4, lower glume with callus, × 8; 5, upper glume, × 8; 6, lower lemma, × 8; 7, upper lemma, × 8; 8, flower, × 8; 9, lodicules, × 8. From *Berhaut* 6649.

8

1. **P. annua** (*Hack.*) *W. D. Clayton* in Kew Bull. 20: 434 (1966); Berhaut, Fl. Sén. ed. 2, 425. *Andropogon annuus* Hack. in Flora 68: 137 (1885). *Parahyparrhenia jaegerana* A. Camus in Bull. Mus. Hist. Nat., sér. 2, 22: 404 (1950); Berhaut, Fl. Sén. ed. 2, 424. *Hyparrhenia jaegerana* (A. Camus) Roberty in Boissiera 9: 109 (1960). *H. djalonica* Jac.-Fél. in J. Agric. Trop. 1: 51 (1954). *H. sulcata* Jac.-Fél. l.c. 54 (1954); Berhaut, Fl. Sén. ed. 2, 424. Loosely tufted annual 0·3–1·6 m. high; shallow pools on sheet ironstone. Very variable, but without clear distinctions between the various forms.
Sen.: Niokolo-Koba (Oct.) *Adam* 15616! 15696! Tambacounda *Berhaut* 3199! Maèl *Berhaut* 3251! Diana Malari (Oct.) *Adam* 18466! **Mali:** Kita *Jaeger* 17! **Guin.:** Madina Tossekré (Oct.) *Adam* 12527! Dalaba (Oct.) *Adames* 392! Timbo (Oct.) *Pobéguin* 1778! Ouassou to Koba (Nov.) *Jac.-Fél.* 7258! Mamou (Nov.) *Chev.* 34600! **S.L.:** Kambia (Aug., Nov.) *Jordan* 666! 964! Kambia to Kukuna (Nov.) *Morton & Gledhill* SL 33! **Iv. C.:** Boundiali to Mankono (Nov.) *Aké Assi* 8296! 8297! **U. Volta:** Baoulé (Oct.) *Kmoch* 123! Ouagadougou to Sikasso (Sept.) *Adam* 15418! Batié to Bouna (Oct.) *Rose Innes* GC 31540! **Ghana:** Burufu (Oct.) *Ankrah* GC 20462! *Hinds* 3770! 5010! Nakpanduri to Yendi (Oct.) *Rose Innes* GC 30734! Also in C. African Rep. and Sudan.

2. **P. perennis** *W. D. Clayton* in Kew Bull. 20: 436 (1966); Berhaut, Fl. Sén. ed. 2, 425. Caespitose grass about 60 cm. high; ironstone outcrops.
Sen.: Niama (Nov.) *Berhaut* 6649! **Guin.:** Conakry (Oct.) *Adam* 12595! **Lib.:** *Warner* 66! Paynesville (Sept.) *van Dillewijn* 82! Duport (Oct.) *van Harten* 172!

129. ELYMANDRA Stapf in F.T.A. 9: 407 (1919); Jacques-Félix in Rev. Bot. Appliq. 30: 169 (1950); Clayton in Kew Bull. 20: 287 (1966).

Racemes solitary, 1-awned; callus of sessile spikelet fulvously bearded 1. *gossweileri*
Racemes paired; callus of sessile spikelet white bearded:
Pedicelled and homogamous spikelets glabrous; upper raceme-base 15–25 mm. long; racemes 2–awned per pair; upper glume of sessile spikelet usually awnless; perennial 2. *androphila*
Pedicelled and homogamous spikelets sparsely hairy to villous; upper raceme-base 3–4 mm. long; upper glume of sessile spikelet with an awn 1–3 cm. long:
Plants annual; racemes 3–5-awned per pair; homogamous pairs 2–10 at the base of each raceme:
Homogamous pairs 7–10 at the base of each raceme, the spikelets 7–9 mm. long; racemes 3–5-awned per pair; fertile lemma bifid to the middle 3. *subulata*
Homogamous pairs 2 at the base of each raceme, the spikelets 10–12 mm. long; racemes 4–awned per pair; fertile lemma shortly bilobed (lobes 1 mm. long) 4. *archaelymandra*
Plants perennial; racemes about 12-awned per pair; homogamous pairs 1 at the base of the lower raceme only .. .. 5. *grallata*

1. **E. gossweileri** (*Stapf*) *W. D. Clayton* in Kew Bull. 20: 288 (1966); Berhaut, Fl. Sén. ed. 2, 424. *Pleiadelphia gossweileri* Stapf in Hook., Ic. Pl. t.3121 (1927). *Themeda gossweileri* (Stapf) Roberty in Boissera 9: 103 (1960). *Elymandra monostachya* Jac.-Fél. in Rev. Bot. Appliq. 30: 170 (1950). Erect annual 2 m. high; shallow sandy soils.
Sen.: Kabrousse, Casamance (Oct.) *Adam* 18295 *bis*! 18374! Niassia (Nov.) *Berhaut* 6648! **Guin.:** Kanea (Oct.) *Adam* 12632! Kindia (Oct.) *Jac.-Fél.* 7193! 7194! **Iv. C.:** Upper Sassandra and Upper Cavally basins *Portères* 238! Also in Angola.

2. **E. androphila** (*Stapf*) *Stapf* in F.T.A. 9: 408 (1919); Jac.-Fél. in Rev. Bot. Appliq. 30: 170 (1950); A. Chev. in Rev. Bot. Appliq. 13: 878 (1933); Berhaut, Fl. Sén. ed. 2, 421; Aké Assi, Contrib. 2: 302. *Andropogon androphilus* Stapf in Journ. de Bot. 19: 103 (1905); Chev. Bot. 715. *Heteropogon androphilus* (Stapf) Roberty in Boissiera 9: 142 (1960). Coarse perennial up to 2·5 m. high; poor sandy soils in uncultivated savanna.
Sen.: Niokolo-Koba *Adam*. **Mali:** Sotuba to Bamako (Sept.) *Adam* 15362! **Guin.:** Soumbalako to Boulivel *Chev.* 18631! Timbo (Oct.) *Pobéguin* 1786! 1798! Kouroussa (Sept.) *Pobéguin* 521! **S.L.:** Mange to Kambia (Oct.) *Adames* 265! Robumbe (Oct.) *Jordan* 800! Dankawali (Nov.) *Glanville* 325! Musaia (Dec.) *Deighton* 4524! Kambaia (Oct.) *Small* 481! **Lib.:** Nimba (Oct.) *Adames* 697! **Iv. C.:** N. Toumodi (Oct.) *Adjanohoun* 337*a*! Aboakouamékro (Nov.) *Aké Assi* 10360! **U. Volta:** Sindou (Sept.) *Adam* 15193! Banfora (Sept.) *Adam* 15215! Bobo-Dioulasso (Oct.) *Koechlin* 6962! Samandeni (Oct.) *Kmoch* 159! Cascade de Toussiana (Oct.) *Scholz* 74*a*! **Ghana:** Gradaw (Oct.) *Rose Innes* GC 31612! Kintampo (Oct.) *Ankrah* GC 20316! Attebubu to Kete Krachi (Oct.) *Rose Innes* GC 30616! Kwahu Tafo to Adawso (Oct.) *Rose Innes* GC 30572! Akaa (Dec.) *Adams* 4713! **Togo Rep.:** Badou to Atakpame (Sept.) *Rose Innes* GC 31357! **N. Nig.:** Keffi to Abuja, Benue Prov. (Nov.) *Freeman* S200! Kontagora (Nov.) *De Leeuw & Magaji* 2000! Sha (Sept.) *Hall* K20! Makwando, Niger Prov. (Nov.) *Ayuba* 73! **W. Cam.:** Nyen, 5,000 ft. (May) *Brunt* 1125! Bambui, 6,300 ft. (Dec.) *Boughey* GC 10387! Also in C. African Rep., Congo and Angola.

3. **E. subulata** *Jac.-Fél.* in Rev. Bot. Appliq. 30: 170 (1950). Annual up to nearly 2 m. high.
Guin.: Seriba (Oct.) *Adam* 12696! Labé (Nov.) *Adames* 412! Timbi Madina (Jan.) *Langdale-Brown* 2553! Kindia (Oct.) *Jac.-Fél.* 7200! Kindia to Conakry (Oct.) *Chev.* 20203! **S.L.:** Batkanu (Nov.) *Jordan* 846! Binle (Oct.) *Adames* 254! **Iv. C.:** Séguéla to Vavoua (Oct.) *Adjanohoun* 285*a*!

4. **E. archaelymandra** (*Jac.-Fél.*) *W. D. Clayton* in Kew Bull. 20: 291 (1966). *Hyparrhenia archaelymandra* Jac.-Fél. in J. Agric. Trop. 1: 48 (1954); Berhaut, Fl. Sén. ed. 2, 424. Annual up to nearly 2 m. high; shallow sandy soils.
Sen.: Niokolo-Koba (Oct.) *Adam* 15504! Tambacounda (Oct.) *Adam* 15489! Ouassadou *Berhaut* 1895! Vélingara (Oct.) *Adam* 18614! Kolda (Oct.) *Adam* 18540! **Guin.:** Tamba to Sambaïlo (Sept.) *Pitot*!

5. **E. grallata** (*Stapf*) *W. D. Clayton* in Kew Bull. 20: 292 (1966). *Hyparrhenia grallata* Stapf in F.T.A. 9: 320 (1919). *H. eylesii* C. E. Hubbard in Kew Bull. 1928: 37. Tufted perennial 1–2 m. high; sandy soils in savanna woodland.
Iv. C.: Sanlo to Kakpin (Nov.) *Aké Assi* 9290! Tanzania to Angola and S.W. Africa.

130. MONOCYMBIUM Stapf in F.T.A. 9: 386 (1919); C. E. Hubbard in Kew Bull. 4: 375 (1949); Jac.-Fél. in Rev. Bot. Appliq. 30: 175 (1950).

Culms erect, caespitose, rarely weak and straggling; leaf-blades linear, 2–5 mm. broad and 5–15 cm. long or more; lower glume of sessile spikelet villous, not spinulose; callus of sessile spikelet longer than broad, ciliate to bearded on base and sides; callus of pedicelled spikelet 0·9 mm. long .. .. 1. *ceresiiforme*

Culms geniculately ascending from a procumbent base, and rooting at the nodes; leaf-blades narrowly lanceolate to lanceolate; lower glume of sessile spikelet glabrous or villous, the margins spinulose towards the tip; callus of sessile spikelet shorter than broad, bearded on the sides but hairless along the obtuse base; callus of pedicelled spikelet 0·3 mm. long:

Leaf-blades narrowly lanceolate, 1–5 cm. long and 2–7 mm. broad, rounded at the base; culms 0·5–1·7 mm. diam. .. .. 2. *deightonii*

Leaf-blades lanceolate, 4·5–9 cm. long and 10–20 mm. broad, amplexicaul at the base; culms 2·5–3 mm. diam. .. .. 3. *lanceolatum*

1. **M. ceresiiforme** (*Nees*) *Stapf* in F.T.A. 9: 387 (1919); A. Chev. in Rev. Bot. Appliq. 13: 877. *Andropogon ceresiiformis* Nees, Fl. Afr. Austr. 109 (1841). *Monocymbium nimbanum* Jac.-Fél. in Rev. Bot. Appliq. 30: 176 (1950). A tufted perennial somewhat fibrous at the base; densely caespitose in southern Africa, but in our region usually 90–120 cm. high with erect leafy culms, these weak and often lodged by heavy rain; shallow stony soils on hillsides.
Guin.: Nzérékoré (Oct.) *Jac.-Fél.* 1930! Mt. Nimba (Sept., Oct.) *Schnell* 1893! *Adam* 6714! Kouroussa (Sept.) *Pobéguin* 511! Macenta to Kérouané (Nov.) *Adam* 6973! **S.L.:** Dankawali (Nov.) *Glanville* 326! Tingi Mts. (Dec.) *Morton & Gledhill* SL 3029! Mt. Loma (Oct., Nov.) *Jaeger* 7893! *Morton* SL 2575! SL 2670! **Lib.:** Nimba (Oct.) *Adames* 650! **Iv. C.:** Mt. Orumbo-Boka, Toumodi (Dec.) *Boughey* A9! Toumodi (Oct.) *Adam* 6582! Bouaké (Oct.) *Adam* 6635! Toumodi to Yamoussokro (Oct.) *Adjanohoun* 327a! Séguélo (Oct.) *Aké Assi* 6602! **U. Volta:** Batié to Gaoua (Oct.) *Rose Innes* GC 31525! Samandeni (Oct.) *Kmoch* 152! Gaoua to Diebougou (Oct.) *Rose Innes* GC 31519! Leo to Ouassa (Oct.) *Rose Innes* GC 31496! Bobo-Dioulasso (Oct.) *Kmoch* 117! **Ghana:** Accra Plains (Sept.) *Ankrah* GC 20011! Maliato (Nov.) *Morton* A3534! Banda (Oct.) *Rose Innes* GC 31587! Damongo to Sawla (Oct.) *Rose Innes* GC 31010! Bawku to Bolgatanga (Oct.) *Rose Innes* GC 31452! **Togo Rep.:** *Baumann* 346! Atacora Mt. Pass, Sokode to Lama Kara (Oct.) *Rose Innes* GC 31396! Sansanné Mango to Dapango (Oct.) *Rose Innes* GC 31421! **Dah.:** Boukombe to Nattitingou (Nov.) *Risopoulos* 1269! **N. Nig.:** Jebba *Barter*! Abinsi *Dalz.* 893! Wuya Ferry, Kaduna R. (Sept.) *Clayton* 366! Kaiama (Oct.) *Ward* 23! Anara F.R. (Oct.) *Keay* FHI 20107! **S. Nig.:** Olokemeji (Nov.) *Onochie* FHI 8111! Akure F.R. (Oct.) *Keay* FHI 22667! Jesse (Sept.) *Butler-Cole* 30! Udi (Nov.) *Jones* 6231! Ogoja Prov. *Rosevear* 17/30a! **W. Cam.:** Bamenda *Dept. of Agric.* 2! Extends eastwards to the Sudan, and southwards through the Congo, Tanzania and Rhodesia into S. Africa. (See Useful Plants.)
[The vegetative characters display a remarkably wide, but apparently continuous, range of variation, while the spikelet characters remain constant; I have, therefore, continued to treat it as a single species for the time being.—W.D.C.]

2. **M. deightonii** *C. E. Hubbard* in Kew Bull. 4: 374 (1949); incl. var. *tonkouii* Jac.-Fél. in Rev. Bot. Appliq. 30: 177 (1950). Procumbent perennial, with flowering culms 30–60 cm. high; spikelet indumentum ranges from glabrous to villous; seepage zones on rock outcrops.
Guin.: Mamou *Adam* 7913! Macenta (Oct.) *Jac.-Fél.* 1314! *Baldwin* 9784! Mt. Yonon (Oct., Nov.) *Adam* 6754! 6801! **S.L.:** Sefadu (Nov., Dec.) *Deighton* 3561! 4666! Kumrabai to Mamila (Nov.) *Deighton* 4942! Picket Hill (Nov.) *T. S. Jones* 208! Sugar Loaf Mt. (Nov.) *T. S. Jones* 265! **Lib.:** Sanokwele (Sept.) *Baldwin* 9525! Soplima, Vonjama Dist. (Nov.) *Baldwin* 10064! Genne-Loffa, Kolahun Dist. (Nov.) *Baldwin* 10095! **Iv. C.:** Mt. Tonkoui, Man *Jac.-Fél.* 1278! Mt. Momi (Oct.) *Aké Assi* 9167!

3. **M. lanceolatum** *C. E. Hubbard* in Kew Bull. 1936: 313. Similar to the preceding, but much more robust; moist, sheltered situations in secondary bush.
Guin.: (Oct.) *Adam* 6762! Mt. Gangan, Kindia *Jac.-Fél.* 272! Benna (Oct.) *Jac.-Fél.* 7184! **S.L.:** Mamuria to Kulufaga (Nov.) *Glanville* 337! Fotane (Nov.) *Glanville* 100! Gbenge Hills (Nov.) *Morton* SL 2893!

131. ANADELPHIA Hack. in Engl., Bot. Jahrb. 6: 240 (1885); F.T.A. 9: 388 (1919); Jacques-Félix in Rev. Bot. Appliq. 30: 177; Clayton in Kew Bull. 20: 275 (1966).

Sessile spikelets 1 in each raceme, embraced by the spatheole:
Upper glume of sessile spikelets awnless (rarely with a short awn-point); lower glume coriaceous:
Pedicelled spikelets present, their callus 2-toothed; sessile spikelets 8–10 mm. long; perennial .. .. 1. *trispiculata*
Pedicelled spikelets absent; lower glume of sessile spikelets 2-toothed at the tip; annuals:
Awn of fertile lemma 4·5–6 cm. long, the column 1–2 times as long as the limb; sessile spikelet 6–7 mm. long (incl. callus) .. .. .. 2. *macrochaeta*
Awn of fertile lemma 10–12 cm. long, the column 3 times as long as the limb; sessile spikelet 10–11 mm. long (incl. callus) .. .. 3. *funerea*
Upper glume of sessile spikelets with an awn 4–9 mm. long; pedicelled spikelets, if present, with a truncate or obtuse callus:
Lower glume of sessile spikelet produced into 2 setae or awns up to 5 mm. long; sessile spikelets 5·5–7 mm. long, the lower glume coriaceous .. .. 4. *trichaeta*
Lower glume of sessile spikelet 2-toothed; sessile spikelets 3·5–5 mm. long, the lower glume firmly chartaceous and finely nerved .. .. .. 5. *trepidaria*
Sessile spikelets 2–6 in each raceme:
Lower glume of the sessile spikelets obtuse at the apex, coriaceous; sessile spikelets 2 per raceme, enclosed by the spatheole:
Perennial; sessile spikelets 5–8 mm. long, including a callus 1·5–2·5 mm. long; pedicelled spikelets 6–9 mm. long; spatheoles 2–3·5 cm. long, green tinged with yellow or brown; panicle containing numerous racemes .. 6. *afzeliana*
Annual; sessile spikelets 10–13 mm. long, including a callus 3–4 mm. long; pedicelled spikelets 15–25 mm. long; spatheoles 4·5–5·5 cm. long, tinged with purple; panicle scanty .. .. 7. *chevalieri*
Lower glume of the sessile spikelets 2-toothed or 2 awned at the apex; raceme exserted from the tip of the spatheole:

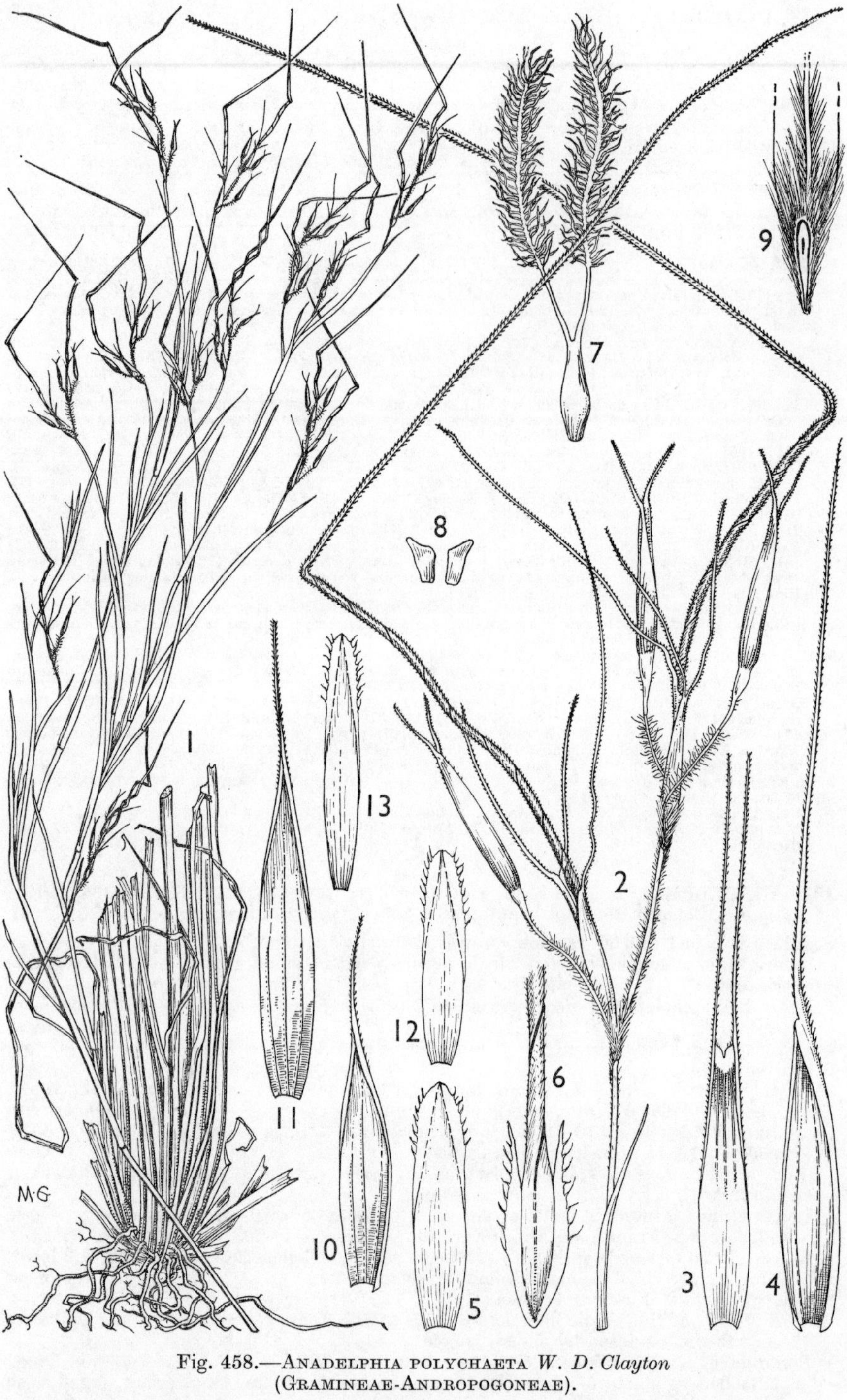

Fig. 458.—ANADELPHIA POLYCHAETA *W. D. Clayton*
(GRAMINEAE-ANDROPOGONEAE).

1, habit, × ⅔. 2, raceme, × 3. Sessile spikelet: 3, lower glume, × 6; 4, upper glume, × 6; 5, lower lemma, × 6; 6, upper lemma, × 6; 7, ovary and stigmas, × 4; 8, lodicules, × 4. 9, callus and base of pedicel, × 16. Pedicelled spikelet: 10, lower glume, × 6; 11, upper glume, × 6; 12, lower lemma, × 6; 13, upper lemma, × 6. From *Adam* 18222.

Perennial; callus of sessile spikelets obtuse, the lower glume firmly chartaceous and finely nerved; leaf-blades up to 6 mm. broad, rarely narrow and involute; sessile spikelets 3–5 per raceme, 4–5 mm. long; awn of fertile lemma 0·7–1·7 cm. long 8. *leptocoma*

Annuals; lower glume of sessile spikelets coriaceous; leaf-blades 1–2 mm. broad:

Sessile spikelets 5·5–7 mm. long, 2 per raceme, the apex of the lower glume drawn out into 2 awnlets 1–5 mm. long; callus slender and pungent; awn of fertile lemma 2–4 cm. long .. .. 9. *polychaeta*

Sessile spikelets 3·5–4·5 mm. long, 2-toothed at the apex; awn of fertile lemma 1·5–2·2 cm. long:

Awn of fertile lemma once geniculate; callus of sessile spikelet slender, narrowly obtuse to pungent, its hairs pallid; leaf-blades 2–8 cm. long; sessile spikelets 4–6 per raceme .. .. 10. *pumila*

Awn of fertile lemma bigeniculate; callus of sessile spikelet oblong and obtuse, its hairs fulvous; leaf-blades 12–30 cm. long; sessile spikelets 2–3 per raceme 11. *bigeniculata*

1. **A. trispiculata** *Stapf* in F.T.A. 9: 398 (1919); A. Chev. in Rev. Bot. Appliq. 13: 877. *Pobeguinea trispiculata* (Stapf) Jac.-Fél. in Rev. Bot. Appliq. 30: 174 (1950). *Hypogynium trispiculatum* (Stapf) Roberty in Boissiera 9: 183 (1960). Stout caespitose perennial about 120 cm. high; panicle stiffly linear, the branches appressed to the main axis; swampy places.
Guin.: Conakry (Oct.) *Adam* 12607! Kanea (Oct.) *Adam* 12633! Kindia *Jac.-Fél.* 257! Baffing R. (Oct.) *Pobéguin* 1787! Fouta Djalon (Oct.) *Adam* 12574! **S.L.:** Mano Bonjema (Jan.) *Adames* 3! Whale R. (Oct.) *Morton* SL 1422! Massa (Sept.) *J. W. D. Fisher* 1765! Toma, Bonthe Is. (Nov.) *Deighton* 2293! **Iv. C.:** Abouabou (Mar.) *Aké Assi* 10008! Moossou (Mar.) *Raynal* 13578! Mbrabo (Jan.) *Hédin*! **Ghana:** Atuabo, Nzima (Jan.) *Thorold* 299! 300! Eikwe (Dec.) *Rose Innes* GC 30548! **Togo Rep.:** Bombouaka (Nov.) *Bille* TG 381!
2. **A. macrochaeta** (*Stapf*) *W. D. Clayton* in Kew Bull. 2: 281 (1966). *Monium macrochaetum* Stapf in F.T.A. 9: 400 (1919); F.W.T.A., ed. 1, 2: 594. *Hypogynium macrochaetum* (Stapf) Roberty in Boissiera 9: 185 (1960). Slender annual 30–60 cm. high; ironstone outcrops.
Guin.: Conakry (Oct.) *Adam* 12618! Mamou (Oct.) *Adam* 12647! Télimélé to Pita (Oct.) *Pitot*! Madina Tossekré (Oct.) *Adam* 12537! Timbo (Oct.) *Pobéguin* 1790!
3. **A. funerea** (*Jac.-Fél.*) *W. D. Clayton* in Kew Bull. 20: 281 (1966). *Monium funereum* Jac.-Fél. in Rev. Bot. Appliq. 30: 186 (1950). Tufted annual about 60 cm. high.
Guin.: Kindia (Sept.) *Jac.-Fél.* 1850!
4. **A. trichaeta** (*Reznik*) *W. D. Clayton* in Kew Bull. 20: 281 (1966). *Monium trichaetum* Reznik in Bull. Mus. Hist. Nat. sér. 2, 4: 1046 (1932); A. Chev. in Rev. Bot. Appliq. 13: 877. *Pobeguinea trichaeta* (Reznik) Jac.-Fél. in Rev. Bot. Appliq. 30: 174 (1950). *Hypogynium trichaetum* (Reznik) Roberty in Boissiera 9: 183 (1960). Slender annual 30–60 cm. high; shallow soils and ironstone outcrops.
Guin.: Conakry (Oct.) *Adam* 12601! Kanea (Dec.) *Adam* 12629! Kindia (Oct.) *Jac.-Fél.* 184! 7198! **S.L.:** Waterloo (Aug.) *Melville & Hooker* 332!
5. **A. trepidaria** (*Stapf*) *Stapf* in F.T.A. 9: 390 (1919). *Andropogon trepidarius* Stapf in Journ. de Bot. 19: 100 (1905). *Monium trepidarium* (Stapf) Jac.-Fél. in Rev. Bot. Appliq. 30: 182 (1950). *M. rufum* Jac.-Fél. l.c. 182 (1950). *M. congestum* Jac.-Fél. l.c. 184 (1950). *M. monianthum* Jac.-Fél. l.c. 184 (1950). *Hypogynium trepidarium* (Stapf) Roberty in Boissiera 9: 186 (1960). Annual 30–60 cm. high on ironstone outcrops.
Guin.: Thiangeul Bori (Oct.) *Adam* 12682! Dalaba (Oct.) *Adames* 394! Labé (Dec.) *Jac.-Fél.* 638! Dinguiraye (Dec.) *Jac.-Fél.* 1472! Kouroussa (Oct.) *Pobéguin* 524!
6. **A. afzeliana** (*Rendle*) *Stapf* in F.T.A. 9: 397 (1919). *Andropogon afzelianus* Rendle in J. Bot. 31: 357 (1893). *A. eriocoleus* K. Schum. (1897). *A. arrectus* Stapf (1905); Chev. Bot. 716. *A. glaucopurpureus* Stapf (1905). *Anadelphia arrecta* (Stapf) Stapf in F.T.A. 9: 396 (1919); F.W.T.A., ed. 1, 2: 592; A Chev. in Rev. Bot. Appliq. 13: 877. *Pobeguinea arrecta* (Stapf) Jac.-Fél. in Rev. Bot. Appliq. 30: 173 (1950); Berhaut, Fl. Sén. ed. 2, 422. *P. afzeliana* (Rendle) Jac.-Fél. l.c. 174 (1950). *Hypogynium arrectum* (Stapf) Roberty in Boissiera 9: 183 (1960). Tufted perennial 1–2 m. high; in savanna.
Sen.: Nioro du Rif (Nov.) *Berhaut* 1697! Niokolo-Koba (Oct.) *Adam* 15900! Kolda to Velingara (Oct.) *Adam* 18554! **Mali:** Zangoula (Oct.) *Farrow* 102*a*! Klela (Oct.) *Farrow* 61! **Port. G.:** Bissau (Nov., Dec.) *Pereira & Correia* 2095! 2128! Comura (Nov.) *Raimundo & Guerra* 155! **Guin.:** Fouta Djalon *Chev.* 18665! Filicoundji (Oct.) *Adam* 12594! Kindia *Jac.-Fél.* 242! Timbo (Oct.) *Pobéguin* 1794! Kouroussa (Sept.) *Pobéguin* 520! **S.L.:** Benekoro (Nov.) *Glanville* 319! Mange to Kambia (Oct.) *Adames* 260! Tabe to Jama (Sept.) *Deighton* 3031! Jigaya (Sept.) *Thomas* 2487! Kasanko (Oct.) *Jordan* 649! **Lib.:** *Warner* 63! 65! Paynesville *Warner* 27! **Iv. C.:** Toumodi to Yamoussokro (Oct.) *Adjanohoun* 329*a*! Bagoué to Boundiali (Dec.) *Adjanohoun* 225*a*! M'Brabo (Jan.) *Hédin*! Bingerville to Eloka (Sept.) *Hédin*! Abouabou (Mar.) *Raynal* 13646! **U. Volta:** Leo to Ouassa (Oct.) *Rose Innes* GC 31493! Bobo-Dioulasso (Oct.) *Koechlin* 6959! **Ghana:** Tamale to Bolgatanga (Oct.) *Rose Innes* GC 31113! Vane (Sept.) *Rose Innes* GC 30901! Tonogo (Nov.) *Morton* A3497! Essiama (Mar.) *Rose Innes* GC 30878! Atuabo (Jan.) *Thorold* 298! **Togo Rep.:** Badou to Atakpame (Sept.) *Rose Innes* GC 31356! **Dah.:** Sémé, Cotonou (Feb.) *Raynal* 13537! **N. Nig.:** Kontagora (Nov.) *Oche* MRK94! Samaru (Oct.-Nov.) *Thatcher* S517! *Freeman* S163! Shendam (Nov.) *Clayton* 1464! William Camp, Jos (Sept.) *Lawlor & Hall* 641! **S. Nig.:** Ogoja Prov. *Rosevear* 12/30*a*! 13/30*a*! Ekoyi, Lagos (Mar., June) *Maitland* 160! *Dalz.* 1317! Lagos (Feb.) *W. MacGregor* 320! Also in the Congos. (See Useful Plants.)
7. **A. chevalieri** *Reznik* in Rev. Bot. Appliq. 14: 199 (1934). *Pobeguinea chevalieri* (Reznik) Jac.-Fél. l.c. 30: 173 (1950). Handsome annual about 1 m. high.
Guin.: Kindia (Oct.) *Jac.-Fél.* 351! 488!
8. **A. leptocoma** (*Trin.*) *Pilger* in Engl., Bot. Jahrb. 54: 284 (1917); F.T.A. 9: 391; Aké Assi, Contrib. 2: 302. *Andropogon leptocomus* Trin. in Mém. Acad. Sci. Petersb. sér. 6, 2: 264 (1832). *A. tenuiflorus* Stapf in Journ. de Bot. 19: 103 (1905). *Anadelphia virgata* Hack. in Engl., Bot. Jahrb. 6: 241 (1885). *A. longifolia* Stapf in F.T.A. 9: 393 (1919); F.W.T.A., ed. 1, 2: 592. *A. tenuifolia* Stapf in F.T.A. 9: 392 (1919); F.W.T.A., ed, 1, 2: 592; Aké Assi, Contrib. 2: 302. *A. pubiglumis* Stapf in F.T.A. 9: 394 (1919); F.W.T.A., ed. 1, 2: 592. *A. triseta* Reznik in Rev. Bot. Appliq. 14: 198 (1934); F.W.T.A. ed. 1, 2: 592; Berhaut, Fl. Sén. ed. 2, 423. *Hypogynium leptocomum* (Trin.) Roberty in Boissiera 9: 188 (1960). Handsome perennial 0·9–1·6 m. high in low-lying savanna, less often on shallow soils and then stunted with narrow leaves.
Mali: Sikasso (Oct.) *Demange* 2744! **Guin.:** Fouta Djalon (Sept.) *Chev.* 18277! 18284! Kindia *Jac.-Fél.* 253! Baffing R. (Oct.) *Pobéguin* 1789 *bis*! Timbo *Pobéguin* 1788! **S.L.:** Northern Prov. (Nov.) *Glanville* 336! Mange to Kambia (Oct.) *Adames* 257! Rokon (Oct.) *Jordan* 566! Kitchom (Jan.) *Deighton* 931! Kent (Nov.) *Deighton* 5631! **Lib.:** Vonjama dist. (Oct.) *Baldwin* 10028! Kolahun dist. (Nov.) *Baldwin* 10082! Bushrod Is. (Aug.) *Baldwin* 13042! Brewersville (Sept.) *Barker* 1428! Monrovia (Nov.) *Linder* 1432! **Iv. C.:** Dabou (Oct., Nov.) *Leeuwenberg* 1939! *Chev.* 17145! 17236! Mont Mofa (Nov.) *Aké Assi*

10368! Toumodi to Dimbokro (Aug.) *Chev.* 22258! **Ghana:** Cape Coast (Nov.) *Hall* 1165! **S. Nig.:** Lagos (Feb.) *W. MacGregor* 57! Brass (Aug.) *Anderson* 1379!

9. **A. polychaeta** *W. D. Clayton* in Kew Bull. 20: 281 (1966); Berhaut, Fl. Sén. ed. 2, 422. Annual, about 60 cm. high; damp soils.
Sen.: Diouloulou (Oct.) *Adam* 18222! Marécage de Bayottes (Aug.) *Berhaut* 6329! **Guin.:** Youkounkoun (Oct.) *Pitot*! Iles Tristão (Nov.) *Jac.-Fél.* 7336!

10. **A. pumila** *Jac.-Fél.* in Rev. Bot. Appliq. 30: 178 (1950). *Hypogynium pumilum* (Jac.-Fél.) Roberty in Boissiera 9: 187 (1960). 15–60 cm. high, on dry sandy soils or hardpan.
Guin.: Gangan Mt. (Oct.) *Schnell* 7486! Kindia (Oct., Nov.) *Jac.-Fél.* 7196! 2075! Friguiagbé (Nov.) *Chillou* 3387! **Ghana:** Sampa to Wenchi (Oct.) *Rose Innes* GC 31637!

11. **A. bigeniculata** *W. D. Clayton* in Kew Bull. 20: 283 (1966). Annual, 30–60 cm. high; on ironstone outcrops.
Guin.: Dalaba (Oct.) *Adames* 390! Kanea (Oct.) *Adam* 12627! Madina Tossekré (Oct.) *Adam* 12530! Télimélé to Pita (Oct.) *Pitot*! **S.L.:** Bumban (Oct.) *Deighton* 3279!

132. VOSSIA Wall. & Griff. in J. Asiat. Soc. Bengal 5: 572 (1836); F.T.A. 9: 41 (1917); *nom. cons.*

Racemes 1–6 or more, subdigitate, up to 25 cm. long; internodes and pedicels setulose; lower glume of both sessile and pedicelled spikelets with a flat herbaceous tail up to 3 cm. long and 2 mm. wide .. .. *cuspidata*

V. cuspidata (*Roxb.*) *Griff.* Not. 3, Index 12 (1851); F.T.A. 9: 41; A. Chev. in Rev. Bot. Appliq. 13: 849; Berhaut, Fl. Sén. ed. 2, 390, 398; including var. *polystachya* Koechl. in Bull. Soc. Bot. Fr. 108: 243 (1961). *Ischaemum cuspidatum* Roxb., Fl. Ind. 1: 325 (1820). *Vossia procera* Wall. & Griff. (1836)—Chev. Bot. 714. Perennial, the spongy culms submerged or floating, rooting from the submerged nodes; in water near the margins of lakes and rivers.
Mali: Dari (Nov) *Farrow* 105*a*! Ouatagouana (Sept.) *Hagerup* 431! Gao (Jan.) *Chev.* 43057! Mopti (Sept.) *Matthes* 1! **S.L.:** Bum-Kittam R. (Sept.) *Fisher* 1770! Bap to Pazehun (Dec.) *Sampson* 54! Subu (Oct.) *Jordan* 600! **U. Volta:** Ouagadougou (Dec.) *Scholz* 195! **Ghana:** Battor (Nov.) *Lawson* VBS 767! **N. Nig.:** Sokoto (Oct.) *Dalz.* 508! Kworre (Sept.) *Palmer* 19! Katagum *Dalz.* 291! Kukawa (Feb.) *Davey* FHI 27168! Kalkala (Oct.) *Gwynn* 103! Throughout tropical Africa and in India (See Useful Plants.)

133. URELYTRUM Hack. in E. & P., Pflanzenfam. 2, 2: 25 (1889); F.T.A. 9: 42 (1917).

Racemes numerous, usually 30 or more, in whorls along the axis of a panicle 30–60 cm. long; leaf-blades up to about 20 mm. broad; sessile spikelet 5–6 mm. long, its lower glume subacute and smooth between the spinously ciliate keels; callus rounded, obtuse; awn 7–15 mm. long, rarely wanting .. .. .. 1. *giganteum*
Racemes 1–9, subdigitate; leaf-blades up to 6 mm. broad:
Lower glume of sessile spikelets conspicuously muricate on the keels:
Annual; racemes solitary; sessile spikelets 5–7 mm. long, glabrous; pedicelled spikelets 3–7 mm. long or almost suppressed; awns 2·5–3·5 cm. long, spirally coiled about the raceme below, the upper part horizontally spreading 2. *annuum*
Perennial; racemes 2–4; sessile spikelets 6–8 mm. long, usually shortly pubescent on the back; pedicelled spikelets more or less reduced; awns 5–7 cm. long, falcate .. .. 3. *muricatum*
Lower glume of sessile spikelets smooth on the keels or spinulose-ciliolate near the tip:
Ligule 4–16 mm. long, adnate to auricles from the top of the sheath:
Sessile spikelets 8–11 mm. long; pedicelled spikelets suppressed, the awn 6·5–10 cm. long, falcate; racemes 1–3; ligule 7–16 mm. long .. .. 4. *auriculatum*
Sessile spikelets 6 mm. long; pedicelled spikelets 5 mm. long, the awn 4·5 cm. long and spirally coiled about the raceme with the upper part spreading horizontally; racemes 1; ligule 4–5 mm. long .. .. 5. *semispirale*
Ligule 1–2 mm. long; pedicelled spikelets more or less reduced; awns falcate:
Racemes 4–9; sessile spikelet 6–7 mm. long, glabrous; awns 2·5–4 cm. long 6. *fasciculatum*
Racemes solitary:
Awn up to 5 cm. long; sessile spikelets 6–8 mm. long, pubescent 7. *gracilius*
Awn 1–2 cm. long; sessile spikelets 8–10 mm. long, glabrous .. 8. *pallidum*

1. **U. giganteum** *Pilger* in Engl., Bot. Jahrb. 34: 125 (1904); F.T.A. 9: 46. *U. thyrsoides* Stapf in F.T.A. 9: 47 (1917); F.W.T.A., ed. 1, 2: 599; A. Chev. in Rev. Bot. Appliq. 13: 850. *Rhytachne gigantea* Stapf (1908). Robust perennial up to 2·5 m. high; swampy places.
N. Nig.: *Ward* 703! Abinsi (Aug.) *Dalz.* 902! Fadan Karshi, Wamba (Oct.) *Kennedy* FHI 8041! Yankari (Sept.) *Oche & Tuley* 1657! **S. Nig.:** Ogoja (Aug.) *Nwanzo* 770! **W. Cam.:** Boku, Donga plain *Brunt* 999! Extends south to the Congo and east to the Sudan. (See Useful Plants.)

2. **U. annuum** *Stapf* in Mém. Soc. Bot. Fr. 2, 8: 99 (1908); F.T.A. 9: 44; Chev. Bot. 714; Berhaut, Fl. Sén. ed. 2, 389. Erect annual about 1·2 m. high; shallow or rocky soils.
Sen.: Maleine Hiani (Oct.) *Adam* 15498! Kanéméré (Sept.) *Fotius* K445! Kounkouba (Oct.) *Diallo* 482! **Guin.:** Soumbalako to Boulivel (Sept.) *Chev.* 18667! Timbo (Oct.) *Pobéguin* 1773! Baffing (Oct.) *Adam* 12664*bis*! **U. Volta:** Gaoua (Oct.) *Rose Innes* GC 31533! **Ghana:** Nakpanduri to Yendi (Oct.) *Rose Innes* GC 30736! Bolgatanga to Tamale (Oct.) *Rose Innes* GC 30237! Bawku to Bolgatanga (Oct.) *Rose Innes* GC 30507! Bawku (Oct.) *Ankrah* GC 20293! **Togo Rep.:** Sansanné Mango to Kande (Oct.) *Rose Innes* GC 31413!

3. **U. muricatum** *C. E. Hubbard* in Kew Bull. 4: 367 (1949). Up to 2·5 m. high; in savanna.
Sen.: Kanéméré (Sept.) *Fotius* K 389! **Mali:** Sikasso (Sept.) *Adam* 15058! Niena (Oct.) *Demange* 3023! **Iv. C.:** Ferkessédougou (Oct.) *Adjanohoun* 199*a*! **U. Volta:** Batié to Gaoua (Oct.) *Rose Innes* GC 31535!

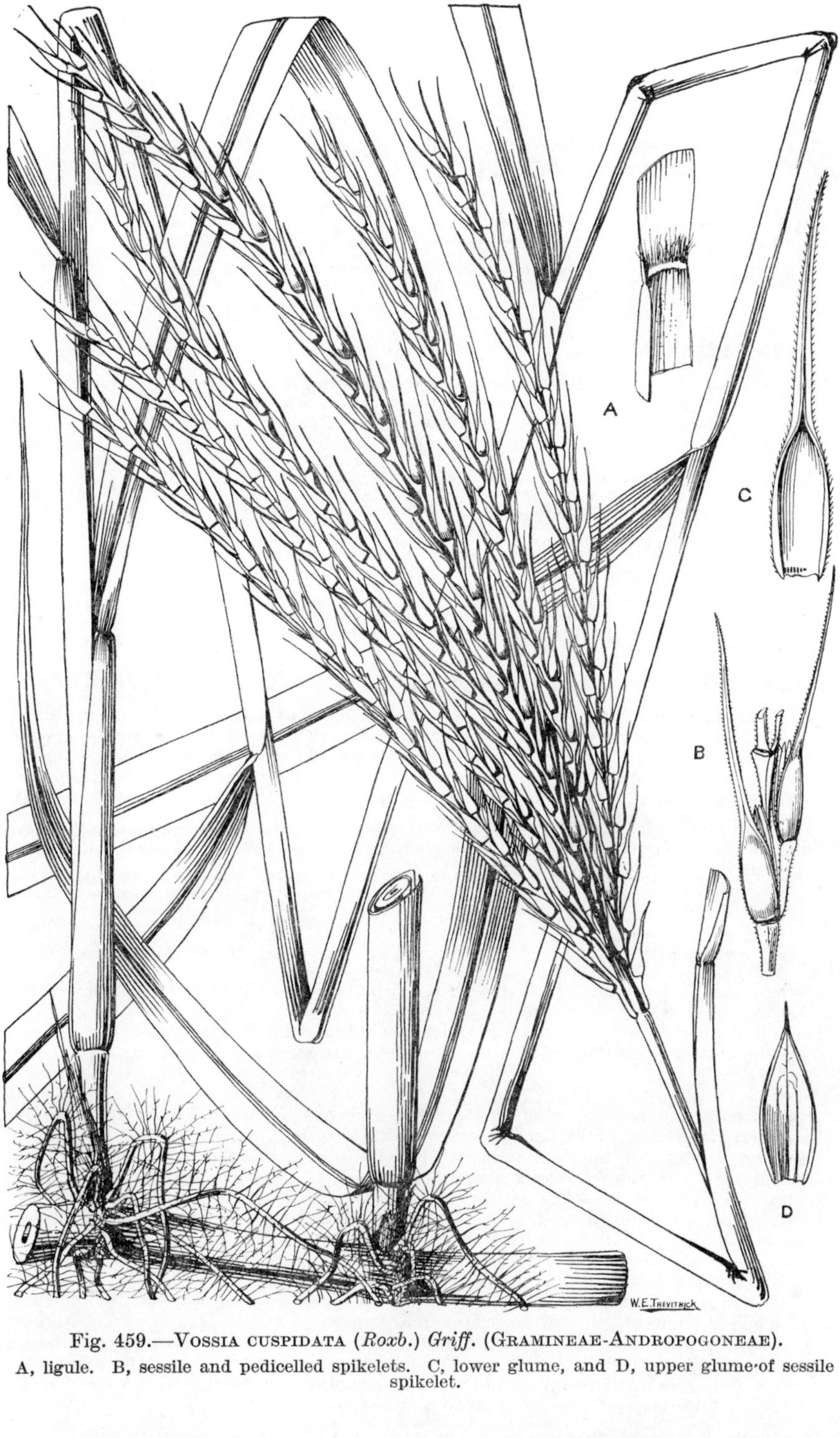

Fig. 459.—Vossia cuspidata (*Roxb.*) *Griff.* (Gramineae-Andropogoneae).
A, ligule. B, sessile and pedicelled spikelets. C, lower glume, and D, upper glume of sessile spikelet.

Varalé (Sept.) *Aké Assi* 6492! Tondolira (Oct.) *Scholz* 89! **Ghana:** Kete Krachi to Yendi (Sept.) *Ankrah* GC 20246! Salaga to Tamale (Oct.) *Rose Innes* GC 30511! Gambaga (Oct.) *Rose Innes* GC 30760! Gambaga to Walewale (Oct.) *Rose Innes* GC 30267! Kpandai to Salaga (Oct.) *Ankrah* GC 20405! **Togo Rep.:** Sokode to Lama Kara (Oct.) *Rose Innes* GC 31390! **N. Nig.:** Gwada, Niger Prov. *Lamb* 3192! Plateau Prov. (Sept.) *Lely* P759! Anara F.R. (Oct.) *Keay* FHI 20135! Samaru, Zaria (Aug., Sept.) *Thatcher* S.466! *Clayton* 1299!

4. **U. auriculatum** *C. E. Hubbard* in Kew Bull. 4: 368 (1949). Perennial, 90–120 cm. high.
N. Nig.: Vodni, Pankshin *Saunders* 36a!

5. **U. semispirale** *W. D. Clayton* in Kew Bull. 20: 257 (1966).
Ghana: Ahamansu (Sept.) *Rose Innes* GC 31325!

6. **U. fasciculatum** *Stapf ex C. E. Hubbard* in Kew Bull. 4: 370 (1949). Caespitose perennial up to nearly 2 m. high.
N. Nig.: Gembu to Donga R., Mambila (Aug.) *De Leeuw* 1804! **W. Cam.:** Jakiri (June) *Brunt* 689! Also in E. Cameroun and C. African Rep.

7. **U. gracilius** *C. E. Hubbard* in Kew Bull. 4: 369 (1949). Perennial about 1 m. high.
N. Nig.: Shika, Zaria *Bumpus* 29!

8. **U. pallidum** *C. E. Hubbard* in Kew Bull. 4: 376 (1949). Perennial up to 2·5 m. high.
Ghana: Yendi (Aug.) *Hinds* 6! Yendi to Gambaga (Sept.) *Ankrah* GC 20255! Kpandai to Salaga (Oct.) *Ankrah* GC 20404!

134. JARDINEA Steud., Syn. Pl. Glum. 1: 360 (1854); F.T.A. 9: 50 (1917).

Racemes numerous, up to 25 cm. long, fasciculate from a short common axis; sessile spikelets 4–7 mm. long, acuminate to apiculate, muricate on sides and back or the smaller spikelets on the sides only; callus truncate; pedicelled spikelets similar, with or without an awn-point up to 2 mm. long .. .. *congoensis*

J. congoensis (*Hack.*) *Franch.* in Bull. Soc. Hist. Nat. Autun 8: 321 (1895); F.T.A. 9: 53; A. Chev. in Rev. Bot. Appliq. 13: 850. *Rhytachne congoensis* Hack. in DC., Monogr. Phan. 6: 277 (1889). Perennial, 1·6–3·6 m. high; best distinguished from awnless forms of *Urelytrum giganteum* by the truncate, square-cornered callus of the sessile spikelets; swampy places.
Ghana: Avedo, Sogakope, Adidome junction (Sept.) *Rose Innes* GC 31280! Kete Krachi to Yendi (Oct.) *Rose Innes* GC 30950! Navrongo to Lawra (Oct.) *Rose Innes* GC 30290! Tumu (Oct.) *Rose Innes* GC 32292! **Togo Rep.:** Lomé *Warnecke* 235! Kande to Lama Kara (Oct.) *Rose Innes* GC 31402! **N. Nig.:** Ilorin (Aug.) *Ward* L104! Badeggi (Sept.) *Clayton* 300! Nupe *Barter* 1381! Zaria (Aug.) *Keay* FHI 28017! Abinsi *Dalz.* 898! **S. Nig.:** Lagos (Feb., Aug.) *Millen* 131! *Dalz.* 1421! *W. MacGregor* 86! Shaki (Aug.) *Latilo & Taylor* FHI 41477! Ogoja to Abakaliki (Aug.) *Nwango* 809! Extends east to the Sudan, and south through the Congo to Angola and Zambia. (See Useful Plants.)

135. LASIURUS Boiss., Diagn. Pl. Nov. Or. sér. 2, 4: 145 (1859); F.T.A. 9: 60 (1917).

Perennial, almost subwoody at the base, with wiry glaucous culms; blades linear, long-attenuate to a setaceous tip, glaucous, glabrous except for a few hairs near the ligule; racemes up to 10 cm. long, densely white villous, the spikelets often three at each node, two sessile and one pedicelled; sessile spikelets 6–9 mm. long, pale green, the lower glume flat, hirsute, emarginate or 2-toothed at the tip .. .. *hirsutus*

L. hirsutus (*Forsk.*) *Boiss.* Diagn. Pl. Nov. Or. sér. 2, 4: 146 (1859); F.T.A. 9: 60; A. Chev. in Rev. Bot. Appliq. 13: 851; De Miré & Gillet in J. Agric. Trop. 3: 733. *Saccharum hirsutum* Forsk., Fl. Aegypt.-Arab. 16 (1775). Up to 90 cm. high; dry soils.
Mali: Bouressa to Tessalit (Sept.) *Rossetti* 194/1959! **Niger:** Agades (Feb.) *Bradley* 3! Aïr (*fide* De Miré & Gillett). Also in N. Africa and eastwards through Iraq, Egypt, Sudan, Somalia and Arabia to India. (See Useful Plants.)

136. ELIONURUS Kunth ex Willd., Sp. Pl. 4: 941 (1806); F.T.A. 9: 62 (1917).

Lower glume of the sessile spikelet with tooth-like warts bearing tufts of hair on the margin:
- Perennial; leaf-sheaths villous to tomentose; blades stiff, glabrescent to villous above, closely ribbed and scaberulous on the veins below; racemes long-exserted 1. *hirtifolius*
- Annuals:
  - Racemes distant, long-exserted; leaf-sheaths glabrous to pilose; blades thin, usually rolled, glabrescent or with a few sparse hairs .. .. .. 2. *elegans*
  - Racemes gathered in bundles, embraced below by the spatheoles 3. *royleanus*

Lower glume of the sessile spikelet merely ciliate:
- Lower glume of both spikelets with awns up to 20 mm. long .. 4. *euchaetus*
- Lower glume of sessile spikelet with awns up to 4 mm. long, the pedicelled spikelet at most apiculate:
  - Culms much branched, the flowering branches 2–3-nate, forming a spathate false panicle; leaf-blades 2–5 mm. broad, often involute:
    - Callus of sessile spikelet 2 mm. long, the lower glume glabrous on the back; hairs fringing glume short, stiff, up to 0·5 mm. long .. .. 5. *platypus*
    - Callus of sessile spikelet 1 mm. long, the lower glume glabrescent to villous; hairs fringing glume usually longer, up to 1 mm. long .. .. 6. *pobeguinii*
  - Culms simple, with a terminal raceme and sometimes 1–2 axillary racemes; leaf-blades filiform, or (in our area) up to 3 mm. broad .. .. 7. *argenteus*

1. **E. hirtifolius** *Hack.* in DC., Monogr. Phan. 6: 341 (1889); F.T.A. 9: 63; A. Chev. in Rev. Bot. Appliq. 13: 851. Tufted perennial, 30–60 cm. high.

Ghana: Burufo (Dec.) *Adams* 4423! Zuarungu (Dec.) *Adams & Akpabla* 4306! Lawra (May) *Morton* GC 7617! **N. Nig.:** Zongon Daji, Kabba Prov. (Nov.) *Clayton* 597! Abinsi (Mar.) *Dalz.* 878! Nupe *Barter* 1176! Birnin Gwari (Mar.) *Meikle* 1345! Samaru, Giwa Dist. (May) *Keay* FHI 25730! **S. Nig.:** Abakaliki (Mar.) *Jackson* 2500*b*! Also in C. African Republic.

2. **E. elegans** *Kunth*, Rev. Gram. 361 (1830); F.T.A. 9: 64; Chev. Bot. 715, and in Rev. Bot. Appliq. 13: 851; Berhaut, Fl. Sén. ed. 2, 396. *Andropogon elegans* (Kunth) Gay ex Steud. (1854). Annual, 30–60 cm. high; dry soils.
Sen.: *Heudelot* 384! Niokolo-Koba (Oct.) *Adam* 15500! Walo (May) *Roger* 66! Bougouni (Sept.) *Adam* 15468! **Gam.:** Sankuli-Kunda (Sept.) *Pirie* 32/33! **Mali:** Sanga *Dieterlen* 126! Dioura (Aug.) *Davey* 044! **Port. G.:** Bafatá (Sept.) *Esp. Santo* 2751! Nova Lamego (Oct.) *Pereira* 3339! Piche (Sept.) *Pereira* 3191! 3242! **Guin.:** Koulikoro *Chev.* 2201! **U. Volta:** Po (Sept.) *Aké Assi* 6440! Bondoukuy (Sept.) *Aké Assi* 6401! Gaoua (Oct.) *Rose Innes* GC 31528! Leo to Po (Oct.) *Rose Innes* GC 31472! Mossi (Aug.) *Chev.* 24758! **Ghana:** Bolgatanga to Tamale (Oct.) *Rose Innes* GC 30235! Lawra (Oct.) *Rose Innes* GC 30295! Jirapa to Lawra (Oct.) *Ankrah* GC 20457! Gambaga (Oct.) *Ankrah* GC 20274! Bolgatanga to Bawku (Oct.) *Rose Innes* GC 31442! **Togo Rep.:** Kande to Sansanné Mango (Oct.) *Rose Innes* GC 31410! **Niger:** Niamey (Oct.) *Hagerup* 494! **N. Nig.:** 110 miles N. of Gusau (Oct.) *De Leeuw & Magaji* 1966! Nupe *Barter* 993! (See Useful Plants.)

3. **E. royleanus** *Nees ex A. Rich.* Tent. Fl. Abyss. 2: 471 (1851); F.T.A. 9: 65; A. Chev. in Rev. Bot. Appliq. 13: 852. Caespitose annual about 15 cm. high; dry soils.
Sen.: *Perrottet* 895! Also in the Cape Verde Is., extending through Uganda, Kenya, Ethiopia and Sudan, to Arabia and India.

4. **E. euchaetus** *Adjanohoun & Clayton* in Adansonia 4: 199 (1964). Tufted perennial 1 m. high.
Iv. C.: Dabakala to Bondoukou (Jan.) *Adjanohoun* 236*a*! 237*a*! Sanlo to Kakpin (Nov.) *Aké Assi* 9279! **U. Volta:** Linguekoro (Nov.) *Scholz* 84!

5. **E. platypus** (*Trin.*) *Hack.* in Bol. Soc. Brot. 3: 135 (1885); F.T.A. 9: 66; Chev. Bot. 715, and in Rev. Bot. Appliq. 13: 852. *Andropogon platypus* Trin. in Mém. Acad. Sci. Petersb. sér. 6, 2: 261 (1832). *A. donianus* Benth., Fl. Nigrit. 570 (1849). *Elionurus pallidus* K. Schum. in Engl., Bot. Jahrb. 24: 326. (1897). Caespitose perennial, 0·6–1·8 m. high, with flattened basal sheaths.
Guin.: Diaguissa *Chev.* 12659! **S.L.:** (May) *Barter*! Freetown (Aug.) *Deighton* 2053! Karina (Feb.) *Glanville* 156! Matotoka (July) *Thomas* 1249! Rokupr (May) *Jordan* 864! **Lib.:** Nimba (Dec.) *Adam* 20365! **Iv. C.:** Danané (June) *Adjanohoun* 230*a*! Kong (Oct.) *Boudet* 3091!

6. **E. pobeguinii** *Stapf* in Journ. de Bot. 19: 99 (1905); F.T.A. 9: 67; A. Chev. in Rev. Bot. Appliq. 13: 852. *E. tenax* Stapf in Kew Bull. 1909: 422; F.T.A. 9: 68; F.W.T.A., ed. 1, 2: 600; A. Chev. in Rev. Bot. Appliq. 13: 852. Tufted perennial up to 2 m. high, with aromatic leaves and silvery racemes.
Guin.: Timbo (Oct.) *Pobéguin* 1772! 1772*bis*! Madina Tossekré (Oct.) *Adam* 12376! Kouroussa (Sept.) *Pobéguin* 517! Kankan *Bardou*! **S.L.:** Mt Loma (Dec.) *Jaeger* 8413! **Iv. C.:** Téhini (Dec.) *Aké Assi* 7511! Kong (Oct.) *Boudet* 3089! **U. Volta:** Samandeni (Sept.) *Kmoch* 112! Gaoua (Oct.) *Rose Innes* GC 31527! Leo to Ouassa (Oct.) *Rose Innes* 31500! Leo to Po (Oct.) *Rose Innes* GC 31072! Banfora to Mangodara (Nov.) *Scholz* 88! **Ghana:** Banda to Bui (Oct.) *Rose Innes* GC 31602! Babile Agric. Sta. (Oct.) *Rose Innes* GC 30296! Navrongo to Tumu (Oct.) *Rose Innes* GC 30781! Sawla (Oct.) *Rose Innes* GC 30666! Bolgatanga to Bawku (Oct.) *Rose Innes* GC 30503! **Togo Rep.:** Sokode to Lama Kara (Oct.) *Rose Innes* GC 31395! **Dah.:** Parakou (Nov.) *Risopoulos* 1288! **N. Nig.:** Gwada, Niger Prov. *Lamb* 3187! Kangimi, Anara F.R. (Oct.) *Keay* FHI 5485! Shika (Oct.) *Clayton* 133! Gindiri, Jos Plateau (Oct.) *Hepper* 1139! Okene (Nov.) *Clayton* 570!

7. **E. argenteus** *Nees* Fl. Afr. Austr. 95 (1841); F.T.A. 9: 70; A. Chev. in Rev. Bot. Appliq. 13: 852. *E. chevalieri* Stapf in Mém. Soc. Bot. Fr. 2, 8: 100 (1908); F.T.A. 9: 70; F.W.T.A., ed. 1, 2: 600; Chev. Bot. 715, and in Rev. Bot. Appliq. 13: 852. Densely caespitose perennial about 60 cm. high.
Guin.: Sangorola (Feb.) *Chev.* 341! **S.L.:** Mt. Loma (Jan.-June) *Jaeger* 8846! 9394! 9417! 9782! Tingi Mts. (Apr.) *Morton & Gledhill* SL 1884! **U. Volta:** Bobo-Dioulasso (June) *Scholz* 102! **N. Nig.:** Abinsi (Mar.) *Dalz.* 866! Minna to Zungeru (Feb.) *Meikle* 1227! Bauchi (Jan.) *Lely* P92! **W. Cam.:** Bali, 4,100 ft. (Apr.) *Brunt* 1054! Gembu, 4,500 ft. (Jan.) *Hepper* 2826! Throughout tropical and S. Africa.

[Specimens from W. Africa, Congo, Uganda and Ethiopia differ from typical *E. argenteus* in their harder basal sheaths and broader leaf-blades; but the two forms intergrade and cannot be reliably separated.

The generic name was spelled *Elyonurus* by Willd., but eventually published by Kunth (1816) as *Elionurus*; the etymology leaves no doubt that Willd. miscopied it. Several of the species are closely allied to, perhaps conspecific with, American species, but the taxonomy of the latter is not clear and it would be premature to attempt a correlation.]

137. LOXODERA Launert in Bol. Soc. Brot. 37: 80 (1963), and in Senck. Biol. 46: 121 (1965).

Pedicelled spikelets awnless .. .. 1. *ledermannii*
Pedicelled spikelets with awns 5–15 mm. long .. .. .. 2. *strigosa*

1. **L. ledermannii** (*Pilger*) *W. D. Clayton ex Launert* in Senck. Biol. 46: 121 (1965); Clayton in Kew Bull. 20: 258 (1966). *Elionurus ledermannii* Pilger in Engl., Bot. Jahrb. 45: 207 (1910). *Lasiurus maitlandii* Stapf & C. E. Hubbard in Kew Bull. 1927: 264. Densely caespitose perennial about 60 cm. high.
Niger: Gaya (Oct.) *Boudet* 5259! **N. Nig.:** 17 miles N. of Kaduna (May) *Keay* FHI 25759! Shaganu (June) *J. Hall* 1283! Also in E. Cameroun and Uganda.

2. **L. strigosa** (*Gledhill*) *W. D. Clayton* in Kew Bull. 23: 295 (1969). *Urelytrum strigosum* Gledhill in Bol. Soc. Brot. sér. 2, 41: 58 (1967). Tufted perennial up to 1 m. high, the basal sheaths becoming fibrous at maturity.
S.L.: Loma Mts. (Feb.) *Gledhill* 369!

138. HACKELOCHLOA O. Ktze., Rev. Gen. 2: 776 (1891).

Leaf-blades linear-lanceolate, rounded and amplexicaul at the base, coarsely setose with tubercle-based hairs, up to 10 cm. long and about 1 cm. broad; racemes solitary, 1–2 cm. long, supported by spatheoles; sessile spikelet 1·5 mm. long, the lower glume pitted and tubercled; pedicelled spikelet ovate, 1·5 mm. long, herbaceous; pedicel fused to the adjacent internode .. .. *granularis*

H. granularis (*Linn.*) *O. Ktze.* Rev. Gen. Pl. 2: 776 (1891); Berhaut, Fl. Sén. ed. 2, 421. *Cenchrus granularis* Linn., Mant. 2, App. 575 (1771). *Manisuris granularis* (Linn.) Linn. f., Nov. Gram. Gen. 40 (1779); F.T.A. 9: 57; Chev. Bot. 715, and in Rev. Appliq. 13: 850. *M. polystachya* P. Beauv. (1804). *Rottboellia granularis* (Linn.) Roberty (1960). Annual, up to 60 cm. high or more; weed of cultivated land.
Gam.: *Saunders* 35! **Mali:** Nyamino to Koulikoro (Oct.) *Chev.* 2280! Labézenga (Sept.) *Hagerup* 438! Dioura (Sept.) *Davey* 074! **Guin.:** *Chillou* 976! Kindia *Jac.-Fél.* 176! Timbo (Oct.) *Pobéguin* 1770!

Bambaya (Oct.) *Jaeger* 2139! **S.L.:** Kortright (Oct.) *Gledhill* 44! Musaia (Dec.) *Deighton* 4429! Kambia (Aug.) *Jordan* 300! Kabala (Sept.) *Thomas* 2262! Pendembu (Oct.) *Glanville* 83! **Iv. C.:** Vavoua to Séguéla (Oct.) *Adjanohoun* 414*a*! Bouaké (July) *Chev.* 22121! Nambonkaha (Nov.) *Leeuwenberg* 2045! Tehini (Aug.) *de Wilde* 688! **U. Volta:** Zabré (Oct.) *Scholz* 128*a*! **Ghana:** Cape Coast *Rose Innes* GC 30016! Achimota (May) *Irvine* 690! Bamboi to Bole (Oct.) *Rose Innes* GC 30648! Gambaga (Oct.) *Vigne* FH 4636! Tamale to Bolgatanga (Oct.) *Rose Innes* GC 30225! **Togo Rep.:** Lomé *Warnecke* 172! Kande (Oct.) *Rose Innes* GC 31406! **N. Nig.:** Lokoja to Kabba (Oct.) *Parsons* 22! Nupe *Barter*! Zaria *Taylor* 5! Gindiri, Plateau (Oct.) *Hepper* 1082! Yola (July) *Dalz.* 273! **S. Nig.:** Ibadan (June) *Jones* FHI 4353! Igboora (Oct.) *Haines* 344! Ogurude (Jan.) *Holland* 272! Enugu (Oct.) *Jones* FHI 6766! Abakaliki (July) *Nwanzo* 766! **W. Cam.:** Gurum, Vogel Peak (Nov.) *Hepper* 1314! Gangumi, Gashaka (Dec.) *Latilo & Daramola* FHI 28798! Throughout the tropics. (See Useful Plants.)

139. OXYRHACHIS Pilger in Not. Bot. Gart. Berl. 11: 655 (1932).

Spikes solitary on the end of the culms, erect, up to 16 cm. long, the spikelets sunk in hollows; internodes of the rhachis 5–10 mm. long, disarticulating very obliquely along a line some 3 mm. long; spikelets lanceolate, 4–6 mm. long, subacute, furnished with 2 glumes and 2 lemmas, the lower barren, the upper bisexual and sometimes accompanied by a tiny palea .. .. *gracillima*

O. gracillima (*Baker*) *C. E. Hubbard* in Hook., Ic. Pl. t. 3454 (1947), incl. subsp. *occidentalis* Gledhill in Bol. Soc. Brot. 41: 61 (1967). *Rottboellia gracillima* Baker in J. Linn. Soc. Bot. 22: 553 (1887). *Oxyrhachis mildbraediana* Pilger in Not. Bot. Gart. Berl. 11: 655 (1932). Caespitose perennial with filiform leaves; marshy soils.

S.L.: Loma Mts. (Jan.) *Jaeger* 9409! 9485! **N. Nig.:** Maisamari, Mambila (Aug.) *De Leeuw* 1746! **W. Cam.:** Jakiri, Bamenda Div. (Feb.) *Hepper* 2069! Also in Tanzania, Zambia and Madagascar.

140. HEMARTHRIA R. Br., Prodr. Fl. Nov. Holl. 207 (1810); F.T.A. 9: 54 (1917).

Racemes solitary to fasciculate, 5–7 cm. long; sessile spikelet linear-oblong to oblong, 5–7 mm. long, glabrous; lower glume laterally 2-keeled, obtuse at the tip; pedicelled spikelet similar, but with the lower glume acute .. .. *altissima*

H. altissima (*Poir.*) *Stapf & C. E. Hubbard* in Kew Bull. 1934: 109; Berhaut, Fl. Sén. ed. 2, 421. *Rottboellia altissima* Poir., Voy. Barb. 2: 105 (1789). *R. fasciculata* Lam., Tab. Encycl. Méth. Bot. 1: 204 (1791). *Hemarthria fasciculata* (Lam.) Kunth, Rev. Gram. 1: 153 (1829); F.T.A. 9: 55; A. Chev. in Rev. Bot. Appliq. 13: 850. *Manisuris altissima* (Poir.) Hitchc. (1934). Perennial up to 1·6 m. high; the culms ascending from a decumbent rooting base; wet places.

Sen.: *fide* Raynal in Adansonia 7: 331 (1967). **Mali:** N. of Mopti (July) *Lean* 48! **N. Nig.:** *Ward* 54! Argungu (July) *Latilo* FHI 62756! Kworre (Sept.) *Palmer* 32! Daura (Aug.) *J. Hall* 467! Chad, Tanzania, south tropical and S. Africa, extending east through Madagascar and Mauritius to India and Burma; also on the shores of the Mediterranean and, probably introduced, in tropical America.

141. ROTTBOELLIA Linn. f., Nov. Gram. Gen. 23 (1779); F.T.A. 9: 72 (1917); *nom. cons.*

Racemes cylindric, 8–15 cm. long, borne in a leafy panicle, the uppermost spikelets barren and forming a slender tail-like appendage; internodes of the rhachis 4–6 mm. long; sessile spikelets a little shorter than the internodes; lower glume coriaceous, ovate, subacute; upper glume boat-shaped, sharply keeled, laterally compressed; pedicelled spikelet as long as the sessile, but green, striate, chartaceous and dorsally compressed .. .. *exaltata*

R. exaltata *Linn. f.* Nov. Gram. Gen. 40, t. 1 (1779); F.T.A. 9: 73; Chev. Bot. 713 and in Rev. Bot. Appliq. 13: 852; Berhaut, Fl. Sén. ed. 2, 392. A tall annual up to 3·5 m. high, often supported by stilt roots, with leaf-blades up to 2·5 cm. broad; lower sheaths painfully hispid with stiff hairs; fallow land, particularly where fertility is high.

Gam.: McCarthy Is. (Sept.) *Pirie* 49/33! Wallikunda (Sept.) *Macluskie* 18! **Guin.:** Dyeke (Oct.) *Baldwin* 9675! Kindia *Jac.-Fél.* 212! Timbo (Oct.) *Pobéguin* 1804! Kouroussa (Nov.) *Pobéguin* 532! **S.L.:** Rokon, Samu (Oct.) *Jordan* 569! Njala (Feb.) *Deighton* 493! Bumbuna (Oct.) *Thomas* 3805! Makump (Oct.) *Glanville* 55! Kambia (Dec.) *Deighton* 859! **Lib.:** Fishtown (July) *Baldwin* 6670! Ganta, Sanokwele Dist. (Oct.) *Harley* 1019! Jabrocca (Dec.) *Baldwin* 10875! **Iv. C.:** Kongasso (Oct.) *Adjanohoun* 387*a*! Man (Oct.) *Adjanohoun* 298*a*! Niapidou *Leeuwenberg* 2427! **U. Volta:** Ouagadougou (Aug., Sept.) *Chev.* 24737! *Scholz* 150! **Ghana:** Achimota (June) *Irvine* 5025! Kpandu (Sept.) *Ankrah* GC 20195! Adawso (May) *Howes* 903! Kumasi (Apr.) *La Danso* 49! Tamale (Dec.) *Williams* 804! **Dah.:** Cotonou to Allada (Oct.) *Risopoulos* 1228! **N. Nig.:** Abinsi (Sept.) *Dalz.* 870! Wamba Div. (Oct.) *Kennedy* 8039! Katagum Dist. *Dalz.* 288! Zaria Dist. *Taylor* 28! Sokoto (Oct.) *Dalz.* 509! **S. Nig.:** Lagos (June, July) *Dalz.* 1315! *Haines* 310! Aguku Dist. *Thomas* 532! Awka (Oct.) *Thomas*! *Jones* 6793! **W. Cam.:** Victoria (Jan.) *Maitland* 900! Nankon, 3,800 ft. (Apr.) *Brunt* 1122! **F. Po:** (Nov.) *T. Vogel* 91! Throughout the Old World tropics, and in the West Indies. (See Useful Plants.)

142. ROBYNSIOCHLOA Jac.-Fél. in J. Agric. Trop. 7: 406 (1960).

Racemes 4–10 cm. long, borne in a scanty leafy panicle; internodes of the rhachis 3–4 mm. long; sessile spikelets about as long as the internodes; lower glume coriaceous, oblong, laterally 2-keeled, the apex truncate, emarginate and winged; upper glume boat-shaped, with a winged keel; pedicelled spikelet 5–7 mm. long, ovate, acute, green, chartaceous, striate .. .. *purpurascens*

R. purpurascens (*Robyns*) *Jac.-Fél.* in J. Agric. Trop. 7: 406 (1960). *Rottboellia purpurascens* Robyns, Fl. Agrost. Congo Belge 1: 66 (1929); F.W.T.A., ed. 1, 2: 602; Jac.-Fél. in Rev. Bot. Appliq. 32: 548; Ballard in Hook., Ic. Pl. t. 3139. Robust swamp grass up to 1·6 m. high, rooting at the lower nodes.

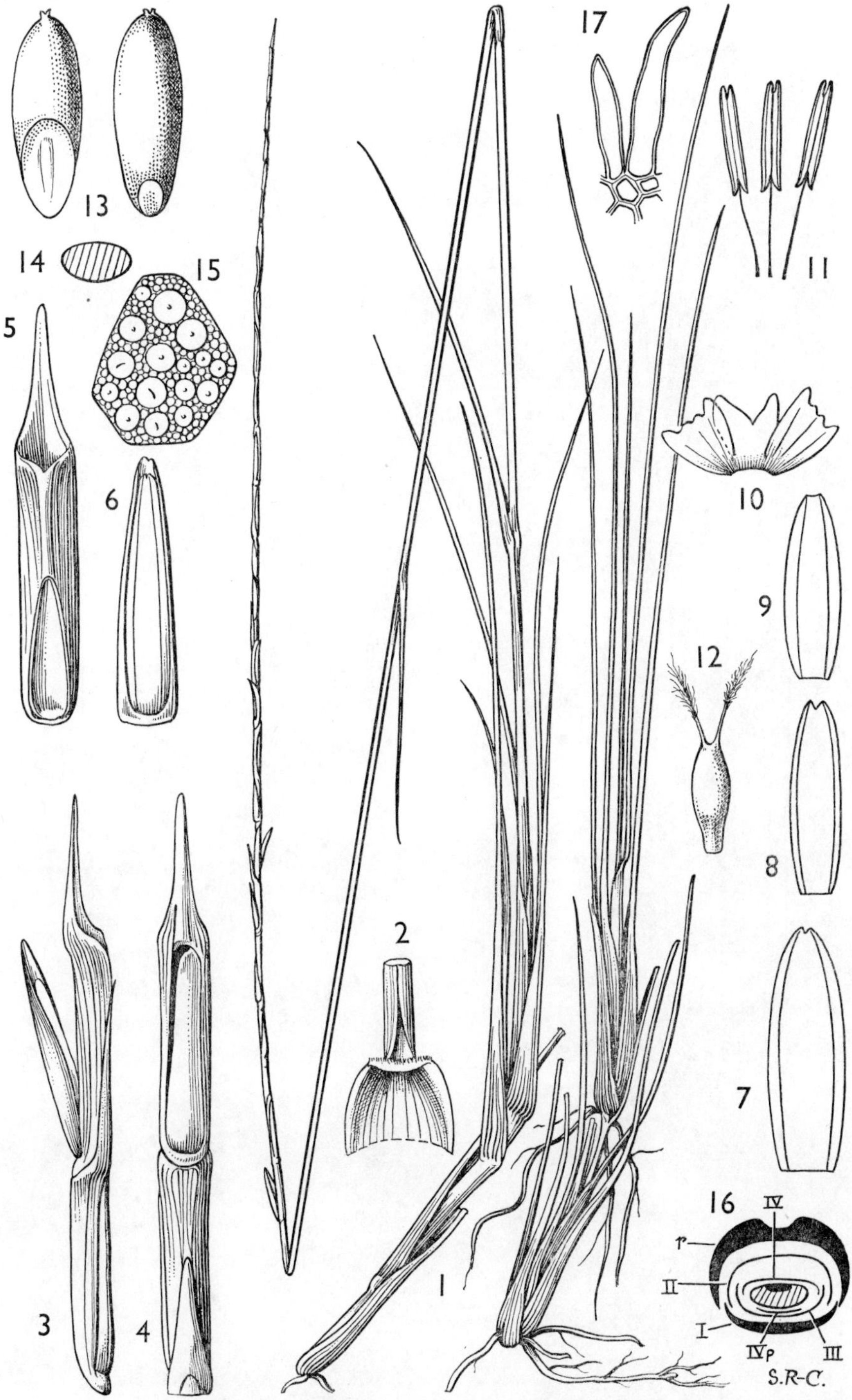

Fig. 460.—OXYRHACHIS GRACILLIMA (*Bak.*) *C. E. Hubbard* (GRAMINEAE-ANDROPOGONEAE).

1, habit, × 1. 2, ligule, × 8. 3 & 4, part of spike, side and front view respectively, × 8. 5, internode of rhachis, × 8. 6, spikelet showing upper glume and margins of lower glume, × 8. 7, upper glume, × 8. 8, lemma of lower floret, × 8. 9, lemma of upper floret, × 8. 10, palea and lodicules, × 20. 11, stamens, × 8. 12, ovary and stigmas, × 20. 13, caryopses, × 12. 14, transverse section of caryopsis, × 12. 15, cell with starch grains, much enlarged. **16**, diagram of spikelet (I = lower glume, II = upper glume, III = lemma of lower floret, IV = lemma of upper floret, IVp = palea of same, r = rhachis). 17, hairs on upper surface of leaf blade, much enlarged.

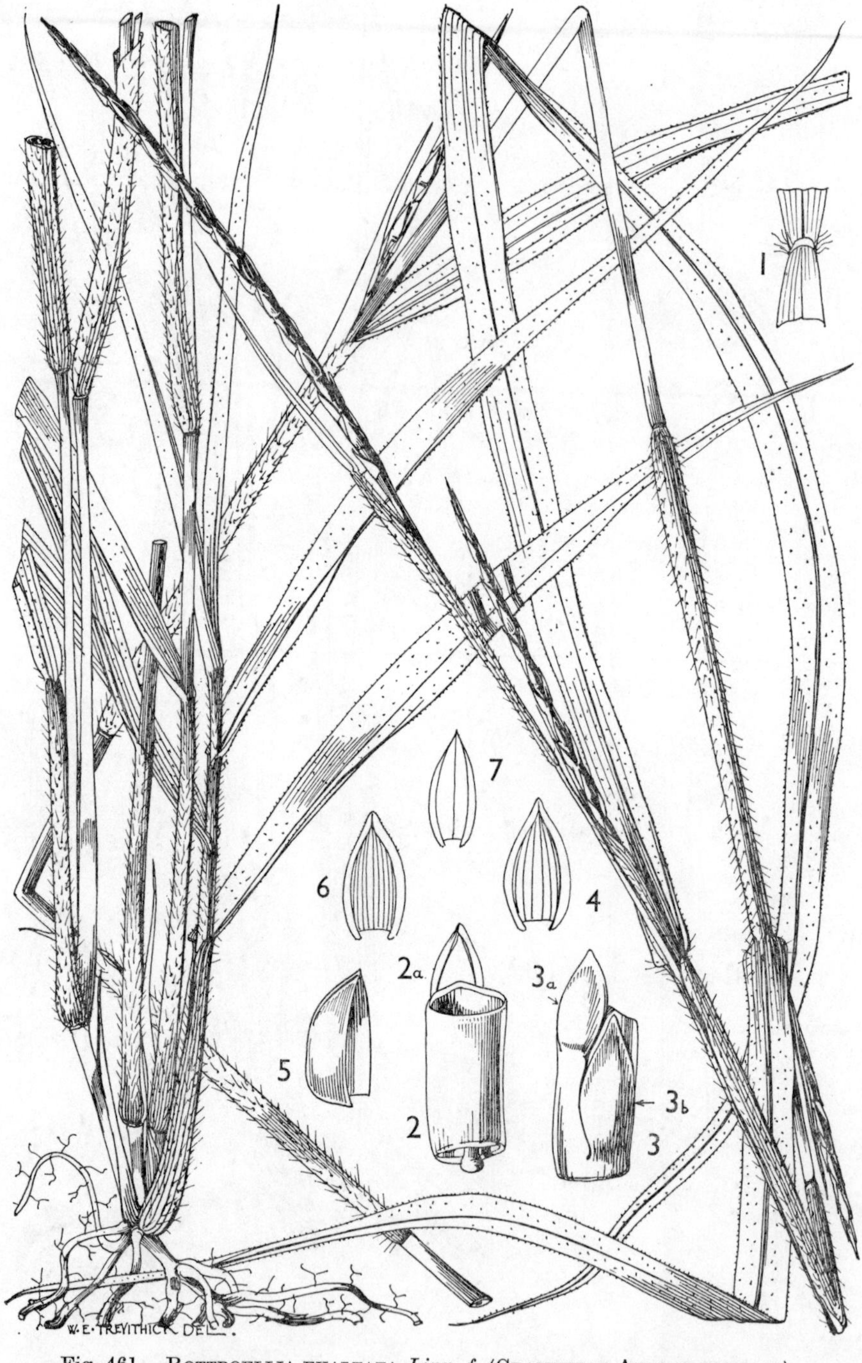

Fig. 461.—ROTTBOELLIA EXALTATA *Linn. f.* (GRAMINEAE-ANDROPOGONEAE).

1, ligule. 2 & 3, internodes: 2a & 3a, pedicelled spikelet. 3b, sessile spikelet. 4, lower glume; 5, upper glume; both of sessile spikelet. 6, lower glume; 7, upper glume; both of pedicelled spikelet.

Guin.: Baffing R. (Oct.) *Pobéguin* 1806! **S.L.:** Bandakor (Sept.) *Fisher* 1768! Rokon, Samu (Oct.) *Jordan* 570! Subu, Nongoba Bullom (Oct.) *Jordan* 598! Bumbe (Apr.) *Glanville* 234! Kambia (Sept.) *Jordan* 307! Also in the Congo.

143. CHASMOPODIUM Stapf in F.T.A. 9: 76 (1917).

Culms 2–4 m. high; leaf-blades 1–4·5 cm. broad, smooth or scaberulous, often with a false petiole at the base; racemes fragile, 5–13 cm. long, the glumes of the terminal spikelet foliaceous and produced into a tail up to 2·5 cm. long; joints and pedicels glabrous or ciliate; lower glume of the sessile spikelet rigidly coriaceous, ovate to oblong, smooth, laterally winged towards the tip and bifid or emarginate at the apex, 20–35-nerved; pedicelled spikelet almost as long as the sessile, its lower glume variously winged at the tip, usually on one side only .. .. .. *caudatum*

C. caudatum (*Hack.*) *Stapf* in F.T.A. 9: 76 (1917); A. Chev. in Rev. Bot. Appliq. 13: 854; Berhaut, Fl. Sén. ed. 2, 392. *Rottboellia caudata* Hack. in DC., Monogr. Phan. 6: 298 (1889). *R. afzelii* Hack. l.c. 300. *R. kerstingii* Pilger in Engl., Bot. Jahrb. 34: 126 (1904). *Chasmopodium afzelii* (Hack.) Stapf in F.T.A. 9: 77 (1917). Annual savanna grass, with a considerable range of variation in both spikelet and vegetative characters.
Guin.: Kindia *Jac.-Fél.* 180! Timbo (Mar.) *Pobéguin* 1805! Macenta (Oct.) *Baldwin* 9839! **S.L.:** (Dec.) *Sc. Elliot* 3938! Bumbuna (Oct.) *Thomas* 3346! Makump (Oct.) *Glanville* 59! Rokupr (Nov.) *Jordan* 691! Kambia (Dec.) *Deighton* 855! **Lib.:** Robertsport (Dec.) *Baldwin* 10888! Vonjama (Oct.) *Baldwin* 9947! **Iv. C.:** Vavoua to Séguéla (Jan.) *Adjanohoun* 238a! Singrobo (Oct.) *Adjanohoun* 357a! Bouna to Bondoukou (Oct.) *Rose Innes* GC 31554! Bondoukou (Dec.) *Aké Assi* 7486! Sorobango (Dec.) *Adjanohoun* 437a! **U. Volta:** Ouaga (Oct.) *Scholz* 164! **Ghana:** Kpandu (Sept.) *Rose Innes* GC 30400! Kafeiro, Hohoe Dist. (Nov.) *St. Cl.-Thompson* 1644! Kintampo to Bamboi (Oct.) *Rose Innes* GC 30621! Banda (Oct.) *Rose Innes* GC 31591! Gambaga to Bawku (Oct.) *Rose Innes* GC 30269! **Togo Rep.:** *Kersting* 237! **N. Nig.:** (Aug.) *Lely* P499! Vodni *Saunders* 16! Naraguta F.R. (July) *Lawlor & Hall* FHI 46531! Onyi F.R. (Oct.) *Jackson* 1! **S. Nig.:** Lagos (Feb.) *W. MacGregor* 168! Ago-Are F.R. (Nov.) *Keay* FHI 37691! Ogoja Prov. *Rosevear* 6/30a! Extends southward through the Congo to Angola; also in the Sudan. (See Useful Plants.)

144. COELORHACHIS Brongn. in Duperr., Voy. Coq. Bot. 64, t. 14 (1831); F.T.A. 9: 78 (1917).

Racemes 5–7 cm. long, with closely overlapping spikelets; lower glume of the sessile spikelet thinly coriaceous, lanceolate-oblong, smooth, 5-nerved with narrow lateral wings produced into a notched or emarginate tip; upper glume broadly curved with a low median keel; pedicelled spikelets a little smaller than the sessile; pedicel with a lanceolate wing-like appendage 1·5 mm. long from the top .. .. *afraurita*

C. afraurita (*Stapf*) *Stapf* in F.T.A. 9: 80 (1917); Berhaut, Fl. Sén. ed. 2, 421. *Rottboellia afraurita* Stapf in Mém. Soc. Bot. Fr. 2, 8: 98 (1908); Chev. Bot. 713. A tall perennial up to nearly 2 m. high, with strongly compressed basal sheaths; marshy places.
Sen.: Niokolo-Koba *Adam* 17507; 17561. *Raynal* 6900. Kédougou (Nov.) *Adam* 19935! **Mali:** Bamako *Chev.* 232. **Guin.:** Konkouré (Mar.) *Pitot* 578! **S.L.:** Giema (Nov.) *Fisher* 64! Kenema (Nov.) *Deighton* 455! Cameron to Badugri (Jan.) *Morton* SL 459! Benekoro & Dankawali (Nov.) *Glanville* 320! **Iv. C.:** Sanlo to Kakpin (Nov.) *Aké Assi* 9286! Vavoua to Séguéla (Jan.) *Adjanohoun* 175a! **N. Nig.:** Maisamari, Mambila (Aug.) *De Leeuw* 1749! **W. Cam.:** Pinyin, 5,500 ft. (Apr.) *Brunt* 1093! Jakiri, Bamenda Div. (Feb.) *Hepper* 2073! Throughout tropical Africa. (See Useful Plants.)

145. RHYTACHNE Desv. in Hamilt., Prodr. Fl. Ind. Occ. 11 (1825); F.T.A. 9: 81 (1917); Clayton in Kew Bull. 20: 258 (1966), and in Kew Bull. 24: 309 (1970).

Plants perennial:
- Pedicelled spikelets similar to the sessile:
  - Spikelets and internodes glabrous; basal sheaths fibrous; lower glume longitudinally striate, and sometimes also transversely rugose; spikelets at most mucronate 1. *glabra*
  - Spikelets and internodes densely pubescent; basal sheaths not fibrous; lower glume smooth; spikelets with an awn 2–2·5 cm. long from their upper glumes 2. *perfecta*
- Pedicelled spikelets suppressed or up to 1 mm. long; basal sheaths not fibrous; spikelets glabrous:
  - Lower glume smooth; rhachis internodes 6–8 mm. long; leaf-blades flat or convolute, up to 4 mm. broad; spikelets awnless; upper glume concealed 3. *megastachya*
  - Lower glume transversely rugose, at least on the keels:
    - Leaf-blades filiform; rhachis internodes 3–5 mm. long; glumes of sessile spikelet and rudimentary pedicelled spikelet with or without awns up to 5 mm. long; upper glume smooth, concealed .. .. 4. *rottboellioides*
    - Leaf-blades flat, basal sheaths compressed and keeled; internodes of rhachis 6–7 mm. long; glumes awnless; upper glume rugose on the keel, visible between the slightly parted pedicel and internode .. .. 5. *furtiva*

Plants annual:
- Rhachis internodes 4–6 mm. long, the pedicel 1–2 mm. longer; lower glume with deep transverse corrugations; glumes of sessile and rudimentary pedicelled spikelets with awns usually 5–7 mm. long .. .. 6. *triaristata*
- Rhachis internodes 2–3 mm. long, the pedicel as long; lower glume inconspicuously rugose; glumes of sessile and rudimentary pedicelled spikelets with awns usually 3–5 mm. long .. .. 7. *gracilis*

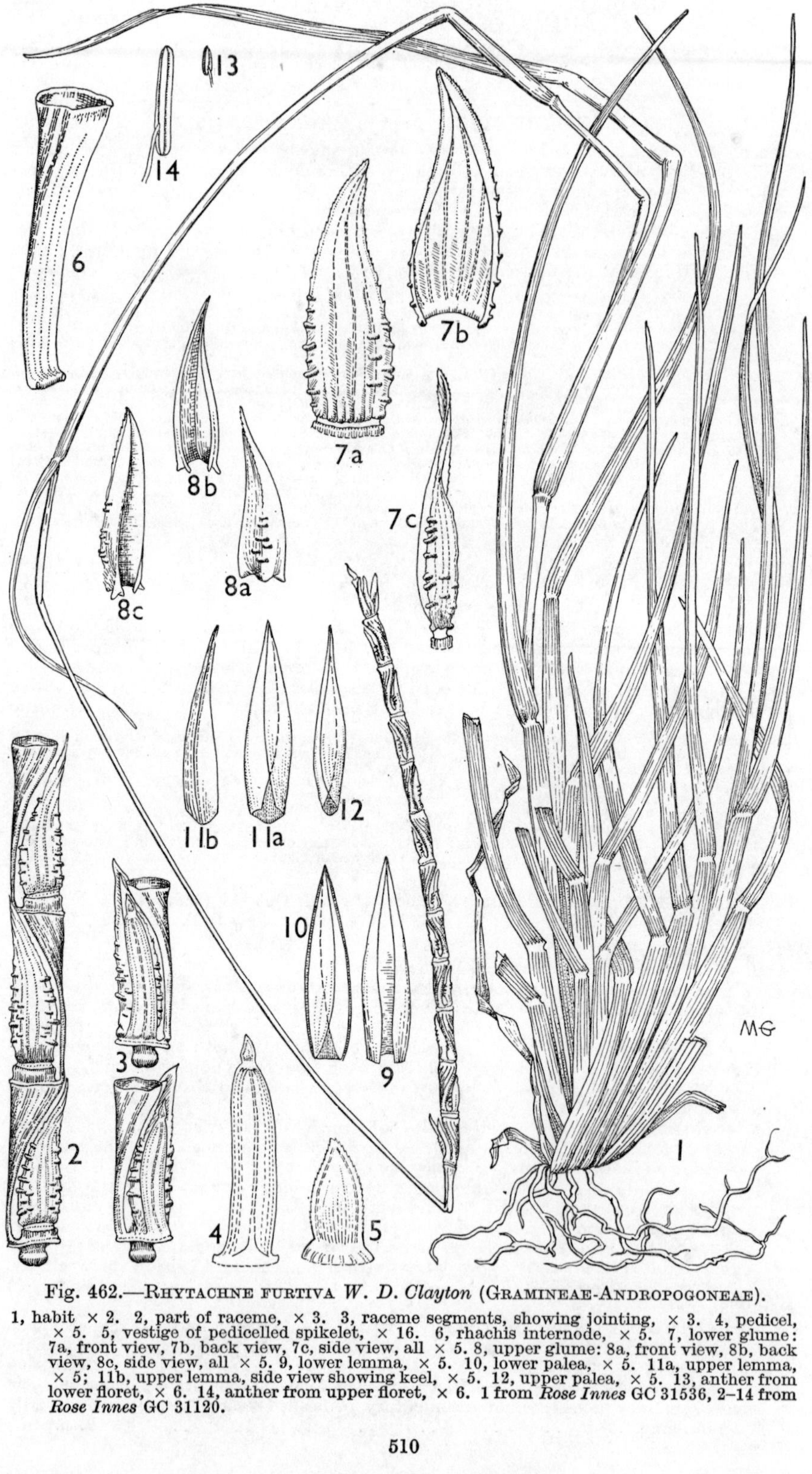

Fig. 462.—RHYTACHNE FURTIVA *W. D. Clayton* (GRAMINEAE-ANDROPOGONEAE).

1, habit × 2. 2, part of raceme, × 3. 3, raceme segments, showing jointing, × 3. 4, pedicel, × 5. 5, vestige of pedicelled spikelet, × 16. 6, rhachis internode, × 5. 7, lower glume: 7a, front view, 7b, back view, 7c, side view, all × 5. 8, upper glume: 8a, front view, 8b, back view, 8c, side view, all × 5. 9, lower lemma, × 5. 10, lower palea, × 5. 11a, upper lemma, × 5; 11b, upper lemma, side view showing keel, × 5. 12, upper palea, × 5. 13, anther from lower floret, × 6. 14, anther from upper floret, × 6. 1 from *Rose Innes* GC 31536, 2–14 from *Rose Innes* GC 31120.

1. **R. glabra** (*Gledhill*) *W. D. Clayton* in Kew Bull. 23: 295 (1969). *Lepargochloa glabra* Gledhill in Bol. Soc. Brot. 40: 65 (1966). Densely tufted perennial 30–45 cm. high; mountain grassland.
Guin.: Mt. Nimba *Schnell* 969. **S.L.:** Tingi Mts. (Apr., Dec.) *Morton & Gledhill* SL 1885! SL 3083! Loma Mts. (Jan.-Apr.) *Morton & Gledhill* SL 1096! *Jaeger* 8845! 9779!

2. **R. perfecta** *Jac.-Fél.* in J. Agric. Trop. 1: 41 (1954). Caespitose plant about 60 cm. high.
Guin.: Massif du Benna *Brun* 7!

3. **R. megastachya** *Jac.-Fél.* in Rev. Bot. Appliq. 32: 552 (1952). Robust perennial about 1 m. high; wet land subject to flooding.
Guin.: (May) *Adam* 5163! Fouta Djalon (Dec.) *Jac.-Fél.* 7432! **S.L.:** Mateboi (July) *Glanville* 283! Rowankiliboli, Mateboi (Nov.) *Jordan* 831! **Ghana:** Babile (Oct.) *Addei* SLUS 1!

4. **R. rottboellioides** *Desv.* in Hamilt., Prodr. Fl. Ind. Occ. 12 (1825); F.T.A. 9: 83; Chev. Bot. 714, and in Rev. Bot. Appliq. 13: 854; Berhaut, Fl. Sén. ed. 2, 389; Aké Assi, Contrib. 2: 303; incl. var. *guineensis* Schnell in Rev. Gen. Bot. 57: 291 (1950). *R. setifolia* K. Schum in Engl., Pfl. Ost-Afr. C: 96 (1895). Densely caespitose grass 30–60 cm. high, typically on shallow soils over ironpan; very variable in length of spikelets and awns, and in degree of corrugation of lower glume.
Sen.: Diégoune, Casamance R. (Sept.) *Adam* 18076! **Mali:** Bamako (Mar.) *Chev.* 44072! **Guin.:** Kindia *Jac.-Fél.* 182! Dalaba (Nov.) *Chev.* 34595! Nzo to Sokonanta (Mar.) *Chev.* 21084! Madina Tossekré (Oct.) *Adam* 12519! Friguiagbé (Sept.) *Chillou* 708! **S.L.:** Mano Bonjema (Apr.) *Deighton* 3714! Gbap (June) *Adames* 58! Wellington (June) *Deighton* 5550! Port Lokko (Apr.) *Sc. Elliot* 5737! Robolou & Mabang (Jan.) *Glanville* 132! **Lib.:** Robertsfield (Mar.) *Harley* 2120! Painesville (Feb.) *Harley* 1860! Monrovia (June) *Baldwin* 5919! Nimba (Sept., Dec.) *Adam* 16460! *Adames* 602! **Iv. C.:** Oumé (Sept.) *Adjanohoun* 255*a*! Toumodi (Oct.) *Adjanohoun* 324*a*! **Ghana:** Essiama (Feb.) *Irvine* 2337! Atuabo (Apr.) *Thorold* 38! Kwahu Tafo to Adawso (Oct.) *Rose Innes* GC 30563! Tamale to Yendi (Oct.) *Rose Innes* GC 30682! Bimbila (Mar.) *Hepper & Morton* A3095! **Dah.:** Sémé (Feb.) *Raynal* 13538! **N. Nig.:** Nupe *Barter* 1383! Bida Div. (Mar.) *Meikle* 1324! Abinsi (July) *Dalz.* 873! Wuya Ferry, Kaduna R. (Sept.) *Clayton* 357! Jos (Dec.) *Haines* 123! **S. Nig.:** Obudu Plateau, 5,000 ft. (Dec., May) *Tuley* 46! 624! **W. Cam.:** Mamfe (Mar.) *Richards* 5227! Bafut-Ngemba F.R., 5,500–7,400 ft. (Feb., Mar.) *Coombe* 214! *Hepper* 2128! High Lava Plateau, 6,500 ft. (Mar.) *Brunt* 1024! Mbiami, 6,000 ft. (June) *Brunt* 750! Throughout tropical Africa. (See Useful Plants.)

5. **R. furtiva** *W. D. Clayton* in Kew Bull. 20: 259 (1966). Caespitose, about 60 cm. high; disturbed places.
U. Volta: Batié to Gaoua (Oct.) *Rose Innes* GC 31536! **Ghana:** Bole to Bamboi (Oct.) *Rose Innes* GC 30651! Kintampo to Tamale (Oct.) *Rose Innes* GC 31120! Damongo to Tamale (Oct.) *Rose Innes* GC 30217!

6. **R. triaristata** (*Steud.*) *Stapf* in F.T.A. 9: 85 (1917); A. Chev. in Rev. Bot. Appliq. 13: 855; Berhaut, Fl. Sén. ed. 2, 389. *Lepturopsis triaristata* Steud., Syn. Pl. Glum. 1: 358 (1854). *Rhytachne triseta* Hack. (1889)—Chev. Bot. 714. *Rottboellia triaristata* (Steud.) Roberty in Boissiera 9: 71 (1960). Slender annual 60 cm. high; shallow soils over rock outcrop, soils disturbed by cultivation, sometimes by muddy stream sides.
Sen.: Kaolak (Nov.) *Berhaut* 2800! Bodi, Simenti (Nov.) *Adam* 15928! **Mali:** Ségou (Sept.) *Chev.* 2329! Marigot Balasoko, Niger Delta (Nov.) *Duong*! Kirilouba, Niger Delta (Nov.) *Duong*! **Guin.:** *Chillou* 757! Timbo (Oct.) *Pobéguin* 1769! Ditinn *Schnell* 7405*bis*! Seriba (Oct.) *Adam* 12699! **S.L.:** Brookfield (Oct.) *Deighton* 2142! Newton (Nov.) *Deighton* 1464! Port Loko (Oct.) *Jordan* 638! Njala (Oct.) *Deighton* 1775! Roboli, Rokupr (Nov.) *Jordan* 171! **Iv. C.:** Séguéla (Oct.) *Aké Assi* 6583! 9697! **U. Volta:** Gaoua to Batié (Oct.) *Rose Innes* GC 31531! Po to Navrongo (Oct.) *Rose Innes* GC 31465! Banankélédaga, Bobo-Dioulasso (Oct.) *Kmoch* 145! Zabré (Oct.) *Scholz* 159! **Ghana:** Yendi (Oct.) *Rose Innes* GC 30690! Tamale (Nov.) *Thorold* 290! Mirigu (Oct.) *Vigne* FH 4645! Burufu (Nov.) *Harris*! Gambaga (Oct.) *Ankrah* GC 20273! **Togo Rep.:** Sansanné Mango (Oct.) *Rose Innes* GC 31416! **N. Nig.:** (Oct.) *Lely*! Kaduna (Oct.) *Thatcher* S542! Kufena Hill, Zaria (Sept.) *Clayton* 1313! Anara F.R. (Oct.) *Keay* FHI 20103! Fodama (Dec.) *Moiser* 178! **S. Nig.:** Enugu to Agbami (May) *Onochie* FHI 35813! **W. Cam.:** Vogel Peak, 3,000 ft. (Nov.) *Hepper* 1415! Also in the Sudan and Zambia. (See Useful Plants.)

7. **R. gracilis** *Stapf* in Journ. de Bot. 19: 98 (1905); F.T.A. 9: 86; A. Chev. in Rev. Bot. Appliq. 13: 855; Berhaut, Fl. Sén. ed. 2, 392, 395. *R. minor* Pilger in Engl., Bot. Jahrb. 54: 280 (1917); F.W.T.A., ed. 1, 2: 601. Slender grass 15–60 cm. high; sandy or shallow soils.
Guin.: Koba (Nov.) *Jac.-Fél.* 7280! Conakry (Oct.) *Adam* 12621! Madina Tossekré (Oct.) *Adam* 12524! Kouroussa (Aug.) *Pobéguin* 494! Friguiagbé (Nov.) *Chillou* 853! **S.L.:** Waterloo (Aug.) *Melville & Hooker* 294! Lungi (Nov.) *Glanville* 103! Kitchom (Jan.) *Deighton* 933! Rolip, Sanda-Tenraron (Sept.) *Jordan* 535! Rokupr (Nov.) *Jordan* 177! **U. Volta:** Ouasso (Nov.) *Jac.-Fél.* 7255! **Iv. C.:** Séguélo (Nov.) *Aké Assi* 9728!

146. ZEA Linn., Sp. Pl. 971 (1753), and Gen. Pl. ed. 5, 419 (1754); F.T.A. 9: 26 (1917).

Leaf-blades linear-lanceolate, up to about 1 m. long and 10 cm. broad or more; male spikelets paired, one sessile or subsessile, the other stalked, in large terminal panicles (the " tassel "); female spikelets in axillary sheathed " ears ", the spikelets paired, both sessile, in longitudinal rows on a thick spongy axis (the " cob "); styles very long, exserted from the ear in a silky tassel (the " silk ") .. .. .. *mays*

Z. mays *Linn.* Sp. Pl. 971 (1753); F.T.A. 9: 26; Chev. Bot. 712, and in Rev. Bot. Appliq. 13: 848; De Miré & Gillet in J. Agric. Trop. 3: 740. The common maize, cultivated throughout the area. Originally from the New World, now introduced to all tropical and warm temperate regions. (See Useful Plants.)

147. COIX Linn., Sp. Pl. 972 (1753), and Gen. Pl. ed. 5, 419 (1754); F.T.A. 9: 27 (1917).

Annual up to about 2 m. high; leaf-blades lanceolate, rounded or subcordate at the base, up to 45 cm. long and 5 cm. broad; male spikelets paired or in threes, one pedicelled, the other(s) sessile, in a raceme 2·5–5 cm. long, subtended by a hard, polished, white or bluish sheath 8–10 mm. long; female inflorescence reduced to a solitary spikelet enclosed within the sheath and forming with it a false fruit .. *lacryma-jobi*

C. lacryma-jobi *Linn.* Sp. Pl. 972 (1753); F.T.A. 9: 27; Mimeur in Rev. Bot. Appliq. 31: 198; Vallayes in Bull. Agric. Congo Belge 39: 247; Chev. Bot. 712, and in Rev. Bot. Appliq. 13: 848; Berhaut, Fl. Sén. ed. 2, 420. " Job's Tears "; stream banks and moist places. The false fruit can vary much in shape and size, but so far only the typical form (var. *lacryma-jobi*) has been found in W. Africa.

Mali: Kawlauba (Oct.) *Rogeon* 217! **Guin.:** *Garret* 4! Mali to Labé (Nov.) *Chev.* 34918! Mafindi to Kabaya (Jan.) *Chev.* 20493! **S.L.:** Yonibana (Oct.) *Thomas* 4165! Binkolo (Aug.) *Thomas* 1733! Tungie (Mar.) *Marmo* 80! Mambolo (Dec.) *Jordan* 740! Mt. Aureol, Freetown (Aug.) *Melville & Hooker* 110! **Lib.:** Kakatown *Whyte*! Jabrocca (Dec.) *Baldwin* 10876! Ganta (Dec.) *Harley* 18! **Iv. C.:** Niapidou (Jan.) *Leeuwenberg* 2629! **Ghana:** Sukum (Sept.) *Ankrah* GC 20072! Kumasi (Mar.) *La Danso* 30! Tafo (July) *Irvine* 4972! **N. Nig.:** Share (Jan.) *Ejiofor* FHI 30801! **S. Nig.:** Umudike (Feb.) *Ariwaodo* 643! Nikrowa Nedo, Iyekuselu (Nov.) *Emwiogbon* FHI 57520! **W. Cam.:** Tiko (Feb.) *Maitland* 978! A native of tropical Asia, and probably introduced into Africa. (See Useful Plants.)

INDEX TO VOLUMES I–III

This Index includes all names and synonyms in Volume III (Parts 1 and 2), as well as keyed-out families and genera in Vols. I and II. The indexes in these volumes should be consulted for species and synonyms.

The volume number is given in roman numerals, followed by the page. Synonyms and misapplied names are shown in *italics*; they may be located by means of the number of the genus and species shown in brackets.